T0262841

THE DYNAMICS OF VEHICLES
ON ROADS AND ON TRACKS

THE DYNAMICS OF VEHICLES ON ROADS AND ON TRACKS

Edited by

Masato Abe

Proceedings of the 18th IAVSD Symposium

held in Kanagawa, Japan

August 24-30, 2003

Supplement to Vehicle System Dynamics, Volume 41

Taylor & Francis
Taylor & Francis Group

Library of Congress Cataloging-in-Publication Data

A Catalogue record for the book is available from the Library of congress

www.tandf.com

ISBN 90 265 1972 9

TABLE OF CONTENTS

PREFACE

The 18th Symposium of the International Association for Vehicle System Dynamics was held at Kanagawa Institute of Technology (KAIT), Atsugi, Kanagawa, Japan August 24-30, 2003. The symposium was hosted by KAIT as one of the memorial events of the 40th anniversary of KAIT.

Though overwhelming numbers of high quality papers were applied in response to the call for papers for the presentation at the symposium, the Scientific Committee accepted 89 papers for the oral presentation and 38 for the poster presentation. Finally 82 papers were presented at the oral sessions and 29 papers at the poster sessions in the symposium. There were five States-of-the-Arts papers presented at the plenary sessions in the symposium and they have been published in a special issue of the journal "Vehicle System Dynamics", Volume 40, Number 1-3, 2003.

We registered 211 distinguished scientists and engineers with 19 accompanied persons from 25 different countries. We had plenary sessions and two road/rail parallel sessions in the morning and three parallel sessions in the afternoon through the symposium week with the technical visits in the Wednesday afternoon. Around two thirds of the papers presented were from road and remainders were from rail.

The IAVSD thanks the companies and organization, Toyota Motor Corporation, Nissan Motor Co., Ltd., Honda R&D Co., Ltd., Hitachi Unisia Automotive, Ltd., Sumitomo Metal Industries, Ltd., Odakyu Electric Railway Co., Ltd. and Suzuki FOUNDATION for their financial supports for the symposium.

I would like to thank all the members of the Board of IAVSD, the members of the Scientific Committee, the local organizing committee members in Japan and all the members of the local operating committee in KAIT for their supports and hard works for a success of the symposium. Finally I would like to thank Kanagawa Institute of Technology for hosting this successful symposium.

I believe that the papers presented at the symposium and published in this supplement issue will contribute to give you informative ideas and encourage you for your further research and development works in the fields of vehicle system dynamics.

October 27, 2003

Prof. Masato Abe
Chairman of the 18th IAVSD Symposium

Vehicle System Dynamics Supplement 41 (2004), p.3-12

Vehicle suspension optimisation for heavy vehicles on deformable ground

S. PARK, A.A. POPOV[1] AND D.J. COLE

SUMMARY

The mathematical prediction of the dynamic behaviour of wheeled off-road heavy vehicles together with the operating limits for vehicles travelling at constant speed in a straight line on deformable ground are investigated. The magnitude and duration of vehicle vibration can increase despite the fact that the deformation of ground acts like a filter for the vertical excitation input. The main result here concerns the conducted optimisation of vehicle suspension parameters for off-road operation by the application of the developed numerical models and procedures. It is shown that higher suspension damping ratio is required as the ground deformability increases in order to minimise vertical vibration.

1   INTRODUCTION

The ride performance of wheeled vehicles travelling in a straight line on a rigid road surface is generally well understood, and the ride behaviour can be predicted accurately by linear models of structural vibration. However, the established modelling techniques and suspension design do not necessarily apply when non-linear ground deformation is involved in the calculation. The non-linearity becomes more significant with the increase of vehicle weight and ground deformability. The stiff suspension of heavy vehicles is another factor that makes difficult the isolation of vibration from roadway input.

Bekker [1] established the area of land locomotion and the theories that he introduced had significant impact on research in vehicle and soil interaction. This interaction is particularly important for the study of off-road vehicles. The relationship between ground pressure and sinkage was investigated, and a non-linear pressure/sinkage equation was derived empirically. Wong [2] conducted extensive field measurements and obtained the various ground parameters by fitting the measured pressure/sinkage data in accordance with the non-linear equation (1). The interaction of wheel and ground was also investigated, and the importance of the

[1] *Address correspondence to:* School of Mechanical, Materials, Manufacturing Engineering and Management, University of Nottingham, University Park, Nottingham NG7 2RD, UK; Tel: +44(0)115 951 3783; Fax: +44(0)115 951 3800; Email: atanas.popov@nottingham.ac.uk.

contact patch was emphasised. The stress due to the wheel sinkage is not uniformly distributed but depended on the magnitude of the sinkage. In the dynamic cases, the position where highest stress occurs moves forward as the speed of vehicle increases. However, the energy transferred to the vehicle suspension is only vertically directed and the magnitude of the energy in vertical direction remains unchanged in steady state conditions. Cole [3] studied extensively heavy vehicle suspension design not only for ride comfort by minimising vehicle body acceleration but also for road-friendly suspension by lowering the dynamic tyre force.

Analysis of the dynamic behaviour of a heavy vehicle during off-road operation is conducted under steady state condition. Section 2 describes a simple and computationally efficient model of ground deformation. Since the ground model is generally non-linear, an iterative procedure in the time domain is employed for the solution of the contact problem of wheel/ground interaction at any position along the vehicle path. The mathematical model of the vehicle and its interaction with the ground are described in section 3. Section 4 presents various responses of the vehicle model while the ground parameters and vehicle suspension parameters vary. The main objective of the work is to investigate the interactive behaviour of heavy vehicle and deformable ground, and to find out optimum suspension parameters depending on ground conditions.

2   MODELLING OF GROUND

During the longitudinal travel of the vehicle, an arbitrary ground profile is the primary input of vertical excitation to the vehicle. When ground deformation is included in the numerical model, the deformation filters the excitation to the vehicle while the level of excitation varies depending on the ground deformability. Therefore, the concept of deformability and surface roughness should be treated in modelling the off-road ground.

When the surface of ground is compressed by a plate, the ground starts to deform non-linearly with a linear increase of the loading, and it collapses when the load reaches a certain magnitude. Bekker [1] obtained the following non-linear equation for the pressure and sinkage relationship

$$p = \left(\frac{k_c}{b} + k_\phi \right) z^n \tag{1}$$

where b is the smaller dimension of the rectangular contact area, k_c is a modulus of deformation due to the cohesive properties of soil and k_ϕ is a modulus related to the frictional component of the ground. Bekker's model is implemented in a lumped-parameter ground model in which the ground properties are represented by a layer of non-linear springs. Random road surface profiles can be generated with statistical properties of road surface profiles suitable for off-road operations.

The finite element method is applied in producing pressure/sinkage data of the various types of grounds. The ground parameters for the FE analyses are taken from the existing literatures, and static forces are applied onto a rectangular plate located on the ground model. Wong [2] showed that Eq. (1) is only applicable to clay type grounds whose characteristics are governed by cohesion. Therefore, four different clays are chosen (very soft clay, soft clay, medium clay and hard clay) that have different levels of deformability. The FE results are fitted using the least-squared curve-fitting and the fitted ground parameters are given in Table 1.

Table 1. Fitted ground parameters using rectangular plates.

Soil Type	n	k_c (kN/m^{n+1})	k_ϕ (kN/m^{n+2})
Very soft day	0.73	16.74	515.7
Soft clay	0.77	189.48	324.88
Medium clay	0.67	105.85	1865
Hard clay	0.62	79	2764.9

2.1 DISCRETE WHEEL/SOIL MODEL

If the ground is compressed by a wheel, the contact surface deforms along the wheel surface. The pressure over the contact patch is not uniform but proportional to the sinkage of the surface. Since Bekker's pressure/sinkage equation is only applicable to the case of uniform pressure, ground can be discretised into small sections, as shown in Figure 1, assuming that uniform pressure applies on each area. Denoting the pressure at the s-th contact area as p_s the ground force at the area is

$$f_s = p_s A_d \qquad (2)$$

where A_d is the elementary area in the discrete soil model. Hence, the sum of the elementary forces gives the total ground force due to sinkage:

$$F_G = f_1 + f_2 + \ldots + f_s + \ldots + f_{m-1} + f_m \qquad (3)$$

where m is the number of discrete elements between wheel and ground. The ground force should be the same as the input force, and this force equilibrium enables the calculation of the sinkage of the wheel using an iterative procedure. The tyre width of a commercial vehicle is assumed to be of 0.3 m. The smaller value of discretisation step gives more accurate results, however, it requires more simulation time. Therefore, after convergence studies a discrete step of 0.01 m is chosen which is sufficiently small to satisfy both accuracy and simulation time.

2.2 GROUND PROFILE

In order to model off-road ground condition, various values of ground surface profiles are necessary as well as ground deformability. The required surface profile is obtained using the value of its power spectral density in the frequency domain. The spectral density is converted into an expression for the height/distance relationship by taking the inverse discrete Fourier transform of the spectral coefficients. Robson [5] proposed that the roughness of ground depended on a function which contained the information of road profile spectral density. The spectral density of the ground surface is found from

$$S(N) = S(N_0) \cdot (N / N_0)^{-\omega} \qquad (4)$$

where ω is the dimensionless profile constant, N is desired wavenumber and reference wavenumber $N_0 = 1/2\pi$. The description of obtaining ground profile using the surface spectrum is presented in detail by Cebon [6] and various road spectra of surface profiles are classified. Mitschke [7] extended the classification of road profiles to off-road conditions as shown in Table 2, and the desired random ground profile can be generated by choosing the range of the spectral density value.

Table 2. Mitschke's [7] off-road surface classification

Ground Profile	Profile constant (ω)	$S(N)$ (m³/cycle)
Good	2.25	199.8×10^{-6}
Medium	2.25	973.9×10^{-6}
Poor	2.14	3782.5×10^{-6}

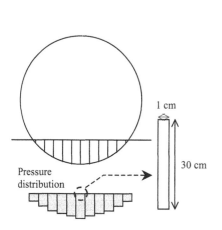

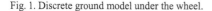

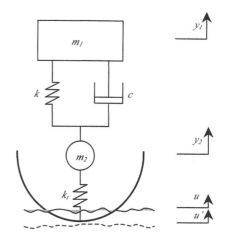

Fig. 1. Discrete ground model under the wheel.   Fig. 2. Circular rigid treadband model.

3   MODELLING OF HEAVY VEHICLES

The quarter-vehicle model is a widely used approach to investigate the vibration behaviour of vehicles due to its simplicity and reasonable accuracy. Typical parameters for the quarter-vehicle model of a heavy vehicle are: sprung mass =3600 kg; unsprung mass = 400 kg; suspension damping = 10 kNs/m; suspension stiffness = 400 kN/m; tyre stiffness = 800 kN/m.

In modelling a vehicle running on paved ground, it can be assumed that the vehicle wheel contacts the ground surface in a point ignoring the deformation of the ground, while the tyre stiffness can be represented by a single spring. However, the influence of ground deformation to the vehicle vibration is significant in off-road heavy vehicles, and the contact area between tyre and ground is not a single point but changes depending on ground deformation and tyre deflection. Therefore, it is essential to develop a new vehicle model considering the extent of contact patch between tyre and ground. By extending the characteristics of contact patch due to ground deformation into the point contact model, a circular rigid treadband contact (CRTC) model is developed as shown in Figure 2. The surface of the tyre is assumed to be rigid material, however, the overall deflection of the tyre is represented by the stiffness of a linear spring between the tyre surface and axle. The deflection of the spring provides the dynamic tyre force. The equation of motion of the CRTC model is

$$m_1\ddot{y}_1 + c(\dot{y}_1 - \dot{y}_2) + k(y_1 - y_2) = 0$$
$$m_2\ddot{y}_2 + c(\dot{y}_2 - \dot{y}_1) + k(y_2 - y_1) + k_t(y_2 - u') = 0 \qquad (6)$$

where u' is deformed ground profile and $k_t(y_2 - u')$ is a force due to both ground sinkage and tyre deflection. Assuming the tyre deflection zero, the second line of the equation of motion is

$$m_2\ddot{y}_2 + c(\dot{y}_2 - \dot{y}_1) + k(y_2 - y_1) = F_G \qquad (7)$$

where F_G is dynamic ground force due to the sinkage of the wheel. For deformable tyre, the tyre force at initial position is obtained from equilibrium with the ground force and vehicle weight: $k_t\, d_{T0} = F_G(d_{G0}) = Mg$ where d_{G0} is initial ground deformation under the wheel centre, k_t is tyre stiffness, d_{T0} initial tyre deflection, M is total vehicle mass and g is gravity. The initial tyre deflection can be obtained since $d_{T0} = Mg/k_t$. The speed of the vehicle is constant in order to make sure the wheel always contacts the ground. The dynamic tyre force under a rigid wheel is found ($F_T = -(F_G - Mg)$) and the dynamic deflection of tyre is obtained ($\Delta = F_T/k_t$). Therefore, a new equation of motion of circular rigid treadband model of the axle can be obtained

$$m_2\ddot{y}_2 + c(\dot{y}_2 - \dot{y}_1) + k(y_2 - y_1) = F_G(d_G + \Delta) - Mg \qquad (8)$$

where the magnitude of $F_G(d_G + \Delta) - Mg$ is the same as $k_t(y_2 - u')$. The radius of the wheel and speed of the vehicle are assumed to be 0.5 m and 5m/s respectively.

4   RESULTS AND DISCUSSION

Simulations were performed by varying not only the road conditions but also the vehicle suspension parameters such as damping and stiffness rates. The initial vehicle parameters are: mass ratio = m_1/m_2 = 3600/400 = 9, stiffness ratio = k_1/k = $800 \times 10^3/400 \times 10^3$ = 2, damping ratio = 0.132 while the vehicle travels with a constant speed of 5m/s on the ground.

The suspension damping ratio of passenger cars usually lie between 0.2 and 0.4 for good ride [8]. However, the same ride quality as for passenger cars cannot be achieved for off-road heavy vehicles, and the ideal damping ratio may lie outside the range for passengers cars because heavy trucks generally have more resonances and are stiffly sprung. In order to support the heavy weight of the vehicle, allowing longer travel on off-road, stiff suspension and large suspension working space are required. Therefore, various damping and stiffness values should be applied in order to find optimum values for the best ride quality and vehicle performance.

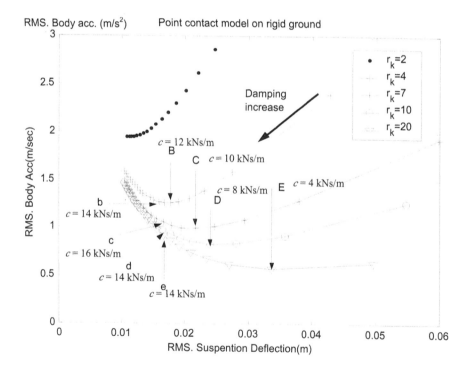

Fig. 3. Body acceleration vs. suspension deflection on non-deformable ground with medium profile.

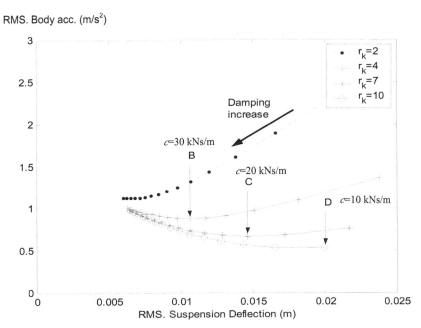

Fig. 4. Responses of the CRTC model on very soft clay with medium profile.

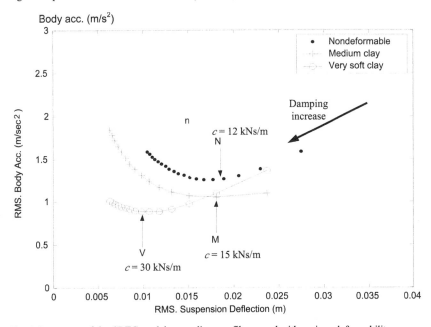

Fig. 5. Responses of the CRTC model on medium profile ground with various deformability.

Figure 3 shows the responses of the point contact (PC) vehicle model on non-deformable ground with medium roughness (Table 2) by varying both suspension stiffness and damping rates simultaneously. High values of the body acceleration are found at lower damping and the stiffness ratio of 2, while the increase of damping rate results in reduction of the body acceleration. The overall body acceleration can be lowered with the increase of stiffness ratio (or decrease of suspension stiffness rate) because more external excitation is absorbed into larger suspension deflection. The best ride quality of the vehicle is achieved by minimising body acceleration, whereas the least road surface damage is achieved by minimising tyre force. Points B, C, D and E show the optimum damping rates (c = 12, 10, 8 and 4 kNs/m) to minimise body acceleration at different stiffness ratio of 4, 7, 10 and 20 respectively. The damping ratios of the points B, C, D and E are 0.2236, 0.2465, 0.2357 and 0.1667 while the points b, c, d and e minimise tyre force at the damping rates of 14, 16, 14 and 14 kNs/m (ζ= 0.2609, 0.3944, 0.4125 and 0.5833) respectively. Some sacrifice of passenger's comfort is unavoidable if the suspension is only tuned to minimise tyre force. At the stiffness ratios of 7, 10 and 20, for example, body acceleration values are increase by 4.8 %, 15.8 % and 33.3 % when the damping rates are tuned for tyre force minimisation. However, nearly no loss of ride comfort is observed at the stiffness ratio of 4.

Figure 4 shows the body acceleration versus suspension deflection for the CRTC model on very soft clay with medium profile. The overall body acceleration values are lower than those of the PC vehicle model, while the body acceleration is minimised at higher damping rate than the PC vehicle model: For the stiffness ratio of 4, 7 and 10 the body acceleration values are minimised at the damping rates of 30 kNs/m (ζ = 0.56), 20 kNs/m (ζ = 0.37) and 10 kNs/m (ζ = 0.19), respectively. When the ground deformation is included in the simulation, the damping rates which minimise body acceleration also minimise tyre force at any stiffness ratio. Therefore, no compromise of body acceleration and tyre force is needed for vehicles on deformable ground.

Higher stiffness ratio (or softer suspension stiffness rate) is required in order to reduce body acceleration as the suspension working space increases. However, the volume of working space cannot exceed a certain design limit (i.e. low platform height of the typical truck does not exceed 0.2 m). Therefore, the maximum stiffness ratio should be chosen as 4 where the initial suspension deflection is 0.17 m. Figure 5 shows the changes of body acceleration for the PC and CRTC models for the stiffness ratio of 4 on medium ground profile. The magnitudes of wheel sinkage in static condition are 6 cm on medium clay and 22 cm on very soft clay. Optimum damping rates increases proportionally to the ground deformability: 12 kNs/m (ζ = 0.22) on non-deformable ground, 15 kNs/m (ζ = 0.28) on medium clay and 30 kNs/m (ζ = 0.56) on very soft clay, giving lower overall body acceleration on more deformable ground. It implies that truck suspension must be tuned differently depending on the road condition on which the vehicle mainly operates; higher damping ratio is necessary as the ground becomes more deformable.

Figure 6 shows the changes of optimum damping rates for the vehicle suspension as the ground profile varies, showing that generally higher damping rate is required as the surface profile becomes rougher with the increase of ground deformability. For good, medium and poor surface profile the optimum damping values vary as follows:

$c_{optimum}$ = 11, 12 and 16 kNs/m on non-deformable ground
$c_{optimum}$ = 15, 15 and 25 kNs/m on medium clay   (static wheel sinkage: 6cm)
$c_{optimum}$ = 20, 20 and 25 kNs/m on soft clay   (static wheel sinkage: 12 cm)
$c_{optimum}$ = 25 , 30 and 25 kNs/m on very soft clay (static wheel sinkage: 22cm)

However, the optimum damping rate does not increase proportionally to the surface profile when the vehicle is simulated on very soft clay because the ground surface exceeds the maximum allowed clearance of the vehicle when the poor surface profile is applied to very soft clay (maximum clearance ≈ radius of the wheel). Moreover, the influence of non-linear ground deformation is significant to the suspension vibration on very soft clay. Therefore, the operating limit for the vehicle is at the ground with very soft clay and medium profile.

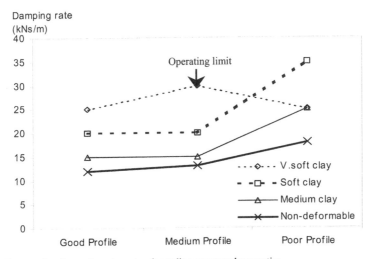

Fig. 6. Change of optimum damping rates depending on ground properties.

By taking non-deformable ground with good profile as reference ground condition, body acceleration is minimised at the damping ratio of 0.21 (c = 12 kNs/m). The optimum damping ratio is increased by 9 % on medium profile ($\zeta_{c=13}$ = 0.24) and by 55 % on poor profile ($\zeta_{c=18}$ = 0.34). If the ground deformation is included, the optimum damping ratios are also increased significantly, by 33% on medium clay ($\zeta_{c=15}$ = 0.28), by 76 % on soft clay ($\zeta_{c=20}$ = 0.37) and by 124% on very soft clay ($\zeta_{c=25}$ = 0.47) for the ground with a good profile.

5   CONCLUSIONS

From the simulation with the CRTC model on various ground conditions the following conclusions can be drawn:

1. Overall body acceleration can be lowered with the increase of stiffness ratio because more external excitation is absorbed into the larger suspension deflection. However, the suspension stiffness ratio should not exceed 4 due to the design limitation of the truck.

2. Ground force of the circular rigid treadband model (or tyre force of the point contact model) and body acceleration are minimised at the same damping rates when the stiffness ratio is 4. Therefore, compromise of neither body acceleration nor ground force (tyre force) is required in tuning off-road truck suspension.

3. Higher damping rate is required as surface profile becomes rougher and ground deformability increases. For vehicles on very soft ground the operating limits are already reached at medium road profile as the ground surface exceeds the maximum clearance of the vehicle.

4. The optimum damping ratio of the heavy vehicle on non-deformable and good profile is 0.21. The optimum damping ratio is increased by 9 % on medium profile and by 55 % on poor profile. The optimum damping ratios are also increased with the increase of ground deformability, by 33% on medium clay (static wheel sinkage = 6 cm), by 76 % on soft clay (sinkage = 12 cm) and by 124% on very soft clay (sinkage = 22 cm).

REFERENCES

1. Bekker, M. G.: Theory of Land Locomotion, Ann Arbor, University of Michigan Press, 1956.
2. Wong, J.Y.: Theory of Ground Vehicles, New York: John Wiley & sons, 3rd edition, 2001.
3. Cole, D.J.: Fundamental Issues in Suspension Design for Heavy Road Vehicles, Vehicle System Dynamics, 35 (4-5), (2001), pp. 319-360.
4. Matlab 5, The MathWorks, Inc., December 1996.
5. Robson, J. D.: Road Surface Description and Vehicle Response, International Journal of Vehicle Design, 1 (1), (1979), pp. 25 – 35.
6. Cebon, D.: Handbook of Vehicle-Road Interaction, Swets and Zeitlinger B.V., the Netherlands, 1999.
7. Mitschke, M..: Dynamik der Kraftfahrzeuge, Springer-Verlag, Berlin, 1972.
8. Chalasani., R. M.: Ride Performance Potential of Active Suspension Systems – Part I: Simplified Analysis Based on a Quarter-Car Model, American society of mechanical engineers symposium on simulation of ground vehicles and transportation systems, ASMES Symposium on Simulation of Ground Vehicles and Transportation Systems, Anaheim, CA, Research publication GMR-5312 December 1986.

Vehicle System Dynamics Supplement 41 (2004), p.13-22

Bridge – friendly truck suspension

MICHAEL VALÁŠEK, JAROMÍR KEJVAL[1], JIŘÍ MÁCA,VÍT ŠMILAUER[2]

SUMMARY

The paper deals with the development of the controlled truck suspension that reduces the loading and deflection of a bridge on which a truck is passing. The trucks moving around the roads and bridges are causing their wear and damage. The mechatronical solution can help by the usage of controlled variable suspension damper that leads to the decrease of road-tyre forces and road damage. Such truck suspensions are called road-friendly truck suspension.

The concept of road-friendliness can be extended into the concept of bridge-friendliness. This paper investigates whether the road-friendly truck suspension is also bridge-friendly and vice versa and even more whether there exists a specifically bridge-friendly truck suspension which reduces the bridge loading beyond that by the road-friendly truck suspension. These concepts are studied within this paper on a complex nonlinear truck suspension model and multi DOF bridge model.

1 INTRODUCTION

The trucks moving around the roads and bridges are causing their wear and damage [1-3]. The mechatronical solution can help by the usage of controlled variable suspension damper that leads to the decrease of road-tyre forces and road damage [2-3]. Such truck suspensions are called road-friendly truck suspension.

The concept of road-friendliness [4] can be extended into the concept of bridge-friendliness. Bridge-friendly truck suspension decreases bridge–tyre forces [10,13]. This paper investigates whether the road-friendly truck suspension is also bridge-friendly and vice versa and even more whether there exists a specifically bridge-friendly truck suspension which reduces the bridge loading. These concepts are studied further on a simple nonlinear quarter car model with flexible pavement [10].

[1] *Address correspondence to*: Prof. Ing. Michael Valášek, DrSc., Ing. Jaromír Kejval, PhD, Czech Technical University in Prague, Faculty of Mechanical Engineering, Department of Mechanics, Karlovo nam. 13, 121 35 Praha 2, Czech Republic, E-mail: valasek@fsik.cvut.cz

[2] *Address correspondence to*: Doc.Ing.Jiří Máca,CSc, Ing. Vít Šmilauer, Czech Technical University in Prague, Faculty of Civil Engineering, Thákurova 7, Prague 6, 160 00, Czech Republic, E-mail: maca@fsv.cvut.cz

2 TRUCK - BRIDGE INTERACTION MODEL AND BRIDGE-FRIENDLINESS

There have been investigated in several projects [1-3] and many papers [5-9] the truck-road interaction and the concept of road-friendliness. From these projects there arises the question how truck suspension designed for road friendliness behaves on the bridge and vice-versa. The results from simplest realistic model of truck-bridge interaction (quarter car on 1DOF bridge) from earlier paper [10] were so promising that new improved model of truck-bridge interaction has been designed. The schema of simulation model of half truck on more DOF bridge is on Fig. 1.

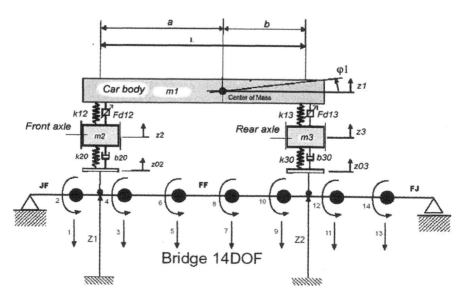

Figure 1. Nonlinear half-car model with multi DOF bridge model interaction

The equations of motion for this half car model with 4DOF are:

$$m_1 \ddot{z}_1 = k_{12} \cdot (z_1 - z_2 - \varphi \cdot a) - 2 \cdot F_{d12} - k_{13} \cdot (z_1 - z_3 - \varphi \cdot b) - 4 \cdot F_{d13}$$
$$I_{01} \ddot{\varphi} = k_{12} \cdot (z_1 - z_2 - \varphi \cdot a) \cdot a + 2 \cdot F_{d12} \cdot a - k_{13} \cdot (z_1 - z_3 - \varphi \cdot b) \cdot b - 4 \cdot F_{d13} \cdot b$$
$$m_2 \ddot{z}_2 = k_{12} \cdot (z_1 - z_2 - \varphi \cdot a) + 2 \cdot F_{d12} - k_{20} \cdot (z_2 - z_{02}(t)) - b_{20} \cdot (\dot{z}_2 - \dot{z}_{02}(t))$$
$$m_3 \ddot{z}_3 = k_{13} \cdot (z_1 - z_3 - \varphi \cdot b) + 4 \cdot F_{d13} - k_{30} \cdot (z_3 - z_{03}(t)) - b_{30} \cdot (\dot{z}_3 - \dot{z}_{03}(t))$$

(1)

$$z_{02}(t) = z_1 + z_r$$
$$z_{03}(t) = z_2 + z_r$$

where m_1 is mass of car body, m_2, m_3 are masses of front and rear axle, k_{12} and k_{13} are stiffness of main springs, k_{20} and k_{30} stiffness of tyres, F_{d12}, F_{d13} are forces of passive or semi-active dampers, zr is road irregularity. All used parameters corresponds to figure (1).

The principles of semi-active control are based on the transformation from the required (active) force F_{act} to the setting of the damping rate $b_{semi-active}$ such that the damping force is nearest to the desired one. For an ideal linear variable shock absorber the damping rate $b_{semi-active}$ is set for the interval /b_{min}, b_{max}/ as

$$b_{semi-active} = \begin{cases} b_{max} & \text{if } b_{max} < b_{act} \\ b_{act} & \text{if } b_{min} < b_{act} < b_{max} \\ b_{min} & \text{if } b_{act} < b_{min} \end{cases} \tag{2}$$

$$b_{act} = \frac{F_{act}}{\left(\dot{z}_1 - \dot{\phi}a - \dot{z}_2\right)} \qquad \text{for the front axle (equation for rear axle is similary)} \tag{3}$$

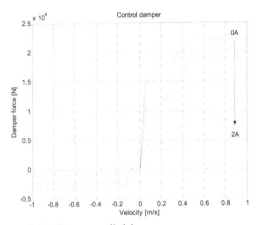

Figure 2. Nonlinear controlled damper

For the car model there is used the model of nonlinear control damper on Fig. 2 with internal dynamics as it is described in [2,4].

In the truck model there were used for front axle two and for rear axle four controlled dampers. Other parameters (masses, stiffness, damping) correspond to full truck model.

The ideal forces F_{d12}, F_{d13} of this element, according to the control law of the nonlinear extended ground-hook [6,7], are

$$F_{d12} = b_{f1}\left(\dot{z}_2 - \dot{z}_{02}\right) - b_{f2}\left(\dot{z}_1 - \dot{\varphi}a\right) - b_{f12}\left(\dot{z}_1 - \dot{\varphi}a - \dot{z}_2\right) + \Delta k_{f10}\left(z_2 - z_{02}\right) - \Delta k_{f12}\left(z_1 - \varphi a - z_2\right)$$
$$F_{d13} = b_{r1}\left(\dot{z}_3 - \dot{z}_{03}\right) - b_{r2}\left(\dot{z}_1 + \dot{\varphi}b\right) - b_{r12}\left(\dot{z}_1 + \dot{\varphi}b - \dot{z}_3\right) + \Delta k_{r10}\left(z_3 - z_{03}\right) - \Delta k_{r12}\left(z_1 + \varphi b - z_3\right)$$

$$(6)$$

where $b_{f1}, b_{f2}, b_{f12}, \Delta k_{f10}$ and Δk_{f12} are state dependent gains of extended ground hook EGH controller for the front axle and $b_{r1}, b_{r2}, b_{r12}, \Delta k_{r10}$ and Δk_{r12} for the rear axle. The tyre-road forces (b20, b30 neglected) for front F_1 and rear F_2 axle are:

$$F_1 \approx k_{20}\left(z_2 - z_{02}\right)$$
$$F_2 \approx k_{30}\left(z_3 - z_{03}\right)$$

$$(7)$$

If the same parameters for front and rear axle are used, there are only 5 parameters for whole half-car model. By changing the parameters b_1, b_2, b_{12}, Δk_{10} and Δk_{12} a variety of modified control laws for the system can be obtained. For the systematic determination of these parameters, the MOPO approach was applied. The multi-objective parameter optimization (MOPO) [2,4] within the environment MATLAB/ SIMULINK with the genetic algorithm toolbox [11] allows one to find a satisfactory compromise among the performance criteria despite the fact that they conflict with each other. The single performance criterion was square root of the time integral of the dynamic tyre force:

$$sumRF_{front\_RMS} = \sqrt{\int_0^t F_1^2 \, dt} \quad , \quad sumRF_{rear\_RMS} = \sqrt{\int_0^t F_2^2 \, dt}$$

$$(8)$$

As the second performance criteria we used integral of truck sprung mass acceleration

$$sumAcc = \sqrt{\int_0^t \left(m1\_acc\right)^2 dt}$$

$$(9)$$

The parameters of the extended ground-hook were originally considered constant for the whole shock absorber velocity interval [2, 4]. Because the characteristics of the shock absorber are nonlinear, the nonlinear Extended Ground-Hook control with state-dependent gains (gain scheduling) is used.

Four subintervals of the relative damper velocity $v = \dot{z}_1 - \dot{\varphi}a - \dot{z}_2$ have been defined as high negative (HN) $v \in (-\infty, -0.131)$ [m/s], low negative (LN) $v \in <-0.131, 0)$ [m/s], low positive (LP) $v \in <0, 0.052)$ [m/s] and high positive (HP) $v \in <0.052, \infty)$

[m/s]. Numerical experiments show that the sensitivity of the criteria (8) to the fictitious damping coefficients (b_1, b_2, b_{12}) is substantially higher than to the fictitious changes in stiffness (Δk_{10}, Δk_{12}). Therefore, only the damping coefficients are subjected to the optimization process. There are $3*4=12$ free parameters to be optimized.

The following set of unevenness patterns has been used as the base of the optimization. It consists of cosine bump with heights of 2 cm and lengths of 1.5m. The velocity of the truck considered for the optimization is 50 [km/s]. The achieved results have been described in several papers [2,4].

Half car model of the truck LIAZ was designed together with FEM model of the bridge. The concept of bridge model is a discrete system with 14DOF (7deviation, 7rotation) [13]. Model is divided into three basic elements: joint-fixation (JF), fixation-fixation (FF) and fixation-joint (FJ) according to Fig. 1. The equation of bridge oscillation is:

$$\mathbf{M}\ddot{r} + \mathbf{B}\dot{r} + \mathbf{K}r = \mathbf{F} \tag{10}$$

where $\mathbf{M}$, $\mathbf{B}$, $\mathbf{K}$ are mass, damping and stiffness matrixes of bridge.

This model interacts with two contact places: for front and rear axle according to Fig. 1. Contact tyre-bridge forces are expanded into all nodes of bridge as the truck moves on the bridge. The bridge model was developed for the following parameters: 20m concrete bridge supported on both ends. Implementation of the interaction between truck suspension model and bridge discrete model is described on the Fig. 4.

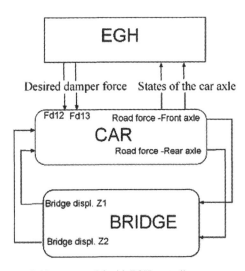

Figure 4. Interaction between bridge -car model with EGH controller

3   INVESTIGATION OF TRUCK-BRIDGE INTERACTION

The developed controlled truck suspensions were tested under the given conditions:

Velocity of the car is 50 km/h, concrete bridge is 20m long, excitation is by cosine bump 2cm/1.5m in the middle of the bridge.

The following integral criteria are used for result evaluation.

$$sumRF_{RMS} = sumRF_{front\_RMS} + sumRF_{rear\_RMS} \tag{11}$$

3.1   PASSIVE DAMPER

Different passive damper (hard, commercial, soft) characteristics were used in basic investigation for possible comparison with another new control methods. The passive damper was modeled as controlled damper with different constant currents.

Constant control current [A]	0A (hard)	1.2A (commercial)	2A (soft)
SumRF_front_RMS	11 424	7 644	9 576
SumRF_rear_RMS	24 181	20 487	30 424
SumRF_RMS	35 605	**28 131**	40 000

Table 1 Comparison of road forces integral criteria for different passive dampers

Time response of tyre-bridge forces for front and rear axle are on the next figure 5.

Static force is removed.

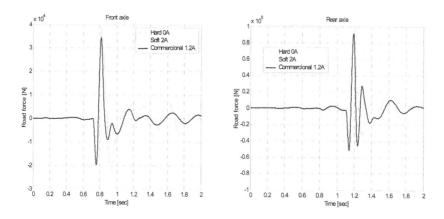

Figure 5. Time response of front and rear axle bridge forces with passive dampers

3.2   CONTROLLED DAMPER

In next part there are compared the simulated behaviour of truck suspensions with different sets of controller gain parameters obtained from design on earlier simplified models and on new half car model with 14DOF bridge model. All comparisons are done on the most complex truck=bridge interaction model. In the earlier papers [2,4] there were determined parameters for the road-friendly truck behaviour. In next part these parameters are called **SADTS parameters** The parameters from paper [10] where the car bridge interaction with simplest model (quarter car + 1DOF bridge) was investigated is called **¼ car+1DOF bridge**

Control method	SumRF_front RMS [N]	SumRF_rear RMS [N]	SumRF_RMS [N]
MOPO SADTS 12const. For road	8 443	19 324	27 767
¼ car+1DOF bridge 12const, 2cm obstacle, T=17ms	7 448	17 883	25 331
½ car+14DOF bridge 12const, 2cm obstacle, T=17ms	6 864	16 976	23 840
½ car+14DOF bridge 24const, 4cm obstacle, T=17ms	6 563	17 103	23 666
½ car+14DOF bridge 12const, 4cm obstacle, T=17ms	8 004	18 606	26 610
½ car+14DOF bridge 12const, 2cm obstacle, T=1ms	**5987**	**15 230**	**21 210**

Table 2 Comparison of bridge forces integral criteria for different control algorithm for semiactive dampers for 2cm cosinus bump in the middle of the bridge.

Using the actual simulation model (called **½ car + 14DOF bridge** there were obtained by genetic algorithms next sets of controller gain parameters with different initial parameters:

- Front and rear axle with the same 12 control parameters for 2cm obstacle with damper time constant 17ms.
- Front and rear axle with different 24 control parameters for 2cm obstacle with damper time constant 17ms.
- Front and rear axle with the same 12 control parameters for bigger obstacle (4cm / 1.5m) with damper time constant 17ms.
- Front and rear axle with the same 12 control parameters for 2cm obstacle for idealistic actuator with 1ms time constant.

The resulting sets of controller gain parameters were again tested on bridge with 2cm obstacle. The values of bridge forces integral criteria are compared in the Table 2.

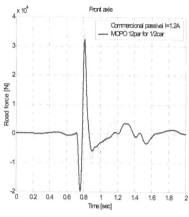 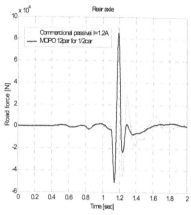

Figure 6. Comparison of time response of front and rear axle bridge forces with for passive and controlled dampers

Comparison of bridge force time response for front and rear axle for commercial passive damper and 12const bridge friendly EHG optimized for ½ car+14DOF bridge with 17ms dynamics constant is on the Figure 6.

The sets of obtained EGH parameters were tested for different conditions of simulation model. Comparison of result is on the next table 3.

Sets of parameters / Test conditions	Passive commercional I = 1.2A For 2cm obstacle	SADTS 12const. For road for 2cm obstacle	¼ car+1DOF bridge 12const for 2cm obstacle	½ car+14DOF bridge – 12const for 2cm obstacle	½ car+14DOF bridge – 12const for 4cm obstacle	½ car+14DOF bridge – 24const for 2cm obstacle
Small obstacle 0.5cm	9 091	8 416	6 808	6 769	7 461	**6 774**
Big obstacle 5cm	96 460	77 070	76 440	79 230	**72 830**	79 770
Obstacle on bridge begin 2cm	27 880	27 780	25 190	24 060	27 180	**23 890**
Light truck loading 10t	28 370	28 040	25 500	24 280	27 190	**24 070**
Heavy truck loading 20t	27 990	27 580	25 140	23 610	26 310	**23 560**
Bad Stochastic Road 2cm	22 000	19 850	19 590	19 190	19 120	**19 090**

Table 3 Comparison of bridge forces integral criteria for different control algorithm and different test conditions (Note: All values correspond to integral criteria sumRF_RMS from front and rear axle together)

The design of truck suspension must take into account two basic criteria: acceleration of sprung mass (comfort) and tire-bridge normal forces (bridge-friendliness). These criteria are conflicting and corresponding Pareto sets were determined. This was done by running several optimization processes by genetic algorithms for weighted criteria function CF:

$$CF = A*sumRF + B*sumAcc \qquad\qquad (12)$$

where A=[0:1], B=[0:1] are weights of particular criteria, sumRF is RMS of normalized normal bridge forces and sumAcc is RMS of normalized truck body acceleration. The optimization processes were run for weights ranging the whole intervals. All investigated cases within genetic algorithm were saved and based on them the boundary curve of Pareto set was determined (Fig. 7). The line at this figure represents the available limits by passive dampers. The shape of boundary of dots area creates the boundary curve of Pareto set of controlled dampers.

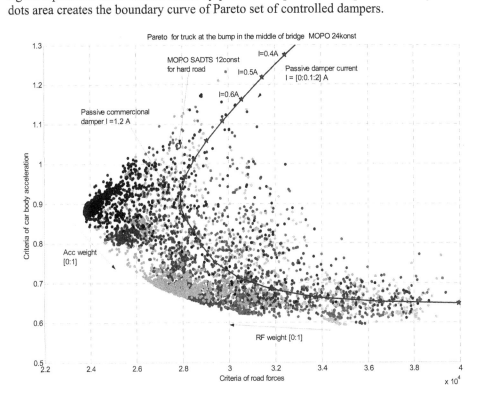

Figure 7. Result Pareto set of bridge forces/ acceleration for passive and controlled dampers (truck on the bridge with bump)

4   CONCLUSION

This paper describes the concept of bridge-friendly truck suspension and has proved that there exists specific bridge-friendly semi-active truck suspension. There have been compared the behaviour of commercial passive, road-friendly and bridge-friendly suspensions within the several important excitation cases on the road and on the bridge. The reduction of bridge loading by the truck has been detected in all cases.

There were designed a special bridge-friendly control of truck suspension. Potential saving of controlled suspension is about 25% of dynamic loading by passive suspension. The behaviour of bridge-friendly truck suspension on the road is excellent. This is valid for shorter bridges (typically concrete bridge with span of 20m, but steel bridge was also successfully tested).   These results could be significantly improved if the dynamics of controllable damper is improved.

The control concept is based on the extended nonlinear ground hook (EGH) [7], the control synthesis is done by design-by-simulation, however the direct nonlinear control synthesis is also considered [12]. The control uses the sensor data only from the truck suspension and thus it is fully feasible.

REFERENCES

1.   *Proceedings of DIVINE Concluding Conference, Rotterdam 1997*
2.   *Kortüm, W., Valasek, M. (Eds.): SADTS Final Technical Report, DLR, Wessling 1998*
3.   *Proceeding of Workshop: Moderne Nutzfahrzeug-Fahrwerke, Potential zur Ladegut- und Fahrbahnschonung, Aachen, March 2000*
4.   *Valášek, M. - Kortüm, W.: Road-friendly Trucks, In: Fryba, L., Naprstek, J. (Eds.): Structural Dynamics-Proc. of Eurodyn '99, Balkema, Rotterdam 1999, pp. 855-860*
5.   *Novak, M. and Valasek, M.: A New Concept of Semi-Active Control of Truck's Suspension, In: Proc. of AVEC 96, Aachen 1996, pp. 141-151*
6.   *Valášek, M. et al.: Extended Ground-Hook -New Concept of Semi-Active Control of Truck's Suspension, Vehicle System Dynamics, 27(1997), 5-6, pp. 289-303*
7.   *Valasek, M. et al.: Development of Semi-Active Road-Friendly Truck Suspensions, Control Engineering Practice 6(1998), pp. 735-744*
8.   *Vaculin, O., Kortüm, W., Schwartz, W.: Analysis and Design of Semi-active Damping in Truck Suspension – Design by Simulation, In: Proc. of AVEC 96, Aachen 1996, pp. 1087-1103*
9.   *Valasek, M., Kortüm, W.: Nonlinear Control of Semi-active Road-friendly Truck Suspension, In: Proc. of AVEC 98, Nagoya 1998, pp. 275-280*
10.   *Valasek, M., Kejval, J., Maca, J.: Control of truck-suspension as bridge-friendly, Grundmann, H., Schueller, G.I. (eds.) Structural Dynamics Eurodyn 02, Balkema, Munich 2002, pp. 1015-1020*
11.   *Houck, Ch.R. et al.: A Genetic Algorithm for Function Optimization, North California State University*
12.   *Valasek, M., Kejval, J.: New Direct Synthesis Of Nonlinear Optimal Control of Semi-Active Suspensions, Proc. of AVEC 2000, Ann Arbor 2000, pp. 691-697*
13.   *Máca, J., Šmilauer, V., Valasek, M.: Dynamic Interaction of Trucks and Bridges, Proc. of AED 2003, Prague*

The effect of nonlinear suspension kinematics on the simulated pitching and rolling dynamic behavior of cars

YUKIO WATANABE AND MICHAEL W. SAYERS [1]

SUMMARY

This paper compares the behavior of two modeling methods for independent suspensions. Both versions are based on the CarSim™ vehicle model, with equations created by the AutoSim™ symbolic multibody code generator. A fully nonlinear model uses the complete kinematical and dynamical multibody equations for the wheel spindle as it moves through the range of the suspension travel. Tables that specify longitudinal movement, lateral movement, steer, inclination, and dive define the 3D vertical motions of the spindle. In the other method, the spindle follows a straight line angled to match the full 3D motion for small displacements. This approach has historically been taken for math models created by hand, to help control the equation complexity. Those motions are combined with nonlinear algebraic corrections for inclination and steer, to obtain proper inputs for the tire model. The fully nonlinear model requires about 20% more overall computation time relative to the simpler "straight-line" model. On a 1.8 GHz PC, the simpler model runs about ten times faster than real-time, and the nonlinear model runs about eight times faster than real time. The models agree closely for moderate levels of braking, acceleration (throttle), and cornering. When the trajectory of the vertical movement of the wheel spindle curves significantly, as it does for most automobiles, the nonlinear model gives better predictions of roll and pitch.

1   INTRODUCTION

Vehicle dynamics simulation programs are now used for many applications, including mechanical design, controller design, replacement of lengthy test procedures, training, and real-time simulation with hardware in the loop (HIL). Model complexity can range from simple models with a few equations to detailed models whose equations of motion cover thousands of lines of computer source code. This paper compares two levels of complexity in the suspension kinematics for math models used to predict how a four-wheeled vehicle system responds to driver controls on a 3D ground surface. The two versions of the model are based on the widely used CarSim™ simulation package [1,2].

External forces acting on a highway vehicle come only from the air and ground, with the ground forces being far more important; hence, the most critical part of a

---

[1] *Address correspondence to:* Mechanical Simulation Corporation, 709 W. Huron Street, Ann Arbor, MI, 48105, USA. Tel.: 734/ 668-9189; Fax: 734/ 668-2877; E-mail: ywata@carsim.com

vehicle math model is the realistic prediction of the tire forces and moments that cause vehicle motions. The significance of tire actions requires that the kinematical input variables (slip angle, longitudinal slip, inclination angle, radius deflection) be determined accurately. With a detailed multibody suspension model with all linkages, bushings, and flexible bodies, these variables are defined directly from the motions of the moving parts. However, the approach taken in historical vehicle models such as HVOSM has been to combine a much simpler dynamic multibody model with quasi-static extensions to obtain the correct inputs for the tire model [3]. If the algebraic equations are based on easily measured steer and inclination angles, then the tire kinematical variables can be predicted fairly accurately without the computational overhead associated with the full multibody equations [4].

Besides determining the kinematical interaction between the tire and ground, the suspension model affects the way tire actions are transferred to the vehicle. The main effect of the force transfer is to cause overall movement of the vehicle system for prediction of braking, acceleration, and handling behavior. Another effect is to generate bounce, roll, and pitch through "jacking," "anti-dive," and "anti-squat" reaction forces and moments [5]. A detailed multibody suspension model properly handles the transfer of working spring/damper forces and non-working reaction forces (at the expense of long computation times). Simpler models attempt to capture the main effects of the force transfer by representing the overall suspension motion without details of the mechanical linkages. For example, the concepts of roll center and pitch center have historically been used to predict how tire forces cause roll and pitch of the sprung mass [6]. Also, models whose equations were developed by hand sometimes explicitly added the effects of jacking and anti-pitch forces [3].

The two vehicle math models described in this paper handle independent suspensions with different modeling methods. The "full nonlinear method" defines the suspension motion as following a curved line through 3D space, with rotation in pitch and roll. Alternatively, a "simple method" defines the suspension motion as following a straight line with no rotation. Rotation effects are added with nonlinear algebraic expressions, such that both models provide the same inputs to the tire model. However, they differ in how the tire forces and moments are transferred to the sprung mass to cause roll and pitch.

2   THE MODELS

The CarSim four-wheeled vehicle dynamics model with independent front and rear suspensions has a sprung mass with six degrees of freedom (DOF) and four unsprung masses, each with one DOF for vertical travel. Each wheel has a spin DOF, and each tire has two DOF for longitudinal and lateral delayed slips. The engine crankshaft rotation adds one DOF, and each brake actuator has one DOF to represent brake fluid dynamics. In total, the model has 27 DOF.

Suspension toe and camber are specified with nonlinear tabular functions of suspension stroke to obtain proper tire kinematical inputs (inclination and slip angle).

Compliance toe and camber effects are handled with linear coefficients and added algebraically to the tire inputs, along with steering control calculated by a steering system model.

The equations for the 3D kinematics and dynamics are complicated. The equations and the programs that solve them were generated by AutoSim™, a symbolic multibody program that combines a multibody formalism [7] with a symbolic algebra language [8] and generates ready-to-compile C code. The equations are always a minimal set of ordinary differential equations (ODEs) that can be solved with almost any numerical integration method. AutoSim was used to make two versions of the CarSim models with different suspension kinematics.

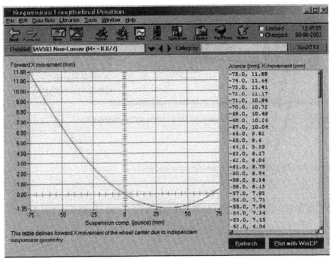

Fig. 1. A CarSim suspension kinematics data screen.

2.1   FULL NONLINEAR MODEL

In the full nonlinear model, each wheel spindle moves vertically in a curved line through 3D space, and rotates. To create the model, the spindle is initially given five DOF. Then, four of the DOF are removed by constraints to match nonlinear input tables that define longitudinal tracking, lateral tracking, camber, and dive (caster angle change) as functions of vertical travel. For example, Fig. 1 shows the user-interface screen for the table used to define longitudinal movement of the wheel center as a function of suspension travel.

The lateral geometry of the full nonlinear model (Fig. 2) defines trajectories of each center of tire contact (CTC) point, whose position is defined by the trajectory of the wheel center in combination with camber angle change. Although this figure shows a double wishbone suspension, the model suits any independent suspension type (McPherson strut, five-link, etc.). Longitudinal geometry is modeled in the same manner, with tables defining longitudinal movement of the wheel center (Fig. 1) and pitching of the wheel spindle as functions of suspension travel.

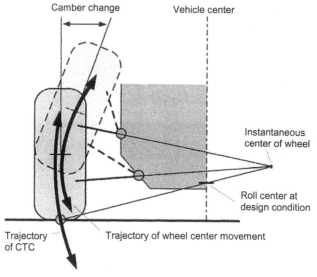

Fig. 2. Lateral geometry of the full nonlinear suspension model.

2.2 SIMPLE MODEL: TRANSLATIONAL JOINT WITH ALGEBRAIC TIRE INPUTS

The simple model attempts to simulate the same behavior as the full nonlinear model with much simpler equations. The wheel spindle in the simple suspension model moves in a straight line, such that the wheel body does not rotate relative to the sprung mass (Fig. 3). The lateral direction of the straight line is set to match the direction at the CTC. Two parameters define this motion: (1) the track width of the suspension at nominal load (L_{tk}), and (2) an equivalent roll-center height (H_{rc}). For each unit of vertical displacement, the wheel moves laterally by $2 \cdot H_{rc}/L_{tk}$. (CarSim uses a roll-center height parameter to maintain compatibility with published datasets. However, the model itself is based on the angled translational motion, which means the roll center moves as the suspensions deflect.)

A similar approach is taken for the longitudinal geometry (Fig. 4). A ratio is defined for the longitudinal movement per unit of vertical movement at the wheel center. Longitudinal tire forces that derive from driveline torques are reacted by shafts connected to the sprung mass through the wheel center, and therefore the angle at the wheel center is needed to transfer drive forces properly to the sprung mass. On the other hand, brake forces are reacted by pads connected to the unsprung mass (the wheel spindle). The proper angle for the suspension motion would be the direction at the CTC. To compensate, a moment is applied to the wheel body based on the difference in angles of the movements at the CTC and wheel center.

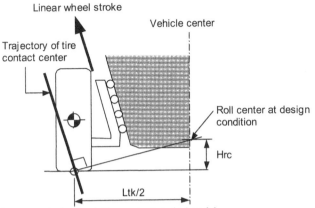

Fig. 3. Lateral geometry in the simple linear wheel stroked model.

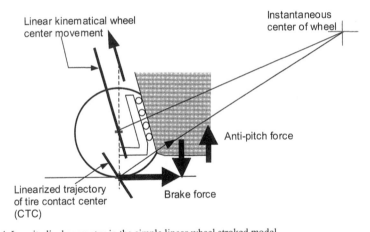

Fig. 4. Longitudinal geometry in the simple linear wheel stroked model.

This simple model is intended to approximate the transfer of tire forces to the sprung mass. However, it is not adequate by itself to provide realistic kinematical inputs to the tire model. Therefore, the lateral slip angle and inclination angles are adjusted algebraically by adding nonlinear toe and camber geometry effects. As with the nonlinear model, compliance steer and inclination are added algebraically to the inputs for the tire model along with the steering system control angle.

3 SIMULATION RESULTS

3.1 VALIDATION TESTING

Preliminary testing was performed in order to validate the full nonlinear suspension model (created more recently than the simple model). A dataset was prepared in which the nonlinear model was given tables to mimic the behavior of the simpler model. For longitudinal effects, a table defined the constant rate of forward movement per unit of vertical travel at the wheel center. A second table defines diving of the wheel as a function of vertical travel, to mimic the rate of forward movement of the CTC as a function of suspension travel. For lateral effects, the nonlinear camber was combined with loaded tire radius to create a nonlinear table that relates lateral position of the wheel center to the suspension stroke, such that the lateral movement of the CTC is the same in the two models.

Nearly identical results were obtained for test runs involving braking, throttle, and handling.

3.2 EXAMPLE VEHICLE DATASET

Suspension measurements typical for a North American mid-size luxury sedan were used with both models to compare behavior when realistic suspension geometric nonlinearities are considered. In this case, each model was set to duplicate the measured suspension geometries as well as possible. For the nonlinear model, the laboratory measurements were tabular, and the data were simply copied and pasted into the CarSim data screens (e.g., Fig. 1). Effective rates were extracted and used for the simpler model to obtain the proper movements of the CTC.

Although it is easy (and standard) to measure the lateral movements of the wheel center in a laboratory test rig, the critical movement needed to represent the transfer of forces and moments from the tire to the sprung mass is where the tire contacts the road at the CTC. It is necessary to combine the lateral movement of the wheel center, the camber changes, and the deflected tire radius as it applied during the suspension test to determine the trajectory at the CTC. Fig. 5 shows that when the suspension properties of the linear model are set to obtain proper movement of the CTC, then the movement of the wheel center does not match. With the example data, the wheel center actually moves in the opposite direction with the two models. (Although this might at first seem like a major error, the position of the wheel center is not a factor in determining how lateral forces are transferred from the CTC to the sprung mass. Results presented later in this paper show that the two models agree closely in predicting lateral response for small changes in suspension travel.)

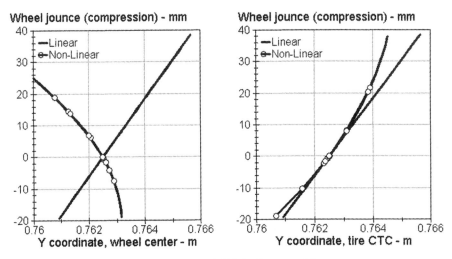

Fig. 5. Lateral movements of the wheel center and CTC for the two models.

3.3   BRAKING BEHAVIOR

A braking input was applied at the initial vehicle speed of 80 km/h on a high-friction surface for the two models, leading to a -0.83g deceleration. Fig. 6 shows that the sprung mass pitch angles for the two models are almost identical. Sprung mass height change (jacking) was also nearly identical between the two models.

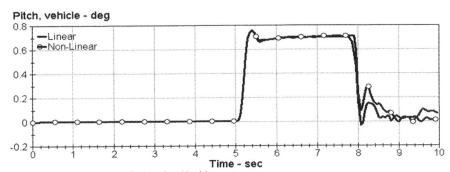

Fig. 6. Body pitch responses for simulated braking test.

3.4   ACCELERATION BEHAVIOR

A full-throttle test was simulated for the two models, leading to nearly identical powertrain behavior, tire forces, and overall speed change. However, Fig. 7 shows that the nonlinear model showed more pitch than the linear one.

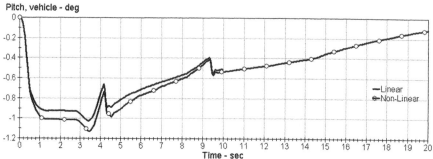

Fig. 7. Body pitch responses for simulated acceleration test.

The reason that the pitch for the two models matched better in braking behavior (Fig. 6) than for acceleration is due to the specific nonlinear property of the example data set. Fig. 1 shows that the measured change in the rear wheel center longitudinal position as a function of jounce approximately follows a straight line for negative suspension travel (extension) as occurs in braking. However, Fig. 1 also shows that the trajectory has significant curvature when going through compression. Thus, the straight-line approximation for the simple models is much more representative for the case of extension (braking) than compression (acceleration).

3.5   UNDERSTEER HANDLING TEST

Fig. 8 shows steering wheel angle as a function of lateral acceleration (A_y) of the two vehicle models during a simulated understeer test in which speed is increased slowly while driving around a circular track (radius: 152.4m = 500 ft). The models agree closely up to A_y of about 0.2 g; for levels above 0.4 g the nonlinear model requires a visibly higher steering control.

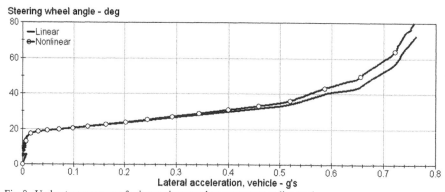

Fig. 8. Understeer responses for increasing speed on a constant-radius track.

3.6 STEP STEER MANEUVER

To illustrate the differences in behavior at high levels of cornering, an extreme step steer input of 135 deg (applied at 270 deg/second) is simulated with the two models, leading to A_y of about 0.8 g. The most significant difference between the models was in the predicted roll angle. Fig. 9 shows the roll angle and sprung-mass height change for the two models. For this test, two variations of the vehicle were simulated: one had no bump stops or auxiliary roll stiffness, and therefore allowed large roll angles; the other had bump stops, and realistic auxiliary roll stiffness that reduces the overall roll angles (labeled as "limit" in the figure). With or without the limits, the nonlinear model shows more roll but less vertical movement (jacking).

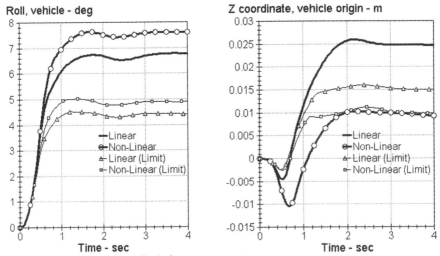

Fig. 9. Roll and height change of body for extreme step steer.

3.7 COMPUTATIONAL EFFICIENCIES

The simpler model runs about 10 times faster than real time on a 1.8 GHz Intel CPU machine. Earlier studies comparing the performance of AutoSim-generated programs with numerical multibody programs showed an improvement of a factor of about 50 [9], implying that a comparable numerical multibody program model would run about five times slower than real time on the same machine.

In this comparison, the multibody kinematical and dynamical equations for the fully nonlinear model were found to require about 40% more computation. However, the linear model also involves considerable computation for subsystems such as the tire model, road geometry, controllers, the powertrain, numerous table lookups, etc. When tested, the nonlinear model ran about eight times faster than real time, about 20% slower than the simpler model.

4   CONCLUSIONS

Two CarSim models were compared that are identical except in the way the wheel spindle moves vertically during suspension deflection. A fully nonlinear version was found to require about 20% more computation relative to a simpler version that assumed straight-line motion. The model with fully nonlinear equations still runs eight times faster than real-time on a 1.8 GHz PC, so the computation considerations are not as much of a factor today as they were years ago when early vehicle math models were developed.

When the trajectory of the vertical movement of the wheel spindle curves significantly, as it does for most automobiles, the nonlinear model gives better predictions of roll and pitch of the sprung mass. Another advantage of the nonlinear model is that when measured suspension data are available from the laboratory, they can be copied and pasted directly into the data screens of a modern graphical user interface (GUI) such as the one in CarSim.

ACKNOWLEDGEMENT

The authors are pleased to acknowledge Dr. Thomas Gillespie for his helpful suggestions and warmest thanks are due to Mr. Mark Gjukich and Mr. Damon Becker for the figures.

REFERENCES

1.   Sayers, M.W., "Vehicle Models for RTS Applications." Vehicle System Dynamics, Vol. 32, No. 4-5, Nov. 1999, pp. 421-438.
2.   Watanabe, Y. and Sayers, M. W., "Extending Vehicle Dynamics Software for Analysis, Design, Control, and Real-Time Testing." Proceeding of 6th International Symposium on Advanced Vehicle Control, AVEC '02, Hiroshima, Japan, September 2002, pp. 407-412.
3.   Jindra, F., "Mathematical Model of Four-Wheeled Vehicle for Hybrid Computer Vehicle Handling Program." U.S. DOT, DOT HS-801-800, NTIS 33658, 1975.
4.   Sayers, M.W. and Fancher, P.S. Jr, "A Hierarchy of Symbolic Computer-Generated Real-Time Vehicle Dynamics Models." Transportation Research Record 1403, National Research Council, Washington, D.C. 1993, pp. 88-97.
5.   Sharp, R.S., "Influences of suspension kinematics on pitching dynamics of cars in longitudinal maneuvering," Vehicle System Dynamics, Vol. 33, 2000, pp. 23-36.
6.   Gillespie, T.D. Fundamentals of Vehicle Dynamics. 1992, Society of Automotive Engineers, Warrendale, PA.
7.   Sayers, M.W., "Symbolic Vector/Dyadic Multibody Formalism for Tree-Topology Systems." Journal of Guidance, Control, and Dynamics, Vol. 14, No. 6, Nov/Dec 1991, 1240-1250.
8.   Sayers, M.W., "Symbolic Computer Language for Multibody Systems." Journal of Guidance, Control, and Dynamics, Vol. 14, No. 6, Nov/Dec 1991, 1153-1163.
9.   Sayers, M.W. and Mousseau, C.W., "Real-time Vehicle Dynamic Simulation Obtained with a Symbolic Multibody Program." Transportation Systems 1990, AMD-Vol 108, American Society of Mechanical Engineers, 1990, pp. 51-58.

Research on high curving performance trucks
- concept and basic characteristics of
active-bogie-steering truck -

AKIRA MATSUMOTO[1], YASUHIRO SATO[1], HIROYUKI OHNO[1], TAKESHI
MIZUMA[1], YOSHIHIRO SUDA[2], YOUHEI MICHITSUJI[2,], MASUHISA
TANIMOTO[3], EIJI MIYAUCHI[3] AND YOSHI SATO[4]

SUMMARY

"Active-bogie-steering (ABS) truck", is one of the methods to realize the compatibility between curving performance and hunting stability. ABS truck has active steering mechanism only between car body and truck frame, and truck flame is steered by actuators toward "radial steering position", which realizes no "bogie angle shortage". Validity tests were carried out on the truck rolling test stand in NTSEL, which can simulate curve-running condition, and we have also carried out numerical simulation using multi-body dynamics simulation program A'GEM. The results of stand tests and numerical simulations show 1) "lateral contact force of outside wheel of leading axle" in curving is produced by "bogie angle shortage" and 2) "bogie angle shortage" is reduced proportionally by bogie-steering actuator force. So active-bogie-steering can reduce the lateral contact force of the outside wheel of leading axle in curving, and it is possible to reduce lateral contact force in sharp less than zero.

1. INTRODUCTION

Railway truck is necessary to have excellent curving performance as well as high speed hunting stability, but generally they are contrary to each other, and curving ability have been sacrificed up to now.

We have carried out theoretical studies and full scale stand tests on various technologies for high curving performance truck, such as "optimized worn tread profile", "independently rotating rear wheels", "asymmetric arrangement of longitudinal stiffness", etc. in order to realize the compatibility between curving performance and hunting stability [1][2].

In this paper we introduce the concept of "active-bogie-steering (ABS) truck" and the results of stand tests and numerical simulation for validation of basic characteristics of the system.

[1] *Address correspondence to:* National Traffic Safety & Environment Laboratory (NTSEL), 7-42-27, Jindaiji-higashimachi, Choufu, Tokyo, 182-0012 JAPAN, TEL./Fax: +81-422-41-4208; E-mail: matsumo@ntsel.go.jp, [2] The University of Tokyo, E-mail: suda@iis.u-tokyo.ac.jp, [3] Sumitomo Metal Technology, tanimoto-msh@sumitomometals.co.jp, [4] Sumitomo Metal

2. CONCEPT OF ACTIVE-BOGIE-STEERING TRUCK

2.1 BASIC MECHANISM

When railway trucks negotiate sharp curve, the rolling radius difference between inside and outside wheels cannot be obtained sufficiently and the attitude of the truck becomes so-called "insufficient steering condition" shown in Fig.1. In that condition the bogie angle of truck is not enough against "radial steering", and large flange contact force is produced in outside wheel of leading axle for the reaction of anti-steering moment of longitudinal creep forces in trailing wheels and lateral creep forces of leading wheels. This large lateral contact force of the outside wheel of the leading axle causes severe wear and sometimes derailment.

Considering these ill-suited phenomena, if the truck frame is steered by actuators toward "radial steering position of the truck", the improvement of curving performance of truck can be realized by simple methods. This is the concept of the ABS truck. The truck has the active steering mechanism only between car body and truck frame and no additional mechanism with axles.

The mechanism of the ABS truck is shown in Fig.2. In this truck the truck frame is steered by actuator toward "radial steering position" and axles are not steered[3].

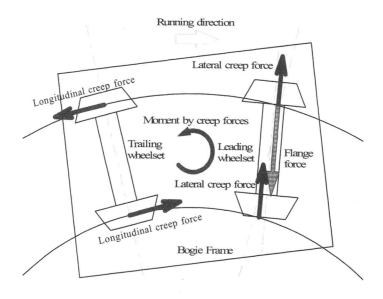

Fig. 1. Attitude of truck running through sharp curve.

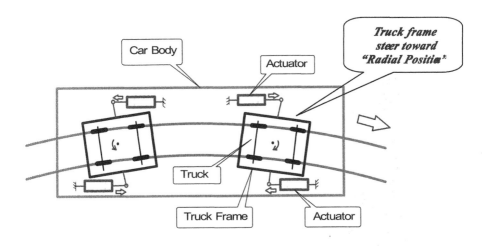

Fig. 2. Basic mechanism of "Active-Bogie-Steering" truck

3. TRUCK ROLLING STAND TEST

3.1 STAND TEST FACILITY

The stand tests are curried out on truck rolling test stand in NTSEL. The test stand was designed mainly to examine the characteristics of trucks of urban rail transits. A great feature of this facility is that, it can examine the truck under curving. Using this facility can simulate half car body condition. This facility has three main functions:
1) an angle corresponding to the rail curvature is given to the rail wheel, i.e. the angle of attack,
2) a difference in rotating speed of the opposing rail wheels (corresponding to difference in rail length of outer and inner rails) is applied,
and
3) lateral force equivalent to excess or deficiency of cant is given to the car body.

Fig.3 shows the photo of the test stand and test truck, Fig.4 shows the layout of the test stand. The present rail wheels have the same profile as 50kgN rail with a 1/40 tie-plate angle. The test truck used in this experiment is a bolster-less bogie designed for subway car. The test truck has wheelsets with worn profile wheels which are used in commercial subway lines. Fig.5 shows rolling radius difference of wheels of test truck, according to this figure its perfect rolling radius is about 350m.

In this experiment we used electrically powered actuators in order to examine basic characteristics of ABS truck. Two actuators are equipped both side of truck as shown in Fig.3.

Fig. 3. Test truck on truck rolling test stand in NTSEL

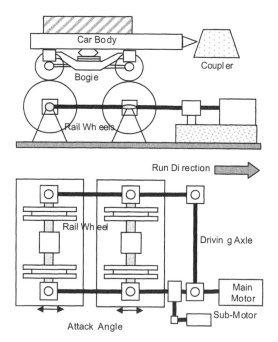

Fig. 4. Layout of truck rolling test stand

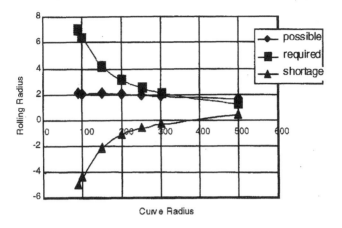

Rolling Radius Difference

Fig. 5. Rolling radius difference of test truck wheel

We examine characteristics of truck curving performance, such as bogie angle, lateral force and attack angle, when truck is steered by actuators in various curving radius.

3.2 Stand test results

The lateral contact force produced between outside rail and the outside wheel of the leading axle is one of the most important indexes of curving performance ("flange force" in Fig.1). The increase of this force causes severe wear of rail and wheel flange, and leads to derailment. So we evaluate the curving performance of the truck by this contact force.

Fig.6 shows the lateral contact force of the outside wheel of the leading axle in curving. Each line shows characteristics of each case of active steering, i.e. 0kN show the case of no active steering. The lateral force increases according to the sharpness of the curve, and in very sharp curve, such as under 200m radius, very large value is observed without bogie-steering, i.e. actuator force = 0kN. Such increase of lateral force is caused by large flange contact force shown in Fig.1, so it is important to examine the shortage of truck bogie angle. Fig.7 shows the relationship between curving radius and bogie angle shortage against "radial steering position" of the truck. According to this figure bogie angle shortage is radically increased in sharp curve

37

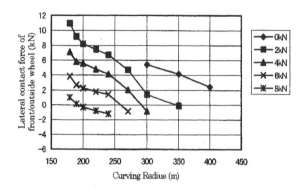

Fig. 6. Lateral contact force of leading axle in curving (stand test results)

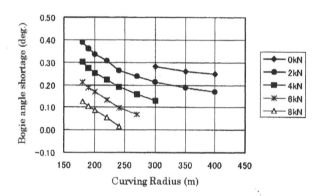

Fig. 7. Bogie angle shortage in curving (stand test results)

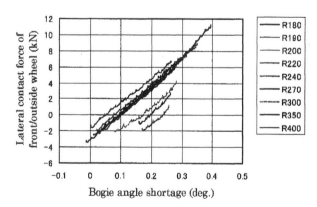

Fig. 8. Relationship between lateral force vs. bogie angle shortage (stand test results)

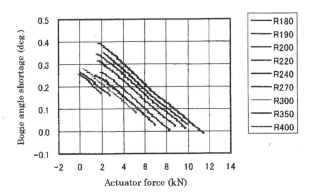

Fig. 9. Reduction of bogie angle shortage by steering actuator (stand test results)

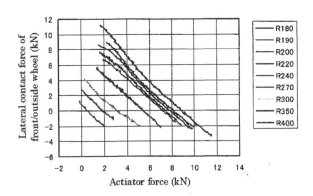

Fig. 10. Reduction of lateral contact force by steering actuator (stand test results)

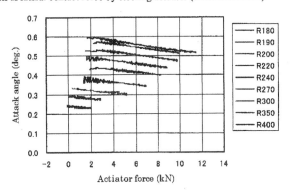

Fig. 11. Reduction of attack angle by steering actuator (stand test results)

Fig.8 shows the relationship between bogie angle shortage and lateral contact force, and this figure shows that the lateral contact force is increased in proportion to bogie angle shortage. Fig.9 shows the relationship between actuator force and bogie angle shortage, and this figure shows that bogie angle shortage is reduced according to the increase of actuator force; i.e. by active-bogie-steering.

From these figures, it is considered that if active bogie steering reduces bogie angle shortage, we can realize the reduction of lateral contact force of the outside wheel of leading axle.

Fig.10 shows that the lateral contact force of the outside wheel of the leading axle is reduced proportionally by actuator force. In this way active-bogie-steering by actuator can realize the reduction of lateral contact force in curving, and by the control of actuator force it is possible to realize "zero lateral contact force" even in sharp curve. To realize "zero lateral contact force", 3kN of actuator force is required at 300mR, 7kN at 240mR and 9kN at 180mR.

Fig.11 shows the reduction of attack angle by bogie-steering, and this figure shows that attack angle is reduced according to the increase of actuator force but that is not clear compared with the reduction of bogie angle shortage. In sharp curve, bogie steering can reduce lateral contact force of the outside wheel but attack angle can't be reduced enough in stand tests.

4.   COMPARISON BETWEEN SIMULATION AND TEST RESULTS

We also carried out numerical simulation by multi-body dynamics program A'GEM as well as stand tests. Fig.12. shows the relationship between lateral contact force of outside wheel of leading axle in various curving radius. According to this figure lateral contact force is reduced by actuator force remarkably. Simulation result agrees very well with stand test results shown in Fig.10 in all-curving radius.

Fig.13 shows the lateral contact force of the outside wheel of the leading axle in curving calculated by numerical simulation. These values also agree with the results of stand tests shown in Fig.6.

Fig.14 shows the reduction of attack angle by bogie-steering from simulation results, and this figure shows that attack angle is reduced according to the increase of actuator force. The reduction of attack angle is more remarkable in simulation results compared with stand test results shown in Fig.11. The reason of difference is now considered.

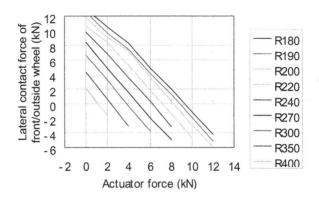

Fig. 12. Reduction of lateral contact force by steering actuator (simulation results)

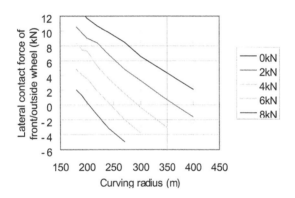

Fig. 13. Reduction of lateral contact force by steering actuator (simulation results)

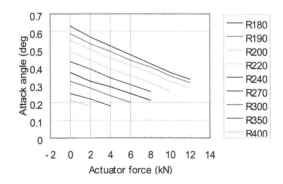

Fig. 14. Reduction of attack angle by steering actuator (simulation results)

5. CONCLUSION

According to the results of stand tests and numerical simulation mentioned above we verified the validity of "active-bogie-steering (ABS) truck". The ABS truck system can remarkably improve the curving performance, for example, reduction of the lateral contact force of the outside wheel of leading axle in the truck. It is one of the effective methods of realizing high curving performance.

For the next step, it is required to develop dynamic steering controller, and it is also important to select optimum actuator, such as electric motor, oil pressure or air pressure and so on. Moreover if ABS truck is joined another high curving performance technology, such as "optimized worn tread profile", higher curving performance can be obtained.

ACKNOWLEDGMENTS

This study is carried out by the assist of financial aid of Metro Cultural Foundation.

REFERENCES

1. Suda, Y.: Improvement of High Speed Stability and Curving Performance by Parameter Control of Trucks for Railway Vehicles Considering Independently Rotating Wheelsets and Unsymmetric Structure, JSME International Journal, Series III, 1990, 33-2, p176

2. Matsumoto A., and et. al : Compatibility of curving performance and hunting stability of railway bogie, Vehicle System Dynamics vol.33 supplement, 1999, p740-748

3. Matsumoto A., Suda Y., Tanimoto M., et. al., Research and development of active-bogie-steering truck: 1st report, JSME No.03-205, International Symposium on Speed-up and Service Technology for Railway and Maglev Systems 2003 (STECH'03), 2003, p15-18

Design and experimental implementation of an active stability system for a high speed bogie

J.T.PEARSON[1], R.M.GOODALL[1], T.X.MEI[2], S.SHEN[2], C.KOSSMANN[3], O.POLACH[3] AND G.HIMMELSTEIN[4]

SUMMARY

This paper describes an experimental active stability system for a high speed railway vehicle. Two models are developed in the study: one for control system design and one for control system performance assessment. A number of possible control schemes are described, these include single axle control schemes and modal control schemes. Results both from comprehensive simulations and for the stability controller functioning on an experimental vehicle during a testing phase are presented.

1. INTRODUCTION

A solid-axle wheelset of a conventional railway vehicle is unstable at all speeds, and the technique used to stabilise this behaviour consists of linking two wheelsets together with a stiff primary yaw suspension via the bogie frame. This stabilises the wheelset but degrades the natural curving behaviour of the wheelset, and this stability-curving trade off is the fundamental problem faced by bogie designers. At higher speeds the stiff primary suspension is insufficient to provide stability and it becomes necessary to add secondary yaw dampers from the bogie to the body. However, these dampers have an unwanted effect of transmitting high frequency vibration to the vehicle body, and this has secondary effects on the design of the vehicle body (stiffness and consequently weight). The dampers themselves are also significantly heavy and are one of the least reliable of the suspension components.

Recently there has been increasing interest in the use of active control technology within the primary suspensions of railway vehicles. This offers the potential freedom to the designer to control stability and steering independently, substantially improving the curving behaviour of the vehicle while allowing higher operational speeds.

The vehicle studied in this paper has a relatively soft primary suspension and no secondary yaw dampers. Removing the secondary yaw dampers offers significant advantages in terms of the vehicle's weight and comfort, however once removed

[1] Department of Electronic and Electrical Engineering Loughborough University, Leicestershire, LE11 3TU, UK Email: J.T.Pearson@lboro.ac.uk,
[2] School of Electronic and Electrical Engineering University of Leeds, Leeds, LS2 9JT, UK
[3] Bombardier Transportation, Winterthur, Switzerland
[4] Bombardier Transportation, Siegen, Germany

stability and consequently high-speed operation are significantly compromised, and the plan is to use active control, both to provide stability and to provide active steering in curves.

This paper focuses upon the design of the active stability control system. Control is applied by means of two electrically driven actuation mechanisms that apply independent yawing actions to each wheelset. Two a.c. servo-motors act through gearboxes, from which steering linkage mechanisms transfer the control action to the wheelsets.

The project has also investigated the integration and design of the active steering control system, which is the subject of another paper [1]. It will utilise the same hardware (actuators, sensors and controller) and will require additional software only.

2. VEHICLE SCHEME AND MATHEMATICAL MODELLING

Two mathematical models of the vehicle have been developed. The first is a relatively simple plan-view half-vehicle linear model containing 14 states. Both yaw and lateral modes have been modelled for the two axles and the bogie frame, but the lateral mode only of the body is modelled. This model is used for developing the controllers, and has been generated using SIMULINK. The second model, which is for simulation and performance predictions, is a non-linear model which incorporates particular features of railway vehicles such as:- the non-linear wheel/rail contact geometry, non-linear characteristics of some suspension components, large displacements and rotations between bodies etc. with high degrees of accuracy. This model contains in excess of 130 states and has been generated using SIMPACK. These models are shown in Figure 1.

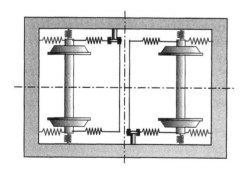 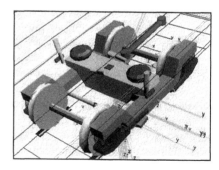

Fig. 1. Control System Design Model (Simulink) and Vehicle Simulation Model (Simpack) (Bogie only is shown.

The parameters, variables and equations for the controller design model are given in the appendix, including data for the electrical actuation components. The models were validated against each other, and were also validated using experimental data obtained from the vehicle. This included both a modal analysis of the bogie and a stability test of the passive vehicle. These models have provided the basis for a study of two different controllers which are described in the following two sections.

3. SINGLE AXLE CONTROL STRATEGIES

A control-oriented analysis of a wheelset has shown that, whereas passive damping does not assist with stability, active damping will stabilise an unconstrained solid-axle wheelset [2]. The approach used in this study is to apply a yaw torque between the bogie frame and the wheelset proportional to the lateral velocity of the wheelset (termed active yaw damping [3]). Figure 2 shows the overall scheme. Although the two axles are treated as being independent clearly there are interactions between the two axles through the bogie frame. However the relatively low primary stiffnesses means that the wheelsets are less strongly-coupled in a passive sense than they would normally be, and analysis has shown that applying the active yaw damping concept separately to each wheelset still provides good stabilisation.

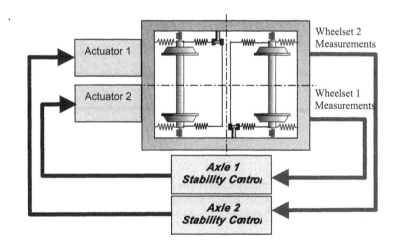

Fig. 2. Single Axle Control Scheme.

The primary design objective for the stability controllers is to provide at least 20% damping across all modes at a vehicle speed of 65m/s and with a conicity of

0.3. A secondary, but nonetheless essential, design goal is to provide at least 5% damping across all modes with a higher conicity of 0.5.

Typical values chosen for the active damping for each wheelset at vehicle speed of 65m/s (234km/h) are:- active yaw damping gain axle 1 204 kNm/ms^{-1}; active yaw damping axle 2 204 $kNm/rad\,s^{-1}$.

4. MODAL CONTROL STRATEGIES

Figure 3 shows the general form of a modal control scheme. The output measurements from the two wheelsets are decomposed to give feedback signals required by the lateral and yaw controllers respectively, and the output signals from the two controllers are then recombined to control the two actuators for the two wheelsets accordingly [4]. This enables the lateral and yaw controllers to be developed individually, at least as far as stability is concerned.

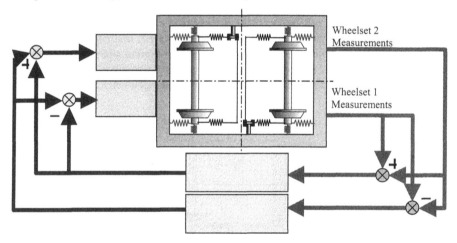

Fig. 3. Modal Control Scheme.

Analysis showed that only marginal improvements in stability could be achieved by having different gains for the lateral and yaw stability control systems. The active yaw damping gain was 260 kNm/ms^{-1}, and the active lateral damping gain was 172 $kNm/rad\,s^{-1}$.

5. SIMULATION STUDIES

Comprehensive simulation studies have been performed of the various control options. These studies have included stability tests and straight track tests using recorded track data. Both the single axle control strategies and the modal control strategies performed equally well during the simulation studies. Figure 4 shows the lateral displacement of the leading wheelset for recorded track data (typical good quality high-speed track), the vehicle speed was 235km/h and the conicity was 0.3.

Figure 5 shows simulation results for the response of the controller to a 1-cosine stability test, at 200km/h and 0.3 conicity and at 230km/h and 0.15 conicity.

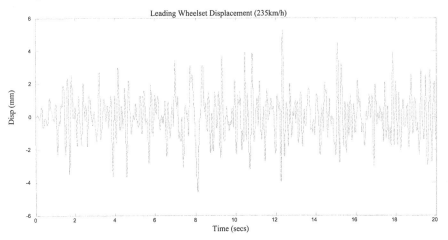

Fig. 4. Simulated Track test at 235km/h

200 km/h 0.3 Conicity   230 km/h 0.15 Conicity

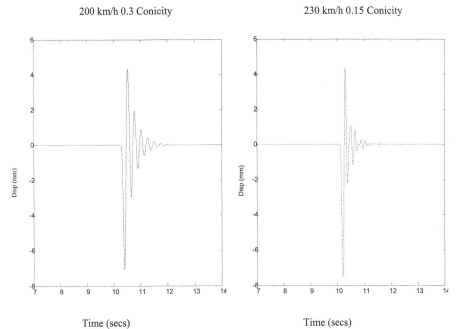

Fig. 5. Simulation Stability Test Result - Leading Wheelset Lateral Displacement

6.   EXPERIMENTAL RESULTS

Following the simulation study a period of testing on an experimental train (Figure 6) has been conducted on a full size roller rig. During this testing phase extensive stability tests and track file tests have been performed and the controller has successfully operated at speeds in excess of 300km/h.

The first test is at the limit of passive stability. Figure 7 shows results for both active control and the passive vehicle at 100km/h, and illustrates clearly the effectiveness of active stabilisation. Beyond this a passive comparison is not possible. Figure 8 shows two typical results for the stability test (a 1-cosine input onto the leading axle with an amplitude of 7.5mm) at speeds of 200km/h and 230km/h. Both results clearly show the stabilising effect of the controller (this vehicle in a passive configuration was unstable at speeds above 100km/h).

Fig. 6. Experimental Bogie.

The results shown are for the single axle control strategies only, although both the modal and single axle strategies were implemented and tested on the experimental vehicle. The performance of the modal strategies proved to be very similar to that of the single axle strategies.

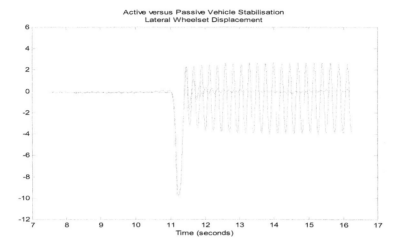

Fig. 7. Rig Stability Test Result - Active and Passive Control.

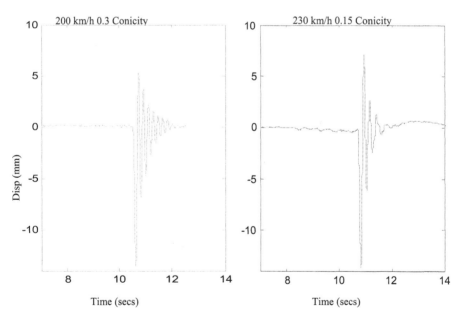

Fig. 8. Rig Stability Test Result - Leading Wheelset Lateral Displacement

7. CONCLUSIONS

This paper describes the implementation of an active stability controller for a high speed railway vehicle. It has shown through simulation and through practical implementation that the lateral velocity signals can be used for the active yaw damping controls scheme in both single axle and modal control formats.

The theoretical and experimental work form the world's first demonstration of the feasibility of a high-speed bogie with active rather than passive stabilisation. Key research challenges are to develop the work into a safe, robust and reliable system that can form the basis for the manufacture of a new generation of actively-controlled bogies, offering substantial improvements in running stability and curving performance.

ACKNOWLEDGEMENTS

The authors wish to acknowledge the support of Bombardier Transportation.

REFERENCES

1. S.Shen, T.X.Mei, R.M.Goodall, J.Pearson and G.Himmelstein: A Study of Active Steering Strategies for a Railway Bogie. IAVSD03, Kanagawa Japan, 2003.
2. Goodall, R. and Li, H., "Solid-axle and independently-rotating wheelsets - a control engineering assessment", Journal of Vehicle System Dynamics, 1999
3. Li, H. and Goodall. R.M., "Modelling and Analysis of a railway wheelset for active control", UK Control 98, Swansea, Sep. (1998)
4. Mei, T.X. and Goodall, R.M. "Wheelset Control Strategies for a 2-Axle Railway Vehicle", Journal of Vehicle System Dynamics. Vol. 33 Supplement, 2000, pp.653-664
5. Pearson, J.T., Goodall, R.M., Mei, T.X., and Himmelstein, G. "Assessment Of Active Stability Control Strategies For A High Speed Bogie" Mechatronic 2002

APPENDIX – Modelling Details

Variables

y_{w1}, y_{w2},	Lateral displacement of leading, trailing wheelset,
y_b, y_v	bogie frame and vehicle body
θ_{w1}, θ_{w2},	Yaw displacement of leading, trailing wheelset,
θ_b, θ_v	bogie frame and vehicle body
V	Vehicle travel speed (83.3 *m/s* or 300 *km/hour*)
R_1, R_2	Radius of the curved track at the leading and trailing wheelsets
θ_{c1}, θ_{c2}	Cant angle of the curved track at the leading and trailing wheelsets (typically 6^0)
y_{t1}, y_{t2}	Track lateral displacement (irregularities)
T_{w1}, T_{w2}	Controlled torque for leading and trailing wheelsets respectively

Parameters

r_0, λ	Wheel radius (0.445 *m*) and conicity (0.3) respectively
m_w, I_w	Wheelset mass (1363 *kg*)and yaw inertia (766 *kg m²*) respectively
m_b, I_b	Bogie frame mass (3447 *kg*)and yaw inertia (3200 *kg m²*) respectively
K_{long}, C_{long}	Longitudinal stiffness (957 *kN/m*) and damping per wheelset (2 *kN s/m*) respectively
K_{lat}, C_{lat}	Lateral stiffness (2.7 *MN/m*) and damping per Axle-box (2 *kN s/m*) respectively
m_v	Vehicle mass (38,600 *kg*)
K_{sc}, C_{sc}	Secondary lateral stiffness (245 *kN/m*) and damping per bogie (20 *kN s/m*) respectively
K_s	Steering linkage stiffness (20 M*N/m*)
l_g	Half gauge of wheelset (0.75m)
l_x	Semi-longitudinal spacing of wheelset (1.225m)
l_y	Semi spacing of longitudinal primary suspension (1.00m)
l_s	Semil-lateral spacing of steering linkages (1.20m)
f_1, f_2	Longitudinal and lateral creepage coefficients (10 and 10 *MN*)
g	Gravity (9.8 m/s²)
Im	Motor moment of inertia (0.00115 kgm²)
Ra	Motor armature resistance (0.112 Ohm)
La	Motor armature inductance (9.04e-4 H)
Kt	Motor torque constant (0.537 Nm/A)
Kv	Motor back emf constant (0.435 V/rad/s)
Cm	Motor-gearbox shaft Damping (0.0084 Nm/rads)
$Ig1$	Gearbox moment of inertia (motor end) (3.864e-4 kgm²)
n	Gear ratio (1/87)
Kg	Gearbox drive stiffness (1.131102e6 Nm/rad)
Cg	Gearbox drive damping (7540.7 Nms/rad)

Equations

$$m_w\,\ddot{y}_{w1} + \left(\frac{2f_{22}}{V} + 2C_{lat}\right)\dot{y}_{w1} + 2K_{lat}y_{w1} - 2f_{22}\theta_{w1} - 2C_{lat}\,\dot{y}_b - 2K_{lat}y_b \cdots$$

$$- 2C_{lat}\,l_x\,\dot{\theta}_g - 2K_{lat}\,l_x\theta_g = m_w\left(\frac{V^2}{R_1} - g\theta_{c1}\right) \tag{1}$$

$$I_w\,\ddot{\theta}_{w1} + \left(\frac{2f_{11}l_g{}^2}{V} + 2C_s\,l_s^2 + 2C_{long}\,l_y^2\right)\dot{\theta}_{w1} + \left(2K_s\,l_s^2 + 2K_{long}\,l_y^2\right)\theta_{w1}\cdots$$

$$+ \frac{2f_{11}\lambda\,l_g}{r_0}\,y_{w1} - \left(2C_s\,l_s^2 + 2C_{long}\,l_y^2\right)\dot{\theta}_b - \left(2K_s\,l_s^2 + 2K_{long}\,l_y^2\right)\theta_b = \cdots \tag{2}$$

$$\frac{2f_{11}l_g{}^2}{R_1} - \frac{2f_{11}\lambda\,l_g}{r_0}\,y_{t1} + T_{w1}$$

$$m_w\,\ddot{y}_{w2} + \left(\frac{2f_{22}}{V} + 2C_{lat}\right)\dot{y}_{w2} + 2K_{lat}y_{w2} - 2f_{22}\theta_{w2} - 2C_{lat}\,\dot{y}_b - 2K_{lat}y_b \cdots$$

$$- 2C_{lat}\,l_x\,\dot{\theta}_g - 2K_{lat}\,l_x\theta_g = m_w\left(\frac{V^2}{R_2} - g\theta_{c2}\right) \tag{3}$$

$$I_w\,\ddot{\theta}_{w2} + \left(\frac{2f_{11}l_g{}^2}{V} + 2C_s\,l_s^2 + 2C_{long}\,l_y^2\right)\dot{\theta}_{w2} + \left(2K_s\,l_s^2 + 2K_{long}\,l_y^2\right)\theta_{w2}\cdots$$

$$+ \frac{2f_{11}\lambda\,l_g}{r_0}\,y_{w2} - \left(2C_s\,l_s^2 + 2C_{long}\,l_y^2\right)\dot{\theta}_b - \left(2K_s\,l_s^2 + 2K_{long}\,l_y^2\right)\theta_b = \cdots \tag{4}$$

$$\frac{2f_{11}l_g{}^2}{R_2} - \frac{2f_{11}\lambda\,l_g}{r_0}\,y_{t2} + T_{w2}$$

$$m_b\,\ddot{y}_b + (2C_s - 4C_{lat})\dot{y}_b + (2K_s + 4K_{lat})y_b - 2C_{lat}\,\dot{y}_{w1} - 2K_{lat}y_{w1} - 2C_{lat}\,\dot{y}_{w2}\cdots$$

$$- 2K_{lat}y_{w2} - 2C_s\dot{y}_v - 2K_s y_v = m_b V^2\left(\frac{1}{2R_1} + \frac{1}{2R_2}\right) - m_b\,g\left(\frac{\theta_{c1}}{2} + \frac{\theta_{c2}}{2}\right) \tag{5}$$

$$I_b\,\ddot{\theta}_b + \left(4C_{lat}\,l_x^2 + 4C_s\,l_s^2 + 2C_{long}\,l_y^2\right)\dot{\theta}_b + \left(4K_{lat}\,l_x^2 + 4K_s\,l_s^2 + 2K_{long}\,l_y^2\right)\theta_g\cdots$$

$$- 2C_{lat}\,l_x\,\dot{y}_{w1} + 2C_{lat}\,l_x\,\dot{y}_{w2} - 2K_{lat}\,l_x y_{w1} + 2K_{lat}\,l_x y_{w2} - \cdots \tag{6}$$

$$\left(2C_s\,l_s^2 + 2C_{long}\,l_y^2\right)\dot{\theta}_{w1} - \left(2C_s\,l_s^2 + 2C_{long}\,l_y^2\right)\dot{\theta}_{w2} - \left(2K_s\,l_s^2 + 2K_{long}\,l_y^2\right)\theta_{w1}\cdots$$

$$- \left(2K_s\,l_s^2 + 2K_{long}\,l_y^2\right)\theta_{w2} = -\left(T_{w1} + T_{w2}\right)$$

$$m_v\,\ddot{y}_v + 2C_{sc}\,\dot{y}_v + 2K_{sc}y_v - 2C_{sc}\,\dot{y}_b - 2K_{sc}y_b = m_v V^2\left(\frac{1}{2R_1} + \frac{1}{2R_2}\right) - m_v\,g\left(\frac{\theta_{c1}}{2} + \frac{\theta_{c2}}{2}\right) \tag{7}$$

Vehicle System Dynamics Supplement 41 (2004), p.53-62 © Taylor & Francis Ltd.

Curving and stability optimisation of locomotive bogies using interconnected wheelsets

O. POLACH[1]

SUMMARY

The paper demonstrates the application of an equivalent axle guidance stiffness in the design of self-steering bogies, as well as the potential of the modular wheelset guidance with optional application of wheelset coupling, in order to optimise the trade-off between curving and stability of locomotive bogies. The sensitivity analysis illustrates the influence of the operating conditions on the self-steering ability.

1   INTRODUCTION

The conflict between curving and stability is a well known challenge concerning railway vehicle dynamics. Bogies with longitudinally soft axle guidance are suitable for curved track, whereas good stability performance can be achieved with stiff axle guidance. Independent of the form of the wheelset guidance and suspension design in the horizontal plane, the stability and curving performance can be described by two stiffness parameters: the shear stiffness and the bending stiffness [1, 2]. In order to achieve optimal curving properties, the bending should be low. However, the vehicle then lacks stability. To achieve the required stability the wheelsets have to be restrained by an increase in shear stiffness. For conventional bogies a limit exists to the shear stiffness that can be provided in relation to the bending stiffness. The trade-off stability/curving in which the bending stiffness must be minimised is restricted.

This limitation can be improved if the wheelsets are connected to each other directly or by a mechanism fitted on the bogie frame. An overview of design options and realised examples can be found in [3, 4]. The application of a cross anchor on the three-piece bogies and the service experience gained is described in [1, 5]. In order to design self steering interconnected wheelsets for locomotives, the transfer of tractive forces between the wheelsets and bogie frame should not influence the

---

[1] *Address correspondence to:* Bombardier Transportation, Zürcherstrasse 41, CH-8401 Winterthur, Switzerland.
Tel.: +41 52 264 1656; Fax: +41 52 264 1101; E-mail: oldrich.polach@ch.transport.bombardier.com

axle guidance. A solution with a wheelset cross coupling with a mechanism fitted on the bogie frame realised on the Locomotive 2000 series [6] was developed by the erstwhile SLM Swiss Locomotive and Machine Works in Winterthur, now a part of Bombardier Transportation. This design solution is in service in Switzerland (SBB Re 460, BLS Re 465), Norway (NSB El 18) and Finland (VR Sr2).

The paper deals with the application of this coupling mechanism in order to optimise curving and stability of newly developed locomotive bogies. A modular wheelset guidance design is presented, which enables utilisation of the bogie performance depending on the service conditions, and the sensitivity of the self-steering to the service conditions is illustrated.

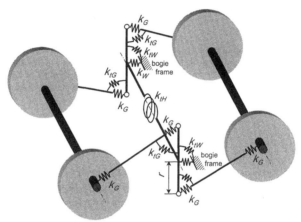

Fig. 1. Scheme of coupling mechanism between the wheelsets.

2   DESIGN ASSESSMENT USING EQUIVALENT AXLE GUIDANCE STIFFNESS

The mechanical scheme of the wheelset coupling analysed in the paper is shown in Fig. 1. The design uses frictionless rubber elements with finite stiffness. The main parameters apply, as follows:

- k_G - radial stiffness of the longitudinal linkage bushing
- k_W - radial stiffness of the bushings between the shaft and bogie frame
- k_{tG} - torsional stiffness of the longitudinal linkage bushing
- k_{tW} - torsional stiffness of the bushings between the shaft and bogie frame
- k_{tH} - torsional stiffness of the coupling shaft
- r - swing arm of coupling shaft.

The longitudinal stiffness of the linkage rod itself is very high, and the body is assumed as being rigid. The bending stiffness of the coupling shaft is not considered in our analysis due to the symmetrical fitting of the bearings (bushings).

The wheelsets, coupled by the mechanism according to Fig. 1, are able to move in the horizontal plane in four eigenmodes, see Fig. 2:

- Shear (lozenging) mode (S)
- Bending (steering) mode (B)
- Tractive (longitudinal in-phase) mode (T)
- Longitudinal anti-phase mode (L).

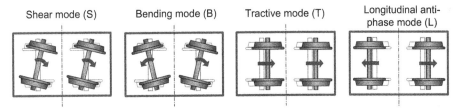

Fig. 2. Wheelset eigenmodes of a two-axle bogie.

When analysing only one eigenmode, the stiffness of the coupling between the wheelsets can be expressed as an equivalent longitudinal stiffness k_{el} between the axle box and bogie frame. Index I indicates the mode, e.g. k_{eB} for bending mode.

The total equivalent axle guidance stiffness includes the longitudinal stiffness k_{Px} and lateral stiffness k_{Py} of primary suspension. The total longitudinal equivalent axle guidance stiffness then comprises

$$k_{xI} = k_{Px} + k_{el} \tag{1}$$

with k_{Px} - longitudinal axle guidance (primary suspension) stiffness.

The equivalent stiffness can be expressed by serial and parallel combinations of the single stiffness parameters shown in Fig. 1. The equivalent stiffness for bending is

$$k_{xB} = k_{Px} + \cfrac{1}{\cfrac{2}{k_G} + \cfrac{1}{\cfrac{k_{tW}}{2 \cdot r^2} + \cfrac{k_{tG}}{r^2}}} \tag{2}$$

The same equivalent stiffness is achieved both for shear and the tractive mode

$$k_{xS} = k_{xT} = k_{Px} + \cfrac{1}{\cfrac{2}{k_G} + \cfrac{2}{k_W}} \tag{3}$$

For the longitudinal anti-phase mode we get

$$k_{xL} = k_{Px} + \cfrac{1}{\cfrac{2}{k_G} + \cfrac{1}{\cfrac{k_{tW}}{2 \cdot r^2} + \cfrac{k_{tG}}{r^2} + \cfrac{k_{tH}}{r^2}}} \qquad (4)$$

To achieve best possible curving performance paired with stable running at high speed independent of the transmission of high tractive forces between the wheelset and bogie frame, the equivalent longitudinal stiffness of the coupling mechanism must be low for bending mode, but high for other eigenmodes. The torsional stiffnesses of the coupling elements k_{tG} and k_{tW} should be as small as possible, and the radial stiffnesses k_G and k_W as large as possible. Friction bearings would fulfil these requirements best, but have proven unsuitable in respect to maintenance during operation. For this reason rubber elements, which provide the closest correspondence with the given requirements, are utilised.

From equation (4) the necessity for high torsional stiffness of the coupling shaft ensues, in order that a greater equivalent stiffness for longitudinal anti-phase mode can be achieved. This requirement must be taken into consideration in the dimensioning of the coupling shaft.

The equivalent stiffness is suitable for parameter analysis during the wheelset guidance design, e.g. see analysis of three-axle steering bogies in [7]. During the remainder of this study the bending stiffness k_{xB} will be applied in order to analyse curving and stability performance of the investigated locomotive bogie. The steering ability and curving properties are related to bending stiffness and can therefore be easily assessed using this equivalent parameter.

3   LOCOMOTIVE BOGIE WITH MODULAR AXLE GUIDANCE

With the development of a new locomotive bogie, the Flexifloat Bogie Family of Bombardier Transportation [8] has been complemented with the proven wheelset coupling mechanism with separation of axle guidance and tractive force transfer as demonstrated by the Locomotive 2000 series. The newly developed modular axle guidance system allows the construction of four axle guidance versions based on specified service conditions:

- longitudinal stiff axle guidance (ST)
- longitudinal soft axle guidance (SO)
- longitudinal very soft axle guidance combined with wheelset coupling shaft (CW)
- longitudinal very soft axle guidance combined with wheelset coupling shaft and dampers of coupling shaft (CWD).

Using the coupling shaft mechanism, the following improvements can be achieved:

- self steering ability and radial adjustment in curves, together with reduction of wheel-rail guiding force and wear

- transfer of tractive force without influencing the axle guidance parameters
- increase of the critical speed to the same range as bogies with stiff axle guidance.

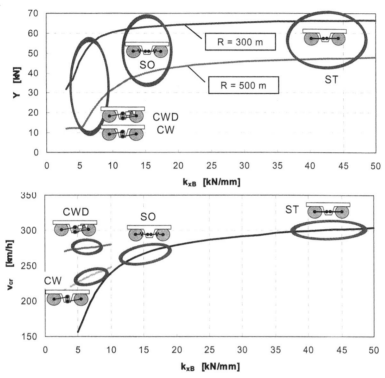

Fig. 3. Guiding force in curve (top diagram) and critical speed (bottom diagram) as function of equivalent longitudinal axle guidance stiffness. Marked areas display recommended equivalent stiffness for the proposed versions of modular axle guidance.

Based on the equivalent longitudinal axle guidance stiffness for the bending mode k_{xB}, the influence on the stability and curving performance of the locomotive with four different axle guidance versions was evaluated using the locomotive model built in to SIMPACK programme. Fig. 3 illustrates the critical speed dependent on the equivalent longitudinal stiffness k_{xB} of the axle guidance. The stability declines with lessening stiffness of the axle guidance. In the case of soft axle guidance with interconnected wheelsets and damping of the coupling shaft, stability is comparable with the stiff axle guidance. The advantage of the soft axle guidance and the coupling of the wheelsets becomes apparent during curve negotiation, see top diagram in Fig. 3. Both the guiding force of the leading wheel and the wear index decrease with sinking stiffness k_{xB}. The version with the coupling shaft mechanism (possibly supplemented with coupling shaft damping) deals excellently with the conflict in objectives between stability and curve negotiation.

4   SENSITIVITY ANALYSIS OF SELF-STEERING

As the radial adjustment of the wheelsets in curves is achieved through creep forces in the contact between wheel and rail, the running characteristic is influenced by the conditions in the wheel-rail contact. The influence of curve radius, cant deficiency, wheel-rail contact geometry, tractive effort and flange lubrication on the self-steering ability of presented locomotive bogie was analysed and a comparison was made between:

- self-steering bogie with longitudinal very soft axle guidance, coupling shaft and dampers of coupling shaft (CWD)
- conventional bogie design with longitudinal stiff axle guidance (ST).

The results presented were calculated using a full non-linear locomotive model in the simulation tool SIMPACK. In order to assess the curving performance, values on the outer wheel of the leading wheelset are presented for

- guiding force Y [kN]
- wear index A_R [Nm/m], given as the sum of products (creep force × creep) in longitudinal and lateral direction.

The **influence of curve radius and cant deficiency** presented in Fig. 4. As can be seen, with decreasing curve radius R the guiding force Y and the wear index A_R increase significantly. The self-steering of the wheelsets demonstrates a good efficiency up to a certain minimum radius. The design presented here for the standard gauge locomotive significantly reduces the guiding force in large, medium and small curve radii. In very small curve radii approx. below 250 m, the creep forces cannot overcome the reaction of the axle guidance, and the guidance force is practically the same for both versions. For self-steering bogies, wear is significantly lower throughout the whole range of the curve radii. With increasing uncompensated lateral acceleration a_{lat} the guiding force and wear index increases. However, the influence of same is significantly smaller than for the curve radius. In the results presented in the following chapters, curve radii $R = 300$ and 500 m and uncompensated lateral acceleration $a_{lat} = 0.98$ m/s^2 are applied as characteristic parameters to assess curving performance.

The **rail profile and inclination** also has an influence on the self-steering ability. Fig. 5 demonstrates the guiding forces between the wheel profile S 1002 and the rail UIC 60. In the case of rail profile S 1002, which is optimised for the rail inclination 1:40, the guiding forces are somewhat higher at a rail inclination of 1:20. However, when the versions with the stiff and very soft axle guidance are compared, the advantages of the self-steering bogie are clearly apparent.

In order to estimate the **influence of rail wear** on self-steering, calculations were carried out on rail profiles which were identified from measurements as characteristic worn rail profiles in curves [9], see Fig. 7. The heavily worn outer rail and small rolling radii difference lead to a reduction in the self-steering ability, see Fig. 6. Even though the effectiveness of the self-steering is significantly lower, the self-steering bogie is still more favourable, particularly with regard to wear.

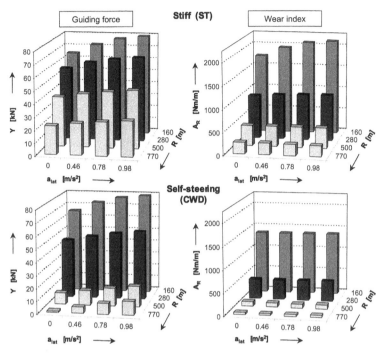

Fig. 4. Comparison of stiff (top) and self-steering (bottom) axle guidance: Guiding force and wear index of the outer leading wheel in a curve in function of lateral acceleration and curve radius.

Fig. 8 illustrates the **influence of tractive effort** on the examined values Y and A_R. With increasing tractive effort the creep forces between wheel and rail reach the saturation point. The longitudinal creep forces incurred by the varying rolling radii difference reach lower values and self-steering is reduced. Nevertheless, the self-steering bogie demonstrates a more favourable curving performance. Wear at full tractive force is mainly caused by tractive creep and is therefore hardly influenced by self-steering. As calculated, the results assume dry rails and favourable adhesion conditions. Should the run be viewed with tractive effort under critical adhesion conditions, the creep force control must be taken into consideration [2, 10], and the calculation methods concerning the creep forces extended, as described in [11, 12].

The radial adjustment of the wheelsets will be slightly reduced by the **influence of the wheel flange lubrication,** so that the guiding force achieves a higher value than without lubrication. However, the wheel flange lubrication definitely has a positive effect on wear. As can be seen in Fig. 9, the wear index demonstrates values which are approx. 5 times lower than without lubrication. It is therefore clear that the utilisation of flange lubrication on self-steering bogies can lead to a further reduction in flange wear if a slight increase in the guiding force can be accepted.

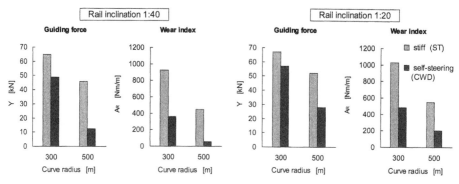

Fig. 5. Influence of rail inclination on the guiding force and wear index.

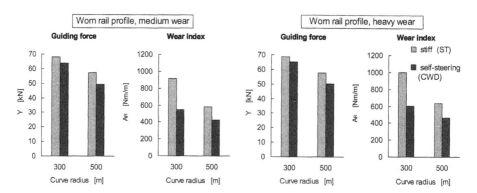

Fig. 6. Influence of worn profile, as shown in Fig. 7, on the guiding force and wear index. Results for new wheel profile – see Fig. 5 (left diagram).

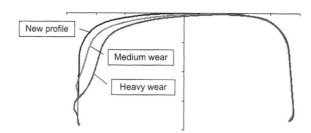

Fig. 7. Worn rail profiles of high rail in curve from [9] as used in the sensitivity analysis.

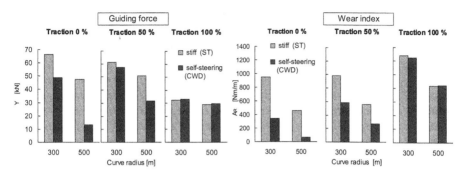

Fig. 8. Influence of tractive effort on curving performance ($a_{lat} = 1.1$ m/s^2).

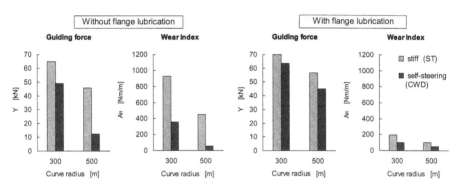

Fig. 9. Influence of flange lubrication on curving performance (wheel-rail friction coefficient: tread 0.4, flange 0.1).

The sensitivity analysis demonstrated that the radial adjustment of the self-steering wheelsets in curves is influenced by various factors. In spite of this, a self-steering bogie with interconnected wheelsets always achieves better running characteristics on curved tracks than a conventional bogie construction with stiff axle guidance. The requirements of the track infrastructure management concerning modern rail vehicles [13] confirm the necessity of radial steering wheelsets. According to Veit [14], a locomotive with coupled self-steering wheelsets only incurs approximately 60% of the annual expenses of the track maintenance in curves having radii between 250 m and 400 m in comparison to other locomotives with stiff axle guidance.

In the case of the design considered here - radial steering design by way of coupling of the wheelsets with a coupling shaft - the reduction of wheel wear has already been proven during service of the locomotives SBB Re 460 and BLS 465 on tracks with a high number of curves. On the Gotthard route in Switzerland, the locomotive SBB Re 460 achieves 3 to 4 times longer running performances between re-profiling of the wheelsets than previous locomotive versions [15].

5 CONCLUSION

The paper demonstrates the application of the equivalent axle guidance stiffness in the design of the radial steering bogies, as well as the potential of the modular axle guidance with optional application of wheelset coupling, in order to optimise the trade-off between curving and stability. The sensitivity analysis demonstrates the influence of operating conditions such as curve radius, rail inclination, rail wear, tractive force and wheel flange lubrication on the self-steering of the wheelsets. Despite a certain dependency on the relative improvement of effective operating conditions, the self-steering bogie generally achieves better characteristics and provides significant potential for savings in connection with maintenance of vehicle and infrastructure when compared with bogies with stiff axle guidance.

REFERENCES

1. Orlova, A., Boronenko, Y., Scheffel, H., Fröhling, R. and Kik, W.: Tuning von Güterwagendrehgestellen durch Radsatzkopplungen. ZEVrail Glasers Annalen 126 (2002), Tagungsband SFT Graz 2002, pp. 200-212.

2. Polach, O.: Optimierung moderner Lok-Drehgestelle durch fahrzeugdynamische Systemanalyse. EI Eisenbahningenieur, No. 7/2002, pp. 50-57.

3. Wickens, A. H.: Steering and stability of the bogie: vehicle dynamics and suspension design. Proceedings of the Institution of Mechanical Engineers, Part F, Journal of Rail and Rapid Transit, Vol. 205 (1991), pp. 109-122.

4. Scheffel, H.: Unconventional bogie designs – their practical basis and historical background. Vehicle System Dynamics, 24 (1995), pp. 497 – 524.

5. Fröhling, R.: Strategies to control wheel profile wear. Proc. of the 17th IAVSD Symposium, Lyngby, Denmark 2001, Vehicle System Dynamics Supplement 37 (2002), pp. 490-501.

6. Gerber, M., Drabek, E. and Müller, R.: Die Lokomotiven 2000 - Serie 460 - der Schweizerischen Bundesbahnen. Schweizer Eisenbahn-Revue, No. 10/1991, pp. 321-376.

7. Ahmadian, M. and Huang, W.: A qualitative analysis of the dynamics of self-steering locomotive trucks. Vehicle System Dynamics 37 (2002), No. 2, pp. 85-127.

8. Altmann, L.: The flexifloat bogie family – from heavy haul to high speed locomotives. In: Extended abstracts of the 5th International Conference on Railway Bogies and Running Gears, Budapest, 24-26 Sept. 2001, pp. 10-12.

9. Jendel, T: Prediction of wheel profile wear – comparisons with field measurements. Wear 253 (2002), pp. 89–99

10. Polach, O: Influence of locomotive tractive effort on the forces between wheel and rail. In: ICTAM, Selected papers from the 20th International Congress of Theoretical and Applied Mechanics held in Chicago, 28 August - 1 September 2000, Vehicle System Dynamics Supplement 35 (2001), pp. 7-22.

11. Polach, O.: Rad-Schiene-Modelle in der Simulation der Fahrzeug- und Antriebsdynamik. Elektrische Bahnen 99 (2001), No. 5, pp. 219-230.

12. Polach, O.: Creep forces in simulations of traction vehicles running on adhesion limit. In: Proceedings of the 6th International Conference on Contact Mechanics and Wear of Rail/Wheel Systems (CM2003) in Gothenburg, Sweden, June 10–13, 2003, Vol. II, pp. 279-285.

13. Junker, K.: Forderungen eines Netzbetreibers an moderne Schienenfahrzeuge. ZEVrail Glasers Annalen 126 (2002), Tagungsband SFT Graz 2002, pp. 144-148.

14. Veit, P.: Einfluss von Triebfahrzeugen auf das Verhalten des Oberbaus. ZEV+DET Glasers Annalen 125 (2001), No. 9-10, pp. 449-454.

15. Müller, R.: Veränderungen von Radlaufflächen im Betriebseinsatz und deren Auswirkungen auf das Fahrzeugverhalten (Teil 1). ZEV+DET Glasers Annalen 122 (1998), No. 11, pp. 675-688.

Vehicle System Dynamics Supplement 41 (2004), p.63-72   © Taylor & Francis Ltd.

Kinematic cross-linking in automotive suspension systems

R. S. SHARP[1] AND MAËLLE DODU

SUMMARY

A relatively new type of kinematically cross-linked automotive suspension system is described and mathematical models are built to predict the behaviour as a function of the geometric design. Three planar model variants are involved. The first allows simulation of the mainly vertical response to load changes, representative of accelerating or braking in straight line motion. The second covers quasi-steady state cornering at constant speed. The third deals with the standard laboratory roll centre test, in which a pure rolling torque is applied to the stationary car body. Mention is also made of a full car handling model incorporating the suspension arrangements but details are not included.

Graphical simulation results and animations of the system are used to demonstrate the behaviour. A systematic approach to choosing the geometry is taken and results are included to illustrate the convergence of the design to a state with very advantageous properties. Small body roll angle, excellent road wheel camber angle control and modest negative jacking with some slight reductions in lateral load transfer, as compared with convention, are apparently possible. The most critical design area is revealed.

1. INTRODUCTION

Automotive suspension systems typically classified as independent often have some cross-dependency through anti-roll bars or torsion beams but an entirely more influential cross-coupling is possible through kinematic cross-linking. A kinematically cross-linked system was employed some years ago on the Fairthorpe Electron car, under the name "Torrix-Bennett Linkage" [1] but it was somewhat limited in its design possibilities. A more versatile arrangement was invented by Walker [2, 3] and is the subject of this paper. The novel arrangement has been used on a successful circuit racing car, the DAX, Fig. 1. As far as the authors are aware, no analysis of the system has been published previously.

Basically, the kinematic cross-linking allows the camber angle of the right hand side road wheel to be influenced by the up / down movements of the left wheel and conversely. This introduces interesting and potentially useful behavioural possibilities. It also implies a need for new analysis and design procedures.

With conventional independent suspension systems, the road wheel has just one suspension degree of freedom relative to the chassis and the constraint system determining the wheel carrier (knuckle) trajectory will invariably be chosen to compromise between the requirements of straight line manoeuvring (acceleration and braking) and cornering. Given that the suspension is sufficiently compliant to deflect significantly in manoeuvring, the symmetric body motion implied by

---

[1] *Address correspondence to*: Department of Electrical and Electronic Engineering, Imperial College London, London SW7 2AZ. Tel.: int+44 207 6272; Fax: int+44 207 6282; E-mail: Robin.Sharp@Imperial.ac.uk

longitudinal accelerations has different implications for the road wheels as compared with those accompanying lateral accelerations. In particular, the camber angle of the road wheel relative to the ground will be affected differently.

Fig. 1 Competition car realisation of kinematically cross-linked suspension (http://www.daxcars.co.uk/start.htm).

Best shear force performance from a pneumatic tyre is obtained when the contact pressure between tread rubber and road is uniform. Tyre tread temperature and wear will also tend to be uniform under such conditions. Control of the road wheel camber angles, as a four wheeled vehicle manoeuvres, is therefore important, especially with racing and sports cars, on which the tyres are likely to be wide and therefore relatively camber sensitive.

Confining the discussion to the lateral properties, pure vertical translation of the road wheels relative to the chassis is ideal for braking and driving, while instantaneous centres on the ground and near to the car centre line (for a conventional system) are, to some extent, ideal for cornering. In this latter case, cambering of the road wheels relative to the chassis will compensate for body roll in cornering, to keep the wheels normal to the road, while no significant jacking will occur [4, 5]. With a conventional suspension system, these two ideals are incompatible.

A diagrammatic representation of the DAX cross-linked suspension is shown in Fig. 2. In the figure, p15, p16, p19 and p20 are chassis points, while p23 and p24 are attached to an extension of the upper wishbone on each side. The upper wishbones are joined to swinging links at p13 and p14 and the angles of the swinging links are controlled by the cross-links from p23 to p22 and from p24 to p21. Spring / damper units act between p11 and p9 on the left side and between p12 and p10 on the right side. Three similar mathematical models of this single end, planar car will be constructed and used to examine the behaviour of the system under different test conditions. A full car model with similar cross-linked suspension at each end has also been built [6] but it will not be described in any detail. In the next section, the basic model building will be described. Then some design issues will be considered. In section 4, results of simulations and animations will be given and discussed and in section 5, conclusions will be drawn.

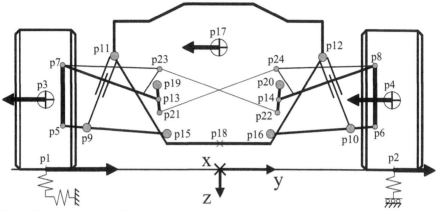

Fig. 2 Diagrammatic cross-linked suspension system in a right turn, showing inertia forces and tyre force reactions in equilibrium – view from rear.

2. SYSTEM MODELLING

Multibody system simulation models in AutoSim™ have been constructed, as illustrated in Fig. 2. AutoSim [7, 8] is written in LISP and is a multibody system modelling suite. Firstly, there is a mechanical system description language. The composer uses commands like (add-body), (add-line-force), (no-movement), (add-out) etc. to create a set of parent / child relationships for the bodies, with a full specification of the freedoms allowed, constraints to remove freedoms already added, the forces and moments acting, the outputs required etc. Loading the complete AutoSim program leads to a "C" simulation file being written. Linking of this file to library files, which are part of the AutoSim suite, enables the creation of an executable simulation program, the output from which can be plotted using WinEP and animated using "Animator"[2].

To build the planar model of the system shown in Fig. 2, the main body (main) is added to the (given) inertial body, N, with lateral and vertical translation and roll freedoms. The various points connected to main are defined and the massless suspension links are added as follows. For the left side, the lower arm is added with x-rotational freedom about its joint with main, p15, and points p5 and p9 are defined

---

[2] WinEP and Animator are freeware packages from the University of Michigan Transportation Research Institute. Copies can be obtained from the author through e-mail.

on it. The wheel is attached to the lower arm at p5 with x-rotational freedom and points p1 and p7 are defined on it. The swinging link is joined to main at p19 with the same x-rotational freedom and points p13 and p21 and similarly, the upper arm is joined to the swinging link at p13, to contain p7 and p23. The longest chain of "children" of main has two bodies, which is the minimum number possible for this system. Constraints are then defined to oblige the upper arm and wheel points at p7 in the nominal configuration to move together in both [ny] and [nz] directions. Best practice is to add the constraints and to thereby allow AutoSim to remove freedoms in the reverse order of the building process. This is a "last in, first out" principle. Using that principle eases considerably AutoSim's task of symbolically eliminating redundant coordinates. The corresponding commands for the right side are then included.

For the cross-linking, a massless link, again with x-rotational freedom, is added to the left swinging link at p21, the point p24 is defined on it and relative movement between this point and the corresponding point on the right upper arm is disallowed. The surplus freedoms are again removed in the reverse order to their additions. Mirror image commands deal with the other cross-link.

Suspension struts, an anti-roll bar, tyre forces, excitation force and outputs are added to complete the model. Three variants of the model are employed for simulating the body bounce response, the quasi-equilibrium cornering response and the standard laboratory roll centre experiment, in which a pure rolling moment is applied to the stationary car chassis. In the first two cases, the wheels are notionally rolling along and they give no position constraint to the motion, while in the third case, they are stationary and they behave like lateral and vertical springs to ground.

For the bounce response, the tyres are represented as the left side one in Fig. 2 but with only a very small lateral tyre stiffness. For the cornering response, the description of Fig. 2 applies directly, with physically representative parameter values, while for the roll centre test, both tyres are modelled as having realistic vertical and lateral stiffnesses as shown on the left side in Fig. 2. Suspension jacking in cornering [4, 5] depends on the sharing of the lateral tyre forces between left and right wheels. In reality, this depends on the tyre characteristics, the load transfer, the steering system design etc. In the cornering model, it is assumed that the lateral forces are shared in proportion to the loads on the tyres. This is an accurate representation when travelling straight ahead, when symmetry demands equal treatment of left and right tyres, and at the rollover limit, when only the outside tyre will give any sideforce. Between these limits, the model is imperfect but reasonable.

3. DESIGN CONSIDERATIONS

The main design objective is to achieve very small road wheel camber angle changes, relative to the road surface, in both longitudinal and lateral manoeuvring. It is regarded as evident that success in this objective will imply that the wheels will camber little relative to the road under any combination of longitudinal and lateral acceleration (latacc). Longitudinal manoeuvring is represented with the planar model by loading and unloading the main body, which situation, being symmetric, is easier to visualise than any involving anti-symmetry. We therefore start with a geometric design which appears to allow the body to move up and down from the nominal state, without the road wheels moving much, either laterally or in camber. This is obtainable only with a near horizontal lower control arm.

To the extent that the upper wishbone inclines upwards towards the outside, the cross-linking needs to ensure that the swinging link moves outwards in bump. Balancing the wishbone inclination and the cross-linking will allow the orbit of the points p7 and p8 to be near vertical in double wheel bump. This basic thinking yields a nominal design, shown in Fig. 3, including notional drive shaft joint locations. Variants of this nominal design will be studied subsequently.

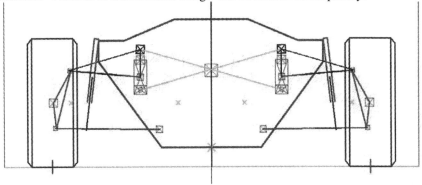

Fig. 3 Nominal system design to yield almost pure wheel vertical translation in double bump.

4 RESULTS

Firstly, it is established that the datum design yields very small camber and track change responses to the body bouncing as a result of a sinusoidal vertical force being applied to the mass centre of main. This implies excellent behaviour in longitudinal manoeuvring, especially in the suspension bump (jounce) direction, this being the more important, since it corresponds with high tyre loading. The results are shown in Fig. 4, confirming that the body translates vertically through a large range with little influence on the wheel camber angles. Fig. 5 shows the corresponding attitude changes in cornering at up to 12m/s/s latacc. The behaviour is not particularly unusual, with a conventional level of body roll and very slight negative jacking. The tyres make excellent contact with the ground through the full range of lateral accelerations.

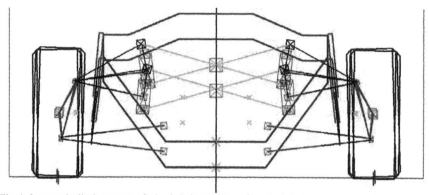

Fig. 4 Symmetric displacements of wheels in bounce test of nominal design.

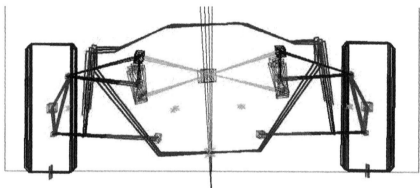

Fig. 5 System configuration in cornering of the nominal design at up to 12m/s/s latacc. Right hand turn; view from the rear.

It is anticipated that the suspension link angles will be much more influential on the behaviour than the lengths, as is the case with conventional systems, so design variations discussed involve only vertical movements of the points, p15 and p16, p23 and p24 and p13 and p14, see Fig. 2. Firstly, the inclination angle of the lower track control arm (lower wishbone) is varied by moving p15 and p16 up or down. These variations show behaviour that would be expected for a conventional suspension, in which raising the joints reduces the roll angle in steady cornering, while causing the body to jack up, Figs 6 and 7. Of these alternative designs, that with $z15 = -0.165$m (1cm up from the original) appears to have a slight advantage over the nominal one. In this case, the mass centre falls a little (negative jacking) as the latacc builds up from zero to 12m/s/s. Subsequently, this design will be regarded as the datum.

Similar systematic variations in the cross-link offset distances are next examined, through moving p23 and p24 upwards. Surprisingly, both a reduction in body roll and an exaggerated lowering of the body mass centre (negative jacking) result but the details are not shown, partly due to lack of space and partly because the designs are surpassed by those to be considered next.

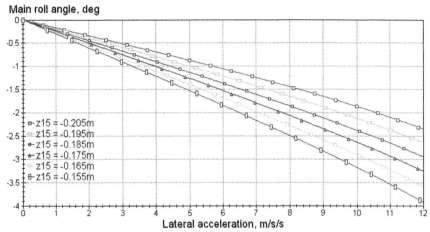

Fig. 6 Quasi-steady right turn simulation showing body roll angle for nominal design and 5 variations on the lower wishbone inclination angle. These negative roll angles are the conventional responses.

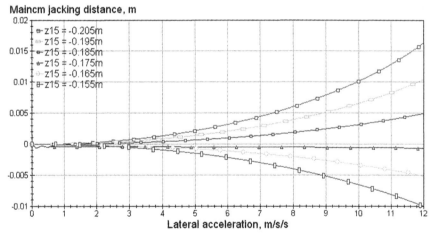

Fig. 7 Quasi-steady turning simulation showing body mass centre jacking for nominal design and 5 variations on the lower wishbone inclination angle.

Keeping the same datum as in the above trials, p13 and p14 are lowered. The region around $z13 = -0.405$m is especially interesting. As p13 and p14 are lowered, body roll is reduced and jacking down increased as the latacc builds but, if the lowering is overdone, the system jacks down quite dramatically at high cornering effort, Figs 8 and 9. Concentrating on the arrangement with $z13 = -0.406$m, as having particularly good properties, it gives a maximum roll angle of 1.2 degrees at

12m/s/s latacc and modest negative jacking. Frozen animations are shown in Fig. 10, where it can be seen how little the body attitude changes and how well the wheel camber angles are controlled. Corresponding to the negative jacking, the tyre load transfer is also contained, the spring forces are little altered from the nominal state and the full suspension travel remains available for dealing with road roughness inputs.

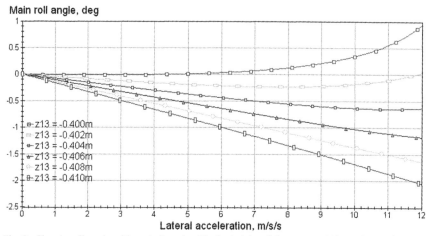

Fig. 8 Chassis roll angle with variations in z13, the height of the upper wishbone inner pivot; quasi-steady turning at up to 12m/s/s.

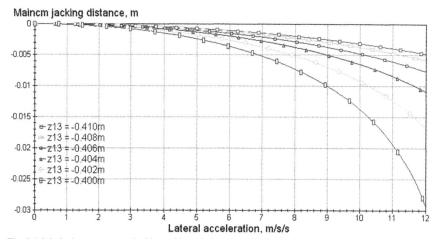

Fig. 9 Main body mass centre jacking with variations in z13; quasi-steady turning at up to 12m/s/s.

Exercising this suspension vertically in double bump confirms that it behaves very much like the original nominal system shown in Fig. 3 in giving very little track or wheel camber changes. If we subject the system to the standard laboratory roll centre test, with the wheels stationary by frictional interaction with the ground and a pure rolling torque applied to the body, the motion is as shown in Fig. 11. The fixed point, which conventionally would be considered the roll centre, is surprisingly high. A conventional system giving this result in the roll centre test would yield an unacceptable amount of jacking up in cornering, quite unlike the kinematically cross-coupled system under study.

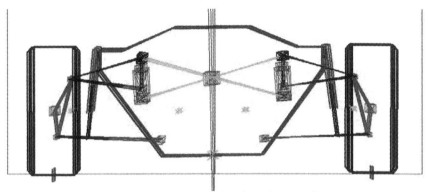

Fig. 10 Quasi-steady turning attitude of cornering car as latacc increases linearly from zero to 12m/s/s. Configuration with z13 = -0.406m.

The full car ride / handling model mentioned earlier has the cross-coupled suspension fitted at each end, is slightly front heavy and has front springs somewhat stiffer than those at the rear. It has a detailed manual steering system and a constant forward speed constraint. It also contains a sophisticated tyre treatment by a Magic Formula based method [8] and has been used to simulate ramp / step steering wheel input responses. It shows essentially the same behaviour as the planar models [6] and reinforces the idea that the simple models properly predict the running behaviour of a real car. This work is to be written up separately.

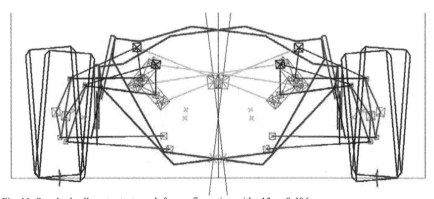

Fig. 11 Standard roll centre test result for configuration with z13 = -0.406m.

5 CONCLUSIONS

This simulation study of the DAX kinematically cross-coupled suspension has involved a systematic approach to the design of such systems and has revealed counter-intuitive behaviour which appears to be practically very advantageous. The additional design freedom which comes with the cross-linking can be used to largely decouple the symmetric, bounce design from the anti-symmetric, roll design. It becomes possible, through the cross-linking, to inhibit body roll while not causing jacking, by geometric design choices. The study has converged on a design with conventional stiffness, which yields very little attitude change in cornering and excellent wheel camber control, while leaving almost all the suspension travel for dealing with road bumps. Corresponding properties cannot be obtained by a usual geometric arrangement. In the most advantageous design region, the behaviour is rather sensitive to certain geometric changes. It is implied that robustness against load variations and reasonably precise manufacturing are needed.

The most critical design region is that involving the height of the upper wishbone inner pivots. In a certain parameter locality, small changes in height, governed by the parameter $z13$, cause large changes in behaviour. The best choice of $z13$ depends on the choices made in respect of other key parameters in the design, notably $z15$ and $z23$. These relate to the height of the lower wishbone inner pivot and the height of the upper wishbone cross-link extension. A full car ride / handling model has shown confirmatory behaviour in responding to ramp step steering control inputs at constant speed but it has not been possible to include the detailed evidence in the paper. Other publications will provide such evidence at a later date.

REFERENCES

1. *(http://www.Fairthorpescc.co.uk).*
2. *Patent: GB2358619: Vehicle Suspension System, Peter John Walker, 01/08/2001 (http://gb.espacenet.com/).*
3. *McBeath, S.and Bamber, J., Camber nectar, Racecar Engineering, 11(6), June 2001, 32-36.*
4. *R. S. Sharp, "Fundamentals of the lateral dynamics of road vehicles", Mechanics for a New Millenium (H. Aref and J. W. Phillips eds), Kluwer Academic Publishers, Dordrecht, 2001, 127-146.*
5. *Sharp, R. S. and Bettella, M., Automotive suspension and chassis interactions, La Dynamique du Véhicule Automobile et Ferroviaire, Proc. SIA International Congress keynote lecture, 6-7 June, 2001, Lyon, 11-21.*
6. *Dodu, M., Analysis, design and performance assessment of the DAX cross-linked suspension system, Cranfield University School of Engineering M. Sc. Thesis, September 2002.*
7. *Anon., AutoSim Reference Manual 2.5+, Mechanical Simulation Corporation, 709 West Huron, Ann Arbor MI (1998) (http://www.trucksim.com).*
8. *Mousseau, C. W., Sayers, M. W. and Fagan, D. J., Symbolic quasi-static and dynamic analyses of complex automobile models, The Dynamics of Vehicles on Roads and on Tracks (G.Sauvage ed.), Proc. 12th IAVSD Symposium, Swets and Zeitlinger, Lisse, 1992, 446-459.*
9. *Sharp, R. S. and Bettella, M., Tyre shear force and moment descriptions by normalisation of parameters and the "Magic Formula", Vehicle System Dynamics, **39**, 2003, 27-56.*

Analysis of Large Vehicle Dynamics for Improvin Roll Stability

SHUICHI TAKANO[1], MASAHIRO SUZUKI[1], MASAO NAGAI[2],
TETSUO TANIGUCHI[3] AND TADASHI HATANO[3]

SUMMARY

Commercial vehicles such as trucks and buses are prone to roll over because of their high centres-of-gravity. There have been static analyses of roll stability of these large vehicles in the past, but not many researches concentrated on dynamic characteristics of their roll motions. In this study, a truck with and without load is used for experiments to determine its roll behaviours and its relationship with the plane motions such as the yaw rate and lateral acceleration in realistic driving conditions. Also, simple vehicle models are applied to simulate these motions without time-consuming model identification processes. It is found that inclusion of the roll motion in the model is necessary but not sufficient for the model to be accurate.

1 INTRODUCTION

In today's society, large-sized vehicles such as trucks and buses play important roles in both passenger and freight transportation. In actual driving, however, the dynamic characteristics of these vehicles are often affected by the amount of the load they carry [1]. Trucks especially carry large and varying amounts of load, which result in their higher centers-of-gravity and therefore greater roll motions compared to passenger cars [2,3]. In the past, there has been static analyses of large vehicles' roll motions [4,5] but not many studies on their dynamic behaviours in realistic load and driving conditions. Many accidents involving large vehicles are caused by rollover from excessive roll motion; these accidents could cause great damages to other motorists and pedestrians [6]. Understanding the dynamics of these vehicles is, therefore, highly important. The first purpose of this study is to examine dynamic characteristics of plane and roll motions of a truck and the effect the load condition has on these ele-

[1]*Address correspondence to*: Division of Mechanical Systems Engineering, The Graduate School of Technology, Tokyo University of Agriculture and Technology, 2-24-16 Naka-cho, Koganei-shi, Tokyo 184-8588, Japan. Tel.: +81-42-388-7090; Fax: +81-42-385-7204; E-mail: y1833201@gc.tuat.ac.jp
[2]Department of Mechanical Systems Engineering, Tokyo University of Agriculture and Technology, 2-24-16 Naka-cho, Koganei-shi, Tokyo 184-8588, Japan
[3]National Traffic Safety and Environment Laboratory, 7-42-27 Jindaijihigashi, Chofu, Tokyo 182-0012, Japan

ments using an experiment vehicle.

The second purpose is to develop a simple mathematical model for describing dynamics of a truck. Experiments involving actual trucks and buses can be prohibitive in terms of cost, time and safety. The models used in past researches, however, have been highly complex, often with over twenty degrees-of-freedom and optimised for advanced simulation software [7,8]. With such models, the process of model identification becomes time-consuming since there are many parameters to be identified. Also, it is difficult to determine the interrelations between each output with a complicated model because there are too many elements to consider. Therefore, vehicle models with two or three degrees-of-freedom are considered in this study.

2   EXPERIMENT VEHICLE

For an experiment vehicle, a 4-ton truck as depicted in Figure 1 is used. To recreate changes in the truck's center-of-gravity height induced by the load condition, a metal frame is installed on the floor of its cargo bay. Weights with total mass of 3,000 kilograms are fixed on the frame to simulate the fully loaded condition; they are removed to represent an empty vehicle. The weights are placed so that they would not affect the stability factor of the vehicle greatly. All experiments are conducted for these two load conditions. Specifications of the vehicle are described in Table 1.

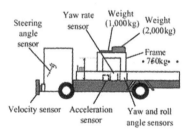

Fig. 1. Experiment vehicle.

3   EXPERIMENTS

To determine the static roll angle of a truck for given lateral acceleration, tilting platform and steady-state turn experiments are performed. In the tilting platform experiment, the vehicle is placed on a platform, and steady lateral acceleration is created by tilting the platform. The resulting lateral acceleration is $g\sin\theta$, θ being the tilt angle. The tilt angle θ is increased by small increments, and the corresponding roll angle is recorded. In the steady-state turn experiment, the lateral acceleration is generated by driving the experiment vehicle at a constant radius of 70.8 metres. The vehicle velocity is increased by small increments, and the lateral acceleration and roll angle are recorded. The above two experiments are carried out for the identical purpose, but greater lateral acceleration can be created with the platform because there is

Table 1. Vehicle parameters and model variables.

	DEFINITION	SYMBOL	UNIT	VALUE W/O LOAD	VALUE W/LOAD
Known parameters	Vehicle mass	m	kg	4660	7660
	Sprung mass	m_s	kg	4660	7660
	Wheelbase	l	m	3.73	3.73
	Front axle tread	d_f	m	1795	1795
	Rear axle tread	d_r	m	1610	1610
	Roll center height	R_c	m	0.4023	0.4023
	Vehicle width	a	m	2.240	2.240
	Vehicle length	b	m	6.820	6.820
	Distance from top of cab to bottom of chassis	c	m	2.023	2.023
	Steering gear ratio	N	-	16.6	16.6
	Gravitational acceleration	g	m/s^2	9.81	9.81
Calculated or identified parameters	Roll moment arm	h_s	m	0.4813	0.9499
	Yaw moment of inertia	I	kg m^2	16001	26364
	Roll moment of inertia	I_ϕ	kg m^2	5333	12854
	Distance from front axle to c.g.	l_f	m	1.716	2.056
	Distance from rear axle to c.g.	l_r	m	2.014	1.674
	Distance from c.g. to top of cab	c_1	m	1.518	0.999
	Distance from c.g. to bottom of chassis	c_2	m	0.505	1.024
	Linear cornering stiffness per front tire	k_f	N/rad	26530	30200
	Linear cornering stiffness per rear tire	k_r	N/rad	39500	60000
	Total roll stiffness	K_ϕ	N m/rad	409675	333356
	Front roll stiffness	$K_{\phi f}$	N m/rad	68279	55559
	Rear roll stiffness	$K_{\phi r}$	N m/rad	341395	277797
	Roll damping coefficient	C_ϕ	N m s/rad	20000	20000
	Constants for nonlinear tire models	P_1	-	-0.7908×10^{-4}	-0.7908×10^{-4}
		P_2	-	3.142	3.142
		P_3	-	-2.532×10^{-4}	-2.532×10^{-4}
		P_4	-	5.519	5.519
Variables	Lateral force per front (rear) tire	$Y_f (Y_r)$	N	-	-
	Load on front (rear) tire	$W_f (W_r)$	N	-	-
	Nonlinear cornering stiffness per front (rear) tire	$k_f' (k_r')$	N/rad	-	-
	Vehicle velocity	V	m/s	-	-
	Front steer angle	δ	deg	-	-
	Side-slip angle	β	deg	-	-
	Yaw rate	r	deg/s	-	-
	Roll angle	ϕ	deg	-	-
	Lateral acceleration	a_{acc}	m/s^2	-	-

no risk of driver injury.

To clarify the relationship between various motions of the vehicle in dynamic condition, a lane change experiment at constant velocity is conducted. The steering wheel angle, yaw rate, roll angle, and lateral acceleration are measured while the experiment vehicle performs a lane change of 3.6 metres while travelling 50 metres. The vehicle speed is set at either 60, 80, or 100 km/h for a vehicle without load, and 60, 80, or 90 km/h for a vehicle with load. Note that the highest speed is lower for the loaded condition because of safety reasons. The manoeuvre is performed three times per condition. An example of the data is shown in Figure 2. For comprehensive comparison of data, the first peak of various variables in this experiment is termed Peak A, and the second peak Peak B as described in Figure 2. The amplitude of these peaks and their output delay with respect to the steering input are examined.

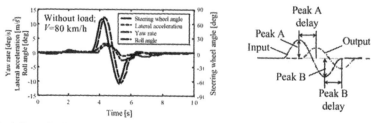

Fig. 3. Example of lane-change maneuver data and definitions of peaks.

Lastly, a random steering input experiment is conducted in order to determine the vehicle's various frequency responses to the steering input. While travelling at constant speed of 80, 100, or 120 km/h, random steering input is entered to the steering wheel while recording the steering wheel angle, yaw rate, roll rate, and lateral acceleration. The frequency responses to the steering angle are obtained by Fourier analysis. Three to four trials per condition are performed, with the results averaged for accuracy.

4 EXPERIMENTAL RESULTS

4.1 Static and Dynamic Roll Angles

Figure 4 displays the roll angle of the experiment vehicle against its lateral acceleration in the tilting platform, steady-state turn, and lane change experiments. The values of Peaks A and B are used for the lane change experiment. In the graph, the vehicle with load displays the roll angle several times larger than the vehicle without load. Also, the data for the vehicle without load largely fall on the same line regardless of whether the experiment is static or dynamic. For the vehicle with load, on the other hand, the data shows a considerably greater slope for the lane change experiment compared to the tilting platform and steady-state turn experiments. This suggests that the driver would experience a larger roll angle than expected in steady-state when

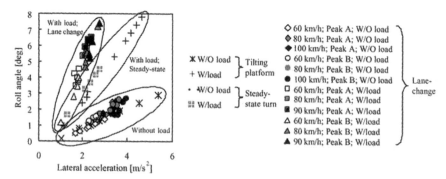

Fig. 4. Roll angle of experiment vehicle for given lateral acceleration.

performing a steering manoeuvre such as a lane change on highway.

4.2 Relationship Between Peak Values in a Lane Change

Figure 5 shows the values of Peaks A and B of the measured output with respect to the steering input. The horizontal axis represents the peak values of the steering wheel angle. The vertical axis marks the peak values of the yaw rate for Figure 5(a), lateral acceleration for Figure 5(b), and roll angle for Figure 5(c), respectively. The peak values for the yaw rate largely fall on the same line regardless of the load condition, indicating that the yaw rate is not significantly affected by the load. Similarly, the data for the lateral acceleration also falls on one line, with slightly pronounced scattering of data for Peak B in the loaded condition. Lastly, the roll angle of the loaded vehicle is three to four times larger than that of the unloaded vehicle, as expected from the previous section. It is also notable that the roll angle data for Peak B in the loaded condition shows some scattering similar to the lateral acceleration.

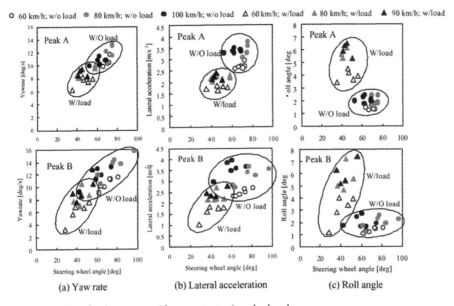

(a) Yaw rate   (b) Lateral acceleration   (c) Roll angle

Fig. 5. Peak value of various output with respect to steering wheel angle.

4.3 Delay with Respect to the Input

Figure 6 shows the delay between a peak of the steering wheel input and a corresponding peak of various output as defined in Figure 3. The yaw rate shows only a small delay with respect to the input regardless of the load condition. On the other hand, both the lateral acceleration and roll angle lag considerably behind the steering angle when loaded, and this delay increases acutely with the vehicle speed. This suggests that roll angle and lateral acceleration would be generated much later than the steering input and that delay depends on the load condition and vehicle velocity.

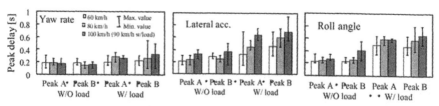

Fig. 6. Delays of various variables with respect to steering wheel angle.

4.4 Frequency Responses

Figure 7 displays the frequency responses of the yaw rate, lateral acceleration, and roll rate, calculated from the results of the random steering input experiment. The load condition does not seem to affect the yaw rate responses. However, the amount of load acutely influences the roll rate, increasing both the gain and phase delay at a wide frequency range. The lateral acceleration is affected by the load as well although to a lesser degree. Its gain decreases with the load, and the phase delay decreases at lower frequencies but increases at higher frequencies.

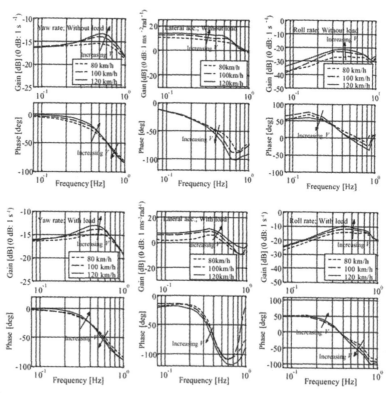

Fig. 7. Frequency response of various experimental output with respect to steering wheel angle (averaged).

5   VEHICLE MODELS

In this section, several mathematical models are considered for simulating the motions of the experiment vehicle.

5.1 Two-DOF "Bicycle" Model

The 2-DOF "bicycle" model is a basic linear model often used to describe dynamics of passenger cars. Figure 8(a) describes this model, and its equations of motion are as below.

Lateral motion:	$mV(\dot\beta + r) = 2Y_f + 2Y_r$	(1)
Yaw motion:	$I\dot r = 2l_f Y_f - 2l_r Y_r$	(2)
Tyre models:	$Y_f = -k_f(\beta + l_f r / V - \delta)$	(3)
	$Y_r = -k_r(\beta - l_r r / V)$	(4)

In Equations (3) and (4), the term in parentheses refers to the side slip angle of the front and rear tyres, respectively. Therefore, the cornering force produced by a tyre in this model is assumed to be linear with respect to its side slip angle.

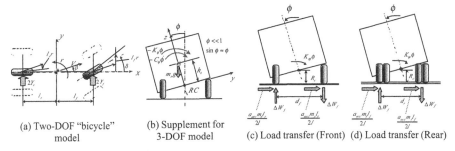

(a) Two-DOF "bicycle" model	(b) Supplement for 3-DOF model	(c) Load transfer (Front) (d) Load transfer (Rear)

Fig. 8. Details of vehicle models.

5.2 Linear 3-DOF Model

To consider the large roll motion of the loaded vehicle, the roll angle θ of the sprung mass as illustrated in Figure 8(b) is incorporated to the 2-DOF model. The equations of motion for the resulting 3-DOF model are as below.

Lateral motion:	$mV(\dot\beta + r) - m_s h_s \ddot\phi = 2Y_f + 2Y_r$	(5)
Yaw motion:	$I\dot r = 2l_f Y_f - 2l_r Y_r$	(6)
Roll motion:	$I_\phi \ddot\phi - m_s h_s V(\dot\beta + r) = (-K_\phi + m_s gh_s)\phi - C_\phi\dot\phi$	(7)

Note that Equation (6) is identical to Equation (2). Equations (3) and (4) are used for the tyre models. Elements such as roll steer and camber thrust are ignored for simplification.

5.3 Nonlinear 3-DOF Model

With real vehicles, the cornering stiffness of a tyre often varies in a nonlinear manner with respect to its vertical load, and the vertical load changes during cornering because of load transfer between the left and right wheels as shown in Figure 8(c) and (d). Since this is very likely to occur with commercial vehicles, a tyre model whose cornering stiffness increases in a nonlinear manner with the vertical load is added to the 3-DOF model mentioned above. By taking moments that acts around the roll center of the sprung and unsprung mass, load transfer can be expressed by the following equations.

$$\Delta W_f = \frac{a_{acc} m_s}{d_f} \left[\frac{h_s}{1 + K_{\phi r}/K_{\phi f} - m_s g h_s/K_{\phi f}} + \frac{l_r}{l} R_c \right] \tag{8}$$

$$\Delta W_r = \frac{a_{acc} m_s}{d_r} \left[\frac{h_s}{1 + K_{\phi f}/K_{\phi r} - m_s g h_s/K_{\phi r}} + \frac{l_f}{l} R_c \right] \tag{9}$$

For the nonlinear tyre model in this study, the relationship between the wheel load and cornering stiffness is expressed by the second degree equations below.

$$k_f' = P_1 W_f^2 + P_2 W_f \tag{10}$$
$$k_r' = P_3 W_r^2 + P_4 W_r \tag{11}$$

where the constants P_1 to P_4 are values to be determined. By taking these equations, total front cornering stiffness, consisting of nonlinear left and right front tyres, can be described by the next equation.

$$2k_f' = P_1(\frac{l_r}{2l} m_s + \Delta W_f)^2 + P_2(\frac{l_r}{2l} m_s + \Delta W_f) + P_1(\frac{l_r}{2l} m_s - \Delta W_f)^2 + P_2(\frac{l_r}{2l} m_s - \Delta W_f) \tag{12}$$

Since the experiment vehicle has double rear tyres, the total rear cornering stiffness is

$$2k_r' = 2 \left[P_3(\frac{l_f}{4l} m_s + \frac{\Delta W_r}{2})^2 + P_4(\frac{l_f}{4l} m_s + \frac{\Delta W_r}{2}) + P_3(\frac{l_f}{4l} m_s - \frac{\Delta W_r}{2})^2 + P_4(\frac{l_f}{4l} m_s - \frac{\Delta W_r}{2}) \right] \tag{13}$$

5.4 Model Identification

To identify the yaw moment of inertia I, the unsprung mass is assumed to be concentrated at the front and rear axles as m_f and m_r, which are determined by measuring the wheel load, and the value of I is calculated as $I = m_f l_f^2 + m_r l_r^2$. The roll moment of inertia I_ϕ is computed by assuming that the sprung mass including the frame is a rectangular parallelepiped with uniform density and the weights a concentrated mass. The roll moment arm h_s is calculated from the manufacturer's specifications and measured dimensions and locations of the frame and mass. The roll stiffness K_ϕ can be derived by matching the following relationship with the results of the tilting platform experiment by varying K_ϕ.

$$K_\phi \phi = m_s h_s (a_{acc} \cos\phi + g\sin\phi) \tag{14}$$

The ratio of the rear roll stiffness $K_{\phi r}$ to front roll stiffness $K_{\phi f}$ is set at five in this study according to typical specifications of trucks in the class. Since the value of roll damping coefficient C_ϕ can be hard to estimate, a value is chosen so that the roll angle would be damped reasonably quickly without making simulation unstable in lane change simulation under highway speed.

The values k_f and k_r for in Equations (3) and (4) are calculated from the stability factor K of the vehicle.

$$K = (V/G_0 N - 1)/V^2 \tag{15}$$

where N is the steering gear ratio of the vehicle, V the vehicle speed, and G_0 the steady-state gain of the yaw rate with respect to the steering wheel angle from Figure 7. Once K is known, the value of either k_f or k_r can be varied in accordance with another definition for K, which is

$$K = -\frac{m}{2l^2}\frac{l_f k_f - l_r k_r}{k_f k_r} \qquad (16)$$

until simulated yaw rate frequency response match the experimental results in Figure
    The cornering stiffness k_f' and k_r' in Equations (10) and (11) can be estimated from the values of k_f and k_r just calculated. Since these equations are second-degree and they both go through the origin, two values of cornering stiffness and corresponding values of vertical loads are necessary in order to estimate them. The cornering stiffness computed for two load conditions provide these two points. Identified parameters for the models are listed in Table 1.

6   EVALUATING THE MODELS

In this section, the models are evaluated by comparing the results of the simulation with the experimental results of the lane-change experiment. Figure 9 shows comparison of simulation and experiment for the lane change manoeuvre at the lowest and highest travel velocities in Chapter 3. For an unloaded vehicle, all the models can simulate the vehicle's plane motions accurately. This implies that the 2-DOF "bicycle" model is sufficient for simulating the unloaded experiment vehicle. For a loaded vehicle, the output of the 2-DOF model deviates considerably from the actual vehicle regardless of the vehicle velocity, especially in lateral acceleration. The plane motions of the linear 3-DOF model matches the experimental values at low velocity, but they also deviates from experimental values at higher speed. The nonlinear 3-DOF model recreates the plane motions of the real vehicle most accurately, with only slight devia-

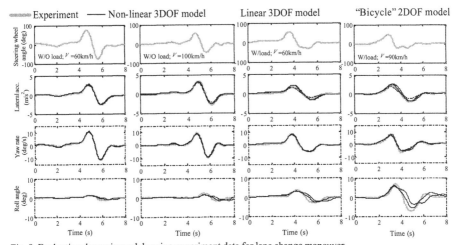

Fig. 9. Evaluating dynamic models using experiment data for lane change maneuver.

tion in yaw rate. It is notable that the roll angle of both 3-DOF models is smaller the experimental value, and its peaks lags behind those of the real vehicle. This may be due to the estimated value of roll damping coefficient C_ϕ used; a more accurate method of determining it may be necessary.

These results imply that inclusion of the roll angle in the model is necessary but not sufficient in order for it to be a reliable representation of a loaded truck at higher speed; the nonlinearity of the tyres due to the load should be considered.

7   CONCLUSIONS

In this study, the following points were clarified for a truck with two different load conditions and its simple mathematical models.

* The vehicle going through a steering manoeuvre could exhibit more pronounced roll motion than that in steady-state condition.
* Larger roll motion induced by the cargo does not greatly affect the magnitude of the vehicle's yaw rate or lateral acceleration in a lane change.
* In a lane change, the roll angle and lateral acceleration lags behind the steering input when the vehicle is fully loaded and travels at high velocity.
* A 2-DOF "bicycle" model becomes inadequate as a mathematical model of the vehicle as the velocity and the load amount increases.
* Inclusion of the roll angle can improve the 2-DOF model at lower speed, but considering the nonlinear characteristics of the tyres is necessary at high speed.
* The assumptions used to simplify the model identification process appear to have no effect on the accuracy of the models' plane motions. However, a method for accurately estimating the roll damping coefficient may be necessary in future.

REFERENCES

1. Nishida, T., Yoshizaki, Y. and Nakabayashi, T.: Influence of road surface on maneuverability and stability of trucks. *Journal of Society of Automotive Engineers of Japan.* Vol. 46, No. 3, 1992, pp.35-41.
2. Nakagawa, K., Nishida, T. and Ohsaki, M.: Influence of roll characteristics on controllability and stability. *Journal of Society of Automotive Engineers of Japan.* Vol. 43, No. 3, 1989, pp.68-73.
3. Higuchi, M., Taneda, K, Shibahata, Y. and Minakawa, M.: A theoretical study of influence of roll on vehicle dynamics. *Proc. 2001 JSAE Annual Congress,* Sapporo, Japan, October 2001. JSAE, 2001, No. 107-01. pp.15-18.
4. Ervin, R.D.: The influence of size and weight variables on the roll stability of heavy duty trucks. *SAE Paper.* 831163, 1983.
5. Winkler, C.B., Karamihas, S.M. and Bogard, S.E.: Roll-stability performance of heavy-vehicle suspensions. *SAE Paper.* 922426, 1992.
6. National Highway Traffic Safety Administration: *2001 Annual Assessment: Motor Vehicle Traffic Crash Fatality and Injury Estimates for 2001,* 2001.
7. Eisele, D.D. and Peng, H.: Vehicle dynamics control with rollover prevention for articulated heavy trucks. *Proc. 5th International Symposium on Advanced Vehicle Control (AVEC 2000).* Hiroshima, Japan, September 2000, pp. 123-30.
8. Ledesma, R.: The effect of tire stiffness parameters on medium-duty truck handling. *SAE Transactions.* Vol. 109, Section 6, 2000, pp. 2168-2177.

An analysis of pitch and bounce motion, requiring high performance of ride comfort

IKUO KUSHIRO[1], EIICH YASUDA[2] AND SHUNICHI DOI[2]

SUMMARY

This paper shows the optimum frequency ratios of the front and rear ends of the car to minimize pitch motion, which can be calculated with the mean square value of pitch. The human sensitivity to pitch and bounce was investigated through sensory evaluation by using a 6-degree-of-freedom simulator. The vehicle design parameters of a two-degree-of-freedom model describe the pitch and bounce motion, and determine the mean square values of pitch and bounce. The mean square values of pitch velocity and bounce velocity with 1/f random input estimate the vehicle frequency and damping settings. There are the optimum front and rear ratios of the frequency and the damping ratio to minimize the mean square value of pitch, but there is no optimum setting to minimize the mean square value of bounce. Higher front and rear ratios of the frequency and the damping ratio are required to achieve minimizing of the mean square value of pitch. The task of passing over a bump takes into account the process of convergence with consideration for human sensitivity. Achieving a high performance of ride comfort consist of, for the most part, finding an optimum front and rear ratio of frequency to minimize the mean square value of pitch. A vehicle with a higher base frequency, and higher base damping ratio or larger dynamic index, requires a higher front and rear ratio of frequency to reduce the mean square value of pitch. The mean square values enable a discussion of the influence of vehicle parameters.

1. INTRODUCTION

Pitch motion is more annoying than bounce motion for passengers. The reduction of pitch motion is advisable for ride comfort. We have empirical knowledge that the rear suspension should have a higher spring rate than the front one in order to reduce pitch motion. Generally, the optimum frequency ratio of the front and rear ends of the car is determined experimentally. Experimental determination prevents sufficient analytical discussion. Achieving high performance of ride comfort requires an evaluation of pitch and bounce, with consideration for human sensitivity. This paper investigates the optimum frequency and damping ratios of the front and rear ends of the car to reduce pitch motion, which can be determined by calculation with the mean square values of pitch and bounce. The influence of the vehicle design parameters on the optimum frequency setting to reduce pitch is also investigated.

[1] *Address correspondence to:* Toyota Motor Corp. Higashifuji Technical Center , 1200, Mishuku, Susono Shizuoka, JAPAN. Tel.: +81-(55) 997-7659; Fax: +81-(55) 997-7872;• ikuo@kushiro.tec.toyota.co.jp
[2] Toyota Central R&D Labs, 41-1 Yokomichi, Nagakute, Nagakute-cho, Aichi-gun, Aichi, JAPAN

2. HUMAN SENSITIVITY TO PITCH AND BOUNCE

The authors investigated sensory evaluation, such as which vehicle motion feels like a smooth ride for passengers, with a 6-degree-of-freedom simulator on which passengers rode. Fig.1 shows representative examples of vehicle motion when passing over a 1 cycle bump at 100km/h. Pitch and bounce are drawn with 80 times the amplitude for easy understanding of the vehicle motion. Case1 felt uncomfortable for passengers due to the large pitch motion, but some passengers also felt the movement was natural. Case2 felt more comfortable for passengers than Case1, but passengers felt some discomfort due to vibrations in the process of convergence.

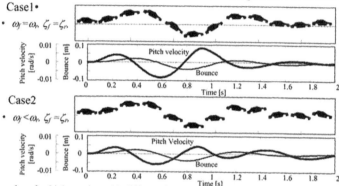

Fig. 1. Examples of vehicle motion with different frequency settings.

On the basis of sensory evaluation, a human discomfort curve was investigated with the complex motions of pitch and bounce under the conditions of an excitation frequency of 1.0, 1.4 Hz, with phases of pitch and bounce as $\pi/2$ radian and various amplitudes. Fig. 2 shows the human discomfort curve, assuming that pitch amplitude multiplied by a wheelbase of 2.78 m is equal to bounce amplitude. Various phases of pitch and bounce excitation were also investigated at the points of a, b, c, and d.

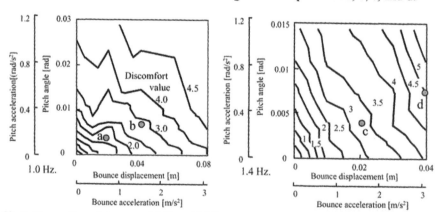

Fig. 2. Human discomfort curve at 1.0 Hz and 1.4 Hz.

The discomfort curve with pitch and bounce shows that pitch motion is about 1.5 times more annoying than bounce motion under a low frequency at 1 Hz. Pitch sensitivity increases at smaller amplitudes. Pitch sensitivity decreases from 1/2 to 1/3 against bounce sensitivity at 1.4 Hz. Humans have a comfort phase of pitch and bounce, and the comfort phase of $\pi/2$, as shown in Fig. 3, feels natural for humans. The direction of initial response of pitch and bounce when passing over a bump is a plus direction in this paper as shown in Fig. 4. Error bars are deviations of each panelist. The vehicle frequency setting would be considered the human sensitivity to pitch and bounce.

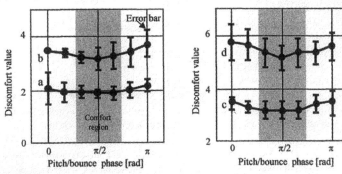

Fig.3. Human discomfort phase of pitch and bounce.

3   VEHICLE MODEL

The vehicle model in Fig. 4 is two-degree-of-freedom system with pitch and bounce, and unstrung mass is ignored in the low frequency response. Rear input is the same as the front one with a time lag of τ that is determined by the wheelbase and vehicle traveling speed.

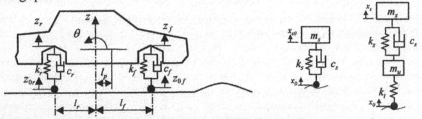

Fig. 4. Vehicle model, quarter model, and quarter model with unsprung mass.

Equations of motion become

$$m\ddot{z}+k_f\left(z_f-z_{0f}\right)+c_f\left(\dot{z}_f-\dot{z}_{0f}\right)+k_r\left(z_r-z_{0r}\right)+c_r\left(\dot{z}_r-\dot{z}_{0r}\right)=0 \tag{1}$$

$$I_y\ddot{\theta}+k_f\left(z_f-z_{0f}\right)l_f+c_f\left(\dot{z}_f-\dot{z}_{0f}\right)l_f-k_r\left(z_r-z_{0r}\right)l_r-c_r\left(\dot{z}_r-\dot{z}_{0r}\right)l_r=0 \tag{2}$$

where $z_f=z+l_f\,\theta,\ z_r=z-l_r\theta.$
The Laplace transformed rear input can be expressed in equation (3).

$$z_{0r} = z_{0f}\,e^{-\tau s} \tag{3}$$

Pitch and bounce transfer functions are expressed in equations (4), (5), and (6) with the front frequency ω_f, the rear frequency ω_r, the front damping ratio ζ_f, the rear damping ratio ζ_r, the wheelbase l, the dynamic index I_{yN}, the front distribution of weight D_{wf}, and the time lag τ under zero initial condition. These transfer functions are also frequency responses of pitch and bounce velocity with step input. Where, $\tau = l/v$, $\omega_f^2 = k_f/m/D_{wf}$, $2\zeta_f\omega_f = c_f/m/D_{wf}$, $\omega_r^2 = k_r/m/(1-D_{wf})$, $2\zeta_r\omega_r = k_r/m/(1-D_{wf})$, $l_f = l(1-D_{wf})$, $l_r = l\,D_{wf}$, $I_y = I_{yN}\,m\,l^2\,D_{wf}(1-D_{wf})$.

The front frequency and the rear frequency are not natural frequencies in this case.

$$\frac{\theta(s)}{z_{0f}(s)} = P_{trn}(s) = \frac{1}{l\,I_{yN}\,Den}\Big[\big(2\zeta_f\omega_f s + \omega_f^2\big)\big(s^2 + 2\zeta_r\omega_r s + \omega_r^2\big)$$

$$- \big(2\zeta_r\omega_r s + \omega_r^2\big)\big(s^2 + 2\zeta_f\omega_f s + \omega_f^2\big)e^{-\tau s}\Big] \tag{4}$$

$$\frac{z(s)}{z_{0f}(s)} = B_{trn}(s) = \frac{1}{Den}\left[D_{wf}\big(2\zeta_f\omega_f s + \omega_f^2\big)\left(s^2 + 2\frac{\zeta_r\omega_r}{I_{yN}}s + \frac{\omega_r^2}{I_{yN}}\right)\right.$$

$$\left. + \big(1 - D_{wf}\big)\big(2\zeta_r\omega_r s + \omega_r^2\big)\left(s^2 + 2\frac{\zeta_f\omega_f}{I_{yN}}s + \frac{\omega_f^2}{I_{yN}}\right)e^{-\tau s}\right] \tag{5}$$

$$Den = s^4 + \frac{2}{I_{yN}}\big[\{D_{wf}I_{yN} + (1-D_{wf})\}\zeta_f\omega_f + \{D_{wf} + I_{yN}(1-D_{wf})\}\zeta_r\omega_r\big]s^3$$

$$+ \frac{1}{I_{yN}}\big[\{D_{wf}I_{yN} + (1-D_{wf})\}\omega_f^2 + \{D_{wf} + I_{yN}(1-D_{wf})\}\omega_r^2 + 4\zeta_f\omega_f\zeta_r\omega_r\big]s^2$$

$$+ \frac{2}{I_{yN}}\big(\omega_f^2\zeta_r\omega_r + \omega_r^2\zeta_f\omega_f\big)s + \frac{1}{I_{yN}}\omega_f^2\omega_r^2 \tag{6}$$

The Laplace transformed time lag is approximated by the first-order Pade polynomial. The pitch and bounce transfer functions are expressed as the polynomial s function.

$$e^{-\tau s} \approx \frac{2 - \tau s}{2 + \tau s} \tag{7}$$

$$P_{trn}(s) = \frac{P_n(s)}{P_d(s)} = \frac{P_0 s^4 + P_1 s^3 + P_2 s^2 + P_3 s + P_4}{a_0 s^5 + a_1 s^4 + a_2 s^3 + a_3 s^2 + a_4 s + a_5} \tag{8}$$

$$B_{trn}(s) = \frac{B_n(s)}{B_d(s)} = \frac{B_0 s^4 + B_1 s^3 + B_2 s^2 + B_3 s + B_4}{a_0 s^5 + a_1 s^4 + a_2 s^3 + a_3 s^2 + a_4 s + a_5} \tag{9}$$

We can calculate the mean square values with the transfer functions to evaluate pitch and bounce velocity under the condition of step input. The pitch mean square value is

$$P_{msv} = \int_{-\infty}^{\infty} |P_{trn}(j\omega)|^2\,d\omega = \int_{-\infty}^{\infty}\frac{P_n(j\omega)P_n(-j\omega)}{P_d(j\omega)P_d(-j\omega)}\,d\omega = -(-1)^s\frac{\pi\cdot\det P_N}{a_0\det P_D} \tag{10}$$

where

$$
\det P_N = \begin{vmatrix} b_0 & b_1 & b_2 & b_3 & b_4 \\ a_0 & a_2 & a_4 & 0 & 0 \\ 0 & a_1 & a_3 & a_5 & 0 \\ 0 & a_0 & a_2 & a_4 & 0 \\ 0 & 0 & a_1 & a_3 & a_5 \end{vmatrix}
\qquad
\det P_D = \begin{vmatrix} a_1 & a_3 & a_5 & 0 & 0 \\ a_0 & a_2 & a_4 & 0 & 0 \\ 0 & a_1 & a_3 & a_5 & 0 \\ 0 & a_0 & a_2 & a_4 & 0 \\ 0 & 0 & a_1 & a_3 & a_5 \end{vmatrix}
$$

$$
P_n(s)P_n(-s) = \left(P_0 s^4 + P_1 s^3 + P_2 s^2 + P_3 s + P_4\right)\left(P_0(-s)^4 + P_1(-s)^3 + P_2(-s)^2 + P_3(-s) + P_4\right)
$$
$$
= b_0 s^8 + b_1 s^6 + b_2 s^4 + b_3 s^2 + b_4
$$

In the same manner, the bounce mean square value B_{msv} is calculated. Evaluating the velocities is appropriate at low frequency with this simple two-degree-of-freedom model in consideration of human sensitivity to pitch and bounce, because pitch sensitivity decreases against bounce at higher frequencies.

The transfer functions of quarter models without unsprung mass and with unsprung mass are equation (11) and equation (12) respectively in Fig. 4, where, $\omega_s^2 = k_s/m_s$, $2\zeta_s \omega_s = c_s/m_s$, $\omega_u^2 = k_t/m_u$

$$
\frac{x_{s0}}{x_0} = \frac{\left(2\zeta_s \omega_s s + \omega_s^2\right)}{\left(s^2 + 2\zeta_s \omega_s s + \omega_s^2\right)} \tag{11}
$$

$$
\frac{x_s}{x_0} = \frac{\left(2\zeta_s \omega_s s + \omega_s^2\right)}{\left(s^2 + 2\zeta_{ds}\omega_{ds}s + \omega_{ds}^2\right)} \frac{\omega_u^2}{\left(s^2 + 2\zeta_{du}\omega_{du}s + \omega_{du}^2\right)} \tag{12}
$$

The unsprung term can be ignored in equation (12) at low frequency. Evaluating equation (11) is equal to evaluating equation (12) in consideration of $\omega_{ds} < \omega_s$, $\zeta_{ds} < \zeta_s$. Because unsprung term makes ω_s and ζ_s reduce. In the same manner, evaluation of the two-degree-of-freedom mode is sufficient.

4. THE MEAN SQUARE VALUES OF PITCH AND BOUNCE

The first evaluation is the mean square values of pitch and bounce assuming road input is 1/f random. The pitch mean square value can determine the optimum front and rear ratios of both the frequency and the damping ratio minimizes the pitch motion, under the conditions of a total spring rate and a total damping which are constant. That is, $k_f + k_r$=constant, and $c_f + c_r$=constant. Other conditions are I_{yN}=0.8, l=2.7, D_{wf}=0.55, v=100km/h.

Fig. 5 shows that there are an optimum front and rear ratios of the frequency and damping ratio to minimize the pitch mean square value. The rear suspension should have a higher frequency and damping ratio than the front in order to reduce the pitch mean square value. There is no optimum front and rear ratios of the frequency and damping ratio to minimize the bounce mean square value. The base frequency ω_0, base damping rate ζ_0, are as follows

$$\omega_0 = \sqrt{\frac{k_f + k_r}{m}} = 2\pi \cdot 1.313 \qquad \zeta_0 = \frac{c_f + c_r}{2\sqrt{(k_f + k_r)m}} = 0.3$$

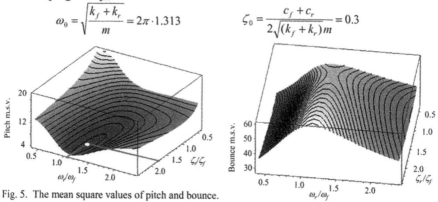

Fig. 5. The mean square values of pitch and bounce.

The vehicle travelling lower speed requires a much higher frequency and damping ratio of the rear than front to reduce pitch motion as shown in Fig. 6.

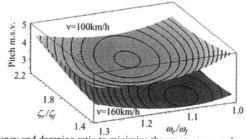

Fig. 6. Optimum frequency and damping ratio to minimize the mean square value of pitch.

5. OPTIMUM FREQUENCY SETTING

The second evaluation is the mean square values with consideration for convergence time assuming bump input. The conditions to reduce the mean square value of pitch are $\zeta_f < \zeta_r$ and $\omega_f < \omega_r$. This case leads to the condition $\zeta_f \omega_f \cdot \zeta_r \omega_r$ and discomfort remains due to vibration with larger pitch and the discomfort phase of pitch and bounce in the process of convergence. Therefore the task of passing over a bump requires the use of another function $R_{\zeta\omega} = \zeta_r \omega_r /(\zeta_f \omega_f) = \tau_f/\tau_r$. Fig. 7 shows a contour plot of the mean square value of pitch in Fig. 5 with the curve of $R_{\zeta\omega}$=constant, and it is sufficient to evaluate the mean square value of pitch with only $R_\omega = \omega_r/\omega_f$. The condition $\zeta_f = \zeta_r$ is adopted to evaluate pitch and bounce motion for simplicity and make R_ω smaller. The frequency or damping ratio to minimize the mean square value of pitch is calculated in Fig. 8. The optimum frequency ratio of the front and rear to minimize the mean square value of pitch is the condition that approximately maximizes the mean square value of bounce. On the other hand, the damping ratio of the front and rear to minimize the pitch mean square value is not the condition that maximizes the bounce mean square value , where $\omega_0 = 2\pi 1.313$, $\zeta_0 = 0.3$, $I_{yN} = 0.8$, $l = 2.7$, $D_{wf} = 0.55$, $v = 100$km/h.

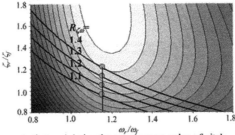

Fig. 7. Optimum frequency ratio to minimize the mean square value of pitch.

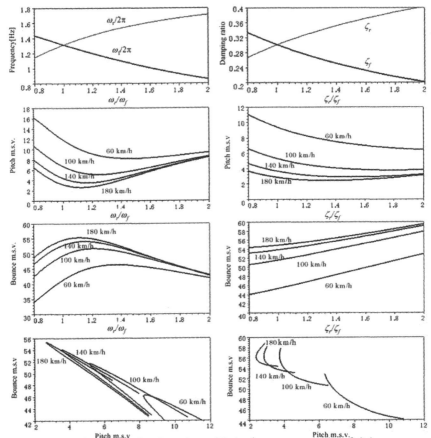

Fig. 8. Optimum frequency or damping ratio to minimize the mean square value of pitch.

The front and rear ratio of frequency to minimize the mean square value of pitch is equivalent to the evaluation of both pitch and bounce in Fig. 9 with consideration for human sensitivity, that is, the pitch is 1.5 times more annoying than bounce. A high performance of ride comfort will be archived to minimize the pitch mean square value.

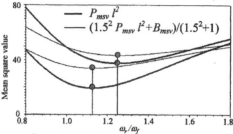

Fig. 9. Optimum frequency ratio to minimize the mean square value of pitch.

The frequency setting to minimize the mean square value of pitch changes according to various conditions and specifications of the vehicle in Fig. 10. Base parameters are $\omega_0 = 2\pi\ 1.313$, $\zeta_0 = 0.3$, $I_{yN} = 0.8$, $l = 2.7$, $D_{wf} = 0.55$, $v = 100$ km/h. A higher base frequency ω_0, higher base damping ratio ζ_0, and longer wheelbase require a higher rear frequency to reduce pitch motion. The wheelbase is treated only with a time lag term in a bounce transfer function. A lower dynamic index needs a rear frequency to reduce pitch motion, but an excessive lower dynamic index will be avoided, because it creates excessive pitch.

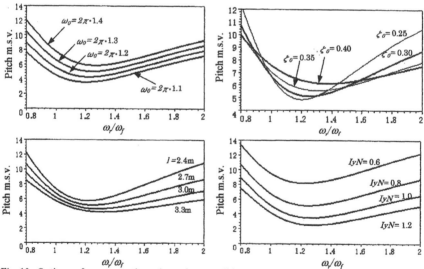

Fig. 10. Optimum frequency ratio under various conditions.

The front distribution of weight D_{wf} only lightly affects the optimum frequency setting to minimize the mean square value of pitch on its own

The front distribution of weight D_{wf} only lightly affects the optimum frequency setting to minimize the mean square value of pitch on its own

The influences of the front distribution of weight D_{wf} and the dynamic index I_{yN} are mutually related. The approximated denominator of the transfer functions becomes equation (13) under the condition that $I_{yN} \approx 1$, where $C_1 = D_{wf} + I_{yN}(1-D_{wf})$, $C_2 = D_{wf} I_{yN} + (1-D_{wf})$. The pitch response is mainly dominated equation (14),

$$Den \approx \left(s^2 + 2\frac{\zeta_f \omega_f}{C_1} s + \frac{\omega_f^2}{C_1} \right)\left(s^2 + 2\frac{\zeta_r \omega_r}{C_2} s + \frac{\omega_r^2}{C_2} \right) \tag{13}$$

$$P_{trn}(s) \approx \frac{1}{I_{yN} Den}\left[\frac{\left(2\zeta_f \omega_f s + \omega_f^2\right)}{\left(s^2 + 2\frac{\zeta_f \omega_f}{C_1} s + \frac{\omega_f^2}{C_1}\right)} - \frac{\left(2\zeta_r \omega_r s + \omega_r^2\right)}{\left(s^2 + 2\frac{\zeta_r \omega_r}{C_2} s + \frac{\omega_r^2}{C_2}\right)} e^{-\tau s} \right] \tag{14}$$

The conditions of $D_{wf} > 0.5$ and $I_{yN} < 1.0$ yield $C_1 > C_2$, then a smaller R_ω is required to reduce the pitch mean square value.

The dynamic index affects the phase of bounce and pitch, which can affect the comfortable ride for a passenger on a seat in another position. The bounce transfer function calculated by equation (15) determines the mean square value of bounce at the point l_p in Fig. 4.

$$\frac{z_{lp}(s)}{z_{0f}(s)} = B_{lp}(s) = B_{trn}(s) + P_{trn}(s) \cdot l_p \tag{15}$$

A smaller dynamic index not only leads to a larger mean square value of pitch, but also to a larger mean square value of bounce at the point of the rear seat as shown in Fig.11, because of the pitch and bounce phase. The small dynamic index makes the pitch mean square value excessive larger, therefore only evaluation of the pitch mean square value is required to obtain the optimum front and rear ratio of frequency in case of $I_{yN} < 1.0$.

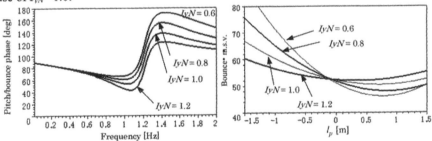

Fig. 11. Phase of pitch and bounce and the mean square value of bounce at the position of lp.

6. EXPERIMENTAL RESULT

A vehicle improved by using evaluation of the mean square values

Fig. 12 shows the experimental result. Mean square values are given by equation (16) The improved vehicle achieved better ride reducing the pitch mean square•value by over 20%, and keeping the bounce mean square value almost equal to the original.

$$P_{ms} = \frac{1}{T} \int_0^T P^2(t)\,dt \qquad (16)$$

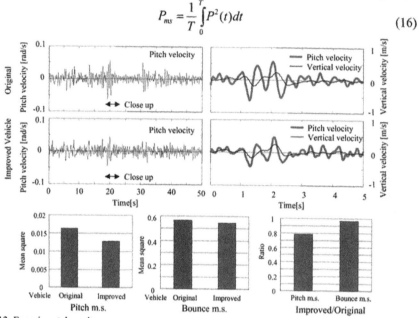

Fig. 12. Experimental result

7. CONCLUSION

The mean square values of pitch and bounce with a simple two-degree-of-freedom model estimate the vehicle frequency and damping setting to achieve a high performance of ride comfort human sensitivity in consideration. Achieving a high performance of ride comfort at low frequencies consist of, for the most part, finding an optimum front and rear ratio of frequency to minimize the mean square value of pitch. Vehicle parameters such as the dynamic index, base frequency and base damping ratio must take into account the frequency setting.

REFERENCES

1. Thomas D. Gillespie: Fundamentals of Vehicle Dynamics. SAE
2. William F. Milliken, Douglas L. Milliken: Race Car Vehicle Dynamics. SAE
3. Phihas Barak: Magic Numbers in Design of Suspensions for Passenger Cars. SAE 911921
4. Eiich Yasuda, Shuichi Doi, Ikuo Kushiro: An Analysis of Human Vibration Sensitivity under Multi-Axial Vibration. JSAE20004447

Directional stability of buses under influence of crosswind gusts

MAGNUS JUHLIN[1]

SUMMARY

Incidents with heavy vehicles have put the issue of directional stability of buses under influence of crosswind gusts into focus. In this work the directional stability of buses when exposed to crosswind gusts has been investigated by using a MBS vehicle model combined with a generalised crosswind gust model and results from static wind tunnel measurements. The simulations carried out show the importance of the aerodynamic properties of buses as well as the importance of a proper weight distribution. It is also shown how important the driver model is in order to get realistic directional deviation in simulations and thereby also how important the driver is in order to keep the directional deviation to a minimum in real life situations.

1.   INTRODUCTION

During recent years, some incidents with heavy vehicles in Sweden have raised doubts in the public society regarding the safety of travelling by bus. Examples of such incidents are buses leaving the road due to crosswind gusts [1] and trucks with front tire explosions [2].

When a vehicle is driven on a road it is exposed to different types of disturbances, e.g. crosswind, uneven roads and the driver's input. These disturbances will affect the direction of the vehicle. In most cases the directional stability is an issue of comfort for the driver and of course also for the passengers in case of a bus. How well the driver is able to correct a deviation from the desired course is a combination of skill and how well the vehicle and driver interacts.

In some situations a crosswind gust disturbance might result in an accident regardless of the driver's corrections, i.e. when the magnitude of deviation is larger than what is allowed by the traffic environment. The reason can be that the directional deviation becomes too large before the driver reacts or that the vehicle does not respond on the driver's effort to control the vehicle. Hence, it is important to keep the lateral and angular deviation from the desired course within certain limits in order to avoid an accident.

[1] *Address correspondence to:* Bus Chassis Development, Scania CV AB, SE-151 87 Södertälje. Tel.: +46 8 553 828 71; Fax: +46 8 553 500 90; E-mail: magnus.juhlin@scania.com

2. THE MODEL

To investigate how different parameters affect the directional stability, a generalised crosswind gust model has been modelled in the MBS-software ADAMS and applied on a parameterised model of a typical 3-axled bus representative for the coach segment.

2.1 THE VEHICLE MODEL

The vehicle model, Figure 1, consists of a double wishbone front axle suspension, a rear axle with two upper and two lower reaction rods, a tag axle also with two upper and two lower reaction rods and a rigid bus body. The steering angles of the tag axle can either be locked or coupled to the steering of the front axle. The tire forces are generated using a magic-formula tire model and the tire data corresponds to existing tires, although the camber forces have not been included.

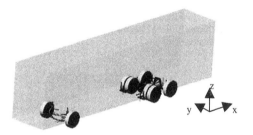

Fig. 1. Graphics of the vehicle model.

In Figure 2 the vehicle response on a single lane change steering wheel input (δ_h) is shown. The unladen bus has a velocity of 60km/h and the excitation frequency is 0.4 Hz. The discrepancy is rather small and the model is therefore considered to be valid.

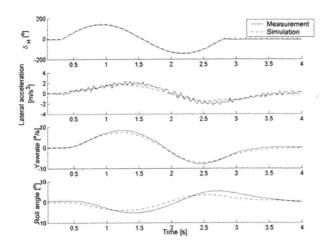

Fig. 2. Single lane change at 60 km/h and 0.4 Hz, unladen case.

2.2 THE GUST MODEL

A generalised crosswind gust model has been proposed by Jarlmark [3,4], where the time lag between the lateral force and yaw moment build up and the amplitude peak at gust entry are incorporated. Wind-tunnel measurements by Chadwick et al. [5] at the Cranfield crosswind facility confirm the proposed crosswind gust model and clarify the physics behind. Chadwick et al. found that a sharp edged box has a yaw moment peak, due to the negative pressure acting on the front half of the body's leeward side, at one body length into the gust with a peak/mean ratio of 2.5. Changing the edges on the box from sharp to a radii corresponding to a radii/height ratio of 0.1 lowers the peak/mean ratio to 1.5. The authors also found that the lateral force has a peak at one and a half body length into the gust with a peak/mean ratio of 1.1. In this work these findings have been used to model the crosswind gust and in figures 3 and 4 the force and moment build-up used in this investigation are shown. On the ordinate-axis the normalised distance travelled from gust entry is shown, where unity equals the bus body length. On the abscissa the normalised force and moments are shown for each direction. The forces and moments are defined as follows: the lateral force (S) is positive in the positive y-direction according to the coordinate system in figure 1, positive yaw moment (YM) in the negative z-direction, positive roll moment (RM) in the negative x-direction, positive lift force (L) in the positive z-direction and positive pitch moment (PM) in the positive y-direction. A ramp build-up is used for the lift force and pitch moment which reaches it's maximum at one body length into the gust.

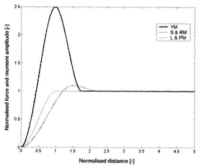
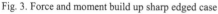

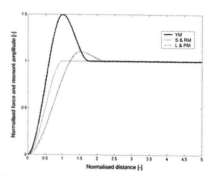

Fig. 3. Force and moment build up sharp edged case

Fig. 4. Force and moment build up radiused edged case

The magnitudes of the static aerodynamic forces and moments are taken from wind tunnel test carried out by the Aeronautical Research Institute of Sweden [6]. These wind tunnel tests were performed with a bus model in scale 1:10. A model of the double-decker bus, with a height of 4 meters and a length of 12 meters, that was involved in the accident investigated in [1] was first tested and later on a generic model, with the possibility to change each edge of the body work from sharp to radiused, was tested. All shapes were tested with a front spoiler and side skirts. A model corresponding to a height of 3 meters was also studied. The results from these tests show the aerodynamic properties of different bus bodies under static conditions.

The most common method when calculating aerodynamic coefficients is to use the frontal projected area as reference area and the wheelbase as reference length. This excludes the possibility to calculate forces and moments for a vehicle with differing length and/or length-wheelbase ratio. Therefore, the aerodynamic coefficients has been recalculated using the projected area in each direction and the total length rather than frontal projected area and wheelbase. In Figure 5 and 6, the recalculated aerodynamic coefficients with indices corresponding to the forces and moments in Figures 3 and 4, are shown as function of the angle of attac.

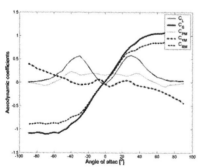

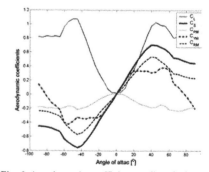

Fig. 5. Aerodynamic coefficients sharp edged.

Fig. 6. Aerodynamic coefficients radiused edged

By combining the results from the static wind tunnel tests with the generalised crosswind gust model, shown in Figure 3 and 4, the aerodynamic loads on a bus due to a crosswind gust can be approximated. This model is valid when the pressure point of the lateral force is located in front of the aerodynamic reference point. To study the effect of the aerodynamic properties on the directional stability the generalised forces and moments due to a gust and the aerodynamic coefficients are imported to ADAMS as spline curves. The aerodynamic forces and moments are applied to the vehicle model, at the same location as the reference point used in the aerodynamic measurements, using a GFORCE element.

2.3 THE DRIVER MODEL

A PID-regulator has been used to correct the directional deviation of the vehicle model. This regulator uses the lateral and angular displacement as error signals and applies a steering motion on the vehicle steering system to reduce the error. Different settings have been used to simulate different driver behaviour and the maximum steering wheel angle (δ_h) has been limited to 275°. In order to keep the desired velocity the vehicle has a cruise control that applies torque at the drive axle.

3. THE SIMULATIONS

Simulations have been carried out using the model described in previous sections with three different vehicle parameter settings according to Table 1. Two different body geometries have been studied, sharp edged and radiused. The vehicle and wind velocity have also been varied and the influence of the driver model on the directional deviation has also been studied.

Table 1. Vehicle parameters.

Vehicle parameters	Unladen	Laden	Rear bias	Units
Vehicle weight	18200	255700	23000	[kg]
C.o.g. position – $\hat{x}$	5.35	5.31	5.70	[m]
C.o.g. position – $\hat{z}$	1.45	1.69	1.62	[m]
Inertia – I	33400	38800	37100	[kgm^2]
Inertia – J	495000	621000	555000	[kgm^2]
Inertia – I	496000	623000	556000	[kgm^2]
Body length	15	15	15	[m]
Body height	3.9	3.9	3.9	[m]
Body width	2.55	2.55	2.55	[m]
Axle distance front to rear	7.175	7.175	7.175	[m]
Axle distance rear to tag	1.5	1.5	1.5	[m]

---

[2] Origo located in intersection of centre of front axle and ground level

3.1 INFLUENCE OF AERODYNAMIC PROPERTIES

In figure 7, the aerodynamic loads on a bus travelling with a velocity ,v_b , of 22 m/s and entering a crosswind gust at t = 1.6 s with a wind velocity, v_g, of 15 m/s perpendicular to the longitudinal direction of the bus, are shown. The direction of forces and moments are defined according to the coordinate system given in Figure 1. As can be seen in Figure 7, there is a large difference in yaw moment for these bus bodies and the pitch moment has opposite signs.

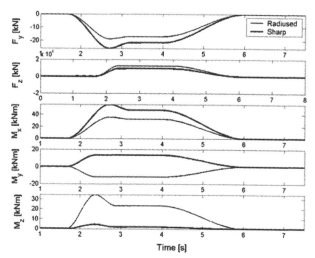

Fig. 7. Aerodynamic loads on radiused and sharp edged bus, v_b = 22 m/s, v_g = 15 m/s.

In Figure 8 the steering wheel angle and vehicle response for an unladen bus, with parameters according to Table 1, affected by the aerodynamic loads shown in Figure 7 are shown.

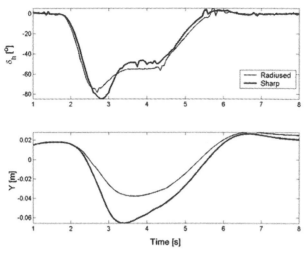

Fig. 8. Steering wheel angle and lateral deviation of radiused and sharp edged bus, v_b = 22 m/s, v_g =15 m/s.

The yaw moment is generally known to have large impact on the directional deviation of a vehicle but in this case the larger lateral force and the positive pitch moment have greater influence and give the sharp edged bus a greater directional deviation than the radiused as can be seen in figure 8. The directional deviation is rather small in this particular case, due to the fact that the wind and vehicle velocity is relatively low and that the driver model is reacting fast, see figure 8.

3.2 INFLUENCE OF WEIGHT DISTRIBUTION

To study the influence of how the bus is loaded on the directional stability three different load cases, with parameters according to table 1 and a radiused bodywork, have been studied. The vehicle velocity, v_b, used is 25 m/s and the wind velocity, v_g, is 25 m/s, gust entry occurs at t = 1.6 s. The difference in directional deviation between these three cases is large as can be seen in figure 9. In the unladen and rear biased cases the driver model reaches the implemented steering wheel angle limit and it is not until the gust starts to fade out at time t = 4 s that the vehicle starts to regain its original direction. In normal traffic the case with rear bias weight distribution can cause a serious accident. If the conditions are favourable it might be possible to avoid an accident in the unladen case but if the road surface friction is reduced, all three cases might cause accidents, regardless of how skilled the driver is.

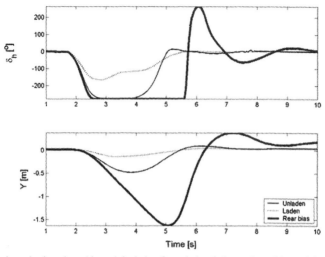

Fig. 9. Steering wheel angle and lateral deviation for unladen, laden and rear bias weight distribution, v_b = 25 m/s, v_g = 25 m/s.

In figure 10 the tire forces in the z-direction are shown. The vertical load on the front right tire is drastically reduced by the aerodynamic forces and the wheel looses contact with the ground in the unladen and rear biased cases.

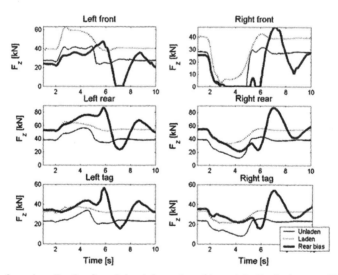

Fig. 10. Tire forces in z-direction for unladen, laden and rear bias weight distribution, v_b = 25m/s, v_g = 25 m/s.

3.3 DRIVER INFLUENCE

In all cases described above the driver model that was used has been tuned to have a fast response in order to show the vehicle's ability to withstand a crosswind gust rather than showing the drivers ability to control the vehicle. Changing the driver model to a driver with slower response greatly affects the directional deviation. In figure 11 the directional deviations of two driver models are shown. The vehicle used in this comparison has a sharp edged body and parameters according to the unladen case in table 1. Driver 1 has fast response while driver 2 has a response time that reminds of a human driver. This shows how important it is to have a realistic driver model when making simulations to predict the directional deviation of a vehicle exposed to a crosswind gust in a real situation. It also illustrates the importance to have skilled drivers that react fast and accurate when driving a bus exposed to crosswind gusts.

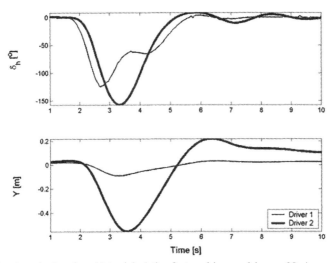

Fig. 11. Steering wheel angle and lateral deviation for two driver models, $v_b = 25$m/s, $v_g = 20$ m/s.

4. CONCLUSIONS

Combining the generalised crosswind gust model with aerodynamic data from static wind tunnel measurements and applying that to a vehicle simulation model provide a tool to investigate the directional deviation of vehicles when exposed to crosswind gusts. The simulations carried out show the importance of aerodynamic properties such as yaw moment, lift and pitch moment. Another property that is important for the directional stability is the weight distribution. This is, however, a parameter that changes under normal operating conditions of a bus and therefore puts a great responsibility on the driver to make sure the vehicle is appropriately laden. The results also show the importance of the reaction time of the driver model and thereby the importance of a representative driver model when simulating vehicles exposed to crosswind gusts.

ACKNOWLEDGEMENT

The author is pleased to acknowledge the financial support of the Swedish National Road Administration. The author also wishes to thank Prof. Annika Stensson, KTH for academic guidance and Dr. Peter Eriksson, Scania for the supervision. The author gratefully acknowledges Per-Åke Torlund, FFA for his help with supplying the aerodynamic data. The author also would like to thank his colleagues at the Bus Chassis Development, Scania for the support with models and measurements.

REFERENCES

1. Swedish Board of Accident Investigation (Statens haverikommision): Investigation of Bus Accident in Sweden (in Swedish). Report RO 2001:04, 2001.
2. The Swedish Association of Haulage Contractors (Svenska åkeriförbundet): Accidents and Incidents with Front Tyres on Heavy Trucks (in Swedish). report 2002:1, 2002.
3. Klasson, J. (Now Jarlmark, J.): Generalised Crosswind Model for Vehicle Simulation Purposes. Supplement to Vehicle System Dynamics, Vol. 35, Proceedings of the 17th IAVSD Symposium, August, 2001, pp. 350-359
4. Jarlmark, J.: Driver-Vehicle Interaction under Influence of Crosswind Gusts. Licentiate thesis, Division of Vehicle Dynamics, KTH, Sweden. ISRN KTH/FKT/L--02/18—SE, 2002.
5. Chadwick, A., Garry, K;. Howell, J.: Transient Aerodynamic Characteristics of Simple Vehicle Shapes by the Measurement of Surface Pressures. SAE 2001-01-0876, 2001
6. Torlund, P-Å.: Experimental Wind Tunnel Investigation of Crosswind Stability of a Double Decker Coach (in Swedish). The Aeronautical Research Institute of Sweden, Report FFA TN 2000-05, 2000.

Vehicle System Dynamics Supplement 41 (2004), p.103-112

State estimation of vehicle handling dynamics using non-linear robust extended adaptive kalman filter

MEDY SATRIA[1] AND MATTHEW C BEST[1]

SUMMARY

This paper considers translational and rotational state estimation of vehicle handling dynamics using a non-linear robust filtering method. The robust method is needed to compensate the modelling errors which are caused by a combination of many factors such as: parameter uncertainties, model simplification and non-linearities; these inevitably compromise the filter performance. A non-linear Robust Extended Adaptive Kalman Filter (REAKF) is proposed, based on a six degree of freedom (6DOF) vehicle model. The model uncertainties are represented using an integral quadratic constraint method. Parameter adaptation is then included to further increase the accuracy of the filter; the tyre longitudinal and cornering stiffnesses are varied, since both of these parameters are central to the vehicle non-linearities. The filter is tested in both simulation and practice using data collected from a real vehicle. Comparisons are also made with the standard Extended Kalman Filter (EKF) and the non-adaptive Robust Extended Kalman Filter (REKF). The combination of evidence from both tests illustrates the performance and stability benefits of the REAKF.

Keywords : vehicle handling dynamics, state estimation, robust Kalman filter, model uncertainties, parameter adaptation

1. INTRODUCTION

State estimation plays an important role in advanced vehicle control design. Indeed, information about vehicle states is essential in designing advanced state feedback control. However, not all of the vehicle states can be measured directly, such as sideslip velocity, and others are relatively expensive to measure directly, such as yaw and roll rates. The use of a state estimator with relatively cheap sensors gives significant benefit to solve this problem. Many studies have been done in both linear [1,3-5] and nonlinear [2,5] filtering for vehicle state estimation. In [3], the use of a linear Kalman filter is examined for a BMW Driver Assistance System, which uses a simple linear 2DOF model to estimate lateral vehicle states (yaw rate and lateral acceleration). More complete vehicle states are estimated in [1] using a linear 3DOF model, introducing a linear robust filter to compensate model uncertainties. However, since these papers deal with linear filtering, the filter performance is only guaranteed at a certain operating point and any admissible uncertainties around it. A nonlinear filter is proposed in [2], by using an extended Kalman filter, both non-adaptive and adaptive, using a nonlinear 3DOF vehicle model. However, this method has shortcomings in terms of model uncertainty compensation, since it still uses a nominal vehicle model. In this paper, a nonlinear robust filter is proposed,

---

[1] Department of Aeronautical and Automotive Engineering, Loughborough University,
Loughborough, Leicestershire LE11 3TU United Kingdom.
Tel : +44 1509 227264 ; Fax : +44 1509 227275 ; E-mail : m.satria@lboro.ac.uk

employing a nonlinear 6DOF vehicle model to give more complete estimation of the states. The method has advantages in that it considers model uncertainty compensation in a more analytical way within the design process. It is based on the work done in [6], which uses an integral quadratic constraint to model structured uncertainties.

2. VEHICLE MODEL

A six degree of freedom (6DOF) vehicle model is developed as the base for the filter design and for the source model, which is used to simulate a 'real' vehicle. The vehicle is assumed to have front wheel drive (FWD) and front wheel steer (FWS). It is also assumed to run on a flat road. A body attached SAE axis system centred at the vehicle centre of mass is employed, as shown in Fig.1, and the dynamics of the eight vehicle states $x = [u, v, w, p, q, r, \varphi, \theta]^T$ is modelled by non-linear state space equations as follows:

$$\dot{u} = 1/M \sum F_x - qw + rv \tag{1}$$

$$\dot{v} = 1/M \sum F_y - ru + pw \tag{2}$$

$$\dot{w} = 1/M \sum F_z - pv + qu \tag{3}$$

$$\begin{bmatrix} \dot{p} \\ \dot{r} \end{bmatrix} = \frac{1}{-I_{xx}I_{zz} + I_{xz}^2} \begin{bmatrix} -I_{zz} & -I_{xz} \\ -I_{xz} & -I_{xx} \end{bmatrix} \begin{bmatrix} \sum M_x - q(I_{zz}r - I_{xz}p) + I_{yy}qr \\ \sum M_z - pI_{yy}q + q(I_{xx}p - I_{xz}r) \end{bmatrix} \tag{4}$$

$$\dot{q} = 1/I_{yy} \left(\sum M_y - r(I_{xx}p - I_{xz}r) + p(I_{zz}r - I_{xz}p) \right) \tag{5}$$

$$\dot{\varphi} = p \quad ; \quad \dot{\theta} = q$$

The system inputs are $u_s = [\delta, \tau]^T$ for the source model and $u_s = [\delta, w_s]^T$ for the filter, which are assumed to be available as deterministically known inputs.

Two accelerometer sensors as the system outputs $y = [s_1, s_2]^T$ are placed at the vehicle centre of mass, oriented longitudinally and laterally as shown in Fig. 2. These are modelled by the following equation:

$$s_1 = \dot{u} + qw - rv$$
$$s_2 = \dot{v} + ru - pw \tag{6}$$

This sensor configuration is chosen because it provides observability for the system, a compulsory requirement for the filter design; it also gives economic benefit since accelerometer sensors are cheap and widely available.

For the filter model, linear tyre force models are used since these reduce the complexity in the filter design, especially in the Jacobian derivation of the model. The full tyre magic formula [8, pp 475-483] is used for the source model to provide 'true' state trajectories and data for parameter identification in the simulation test.

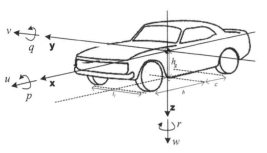

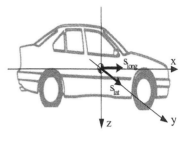

Fig. 1. Vehicle axis system. Fig. 2. Sensor placement and orientation.

Table 1. Model Nomenclature.

States (x)		Parameters	
u	Forward velocity (m/s)	M	Vehicle mass (1475 kg)
v	Sideslip velocity (m/s)	I_{xx}	Roll moment of inertia (382 kgm²)
w	Vertical velocity (m/s)	I_{yy}	Pitch moment of inertia (2132 kgm²)
p	Roll rate (rad/s)	I_{zz}	Yaw moment of inertia (2222 kgm²)
q	Pitch rate (rad/s)	I_{xz}	Roll/yaw product of inertia (80 kgm²)
r	Yaw rate (rad/s)	κ_x	Tyre longitudinal slip rate (62289 N)
φ	Roll angle (rad)	C_{af}	Front tyre cornering stiffness (66851 N/rad)
θ	Pitch angle (rad)	C_{ar}	Rear tyre cornering stiffness (54761 N/rad)
Source Model Inputs (uₛ)		Filter Inputs (uₛ)	
δ	Steer angle (rad)	δ	Steer angle (rad)
τ	Driven wheel torque (Nm)	w_s	Driven wheel speed (m/s)

3. ROBUST EXTENDED KALMAN FILTER

Robust extended Kalman filter (REKF) is a nonlinear filter which takes into account model uncertainty in its design process. Unlike the standard extended Kalman filter (EKF) which only based on the nominal model, REKF is based on the nominal model plus admissible uncertainty model in the design process. This gives more accurate way to compensate the model uncertainty than that of EKF.

Consider a vehicle model with uncertainty represented by a non-linear state space equation as follow:

$$\dot{x} = f(x,u) + D_1\Delta_1(t)K(x,u) + D_2 w_o \qquad (7)$$
$$y = g(x,u) + \Delta_2(t)K(x,u) + v_o$$

where $\Delta(t) = [\Delta_1(t) \ \Delta_2(t)]^T$ is a matrix of uncertain parameters satisfying the norm

bound $\left\| \begin{bmatrix} Q_1^{1/2}\Delta_1(t) \\ R_1^{1/2}\Delta_2(t) \end{bmatrix} \right\| \le 1$, and $w_o(t)$ and $v_o(t)$ are noise signals satisfying the bound:

$$\Phi(x(0)) + \int_0^s \left[w_o(t)'Q_2 w_o(t) + v_o(t)'R_2 v_o(t)\right]dt \le d \qquad (8)$$

The influence of the uncertainty in the model is subject to the following integral quadratic constraint (ICQ):

$$(x(0)-x_o)'N(x(0)-x_o)+1/2\int[w(t)'Qw(t)+v(t)'Rv(t)]dt \le d+1/2\int K(x,u)'K(x,u)dt \qquad (9)$$

where $Q = \begin{bmatrix} Q_1 & 0 \\ 0 & Q_2 \end{bmatrix}$, $R = \begin{bmatrix} R_1 & 0 \\ 0 & R_2 \end{bmatrix}$, $w(t) = \begin{bmatrix} \Delta_1(t)K(x,u) \\ w_o(t) \end{bmatrix}$ and $v(t) = \begin{bmatrix} \Delta_2(t)K(x,u) \\ v_o(t) \end{bmatrix}$.

The uncertainties which satisfy this ICQ are called *admissible uncertainties*, in which the filter performance is guaranteed. The robust extended Kalman filter then can be derived with the following state space equation:

$$\dot{\hat{x}} = f(\hat{x},u) + P[\nabla_x g(\hat{x},u)'R(y - g(\hat{x},u)) + \nabla_x K(\hat{x},u)'K(\hat{x},u)] \ ; \ \hat{x}(0) = x_o \qquad (10)$$

where P is the solution of the following Riccati differential equation (RDE):

$$\dot{P} = P\nabla_x f(\hat{x},u) + \nabla_x f(\hat{x},u)P + DQ^{-1}D' + P[\nabla_x K(\hat{x},u)'\nabla_x K(\hat{x},u) - \nabla_x g(\hat{x},u)'R\nabla_x g(\hat{x},u)]P \qquad (11)$$
$$P(0) = N^{-1} \ ; \quad D = [D_1 \ D_2]$$

Further detail of the filter derivation can be seen in [6]. The REKF (10) is actually a general form of EKF; if the structured uncertainty model $K(x,u)$ is set to zero, then REKF (10) and RDE (11) revert to the standard EKF formula.

The structured uncertainty model of the vehicle is constructed by direct extraction from the vehicle model. The parameters of the vehicle are assumed to have uncertainty within a reasonable bound. Extracting the uncertainty part from the vehicle model then yields the structured uncertainty model. In this study, the main emphasis is given to uncertainty in the tyre longitudinal and cornering stiffnesses, because these parameters have a major influence on the model error between filter model and the actual/source model. Further unstructured modeling errors are then imposed by adding process and measurement noise, which is accounted for by tuning Q_2 and R_2 matrices.

Consider nominal longitudinal and lateral tyre forces which are given by the following linear relationship:

$$F_{xi}^{nom} = \kappa_x S_{ri} \ ; \ F_{yfi}^{nom} = C_{af}\alpha_{fi} \ ; \ F_{yri}^{nom} = C_{ar}\alpha_{ri} \qquad (12)$$
$$i = 1,2 \ ; \ 1:\text{left} \ , \ 2:\text{right}$$

where S_r is the longitudinal slip ratio at the front tyres, and α_f and α_r are front and rear tyre sideslip angle. The errors between true tyre forces and nominal tyre forces can then be expressed by Equation (13).

$$\Delta_{Fxi} = F_{xi}^{true} - F_{xi}^{nom} \;;\; \Delta_{Fyfi} = F_{yfi}^{true} - F_{yfi}^{nom} \;;\; \Delta_{Fyri} = F_{yri}^{true} - F_{yri}^{nom} \tag{13}$$
$$i = 1,2 \;;\; 1:\text{left} \;,\; 2:\text{right}$$

In this study, a linear approximation is used to model the tyre force errors in Equation (13), yielding the following equation:

$$\Delta_{Fxi} = \hat{k}_{xi} S_{ri} \;;\; \Delta_{Fyfi} = \hat{c}_{afi} \alpha_{fi} \;;\; \Delta_{Fyri} = \hat{c}_{ari} \alpha_{ri} \tag{14}$$
$$i = 1,2 \;;\; 1:\text{left} \;,\; 2:\text{right}$$

where $\hat{k}_{xi}$, $\hat{c}_{afj}$ and $\hat{c}_{ark}$ are uncertainty gains which are obtained by least square identification. Substituting Equation (14) into Equation (1) – (5) results in a structured uncertainty equation as follow:

$$K(x,u) = \left[1/M\sum_{i=1}^{2}\Delta_{Fxi} \;\; 1/M\sum_{i=1}^{4}\Delta_{Fyi} \;\; 1/M\sum_{i=1}^{4}\Delta_{Fzi} \;\; \frac{-I_{zz}\sum\Delta M_x - I_{xz}\sum\Delta M_z}{-I_{xx}I_{zz}+I_{xz}^2} \;\; \frac{\sum\Delta M_y}{I_{yy}} \;\; \frac{-I_{xz}\sum\Delta M_x - I_{xx}\sum\Delta M_z}{-I_{xx}I_{zz}+I_{xz}^2} \right]^T \tag{15}$$

The bound of the uncertainties can be tuned by tuning the Q and R matrices or the uncertainty gains in Equation (14), to give a reasonable range of uncertainty.

To further increase the robustness of the filter, the nominal tyre longitudinal and cornering stiffnesses (κ_x, C_{af} & C_{ar}) are then adapted, to form a robust extended adaptive Kalman filter (REAKF). The adaptation law is derived based on tyre magic formula [8, pp 475-483], whereby :

$$\hat{\kappa}_{xi} = \frac{\mu_{xi} Z_{fi} R_i}{\sqrt{S_{ri}^2 + \mu_i^2 \tan^2 \alpha_{fi}}} \tag{16}$$

$$\hat{C}_{afi} = \frac{\mu_{yi} Z_{fi} \mu_i R_i}{\sqrt{S_{ri}^2 + \mu_i^2 \tan^2 \alpha_{fi}}} \tag{17}$$

$$\hat{C}_{ari} = \frac{\mu_{yri} Z_{ri} \overline{F}_{yi}}{\alpha_{ri}} \tag{18}$$

$$i=1,2 \;;\; 1:\text{left} \;,\; 2:\text{right}$$

where R is normalised resultant tyre force at front wheel, $\overline{F}_y$ is normalised lateral tyre force at rear wheel, μ_x is longitudinal friction coefficient, μ_y is lateral friction coefficient, μ is lateral/longitudinal friction coefficient, and Z_f and Z_r are loads at front and rear wheels. R is formulated using combined-slip tyre magic formula, whereas $\overline{F}_y$ is formulated using pure-slip tyre magic formula.

4. SIMULATION TEST

The simulation is carried out using a source model acting as the 'true' vehicle. The vehicle has an initial forward velocity of 25 m/s. The source model is then excited by random steer angle and torque inputs to generate the data for parameter identification and noise generation. The inputs are chosen with a view to feasible vehicle tests, to induce a wide range of operating condition, within the achievable frequency response of a test driver :

$$\delta(t) = \delta_c(t) + N(0, \sigma_\delta^2)$$
$$\tau(t) = \tau_c(t) + N(0, \sigma_\tau^2)$$

(19)

bandlimited $0 - 5$Hz, $\sigma_\delta = 0.0349$ rad ($2°$), $\sigma_\tau = 200$ Nm

where $\delta_c(t)$ and $\tau_c(t)$ are piecewise continuous inputs as shown in Fig. 3; these are needed to widen the operating condition range since these inputs give a broad state dynamics at various steer angle (lateral dynamics) and wheel speed (longitudinal dynamics), thus give more accurate parameter identification. Parameter identifications for κ_x, C_{af}, C_{ar}, $\hat{k}_{xi}$, $\hat{c}_{afj}$ and $\hat{c}_{ark}$ are then made by least square regression of this data for 25 second data batch.

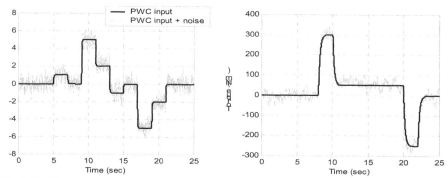

Fig. 3. Handling input test.

To simulate the external process and measurement noises $w_o(t)$ and $v_o(t)$, additional external noises are added to handling inputs (17) as follow:

$$\delta_o(t) = N(0, \sigma_{\delta o}^2) \quad ; \quad \tau_o(t) = N(0, \sigma_{\tau o}^2)$$

(20)

$$\sigma_{\delta o} = 0.0017 \text{ rad } (0.1°), \ \sigma_{\tau o} = 5 \text{ Nm}$$

These noises simulate noises that disturb steer angle and torque/wheel speed sensors, which influence the vehicle state derivative and sensor models, yield process and measurement noises $w_o(t)$ and $v_o(t)$. The Q_2 and R_2 matrices are then constructed from this data such that these satisfy ICQ (8).

The filters are then tested under handling inputs $\delta_c(t)$ and $\tau_c(t)$ plus the external noises (20), as shown in Fig. 4 (h) and (i); these are chosen to illustrate transient and steady state dynamic effects. The estimation result against the source model state trajectories is shown in Fig. 4 (a) – (g). Forward velocity estimation (Fig. 4(a)) is good for all filters; the filters don't need to do a 'hard' job since most of forward velocity information is given by wheel speed sensors. Sideslip velocity (Fig. 4(b)) is poorly estimated by EKF and REKF with steady state errors in every cornering; both EKF and REKF don't give appropriate compensation of lateral tyre force error. REAKF gives extremely good estimation of sideslip velocity, which means that parameter adaptation gives significant enhancement of tyre force error compensation. Fig. 4 (j) – (l) show the adaptation done by REAKF, these show the effect of load transfer in the adaptation process. For example under high positive steer angle between 9 and 11 seconds, the instantaneous cornering stiffness at the left wheel is much greater than that of at the right one, which mean that there is more load at the outer side (left side) than that of the inner side (right side) which cause greater tyre forces at the left side than those of the right side, thus result in greater instantaneous cornering stiffness at the left side than that of the right side.

Roll and pitch rate (Fig. 4(c)-(d)) are not well estimated by EKF, due to noise sensitivity. However, the plot shows that transients are well tracked, and the related roll and pitch angle estimation (Fig. 4(f)-(g)) are consistently good. REKF can reduce these errors, although is not great for pitch rate. Both EKF and REKF give a good estimation for yaw rate (Fig 4(h)), with small steady state error. The effect of external noises is well compensated by all filters, as can be seen in the 'clean' signal in the plot. However, the external noises cause excessive parameter adaptation of the REAKF at small angle steer, which mean that the REAKF is more susceptible to noise in small steer angle than in large one.

Table 2 shows the percentage of RMS errors against the true states. The REKF generally gives a better estimation than that of EKF, except in forward and sideslip velocity estimations are worse, but these are not really great. Significant improvement is given in roll rate estimation, with consistent roll angle improvement. It means that the inclusion of structured uncertainty model in the REKF gives performance enhancement than that of standard nominal EKF. Finally, the REAKF has the best performance among the others with smallest RMS errors.

Table 2. Percentage of estimation errors with respect to true states.

%	u	v	p	q	r	φ	θ
EKF	0.3326	42.1741	39.2002	35.0511	11.9527	8.3227	15.6015
REKF	0.3690	49.0700	8.7221	28.4620	8.0829	3.2512	14.3845
REAKF	0.0196	6.4961	5.1217	4.7092	2.4291	0.7067	1.0140

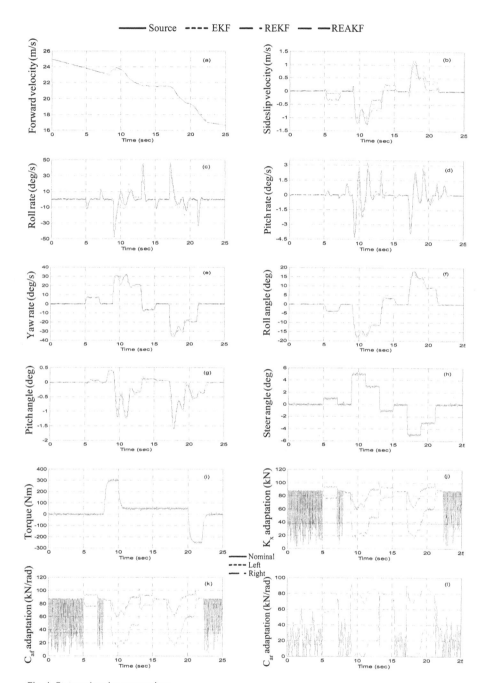

Fig. 4. State estimation comparison.

5.   EXPERIMENT TEST

A preliminary experiment test is done using test data collected from a vehicle (Ford Mondeo) instrumented with steer angle sensor, wheel speed sensors, body-centred triaxial accelerometers, and yaw and roll rate sensors. To obtain data for parameter identification, the vehicle is excited by random steer and wheel speed inputs, sampled as 200 Hz for 10 seconds, as shown in Fig. 5. Filter parameters are then identified by least square regression of this data batch.

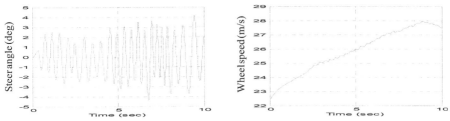

Fig. 5. Random input sensor data.

The filters are then tested using J-turn handling inputs as shown in Fig. 6 (a) and (b), the estimation result against the data from sensors is shown in Fig. 6 (c)-(f). Roll rate estimation (Fig. 6(c)) is not really good for all filters, with high RMS error as shown in Table 3. A high transient error occur at t=12 seconds when the vehicle make cornering. Suspension parameter inaccuracy is likely to be the major cause.

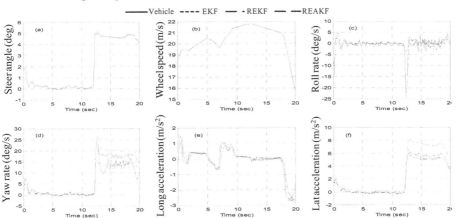

Fig. 6. Estimation result for all filters against sensor data.

Table 3. Percentage of estimation errors with respect to true states.

%	p	r	s_1	s_2
EKF	365.6776	46.5128	37.8277	43.0379
REKF	369.1837	29.4363	22.8538	10.4829
REAKF	326.3010	23.5206	32.7912	3.9410

Yaw rate (Fig. 6(d)) is estimated better than roll rate, however there still steady state error while cornering at t>12 seconds. Longitudinal acceleration (Fig. 6(e)) is relatively well estimated by all filters, as again the information about longitudinal dynamics has been provided very well by wheel speed sensors. Lateral acceleration (Fig. 6(f)) is well estimated by REKF and REAKF, but is bad estimated by EKF with high steady state error while cornering. REAKF generally give the best estimation among the others, however further model validation is needed to increase the accuracy of this filter.

The other states have not been validated due to sensor limitation, however this study has given insight of the implementation issue. Further on-line and real-time test is needed to test the filter for real-time embedded filter implementation.

6.   CONCLUSIONS

The extended adaptive robust Kalman filter has been examined, and it shows enhancement in the state estimation compare with the standard EKF and non-adaptive REKF. The combination of robust method and parameter adaptation gives significant benefit in performance enhancement. Parameter adaptation gives greater contribution in the performance improvement than robust method. Further uncertainty model algorithm formulation is needed such that it gives greater contribution in performance enhancement.

REFERENCES

1.   Satria, M. and Best, M. C. : Comparison between Kalman Filter and Robust Filter for Vehicle Handling Dynamics State Estimation. Proceedings of The 2002 SAE World Congress, Detroit-Michigan, USA, March 2002, SAE Paper No 2002-01-1185, pp. 1-10.
2.   Best, M. C., Gordon, T. J. and Dixon, P. J. : An Extended Adaptive Kalman Filter for Real-time State Estimation of Vehicle Handling Dynamics. Vehicle System Dynamics 34(2000), pp. 57-75.
3.   Best, M. C. and Gordon, T. J. : Real Time State Estimation of Vehicle Handling Dynamics Using an Adaptive Kalman Filter. Proceedings of The 4th International Symposium on Advanced Vehicle Control (AVEC), Nagoya, Japan, September 1998, pp. 183-188.
4.   Venhovens, P. J. TH. and Naab, Karl. : Vehicle Dynamics Estimation Using Kalman Filters. Vehicle System Dynamics 32 (1999), pp. 171-184.
5.   Kienecke, U. and Daiβ, A. : Observation of Lateral Vehicle Dynamics. Control Engineering Practice 5(8)(1997), pp. 1145-1150.
6.   James, M. R. and Petersen, I. R. : Nonlinear State Estimation for Uncertain Systems with an Integral Constraint. IEEE Transactions on Signal Processing 46(11)(1998), pp.2926-2937.
7.   Gelb, A. : Applied Optimal Estimation. MIT Press, Cambridge Mass. 1974.
8.   Milliken, W. F. and Milliken, D. L. : Race Car Vehicle Dynamics. SAE International, 1995.

Vehicle System Dynamics Supplement 41 (2004), p.113-122   © Taylor & Francis Ltd.

Effect of Asymmetric Brake Shoe Force Application on Wagon Curving Performance

Y. HANDOKO, F. XIA, M. DHANASEKAR[1]

SUMMARY

This paper reports the effect of asymmetric brake force on wheelsets to the curving performance of freight wagons fitted with three-piece bogies. For the purpose of study a wagon model containing eleven rigid body elements was set up. The non-linear characteristics caused by friction and clearances between critical rigid elements are considered in the modelling. The effects of yaw torque (positive and negative) induced by asymmetric brake forces, the wheel profiles, and the friction coefficients on the curving performance of the wagons are examined. The results show that the negative yaw torque potentially deteriorates the curving performance as measured by the increase in the angle of attack and lateral force as well as lateral to vertical force (L/V) ratio.

1   INTRODUCTION

When trains negotiate downhill grades on either tangent or curved tracks under normal operating condition, brakes are commonly applied. Brake application either helps maintain the speed at constant level or reduces the speed until the train stops.

Only limited information is found on the effect of braking to the curving performance of wagons in the literature. Lixin and Haitao [1] reported that in general brake forces adversely affect the lateral and vertical dynamics of wagons. Recently, Berghuvud [2] has investigated the effect of symmetric braking on the curving performance of wagons containing three-piece bogies. In practice, however, asymmetric brake shoe normal force within a single wheelset could be applied especially in wagons equipped with one-side push brake shoe arrangement. This paper, therefore, examines the effect of asymmetric brake shoe force application on the curving performance of freight wagons containing three-piece bogies.

[1]*Address correspondence to:* Centre for Railway Engineering, Central Queensland University, Building 70, Rockhampton, QLD 4702, Australia.   Ph. +61 7 4930 9677; Fax. +61 7 4930 6984; email: m.dhanasekar@cqu.edu.au

1.1 Description of the wagon-bogie system

A wagon with standard three-piece bogie is considered. This type of bogie is widely used in freight wagons around the world. It has a very simple construction consisting of two side frames, one bolster, and two wheelsets. The suspension damping is provided by the coulomb friction of the wedges placed between the bolsters and side frames. Typical configuration of the three-piece bogie is shown in Fig.1.

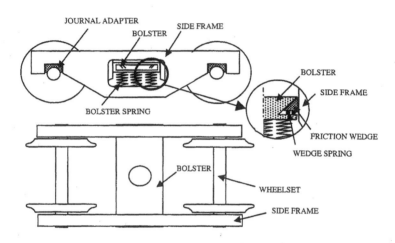

Fig. 1. Typical configuration of three-piece bogie (constant damping type)

1.1. Brake forces distribution to the wheels

The bogie is equipped with one-side push brake shoe arrangement. The braking force produced by the brake cylinder that is mounted at the wagon underframe is distributed to the wheels through a mechanical link arrangement (brake rigging) as shown in Fig.2. The link consists of rods and levers suspended from the underframe and bogies and linked with pins and bushes.

The brake rigging requires careful setting up and regular adjustment to ensure forces are evenly distributed to all wheels. It can be seen from the rigging diagram that any bad adjustment of the brake rigging could lead to uneven distribution of braking forces to each wheel. Such situation can occur when the centre-pin on rod AB is slightly off-centred or if the fixed-end pin in the bolster is disoriented. The uneven distribution of the braking force to wheels may also occur during curve negotiation if the bogie deforms in shear (warping) mode.

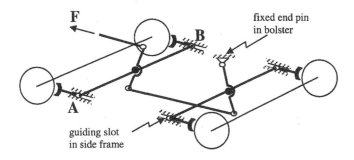

Fig. 2. Brake rigging arrangement

2 WAGON-TRACK INTERACTION UNDER BRAKING

Effect of braking torque application to the curving performance of a railway wagon in terms of wheel-rail interaction forces is explained in [2] and [3]. It is shown that the braking torque could reduce the steering torque of the wheelset. This concept is explained in Fig.3 where a wheelset free body as it moves along a curved track is shown.

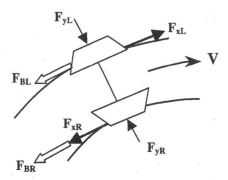

Fig. 3. Curving diagram of a wheelset

While negotiating a curve, the rolling radius of the outer wheel becomes larger than the inner wheel leading to the generation of longitudinal creep forces F_{xR} and F_{xL}, where in general $F_{xR} \neq F_{xL}$. These creep forces produce steering torque that guides the wheelset to follow the curve appropriately. Application of brake produces additional longitudinal force F_{BR} and F_{BL} on the contact point which produces a steering moment as shown in equation (1).

$$M = a\left[\left(F_{xL} - F_{BL}\right) + \left(F_{xR} + F_{BR}\right)\right],$$ (1)

where a is the semi distance between the contact points.

However, the resultant of total longitudinal creep force and lateral creep force in the contact point cannot exceed the maximum frictional force between the wheel and the rail. The resultant creep forces are calculated as shown in equation (2).

$$F_{CR} = \sqrt{\left(F_{xR} + F_{BR}\right)^2 + F_{yR}^2} \quad ; \quad F_{CL} = \sqrt{\left(F_{xL} - F_{BL}\right)^2 + F_{yL}^2}$$ (2)

where F_{CR} and F_{CL} are resultant creep force in the right and left rail respectively.

If F_{CR} and F_{CL} exceed the creep condition, it becomes saturated leading to reduction in the longitudinal creep force. The direction of longitudinal creep force and additional force due to braking is the same on the inner wheel, so the reduction in longitudinal force is larger than that on the outer wheel. Because of this reduction of longitudinal force, the steering moment of the wheel reduces. This is the main mechanism of steering torque reduction due to braking.

For the three-piece bogie, the presence of friction and longitudinal clearance makes the operating condition more complex. The friction characteristic causes no change in the displacements of the wheelset relative to the bogie frame until the static friction in the suspension is overcome.

3   PRINCIPLE OF MODELLING OF THE SYSTEM

The dynamics of wagons containing three-piece bogies running on a downhill grade is modelled. Throughout the journey, brake is applied to keep the velocity constant. The principle of modelling is shown in Fig.4. Essentially instead of modifying the track geometry data, the static gravitational force components $mg \sin \alpha$ and $mg(1 - \cos \alpha)$ (α is the slope in longitudinal direction) are added to the wagon model.

The effect of braking is modelled by simultaneously applying the brake pitch torque (T_b in Fig.4) to each wheelset. Friction coefficient between the shoe and the wheel is assumed to remain constant, thus the brake shoe normal force corresponding to the braking torque is easily calculated. Yaw torque is applied to the leading wheelset of the front bogie to represent the asymmetric braking forces.

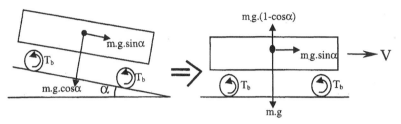

Fig.4. Modelling principle of wagon running on downhill slope with braking applied.

The equation of motion of the wagons system can be written:

$$[M]\{\ddot{q}\} + [C]\{\dot{q}\} + [K]\{q\} = \{Q_{cr}\} + \{Q_{ir}\} + \{Q_e\}, \tag{3}$$

where $\{q\}$ is displacement vector and $[M],[C],[K]$ are matrices representing the mass, damping and stiffness of the system. The vectors $\{Q_{cr}\}$, $\{Q_{ir}\}$ and $\{Q_e\}$ represent the generalised forces due to wheel-rail interaction, rail irregularity, and external forces applied to the system. The braking torque and the yaw torque are components of vector $\{Q_e\}$.

This model was established in VAMPIRE software using 11 rigid bodies defined by 62 DOF that excluded the wheelset pitch. The pitch degree of freedom of a wheelset is a special case, being a function of the forward velocity of the wheelset. While calculating the wheel/rail force, the wheelset pitch inertia was assumed to be negligible relative to the creep forces [6]. This means that at each time step the pitch torques due to creep forces and externally applied braking torque must balance exactly. This leads to the wheelset pitch displacement not being explicitly calculated. The effect of the wheelset pitch especially during the application of braking, forms a part of ongoing research at the Centre of Railway Engineering and is not included in this paper.

The non-linear characteristics caused by friction and clearances of the suspension system were considered. Surface friction between various contacting components in the wagon model was given due consideration. For example, the friction between side frame and (a) axle box adapter (b) bolster split wedge were considered. Where sensitivity of friction coefficient was examined, the values were changed from 0.10 to 0.35. In all other cases a constant value of 0.3 was used. Full nonlinear creep theory was used to calculate the creep force between the wheel and the rail.

The curved track was modelled with perfect equilibrium cant (ie, no cant deficiency was considered). Transition curves were attached to entry and exit ends of the main curve. The wagon was kept at a constant velocity of 60 km/h both on the tangent track and on the curved segment of the downgrade.

To study the effect of wheel geometry, worn wheel (hollow wheel) profile as shown in Fig.5 is used in the simulation. This hollow wheel profile is known to adversely affect the curving performance of wagons [4,5].

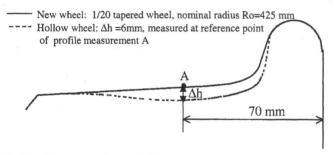

Fig.5. Profiles of new wheel and hollow wheel.

The wagon input data is based on a freight wagon operated in Queensland that runs on 1067 mm gauge. A detailed data of similar wagon is reported by Sun and Dhanasekar [7]. Table 1 shows the parameters investigated in this simulation. The number of analyses carried out using combinations of various parameters shown in table 1 was 180. Each analysis typically took 40 secs on a Pentium IV PC.

Table 1.The parameters investigated.

No	Parameters	Value
1	Yaw torque (kN.m)	-6.0 to 6.0 in step of 1.2
2	Curving radius (m)	600, 300, 150
3	Wheel profiles	New and worn (hollow wheel)
4	Friction coefficient between side frame and axle box	0.10, 0.25, 0.35

4 SIMULATION RESULTS

Angle of attack (AoA) and track forces were output from the analyses. As only perfect cant was considered the centrifugal force was eliminated as shown in Fig 6. This figure shows the lateral acceleration of the leading wheelset being negligibly small. With the elimination of the centrifugal forces, the lateral track forces obtained were mainly due to the wagon-track interaction. A typical lateral track force time series obtained at the outer rail when the wagon runs on a 300 m radius curved track with asymmetric braking is shown in Fig 7. The graph shows that in the main curve the lateral track force has a relatively stable value.

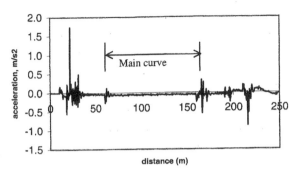

Fig.6. Lateral acceleration of leading wheelset

For each braking case the mean value of the lateral force on the outer rail and the mean value of the AoA of the leading wheelset in the main curve were evaluated. The results were considered representative 10 m after the wagon entered and 10 m before it exited the main curve. Only a few important results are presented in this section.

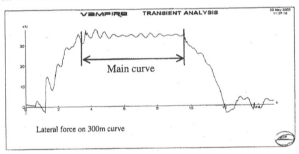

Fig.7. Typical result of simulation in time domain

Two types of asymmetric shoe force applications were examined – the first producing positive yaw torque and the second producing negative yaw torque. Figs 8(a) and 8(b) show the effect of these yaw torque application to the angle of attack whilst negotiating 300 m radius curved track with wheels of new and hollow profiles respectively. Solid lines in the figures show the effect of negative yaw torque and the dotted lines show the effect of positive yaw torque. These curves are shown for a selected set of three values of coefficients of friction between side frame and axle box (μ=0.10, 0.25 and 0.35).

It is clear from Fig.8(a), for wagons containing new wheels, that an increase in negative yaw torque increases the AoA for all levels of the coefficient of friction. The coefficient of friction affects the AoA significantly under no yaw torque (symmetric braking) compared to increased level of negative yaw torque. Under increased level of positive yaw torque a small reduction in AoA is observed for all values of coefficients of friction.

Fig.8(b) shows that the AoA is significantly larger for the wagons containing the worn wheels. In spite of this difference, the trend in the variation of the AoA to the increase in negative and positive yaw torque is similar to that of the result shown in Fig.8(a) except for the effect of the coefficient of friction. The worn wheels with lower friction produce increasing AoA, whilst the new wheels with higher friction produce increasing AoA. This phenomenon can be explained as follows: The negative conicity of worn wheels leads the wheelset to steer against the curvature of the track. Lower friction coefficient between the axle box adapter and the side frame provides smaller resistance to this movement resulting in higher AoA. A reverse situation occurs for the wheelsets containing new wheels as the new wheels steer along the curve properly. In this case, lower friction coefficient obviously lead to smaller AoA.

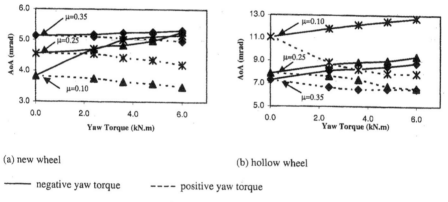

(a) new wheel

(b) hollow wheel

———— negative yaw torque   ‑ ‑ ‑ ‑ positive yaw torque

Fig.8. AoA as function of yaw torque at 300 m radius curved track

Figs. 9(a) and 9(b) show the effect of the yaw torque to the lateral force on the outer rail for the wheelset containing new and worn profiles respectively. As these lateral forces are derivatives of the AoA, the general trend of the variation of lateral forces to yaw torque is very similar to that of the AoA to yaw torque. The lateral track force increases with the negative yaw torque and decreases with the positive yaw torque. The lateral track force produced by the hollow wheel is much larger compared to that produced by the new wheel due to the larger AoA and the difference in wheel-rail contact (two point contact occurs when the wheel is hollow, whilst for the new wheel profile single point contact occurs).

Running the model on smaller or larger curving radius does not change the trend of AoA or the lateral track force to the yaw torque. Generally, smaller curving radius produces larger AoA while larger curving radius produces smaller AoA. Curving radius also affects the lateral forces similarly.

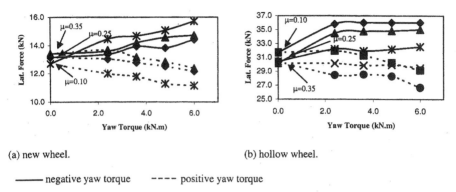

(a) new wheel.   (b) hollow wheel.

——— negative yaw torque   - - - - positive yaw torque

Fig.9. Lateral force on outer rail as function of yaw at 300 m radius curved track.

Safety against derailment is normally examined from the lateral to vertical track force ratio (L/V ratio). Fig.10 exhibits a typical L/V ratio time series of the leading wheelset obtained on a 150 m curve radius track with worn wheel and asymmetric braking applied. The maximum value above 0.6 was recorded for a duration of about 0.8 second just after the wheelset exited the main curve to the transition curve.

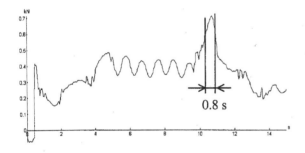

Fig.10. Typical L/V ratio in time series

The maximum L/V ratio for each case is presented in Table 2. These values correspond to the maximum negative yaw torque of 6 kN.m.

Table 2. Maximum L/V ratio

Curve Radius	Max L/V (New Wheel)	Max L/V (Hollow Wheel)
600 m	0.18	0.43
300 m	0.30	0.46
150 m	0.58	0.71

It is seen from Table 2 that reduction in track curving radius increases the maximum value of the L/V ratio for wagons containing either new or hollow wheels. Generally axles containing hollow wheels have produced higher maximum L/V ratios.

5   CONCLUSION

The effect of asymmetric brake shoe force on the curving performance of wagons containing three-piece bogies has been reported. The results show that the asymmetric braking induced negative yaw torque adversely affects the angle of attack and hence the lateral track force. The friction coefficient between the bogie side beam and the axle box is an important factor that affects the degree of braking force influence on the curving performance of wagons. It was shown that the hollow wheels significantly increase the AoA, in particular when the coefficient of friction between the axle box and the side frame is lower. The hollow wheel also increases the L/V ratio during curving.

ACKNOWLEDGEMENT

This work is a part of an ongoing research on the effect of brake force application to the wagon and bogie dynamics jointly funded by the Centre for Railway Engineering (CRE), Central Queensland University (CQU) Australia and Indonesian Railways Industry (PT INKA).

REFERENCES

[1]   Berghuvud. A, *Freight Curving Performance in Braked Condition*, Proceeding of the Institution of Mechanical Engineers, Part F Journal of Rail & Rapid Transit, V216, 2002, pp23-29.

[2]   Suda Y and Grencik J, *The mechanism of enhanced curving performance of unsymmetri suspension trucs under acting traction/brake torque,* Proceeding of the the 14th IAVSD symposium, University of Michigan, 1995, pp 629-640.

[3]   Qian Lixin and Cheng Haitao, *Threedimension dynamic response of car in heavy haul train during braking mode,* Proceeding of the 7th International Heavy Haul Conference, 2001, pp 231-238.

[4]   S Mace, R Pena, N Wilson and D DiBrito, *Effect of Wheel-rail contact geometry on wheelset steering forces,* Wear 191,1995, pp 204-209.

[5]   Kevin J. Sawley, *Wheel/Rail profile maintenance,* Proceeding of WCCR conference, Koln, Germany, 2001.

[6]   AEA Technology Rail, *VAMPIRE Software Manual,* 2003

[7]   Sun Y.Q, Dhanasekar.M, Roach D, *A three-dimensional model for the lateral and vertical dynamics of wagon-track systems,* Proceeding of the Institution of Mechanical Engineers, Part F Journal of Rail & Rapid Transit, V217 Iss.1, 2003, pp31-45.

Vehicle System Dynamics Supplement 41 (2004), p.123-132   © Taylor & Francis Ltd.

On the interaction and integration of wheelset control and traction system

T. X. MEI [1] AND J. W. LU [1]

SUMMARY

This paper carries out a fundamental study of integrating the control of railway wheelset with the provision of traction. Three key elements where critical interactions are likely to occur are considered. One is the traction subsystem including a traction motor and the transmission dynamics; the second is the active control with real dynamics of an electro-mechanical actuator; and finally the flexibility of the axle is also considered. A number of active control strategies and actuation configurations are investigated to study the effect of the interactions on different control methods. The stability of the system as a whole is analysed and performances are assessed.

1. INTRODUCTION

Recent research has shown that wheelsets may be controlled actively in a manner that the creep force at the wheel-rail contact is significantly reduced [1]. It has been recognised that active steering can provide the critical stability control for railway vehicles and at the same time offer almost ideal curving performances - something extremely difficult for the conventional passive solutions. Different active steering strategies and configurations have been proposed and their performances assessed [1]. Experimental work is also being carried out to prove the effectiveness of active approaches [2]. On the other hand modern railway vehicles have been moving towards multiple units and distributed traction systems for increased availability of tractive and braking effort. However the delivery of tractive effort and the stability control of railway wheelset are both achieved via the contact force at the wheel-rail interface, and interactions between the two subsystems will be difficult to avoid.

Whilst the use of passive suspensions is still the only means of stabilisation for modern railway vehicles, this is less a problem as passive elements are simpler and easier to deal with compared with active controls and more importantly there is no additional energy supplied into the system. This is not to say that there is no adverse effect of the interactions, and in fact some studies have been carried out to assess the

---

[1] School of Electronic and Electrical Engineering, The University of Leeds, Leeds, LS2 9JT, UK.
Tel: 44(0) 113 343 2066, Fax: 44 (0) 113 343 2032, E-mail: t.x.mei@ee.leeds.ac.uk

dynamic behaviour of rail vehicles under the influence of traction, e.g. a study was carried out to assess the performance of various vehicle configurations during curve negotiation when traction or braking torque is applied [3].

However, once the active steering is used as the primary control of wheelset dynamics for its obvious advantages, some more fundamental issues must be addressed and it is essential to establish the necessary understanding that will enable effective design of an integrated control system with optimised solutions for the provision of traction in actively steered railway vehicles of the future. Some studies have explored the possibility to actively steering independently-rotating wheels using separately controlled traction motors [4-6]. A paper represented at IAVSD'01 in Copenhagen has reported some initial findings of a systems study [7]. This paper is to report further progress made since then, mainly concentrating on dynamic interactions between the two sub-systems and the effect of different active control approaches and configurations on the overall stability as well as performance.

2. CONFIGURATION AND MODELLING

The study is mainly focussed on a single railway wheelset, where primary/critical interactions between the active steering and traction sub-systems take place. On a complete vehicle, other components such as bogie and body frames will also have some influence via suspension connections. But those are not considered essential and will therefore only be included at a later stage of the study. Figure 1 gives a simplified plan view diagram of the configuration used in the study.

Two wheelset configurations are considered – the solid-axle wheelset and the independently rotating wheelset. For the former, the axle flexibility is considered in the study, and the stiffness is set at a value for the relative rotational mode of the wheels of 40-60Hz, a level of typical axles. The use of a flexible axle is both realistic and necessary for the study, because otherwise some interactions between the longitudinal motion and other wheelset modes may be unrealistically reduced. The wheelset is connected to a frame (or ground) in the lateral direction via typical primary suspensions. Also mounted on the axle are a traction motor for the provision of tractive effort and an actuator for the implementation of active control.

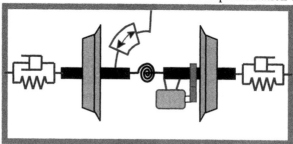

Figure 1. Wheelset configuration.

All motions of the wheelset relevant to the study are considered in the modelling, which are the lateral, yaw and longitudinal motions of the wheelset as well as the rotational motion of the wheels. Passive components (the springs and dampers) are considered to be linear. A linearised creep law is used for stability/eigenvalue analysis and initial performance assessment – a standard approach commonly used for the modelling of the wheel-rail contact mechanics in many studies [8-10]. For more detailed studies, the effect of reduced creep coefficients due to traction and that of non-linear wheel/rail profiles should also be included in simulations [7].

The traction subsystem consists of a DC motor and the traction transmission dynamics. Modern railway vehicles are typically equipped with AC motors for traction. However advances in high power switching devices as well as motor control methods have enabled an induction motor to behave very similarly to a separately excited DC motor in the range of frequencies of interest for this application. Therefore the complexity of the AC traction motor and its associated power electronics and control is substituted in this work by a DC motor with the separate excitation control of the torque and flux producing currents, with the reasonable expectation that no significant difference will be introduced in the dynamic behaviour of the wheelset.

The dynamics of a typical electro-mechanical actuator are modelled for the implementation of active steering [2]. The actuator is normally placed such that it is to provide a net yaw torque for controlling the wheelset, as illustrated in Figure 1. Studies have suggested that this is less demanding on the actuator than the use of an actuator in the lateral direction and also the arrangement does not have any adverse effect on the ride quality [1], but the lateral actuation is nevertheless included in the study to investigate if it would have a different effect in terms of interaction. When passive stabilisation is considered, yaw stiffness is used to replace the actuator.

A list of relevant parameters used in the study is given in Table 1.

Table 1. A list of parameters.

Friction coefficient	0.05 - 0.4
Semi space of wheelset	0.7 m
Lateral stiffness	511 kN/m
Lateral damping	37 $kN s/m$
Lateral creep coefficient	10 MN (nominal)
Longitudinal creep coefficient	10 MN (nominal)
Motor armature resistance, and inductance	0.04 Ohm; 0.4 mH
Motor rotor/gearbox inertia, and gear ratio	11 $kg\ m^2$ (nominal); 5
Motor (machine) constant	4.28 Nm/A
Travel speed	83.3 m/s
Wheel conicity	0.2 (nominal)
Wheelset mass, yaw inertia, rotational inertia	1250 kg; 600 $kg\ m^2$; 35 $kg\ m^2$
Wheel radius	0.45 m
Yaw stiffness (passive only)	4.7 $MN s/rad$

3.   CONTROL STRATEGIES

There have recently been a number of studies on the control strategies for the active steering of a railway wheelset, and a comprehensive review of configurations, control options and design methodologies is given in [1]. In this study, several control strategies are considered in order to investigate their effect on the overall stability and performance especially in the presence of a traction subsystem. The basic concepts of the selected controllers are highlighted below and references are provided for further particulars. Control techniques for traction and actuator sub-systems are well established and a brief description of the controllers used in the study is given.

Control of Solid Axle Wheelset. A solid axle wheelset can be stabilised using the conventional *Yaw stiffness stabilisation* (C1), which can be provided either passively or actively. It can be controlled by *Active yaw damping* (C2), where a yaw torque is applied to be proportional to the lateral velocity of the wheelset [2, 8]. In some cases a phase lead compensator is added to the active yaw damping for the improvement of stability margins [9], which leads to *Active yaw damping with phase compensation* (C3). A full state feedback controller may be designed using a yaw actuator [10] - *Active yaw optimal control* (C4), or a lateral actuator. For the latter configuration, *Active lateral damping* (a lateral force proportional to the yaw velocity of the wheelset) is also effective [2].

Control of Independently-Rotating Wheelset. An instability problem associated with the independently-rotating wheelset has been observed experimentally [11], and later explained in a theoretical analysis [8]. This instability has been found to be easier to control than the kinematic mode of the solid-axle wheelset. Control methods such as *yaw damping* (CI1) and *active yaw moment control* (CI2, - yaw torque proportional to the lateral acceleration of the wheelset) have been shown to provide a high level of damping, but the passive yaw stiffness has a very limited effect [12].

Control of Traction Motor. As mentioned earlier, a separately-excited dc motor for traction is used in the study for simplicity. Classical PI controllers are used and tuned to give a response time of motor torque in the range of 10-20ms. The slip/slide protection, which is usually a part of the traction control, is not included because the study is focussed on the dynamic characteristics of the system before a slip/slide occurs with the ultimate goal of preventing it from happening.

Control of Actuator. There are two different approaches to control the electro-mechanical actuator used to implement the active steering. One is to control the motor current, which is very simple and convenient to achieve. But the drawback is that there is no control on the actuator mode due to the effective moment of inertia of the motor/gearbox and the stiffness in the linkage between the actuator and the wheelset axle, which can compromise the effectiveness of active steering. The other approach is to control the actuator torque delivered at the wheelset, which would definitely give an improved response but at the expense of extra sensors.

4. SIMULATIONS AND ASSESSMENTS

Gains and parameters of the control strategies selected above are tuned to provide the best stability margins possible at the speed of 300 *km/h*.

4.1 SOLID-AXLE WHEELSET

An initial assessment is carried out by studying the effect of those controllers on each of the wheelset modes for the axle being absolute rigid (ideal case) or slightly flexible (more realistic case). Table 2 lists all frequencies and damping levels of three wheelset modes - kinematic, high frequency and torsional. The high frequency mode is associated with the creep and wheelset mass/inertia. In normal circumstances, the mode is well damped and its frequency is quite high – hence the term high frequency mode. The frequency decrease as the travelling speed increases.

Table 2: Eigenvalues of the wheelset with different controllers (V=300 km/h).

Axle	Control Options	Kinematic mode Frequency (Hz)	Damping (%)	High freq. mode Frequency (Hz)	Damping (%)	Torsional mode Frequency (Hz)	Damping (%)
	No	9.1	-24	36	97	N/A	
	C1	9.8	3.0	34	96	N/A	
Solid	C2	12.3	13	15; 46	100	N/A	
	C3	6.4	33	38	33	N/A	
	C4	9.3	23	36	97	N/A	
	C5	15.5	21	9; 50	100	N/A	
	No	9.2	-25	39; 89	100	51	51
	C1	9.6	4.0	41; 86	100	49	49
60Hz	C2	12.0	21	25; 91	100	48	56
	C3	6.3	29	45	13	63	78
	C4	8.7	23	39; 94	100	52	46
	C5	16.8	15	9.2; 96	100	56	59
	No	9.3	-27	36; 122	100	29	37
	C1	9.8	5.0	36; 124	100	30	22
40Hz	C2	11.0	30	35; 122	100	26	22
	C3	6.1	25	39	-0.2	46	98
	C4	8.2	22	36; 122	100	34	30
	C5	17.8	-0.4	9.6; 124	100	30	70

For the rigid axle, it is clear that the active yaw damping (C2) has a stabilising effect on both kinematic and high frequency modes of the wheelset. If a phase lead compensator is also used (C3), which is necessary in some cases to improve the damping ratio of the kinematic mode, the high frequency mode is pushed towards the imaginary axis. The passive stabilisation and the optimal control stabilise the kinematic mode, but the effect on the high frequency mode is marginal. An examination of the optimal control gains reveals that it relies heavily on a feedback of yaw movement – a key reason why the effect of two strategies (C2, C4) are similar. The active lateral damping (C5) can be used to stabilise the kinematic mode with no adverse impact on the other two modes, but it appears to significantly increase the frequency of the kinematic mode.

When the axle flexibility is considered, the active yaw damping with a phase lead compensator and the active lateral damping are affected the most. For C3, the high frequency mode appears to be moving even further towards the unstable region, although the torsional mode becomes more stable. For C5, the axle flexibility has a de-stabilising effort on the kinematic mode.

The main source of interaction between the traction and active steering sub-systems is at/via the wheel-rail interface. Creep coefficients are known to reduce with the increase of creep, especially in poor contact conditions when the friction coefficient is low. In the design for stabilisation, the effect of reduced creep coefficient is often examined using expected variations. However the application of tractive effort can make the problem much worse as it can significantly increase the creep in the longitudinal direction (upto 10% of the normal force or even more in braking). Therefore it is necessary to extend the study into where the creep is near the point of traction slip. Figures 2 and 3 compare how the kinematic and high frequency modes are affected. In Figure 2, the kinematic mode of an uncontrolled wheelset is unstable as expected and the frequency of the mode decreases as the effective creep coefficient decreases. The passive stabilisation (C1) appears to be reasonably robust against the variation, and it remains stable for the entire range examined. The active yaw damping with phase compensation (C3) and the active optimal control (C4) are shown to provide even better robustness for this mode, with a high level of damping achieved throughout. However the two (pure) active damping techniques (C2 & C5) are problematic and the mode becomes unstable when the creep coefficient is reduced to around 60-70% of the nominal values.

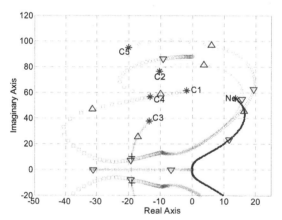

Figure 2. Migration of kinematic mode with creep coefficient.

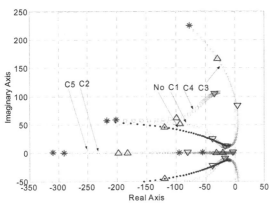

Figure 3. Migration of high frequency mode with creep coefficient.

As shown in Figure 3, the high frequency mode is well damped for the uncontrolled wheelset where the damping is mainly from the creep forces. Most control strategies either do not interfere significantly with the mode (C1, C4) or tend to improve the damping (C2, C5). The only exception is the controller C3, and the additional phase compensator appears to have a destabilising effect on this particular mode. This instability is made worse by a softer axle, as indicated in Table 2. In addition, the frequency of the mode decreases with the effective creep coefficients and it can be as low as that of the kinematic mode at high speeds. A more detailed study is needed to establish its implications on both system stability and noise generation at the wheel-rail interface. The high frequency mode has been omitted in many studies for vehicle dynamics, which is probably justified for conventional vehicles as the passive stabilisation has little effect on it as demonstrated in Figure 3. At lower speeds, this is less a concern as the frequency of the mode will be too high (in high tens or even hundreds) to be significantly affected by the control schemes, which must be implemented via some form of a mechanical device.

The extra moment of inertia attached to a wheelset due to traction equipment reduces the frequencies of the high frequency and torsional motions in general, the extent of which is dependent on the control strategies. The wheelset kinematic mode is not affected, which is expected. The presence of an extra mass tends to help and stabilise the high frequency mode which is made less stable by the active yaw damping with phase compensation (C3), in particular when the axle flexibility is taken into account. The effect of the mass on the torsional motion is not critical as this mode is well damped. Figure 4 shows how different modes of the wheelset with the controller C3 are affected by the variation of traction mass/inertia.

When the full dynamics of the traction and the actuator for active control are considered, interactions between different modes (the traction, the axle and the actuator modes) are also evident if the frequencies of those modes are close to one another. However the effect of the interactions is largely transient, e.g. it is found that the stiffness in the transmission dynamics of the traction sub-system has an adverse effect on the transient performance of the wheelset. When an inappropriate stiffness is used, the traction dynamics may interfere with the wheelset kinematical mode and the phenomenon of a beating between two modes with similar frequencies is observed.

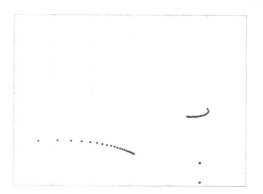

Figure 4. Migration of wheelset modes with traction mass/inertia.

4.2 INDEPENDENTLY-ROTATING WHEELSET

The instability of an independently-rotating wheelset (IRW) is caused due to a combination of the conicity and the need of a longitudinal creep to rotate the wheels. Although the unstable mode is still referred to as the 'kinematic' mode for convenience, the nature is quite different from that of a solid axle wheelset. The required creep is relatively small and hence the instability is not as severe – a high level of damping can be readily achieved either passively or actively [12].

Consequently the control effort to stabilise the mode is much less demanding, and the control strategies tend to be very robust against parameter variations such as reduced effective creep coefficients, as shown in Figures 5, or increased rotation/yaw inertia due to traction. The study has also revealed that the stability of the high frequency mode of an independently-rotating wheelset is not greatly affected by above mentioned variations. For the uncontrolled and controlled (with CI1 or CI2) IRW, this mode remains over-damped in all cases although its frequency varies in a similar manner as that of the solid-axle wheelset.

However care must be taken for a potential problem associated with the actuator. The small longitudinal creep forces at the wheel-rail interface are the main reason for the use of IRWs, but on the other hand it may not be sufficient to provide adequate damping for the actuator. The study has found that it is more likely that an actuator used for IRW control to become unstable than that for the solid-axle wheelset. A local damping for the actuator mode may be necessary, which is straight-forward to implement.

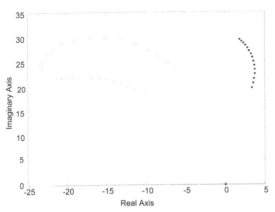

Figure 5. Migration of 'kinematic' mode (IRW) with creep coefficient.

5. CONCLUSIONS AND FURTHER WORK

This paper has presented a detailed systems study on the interactions between two subsystems - the active control of a railway wheelset and traction control. Two wheelset configurations have been used in the study – the conventional solid-axle wheelset and the independently-rotating wheelset. In general, it has been found that the interactions associated with the solid-axle wheelset can be profound. Some active control strategies appear to be less robust in stabilising the kinematic mode in large creep conditions, whereas others may potentially cause instability or poor performance at other frequencies, e.g. that of the high frequency mode. The flexibility of the wheelset axle has been shown to play a significant role in the effectiveness of active controls. The torsional mode associated with the flexibility is stable, but other modes of the wheelset can become unstable if a soft axle is used. On the other hand, the active control for the independently-rotating wheelset has been shown to maintain the stability under the influence of traction without serious problem, largely because the wheelset is much easier to stabilise in the first place.

Although various studies including this one have shown the independently-rotating wheels to be advantageous in many aspects once active steering is used, the solid-axle wheelset is still preferred by rail industry for its mechanical robustness - in short term at least. A systems solution to overcome the problems of the solid-axle wheelset revealed in this study may lie with a combination of two measures. At the wheelset level, a more robust control strategy is necessary to take into account of all modes that can potentially become unstable. At the system level, a supervisory control is required to co-ordinate the active control with the application of tractive effort such that the interactions can be minimised, in particular in poor wheel-rail contact conditions.

ACKNOWLEDGMENT

Authors wish to acknowledge the support of the UK research council EPSRC for funding the project GR/R51636, which made this study possible.

REFERENCES

1. Mei T.X. and Goodall R.M. "Recent development in active steering of railway vehicles", Journal of Vehicle System Dynamics (accepted for publication).
2. Pearson J.T., Goodall R.M. Mei T.X., Shuiwen S, Kossmann C., Polach O. and Himmelstein G. "Design and experimental implementation of an active stability system for a high speed bogie", 18th IAVSD Symposium: Dynamics of Vehicles on Roads and Tracks, Japan, 2003
3. Suda Y. and Grencik J. "Influence of traction/brake force on steering ability of railway trucks", Trans. Japan Society of Mechanical Engineers, Part C, v62, Apr 1996, pp.1361-1366.
4. Mei T.X., Li H., Goodall R.M. and Wickens A.H. "Dynamics and control assessment of rail vehicles using permanent magnet wheel motors", Vehicle System Dynamics, Vol. 37 Supplement, 2002, pp 326 – 337.
5. Gretzschel, M and Bose, L "A new concept for integrated guidance and drive of railway running gears", 1st IFAC Conference on Mechatronic Systems, Vol. 1, pp 265-270, Darmstadt, Germany, September, 2000
6. Liang B, Iwnicki S.D. and Swift F.J. "Simulation of the behaviour of a railway vehicle with independently-driven wheels", 2nd IFAC Conference on Mechatronic Systems, Vol. 1, pp 819-823, Berkeley, USA, December, 2002
7. Mei T.X. Goodall R.M. and Wickens A.H. "A Systems approach for wheelset and traction control", Vehicle System Dynamics, Vol. 37 Supplement, 2002, pp. 257 – 266
8. Goodall, R.M. and Li, H. "Solid axle and independently-rotating railway wheelsets - a control engineering assessment ", Vehicle System Dynamics, Vol. 33, 2000, pp.57-67
9. Mei T.X. and Goodall R.M. "Modal control for active steering of railway vehicles with solid axle wheelsets", Vehicle System Dynamics, 34/1, 2000, pp. 25-41
10. Mei T.X. and Goodall R.M. "LQG solution for active steering of solid axle railway vehicles", IEE Proceedings - Control Theory and Applications, Vol. 147, 2000, pp.111-117.
11. Eickhoff, B.M., 1991, "The application of independently rotating wheels to railway vehicles", IMechE Proc. (Part F), Vol. 205, 1991, pp.43-54.
12. Mei, T.X. and Goodall, R.M. "Practical Strategies for Controlling Railway Wheelsets With Independently Rotating Wheels", ASME, Journal of Dynamic Systems, Measurement and Control, Vol. 125, September. 2003

Equivalent conicity and curve radius influence on dynamical performance of unconventional bogies. comparison analysis

JAVIER SANTAMARIA AND ERNESTO G. VADILLO[1]

SUMMARY

In this paper the dynamic behaviour of several unconventional bogies is compared. The study takes into account the radius of the curve and the maximum level of wear allowed to the wheels. Ranges of conicities and curve radius in which each bogie is advantageous are studied, for both high-speed and urban transit vehicles. Simulations have been carried out using an in house software that makes it possible to solve the wheel-rail contact problem in 3D, and to simulate accurately the negotiation of very sharp curves.

1. INTRODUCTION

Unconventional bogies arise in order to improve curve negotiation of railway vehicles, forcing the wheelsets to adopt a radial position on curve. Several papers show the advantages of such bogies under certain conditions. The aim of this paper is to analyse and compare the curving behaviour of several unconventional bogies, taking into account the curve radius and the maximum level of wear allowed to the wheels. This condition is considered in this work using as a parameter the equivalent conicity of the wheels. As wheel profiles wear out, the equivalent conicity increases reducing dramatically the vehicle stability. It will be necessary in these cases to increase primary suspension stiffness in order to retrieve initial stability. Such stiffness increase, when necessary, has a negative influence on curving behaviour, and therefore an equivalent conicity should be included as a parameter affecting curving response.

Five different bogie configurations have been studied in this paper: A) Two-axle conventional bogie; B) Radial bogie, with elastic connections between

---

[1] *Address correspondence to:* Department of Mechanical Engineering, University of the Basque Country, Bilbao, 48013 SPAIN. Tel.: +34 946014230; Fax: +34 946014215; E-mail: impsamaj@bi.ehu.es

wheelsets (Fig. 1 left); C) Conventional bogie in which axle boxes of the same side of the wheelsets are linked to an equalising bar articulated on bogie frame (Fig. 1 right). In this way, angles of attack of both wheelsets tend to be opposite and radial to the curve; D) and E) bogies are B and C bogies respectively with independently rotating wheels (IRW), only in the trailing axle [1]. Two different passenger services have been analyzed, High Speed and Urban Transit.

Fig. 1. On the left, radial bogie with elastic connections between the wheelsets (config. B) and on the right, bogie with longitudinal connections articulated on bogie frame (config. C).

An in house software [2] has been used to carry out all the necessary calculations. Three types of analysis can be made with this software: a) Stability linear analysis, obtaining critical speeds on straight track, b) Non-linear steady-state analysis on curves, obtaining equilibrium position, and c) Dynamic simulations, in which the non-linear equations of motion are numerically integrated.

The wheel-rail contact problem is solved in 3D. This allows taking into account the influence of the angle of attack on contact parameters, and in this way, it is possible to study the vehicle response when negotiating very sharp curves on urban transit (R<30m). Figure 2 shows three different results obtained for wear index of a bogie, when negotiating several curves, depending on the method of calculation chosen. It is very usual to obtain the wheel-rail contact point solving a 2D problem. In this way, the angle of attack is not taken into account, but it is assumed that its influence is very low. Another way to solve the problem is to linearize the wheel and rail profiles around the contact point, and estimate the influence that the angle of attack has on the location of this point. As seen in Figure 2, linear estimation gives excellent results for large and medium curve radii. However, differences with 3D analysis become more significant when very sharp curves are computed (approximately radius of 15 m). Regarding the 2D method, it is clear that is not valid if sharp curves of radius below 125 m. have to be studied.

In order to obtain creep forces, FASTSIM [3] algorithm has been used, both in quasiestatic and dynamic analysis. There are no simplifications regarding low angles or secondary accelerations in the formulation of the equations of motion and equilibrium.

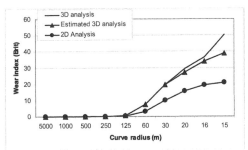

Fig. 2. Wear index differences depending on the type of method used to solve wheel-rail contact problem.

2. METHODOLOGY AND PARAMETERS DESCRIPTION

For each configuration, linearized stability analyses are computed, and primary suspension is optimized according to the conicity value adopted. Generally, the aim of the analysis is to determine a longitudinal stiffness as low as possible compatible with stability criteria and the conicity adopted each time. A low value of longitudinal stiffness allows the bogie to negotiate curves more easily, because each wheelset has more freedom to become radial to the curve.

Once the primary suspension is defined, an exhaustive curving performance is studied for each bogie configuration, each conicity and each curve radius. In order to compare the curving behaviour of the bogies and to determine which configuration is advantageous against the others, it is necessary to compute representative parameters of curve negotiation. In this paper, four different parameters have been considered:

a) Wear index: This is the most important index, due to its direct influence in costs of maintenance, safety and comfort levels. In this work, wear index is computed according to the expression [4]: $W = T^*\gamma / A$ (N/m^2), where $T^*\gamma$ is the scalar product of friction force and the creepage and A is the contact area. Another way to calculate wear index is according to [5] as $W = 0.005\ T^*\gamma$ for $T^*\gamma < 160$ N and $W = 0.025\ T^*\gamma$ for $T^*\gamma > 160$ N. Actually the results are very similar for both cases.

b) Lateral force distribution among wheelsets: When negotiating a curve, a total lateral force appears in each wheelset in order to balance the whole centrifugal force of the bogie. The optimum situation will occur when both lateral forces, the leading wheelset's one ($H1$) and the trailing wheelset's one (Ht) are balanced ($H1$-Ht=0). In this way, the maximum value will be the lowest possible, and the difference with Prud d'Homme limit will be maximum. Lateral forces difference let us estimate the level of aggression of the vehicle to the track.

c) Angle of attack: In general, as the curve radius becomes smaller, the angle of attack of the leading wheelset increases. This is unfavourable, since it increases the risk of derailment, the wear index and the acoustic emissions. It is intended to obtain as low a value as possible.

d) Risk of derailment: It is estimated using Nadal coefficient, Y/Q, evaluated in the outer wheel of the leading wheelset.

3. URBAN VEHICLES

3.1 LINEARIZED ANALYSIS

In general, urban transit bogies reach maximum speeds of 80 Km/h, so it will be a necessary condition that the critical speed is above 80 Km/h, with a minimal damping ratio of 10%. Curve radii that can appear in this case are very small. The analysis has taken into account curve radii from 5000m to 15m. Conicities analysed vary from 0.05 (which simulates new wheel profiles) to 0.40 (which simulates extremely worn wheel profiles).

Once the vehicle model is linearized, stability analyses are made in order to obtain critical velocity for a wide range of siffnesses K_x (longitudinal primary suspension stiffness) and K_y (lateral primary suspension stiffness). The maximum stiffness value considered has been 2×10^7 N/m in all cases. In this way, stability maps are calculated for each configuration and for each equivalent conicity studied. Through the analysis of these maps, it is intended to determine the optimum couple of K_x and K_y for a good curving behaviour, among all the possible values that achieve a critical speed higher than 80 Km/h. Table 1 summarizes the couples chosen for each bogie configuration and conicity.

Table 1. Chosen couples of longitudinal stiffness (K_x) and lateral stiffness (K_y) for each configuration and conicity value for urban service (N/m).

Conicity	Config. A	Config. B	Config. C	Config. D	Config. E
0.05	$1.1 \times 10^6 / 2 \times 10^7$	$7.3 \times 10^5 / 5 \times 10^5$	$7.5 \times 10^5 / 2 \times 10^7$	$9 \times 10^6 / 1 \times 10^6$	$3 \times 10^6 / 1 \times 10^7$
0.10	$1.7 \times 10^6 / 2 \times 10^7$	$1.1 \times 10^6 / 5 \times 10^5$	$1.2 \times 10^6 / 2 \times 10^7$	$1 \times 10^7 / 1 \times 10^5$	$3 \times 10^6 / 1 \times 10^7$
0.15	$2.3 \times 10^6 / 2 \times 10^7$	$1.4 \times 10^6 / 4 \times 10^5$	$1.6 \times 10^6 / 2 \times 10^7$	$2 \times 10^7 / 1 \times 10^5$	$3 \times 10^6 / 1 \times 10^7$
0.20	$2.9 \times 10^6 / 2 \times 10^7$	$1.6 \times 10^6 / 4 \times 10^5$	$1.9 \times 10^6 / 2 \times 10^7$	$2 \times 10^7 / 2 \times 10^5$	$3 \times 10^6 / 1 \times 10^7$
0.25	$3.4 \times 10^6 / 2 \times 10^7$	$1.9 \times 10^6 / 4 \times 10^5$	$2.3 \times 10^6 / 2 \times 10^7$	$2 \times 10^7 / 2 \times 10^5$	$3 \times 10^6 / 1 \times 10^7$
0.30	$4 \times 10^6 / 2 \times 10^7$	$2.0 \times 10^6 / 6 \times 10^5$	$2.6 \times 10^6 / 2 \times 10^7$	$2 \times 10^7 / 2 \times 10^5$	$3 \times 10^6 / 1 \times 10^7$
0.35	$4.5 \times 10^6 / 2 \times 10^7$	$2.3 \times 10^6 / 1 \times 10^6$	$2.9 \times 10^6 / 2 \times 10^7$	$2 \times 10^7 / 2 \times 10^5$	$3 \times 10^6 / 1 \times 10^7$
0.40	$5.1 \times 10^6 / 2 \times 10^7$	$2.4 \times 10^6 / 8 \times 10^5$	$2.9 \times 10^6 / 2 \times 10^7$	$2 \times 10^7 / 2 \times 10^5$	$3 \times 10^6 / 1 \times 10^7$

In the case of a conventional bogie designed for an equivalent conicity of 0.05, and negotiating a curve with a radius of 30 m, the curving behaviour according to the lateral stiffness K_y for three parameters: wear index, angle of attack and risk of derailment are represented in Figure 3b. Each K_y value has its corresponding K_x value that provide the required critical speed of 80 Km/h (Fig. 3a). In config. A, B and C, as seen in Fig. 3b, the lower K_x (larger K_y), the better curving performance.

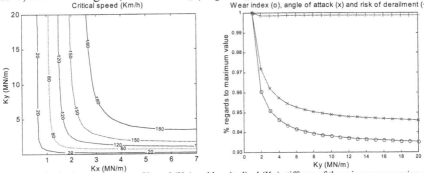

Fig 3. Critical velocity as a function of lateral (Ky) and longitudinal (Kx) stiffness of the primary suspension of an urban conventional bogie with 0.05 equivalent conicity (left), and influence that chosen stiffnesses have on curving behaviour for the same vehicle (right).

Regards to config. D, analysing the curving performance for a representative curve radius and for several values of K_x and K_y, one can see that, on the contrary, wear index and the other parameters decrease very lightly as K_x increases (Fig. 4). The differences are only about 6%. Then, the couples of stiffnesses values are chosen in order to have a good stability, and if possible a high K_x value at the same time, as K_y hardly has influence. Config. E shows a similar behaviour, again with very small differences, and couples of K_x and K_y will be chosen mainly in order to increase stability.

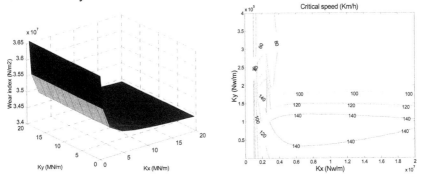

Fig 4. Wear index (N/m^2) according to the primary suspension stiffness of config. D in a 30 m. radius curve (left) and stability map (right) for an equivalent conicity of 0.10.

3.2 CURVE STEADY-STATE ANALYSIS

At this point, the curving performance of all the vehicle configurations described has to be studied, for a wide range of curve radius and conicities. Once all the primary suspension values are determined for each configuration and conicity, steady-state analyses are computed with curve radii from 5000 to 15m.

The best bogie configurations depending on curve radius and conicity are represented in the following pictures, regarding wear index (Fig 5a) and risk of derailment (Fig 5b). Graphs regarding H1-Ht and angle of attack are similar to Figure 5a. It is clear that if a conventional bogie is designed with an optimized primary suspension, its performance when low conicities and not very sharp curves are analysed is not only as good as unconventional's ones, but even much better (curve radius of 150 m approximately), as proved in [6]. Only if the risk of derailment is studied, config. B is the best option in this area, nevertheless the differences being little significant.

Config B also arises as the best option in wear index when curve radius becomes lower than 150 m, specially with high conicities. With extremely small radii, config. D appears as the advantageous option for all type of conicities. It can be seen that in urban service, there are no significant areas where config. C is the best option. On the other hand, config. E only shows improvement in large curves, and if the wheels are allowed to wear out much.

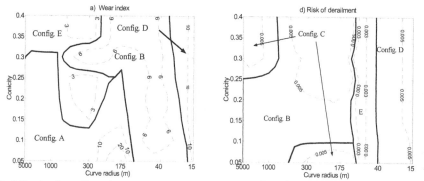

Fig 5. Best bogie configuration according to a) wear index (N/m^2 x 10^5) and b) risk of derailment. Dotted lines show the improvement between the best configuration and the second best configuration in absolute value for both cases.

Figures 6 and 7 show a more specific analysis of configurations B and D, comparing them directly to the conventional bogie. They show the ranges of curve radii and conicities in which their response is better than conventional bogie's one. As said, radial bogies improve curve response in sharp curves, and it can be observed with regards to wear index that config. B is advantageous in a larger area than config. D, but when radii are large config. D behaviour is better, more clearly shown by Hl-Ht graphs. For this reason, both of them can be a good option if very sharp curves are likely to appear in the railway line.

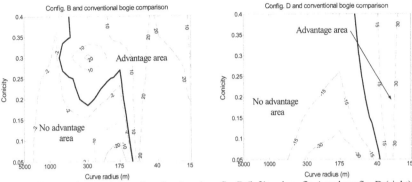

Fig. 6. Differences in wear index of config. A and config. B (left) and config A and config. D (right) in N/m² x 10⁵. Dotted lines show the difference in absolute value.

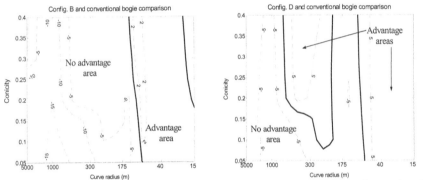

Fig. 7. Differences in lateral forces distribution of config. A and config. B (left) and config A and config. D (right) in KN. Dotted lines show the difference in absolute value.

4.   HIGH-SPEED VEHICLES

4.1 LINEARIZED ANALYSIS

High-speed bogies reach maximum speeds between 200 and 350 Km/h, so it will be necessary to achieve a critical speed of at least 350 Km/h with a damping ratio of 0%, because anti-yaw dampers are usually installed, improving stability. In this case curve radii are generally very large, of approximately 3000 m., although at some points of the railway line radii could be smaller, being necessary to negotiate them at lower speeds. The analysis has taken into account conicities from 0.05 (new wheel profiles) to 0.35 (worn wheel profiles), and curve radii from 5000 m. to 180 m.

In the same way as shown in urban vehicles, Table 2 summarizes the chosen couples K_x/K_y for each case.

Table 1. Chosen couples of longitudinal stiffness (K_x) and lateral stiffness (K_y) for each configuration and conicity value for high-speed service (N/m).

Conicity	Config. A	Config. B	Config. C	Config. D	Config. E
0.05	$2.1x10^6/2x10^7$	$1.2x10^6/2x10^6$	$1.1x10^6/1x10^6$	---	$1.5x10^6/2x10^7$
0.10	$3.8x10^6/2x10^7$	$2.0x10^6/1x10^6$	$1.4x10^6/1x10^6$	---	$1.0x10^6/1x10^7$
0.15	$4.0x10^6/2x10^7$	$2.7x10^6/1x10^6$	$1.8x10^6/2x10^6$	$4.0x10^6/1x10^6$	$1.3x10^6/1x10^7$
0.20	$6.4x10^7/2x10^7$	$3.2x10^6/2.5x10^6$	$2.7x10^6/2x10^6$	$2.5x10^6/2x10^6$	$1.5x10^6/8x10^6$
0.25	$7.0x10^7/2x10^7$	$3.6x10^6/3.5x10^6$	$3.0x10^6/2x10^6$	$3.5x10^6/2.5x10^6$	$1.8x10^6/6x10^6$
0.30	$8.0x10^7/2x10^7$	$4.1x10^6/4x10^6$	$3.3x10^6/2x10^6$	$4.0x10^6/3x10^6$	$2.1x10^6/6x10^6$
0.35	$9.0x10^7/2x10^7$	$4.5x10^6/4x10^6$	$3.5x10^6/2x10^6$	$5.0x10^6/5x10^6$	$2.3x10^6/2x10^6$

Config. D is not able to achieve a critical speed of 350 Km/h for low conicities, disregarding it for subsequent studies. Stiffnesses selected for config. E reaches values up to 1000 Km/h, proving to be a very stable vehicle on straight track. For A, B and C configurations, a critical speed of 350 Km/h is obtained with a damping ratio of 0%.

4.2 CURVE STEADY-STATE ANALYSIS

Figure 8 shows vehicle responses in terms of wear index and angle of attack against curve radius, for the case of a 0.05 conicity. Config. C minimizes these parameters for a wide range of curve radii, while config. E shows a very deficient response in sharp curves. For this reason, config. E will also be disregarded unless there are only large curves in the railway line.

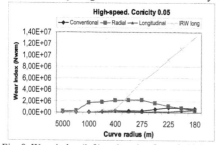 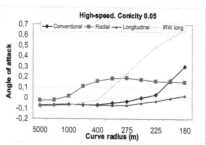

Fig. 8. Wear index (left) and angle of attack (right) versus curve radius, according to four of the analysed bogie configurations (conicity 0.05).

The following graphs show which bogie configurations (A, B or C) is the most advantageous in terms of wear index (Fig. 9a) and lateral forces distribution (Fig.

9b). Differences between configurations in terms of risk of derailment and angle of attack are not significant. Config. E, initially disregarded for its bad behaviour in medium and sharp curves, however shows excellent results in the distribution of lateral forces when large curves are negotiated.

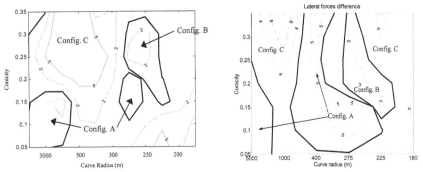

Fig. 9. Best bogie configuration according to a) wear index (N/m² x 10⁵), and b) lateral forces distribution (KN). Dotted lines show the improvement between the best configuration and the second best configuration in absolute value.

It can be seen in these figures that conventional bogie is good enough when curves have large radius and wheel profiles are not allowed to wear in excess. In other cases, config. C arises as the best option. Regarding the Hl-Ht parameter, conventional bogies are also advantageous in curves of medium radius, while config. C is still the best option for a wide range of radii, specially with high conicities.

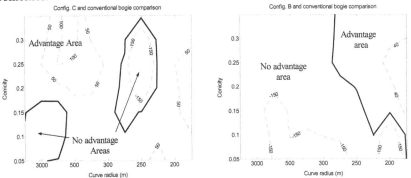

Fig. 10. Differences in wear index of config. A and config. C (left) and config A and config. B (right) in N/m² x 10⁴. Dotted lines show the difference in absolute value.

Figure 10 shows the improvement obtained with config. C (Fig. 10a) and config. B (Fig. 10b) in comparison to the conventional bogie, in terms of wear index. It can be seen that config. C is to be preferred for a wide range of conicities and curve radii, being the improvement more important if high conicities and large curves are considered. On the other hand, it can be observed that configuration B is advantageous in sharp curves in comparison with the conventional bogie, specially with high conicities. When conicities are small (wheel profiles not worn), config. B hardly shows advantageous areas. Moreover, for a wide range of curve radii, not very small (larger than approximately 300 m), its behaviour is much worse than the conventional bogie's one. For this reason, the election must be between conventional and config. C bogies.

5. CONCLUSIONS

In the analysis presented in this paper, it can be concluded that a conventional bogie adequately optimized is good enough or even better than unconventional bogies, when the curve radii are not very short, and the wear level of the wheels profiles is not high. This is more significant in urban service rather than in high-speed vehicles. Nevertheless, if urban vehicles are to negotiate very sharp curves, radial bogies B and D are to be preferred. When high-speed vehicles are analysed, config. C offers a good curving performance for a wide range of conditions. Advantages compared to the conventional bogie are more significant if high conicities or small curve radius are simulated. Config. B does not improve in general conventional bogie response, and when conicity is small, its performance is significantly worse.

6. ACKNOWLEDGEMENTS

This work was partly supported by the Basque Government.

REFERENCES

1. Dukkipati, R. V., Narayanaswamy, S. and Osman M.O.M.: Independently Rotating Wheel Systems for Railway Vehicles – A State of the Art Review. Vehicle System Dynamics, 21 (1992) pp 297-330.
2. Santamaria, J., Vadillo, E.G., Santamaria, L. and Reguart, O.: Dinatren: una nueva herramienta para la simulacion de vehiculos ferroviarios. Anales de Ingenieria Mecanica, (2000) pp 225-232.
3. Kalker, J.J.: Three-Dimensional Elastic Bodies in Rolling Contact. Kluwer Academic Publishers, (1990).
4. Clayton, P., Allery, M.B.P. and Bolton, P.J.: Surface Damage Phenomena in Rails. Proceedings of the Internationsl Symposium of Contact Mechanics and Wear of Rail / Wheel Systems hel at the Univ. of British Columbia, Vancouver 1982.
5. Group of British Rail Research, London: Introduction to Railway Vehicle Dynamics. Traction and Rolling Stock. BRR. 1996
6. García, J.F., Olaizola, X., Martín, L.M. and Giménez, J.G.: Theoretical Comparison Between Different Configurations of Radial and Conventional Bogies. Vehicle System Dynamics 33 (2000), pp 233-259.

Vehicle System Dynamics Supplement 41 (2004), p.143-152   © Taylor & Francis Ltd.

Analysis of Active Suspension Systems with Hydraulic Actuators

XIAOMING SHEN[1] and HUEI PENG[2]

SUMMARY

Most of the existing active suspension studies assume that an ideal force actuator exists and will carry out the commanded force accurately. In reality, due to the interaction between the hydraulic actuator and suspension system, the lightly damped modes of the plant will generate lightly damped zeros (LDZ) for the servo-loop system, which confines the closed-loop performance to a low frequency range. Converting the active suspension problem formulation into a displacement control problem, as suggested by several recent papers, will not solve this problem. Four candidate remedies to reduce the effect of the LDZ are studied and their potentials and limitations are discussed.

1 INTRODUCTION

Active Suspensions (AS) have been widely studied over the last 20 years, with hundreds of papers published (see [1] for an extensive review). Most of the published results focus on the main-loop designs, i.e., on the computation of the desired control force, as a function of vehicle states and the road disturbance. It is commonly assumed that the commanded force will be produced accurately. Simulations of these main-loop designs were frequently done without considering actuator dynamics, or with highly simplified sub-loop dynamics.

In reality, actuator dynamics can be quite complicated, and interaction between the actuator and the vehicle suspension cannot be ignored. This is especially true for hydraulic actuators, which remain to be one of the most viable choices due to their high power-to-weight ratio, low cost and the fact force can be generated over a prolonged period of time without overheating. However, hydraulic actuators also have several adverse attributes: they are nonlinear and their force generation capabilities are highly coupled to the vehicle body motions [2]. Few experimental verification on AS algorithms have been reported and all are limited to bandwidth of 2-4 Hz [3][4][5].

In other applications such as durability tests, hydraulic actuators have achieved bandwidth of 50Hz or higher successfully. However, in these applications, the closed-loop systems are usually designed to track a displacement signal, rather than a force signal. Some researchers thus speculate that for hydraulic actuators, "force tracking" is harder than "displacement tracking". Several papers have been published to recast the force control problem in AS designs to a displacement control problem [6][7].

But do we really have a fundamental limit in using hydraulic actuators for force control in general? Or is it an application-specific phenomenon? For example, AS systems have to settle for a smaller actuator, while for durability test rigs more powerful actuators and higher supply pressures can be used. AS actuators are commonly placed in parallel with existing spring/damper, while the durability test actuators usually are connected in series with

---

[1] Graduate student, School of Mechanical and Power Energy Engineering, Shanghai Jiao Tong University, China
[2] Corresponding author, Associate Professor, Department of Mechanical Engineering, University of Michigan, Ann Arbor, MI 48109-2125, 734-936-0352, hpeng@umich.edu

vehicle suspensions. Were AS researchers attacking the wrong problem by solving a force tracking problem? If there is a fundamental issue related to force control, can we avoid this fundamental limit by formulating a displacement control problem? Are there other remedies? The main purpose of this paper is to provide answers to these questions.

First, we compare the AS servo-loop control problem with that of durability test rigs. For both AS and durability test rigs, linearized equations are used so that transfer functions and frequency response can be used for the analysis. Since the lightly-damped zeros (LDZ) of the closed-loop system are the main source of performance limit [7], we will solve the closed-loop transfer functions of the two tracking problems (both force and displacement) analytically. The poles and zeros of the closed-loop transfer functions will then be analyzed. By doing so, the fundamental limits imposed by the LDZ will be clearly understood. It is shown that the displacement control problem also has its own pair of LDZ. Furthermore, while the natural frequency of the displacement control LDZ is a little higher, their damping ratio is lower. Therefore, switching to a displacement control problem is not a complete answer.

Subsequently, remedies to reduce the adverse effect of the LDZ are studied. Four candidate approaches are analyzed—new actuator; suspension parameter optimization; add-on mode such as vibration absorbers; and advanced control algorithms. Analysis and simulation results are presented to show the effect of the proposed remedies.

2 MODEL DEVELOPMENT

2.1 Quarter-Car Active Suspension Model with a Hydraulic Actuator

In this section, we will present the model of a quarter-car AS system equipped with a hydraulic actuator. The governing equations of the quarter-car suspension system (Fig. 1) are presented. The two degrees of freedom (DOF) are the vertical motions of the sprung mass (m_s) and unsprung mass (m_{us}), respectively. This 2 DOF system has one external input, $\dot{z}_0$, which is the rate of change of road surface elevation. A force (f) is applied by the active suspension actuator on m_s and m_{us} . The state space model is:

$$\frac{d}{dt}\begin{bmatrix} z_{us} - z_0 \\ \dot{z}_{us} \\ z_s - z_{us} \\ \dot{z}_s \end{bmatrix} = \begin{bmatrix} 0 & 1 & 0 & 0 \\ \dfrac{k_{us}}{m_{us}} & -\dfrac{c_s+c_{us}}{m_{us}} & \dfrac{k_s}{m_{us}} & \dfrac{c_s}{m_{us}} \\ 0 & -1 & 0 & 1 \\ 0 & \dfrac{c_s}{m_s} & -\dfrac{k_s}{m_s} & -\dfrac{c_s}{m_s} \end{bmatrix} \begin{bmatrix} z_{us} - z_0 \\ \dot{z}_{us} \\ z_s - z_{us} \\ \dot{z}_s \end{bmatrix} + \begin{bmatrix} 0 \\ \dfrac{1}{m_{us}} \\ 0 \\ \dfrac{1}{m_s} \end{bmatrix} f + \begin{bmatrix} -1 \\ \dfrac{c_{us}}{m_{us}} \\ 0 \\ 0 \end{bmatrix} \dot{z}_0 \qquad (1)$$

The transfer function from actuator force to suspension stroke is then

$$G_p = \frac{z_s - z_{us}}{f} = \frac{(m_s + m_{us})s^2 + c_{us}s + k_{us}}{m_s m_{us}s^4 + (c_s m_{us} + m_s c_s + m_s c_{us})s^3 + (m_{us}k_s + c_s c_{us} + k_s m_s + m_s k_{us})s^2 + (k_{us}c_s + k_s c_{us})s + k_s k_{us}} \qquad (2)$$

By using standard servo-valve dynamic equations, the hydraulic system is described by:

$$\dot{f} = \frac{\sqrt{2}A_p \cdot \beta \cdot k_{xd}}{V} \cdot x_{sp} \cdot signsqrt(P_s - sign(x_{sp})f / A_p) + \frac{2 \cdot A_p^2 \cdot \beta}{V} \cdot (\dot{z}_{us} - \dot{z}_s)$$

$$\dot{x}_{sp} = \frac{1}{\tau}(-x_{sp} + k_{pv}i_{sv}) \qquad (3)$$

where A_p is the piston area, β is the fluid bulk modulus, k_{xd} is the orifice flow coefficient, x_{sp} is the servo valve displacement, $signsqrt(y) \equiv sign(y) \cdot \sqrt{|y|}$, P_s is the supply pressure, V is the cylinder chamber volume, and k_{sv} is the valve gain. Eq.(3) is then linearized, which enables us to do a more in-depth analysis:

$$2C_x x_{sv} + 2A_p(\dot{z}_{us} - \dot{z}_s) = \frac{V}{\beta A_p} f$$ (4)

$$\tau \dot{x}_{sv} + x_{sv} = K_v i_{sv}$$

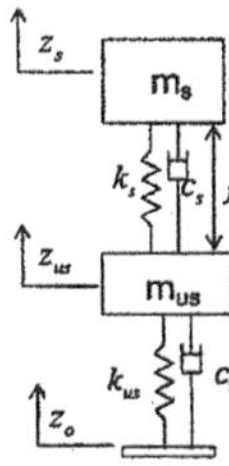

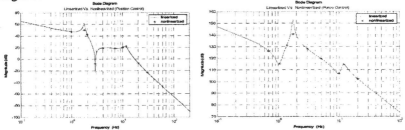

Fig. 1. Quarter-car active suspension model

In Fig. 2, the frequency response of the linear model is plotted against sinusoidal simulation results of the nonlinear system at input magnitude of 7mA (~400N force at 1Hz). It can be seen that Eq.(4) is a good approximation of Eq.(3) under reasonable input magnitude.

Fig. 2. Validation of the linearization results (suspension stroke and force).

Using the linear approximation of Eq.(4), the dynamics of the overall plant with servo valve current i_{sv} as the control input, has the following form

$$\frac{d}{dt}\begin{bmatrix} z_{us} - z_0 \\ \dot{z}_{us} \\ z_s - z_{us} \\ \dot{z}_s \\ f \\ x_{sv} \end{bmatrix} = \begin{bmatrix} 0 & 1 & 0 & 0 & 0 & 0 \\ \frac{k_{us}}{m_{us}} & \frac{c_s + c_{us}}{m_{us}} & \frac{k_s}{m_{us}} & \frac{c_s}{m_{us}} & \frac{-1}{m_{us}} & 0 \\ 0 & -1 & 0 & 1 & 0 & 0 \\ 0 & \frac{c_s}{m_s} & \frac{k_s}{m_s} & \frac{c_s}{m_s} & \frac{1}{m_s} & 0 \\ 0 & \frac{2\beta A_p^2}{V} & 0 & -\frac{2\beta A_p^2}{V} & 0 & \frac{2c_x \beta A_p}{V} \\ 0 & 0 & 0 & 0 & 0 & -\frac{1}{\tau} \end{bmatrix} \begin{bmatrix} z_{us} - z_0 \\ \dot{z}_{us} \\ z_s - z_{us} \\ \dot{z}_s \\ f \\ x_{sv} \end{bmatrix} + \begin{bmatrix} 0 \\ 0 \\ 0 \\ 0 \\ 0 \\ \frac{k_v}{\tau} \end{bmatrix} i_{sv} + \begin{bmatrix} -1 \\ \frac{c_{us}}{m_{us}} \\ 0 \\ 0 \\ 0 \\ 0 \end{bmatrix} \dot{z}_0$$ (5)

145

2.2 Active Suspension—Force Control Problem

Given Eq.(5), the transfer function from i_{sv} to f can be obtained:

$$G_{fi} = \frac{f}{i_{sv}} = 2C_x \beta A_p K_v \frac{m_a m_s s^4 + (c_{ua} m_s + c_s m_s + m_a c_s)s^3 + (k_{ua} m_s + k_s m_s + c_{ua} c_s + m_a k_s)s^2 + (k_{ua} c_s + c_{ua} k_s)s + k_{ua} k_s}{D} \quad (6)$$

where the 6th-order denominator polynomial is

$$D = s(s\tau+1)[m_a v m_s s^4 + (c_{ua} v m_s + c_s v m_s + m_a v c_s)s^3 + (2m_a \beta A_p^2 + k_s v m_s + c_{ua} v c_s + m_a k_s v + 2\beta A_p^2 m_s + k_{ua} v m_s)s^2$$
$$+ (2c_{ua}\beta A_p^2 + c_{ua} k_s v + k_{ua} v c_s)s + 2k_{ua}\beta A_p^2 + k_{ua} k_s v] \quad (7)$$

Similarly, the transfer function from i_{sv} to suspension displacement, $z_s - z_{us}$, is

$$G_{zi} = \frac{z_s - z_{us}}{i_{sv}} = \frac{2C_x \beta A_p K_v [(m_s + m_{us})s^2 + c_{us}s + k_{us}]}{D} \quad (8)$$

Zhang and Alleyne [7] pointed out that the numerator of the transfer function shown in Eq.(6) is the same as the denominator polynomial of the quarter-car suspension plant shown in Eq.(2). This is not a coincidence, and can be explained by formulating a block diagram of the augmented system. Fig. 3 shows a decomposed form of Eq.(5). The "feedback" block $H(s)$ does not represent a feedback control action. Rather, it is an inherent coupling between suspension motion and hydraulic force generation. From Fig. 3 and Eqs.(4)(5), it can be seen that $G_p(s)$ is the suspension dynamics shown in Eq.(2). The force generation dynamics, Eq.(4), can be written as $f = G_a i_{sv} - G_a Hy$, where

$$G_a = \frac{2\beta A_p C_x K_v}{V} \frac{1}{\tau s + 1} \quad \text{and} \quad H = \frac{A_p s(\tau s + 1)}{C_x K_v}.$$

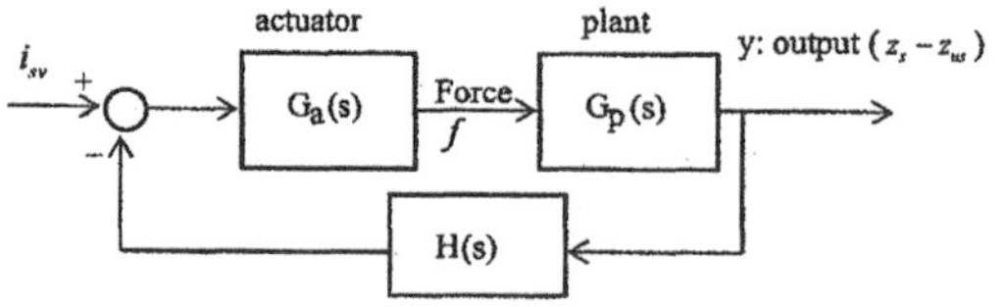

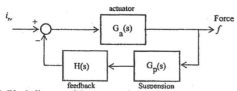

Fig. 3. Block diagram of the suspension displacement sub-loop.

When an engineer is designing the servo-loop control algorithm to ensure proper force generation from the hydraulic actuator, he/she is facing a dynamic system shown in Fig. 4. Again, the two blocks shown in the "feedback loop" were not from a feedback control algorithm, but rather is a physical feedback the control designer has to deal with.

Fig. 4. Block diagram of the actuator force sub-loop.

Given Fig. 4, the "plant' to be controlled is:

$$G_{fi} = \frac{f}{i_{sv}} = \frac{G_a}{1 + G_a G_p H} \qquad (9)$$

Obviously, the poles of the active suspension plant (G_p) become zeros of G_{fi}. From Eq.(6), we can see that this is indeed the case. Why is this a concern? Because the active suspension plant (shown in Eq.(2)) includes two under-damped modes (at around 1Hz and 10Hz). The small damping of the tire mode (~0.01-0.03) and the suspension mode (~0.02-0.3?)implies that the plant of the force servo loop (G_{fi} shown in Eq.(9)) will have a pair of under-damped zeros at around 1Hz. For feedback-only control algorithms, the under-damped zeros will remain to be zeros of the closed-loop system. Therefore, the closed-loop bandwidth will be limited by this pair of lightly damped zeros.

What if we use other control configurations? After all, nonlinear algorithms based on the cancellation of suspension-actuator interactions have been proposed and a somewhat higher closed-loop tracking performance has been reported [8][9]. Let's refrain from analyzing specific control algorithms, and instead examine this problem from a more fundamental viewpoint. Assuming that we know little about the servo-loop control algorithm, except that it is "doing a good job". Based on the "perfect control" analysis [10][11], the "well-behaved" control algorithm will generate an inverse of the plant. In other words, a pair of lightly damped zeros (LDZ) demand much larger control actions at their resonant frequency for acceptable control performance. This is theoretically possible but usually difficult to realize in practice.

From the above analysis, and the fact that hydraulic actuators have been used in displacement-tracking applications such as durability test rigs up to 50-80 Hz, it is tempting to claim that reformulating the AS sub-loop problem (from force control) into a displacement tracking problem will be the answer. As mentioned in Section 1, several prior publications based on this belief were published [6][7][12], and two ways to recast the AS servo-loop into displacement tracking problems were proposed.

2.3 Active Suspension—Displacement Control Problem

In this section, by following the same procedure shown in Section 2.2, we examine whether the performance limitation imposed by the LDZ can be removed in the displacement control problem. The system configuration is still the one shown in Fig. 1, and the control problem is to manipulate the servo-valve current so that a desired suspension displacement is followed. The plant of this control problem is shown in Fig. 3. Its transfer function is:

$$G_{zi} = \frac{z_s - z_{us}}{i_{sv}} = \frac{G_a G_p}{1 + G_a G_p H} \qquad (10)$$

Obviously, the zeros of the suspension plant (G_p) are also zeros of G_{zi}. This fact is confirmed by examining Eqs. (2) and (8). From Eq.(8), it is obvious that the small tire damping creates a pair of LDZ for G_{zi}. The resonant frequency of these zeros is at around 3Hz (vs. 1Hz for the force loop). In other words, the limitation for tracking performance (as analyzed by the "perfect control" analysis) is somewhat relieved, but not by much. In fact, while the resonant frequency of the LDZ of G_{zi} is higher, their damping ratio is

147

much lower. This fundamental performance limit arises from low tire damping, and thus design changes such as using a larger and more powerful hydraulic actuator or a faster servo valve will not help.

3. DURABILITY TEST RIG

Before we discuss possible remedies to address the negative impact imposed by the LDZ, let's first examine why this has not been a problem for other hydraulic applications such as durability test rigs. In Fig. 5, a simplified one degree-of-freedom durability test rig is shown. The mass-spring-damper system shown on the top represents the dominate mode of the tested system. The weight of the ground plate plus the tire(s) is denoted as m_{us}. It is straightforward to find the following two transfer functions

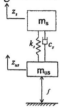

Fig. 5. Durability Test Rig model.

$$G_{p2} = \frac{z_{us}}{f} = \frac{m_s s^2 + c_s s + k_s}{s^2[m_s m_{us} s^2 + (c_s m_{us} + m_s c_s)s + m_s k_s + m_{us} k_s]} \quad (11)$$

$$G_{zi2} = \frac{z_{us}}{i_{sv}} = \frac{2C_x \beta A_p K_v (m_s s^2 + c_s s + k_s)}{D'} \quad (12)$$

where $D' = s(s\tau + 1)[m_{us} vm_s s^4 + (c_s vm_s + m_{us} vc_s)s^3 + (k_s vm_s + m_{us} k_s v + 2\beta A_p^2 m_s)s^2 + 2c_s \beta A_p^2 s + 2k_s \beta A_p^2]$.

The zeros of the plant shown in Eq.(11) are again zeros of the displacement servo-loop, shown in Eq.(12). Since the damping ratio of these zeros is higher than those of Eq.(8) (~0.3 vs. ~0.02), the imposed limitation on tracking performance is much easier to deal with.

4. POSSIBLE REMEDIES

In this section, we will discuss possible remedies (design changes) to address the adverse effects of LDZ on AS systems. These changes aim to eliminate the LDZ, or to increase their damping ratio. Both force and displacement problems are discussed. Here we are not trying to distinguish whether the target application is for narrow band (< 3Hz) or wide band (>10Hz). Rather, we will focus on the influence of the working principle of a force-producing actuator on the servo-loop performance. It should be pointed out that while LDZ problem can be addressed by pole/zero cancellation, this method is not considered because of its poor robustness in real implementations.

4.1 Alternative actuator system

The results presented in this paper so far assume a hydraulic actuator is used. As pointed out earlier, the LDZ arise from the suspension plant (shown in Eq.(2)). Therefore, modification of the hydraulic actuator system (higher supply pressure, larger cylinders, faster servo valves) will not be effective.

When an electromagnetic, a pneumatic or a piezoelectric actuator is used, a "physical feedback" (the H(s) block in Fig. 3) still exists, although in different forms. For example, in the voltage control mode, a DC motor will have back electromotive force, and for pneumatic actuators, volume/pressure coupling effect exists. The fact such physical feedbacks exist implies that LDZ are present and pose a problem for the control design.

When an actuator without physical feedback is used, the servo-loop is no longer in the form shown in Fig. 3, and the analysis results discussed in the previous two sections need to be modified. An example of force actuators that do not exhibit a physical feedback path is electromagnetic motors running under current drive mode. Since the motor current is supplied regardless of the motor speed, the servo-loop control system is independent of the suspension plant, which is depicted in Fig. 6. Obviously, since $G_p(s)$ is outside of the servo-loop, the LDZ no longer exists for the force sub-loop. For the displacement tracking problem, however, LDZ still exist because $G_p(s)$ is inside the servo-loop.

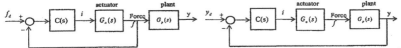

Fig. 6. Block diagrams of the force sub-loop and displacement sub-loop for a current-mode drive electric actuator.

4.2 Effect of modified suspension parameters

In this sub-section, we will examine the effect of modified suspension parameters on the characteristics of the LDZ. In particular, we will focus on their effect on the damping ratio of the LDZ. We assume that the sprung mass is given and cannot be assigned. The nominal values of the suspension parameters are shown in Table 1.

Table 1. Nominal values of the suspension parameters.

Parameter	Meaning	Value	Parameter	Meaning	Value
m_s	Sprung mass	$253kg$	c_{us0}	Tire damping	$10N/m/sec$
m_{us0}	Unsprung mass	$26kg$	k_{s0}	Suspension stiffness	$12000N/m$
c_{s0}	Suspension damping	$348.5N/m/sec$	k_{us0}	Tire stiffness	$90000N/m$

4.2.1 Displacement control problem

Only three parameters will influence the LDZ: m_{us}, c_{us} and k_{us} (see Eq.(2)). Apparently, the most important parameter is the tire damping c_{us}. Increasing c_{us} will increase the damping ratio of the LDZ. However, since this will also increase rolling resistance, it is quite obvious little room is left for the designer.

4.2.2 Force control problem

Five parameters will influence the LDZ (numerator of Eq.(6)): c_s, k_s, m_{us}, c_{us} and k_{us}. Note that there are two pairs of LDZ. We varied all five parameters within a range of their nominal values, depending on our judgment of what could be achieved in practice. It was found that perturbing suspension damping (c_s) or suspension stiffness (k_s) are most effective. The effect of c_s and k_s on the smaller of the two damping ratios is shown in Fig. 7. The effect of c_s is not monotonic because the faster mode becomes less damped than the slower mode at high frequency. We can increase LDZ damping by either increasing c_s or decreasing k_s. For example, if we increase c_s by a factor of 7, the damping ratio will become 0.6, high enough that the zeros no longer pose a problem for the control design. Smaller improvement can be achieved by reducing k_s. Changing these two parameters needs to be done carefully by weighing against their adverse effect—reduced actuator energy efficiency and increased rattle space, respectively.

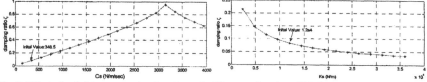

Fig. 7. Effect of perturbing suspension damping and stiffness on damping of LDZ.

To study the effect of c_s, an AS efficiency is defined as $\eta = 1 - E_{wasted} / E_{total}$, where the total input energy E_{total} and the energy absorbed by c_s, E_{wasted} are

$$E_{total}(t) = \int_0^t f \cdot (\dot{z}_s - \dot{z}_{us}) d\tau, \quad f \cdot (\dot{z}_s - \dot{z}_{us}) > 0$$

$$E_{wasted} = \int_0^t \min\{f \cdot (\dot{z}_s - \dot{z}_{us}), c_s(\dot{z}_s - \dot{z}_{us})^2\} d\tau, \quad f \cdot (\dot{z}_s - \dot{z}_{us}) > 0 \tag{13}$$

The force f used is commanded from a main-loop LQ algorithm, where the control gains are re-computed for each perturbed c_s. The relationship between c_s and η is shown in Fig. 8(a). Increasing c_s results in an almost monotonic reduction in η. If we increase c_s by a factor of 7, η will reduce from 60% to 20%. This may be unacceptable because power consumption and cooling requirement. The most obvious adverse effect of a softer suspension (smaller k_s) is increased rattle space (shown in Fig. 8(b)) and body leaning during cornering. In other words, in practice there is limited room for a control engineer to improve the damping of LDZ by reducing k_s.

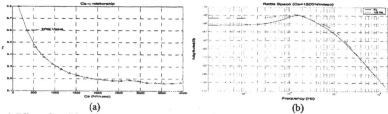

(a)   (b)

Fig. 8. Effect of modified suspension damping and stiffness.

4.3 Vibration absorber

A vibration absorber (VA) refers to a single degree of freedom mass-spring oscillator with low (no) damping. The basic idea for using a vibration absorber is that the mode associated with the additional "energy storage buffer" creates a pair of zeros at the designed frequency. Here we introduce a lightly-damped vibration absorber attached to the unsprung mass (see Fig. 9). The suspension dynamics become

$$G_p = \frac{z_s - z_{us}}{f} = \frac{m_a(m_{us} + m_s)s^4 + (m_a c_{us} + m_{us} c_a + c_a m_a + c_a m_s)s^3}{d} + \frac{(k_a m_a + k_a m_s + m_{us} k_a + k_{us} m_a + c_{us} c_a)s^2 + (c_{us} k_a + k_{us} c_a)s + k_{us} k_a}{d}$$

$$\approx_{m_a \ll m_s} \frac{(m_a s^2 + c_a s + k_a)[(m_{us} + m_s)s^2 + c_{us} s + k_{us}]}{d} \tag{14}$$

where, by using the facts $m_a \ll m_s$, $c_a \ll c_s$ and $k_a \ll k_{us}$

$$d \approx (k_a + c_a s + m_a s^2)(m_s s^2 + c_s s + k_s)(m_{us} s^2 + c_{us} s + k_{us}) + (k_a + c_a s + m_a s^2)(c_s m_s s^3 + k_s m_s s^2)$$

$$= (k_a + c_a s + m_a s^2)[(m_s s^2 + c_s s + k_s)(m_{us} s^2 + c_{us} s + k_{us}) + k_s m_s s^2 + c_s m_s s^3] \tag{15}$$

Notice that both the numerator and denominator approximately contain a vibration mode $(k_a + c_a s + m_a s^2)$, which results in a near-pole-zero cancellation for G_{fi} and G_{zi}. Therefore, the effect of VA is localized. Simulations in Fig. 10 show that the vibration absorber is not effective for the force control loop. For the displacement control problem, the vibration absorber can reduce the adverse effect of LDZ by almost 30dB (at 3Hz), by properly tuning the VA damping c_a.

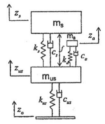

Fig. 9. Quarter-car suspension model with vibration absorber.

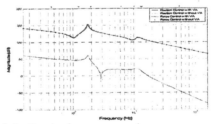

Fig. 10. Frequency response of vibration absorber.

4.4 Advanced control algorithms

By employing advanced control strategies, we can cancel the physical feedback introduced by the hydraulic actuator (or other actuators). Therefore, the effect of this remedy is similar to introducing a non-feedback actuator discussed in Section 4.1. We can even state that the current-mode motor control concept is a special case of the advanced control based algorithm. Inside the current-mode drive is a feedback law that cancels the back electromotive force. This "using feedback control to cancel physical feedback" idea can be applied to other actuator systems. For hydraulic actuators, we expect this to be much harder to achieve because the large bulk modulus of hydraulic fluids represents a rigid coupling. Fast sensing of hydraulic pressure and subsequent manipulation of pumps/valves to react to pressure increase/drop is harder compared with current-mode motor drive. Therefore, this remedy may not work well with hydraulic actuators. No matter which actuator system is used, the servo-loop control algorithm and hardware need to be designed to work with the surge in control signal magnitude at the resonant frequency of LDZ.

5. CONCLUSION

The inherited challenge and possible remedies of the servo-loop control design for active suspension systems are presented. It is shown that for both force servo-loop and displacement servo-loop, we have lightly damped zeros (LDZ) in the "plant", which impose performance limit in practical implementations. It was found that recasting the active suspension servo-loop into a displacement control problem does not provide significant benefits. Four possible remedies were studied—different actuation systems, modified suspension parameters, vibration absorbers, and advanced control algorithms. It was found these remedies can either completely eliminate the LDZ, or moderately increase their damping ratio so that the imposed performance limit is reduced.

REFERENCES

1. Hrovat, D., (1997) "Survey of Advanced Suspension Developments and Related Optimal Control Applications," Automatica, v 33 n 10, pp.1781-1817.
2. Alleyne, A., Liu, R., (1999) "On the limitations of force tracking control for hydraulic servo systems," Journal of Dynamic Systems, Measurement and Control, Trans. of the ASME, Vol. 121, n. 2, 1999, pp. 184-190.
3. Alleyne, A., Hedrick, J.K., (1995) "Nonlinear Adaptive Control of Active Suspension," IEEE Trans. on Control Systems Technology, Vol. 3, n. 1, March, pp. 94-101.
4. Goran, M. B., Bachrach, B. I., Smith, R. E., (1992) "The Design and Development of a Broad Bandwidth Active Suspension Concept Car," ImechE 1992, pp. 231-252.
5. Rajamani, R., Hedrick, J.K., (1994) "Performance of Active Automotive Suspensions with Hydraulic Actuators: Theory and Experiment," Proc. of the American Control Conference, June 1994, pp. 1214-1218.
6. Peng, H., Stratheam, R., Ulsoy, A.G., "A Novel Active Suspension Design Technique--Simulation and Experimental Results," Proceedings of the 1997 American Control Conference, Albuquerque, New Mexico, June 4-6, 1997.
7. Zhang, Y., Alleyne, A., "A Practical and Effective Approach to Active Suspension Control," Proceedings of the 6th International Symposium on Advanced Vehicle Control, Hiroshima, Japan, September 2002.
8. Chantranuwathana, S. and Peng, H., "Practical Adaptive Robust controller for Active Suspensions," Proceedings of the 2000 ASME International Congress and Exposition, Orlando, Florida.
9. Chantranuwathana, S., Adaptive Robust Force Control For Vehicle Active Suspensions, Ph.D. Dissertation, University of Michigan, 2001.
10. Skogestad, S. and Postlethwaite, I., Multivariable Feedback Control—Analysis and Design, Wiley, 1996.
11. Morari, M., "Design of Resilient Processing Plants—III, A General Framework for the Assessment of Dynamic Resilience," Chemical Engineering Science, Vol.38, No.11, pp.1881-1891, 1983.
12. Zhang, Yisheng, Alleyne, A., "A New Approach to Half-Car Active Suspension Control," Proceedings of the 2003 American Control Conference, Denver, Colorado.

Control of Vehicle Suspension Using Neural Network

WAKAE KOZUKUE AND HIDEYUKI MIYAJI[1]

SUMMARY

Holographic neural network, which is a new algorithm of neural network, is applied to the control of a vehicle suspension. For the simplified suspension model, using the feedback gain obtained from the optimal control theory, the equation of motion is solved by Runge-Kutta method. The displacement, velocity, and control force obtained are adopted as training data for neural network. In the simulation the road displacement is assumed to be sinusoidal and its frequency is changed from 1.0Hz to 4.0Hz by 0.1Hz intervals. As the trained network is tested, the results for the road frequency not contained in the training data agree well with the results obtained from the optimal control theory.

1 INTRODUCTION

Recently, with the development of highways, long time and long way driving has become frequent. According to this, a vehicle suspension which achieves good drive feelings and eases drivers' fatigue is strongly desired. As for the control of vehicle suspensions, passive systems which control the spring coefficient and damper coefficient according to a shape of a road has become popular due to the development of electronics and sensor technology. Moreover, the research on active suspension, which realizes the improvement of performance, has been performed [1].

Active suspensions have the following two features.

(1) Supplying energy and controlling it constantly

---

[1] Address correspondence to: Department of Mechanical Engineering, Kanagawa Institute of Technology, 1030 Shimo-ogino,Atsugi-shi, 243-0292 Japan. Tel.:(046)291-3192;Fax:(046)242-8735; E-mail:kozukue@me.kanagawa-it.ac.jp

(2) Having various sensors and operating devices for signals and providing force as a function of these variables

Now, in order to achieve the above-mentioned systems a hydraulic control system is advantageous and recently an active suspension which has a hydraulic actuator in it and controls a car body actively has been mounted on some commercial vehicles. In order to realize the performance the application of control logic based on classical or modern control theory is required. As for the modern control theory the design example of active control for seat suspension of truck based on H∞ control theory was developed[3]. However, even if the control utilizing various theories is possible, many expensive sensors are required and a complicated hydraulic system has to be mounted for active suspensions, so its application is limited to expensive vehicles.

In this study we consider that the optimal regulator, which returns systems back to an equilibrium state, is effective for suppressing up and down vibrations of a body, so the optimal regulator theory is used as the basic control theory. In the optimal regulator theory there is the assumption that the value of all state variables can be directly measured. However, even if state variables are not directly measured, there is the possibility that the number of necessary sensors could be saved by estimating the value of unknown state variables using neural network (NN). Observer is one of the methods to estimate the value of not directly measured state variables in order to save sensors. However, in this study we do not use observer but obtain the value of unknown state variables and control force by NN.

As for the first step to achieve the above-mentioned purpose to save sensors, we investigate whether it is possible to interpolate the result of the optimal control theory by NN. Finally it is shown that the interpolation can be done successfully and the accuracy of interpolation by using HNN is excellent.

2 MODELING OF A VEHICLE CONTAINING A SUSPENSION

The vehicle model containing a suspension used by the optimal control theory is shown in Fig.1. Objective model is a half model of a suspension of a bus.

The various parameters in Fig.1 are shown as follows.

m_1 : mass of axle m_2 : mass of a body k_1 : a spring coefficient of a tire

k_2 : a spring coefficient of a suspension c_1 : a damping coefficient of a tire

c_2 : a damping coefficient of a suspension x_1 : displacement under spring

x_2 : displacement over spring x_0 : road displacement

u : control force of a suspension

2.1   The equation of motion

The equation of motion for two degrees of freedom system shown in Fig.1 is represented as follows.

$$m_1\ddot{x}_1 = c_2(\dot{x}_2 - \dot{x}_1) + k_2(x_2 - x_1) - c_1(\dot{x}_1 - \dot{x}_0) - k_1(x_1 - x_0) - u \qquad (1)$$
$$m_2\ddot{x}_2 = -c_2(\dot{x}_2 - \dot{x}_1) - k_2(x_2 - x_1) + u \qquad (2)$$

2.2   Road displacement

As for road displacement, there are step offset or irregular up and down, but in this study a sinusoidal displacement is assumed for simplicity. Namely the following expression is assumed.

$$x_0 = A\sin(\omega t) \qquad (3)$$

Here the amplitude $A = 0.1(m)$ and ω is changed from 2.0π to 8.0π.

2.3   State equation

From the equation of motion shown in equation (1) and (2) the following state equation is obtained.

$$\dot{x}(t) = Ax(t) + Bu(t) + Dw(t) \qquad (4)$$
$$\dot{x}_2(t) = Cx(t) \qquad (5)$$
$$x(t) = [x_1 \quad x_2 \quad \dot{x}_1 \quad \dot{x}_2], w(t) = [x_0 \quad \dot{x}_0]^T \qquad (6)$$

Here the matrices A, B, C, D are as follows.

$$A = \begin{bmatrix} 0 & 0 & 1 & 0 \\ 0 & 0 & 0 & 1 \\ -\dfrac{k_1+k_2}{m_1} & \dfrac{k_2}{m_1} & -\dfrac{c_1+c_2}{m_1} & \dfrac{c_2}{m_1} \\ \dfrac{k_2}{m_1} & -\dfrac{k_2}{m_2} & \dfrac{c_2}{m_2} & -\dfrac{c_2}{m_2} \end{bmatrix} \qquad (7)$$

155

$$B = \begin{bmatrix} 0 & 0 & -\dfrac{1}{m_1} & \dfrac{1}{m_2} \end{bmatrix}^T \tag{8}$$

$$C = \begin{bmatrix} 0 & 0 & 0 & 1 \end{bmatrix} \tag{9}$$

$$D = \begin{bmatrix} 0 & 0 \\ 0 & 0 \\ \dfrac{k_1}{m_1} & \dfrac{c_1}{m_1} \\ 0 & 0 \end{bmatrix} \tag{10}$$

3   THE APPLICATION OF THE OPTIMAL CONTROL THEORY

3.1   Setup of estimation function

Here the objective function is selected as the addition of the square of the five variables multiplied by the proper weighting coefficient. The four variables are the absolute displacement of axle x_1, the absolute velocity of axle $\dot{x}_1$, the absolute displacement of a body x_2, and the absolute velocity of a body $\dot{x}_2$. In order for the control energy not to become too large the estimation function includes the square of the control input u multiplied by proper weighting coefficient. According to this the estimation function is set up as the following.

$$J = \int_0^\infty (q_1 x_1^2 + q_2 x_2^2 + q_3 \dot{x}_1^2 + q_4 \dot{x}_2^2 + ru^2)dt \tag{11}$$

By rewriting this using state vector Y the following the quadratic estimation function is obtained.

$$J = \int_0^\infty (Y^T QY + ru^2)dt \tag{12}$$

Here Q is the symmetric matrix having the weighting coefficient as its elements and r represents the weighting coefficient for control force.

3.2   The optimal control theory

The optimal control rule u^0 which minimizes the estimation function represented by the equation (12) is shown as follows.

$$u^0 = -FY \tag{13}$$

$$F = r^{-1}B^T P \tag{14}$$

Here P is the unique positive definite solution of the Riccati algebraic equation.

3.3 The optimal control theory for discrete time system

In order to formulate this control system for discrete time system the evaluation function of a continuous system is discretized and the optimal control rule is obtained based on the optimal control theory for discrete time system.

The evaluation function of a discrete time system equivalent to the evaluation function of the continuous system represented by the equation (12) is shown as follows.

$$J = \sum_{i=0}^{\infty} [Y(i)^T Q_d Y(i) + 2Y(i)^T S_d u(i) + r_d u(i)^2] \tag{15}$$

Here Q_d, S_d, r_d are represented as follows by using a sampling time T, weight coefficient Q,r of continuous system, and A, B of the state space equation.

$$Q_d = \int_0^T [\Phi^T Q \Phi] d\tau \qquad\qquad S_d = \int_0^T [\Gamma^T Q \Phi] d\tau \tag{16}$$

$$r_d = \int_0^T [\Gamma^T Q \Gamma + r_0] d\tau \tag{17}$$

$$\Phi(\tau) = \exp(A\tau) \qquad\qquad \Gamma(\tau) = \int_0^T [\exp(At)] dt \cdot B \tag{18}$$

The control rule which minimizes the equation (15), namely the evaluation function of discrete time system, is obtained as follows.

$$u^0(i) = -F_d Y(i) \tag{19}$$

$$F_d = r_d^{-1} S_d + (r_d + b_a^T M b_a)^{-1} b_a^T M (A_d - r_d^{-1} B_d S_d) \tag{20}$$

Here the matrix M is the positive definite symmetric solution of the discrete Ricatti equation.

4 THE OUTLINE OF HOLOGRAPHIC NEURAL NETWORK[2]

HNN was developed by Sutherland [2] and its most prominent feature is that input data and output data have the linear relationship by transforming both onto the complex plane uniformly. In HNN the number of the neuron is one and the construction of NN is equivalent to obtaining the transfer function between the input and the output. Therefore the time for convergence can be reduced sufficiently.

For learning l −class m-dimensional input vector s and n-dimensional output vector r are assumed to be used. Each element of input and output vector is transformed on to the complex plane by nonlinear function.

$$f(s_{hk}) = \lambda_{hk}e^{i\theta_{hk}} \qquad g(r_{jk}) = \gamma_{jk}e^{i\phi_{jk}} \tag{21}$$

Here i represents imaginary unit and θ_{hk}, ϕ_{jk} are the phase angle by using mapping function such as Sigmoid function and have the value in the interval $[0, 2\pi)$. $\lambda_{hk}, \gamma_{jk}$ represent the probability of occurrence in the corresponding phase angle region of input and output data and have the value in the interval $[0, 1]$. The following input matrix [S] and the teacher matrix [T] are obtained by manipulating the above equation (21).

$$[S] = \begin{bmatrix} \lambda_{11}e^{i\theta_{11}} & \lambda_{12}e^{i\theta_{12}} & \cdots & \lambda_{1m}e^{i\theta_{1m}} \\ \lambda_{21}e^{i\theta_{21}} & \lambda_{22}e^{i\theta_{22}} & \cdots & \lambda_{2m}e^{i\theta_{2m}} \\ \vdots & \vdots & & \vdots \\ \lambda_{l1}e^{i\theta_{l1}} & \lambda_{l2}e^{a_{l2}} & \cdots & \lambda_{lm}e^{i\theta_{lm}} \end{bmatrix} \tag{22}$$

$$[T] = \begin{bmatrix} \gamma_{11}e^{i\phi_{11}} & \gamma_{12}e^{i\phi_{12}} & \cdots & \gamma_{1m}e^{i\phi_{1m}} \\ \gamma_{21}e^{i\phi_{21}} & \gamma_{22}e^{i\phi_{22}} & \cdots & \gamma_{2m}e^{i\phi_{2m}} \\ \vdots & \vdots & & \vdots \\ \gamma_{l1}e^{i\phi_{l1}} & \gamma_{l2}e^{\phi_{l2}} & \cdots & \gamma_{lm}e^{i\phi_{lm}} \end{bmatrix} \tag{23}$$

The output matrix [A] is represented as follows by using the transfer matrix [X] .

$$[A] = [S] \bullet [X] \tag{24}$$

[X] is determined by minimizing the following difference between [A] and [T].

$$E_{rr} = ([A]-[T])^{H} \bullet ([A]-[T]) \tag{25}$$

Here H represents complex conjugate. From this condition the following expression is obtained.

$$[X] = ([S]^{H} \bullet [S])^{-1} \bullet [S]^{H} \bullet [T] \tag{26}$$

Since [S] and [T] are known [X] is obtained from the equation (26) directly. However, when Gauss-Jordan elimination method is used to obtain an inverse matrix of n-order the number of operation becomes n^{3}. Since [S] is high order the huge amount of calculation time is required. Thus, in order to avoid this the following iterative learning is used to obtain [X].

$$[X]_{i} = \frac{1}{E}[S]^{H} \bullet [T] \tag{27}$$

$$[X]_{i+1} = [X]_{i} + [S]^{H} \bullet ([T] - \frac{1}{E}[S] \bullet [X]_{i}) \tag{28}$$

Here E is a parameter which normalizes the norm of [X].

5 THE CONSTRUCTION OF LEARNING DATA FOR NN

Feed back gain is obtained using optimal control rule as described in 3. The equation of motion is solved by fourth order Runge-Kutta method using the above feed back gain. The values of parameters used are shown below. These values are for the half model of the suspension for a bus.

$$k_{1} = 1.8 \times 10^{5}[N/m] \qquad c_{1} = 0.56 \times 10^{4}[Ns/m] \qquad m_{1} = 77[kg]$$
$$k_{2} = 2.2 \times 10^{5}[N/m] \qquad c_{2} = 2.044 \times 10^{4}[Ns/m] \qquad m_{2} = 5950[kg]$$

The frequency of road displacement is set up as follows.

$$f = 1.0(Hz) - 4.0(Hz)$$

At first the sinusoidal road displacement is given for five seconds and the absolute velocity, absolute displacement, and control force are calculated for each sampling time 0.01 second. The weighting coefficients for optimal control are determined so that the velocity of the body is decreased by 20% with control at road frequency 2.5Hz.

The above obtained data are learned by using NN as follows. Namely, the following six state variables and control force at certain time k are use as input.

$$x_0(k), x_1(k), x_2(k), \dot{x}_0(k), \dot{x}_1(k), \dot{x}_2(k), u(k)$$

The following three state variables at the next time step k+1 is used as output.

$$x_2(k+1), \dot{x}_2(k+1), u(k+1)$$

This corresponds to the prediction of step k+1 from step k.

Actually the data for thirty one kinds of road frequencies for 1.0 Hz to 4.0Hz changed by 0.1Hz obtained from the above numerical simulation is edited as one learning data. For editing the data for two periods are extracted based on the body velocity $\dot{x}_2$ and these are synthesized according to the road frequencies.

6 ANALYSIS RESULTS BY HNN

At first the passive suspension which dose not do control is analyzed and next the active suspension is analyzed. The time history of the velocity of the body obtained from this analysis is shown in Fig.2. The data obtained from these analysis is learned by HNN and the time history of the absolute velocity of the body $\dot{x}_2$, the absolute displacement x_2, and the control force u are calculated and compared with those results obtained from the corresponding optimal control. As for the testing data given to N.N. the appropriate frequency road displacement not used in learning is chosen among the road displacement frequencies of 1.0 Hz to 4.0 Hz.

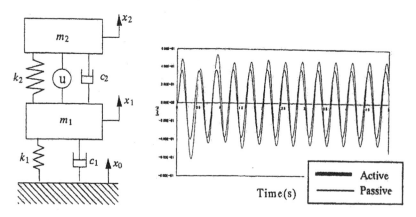

Fig.1. Suspension and vehicle Model. Fig.2 .Absolute velocity of the body by passive and active control.

160

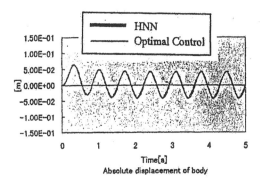

Fig.3. Absolute displacement of the body for active control (road frequency 1.45Hz).

In Fig.3 the displacement of the body in the active control state is shown for the road frequency 1.45Hz. In this figure the value obtained by optimal control and the value obtained by NN agree well. From this fact it is seen that the value of the state variable can be obtained for not learned frequency of the road displacement

7 CONCLUSIONS

Holographic neural network (HNN), which is a new algorithm of neural network, is applied to the active control of vehicle suspension. For HNN, the possibility of accurately interpolating the state variables and control force obtained from optimal control theory is shown by setting up learning data carefully.

REFERENCES

1. Fukushima, N., Irie,N., Akatsu, Y.,Sato, M., and Takahashi, T. "Vehicle Vibration Control by Hydraulic Active Suspension" JSME Trans.C.57-535(1991-3)pp. 722 - 736.(in Japanese)

2. J.G.Sutherland, "The Holographic Model of Memory, Learning and Expression", International Journal of Neural Systems, 1990, Vol.1 No.3, pp. 259 -267.

3. Shimogo, T., Oshinoya, Y., and Shinjo, H. "Active Suspension of Truck Seat", JSME Trans. C. 62-600(1996-8)   pp. 3132 - 3138. (in Japanese)

Vehicle System Dynamics Supplement 41 (2004), p.162-171 © Taylor & Francis Ltd.

Ride comfort enhancement considering car body aerodynamics using actively controlled spoilers

LODEWIJKS G., NUTTALL A.J.G., MAZZALI C., MECHERONI R. & SAVKOOR A.R.[1]

SUMMARY

The paper studies the application of aerodynamic actuators based on actively controlled spoilers mounted on the car body for enhancing the ride comfort of automobiles running on randomly uneven road surfaces. In particular this work extends previous analysis developed by one of the authors by considering the effect of aerodynamic forces induced by the pitch and bounce motions of the vehicle body itself. The design aspects of a wing actuator and the control law for optimising its performance is addressed. The control design considers the non-linearity due to saturation based on the SLQR approach using the stochastic linearisation technique. It is shown that under motorway driving conditions significant improvements in ride performance of automobiles can be achieved with the aid of actively controlled rooftop spoilers having a sufficiently fast response.

1   INTRODUCTION

The development of conventionally designed passive suspension of automobiles has reached a point where the law of diminishing returns begins to apply. Research in vehicle dynamics during the last two decades has firmly established that further improvements in ride quality can be achieved by incorporating actively controlled elements in the vehicle suspension system. The basic concept of a "sky-hook damper" that generates force acting on the vehicle body proportional to the absolute velocity of the body was proposed by Karnopp et al. [1]. This opened up the way to improve ride comfort by introducing such force with the aid of active suspension elements. However, in practice, the design of active suspension generally increases both the cost and complexity of chassis design and furthermore controlling body motion with suspension forces does not mitigate the strongly restrictive trade-off necessary between road holding and ride comfort. The restriction arises because suspension forces act reciprocally on the suspended and the unsuspended masses of the vehicle. The fact that suspension actuators, irrespective of the control algorithm employed, are ineffective in influencing body acceleration in the region of the wheel-hop frequency has been proved mathematically by Hedrick et al. [2].

A relatively novel approach that was suggested recently by the last author [3] aims to improve ride comfort without requiring any modification of a conventionally

---

[1] *Address correspondence to:* Transportation and Logistics Technology, Delft University of Technology, Mekelweg 2, 2628 CD, Delft, The Netherlands. E-mail: A.J.G.Nuttall@wbmt.tudelft.nl

well designed passive suspension. Instead of calling upon the suspension to deliver the control force this approach rests on the aerodynamic forces generated by a wing or spoiler mounted directly on the car body. This method achieves generation of real sky-hook forces, primarily lift force on the wing that is controlled actively by changing its angle of attack by means of an actuator. The control goal is to reduce in-plane chassis vibrations, mainly the bounce and pitch modes of body motion without causing any degradation of the road holding quality of the vehicle. A comparative study [4] performed to assess the achievable improvements in ride performance using suspension actuator design with that achievable using two aerodynamic actuators mounted on the roof clearly showed the superior design performance with the latter design configuration. These studies of aerodynamic actuator design did not consider the influence of aerodynamic forces induced by the pitch and bounce motion of the vehicle body. Donicelli, Mastinu and Gobbi [5] have recently presented that a thorough treatment of that influence that will be followed in this paper.

2   MODELLING OF IN-PLANE HALF-CAR & ROAD PROFILE

2.1   THE HALF-CAR MODEL

The half-car vehicle model (see Fig. 1) comprises of a sprung mass representing the vehicle body and an unsprung mass representing the axle and the wheel assembly. The in-plane motion of the vehicle body is defined by two degrees of freedom; namely the vertical and the pitching motion about its centre of mass. The vertical motion of the front and rear unsprung masses defines the wheel hop motion.

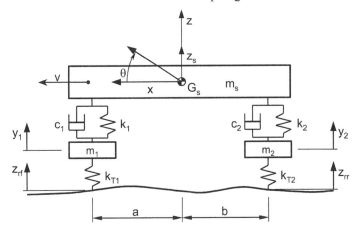

Fig. 1 The half car model running on an uneven road.

2.2 IN-PLANE DYNAMICS OF HALF-CAR EXCITED BY THE ROAD UNEVENNESS

The equation of motion of the vehicle system reads

$$[M]\ddot{x} + [C]\dot{x} + [K]x = F_r \qquad (1)$$

Where x is the state vector and F_r represents a vector of the road excitation input.

$$x = [z_s \quad \theta \quad y_1 \quad y_2]^T \qquad F_r = [0 \quad 0 \quad k_{T1}z_{rf} \quad k_{T2}z_{rf}]^T$$

The matrices in equation (1) are:

$$[M] = \begin{bmatrix} m_s & 0 & 0 & 0 \\ 0 & J_y & 0 & 0 \\ 0 & 0 & m_1 & 0 \\ 0 & 0 & 0 & m_2 \end{bmatrix} \quad [C] = \begin{bmatrix} c_1+c_2 & c_1a-c_2b & -c_1 & -c_2 \\ c_1a-c_2b & c_1a^2+c_2b^2 & -c_1a & c_2b \\ -c_1 & -c_1a & c_1 & 0 \\ -c_2 & c_2b & 0 & c_2 \end{bmatrix}$$

$$[K] = \begin{bmatrix} k_1+k_2 & k_1a-k_2b & -k_1 & -k_2 \\ k_1a-k_2b & k_1a^2+k_2b^2 & -k_1a & k_2b \\ -k_1 & -k_1 \cdot a & k_1+k_{T1} & 0 \\ -k_2 & k_2 \cdot b & 0 & k_2+k_{T2} \end{bmatrix}$$

The parameter values of a typical passenger car are summarised in Table 1 below.

Table 1: Model Parameter Values.

Parameters	Symbol	Unit	Value
Suspended mass	ms	kg	600
Pitching moment of inertia	J	kgm^2	750
Front unsuspended mass	m1	kg	35
Rear unsuspended mass	m2	kg	30
Front damping coefficient	C1	Ns/m	1200
Rear damping coefficient	C2	Ns/m	1500
Front spring stiffness coefficient	K1	N/m	12000
Rear spring stiffness coefficient	K2	N/m	15000
Front tyre stiffness coefficient	p1	N/m	200000
Rear tyre stiffness coefficient	p2	N/m	200000
Distance between front axle and centre of gravity	a	m	1
Distance between rear axle and centre of gravity	b	m	1.5

2.3 ROAD PROFILE MODEL

The height of road surface profile above its mean level z_0 will be modelled as a filtered white noise process [6] described by a first order stochastic differential equation

$$\dot{z}_0 = avz_0 + w \qquad (2)$$

Where w is a white noise of variance $2\,av\sigma^2$, v is the speed of the vehicle, and the constants a and σ define the characteristics of road surface profile. Typical values for an asphalt motorway road are a = 0.15 rad/m and σ = 3.3 mm and v ≈ 40 m/s.

The above parameter values give a road that varies effectively up to about 1 Hz. Furthermore, we assume that the road variation experienced by the rear tyre is a delayed version of that in front. For a wheelbase of 2.5 m and a speed of 40 m/s, this delay is 62.5 ms. For the purpose of controller design, this delay element is modelled by a fifth order Padé approximation.

Assuming that the vehicle is a linear and time invariant system, it is convenient to represent the equations of motion in the state space form

$$x = [A]x + \{G\}w$$
$$y = [C]x \tag{3}$$

Where x is the state vector, y is the output vector or observation vector while the matrices A, G and C determine the relationships between the state input and output.

Introducing the state vector including the Padé filter states defined by:

$$\dot{x}_p = [A_p]x_p + [B_p]z_{rf}\ ,\ y_p = [C_p]x_p + [D_p]z_{rf}$$

The system has globally 14 states, 8 coming from the vehicle, 1 from the road filter and 5 from the Padè time delay approximation of the rear input. Hence,

$$x = \begin{bmatrix} z_s & \theta & y_1 & y_2 & \dot{z}_s & \dot{\theta} & \dot{y}_1 & \dot{y}_2 & |z_{rf}| & \ddddot{z}_{rr} & \dddot{z}_{rr} & \ddot{z}_{rr} & \dot{z}_{rr} & z_{rr} \end{bmatrix}$$

3   AERODYNAMIC EXCITATION DUE TO BODY MOTION

In addition to the excitation from the road input, one needs to consider additional excitation from the non-stationary aerodynamic field forces resulting from the motion of the vehicle body. The modelling of the variation of aerodynamic forces developed in this section is on the line of the treatment in [5].

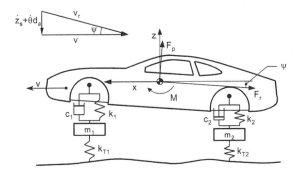

Fig. 2 Characteristic angles and aerodynamic forces and moment on vehicle body.

For modelling purposes, the two resultant forces (lift and drag) applied at the gravity centre and a pitching moment around the y-axis, as shown in Fig.2, replace the distributed pressure and viscous traction acting over the surfaces of the vehicle body. The study is not concerned with the aerodynamic forces arising in stationary state cruising without body oscillations on an ideally flat road. The aerodynamic forces and moment depend on the running speed of the vehicle, the body shape, the spatial attitude and distance from the ground and on the bounce and pitch velocities.

Considering the in-plane motion of the half car subjected only to excitation originating from aerodynamic inputs, the equation of motion in the matrix form is:

$$[M]\ddot{x} + [C]\dot{x} + [K]x = F_{aero}$$

$$\text{With } x = [z_s \quad \theta \quad y_1 \quad y_2]^T$$

$$\text{and } F_r = [F_{Lift} \quad M \quad 0 \quad 0]^T \text{ where } F_{Lift} = F_p \cos\psi - F_r \sin\psi$$

$$F_p = -\frac{1}{2}\rho S V_r^2 C_z(\alpha, z_s), \quad F_r = \frac{1}{2}\rho S V_r^2 C_D(\alpha, z_s), \quad M = \frac{1}{2}\rho S V_r^2 C_\alpha(\alpha, z_s)$$

The angle ψ is the angle between the wind speed V and Vr. Here V is the absolute velocity while Vr is the relative velocity that derives from the pitch and bounce velocities respectively $(\dot{\theta}), (\dot{z}_s)$. The angle of attack of the wind $\alpha = \psi - \theta$ is a function of variables describing the motion of the body $\alpha = f(\dot{z}_s, \theta, \dot{\theta})$. Since the attention is focused only upon the oscillatory body motion it is useful to linearise the expressions for C_z, C_α and C_D referring to the steady state and to change the set of co-ordinates. Denoting $\overline{x}$ to represent the steady state condition and $\underline{x}$ as the variation about the steady state such that is $x = \underline{x} - \overline{x}$, the equation for the variation of motion becomes:

$$[M]\ddot{\underline{x}} + [C]\dot{\underline{x}} + [K]\underline{x} + [K]\overline{x} = [\overline{F}_{aero}] + [\Delta F_{aero}] \tag{4}$$

$$[K]\overline{x} = [\overline{F}_{aero}]$$
$$[M]\ddot{\underline{x}} + [C]\dot{\underline{x}} + [K]\underline{x} = [\Delta F_{aero}] \tag{5}$$

$$[\Delta F_{aero}] = [C_f]\dot{\underline{x}} + [K_f]\underline{x} \tag{6}$$

Finally rearranging the equations the equation reads:

$$[M]\ddot{\underline{x}} + ([C] - [C_f])\dot{\underline{x}} + ([K] - [K_f])\underline{x} = 0 \tag{7}$$

From the above equation it is clear that the aerodynamic forces produce a change in the overall stiffness and damping matrices. In particular it is interesting to note that at sufficiently high speed the elements of matrices $[K_f]$ and $[C_f]$ may become of the same order of magnitude of the corresponding elements of matrix $[K]$ and $[C]$. The present analysis indicates that with increasing running speed the total stiffness tends to increase while the total damping tends to decrease. The combined influence of disturbances due to the aerodynamic effects and the road irregularities on the ride dynamics of the vehicle may be studied by inserting the equations in this section into the equation of motion (3) for the passive vehicle. The influence of considering the aerodynamic disturbance induced by the motion of the body on the ride performance of the passive vehicle is summarised in figure 3.

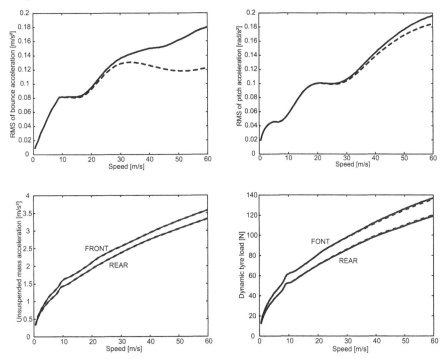

Fig. 3 Comparison of ride comfort and road holding of a (passive) vehicle running on motorway with (solid) and without (dotted) taking into account the effect of aerodynamic forces induced by the motion of the vehicle body.

It is evident aerodynamic forces induced by the motion of the vehicle body adversely affect ride comfort (body vertical and pitch acceleration) The effect on body acceleration level becomes appreciable at motorway speeds of over 35 to 40 m/s as observed in [5]. Interestingly, it is seen that the bounce mode is more severely affected than the pitch mode at higher running speeds of the vehicle.

4   DESIGN AND POSITIONING OF THE WING

4.1   WING DESIGN CONSIDERATIONS

A reasonably symmetrically profiled wing, a standard NACA 0012 with a chord of 0.3 m will be considered, with its camber line coinciding with the chord line. The aerodynamic problem of choosing an appropriate wing profile that requires some wind tunnel and field experimentation with a prototype mounted on the car roof is beyond the scope of this preliminary study. Similarly the ideal mounting height above the roof for minimising interference with the airflow around car body is not considered. Considering vehicle speeds in the range 10 to 60 m/s, the Reynolds and Mach numbers for air (assuming the density ρ = 1,295 Kg/m3 and viscosity v = 1,7894*10^{-5} Kg/(s*m)), are in the range: 3,2*105 <Re<8,7*105 and 0,05<Ma<0,12. It is assumed that the lift coefficient C_l is linearly dependent on angle of attack α in the range (-15° to 15°) where stalling is not likely. The drag coefficient for such profile is small $\left(C_D \approx C_L/100\right)$ and in the velocity range of interest it is well justified to ignore the wing drag.

4.2   OPTIMAL MOUNTING POSITION FOR THE WING

Intuitively it is evident that optimal reduction of both pitch and vertical motions ideally would require the installation of two aerodynamic actuators mounted on the body. Using a single actuator implies that some sort compromise between the performances of pitch and vertical modes becomes inevitable depending upon the mounting position of the actuator with respect to the centre of mass of the vehicle. The present choice for actuator mounting (1 meter behind the CG) has been made considering that pitch reduction merits a higher priority than the reduction of bounce. It turns out as expected that aerodynamic actuators have no significant influence on the dynamic tyre loads irrespective of their mounting position.

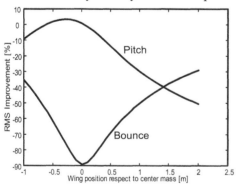

Fig. 4 Influence of actuator mounting position on bounce, pitch.

5   ACTUATOR CONTROL DESIGN

To improve ride comfort an external force is added to the vehicle model by means of a skyhook actuator. This actuator consists of a wing mounted on the vehicle body. Changing the wing's angle of attack influences the force that the wing produces. The wing has been added to the half car model for an analysis of an appropriate controller. The dynamics of the actuator actually changing the wing position have also been included, making it possible to investigate the influence of the bandwidth of the wing rotation unit on system performance. For this initial investigation the actuator is approximated as a first order system.

To obtain the desired suppression of the body accelerations due to road irregularities a feedback controller is required. However in the closed loop situation saturation of the input signal will occur, because the wing angle is limited to prevent the wing from stalling. Due to this non-linear actuator behaviour a LQR theory for systems with saturating actuators is adopted to obtain a linearised system with full state feedback as presented in [7]. In this theory referred to as SLQR, where S stands for saturation, stochastic linearisation replaces the saturation by a gain as function of the variance of the input signal. With the aid of stochastic linearisation and the Lagrange multiplier technique solutions of the SLQR problem are found using the standard Ricatti and Lyapunov equations. These standard equations are coupled to two transcendental equations that characterise the variance of the saturation of the input signal and the Lagrange multiplier associated with the constrained minimisation problem.

Following the algorithm described in [7], the stochastic linearised version of the system dynamics can now be described by:

$$
\begin{aligned}
\dot{\hat{x}} &= [A]\hat{x} + \{B_1\}w + \{B_2\}N\hat{u} \\
\hat{z} &= [C_1]\hat{x} + \{D_{12}\}\hat{u} \\
\hat{y} &= [C_2]\hat{x} + \{D_{21}\}\hat{u}
\end{aligned}
\tag{8}
$$

Where x is the state vector, w the noise input, u the wing angle, N the gain resulting from the stochastic linearisation, y the measured output vector and z denotes controlled outputs of interest after application of a weighting filter. In this case the outputs of interest are the comfort indexes or the body and pitch accelerations. The regulator resulting from the SLQR algorithm operates a feedback such that the controller input $u = -\{K\}^T x$ minimises the variance of the controlled output z.

6   RESULTS OF RIDE PERFORMANCE

The results for the standard deviations of vertical and pitch accelerations of the body are presented as Pareto curves. Clearly it is not possible to find an overall optimum set of weights that can simultaneously minimise both pitch and bounce accelerations. With the aid of the Pareto curves a vehicle designer will have to select a suitable set of weights that yield the desired compromise between the two facets of ride. The weights are normalised to generate comparable numbers. Figures 5 and 6 show the Pareto curves for different roads types. These curves have been attained with two different bandwidths (10 and 50 Hz) of the wing actuator with the wing mounted 1 meter behind the CG to minimise again the earlier mentioned preference towards pitch comfort. Figure 5 shows the influence of the actuator bandwidth on system performance at a motorway cruising speed of 30 m/s.

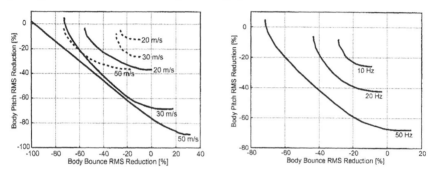

Fig. 5 Pareto curves for motorway with (left) different speeds and actuator bandwidth 10 Hz (dotted) and 50 Hz (solid) and (right) for set speed of 30 m/s and different actuator bandwidths.

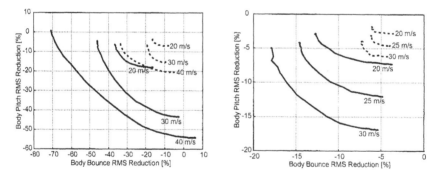

Fig. 6 Pareto curves for (left) secondary and (right) rural road with actuator bandwidth 10 Hz (dotted) and 50 Hz (solid).

In all the considered road conditions a reduction of the pitch and bounce acceleration can be achieved. However, improvement in pitch performance is generally accompanied by a worsening in bounce performance. For running speeds typical on motorways the most significant improvement may be realised on the smooth asphalt motorway but on the relatively rough rural road with lower running speeds the improvement is only marginal. On all the three roads considered it is clear that the higher the running speed, the larger the force generated by the wing becomes and consequently the greater is the improvement achievable in ride comfort.

The effect of the actuator dynamics on the ride comfort performance can be seen from the Pareto curves in figures 5 and 6. Clearly, a slow responding wing actuator seriously jeopardises the overall system performance. From the Pareto curves it can be concluded that the bandwidth of the actuator needs to be significantly higher than 10 Hz to achieve substantial improvements in ride comfort.

7   CONCLUSION

An actively controlled wing mounted on a car has the possibility of enhancing ride comfort without influencing road holding performance. Therefore the active wing device solves the conflict between ride comfort and road holding. The achievable ride comfort improvement increases progressively with increasing vehicle speed. However the bandwidth of the wing actuator also plays an important role in achieving this performance. The design of the actuator rotating the wing will require special attention to attain a high bandwidth and good overall system performance. Furthermore the use of a single wing will always be accompanied by a trade off between pitch and bounce enhancements. An extra active wing could possibly solve this conflict.

The wing counters the motion caused by irregularities of the road surface. Aerodynamic forces resulting from body motion have to be included in ride comfort analysis, because at high speeds these forces have an adverse effect on ride comfort but not on road holding.

REFERENCES

1.   Karnopp, D, Crosby, M.J. and Harwood, R.A.: Vibration control using semi-active force generators, J. Engineering for industry (ASME), 1974, p. 619-626.

2.   Yue, C., Butsuen, T. and Hedrick, J.K.: Alternative control laws for automotive active suspensions, Trans. ASME, J. Dynamic Systems, Measurement and Control, vol. 11, June 1989, p. 286-291.

3.   Savkoor, A.R. and Happel, H.: Aerodynamic vehicle ride control with active spoilers, Proc. Intern. Symposium on Advanced Vehicle Control (AVEC '96), Aachen, Germany, 1996, p. 647-681.

4.   Savkoor, A.R. and Chou, C.T.: Control strategies for improving vehicle ride, Proc.- Vol. 3, Fourth Intern. Conf. On Motion & Vibration Control (Movic) in Zurich Switzerland, 1998, p. 857-862.

5.   Donicelli, C., Mastinu, G. and Gobbi, M.: Aerodynamic effects on ride comfort and road holding of automobiles, Vehicle System Dynamics Supplement 25, 1996, p.99-125.

6.   Venhovens, P.J.Th.: Optimal control of vehicle suspensions, PhD thesis, Delft University of Technology, The Netherlands, 1994.

7.   Gökçek, C, Kabamba, P.T., Meerkov, S.M.: An LQR/LQG theory for systems with saturating actuators, IEEE Transactions on automatic control, vol. 46, no. 10, 2001, p. 1529-1542.

Vehicle System Dynamics Supplement 41 (2004), p.172-181   © Taylor & Francis Ltd.

Possibilities and limitations with distributed damping in heavy vehicles

PETER HOLEN[1] AND BORIS THORVALD

SUMMARY

This paper investigates passive damping performance in heavy vehicles. The objective is comparison of conventional individual dampers and system with modally distributed damping. An analytical model with 7 DOF is first used to derive theoretical damping expressions. Considered degrees of freedom are for sprung mass; bounce, pitch, roll and for unsprung mass; bounce and roll. Vehicle simulations are then performed with a 4x2-tractor semitrailer combination. Load cases of both ride and handling character are used for damping system evaluation and comparison. The vehicle model with conventional individual dampers is validated against measurement data.

1. INTRODUCTION

Heavy vehicles are sold for commercial use. In comparison to cars where new technology may be introduced on a feature basis, more expensive technology solutions are not easily introduced on heavy vehicles if they do not result in a corresponding cost reduction for the customer. The primary function of a heavy vehicle - to transport as large amount of goods as cheap and fast as possible, is still the main sales argument. As a consequence components in e.g. truck chassis suspension are chosen to meet cost and reliability demands.

In a truck, chassis suspension, damping is commonly included by passive hydraulic dampers which are cost efficient and provide acceptable performance. Damper settings are usually derived by manually tuning the dampers at their given position, while driving various load cases. Since conflicts often occur between optimal settings for different load cases, the final setting becomes a compromise. To decrease trade-off effects various technical improvements have been developed. Individual damper force may typically be a function of not just displacement velocity but also displacement amplitude, acceleration and vehicle load [1,2]. Even better adjustment possibilities are obtained using sophisticated systems as with electronic semi-active damping control. Such systems are utilised in high-end luxury cars today, but will on cost sensitive segments like heavy trucks probably not be seen in large-scale production for several years.

[1] *Address correspondence to*: Scania CV AB, Department Vehicle Modularization, SE-151 87 Södertälje, Sweden. Tel: +46 (0)8 553 806 74 Fax: +46 (0)8 553 813 95; E-mail: peter.holen@scania.com

The main drawback with solutions that rely only on technical refinement of the individual damper, here referred to as corner damping, is lack of system approach. Damping of sprung mass eigen modes (roll, pitch and bounce) and axle eigen modes (roll and bounce) is determined by damper characteristic and geometric position of each individual damper. This result in restrictions when trying to obtain specific damping for different eigen modes. Since perpendicular distance from the damper to respective mode rotation centres plays a key role [3], it is with this approach important to also optimise geometric positions. Positioning of the dampers is however often fixed early in design process and may due to packaging aspects not easily be altered.

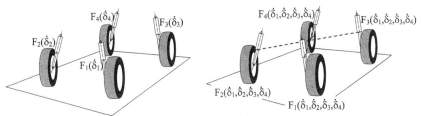

Figure 1. Systems with individual dampers (left) and with interconnected dampers (right).

With a system approach, here referred to as modal damping, the restriction imposed by geometry may be reduced. This improvement is usually associated with electronic semi-active damping control. However, recent technical trends in suspension design focus on more cost efficient passive systems, where the damping is distributed modally by connection of dampers [4,5]. Dampers are usually connected hydraulically, but other solutions are possible. The connected dampers form a system such that individual damper force is a function of all damper displacement velocities, fig. 1.

2. ANALYTICAL MODEL

An analytical model of a two-axle tractor is initially used to calculate mode shapes and corresponding relative dampings, furthermore to derive expressions for distribute damping. The tractor model consists of three rigid bodies with totally 7 DOF; bounce d and roll DOF for the two axles and bounce, pitch and roll DOF for the sprung mass, fig. 2. Both the chassis suspension and the tires are modeled by linear springs and dampers.

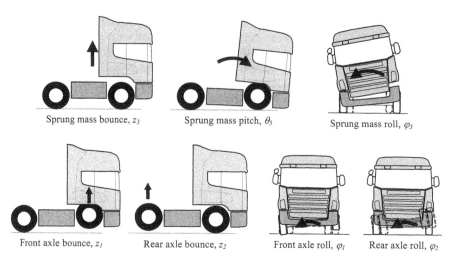

Figure 2. Schematic picture of 7 DOF tractor model.

2.1 MODAL ANALYSIS

Parameter values taken from a real tractor are applied to the 7 DOF model and modal analysis is then performed giving result according to table 1. Even though each mode shape is characterized as a specific displacement it often also consists of small displacements in other directions. The last column in table 1, denoted "Mode shape product", shows the inner product between the calculated mode shape and a mode shape with unit displacement in just one coordinate (zero elsewhere). A value close to one thus indicates close similarity to unit displacement in corresponding DOF.

Table 1. Result from modal analysis of 7 DOF tractor model.

Mode no.	Mode shape character	Frequency [Hz]	Damping [-]	Mode shape product
1	Sprung mass bounce	1.18	0.19	0.97
2	Sprung mass pitch	1.22	0.19	0.86
3	Sprung mass roll	1.84	0.06	0.97
4	Front axle bounce	8.92	0.27	1.00
5	Rear axle bounce	9.06	0.22	1.00
6	Front axle roll	9.28	0.08	1.00
7	Rear axle roll	10.04	0.06	1.00

It may be observed that the relative damping varies between 0.06-0.19 for sprung mass modes and between 0.06-0.27 for un sprung modes.

2.2   DERIVATION OF DISTRIBUTED DAMPING MATRIX

At each wheel in the 7 DOF tractor model, a damper is attached between the axle and the chassis. The relative damper displacements may be described as functions of the chosen DOF, fig. 2, which also are very similar to the seven eigen modes. Both relative damper displacements, δ, and corresponding external forces on the damper, F, are defined positive for damper extension.

Front left :
$$\delta_1 = z_3 - z_1 + (\varphi_3 - \varphi_1)\frac{t_{d1}}{2} - \lambda L \theta_3 \tag{1}$$

Front right :
$$\delta_2 = z_3 - z_1 - (\varphi_3 - \varphi_1)\frac{t_{d1}}{2} - \lambda L \theta_3 \tag{2}$$

Rear left :
$$\delta_3 = z_3 - z_2 + (\varphi_3 - \varphi_2)\frac{t_{d2}}{2} + (1-\lambda)L \theta_3 \tag{3}$$

Rear right
$$\delta_4 = z_3 - z_2 - (\varphi_3 - \varphi_2)\frac{t_{d2}}{2} + (1-\lambda)L \theta_3 \tag{4}$$

Where :

t_{d1}	is front axle damper distance
t_{d2}	is rear axle damper distance
L	is vehicle wheel base
λ	is relative position of chassis pitch centre

If only damper relative displacements are used (no additional sensors) there are only four inputs to control damping of the seven modes. This meaning, that nor is it possible to distinguish or specify damping individually for all seven modes. The following *estimations* of mode displacements may though be utilized:

Front axle bounce :
$$z_1 = -\frac{(\delta_1 + \delta_2)}{2} \tag{5}$$

Front axle roll:
$$\varphi_1 = \frac{(\delta_2 - \delta_1)}{t_{d1}} \tag{6}$$

Rear axle bounce:
$$z_2 = -\frac{(\delta_3 + \delta_4)}{2} \tag{7}$$

Rear axle roll:
$$\varphi_2 = \frac{(\delta_4 - \delta_3)}{t_{d2}} \tag{8}$$

Sprung mass bounce:
$$z_3 = \frac{(\delta_1 + \delta_2)}{2}(1-\lambda) + \frac{(\delta_3 + \delta_4)}{2}\lambda \tag{9}$$

Sprung mass pitch:

$$\theta_3 = -\frac{(\delta_1 + \delta_2)}{2L} + \frac{(\delta_3 + \delta_4)}{2L} \tag{10}$$

Sprung mass roll:

$$\varphi_3 = \frac{(\delta_1 - \delta_2)}{t_{d1}}(1 - \kappa) + \frac{(\delta_3 - \delta_4)}{t_{d2}}\kappa \tag{11}$$

In the expression for sprung mass roll angle estimation, eq. (11), a factor κ is introduced to enable different weighting of relative front and rear roll angle. With the objective to investigate different damping for sprung mass modes eq. (9-11) are selected. Since there are totally 4 DOF for the axles relative to the chassis a fourth equation, corresponding to axle crossing, is created by subtracting eq. (6) from (8).

Axle crossing:

$$\chi = \frac{(\delta_1 - \delta_2)}{t_{d1}} - \frac{(\delta_3 - \delta_4)}{t_{d2}} \tag{12}$$

Eq. (9-12) written on matrix form, where subscripts for sprung mass modes have been omitted to emphasize estimation, thus become:

$$
\begin{bmatrix} z \\ \theta \\ \varphi \\ \chi \end{bmatrix} =
\begin{bmatrix}
(1-\lambda)/2 & (1-\lambda)/2 & \lambda/2 & \lambda/2 \\
-1/2L & -1/2L & 1/2L & 1/2L \\
(1-\kappa)/t_{d1} & -(1-\kappa)/t_{d1} & \kappa/t_{d2} & -\kappa/t_{d2} \\
1/t_{d1} & -1/t_{d1} & -1/t_{d2} & 1/t_{d2}
\end{bmatrix}
\begin{bmatrix} \delta_1 \\ \delta_2 \\ \delta_3 \\ \delta_4 \end{bmatrix} = [\Psi]^{-1}\,\overline{\delta} \tag{13}
$$

If a damping coefficient is defined for each selected modal coordinate the corresponding generalized forces become:

$$
\begin{bmatrix} F_z \\ M_\theta \\ M_\varphi \\ M_\chi \end{bmatrix} =
\begin{bmatrix}
c & 0 & 0 & 0 \\
0 & c_\theta & 0 & 0 \\
0 & 0 & c_\varphi & 0 \\
0 & 0 & 0 & c_\chi
\end{bmatrix}
\begin{bmatrix} \dot{z} \\ \dot{\theta} \\ \dot{\varphi} \\ \dot{\chi} \end{bmatrix} = [C]_q\,\dot{q} \tag{14}
$$

Changing coordinates to damper relative displacements by inserting the time derivative of eq. (13) in (14) gives:

$$\overline{F}_q = [C]_q[\Psi]^{-1}\dot{\overline{\delta}} \tag{15}$$

Transforming the generalized force by multiplying eq. (15) from the left with the transpose inverse transformation matrix then gives the damping matrix for relative damper displacements.

$$\left([\Psi]^{-1}\right)^{T} \overline{F}_{q} = \left([\Psi]^{-1}\right)^{T}[C]_{q}[\Psi]^{-1}\dot{\delta} = [C]_{\delta}\dot{\delta}$$

$$\overline{F}_{\delta} = [C]_{\delta}\dot{\overline{\delta}}$$

Where

$$[C]_{\delta} = \begin{bmatrix} c_{11} & c_{12} & c_{13} & c_{14} \\ c_{21} & c_{22} & c_{23} & c_{24} \\ c_{31} & c_{32} & c_{33} & c_{34} \\ c_{41} & c_{42} & c_{43} & c_{44} \end{bmatrix}$$

$$c_{11} = c_{22} = \left(\frac{c(1-\lambda)^2}{4} + \frac{c_{\theta}}{4L^2} + \frac{c_{\varphi}(1-\kappa)^2}{t_{d1}^2} + \frac{c_{\chi}}{t_{d1}^2} \right)$$

$$c_{12} = c_{21} = \left(\frac{c(1-\lambda)^2}{4} + \frac{c_{\theta}}{4L^2} - \frac{c_{\varphi}(1-\kappa)^2}{t_{d1}^2} - \frac{c_{\chi}}{t_{d1}^2} \right)$$

$$c_{13} = c_{31} = c_{24} = c_{42} = \left(\frac{c\lambda(1-\lambda)}{4} - \frac{c_{\theta}}{4L^2} + \frac{c_{\varphi}\kappa(1-\kappa)}{t_{d1}t_{d2}} - \frac{c_{\chi}}{t_{d1}t_{d2}} \right)$$

$$c_{14} = c_{41} = c_{23} = c_{32} = \left(\frac{c\lambda(1-\lambda)}{4} - \frac{c_{\theta}}{4L^2} - \frac{c_{\varphi}\kappa(1-\kappa)}{t_{d1}t_{d2}} + \frac{c_{\chi}}{t_{d1}t_{d2}} \right)$$

$$c_{33} = c_{44} = \left(\frac{c\lambda^2}{4} + \frac{c_{\theta}}{4L^2} + \frac{c_{\varphi}\kappa^2}{t_{d2}^2} + \frac{c_{\chi}}{t_{d2}^2} \right)$$

$$c_{34} = c_{43} = \left(\frac{c\lambda^2}{4} + \frac{c_{\theta}}{4L^2} - \frac{c_{\varphi}\kappa^2}{t_{d2}^2} - \frac{c_{\chi}}{t_{d2}^2} \right)$$

3. SIMULATION MODEL

A non-linear MBS model of a 36100-kg tractor semi trailer combination is used to evaluate damping performance of systems with corner and modal damping. The model, with main geometry according to fig. 3, is composed of several sub models such as frame, cab, drive train, axles, wheels, fifth wheel and semi trailer (totally 49 DOF). Model parameter values are taken from a real tractor. The axle suspensions are of so-called functional type, where the suspension behaviour is represented rather than exact geometrical layout.

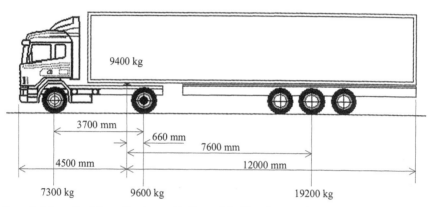

Figure 3. Schematic picture of tractor semi trailer model with main properties.

The tractor chassis suspension is equipped with two dampers at both the front and rear axle. For corner damping, the damper forces are individually determined by each damper displacement velocity and corresponding damping coefficient. To ensure reliability of the vehicle model, simulations of typical vehicle behaviour is for this case validated against test track measurements.

For case of modal damping, the tractor chassis damper forces are determined by the damping matrix in eq. (18) combined with damper displacement velocities. A restriction is however imposed on the damping force in the individual damper so that it always requires a local displacement velocity to generate a force (with the same sign), otherwise the force is set to zero. This way the implemented modal dampingsystem requires no actuator force.

4. SIMULATIONS

Vehicle simulations with the validated MBS model are performed to evaluate corner and modal damping strategy. Load cases representing typical driving conditions of both ride and handling character are used.

If the values on bounce, roll, pitch and warp damping in eq. (14) are chosen with relations according to eq. (25-27) corner damping will result.

$$c_\theta = \lambda(1-\lambda)L^2 c \tag{25}$$

$$c_\varphi = \left((1-\lambda)t_{d1}^2 + \lambda t_{d2}^2\right)\frac{c}{4} \tag{26}$$

$$c_\chi = \kappa(1-\kappa)c_\varphi \tag{27}$$

In the simulations bounce, roll, pitch and warp damping are each varied one at the time from parameter values corresponding to original corner damping. The interval used corresponds to relative damping between 0.05-0.50 for respective mode shape. Values on λ and κ are furthermore chosen as rear bounce and rear roll damping fraction of total damping, for the case nominal corner damping.

4.1   LOAD CASES

Load cases of both ride and handling character are used in the simulations. Four of the road excitation types are illustrated in fig. 4. The ISO road [6] is used with uncorrelated left and right track. Influence from steering wheel excitation is investigated with a single lane change maneuver. For all load cases vehicle velocity is 20 m/s.

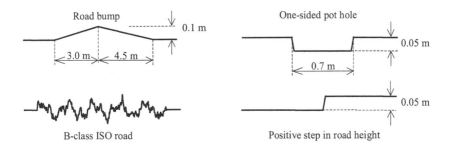

Figure 4. Road excitation load cases used in vehicle simulations.

4.2   RESULT

Due to space limitations only results from two load cases are possible to show.

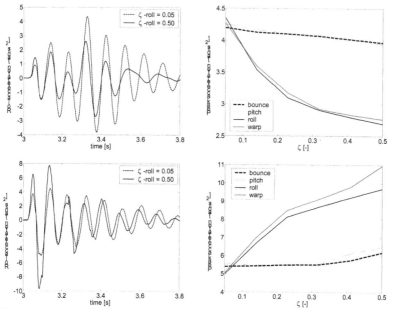

Figure 5. Damping effect on cab (upper) and chassis (lower) roll acceleration for load case one sided pothole.

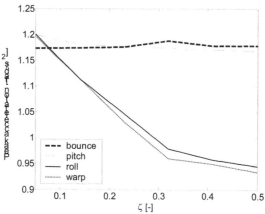

Figure 6. Damping effect on cab peak roll acceleration during single lane change maneuver

Simulation over one sided pot hole show that cab peak acceleration benefits from increased warp and roll damping while chassis peak acceleration is increased, fig. 5.

In general simulations show the following result for the different load cases. For the ISO road, studying PSD of cab, chassis and axle accelerations, damping increase for all four modes reduces amplitudes. Driver comfort during double sided road bump passage and lane change maneuver benefit from increased pitch and roll damping respectively. For positive step in road height and one sided pot hole, damping is limited by what peak acceleration that can be accepted.

5. CONCLUSIONS

Even though modal damping lifts restrictions imposed by geometry, thus enabling a more free choice for sprung mass mode damping, some problems still remain e.g. the compromise between initial impact peak and resulting length of oscillation when driving over a pothole. Low damping values should be used to minimize the initial impact peak while high values are needed to reasonable quick damp out the resulting oscillations.

Modal damping allows sprung mass relative roll damping to be increased from the low nominal value 0.06, without affecting bounce and pitch damping, to improve roll stability during lane change maneuvers, fig.6.

In performed simulations a value of $\lambda = 0.60$, corresponding to rear bounce damping fraction of total damping, was used. When studying effect of bounce and pitch damping no distinct uncoupling could be observed. A reason to this could be that when coupled to a semi trailer, tractor pitch centre is located at the fifth wheel, corresponding to $\lambda = 0.82$.

ACKNOWLEDGEMENTS

The authors are pleased to acknowledge the financial support of VINNOVA (The Swedish Agency for Innovation Systems) and Scania CV AB.

REFERENCES

1. Fukushima, N; Hidaka, K; Iwata, K; Optimum Characteristics of Automotive Shock Absorbers under Various Driving Conditions and Road Surfaces, JSAE Review, March, 1983.

2. Duym, S.; Lemmens, L.; Frequency Dependent Damper using a Compressible Damping Medium; TENNECO Automotive, Monroe European Technical Center, Belgium.

3. Holen, P; Thorvald, B; Aspects on Roll and Bounce Damping for Heavy Vehicles, SAE Technical Paper, 2002-01-3060, 2002.

4. Fontdecaba i Buj, J; Integral Suspension System for Motor Vehicles Based on Passive Components, SAE Technical Paper, 2002-01-3105, 2002.

5. Furihata, K; Toyofuku, K; Sonehara, T; Takahashi, A; Kitagawa, A; Sanada, K; A Study of an Active Suspension System with Modal Control Algorithm, JSAE Review, Vol. 14, No. 4, 1993.

6. ISO8608:1995(E); Mechanical vibration – Road surface profiles – Reporting of measured data, International Organisation for Standardisation, Schweiz, 1995.

Vehicle System Dynamics Supplement 41 (2004), p.182-191

Neuromuscular dynamics and the vehicle steering task

ANDREW PICK AND DAVID COLE[1]

SUMMARY

The paper describes part of a research programme aimed at understanding and modelling the role of steering torque feedback in the vehicle-driver system. Measurement and data analysis procedures for identifying the passive properties (inertia, damping and stiffness) of the driver's arms are presented. A linearised model is found to be satisfactory. In addition, the adaptive properties of the driver's arm dynamics are identified. In particular the driver is able to stiffen the arms by co-activating opposing muscle pairs. By measuring muscle activity using electromyography the muscles involved in the steering task have been identified and muscle co-activation during a simulated driving task has been observed.

1.   INTRODUCTION

Active steering devices, such as electrically assisted power steering and steer by wire, offer great freedom in the control of the torque fed back to the driver. In order to maximise the benefit offered by active steering devices, it is necessary to understand the role of steering torque feedback in the closed-loop behaviour of the driver-vehicle system.

The importance of torque feedback in driver steering control has been recognized for many years. In 1964 Segel [1] investigated the influence of steering torque gradient through a series of subjective evaluations. The results indicated that steering torque has a first order effect on driver perception of handling quality. Other subjective studies rate steering feel and torque feedback as some of the most important cues to the driver, preceded only by visual information [2]. Despite the acknowledged importance of torque feedback, to date most driver steering control models have simplified the situation by assuming that steering control is achieved only through the application of instantaneous and precise steer angles [3, 4].

The work described in this paper is part of a research programme aimed at establishing a driver steering control model capable of predicting subjective evaluations of steering quality. A key part of the research is to develop a dynamic model of the driver's arms, including the neuromuscular system. Neuromuscular models are reviewed briefly in section 2. Sections 3 and 4 describe the equipment and data analysis for measuring the dynamic properties of the arms. Section 5 and 6 describe equipment and experiments for measuring neuromuscular activity during steering tasks. Conclusions are given in section 7.

---

[1] *Address correspondence to:* Department of Engineering, University of Cambridge, Trumpington Street, Cambridge, CB2 1PZ, UK. Tel: +44 1223 332600; Fax: +44 1223 332662; E-mail: djc13@eng.cam.ac.uk

2.   NEUROMUSCULAR MODELLING

Human control of aircraft joysticks has been the subject of past research [5, 6]. From this work, model structures of the driver's neuromuscular dynamics involved in the vehicle steering task have been proposed [7, 8]. The essential components of such models include the driver's arm dynamics such as the inertia, damping and stiffness, as well as sensory feedback representing the driver's position and force sensing capability (Fig. 1). The feedback paths act at a spinal level, through the muscle spindle and Golgi tendon organs where the time delays are small, and at higher cognitive levels with larger time delays.

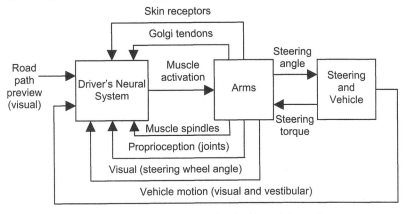

Fig. 1. Proposed model of neuromuscular dynamics involved in vehicle steering.

Whilst the importance of the combined steering and arm dynamics has been recognised, until now the lack of experimental data collected for the driving task has meant that data from aircraft pilots and joystick control have been assumed. Additionally the adaptive capability of the driver's arm dynamics has been widely neglected.

The properties of the muscles are key to understanding arm dynamics. Hill's model [9], as cited by [10], still forms the basis of many muscle models. The model consists of three components: a parallel elastic component (PEC) which represents the elasticity of the passive muscle and ligaments; an active force producing contractile component (CC); and a series elastic component (SEC) arising from tendon compliance (Fig. 2). The total muscle force, F_T, is a non-linear function of muscle length, velocity, tension and activation.

For aircraft joystick control successful attempts have been made at linearising the dynamics by representing the muscle components as springs and dampers connected to the inertia of the arm [5]. The aim of the work described in sections 3 and 4 of this paper is to establish the validity of such a linearised model for representing the muscle properties involved in the steering task.

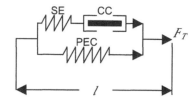

Fig. 2. Components of Hill type muscle model.

3.   HARDWARE FOR DRIVER IN THE LOOP EXPERIMENTS

To collect experimental data for the research, a fixed-base driving simulator with a torque feedback steering wheel has been developed (Fig. 3). The steering hardware consists of a modified column-drive electric power steering (EPS) unit from a small passenger car, which is controlled using Matlab's xPC toolbox. Incorporated in the EPS unit is a motor torque controller with bandwidth in excess of 500Hz. Sensors on the EPS unit send information on column angle, velocity and torque back to the xPC target computer for logging.

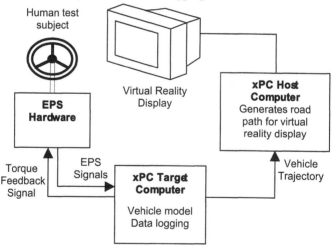

Fig. 3. EPS hardware controllers, signals and sensors.

The EPS steer angle signal is used as the input to a linear 2 DOF yaw/sideslip vehicle model running on the xPC target computer in real time. The model generates a steering torque feedback signal and sends it to the steering hardware. The vehicle trajectory data is sent to the xPC host computer to generate a visual display using Matlab's Virtual Reality toolbox. To represent a conventional passive steering system, the steering torque feedback is calculated as a weighted sum of steer angle, steer velocity and front tyre slip angle signals (Fig. 4). By changing the weights the steering feel can be varied.

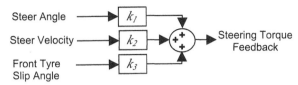

Fig. 4. Steering torque feedback controller.

4.   STEERING AND ARM DYNAMICS

The EPS hardware was used in isolation from the virtual reality display and vehicle model as a dynamic shaker in order to identify the passive properties of the driver's arms. The experiments involved the test subject holding the steering wheel with both hands whilst a random torque demand signal was sent to the EPS motor in the form of a pseudo random binary sequence (PRBS). The test subject was asked to hold the steering wheel with just enough force to prevent their hands from slipping as the wheel moved, but not to actively resist the wheel's movement in order that only the passive arm dynamics where measured.

A PRBS signal consists of a random sequence of logic levels 0 or 1 and has properties similar to those of band limited white noise. However, unlike white noise, the peak amplitude variation of a PRBS signal is known (0 or 1). Hence the signal can be scaled to prevent saturation of actuators or sensors. An 11-bit PRBS test signal was used, with 0.01s time interval, generated from a shift register [11, 12]. The signal was scaled and offset to one of three levels, +/-3Nm,+/-5Nm and +/-7Nm (Fig. 5). The resulting signal was low pass filtered at 50Hz, using a fourth order Butterworth filter, before being sent to the EPS motor. In each test the PRBS signal was repeated four times and the results averaged to improve the signal to noise level.

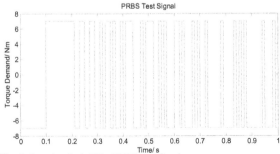

Fig. 5. Sample of +/- 7Nm PRBS test signal.

The measured transfer function and coherence between motor demand torque and column angle was calculated using the cross spectral method [13, 14]. It was found that the coherence could be improved by amplifying the excitation at certain frequencies where the column angle signal showed attenuation (arising from the dynamic response of the driver's arms and steering system). The good coherence between 0-30Hz indicates that a linear transfer function can be fitted to the data and a simple model structure (Equation (1)) is sufficient to represent the measured response (Fig. 6). The time delay is included to model sensor delays.

$$\frac{\theta_{col}(\omega)}{T_{dem}(\omega)} = \frac{15.7(s^2 + 33.1s + 1180)}{(s + 61.7)(s + 56)(s^2 + 4.2s + 12)} e^{-0.012s} \tag{1}$$

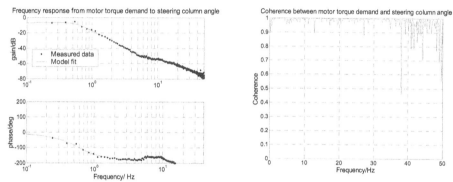

Fig. 6. Frequency response of model and measured data (+/-7Nm PRBS excitation).

The structure of the transfer function fitted to the data is consistent with a two degree of freedom model of the driver and EPS system. The two degrees of freedom are column angle, θ_{col}, and steering wheel angle, θ_{sw}, and arise because of compliance at the torque sensor, K_{sen}, in an otherwise rigid steering column. The driver adds additional inertia, J_{dr}, damping, B_{dr}, and a restraining stiffness, K_{dr}, to the steering wheel. Additional parameters are the inertia of the motor, J_{mot}, inertia of the steering wheel, J_{sw}, and viscous damping terms B_{mot} and B_{sw} to represent friction in the motor and steering wheel bearings respectively (Fig. 7). The torque generated by the motor is assumed to match the demand torque, T_{dem}, due to the high bandwidth motor torque controller.

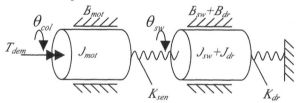

Fig. 7. Lumped parameter model of the driver's arm and steering dynamics.

The equations of motion for the model of the driver's arms and EPS hardware are:

$$J_{mot}\ddot{\theta}_{col} + B_{mot}\dot{\theta}_{col} + K_{sen}(\theta_{col} - \theta_{sw}) = T_{dem}$$

$$(J_{sw} + J_{dr})\ddot{\theta}_{sw} + (B_{sw} + B_{dr})\dot{\theta}_{sw} + K_{sen}(\theta_{sw} - \theta_{col}) + K_{dr}\theta_{sw} = 0$$

(2)

Having established a suitable structure for modelling the EPS and driver's arm dynamics, parameter identification was carried out. The mechanical properties of the EPS hardware had been previously identified and only the values for the driver's arm dynamics needed to be found. A frequency domain fitting procedure was carried out that minimised the phasor error between the experimentally measured frequency response and the model frequency response across the frequency range of interest. The errors were normalised by the magnitude of the experimentally measured frequency response (equation 3).

$$Error = \sum_{\omega} \sqrt{\frac{\text{Re}(H_{exp}(\omega) - H_{mod}(\omega))^2 + \text{Im}(H_{exp}(\omega) - H_{mod}(\omega))^2}{\text{Re}(H_{exp}(\omega))^2 + \text{Im}(H_{exp}(\omega))^2}}$$

(3)

$H_{exp}(\omega)$ -Experimentally measured frequency response.

$H_{mod}(\omega)$ -Model frequency response.

The adaptive nature of the neuromuscular system was also investigated. By co-activating opposing muscles it is possible to pretension and stiffen muscle groups [15]. The stiffening arises from the non-linear properties of the muscle. The frequency response of the driver's arms was measured in this stiffened state and values for the inertia, damping and stiffness were identified for comparison with the values identified for the relaxed muscles (Table 1). The identified parameters where not found to vary significantly with variations in excitation amplitude.

Table 1. Arm dynamic properties identified with +/-7Nm PRBS excitation.

Muscle state	Inertia: J_{dr}/ kgm^2	Damping: B_{dr}/ Nms/rad	Stiffness: K_{dr}/ Nm/rad
Relaxed muscle	0.105	0.72	3.2
Stiffened muscle	0.103	1.81	59.4

The frequency response of the driver arm dynamics alone (equation 4) is plotted (Fig. 8) using the parameter values given in Table 1. It can be seen that by stiffening the muscles the driver potentially increases the bandwidth of control of the steering system. However the increased stiffness requires increased muscle activation and increases the physical workload on the driver.

$$\frac{\theta_{sw}(s)}{T_{sw}(s)} = \frac{1}{J_{dr}s^2 + B_{dr}s + K_{dr}}$$

(4)

where: T_{sw} is the torque from the steering wheel to driver.

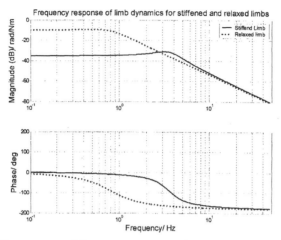

Fig. 8. Model fit to measured frequency response between steering angle and steering torque of the driver's arms.

5   DRIVER MUSCLE ACTIVITY IN THE STEERING TASK

The experiments described above show the adaptive ability of the driver. By co-activating opposing muscles, a considerable increase in the effective stiffness of the arm can be obtained, thus modifying the dynamic properties of the arm. If co-activation is a significant factor in driver steering control behaviour, the design of any steering torque feedback should take this into account. The objective of the work described in this section and section 6 is to establish experimentally the presence of co-activation in typical steering tasks.

Muscle activity was measured using electromyography (EMG). The EMG instrumentation consists of skin electrodes, an amplifier and an optical isolator. A pair of EMG electrodes is attached to the skin in the region of the muscle of interest. The EMG amplifier has high gain and input impedance and is used to measure the small voltages picked up by the electrodes when the muscle contracts (Fig. 9).

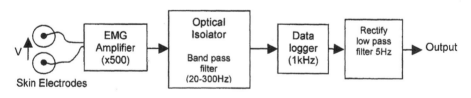

Fig. 9. EMG instrumentation for measuring muscle activity.

When rectified and then smoothed, by low pass filtering, the EMG signal is linearly proportional to muscle force. By simultaneously measuring the applied force on the steering wheel and EMG signal the contribution of key muscles to steering torque was established. Linear regression between the smoothed rectified EMG signal and force on the steering wheel was carried out (Fig. 10). The regression line is only fitted to data points where the muscle is active as muscles only act in tension. The anterior deltoid and sternal portion of the pectoral muscles where found to be strongly correlated with steering torque. Activity was seen in the anterior deltoid when the driver's arm pushes upwards on the steering wheel, and activity in the sternal portion of the pectoral occurs when the driver pulls down on the steering wheel.

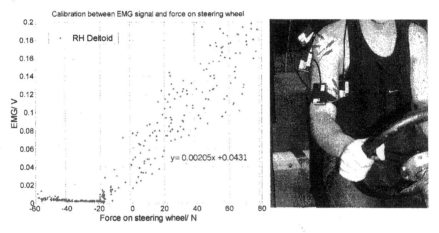

Fig. 10. Calibration of EMG signal to force on steering wheel (left) using EMG instrumentation (right).

6 DRIVING SIMULATOR TESTS

The extent to which the driver adapts the neuromuscular system to the vehicle dynamics has been investigated using the driving simulator described in section 3. Of particular interest is the way in which drivers might use muscle co-activation to stiffen their arms to compensate for changes in the vehicle response, for example with speed, or before initiating a demanding manoeuvre. Several different driving scenarios have been implemented and investigated; one example is a double lane change.

EMG signals were measured and used to predict the steering torque. It was found that steering torque can be predicted quite well using the smoothed rectified EMG signals and the coefficients of the regression lines identified in section 5 (Fig. 11a). The contribution to steering torque from muscles that produce positive and negative torques on the steering was calculated. Simultaneous activity in the muscles that produce positive torques and those that produce negative torques on the steering wheel indicate that the driver is stiffening the limb through co-activation. Steering torque and EMG signals where measured from a driver during a simulated double lane change manoeuvre. Co-activation from opposing muscles can be seen (Fig. 11b), indicating an increase in stiffness of the driver's arms. The co-activation occurs during the transient motion of the vehicle following the double-lane change. It is thought that in this case, the driver uses co-activation to minimise the displacement response of the arms arising from the transient steering torque feedback. Work is underway to identify other conditions for co-activation, and to develop a model of the driver's neural system that correctly predicts the occurrence of co-activation.

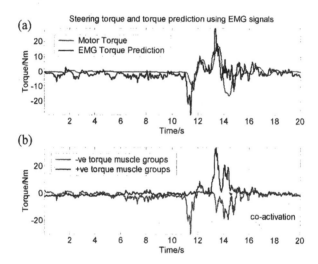

Fig. 11. (a) Steering torque and EMG steering torque prediction during lane change.
(b) Contribution of opposing muscle groups to steering torque.

7 CONCLUSIONS

- The role of steering torque feedback in the dynamics of the vehicle-driver system is poorly understood.
- Research is underway to develop a mathematical model capable of predicting subjective evaluations of steering feel.
- Measurement and data analysis procedures for identifying the passive properties of the driver's arms have been developed. A linearised model appears to be appropriate.

- The significance of muscle co-activation in adapting the stiffness and damping of the driver's arms has been highlighted.
- Electromyography has been used successfully to identify the key muscles involved in the steering task, and to identify the occurrence of co-activation.
- Further work is underway to understand how the driver uses the sensed information on vehicle and steering response to generate the muscle activation signals.

ACKNOWLEDGEMENT

The authors wish to thank the Engineering and Physical Sciences Research Council (EPSRC) and TRW Conekt for their ongoing support of this research.

REFERENCES

1. Segel, L.: Investigation of Handling as Implemented by Variable Steering Automobile. *Human Factors*, Aug (1964), pp. 333-341.
2. Gordon, D.A.: Experimental Isolation of Drivers' Visual Input. *Public Roads*, 33 (1966), pp. 53-68.
3. MacAdam, C.C.: Application of an Optimal Preview Control for Simulation of Closed-Loop Automobile Driving. *IEEE Transactions on Systems, Man and Cybernetics*, 11 (June), (1981).
4. Sharp, R.S. and Valtetsiotis, V.: Optimal Preview Car Steering Control. *Vehicle System Dynamics Supplement*, 35 (2001), pp. 101-117.
5. Paassen, R.V.: A Model of the Arm's Neuromuscular System for Manual Control. *Proc. IFAC Analysis, Design and Evaluation of Man-Machine Systems*, Cambridge, USA 1995.
6. Magdaleno, R.E. and McRuer, D.T.: Experimental Validation and Analytical Elaboration For Models of the Pilot's Neuromuscular Subsystem in Tracking Tasks. NASA Contractor Report CR1757, 1971.
7. Yuhara, N., Horiuchi, S., Iljima, T., Shimizu, K. and Asanuma, N.: An Advanced Steering System with Active Kineshetic Feedback for Handling Qualities Improvement. *Vehicle System Dynamics*, 27 (1997), pp. 327-355.
8. Hess, R.A.: A Model of Driver Steering Control Behaviour for use in Assessing Vehicle Handling Qualities. *Journal of Dynamic Systems, Measurement and Control, Transactions of ASME*, 115 (1993), pp. 456-64.
9. Hill, A.V.: The Heat of Shortening and Dynamic Constants of Muscle. *Proceedings of the Royal Society London Series B*, 76 (1938), pp. 136-195.
10. Winter, D.A.: *Biomechanics and motor control of human movement*. 2nd ed. New York 1990. John Wiley and Sons, Inc.
11. Dutton, K., Thompson, S. and Barraclough, B.: *The Art of Control Engineering*. Harlow, UK. 1997. Prentice and Hall.
12. Horowitz, P. and Hill, W.: *The art of electronics*. 2nd ed, Cambridge, 1989.
13. Ljung, L.: *System Identification Theory for the User*. 2nd ed, London, 1999. Prentice and Hall.
14. Newland, D.E.: *An Introduction to Random Vibrations, Spectral and Wavelet Analysis*. 3rd ed, Harlow, UK, 1993. Longman Scientific & Technical.
15. Hogan, N.: Adaptive Control of Mechanical Impedance by Coactivation of Antagonist Muscles. *IEEE Transactions on Automatic Control*, 29(8) (1984), pp. 681-690.

Proposal on Variable Gear Ratio Characteristics of Cable-type EPS with Active Torque Control in Over-steer Situation

KATSUHIRO SAKAI   ATSUHIKO YONEDA   YASUO SHIMIZU[1]

SUMMARY

This paper proposes the new cable-type EPS system with an active steering torque control using the driving diagram. This diagram expresses a state of vehicle motion when a driver steers, using the front wheel angle and difference between front wheel side slip angle and rear wheel side slip angle. The torque control applies the steering torque to lead the vehicle motion for a stable state utilizing the driving diagram. As simulation results, effectiveness of corrective steering assisted by the active controlled steering torque in this EPS system for recovery from spinning on a slippery road surface has been clarified.

1 INTRODUCTION

In conventional automotive steering systems, steering ratios are designed to be almost constant. For example, input to a steering wheel is transmitted to front wheels by a constant gear ratio mechanism in a steering gear box (e.g. rack and pinion gears). In addition, in order to suppress variation of the gear ratio by making a torsional rigidity high, torsion between the steering wheel and the gearbox is reduced by connecting those parts using steel shafts and universal joints which have high torsional rigidity.

Since the steering ratio is generally maintained constant in this way, the yaw rate gain of a vehicle increases with the increase in the vehicle velocity, and decreases with the increase in the lateral acceleration. For this reason, a driver needs to be adequately experienced in maneuvering the vehicle in order to compensate these yaw rate gain characteristics.

In order to solve this problem, we have developed a steering system in which the gear ratio in the steering mechanism increases with the increase in the vehicle velocity and decreases with the increase in the steering wheel angle. This system enables the yaw rate gain to be almost constant and can improve the controllability of the vehicle significantly[1].

Incidentally, the steering ratio can be varied not only by varying the gear ratio of the steering mechanism but also by the torsion of the steering system caused by setting the torsional rigidity at a low value. This concept is called the "reduction stiffness concept" in the aeronautical field, and the aeroplane based on this concept was produced 60 years ago. This had an elevator control system in which the stiffness was reduced significantly and the gear ratio was designed to obtain a desirable response for the low airspeed condition. Using this control system, the movement of the control stick could be suitably increased at high speed without degrading the response at low speed. Consequently, since the aeroplane response at the each air speed could de matched to pilot characteristics, the flying quality was significantly improved[2].

However, when this concept is applied to a automotive steering system, the problem of increase in phase lag occurs, since the torsional rigidity of the steering system is set at a low value[3]. To solve this problem, we have proposed a new construction for EPS (Electric Power Steering) and the cable-type EPS applied this construction has been developed. This EPS makes use of flexible cables to connect the steering wheel and the gearbox. The flexibility of the cable enables various layouts of steering system components and realizes new designs of cabin space[4]. On the other hand, since such flexibility is caused by the low torsional rigidity

[1]*Address correspondence to:* Honda R&D Co.,Ltd. Tochigi R&D Center, 4630 Shimotakanezawa, Haga-machi, Haga-gun, Tochigi, 321-3393 JAPAN. Tel.: +81-28-677-3311; Fax: +81-28-677-7510; E-mail: katsuhiro_sakai@n.t.rd.honda.co.jp

of the cable, the new construction in which a torque sensor is placed directly below the steering wheel is applied in order to improve a phase characteristic of the cable-type EPS. We have clarified that this construction enables the phase characteristic to be equivalent to that of a conventional EPS with shaft[5][6]. Accordingly, we have investigate effects of variable gear ratio characteristics in the cable-type EPS applied the reduction stiffness concept. The investigation results have been confirmed that the cable-type EPS can improve yaw rate gain characteristics which depend on the vehicle velocity and the lateral acceleration by its variable gear ratio characteristics according to the vehicle velocity and the steering wheel angle, in comparison with conventional EPS[7].

Furthermore, in recent years, active controls of steering gear ratio and steering torque have been positively studied from a view point of active safety. For example, in a urgent situation (e.g. variation of the road friction coefficient μ in turning, braking on a μ-split road or a wind disturbance), the steering gear ratio control to reduce the gear ratio for quick steering[1][8] and the steering torque control according to the vehicle yaw rate[9] or based on a driving diagram[10] are effective for vehicle stability. In addition, since it is known that the steering reaction time of a driver increases to about 0.6[sec] in the urgent situation in comparison with ordinary reaction time (about 0.2[sec])[11], these controls can also compensate the increase in the reaction time, and will contribute to improvement of the active safety.

In this paper, we propose a new cable-type EPS system applied a active steering torque control based on the driving diagram, and investigate effectiveness of the active controlled steering torque to avoid a urgent state of vehicle motion in over-steer situation. As results of simulations, it has been confirmed that this EPS system improves ease of recovery from over-steer situation, because of corrective steering assisted by the controlled steering torque.

2 NOMENCLATURE

Nomenclatures used in this paper are listed in Table 1.

3 CABLE-TYPE EPS

3.1 Model of Cable-type EPS

This EPS places a torque sensor directly below the steering wheel as shown in Fig. 1. Based on output of this sensor $T_S = K_{TS}(\delta_H - \theta_C)$ and the assist gain G_A, a control unit drives the motor to provide the assist torque $T_A = G_A \cdot T_S$ on the steering wheel axis in order to reduces the driver's steering torque T_H. Equations of motion of the cable-type EPS are as follows.

Table 1. Nomenclatures

a_y : lateral acceleration	K_G : torsional rigidity (G/BOX mount)	$W_{f,r}$: weight on tire
$CF_{f,r}$: cornering force	K_{TS} : torsional rigidity (torque sensor)	β : side slip angle of vehicle
C_C : damping coefficient (column)	ℓ : wheel base($= \ell_f + \ell_r$)	$\beta_{f,r}$: side slip angle at wheel
C_M : damping coefficient (motor)	$\ell_{f,r}$: distance from C.G to wheel axle	β_{fr} : difference between β_f and β_r
C_W : damping coefficient (front wheels)	m : vehicle mass	δ_H : steering wheel angle
G_1 : controlled torque gain for β_{fr}	n_G : gear ratio of a steering mechanism	δ_f : front wheel angle
G_2 : controlled torque gain for γ	n_M : EPS reduction gear ratio	δ_{fct} : counter steering by T_C
G_A : EPS assist gain	n_S : steering ratio	γ : vehicle yaw rate
G_γ : vehicle yaw rate gain	T_A : EPS assist torque	γ_{inc} : increase in γ
I : vehicle yawing moment of inertia	T_C : controlled steering torque	γ_{max} : maximum of γ
I_C : moment of inertia (column)	T_H : steering torque	μ : coefficient of friction
I_H : moment of inertia (steering wheel)	T_S : sensing torque	θ_C : rotation angle of column
I_M : moment of inertia (motor)	T_W : self aligning torque around kingpin	θ_M : rotation angle of motor
I_W : moment of inertia (front wheels)	t : time	θ_P : pinion angle
K_D : proportional gain of a driver	t_{ac} : time across the line $\delta=0$	τ : reaction time of driver
K_C : torsional rigidity (column)	t_{set} : settling time of γ	ξ : trail
$K_{f,r}$: cornering stiffness of tire	V : vehicle velocity	f,r : front and rear

$$I_H \ddot{\delta}_H + K_{TS}\left(\delta_H - \theta_C\right) = T_H \tag{1}$$

$$I_C \ddot{\theta}_C + C_C \ddot{\theta}_C + K_{TS}\left(\theta_C - \delta_H\right) + K_C\left(\theta_C - \theta_P\right) = 0 \tag{2}$$

$$n_M^2 I_M \ddot{\theta}_P + n_M^2 C_M \ddot{\theta}_P + K_C\left(\theta_P - \theta_C\right) + \frac{1}{n_G} K_G\left(\theta_P/n_G - \delta_f\right) = T_A \tag{3}$$

$$I_W \ddot{\delta}_f + C_W \ddot{\delta}_f + K_G\left(\delta_f - \theta_P/n_G\right) = T_W \tag{4}$$

$$\theta_P = \theta_M/n_M \tag{5}$$

$$T_A = G_A T_S = G_A K_{TS}\left(\delta_H - \theta_C\right) \tag{6}$$

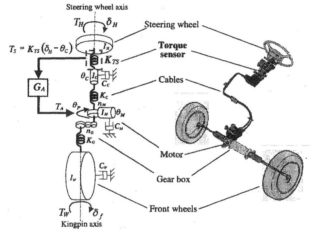

Fig. 1. Model of cable-type EPS system.

3.2 Steering Ratio Characteristics Caused by Torsional Rigidity

Using Equations (1) to (6), the front wheel angle δ_f is obtained as[7]

$$\delta_f = \frac{\delta_H}{n_G} - \frac{K_{TS} + K_C}{n_G K_{TS} K_C\left(1 + G_A\right)} \cdot \frac{T_W}{n_G} \tag{7}$$

The first term in Equation (7) is determined from the gear ratio of the steering mechanism n_G and the steering wheel angle δ_H. The second term is the decrease caused by torsion of the steering system depending on the self-aligning torque around the kingpin axis T_W. Therefore, even if δ_H is unchanging, δ_f is varied by T_W, and this variation of δ_f increases with lowering the torsional rigidity of the cable K_C.

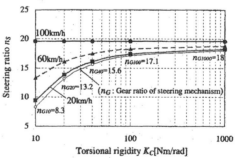

Fig. 2. Steering ratio characteristics with K_C.

Figure 2 shows the relation between the steering ratio n_S ($= \delta_H / \delta_f$) and K_C[7]. Utilizing the variation of δ_f caused by T_W and K_C mentioned above, n_G can be set at a low value(i.e. a quick gear ratio) when a value of K_C is low, without a change of n_S at a high vehicle velocity. In addition, n_S at a low velocity can be varied to low because T_W decreases with decreasing the vehicle velocity V.

4 ACTIVE STEERINGTORQUE CONTROL BASED ON DRIVING DIAGRAM

4.1 Driving Diagram

The driving diagram (Fig. 3) express a state of vehicle motion when a driver steers, using δ_f (the front wheel angle) and β_{fr} (difference between front wheel side slip angle β_f and rear wheel side slip angle β_r). Therefore, this diagram can be utilized in the active steering torque control which lead the vehicle motion to a stable state[10].

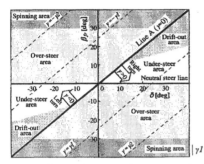 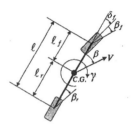

Fig. 3. Driving diagram (in a case of a constant velocity state)   Fig. 4. Vehicle model

A relation between δ_f and β_{fr} is obtained as the following equation, using a vehicle model as shown in Fig. 4.

$$\beta_f = -\beta - \frac{\ell_f}{V}\gamma + \delta_f \tag{8}$$

$$\beta_r = -\beta + \frac{\ell_r}{V}\gamma \tag{9}$$

$$\therefore \ \beta_{fr} = \beta_f - \beta_r = -\frac{\ell}{V}\gamma + \delta_f \tag{10}$$

4.2 Driving Locus in Driving Diagram Obtained by Actual vehicle Test

4.2.1 Under-steer Area ($\beta_{fr} \cdot \gamma > 0$)

Assuming the yaw rate gain $G_\gamma (= \gamma / \delta_f)$ is almost constant in a constant velocity condition, Equation (10) is modified as follows.

$$\beta_{fr} = \left(1 - \frac{\ell}{V}G_\gamma\right)\delta_f = G_U \cdot \delta_f \tag{11}$$

From Equation(11), β_{fr} is proportional to δ_f according to the gradient G_U. Figure 5(a) shows the driving locus of the single lane change task.

4.2.2 Over-steer Area ($\beta_{fr} \cdot \gamma < 0$)

When the rear wheels are locked in turning, the yaw rate γ sharply increases and the vehicle goes into spinning. From Equation (10), $|-\beta_{fr}|$ increases with increasing γ, even if δ_f is unchanging. A test result is shown in Fig. 5(b).

Figure 5(c) shows recovery from spinning by driver's counter-steer behavior. After a state of the vehicle motion enters the over-steer area, the increase in $|-\beta_{fr}|$ is suppressed and γ is leaded to zero (approach Line A) by steering δ_f in the counter (opposite) direction. This steering behavior can avoid spinning.

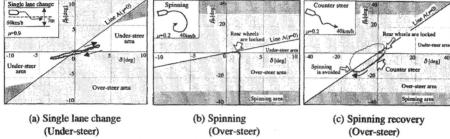

(a) Single lane change	(b) Spinning	(c) Spinning recovery
(Under-steer)	(Over-steer)	(Over-steer)

Fig. 5. Driving Loci obtained by actual vehicle tests.

4.3 Steering Torque Control with $\beta_{fr} + \gamma$

The steering torque is controlled in order to lead vehicle motion to a stable state. In over-steer situation, since the front wheel angle δ_f should be steered to avoid spinning as counter-steer mentioned above, the controlled steering torque is applied in the direction to return β_{fr} and γ to zero. From equation (10), δ_f can be expressed as follows.

$$\delta_f = \beta_{fr} + \frac{\ell}{V}\gamma \tag{12}$$

In order to obtain δ_f which reduce β_{fr} and γ toward zero, the controlled steering torque T_C is calculated as follows.

$$T_C = G_1 \cdot \beta_{fr} + G_2 \cdot \gamma \tag{13}$$

Where, G_1 and G_2 are the controlled torque gains ($G_1 < 0$, $G_2 > 0$).

Therefore, the EPS assist torque T_A is provided by subtracting T_C from $G_A \cdot T_S$. Namely, it can be obtained from Equation (6) and (13) as,

$$T_A = G_A \cdot T_S - T_C = G_A \cdot T_S - G_1 \cdot \beta_{fr} - G_2 \cdot \gamma \tag{6'}$$

5   EFFECT OF ACTIVE TORQUE CONTROL IN OVER-STEER SITUATION

5.1   Effect on the EPS system

Since the EPS assist torque T_A is provided as Equation (6)', the steering torque T_H is obtained as follows from Equations (1) - (5) and (6)'.

$$T_H = \frac{T_W}{n_G} - T_A = \frac{T_W}{n_G} - (G_A \cdot T_S - T_C)$$

$$= \frac{1}{(1+G_A)}\left(\frac{T_W}{n_G} + T_C\right) = \frac{T_W}{n_G(1+G_A)} + G_1^* \beta_{fr} + G_2^* \gamma \tag{14}$$

$$\left(G_1^* = \frac{G_1}{(1+G_A)} , \quad G_2^* = \frac{G_2}{(1+G_A)} \right)$$

Equation (14) shows that the controlled steering torque T_C is provided in the direction to increase self-aligning torque T_W when γ and $|-\beta_{fr}|$ increase in over-steer situation, so it can assist the driver's counter steering behavior which reduce γ and $|-\beta_{fr}|$. In addition, the front wheel angle δ_f is obtained as follows.

$$\delta_f = \frac{\delta_H}{n_G} - \frac{K_{TS} + K_C}{n_G K_{TS} K_C (1 + G_A)} \left(\frac{T_W}{n_G} + T_C \right)$$

$$= \frac{\delta_H}{n_G} - \frac{K_{TS} + K_C}{n_G K_{TS} K_C} \left(\frac{T_W}{n_G(1 + G_A)} + G_1^* \beta_{fr} + G_2^* \gamma \right) \qquad (7)'$$

Equation (7)' means that T_C can steer the front wheel in the direction to reduce γ and $|-\beta_{fr}|$, such as the driver's counter-steer behavior. Particularly, it is effective when the torsional rigidity of the cable K_C is set at a low value.

5.2 Performance of Vehicle with Cable-type EPS for Spin Avoidance

Based on a relation between the steering rate which is required to avoid spinning and the reaction time (time delay) until beginning of steering behavior after the vehicle yaw rate γ increases[1], vehicle performance of spin avoidance on a slippery road surface is examined by simulation.

5.2.1 Vehicle Model

A model which is considered a nonlinear characteristic of the cornering force[12] is used in the simulation. It is expressed as

$$mV(\dot{\beta} + \gamma) = 2CF_f + 2CF_r \qquad (15)$$
$$I\dot{\gamma} = 2\ell_f CF_f - 2\ell_r CF_r \qquad (16)$$
$$T_W = 2CF_f \cdot \xi \qquad (17)$$
$$CF_f = K_f \beta_f - \frac{K_f^2}{4\mu W_f} \beta_f^2 \qquad (18)$$
$$CF_r = K_r \beta_r - \frac{K_r^2}{4\mu W_r} \beta_r^2 \qquad (19)$$

Where, β_f and β_r are given from Equations (8) and (9).

5.2.2 Simulation Condition

A driving situation as shown in Fig. 6 is assumed in the simulation. Namely,
①The vehicle turns at the velocity $V=50$[km/h] and the lateral acceleration $a_y=0.15$[G] under slippery road surface condition (friction coefficient $\mu=0.3$).
②μ at rear tyres suddenly decreases from 0.3 to 0.1 for one second. This causes the vehicle to spin (go into over-steer area).
③A counter steering is begun to avoid spin, after the reaction time (time delay) τ.
④A ramp input of the steering wheel angle δ_H (i.e. a constant steering rate input) is applied as the counter steering.
⑤The spin is avoided by the steering input if γ can be leaded to zero (on Line A). In this simulation, the spin avoidance time t_{av} which is time between ② and ⑤ as shown Fig.6 is set at two seconds.

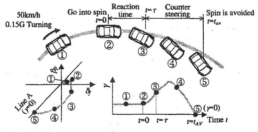

Fig. 6. Driving situation in simulation.

Four EPS systems listed in Table 2 are compared. Each mechanical gear ratio n_G is set based on Fig. 2. In addition, since the steering torque T_H can be obtained as Equation (14), the assist gain G_A and the controlled torque gains G_1, G_2 are adjusted according to n_G, in order that T_H remains unchanged for each n_G.

Table 2. Compared EPS systems

	Rigidity of cable K_C[Nm/rad]	Mechanical gear ratio n_G	Assist gain G_A	Controlled torque gain	
				G_1 [Nm/rad] ($G_1{}^*$=50)	G_2 [Nms/rad] ($G_2{}^*$=35)
1. Conventional (without T_C)	1000 (equivalent to shaft's)	18	1.5	0	0
2. Conv. with T_C	↑	↑	↑	125	87.5
3. Cable40 with T_C	40	15.6	1.9	145	102
4. Cable10 with T_C	10	8.3	4.2	260	182

5.2.3 Simulation Result

Varying the the reaction time τ, the steering rate which can lead the yaw rate γ to zero is calculated for each τ. Results are shown in Fig. 7. The controlled steering torque T_C can reduce the steering rate which is needed to avoid spinning, in comparison to the case without T_C. In addition, this is remarkably effective when the torsional rigidity of cable K_C is set at a low value. For example, in the case of "Conventional", i.e. without T_C and K_C =1000[Nm/rad], when τ is 0.6[sec], the steering rate needs to be at least 430[deg/s]. On the other hand, in the case of "Cable40 with T_C" (K_C=40[Nm/rad]), the needed steering rate reduces until 280[deg/s], furthermore, in the case of "Cable10 with T_C" (K_C=10[Nm/rad]), that is only 20[deg/s].

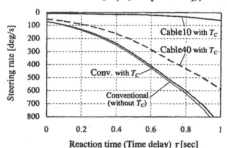

Fig. 7. Steering rate to avoid spinning.

Figures 8(a), 8(b), and 8(c) show time charts of the steering wheel angle δ_H, the front wheel angle δ_f, and the yaw rate γ, when τ is 0.6[sec], respectively. Even if δ_H remains unchanged until t=0.6[sec], δ_f is steered in the counter direction by the controlled steering torque T_C provided to reduce $|-\beta_f|$ and γ as shown in Fig. 9. Because of this counter steering by T_C, increase in γ can be suppressed (Fig. 8(c)).

Figure 10 shows the torsional rigidity of cable K_C versus δ_{fcs} : the angle of the counter steering by T_C defined as the difference between δ_f at t=0[sec] and δ_f at t=0.6[sec], and γ_{inc} : the amount of the increase in γ defined as the difference between γ at t=0[sec] and γ at t=0.6[sec]. The lower K_C makes the counter steering δ_{fcs} be remarkably large, and this significantly reduces γ_{inc}. For example, γ_{inc} at K_C =10[Nm/rad] becomes 1/4 in comparison with γ_{inc} at K_C =1000[Nm/rad].

From the examinations in this section, it has been confirmed that the cable-type EPS set at low torsional rigidity can reduce the steering rate which is required to lead γ to zero, by the counter steering of T_C. In addition, the low steering ratio also reduce the required steering rate[1]. Consequently, these effects of the cable-type EPS can improve performance of spin avoidance on the slippery road surface.

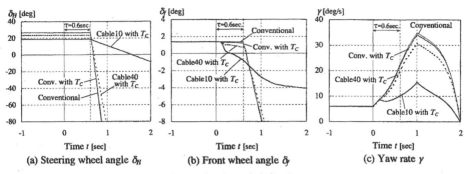

(a) Steering wheel angle δ_H (b) Front wheel angle δ_f (c) Yaw rate γ

Fig. 8. Time charts in the case that the reaction time τ is 0.6[sec].

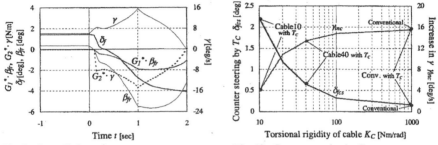

Fig. 9. Controlled steering torque
in the case of "Cable10 with T_C".

Fig. 10. Counter steering by T_C
and Increase in γ at t=0.6 [sec]

5.3 Confirmation on Ease of Recovery from Spinning(Over-steer Situation)

For recovery from spinning, it is necessary that not only γ but also β_{fr} is returned to zero. In this section, ease of the spinning recovery is examined by simulation using a driver-vehicle system model. In order to clarify effect of the active steering torque control, a driver model which outputs the steering torque T_H is used.

5.3.1 Driver-Vehicle System Model

A model of driver-vehicle system used in the simulation is shown in Fig. 11[13]. It is assumed that the driver steers according to the yaw rate γ in over-steer situation, in order to recover from spinning. An EPS model and a vehicle model are the same with those in section 5.2.

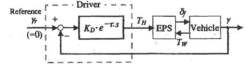

Fig. 11. Driver-vehicle system model

5.3.2 Simulation Condition

Conditions are the same with those in section 5.2 except the steering input. In this section, the steering input as the counter steering for recovery from spinning is given by the driver model.

The reaction time τ which is time until the driver begins the counter steering behavior is set at 0.6[sec]. After that, τ is set at 0.2[sec] as ordinary time delay of the driver.

The controlled steering torque T_C is assumed to be provided until spinning is recovered.

5.3.3 Simulation Result

Figures 12 and 13 show time charts and Figure 14 shows a driving diagram of the results. Even if a state of the vehicle motion goes into spinning in the cases of "Conventional"(without T_C), in the case with T_C, that can recover from spinning by the steering behavior assisted by T_C, because T_C can make the beginning of the driver's counter steering behavior early. Namely, it is equivalent to reducing the reaction time τ. T_C is provided at the same time as a state of the vehicle motion goes into the over-steer area, and assists the driver's counter steering. When γ increases, it is suppressed by the controlled torque $G_2^* \cdot \gamma$, and when γ approaches zero, the controlled torque $G_1^* \cdot \beta_{fr}$ leads the driver's steering behavior in the direction to reduce β_{fr}.

Figure 15 shows the torsional rigidity of cable K_C versus t_{ac} : time until the steering wheel angle δ_H or the front wheel angle δ_f crosses the line of zero. By the assistance of T_C, t_{ac} of δ_H significantly reduces and is almost constant for a value of K_C. t_{ac} of the "Conventional" case is 0.73[sec], t_{ac} of the case with T_C are about 0.38[sec]. On the other hand, t_{ac} of δ_f can be reduced with lowering a value of K_C, because of the counter steering of δ_f by T_C regardless of δ_H as mentioned in section 5.2. t_{ac} of δ_f at K_C =10[Nm/rad] further reduces to 0.23[sec].

Figure 16 shows K_C versus t_{set} : the settling time of γ defined as time required to converge into $|\gamma|<0.5$[deg/s] and γ_{inc} : the maximum of γ. Both t_{set} and γ_{inc} can be reduced when K_C is set a low value, because the steering behavior by T_C is more effective.

The results of the examinations in this section have confirmed that corrective steering, i.e. the early counter steering and the steering behavior to reduce γ and β_{fr} to zero, can be effectively assisted by T_C in the cable-type EPS. Consequently, the cable-type EPS can improve ease of recovery from spinning, applying the torque control in over-steer situation.

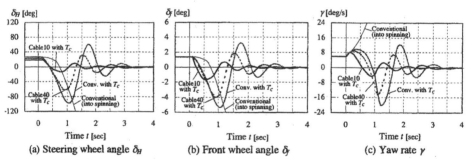

(a) Steering wheel angle δ_H (b) Front wheel angle δ_f (c) Yaw rate γ

Fig. 12. Time charts of the simulation results.

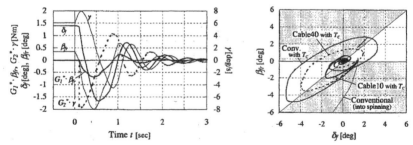

Fig. 13. Controlled torque in the "Cable10 with T_C"case.   Fig. 14. Driving diagram

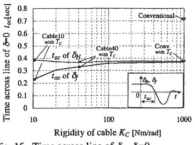

Fig. 15. Time across line of δ_H, $\delta_f = 0$

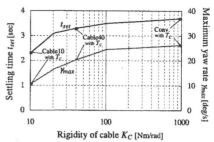

Fig. 16. Settling time and maximum of yaw rate

6 CONCLUSION

We have proposed the new cable-type EPS system applied active steering torque control based on the driving diagram. As the results of theoretical examinations, it has been confirmed that this EPS system can improve ease of recovery from spinning on the slippery road surface.

REFERENCES

1. Shimizu, Y., Kawai, T., Yuzuriha, J. : Improvement in driver-vehicle system performance by varying steering gain with vehicle speed and steering angle :VGS (Variable Gear-ratio Steering system). *SAE International Congress & Exposition*, No.99 PC-480, 1999.
2. Horikoshi, J. : RESEARCH ON THE IMPROVEMENT OF FLYING QUALITIES OF PILOTED AIRPLANES - REDUCTION STIFFNESS CONCEPT APPLIED TO ELEVATOR CONTROL SYSTEM. *Doctoral Thesis*, University of Tokyo, 0599, 1965.
3. Nishii, K., Higuchi, T., : Effect of Component Factors on Controllability and Stability of Vehicles. *Journal of JSAE*, Vol.26, No.7, 1972, pp. 773-783
4. Suyama, K. : An Examination of EPS using Cables to Connect the Steering Wheel and Gearbox. *SAE Paper*, 2003-01-0584, 2003.
5. Sakai, K., Yoneda, A., Shimizu,Y. : A Study on Control Performance of Vehicle using Electric Power Steering(EPS) with Low Torsional Rigidity. *Proceedings of TRANSLOG 2001*, 2001, pp. 397-400
6. Sakai, K., Yoneda, A., Shimizu,Y. : Improvement in Control Performance of Driver-Vehicle System with EPS using Cables to Connect the Steering Wheel and Gearbox. *Proceedings of AVEC '02*, 2002, pp. 641-646.
7. Sakai, K., Yoneda, A., Shimizu, Y. : A Theoretical Study on Variable Gear-ratio Characteristics of EPS(Electric Power Steering) applied Cables to Connect Steering Wheel and Gearbox. *Proceedings of TRANSLOG 2002*, 2002, pp. 357-360.
8. Kojo, T., Suzumura, M., Fukui, K.,Sugawara, T., Matsuda, M., Kawamuro, J. : Development of Front Steering Control System. *Proceedings of AVEC '02*, 2002, pp. 33-38.
9. Yuhara, N., Horiuchi, S., Asanuma, N., Nishi, Y., Kohata, T. : Improvement of Vehicle Handling Quality Through Active Control of Steering Reaction Torque. *Proceedings of AVEC '92*, 923073, 1992
10. Yamawaki, S., Shimizu, Y. : Effect of Steering Torque Control Using a Driving Diagram. HONDA R&D Technical Review, Vol.13, No.2, 2001, pp. 99-106.
11. Nagiri, S., Doi, S., Takei, K., Mizuno, M. : Experimental Analysis of Avoidance Maneuver Using a Driving Simulator. *Proceedings of JSAE 941 1994-5*, 1994, pp. 17 - 20.
12. Abe, M. : *Vehicle Dynamics and Control*. Sankaido, Tokyo, 1992.
13. Shimizu, Y. : Applications of Control Method in the Advance Steering System. No. 04-03 JSAE SYMPOSIUM, New Techno logies in Active Safety, 20003.

APPENDIX

Table 3. Parameter values used in the simulation.

Vehicle		ξ : 0.05[m]		C_M : 4.7×10^{-3}[Nms/rad]
m	:1060[kg]	EPS System		C_W : 1.5×10^2[Nms/rad]
I	:1800[kgm^2]	I_H : 6×10^{-2}[kgm^2]		K_{TS} :120[Nm/rad]
K_f	:40000[N/rad]	I_C : 2×10^{-4}[kgm^2]		K_G : 1.3×10^5[Nm/rad]
K_r	:40000[N/rad]	I_M : 3×10^{-4}[kgm^2]		n_M :17
ℓ_f	:0.9[m]	I_W : 1.3[kgm^2]		Driver
ℓ_r	:1.6[m]	C_C : 0.68[Nms/rad]		K_D : 7×10^{-2}[Nms/deg]

Vehicle System Dynamics Supplement 41 (2004), p.202-211   © Taylor & Francis Ltd.

A Study on the Relationship between Vehicle Behavior and Steering Wheel Torque on Steer by Wire Vehicles

MASAYA SEGAWA, SHUUJI KIMURA,
TOMOYASU KADA, SHIROU NAKANO [1]

SUMMARY

It was reported that Steer-By-Wire (SBW) system, which eliminates driver interference, could introduce the active front steering control easier than that with a conventional steering system. But there have been few reports on studies about reactive torque control strategy for the steering wheel. This research on reactive torque control strategy was carried out in order to improve the driver usability and manoeuvrability of SBW system. It was found that the relation between what drivers feel by reactive torque when turning the steering wheel and what they feel from vehicle behavior could be improved by means of applying newly developed reactive torque that incorporates vehicle dynamics parameter.

1   INTRODUCTION

Present steering systems are the parallel link type, wherein the independent right and left axles of the front wheel are supported by the suspension, and the knuckle arms are connected by the tie rods. The steering wheel is used for the interface with the driver. Disturbance from the road is suppressed and steering wheel torque is reduced because the reduction ratio is enlarged by use of a steering wheel.

After steering systems such as the rack-and-pinion type and ball-screw type began to be mass-produced, hydraulic power steering, which has a lighter steering performance, was developed in 1926. As the demand for lighter steering feeling and a safety function in power steering systems increased along with a rapid expansion of the market for power steering systems, the vehicle-speed-sensitive type hydraulic power steering system was developed. Afterward, electric power steering with steering assistance power that is electronically controlled with an electric motor came to be mass-produced because of its advantages in regard to environmental preservation and energy conservation. However, intelligent steering function having integrated control with other chassis and drive systems is required in addition to independent steering systems. The Steer By Wire (SBW) system has been developed as a steering system that achieves this function.

[1] *Address correspondence to:* R&D centre, Koyo-seiko Co., Ltd., No.333 Toichi-cho, Kashihara-shi, Nara JAPAN. Tel.: +81(744) 29-7043; Fax: +81(744) 29-7049; E-mail: masaya_segawa@koyo-seiko.co.jp

Since the operation of the steering gear is electronically controlled in the SBW system, it is possible for safe and comfortable vehicle motion control to be automated. The relationship between the system, vehicle and driver will be discussed later along with the automation of the steering function. Since there is no mechanical linkage between the steering wheel and steering gear in a SBW system, it is easy to avoid interference between the steering system and driver operating active front steering control, which was difficult with a conventional steering system. It has been confirmed in a full braking test on a μ-split road that the stabilization performance of the vehicle by the SBW system is better than that of other practical active stability control systems using the distribution of driving and braking force.[1][2]

In conventional steering systems, the amount of steering wheel torque and its value or continuousness has been focused on steering design technique to improve the usability of steering systems equipped with a steering wheel. Therefore, steering wheel torque and steering wheel angle in operation of the steering wheel have not been designed taking into account the driver's characteristics because it is a precondition that reactive torque from the road surface is transmitted to the driver through the mechanical linkage in conventional steering system. On the other hand, since there is no mechanical linkage between the steering wheel and steering gear in SBW, the degree of freedom of the design rises greatly.

In a conventional steering system, the torque characteristic and manoeuvrability have been influenced by elements such as suspension geometry and road surface conditions, which are not related to the steering system. In SBW, it is possible that fine manoeuvrability and torque characteristics can be applied to the driver by controlling information appropriately. The feasibility of intelligent usability in SBW different from in the conventional system was studied. The relation between various elements concerning the vehicle and its relation with the driver, such as between the driver and vehicle behavior, should be clarified. A steering system as a man-machine interface that feeds back the physical characteristics of the vehicle and environmental characteristics of the road to the driver more linearly was researched.

This paper describes approaches and technical issues to improve the usability by means of the reactive torque control as an introduction stage from among our research on SBW.

2   STEER BY WIRE SYSTEM

Figure 1 shows the construction of SBW system used in this study.
Steering manoeuvring is detected by means of a steering wheel angle sensor and a torque sensor, and this information is transferred to a controller, which uses this and other sensor information to control front wheel angle via a steering actuator. A rack coaxial type electric power steering system is used as the steering actuator.

The characteristics of the input device, which is the interface between the driver and the steering system, have much influence on the driver usability of SBW. It is considered that the shape of the input device and the characteristics of the torque for manoeuvring that input device are important among those characteristics. A steering wheel, which is expected to be accepted easily by driver, is used as the input device in order to promote the market expansion of SBW. A motor is connected to the steering wheel to load the torque to the steering wheel manoeuvring.

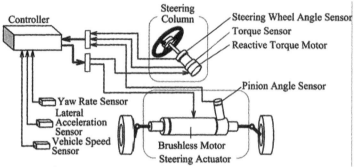

Fig. 1. System construction.

3 SIMULATE REACTIVE TORQUE

3.1 Reactive torque

When driver inputs steering wheel torque via steering wheel, and actual steering wheel angle δ_h is generated. This steering manoeuvring is transmitted to the steering gear via mechanical linkage, and the front wheel is turned through the reduction mechanism. Vehicle behaviour is generated by the force between the tires and road surface. This force transmitted to driver is considered as reaction when driver turns the steering wheel. This reactive torque is one of the most important pieces of information that driver uses in driving. Generally, the relationship between the steering wheel torque and vehicle dynamics concerning front wheel steering system is defined in this equation.[3]

$$Ks(\delta_h - N\delta) = 2\xi K_f\left(\beta + \frac{l_f}{V}\gamma - \delta\right) \qquad (1)$$

In this equation, the left-hand-side shows steering wheel torque of the dynamics model that has a spring between steering input and steering output. The right-hand-side shows the recovery moment of the front wheel. In a conventional vehicle, reactive torque is generated corresponding to the recovery moment in the linear area of the front tires. In conventional power steering, a part of reactive torque is loaded to driver.

Because SBW has no mechanical linkage between the steering wheel and the steering gear, the torque from the front wheels isn't transmitted to driver as in a conventional steering system. As the mechanical linkage is removed, SBW can overcome these problems of conventional steering systems, such as the kickback torque from the road surface, and the decline of the manoeuvrability caused by the unbalance of the force among the four wheels. Driver can benefit from solutions to those problems. But, the road information, which is necessary for driver, cannot be transmitted to the steering wheel because that mechanical linkage was removed. The artificial torque based on the relationship given in equation (1) was generated as reactive torque to improve the driver usability by means of actuators such as a motor mentioned in the section 2. As a preliminary study, the steering feeling in a conventional vehicle was tried to realize by means of reactive torque control.[4]

3.2   Simulation of conventional feeling

The steering wheel angle δ_h was only applied as the driver's input signal because it was considered that the driver evaluates steering feeling by the relationship between reactive torque transmitted via a steering wheel and steering wheel angle. The reactive torque control is regarded as a closed loop between driver and a steering wheel. In order to realize the force shown in equation (1) as much as possible, the reactive torque to driver was calculated based on formula (2), which included as variables steering wheel angle δ_h, cornering force coefficient $K_c(V)$ and proportional gain K_T. The block diagram of this reactive torque control is shown in Figure 2.

$$T_r = K_T \cdot K_c(V) \cdot \delta_h \tag{2}$$

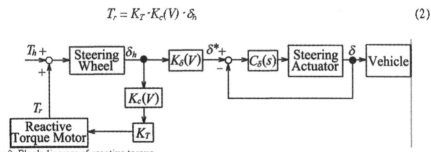

Fig.2. Block diagram of reactive torque.

Concerning front wheel angle control, the target front wheel angle δ^* was determined on the basis of steering wheel angle δ_h and gear ratio $K_\delta(V)$ set as a linear function of vehicle speed V. Actual front wheel angle δ was generated with steering actuator feedback control.

This control was installed in the actual vehicle, and the steering feeling of the SBW vehicle was evaluated by comparing to a conventional vehicle in slalom testing. Test conditions were as follows: Maintained vehicle speed at 13.9m/s on a dry asphalt road, steering wheel angle amplitude peak to peak 3.0rad, and a steering cycle of 0.5Hz. Figure 3 shows the relationship between steering wheel angle and steering wheel torque.

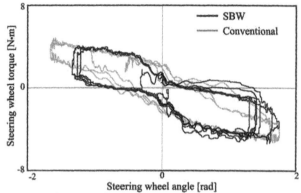

Fig.3. Reactive torque simulating conventional feeling.

As shown in Figure 3, the relationship between steering wheel angle and steering wheel torque equivalent to that of a conventional vehicle was obtained, when driver turns to the direction away from centre.

Evaluation of the SBW manoeuvrability was carried out using the driving simulator shown in Figure 4, where test driver maintained vehicle speed at 22.2m/s on a dry asphalt road, and drove through a slalom course with pylons placed at 30m intervals. The result of that test is shown in Figure 5. The fluctuation of steering wheel torque was appeared, when the vehicle passed by a pylon. It seemed that the driver turned the steering wheel to compensate the deviation between the actual vehicle behaviour and the expected vehicle behaviour against steering wheel manoeuvring.

Fig.4. Driving simulator.

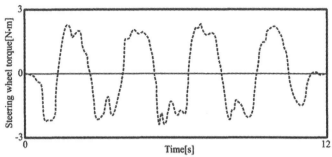

Fig.5. Test result on driving simulator.

In conventional vehicles, driver operates phase lead manoeuvring by controlling front wheel steering angle according to the vehicle dynamics in order to compensate the phase delay of the steering input system. Driver had to control the dynamic characteristics of man-machine system to be constant, and that manoeuvring was the workload to driver. [5][6] It is not possible to overcome this problem with a conventional steering system.

Newly developed control logics of the reactive torque for SBW, which were different from the reproduction of the conventional feeling, should be recommended. If driver can control vehicle behaviour directly as their intention, it will be expected that the workload of phase lead manoeuvring will be reduced and the manoeuvrability will be improved.

4   REACTIVE TORQUE FOR SBW

4.1   Reactive torque control indicating vehicle behaviour

In the previous section, it was clarified that the characteristics of reactive torque are not effective to improve the usability of SBW from the relationship between driver and the input device. When the logic of the reactive torque control is designed, it is necessary to take into account functions that influence vehicle behaviour rather than the realization of the self aligning torque. Those functions cannot be provided by a conventional steering system, such as the value of the variable yaw rate gain against driver's characteristic, and the reduction of driver workload of the feed forward front wheel angle control that a driver does in a conventional vehicle to compensate the vehicle dynamics against driver's intention. From the above, it is necessary for the reactive torque control logic to be designed based on the correlation between the driver, the input device and the vehicle as shown in Figure 6.

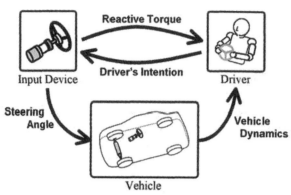

Fig.6. Concept of the reactive torque control for SBW vehicle.

As a part of the research directed at improving vehicle stability control in SBW systems, the reactive torque control algorithm whereby vehicle behaviour was communicated to driver as steering wheel angle was developed.[7] This control block diagram as shown in Figure 7, vehicle behaviour feedback T_{r2} was integrated into reactive torque T_{r1}, which was calculated from steering wheel angle. The steering wheel angle δ_h was made to follow the target steering wheel angle δ_h^* calculated from vehicle behaviour.

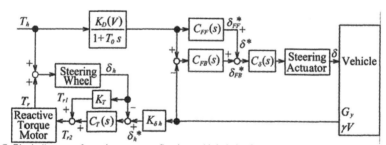

Fig.7. Block diagram of reactive torque reflecting vehicle behaviour.

To investigate the effect of this control logic, the steering manoeuvring with this reactive torque control was evaluated using driving simulator. Test conditions were the same as section 3.2.

As shown in Figure 8, the fluctuation of steering wheel torque, which appeared in the case of the conventional reactive torque control, was reduced. It is considered that the driver's feeling of vehicle behaviour corresponded to the actual vehicle behaviour. It was confirmed that the driver workload was reduced, compared to realization of conventional feeling.

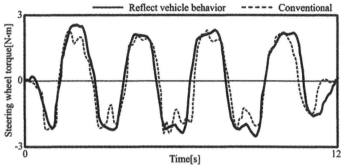

Fig.8. Comparison of control logics using driving simulator.

This reactive torque control was a part of the system whose purpose was vehicle stability control. And the feedback reactive torque was generated corresponding to the difference between the driver's input steering wheel angle and the target steering wheel angle calculated from the vehicle behaviour.

When the deviation between the driver's input angle and the target angle from the vehicle behaviour occurred, the driver could drive more safely, because of reactive torque to compensate that deviation. When there was no difference between the driver's input and the vehicle behaviour, the reactive torque corresponding to the vehicle behaviour was not generated; only the reactive torque corresponding to the steering wheel angle was generated. Because driver could not always sense the vehicle behaviour through the reactive torque, there were those who preferred further linearity between the reactive torque felt via steering wheel and the vehicle behaviour felt by drivers when they turned steering wheel.

4.2　Reactive torque control of vehicle dynamics parameter feedback

The control logic as shown in the Figure 9 was investigated as to whether the reactive torque could be corresponded more clearly to the vehicle behaviour recognized by driver in turning steering wheel. In this control logic, the vehicle behaviour was reflected directly into the reactive torque, not adding the factor of vehicle behaviour to the reactive torque determined by steering wheel angle as shown in the section 4.1.

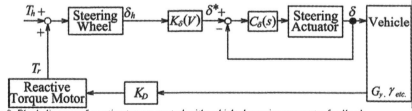

Fig.9. Block diagram of reactive torque control with vehicle dynamics parameter feedback.

Table 1 shows the feedback parameters concerning the vehicle behaviour used in this study. The parameters concerning yaw rate and lateral acceleration were used as the feedback parameters to determine the reactive torque. Although there were many parameters concerning vehicle behaviour, the parameters that seemed to be closely related to steering wheel manoeuvring were chosen.

Table 1. Feedback parameters for reactive torque control.

	Yaw rate : γ	Lateral acceleration : Gy	$k_1V\gamma + k_2Gy$ $(k_1 + k_2 = 1)$
-	I	II	III
$,\dot{\gamma}$	IV	V	VI
$,\dot{Gy}$	VII	VIII	IX

$$※IV = \gamma + \quad ,\dot{\gamma}$$

In our future reports, the relationship between those parameters and the driver's characteristic will be identified from the viewpoint of usability, and the quantification analysis of the reactive torque will be carried out using an actual vehicle.

Further more, the reactive torque control logic considering the driver's characteristic and the mechanical characteristics of the input device will be deeply examined and investigated so to indicate at most linearity between the reactive torque and the vehicle behaviour.

In future reports, the relationship between those parameters and the driver's characteristic will be identified from the viewpoint of usability, and the quantification analysis of the reactive torque will be carried out using an actual vehicle.

The reactive torque control logic considering the driver's characteristic and the mechanical characteristics of the input device will be examined and investigated as to whether the variable transition characteristics from the steering wheel to the front wheel will be transmitted to the driver comfortably.

5 CONCLUSION

It is expected that SBW will improve and reduce the driver's workload and safety, by compensate driver's phase lead manoeuvring. As a preliminary examination, the effect to the reactive torque was investigated when vehicle behaviour parameters feedback were applied to that control logic.

In the future, the relationship between the driver's characteristics and reactive torque including mechanical characteristics of the input device will be identified, and the proper relationship between vehicle behaviour and reactive torque will be more investigated as well.

The final approach of our development is to combine the results of these researches in order to create the reactive torque control logic to improve the driving ability and their safety. And that control logic will be expected to enhance the robustness against vehicle dynamics and the flexibility to traffic environment.

It is expected that driver will be benefit from the effect of direct behaviour control like the fly-by-wire system by means of the improved reactive torque control for SBW vehicle.

REFERENCES

1. Segawa, M., Nishizaki, K. and Nakano, S.: A STUDY OF VEHICLE STABILITY CONTROL BY STEER BY WIRE SYSTEM. Proc. Int. Conf. On AVEC, 2000

2. Nishizaki, K., Hayama, R., Nakano, S. and Kato, K.: Active Stability Control Strategy for Seer By Wire System during Sudden Braking. CD-ROM of the 7th congress on ITS, Turin, 2000

3. Abe, M. :VEHICLE DYNAMICS AND CONTROL (in Japanese). pp.93-97, The Co., Ltd. Sankaidou publication.

4. Segawa, M., Kimura, S., Tomoyasu K., and Nakano, S.: A Study of Reactive Torque Control for Steer By Wire System. Proc. Int. Conf. On AVEC, 2002

5. McDonnell, J. D.: Pilot Rating Techniques for the Estimation and Evaluation of Handling Qualities, AFFDL-TR-68-76, 1968.

6. Horiuchi, S.: A Brief Review of Human Operator Models and Their Application to Handling Qualities Evaluation. Journal of JSAE (in Japanese with English summary). Vol. 45, No. 3(1991), pp. 5-11.

7. Nakano, S., Nishizaki, K., Nishihara, O., and Kumamoto, H.: Steering Control Strategies for the Steer-by-Wire System (Second Report). Trans. of JSAE (in Japanese with English summary). Vol. 33, No. 3(2002), pp. 121-126.

Vehicle System Dynamics Supplement 41 (2004), p.212-221   © Taylor & Francis Ltd.

The Dynamics of the Three-Piece-Freight Truck

FUJIE XIA * and HANS TRUE †

SUMMARY

The paper provides a model to describe the dynamical performances of *the three-piece-fright truck*. The model has 19 rigid bodies and 81 degrees of freedom. The focus is put on the two-dimensional dry friction on the surfaces of the wedges in the lateral and vertical directions, and on the contact surfaces between the side frames and the adapters(wheelsets). The *stick- slip motion* between the corresponding components caused by the friction and the *structure varying systems* occurring together with the stick-slip motion are considered. The *friction direction angle*, the *switch condition* and the *acting friction force* are used in the model. The critical speed is determined through Hopf bifurcation analysis. The phase diagram and first return map are used to verify the chaotic motion of the systems. The dynamic responses and the wheel/rail contact forces to the irregularity track are provided.

1 INTRODUCTION

For the three-piece-freight truck the one obvious structural characteristic is that almost all contacting surfaces among the components are direct contacts, in other words, the interconnections between the components are realized through *motion pairs* such as *sphere joints* and *plane slide pairs*. Comparing with the counterpart of a passenger car, the motion pairs are replaced by suspension elements and hydraulic or air dampers, such that the stability of the passenger car is improved greatly. From a mathematical point of view the multibody system such as the passenger car can be modelled with the general theory of multibody systems dynamics without principled difficulty because the dynamical system of which is sufficient smooth.

Unfortunately, up to now, the dynamic performances of the three-piece-freight-truck have not been thoroughly understood by the dynamical experts. Many investigation efforts have been carried out less or more on this problem by investigators[1][2][3][4][5][6]. But not any model including all the concepts

---

*Rail CRC, Faculty of Engineering and Physical Systems, Central Queensland University, North Rockhampton, QLD 4702, Australia. Email: f.xia@cqu.edu.au

†Informatics and Mathematical Modelling, the Technical University of Denmark, Richard Petersens Plads, Building 321, DK-2800, Lyngby, Denmark. Email: ht@imm.dtu.dk

of the *two-dimensional friction*, the *stick-slip motion* and the *structure varying systems*. The main reason is that there is no suitable way to deal with these basic concepts.

The model is definitely nonlinear resulting from the nonlinear kinematic and dynamical constraints between the wheels and the rails, the nonlinear suspensions and the nonlinear dry friction damping. For low running speed of the truck the nonlinearities of the kinematic and dynamical constraints could be linearized, but the very strong nonlinear suspensions and the dry friction damping can not be linearized at all. The motion of the bolsters are at least two dimensional in the lateral and the vertical directions, so the friction on the surfaces of a wedge should be treated as *two-dimensional friction*, and the same holds for the dry friction on the surfaces of an adapter. For the motion with dry friction there exist two motion states: *Stick motion* and *slip motion*, which lead to a discontinuity in the dynamical system and cause the collapse of state space. Consequently, it changes the degrees of freedom of the system which leads to the structure varying system.

Because the design clearances between the car body and the side supports on the bolsters are necessary such that leads to the side supports acting as dead-band springs. The clearance in the assembly of the wedge damper systems cause the relative yaw motion of the bolster with respect to the side frame to rotate about the truck center line and to assume a state of *warping*. Also the assembly clearances between the side frame and the adapter both in longitudinal and lateral directions produce another dead-band spring force.

F.Xia[7][8][9] introduces the concept of friction direction angle which is used to determine the friction force components in the two orthogonal directions. For the stick-slip motion and the corresponding structure varying systems the switch conditions and the acting friction forces are provided to deal with them.

2   MODELLING OF THE WAGON WITH TWO THREE-PIECE-FREIGHT TRUCKS

The model of the wagon with two three-piece-freight trucks has nineteen principal components. These are the car body, two truck bolsters, eight wedges, four side frames, and four wheelsets. The degrees of freedom of the system are given in Table 1, where the total degrees of freedom is 81. But not all the degrees of freedom are independent and there exist some relations due to the joint connections between the car body and the bolsters, the sliding contact between the bolsters and the wedges, and the side frames and the adapters(wheelsets).

There are 41 independent generalized variables and 40 dependent variables. There is no general way to select the optimal generalized coordinates for a complicated multibody system. We select the 41 independent variables as

$$
\begin{aligned}
\mathbf{x}_i = \quad & [\phi_o \quad v_{b1} \quad w_{b1} \quad \phi_{b1} \quad v_{b2} \quad w_{b2} \quad \phi_{b2} \quad \psi_{b1} \quad v_{f1} \quad \psi_{f1} \quad v_{f2} \\
& \psi_{f2} \quad u_{w1} \quad v_{w1} \quad \psi_{w1} \quad u_{w2} \quad v_{w2} \quad \psi_{w2} \quad \psi_{b2} \quad v_{f3} \quad \psi_{f3} \qquad (1)
\end{aligned}
$$

Table 1: *Degrees of freedom of the wagon with two three-piece-freight trucks*

Components	Long.	Lat.	Vert.	Roll	Pitch	Yaw
Front truck leading wheelset	u_1	v_1	w_1	ϕ_1	χ_1	ψ_1
Front truck trailing wheelset	u_2	v_2	w_2	ϕ_2	χ_2	ψ_2
Rear truck leading wheelset	u_3	v_3	w_3	ϕ_3	χ_3	ψ_3
Rear truck trailing wheelset	u_4	v_4	w_4	ϕ_4	χ_4	ψ_4
Front truck left side frame	u_{f1}	v_{f1}	w_{f1}	—	χ_{f1}	ψ_{f1}
Front truck right side frame	u_{f2}	v_{f2}	w_{f2}	—	χ_{f2}	ψ_{f2}
Rear truck left side frame	u_{f3}	v_{f3}	w_{f3}	—	χ_{f3}	ψ_{f3}
Rear truck right side frame	u_{f4}	v_{f4}	w_{f4}	—	χ_{f4}	ψ_{f4}
Front truck left wedges	u_{d1}	v_{d1}	w_{d1}	—	—	—
	u_{d2}	v_{d2}	w_{d2}	—	—	—
Front truck right wedges	u_{d3}	v_{d3}	w_{d3}	—	—	—
	u_{d4}	v_{d4}	w_{d4}	—	—	—
Rear truck left wedges	u_{d5}	v_{d5}	w_{d5}	—	—	—
	u_{d6}	v_{d6}	w_{d6}	—	—	—
Rear truck right wedges	u_{d7}	v_{d7}	w_{d7}	—	—	—
	u_{d8}	v_{d8}	w_{d8}	—	—	—
Front truck bolster	—	v_{b1}	w_{b1}	ϕ_{b1}	—	ψ_{b1}
Rear truck bolster	—	v_{b2}	w_{b2}	ϕ_{b1}	—	ψ_{b1}
Car body	—	v_o	w_o	ϕ_o	χ_o	ψ_o

$$v_{f4} \quad \psi_{f4} \quad u_{w3} \quad v_{w3} \quad \psi_{w3} \quad u_{w4} \quad v_{w4} \quad \psi_{w4} \quad w_{w1} \quad \phi_{w1}$$

$$w_{w2} \quad \phi_{w2} \quad w_{w3} \quad \phi_{w3} \quad w_{w4} \quad \phi_{w4} \quad \chi_{w1} \quad \chi_{w2} \quad \chi_{w3} \quad \chi_{w4}]^T$$

and the other 40 variables in Table 1 are dependent variables.

The interconnections exist between the car body and the two bolsters via two center plates; the bolsters and the wedges, the wedges and the side frames and the side frames and the adapters through sliding contacts. Wheel/rail kinematic constraints are determined on-line considering the effects of both lateral displacement and yaw of the wheelset. The Hertzian springs are used between wheelsets and rails contacts for determining normal wheel loads and SHE formulae are used to calculate the non-linear creep forces. For the kinematic constraint of the center plate which is modelled as a sphere joint, the translation motions of the bolster and car body is identical in the longitudinal, lateral, vertical directions. In the pitch and yaw rotations they are controlled by kinematic constraints. For the wedge and the bolster, the relations are detailed described by F.Xia[8]. The yaw of the bolster and the longitudinal translation of the side frame are not independent of each other. The vertical displacement and the pitch of the frame are related to the motion of corresponding wheelsets.

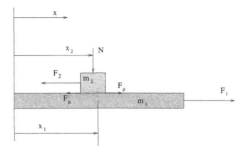

Figure 1: *A moving body on a moving plate both which are subjected to excitation*

3 THE SYSTEM ANALYSIS

The dynamic equations of the system can be written as[8]

$$\begin{bmatrix} \mathbf{A}_a & \mathbf{M}_{dep}\mathbf{J} \\ \mathbf{A}_b & \mathbf{M}_{ind} \end{bmatrix} \begin{bmatrix} \mathbf{F}_{ev} \\ \ddot{\mathbf{x}}_i \end{bmatrix} = \begin{bmatrix} \mathbf{F}_{dep} \\ \mathbf{F}_{ind} \end{bmatrix} \tag{2}$$

where $\mathbf{M}_{dep}$ is the mass matrix related to the dependent variables, and the $\mathbf{M}_{ind}$ denotes the mass matrix related to the independent variables; $\mathbf{J}$ is the matrix related to the constraints. $\mathbf{F}_{ev}$ denotes the unknown force vector. $\mathbf{F}_{dep}$ denotes the forces related to the dependent variables and $\mathbf{F}_{ind}$ denotes the forces related to the independent variables. The matrices $\mathbf{A}_a$ and $\mathbf{A}_b$, where their entries are a function of the friction direction angles θ, friction coefficients μ and the construction parameters p, can be written as

$$\mathbf{A}_{a,b} = f(\theta, \quad \mu_b, \quad \mu_d, \quad \mu_f, \quad p). \tag{3}$$

If the purpose of the simulation is only focussed on the slip motion then the friction direction angle is uniquely determined by the corresponding relative velocities alone. In that case the entries of the matrices $\mathbf{A}_a$ and $\mathbf{A}_b$ are determined by the state space variables. That means in this case the friction force terms in matrix $\mathbf{A}_a$ and $\mathbf{A}_b$ can be determined by the normal force times friction coefficient, i.e.

$$F_{\mu t} = \mu N. \tag{4}$$

The velocity-dependent friction coefficient μ can be described by the hyperbolic secant function as[6][8].

$$\mu(V_r) = \mu_s sech(\alpha|V_r|) + \mu_k(1 - sech(\alpha|V_r|)). \tag{5}$$

By the selection of different values of the parameter α the formula yields different steepness of the continuous curve that describes the change from the static to the kinematic friction coefficient.

In the case of stick motion the relation (4) does not hold. In that case switch conditions are needed to determine the acting friction forces. As an example, Figure 1 shows two relative moving bodies, in the stick motion the friction force can be determined by[8]

$$F_{\mu t} = \frac{1}{m_1 + m_2}[m_2 F_1 + m_1 F_2].$$ (6)

For a side frame contacting wheelsets at two points with stick-slip motion the case becomes rather complicated, which is described in detail in[8].

4 CASE STUDIES

4.1 The critical speed

There are two different critical speeds of vehicles on railways: The *linear critical speed* and the *nonlinear critical speed*. The line critical speed means the velocity at which the stationary solution loses its stability through a Hopf bifurcation. The linear critical speed usually gives a higher critical speed than the measured one[10]. The nonlinear critical speed is the lowest speed for which a periodical motion exists and it will be determined by the bifurcation analysis. Normally the critical speed means the nonlinear critical speed in the railway vehicle system dynamics.

Figure 2 shows the bifurcation diagram for the determination of the critical speed. From the figure we see that the motion of the wheelset 1 changes from a stable steady motion to a periodic motion and then jumps into a chaotic motion when the speed increases. Otherwise, when decreasing from high speed to low speed then the motion changes from chaotic to steady without the periodic motion state. The linear critical speed of the three-piece-freight-truck is about 28.1 m/s and the nonlinear critical speed is about 20.5 m/s for the empty car.

4.2 An investigation of the chaotic motion

In the study of dynamical systems the tools that are often used are: a phase space, a Poincaré map(or Lorenz first return map), the power spectra and the Lyapunov exponents. They provide information about the dynamics of the system for specific values of the parameters, e.g. the speed V in railway vehicle systems.

For the speed $29m/s$ the lateral displacement of the four whelssets is shown in left of Figure 3. The effects of the impacts between the side frames and the adapters and as well as the friction forces on the surfaces of the wedge and the adapters may be the main facts to make the chaotic motion special.

The phase diagrams of the wheelsets 1 and 2 are shown in right of Figure 3. The first return maps of the lateral displacements of wheelsets number 1 and 2 are shown in the Figure 4.

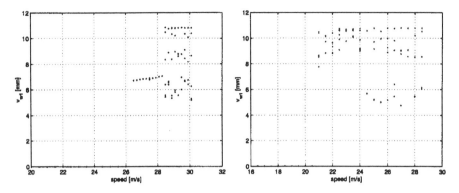

Figure 2: *Left: The bifurcation diagram of the three-piece-freight-truck with speed from 20 m/s to 30 m/s. Right: The bifurcation diagram of the three-piece-freight-truck with speed from 28.5 m/s to 16 m/s.*

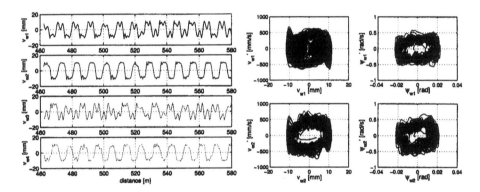

Figure 3: *Left: The lateral displacements of the wheelsets with the speed 29m/s. Right: The phase diagrams of the wheelsets 1 and 2 with the speed 29m/s.*

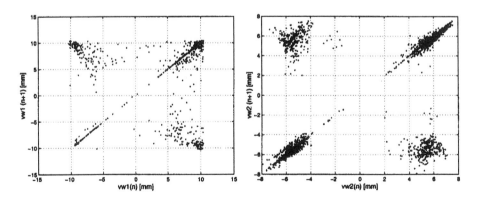

Figure 4: *The first return maps of the lateral displacements of wheelsets number 1 (left) and 2 (right) with the speed 29 m/s for the empty car.*

217

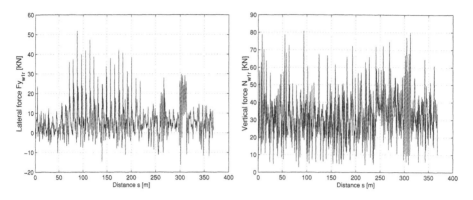

Figure 5: *The measured lateral and vertical forces between the right wheel/rail.*

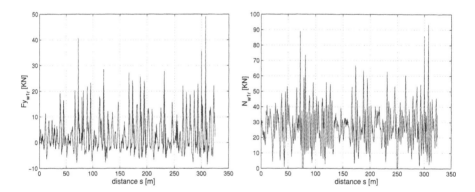

Figure 6: *The simulated lateral and vertical forces between the right wheel/rail.*

diagram, the lateral displacement curves of the wheelsets, the phase diagrams and the first return maps we can definitely claim that the motion of the three-piece-freight-truck is chaotic for the running speed 29 m/s. We can also find that the chaotic motion of the three-piece-freight-truck is obviously different from the chaotic motion of the Cooperrider's passenger car model[10] since the chaotic motion is not a small perturbation on a dominating periodic motion.

4.3   The responses of the system to an irregular track

We use simulated track irregularities from Power Spectral Density(PSD) provided by the Academy of China Railway Sciences [8]. Unfortunately there is no measured data of the irregular track corresponding directly to the track where the wheel/rail contact forces of the three-piece-freight-truck are measured. The parameters of the three-piece-freight-truck are shown in[8]. The running speed used in the simulation is $78km/h$. The measured lateral and vertical wheel/rail contact forces on the right rail are shown in Figure 5.

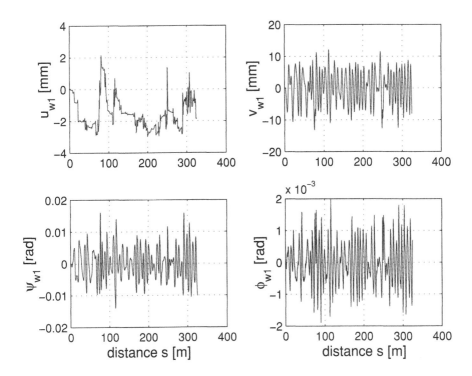

Figure 7: *The parasitic motion of the front wheelset. u, v denotes the longitu-dinal and lateral displacements; ϕ, ψ denotes the roll and yaw rotations of the wheelset.*

The numerical results of the lateral and vertical forces between the right wheel/rail are shown in Figure 6. We can find that the range of the lateral and vertical contact forces between the numerical and the measured results are similar. The difference between the numerical and measured results is probably due to the fact that the new truck is used in the numerical simulation but the used truck is used in the measurement.

Figure 7 shows the displacements of the front wheelset. We can find that the longitudinal displacement of the wheelset is about $2mm$ so it should not be neglected for the modelling of the dynamics of the three-piece-freight-truck.

5 CONCLUSIONS AND REMARKS

The dynamical system of the three-piece-freight truck is a strong non-linear non-smooth system because of the kinematical and dynamical constraints be-tween the wheel and the rails and the dry friction on the surfaces of the wedges and adapters. The two-dimensional dry friction will cause the stick-slip motion between the contacting surfaces that will leads to the change of the degrees of freedom of the system. To understand thoroughly the dynamics of the three-piece-freight truck the structure varying systems caused by stick-slip motion which is related to the dry friction should be included.

To describe the components of the friction force vector on a plane is important for the stick-slip motion analysis, especially for a complex system. The friction direction angle is an effective way to determine the friction force components in orthogonal directions both for the motion in slip and stick states.

In the stick motion state the friction force is not equal to zero unless the all external forces are zero. Since the stick-slip motion exists in the system with friction the structure varying system will takes place which will cause the collapse of the state space. In order to deal with the structure varying systems the acting friction force should be determined by means of the switch condition and then the structure varying system changes its degrees of freedom automatically. In other words, the non-smooth system can be automatically switch into a piece-wise differentiable system.

The linear critical speed of the three-piece-freight-truck is about $28.5m/s$ and the nonlinear critical speed is about $20.5m/s$. The motion of the three-piece-freight-truck is a chaotic motion for a certain speed range.

The normal forces and the corresponding friction forces on the two surfaces of the wedge are asymmetric due to the tractive effort. It must be included in the simulation of the dynamics of the three-piece-freight-truck. The friction forces on the surfaces of the wedges can be used to evaluate the wear state of the wedge damper.

The parasitic motion of the wheelset along the longitudinal direction influences as well the creep forces as the contact state between the frame and the wheelsets. Therefore it should be included in the dynamics analysis of the three-piece-freight-truck.

In the stable motion state at low speed, the motions of the wheelsets are different from zero because of the effect of the stick-slip motion produced by the dry friction. In the present investigation the clearances between the frame and the adapter are the same for left/right sides. If they are asymmetrical then the amplitude of the lateral displacement of the wheelset in the steady motion may increase more or less.

From the bifurcation diagram, the phase diagram and the first return map of the lateral motions of the wheelsets we have concluded that the motion of the three-piece-freight-truck is chaotic motion for a running speed. And more the chaotic motion of the three-piece-freight-truck is obviously different from the chaotic motion of the Cooperrider's passenger car model[10] in that the chaotic motion is not a small perturbation on a periodic motion.

ACKNOWLEDGEMENTS

The work was mainly performed in the Informatics and Mathematical Modelling of the Technical University of Denmark and obtained the financial support of the DSB(Danish State Railways), the Danish Research Agency, the Technical University of Denmark. The Cooperative Research Centre for Railway Engineering and Technologies of Australia(Rail CRC) and Faculty of Engineering and Physical Systems of Central Queesland University also provide many assistances for the presentation. Thanks are also due to Professor Per Grove for his helps with numerical methods and Professor Dudley Roach and Professor Peter Wolfs for their supports in many ways.

REFERENCES

1. A.B. Kaiser *et. al.*, Modeling and Dynamics of Friction Wedge Dampers in Railroad Freight Trucks, *Vehicle System Dynamics*, 38(1):55-82, 2002.
2. R.D. Fröhling, The Influence of Friction Wedges on the Dynamic Performance of Three-Piece Self-steering Bogies. Proceedings of the 4th International Conference on Railway Bogies and Running Gears, Budapest, Hungary, pp.95-103, 1998.
3. C. Cole and M. McCanachan, Snubber Stiction in Three Piece Bogies, Proceedings of the 5th International Conference on Railway Bogies and Running Gears, Budapest, Hungary, 2001.
4. R.F. Harder, Dynamic Modeling and Simulation of Three-Piece North American Freight Vehicle Sunspensions with Non-linear Frictional Behaviour Using ADAMS/Rail, 5th ADAMS/Rail User's Conference, Haarlem, The Netherlands, 2000.
5. N. Bosso, A. Gugliotta and A. Somá, Simulation of a freight bogie with friction dampers, 5th ADAMS/Rail User's Conference, Haarlem, The Netherlands, 2000.
6. H. True and R. Asmund, The Dynamics of a Railway Frieght Wagon Wheelset with Dry Friction Damping, *Vehicle System Dynamics*, 38:149-163, 2002.
7. F. Xia, Modelling of a Two-Dimensional Coulomb Friction Oscillator, *Journal of Sound and Vibvation*, 265(5):1063-1074, 2003.
8. F. Xia, The Dynamics of the Three-Piece-Freight Truck, Ph.D. Dissertation, Informatics and Mathematical Modelling, the Technical University of Denmark, 2002.
9. F. Xia and H. True, On the Dynamics of the Three-Piece-Freight Truck, ASME RTD 2003-1660, Proceedings of the 2003 IEEE/ASME Joint Rail Conference, April 22-24, 2003, Chicago, lllinois, pp.149-159.
10. H. True, On the Theory of Nonlinear Dynamics and its Application in Vehicle Systems Dynamics, *Vehicle System Dynamics*, 31:393-421, 1999.

Vehicle System Dynamics Supplement 41 (2004), p.222-231 © Taylor & Francis Ltd.

Analysis and method of the analysis of non-linear lateral stability of railway vehicles in curved track

KRZYSZTOF ZBOINSKI AND MIROSLAW DUSZA[1]

SUMMARY

The paper is a contribution into the question of stability in curves and into breaking down the traditional approach to it. It shows that considering stability in curves is reasonable, justifiable and worthwhile. Formal shape of the original method of stability analysis in curved track is presented. Straight track case is just a circular curve of infinite radius here. The method is based on results of numerical simulation. Authors use such results to verify and present corresponding procedure applied while investigating stability in curves. Results of simulation enabled to show how the problem looks for a particular vehicle as well as to analyse stability for objects' variants. The analysis includes influence of angle of attack, cant deficiency and excess as well as suspension parameters on vehicle's stability properties and also comparison of stability for 2-axle vehicle and a bogie of 4-axle one.

1. INTRODUCTION

A great number of works exists that treat railway vehicle motion in curved track dynamically. However, traditional approach treating just straight track motion dynamically and restraining to quasi-static case in circular curve still keeps strong. Consequently the problem of stability confines to straight track dynamics in many references. Despite common use of numerical simulation to study non-linear stability in straight track, people rather do not discuss it in curved track explicitly.

The case of straight track manifests that stability problem of non-linear model deals with stable and unstable limit cycles [1] and also problems of bifurcation and chaos [2,3]. On the other hand the author showed in [4,5] and it is visible in Figures 1 and 2 that talking about and expecting limit cycles in circular curves is justifiable. This is due to the same nature of the phenomena and the same subcritical Hopf bifurcation velocity v_n (called also critical velocity of non-linear system) [1] in straight and circular track. The amplitude independence of initial conditions was shown in [5], too. Also lengths of the circular sections appeared to be sufficient. However one should not expect limit cycles in transition curve due to its short length as well as radius and cant change. Despite it we preserve an interest in this case, too.

In this circumstances the task to build and test the method of stability analysis including circular track, which is based on simulation seems to be worthwhile.

[1]*Address correspondence to:* Warsaw University of Technology, Faculty of Transport, Koszykowa 75, 00-662 Warsaw, POLAND. Tel. +48 (22) 660-77-07; Fax: +48 (22) 621-56-87; E-mail: kzb@it.pw.edu.pl

2. RAILWAY VEHICLE-TRACK SYSTEM MODELLING, SIMULATION SOFTWARE AND OBJECTS USED

We used a generalised approach to the modelling as described by us in [4,6,7]. Single model of discrete vehicle-track system describes dynamics in general case of 3-dimensional track of any shape (transition curve). Circular curve and straight track are just particular sub-cases of it. The distinguishing feature is taking all inertia forces into account in the models.

This approach was implemented numerically both in the traditional software representing a particular vehicle [5] and in the more versatile software enabling to build equations of motion numerically [7]. Non-linear contact geometry and forces are regarded. Real wheel/rail profiles were used. Creep forces are calculated by FASTSIM programme. Contact parameter tables introduce the geometry. The tables were in principle generated with use of RSGEO by ArgeCare. Depending on the need we used single table for yaw angle $\psi=0$ or set of tables for the ψ range.

We used two objects while doing the simulation. One is 2-axle laden HSFV1 freight car and the other is bogie of 4-axle freight car. Interaction between the bogie and vehicle body is modelled just through application of a half of the body weight. In addition, several parameter variants of HSFV1 were considered (see section 4). Structure of the bogie is the same as that of HSFV1 car. It is shown in [4,6]. The car parameters are given there, too. Parameters of the bogie are the same as for 25TN bogie given in [4]. The only difference is $k_{zx}=2615$ kN/m and $c_{zx}=52,2$ kNs/m.

The discrete track model is flexible both vertically and laterally [5].

3. FORMAL FORM OF THE METHOD OF STABILITY ANALYSIS

In general the method bases on results of simulation for the range of velocities from v_n (defined in section 1) to v_d (derailment velocity [1]). Figures 1 and 2 are sample of such results for v_n. They represent 2-axle car behaviour on the route including in

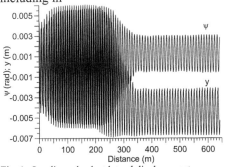 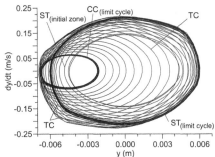

Fig. 1. Leading wheelset lateral displacements y and yaw angle ψ for velocity v_n=45.3 m/s.

Fig2. Phase-plane representation of the limit cycles shown in Figure 1.

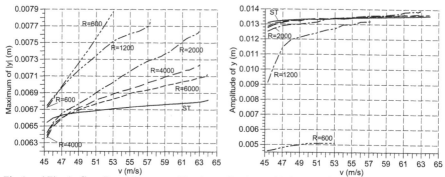

Fig. 3 and Fig. 4. **Case I**- stability map of 2-axle car fitted out with {BR-P10/UIC60} wheel/rail pair (no account taken of angle of attack influence).

succession straight track **ST**, transition curve **TC** and circular curve **CC**. The system is fitted out with BR-P10/UIC60 wheel/rail pair. Curve radius $R=600$ m and track cant $H=0,16$ m. Results analogous to shown in Figures 1 and 2 but for the range of radii from small up to large (e.g. from $R=300$ m to $R=10000$ m) and straight track ($R=\infty$) and velocities from v_n to v_d constitute data to elaborate graphs in type of Figure 3. It represents change in maximum of lateral displacement absolute value $|y|$ for circular sections of different radii and straight track versus velocity. Note that the last case corresponds to well-known limit cycle stability plot for a straight track, e.g. [1]. Despite similarity for straight and circular track sections Figure 3 gives no enough information on vehicle behaviour in curves. This results from non-symmetry that increases with the radius decrease. That is why we supplied it with Figure 4. It shows change in the limit cycle's amplitude of y, corresponding to cases in Figure 3.

Figure 5 represents formally a descriptive idea of the method. Figure 6 is its brief algorithm. There are some numbers given in Figures 5 and 6. For instance: length of a route, initial conditions, velocity intervals etc. They are not fixed values but just suggestions based on our experience. So they can be changed whenever necessary.

4. STABILITY ANALYSIS OF VARIOUS CASES IN CURVED TRACK

Couples of figures analogous to Figures 3 and 4 create maps of vehicle stability for curve radii from small ones to infinite one. If results start with $R=600$ m this means that generation of the plots for smaller radii (e.g. $R=300$ m) was not possible. Another words the derailment occurred prior to the critical velocity v_n was reached.

Before we start detailed discussion of the particular cases let us make some general remarks. So, for vehicle velocities v smaller than v_n the system is generally asymptotically stable in whole range of the R. It can happen for sufficiently small initial conditions and $v > v_n$ and $v < v_c$ (critical velocity of linear system, e.g. [1])

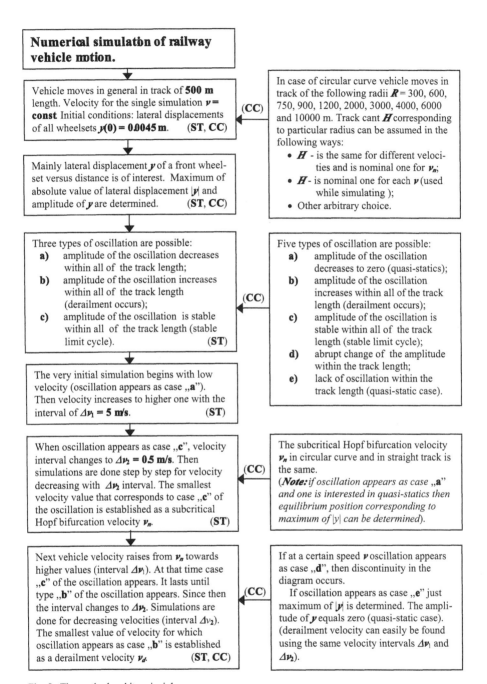

Numerical simulation of railway vehicle motion.

Vehicle moves in general in track of **500 m** length. Velocity for the single simulation $v =$ **const** Initial conditions: lateral displacements of all wheelsets $y(0) = 0.0045\,\text{m}$. **(ST, CC)**

In case of circular curve vehicle moves in track of the following radii $R = 300, 600, 750, 900, 1200, 2000, 3000, 4000, 6000$ and 10000 m. Track cant H corresponding to particular radius can be assumed in the following ways:

- H - is the same for different velocities and is nominal one for v_n;
- H - is nominal one for each v (used while simulating);
- Other arbitrary choice.

(CC)

Mainly lateral displacement y of a front wheelset versus distance is of interest. Maximum of absolute value of lateral displacement $|y|$ and amplitude of y are determined. **(ST, CC)**

Three types of oscillation are possible:
- **a)** amplitude of the oscillation decreases within all of the track length;
- **b)** amplitude of the oscillation increases within all of the track length (derailment occurs);
- **c)** amplitude of the oscillation is stable within all of the track length (stable limit cycle). **(ST)**

Five types of oscillation are possible:
- **a)** amplitude of the oscillation decreases to zero (quasi-statics);
- **b)** amplitude of the oscillation increases within all of the track length (derailment occurs);
- **c)** amplitude of the oscillation is stable within all of the track length (stable limit cycle);
- **d)** abrupt change of the amplitude within the track length;
- **e)** lack of oscillation within the track length (quasi-static case).

(CC)

The very initial simulation begins with low velocity (oscillation appears as case „a"). Then velocity increases to higher one with the interval of $\Delta v_1 = 5\,\text{m/s}$. **(ST)**

When oscillation appears as case „c", velocity interval changes to $\Delta v_2 = 0.5\,\text{m/s}$. Then simulations are done step by step for velocity decreasing with Δv_2 interval. The smallest velocity value that corresponds to case „c" of the oscillation is established as a subcritical Hopf bifurcation velocity v_n. **(ST)**

The subcritical Hopf bifurcation velocity v_n in circular curve and in straight track is the same.
(*Note:* if oscillation appears as case „a" and one is interested in quasi-statics then equilibrium position corresponding to maximum of $|y|$ can be determined).

(CC)

Next vehicle velocity raises from v_n towards higher values (interval Δv_1). At that time case „c" of the oscillation appears. It lasts until type „b" of the oscillation appears. Since then the interval changes to Δv_2. Simulations are done for decreasing velocities (interval Δv_2). The smallest value of velocity for which oscillation appears as case „b" is established as a derailment velocity v_d. **(ST, CC)**

If at a certain speed v oscillation appears as case „d", then discontinuity in the diagram occurs.
If oscillation appears as case „e" just maximum of $|y|$ is determined. The amplitude of y equals zero (quasi-static case). (derailment velocity can easily be found using the same velocity intervals Δv_1 and Δv_2).

(CC)

Fig. 5. The method and its principle.

that asymptotical stability also occurs. This is due to so called tough character of limit cycle excitation. This is discussed in [1] for straight track and in [5] for circular sections. Excluding just described case there exists stable limit cycle between v_n and v_d within whole range of R. For $v > v_d$ oscillations go towards infinity. For $v < v_d$ but in greater range of amplitudes there may theoretically exist unstable limit cycle, too.

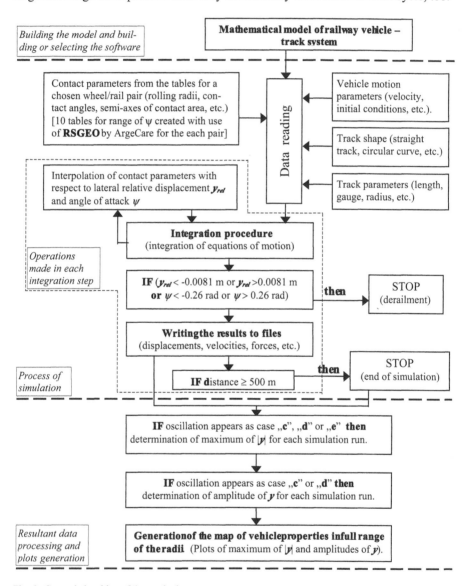

Fig. 6. General algorithm of the method

4.1 INTRODUCTION OF THE CASES AND CORRESPONDING PLOTS PRESENTATION

Meaning of figure couples within rest of the paper is the same as that for Figures 3 and 4 in section 3. **Cases I, II** and **III** (Figs. 3, 4, 7-10) are results for the same 2-axle HSFV1 freight car of nominal parameters (see section 2). Superelevation is nominal one (balance ensured between components of gravity and centrifugal forces) for $v=45,3$ m/s that equals v_n in **case I** If for small R nominal superelevation exceeded maximum cant used in practice ($H_{max}=0.160$ m) this maximum was applied. Single table of parameters for $\psi=0$ represents contact geometry. Difference between the cases lies in different wheel/rail pairs (specified in the figure captions). **Case I** corresponds to British wheels and European track, **case II** to Russian wheels and track and **case III** to German wheels and European/German track.

Case IV (Figs. 11, 12) differs from **case III** just in contact geometry representation. This time, and in all farther cases, it is represented by set of the tables for a range of ψ. This way account was taken of the angle of attack influence.

Cases V and **VI** (Figs. 13-16) differ from **case IV** just in the changes of suspension parameters. In **case V** longitudinal stiffness between wheelsets (axleboxes) and car body is 10 times bigger ($k_{zx}=2067$ kN/m) than its nominal value (no longitudinal motion between wheelset centre and car body is allowed, however). In **case VI** lack of the lateral dumping between wheelsets and car body exists ($c_{zy}=0$).

Case VII (Figs. 17, 18) represents the results for nominal superelevation in any case, i.e. for any velocity and curve radius. This means that track cant for smaller radii can exceed the maximum used in practice by a few times. Except the cant change this case is identical with **case IV**.

Case VIII (Figs. 19, 20) is again analogous to **case IV**. The difference lies in the object. Namely, the bogie of 4-axle freight car is of interest, here.

4.2 ANALYSIS OF THE INTRODUCED CASES

Each of the figure pairs separately is a source of valuable information. Amount of it rises however if one compares analysed cases with each other. Here we are rather interested in such comparisons than in analysis of the individual couples itself.

Choose **case I** to illustrate briefly a kind of the information one can obtain from the analysis of the single couple. One can notice qualitative similarity in full range of the radii (Figs. 3, 4). However, some features could be a surprise. For instance, the maximum displacements are consistently bigger for smaller radii just for velocities far enough from v_n or the derailment velocities v_d are reached at different maximum displacements. One might be also surprised that amplitudes for small radii are so small then they increase and next decrease slightly for straight track (Fig. 4).

Let us proceed to the particular cases comparison. **Cases I, II** and **III** manifest strong influence of wheel/rail pair on stability in curved and straight track. Both the maximums of $|y|$ and the amplitudes of y differ qualitatively and quantitatively. Also critical velocities v_n and derailment velocities v_d for particular radii R vary from each

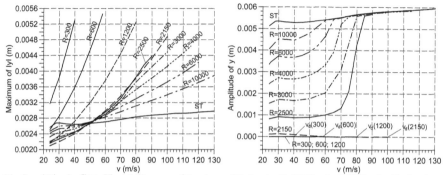

Fig. 7 and Fig. 8. Case II- stability map of 2-axle car fitted out with {SZD wheel/R65} wheel/rail pair (no account taken of angle of attack influence).

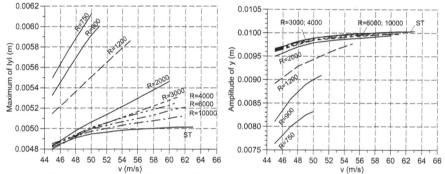

Fig. 9 and Fig. 10. Case III- stability map of 2-axle car fitted out with {S1002/UIC60} wheel/rail pair (no account taken of angle of attack influence).

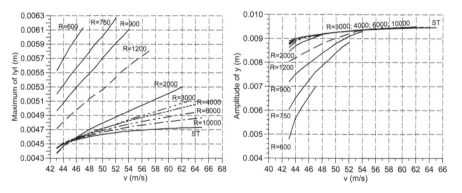

Fig. 11 and Fig. 12. **Case IV** - stability map of 2-axle car fitted out with {S1002/UIC60} wheel/rail pair (account taken of angle of attack influence).

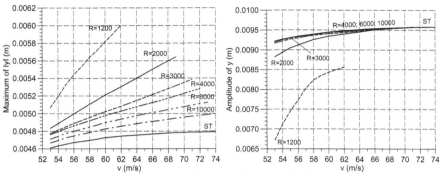

Fig. 13 and Fig. 14. **Case V** - stability map of 2-axle car fitted out with {S1002/UIC60]} wheel/rail pair and for longitudinal stiffness increased 10 times (account taken of angle of attack influence).

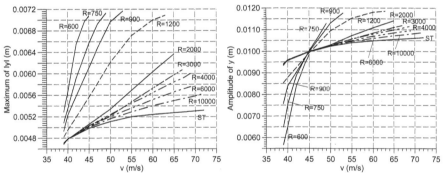

Fig. 15 and Fig. 16. **Case VI** - stability map of 2-axle car fitted out with {S1002/UIC60} wheel/rail pair and for lack of lateral damping (account taken of angle of attack influence).

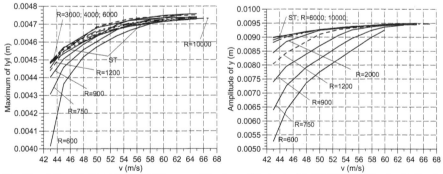

Fig. 17 and Fig. 18. **Case VII** - stability map of 2-axle car fitted out with {S1002/UIC60} wheel/rail pair and for nominal track cant (account taken of angle of attack influence).

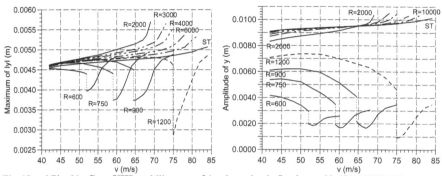

Fig. 19 and Fig. 20. **Case VIII**- stability map of 4-axle car bogie fitted out with {S1002/UIC60} wheel/rail pair (account taken of angle of attack influence).

other. Note significant qualitative similarity of **cases I** and **III** on the one hand and their deep qualitative and quantitative discrepancy with **case II** on the other hand. **Case II** reviles how wide the range between v_n and v_d can be, especially for large radii R. The biggest values of $|y|$ and y amplitude in **case I** are caused by British wheelset suited for 3mm smaller track gauge than European continental gauge used.

Comparison of **case III** to **IV** let us conclude about influence of accuracy of angle of attack determination on the obtained stability results. Generally, in terms of value of quantities in the figures, which correspond with the same velocity this influence is rather small. Nevertheless, neglect of angle of attack real values while interpolating contact parameters leads to higher values of v_n, lower of v_d and higher the smallest value of R for which a limit cycle occurs.

Comparison of **case IV** to **V** and to **VI** enables to study the influence of two suspension parameters. The rise of longitudinal stiffness leads to noticeable increase of v_n and v_d in whole range of the radii R. Another consequence is much higher the smallest value of R for which a limit cycle occurs. Lack of lateral damping causes slight decrease in v_n as well as noticeable increase in v_d and maximums of $|y|$ for the same velocities in whole range of the radii. Additionally change in the amplitudes is quite untypical (Fig. 16). Namely, the amplitudes are higher for small radii than for big ones for velocities above the plots' node (v=45,3 m/s). Note that for this range of velocity a cant deficiency exists while for the rest not necessarily.

Comparison of **case IV** with **VII** shows great influence of track cant value on the stability in curved track. Its nominal value in any case causes that maximums of $|y|$ are close to each other and to straight track values. Thus they are generally smaller. The amplitudes are qualitatively similar to unbalanced case but v_d increased much.

Comparison of **case IV** and **VIII** demonstrates that difference resulting from different types of vehicle can be vital. Besides obvious quantitative difference the main qualitative one is two different bogie behaviours for small and big radii. The next one are discontinuities for the small radii. These features probably arose from smaller mass and other inertia parameters of the single bogie compared to 2-axle car.

4.3 DISCUSSION OF DIFFICULTIES FACED WHILE DOING THE ANALYSES

It is obvious that some results of basic simulations (in type of Fig. 1) might be and were confusing indeed, when you realise that we needed 80 of them on an average in order to elaborate the single case. Example of such results are discontinuities described in **case VIII** Most likely a bifurcation occurs, here. Another similar example is **case II**. The solutions for radii smaller than R=2150 m bifurcate to quasi-static ones. So, their amplitudes equal 0 (Fig. 8). The intermediate case of R=2150 m where limit cycle exists for smaller velocities and qausi-static behaviour appears for bigger ones explains it but separately can be additional interpretation difficulty. Also more typical cases can be troublesome, especially when determining v_d. In fact in case of just slightly increasing amplitude but without the derailment within tested section you never know if you see reality or just effect of calculation inaccuracy. Similarly, when you search for v_n you should beware of too small initial conditions.

In all these sometimes frustrating cases patience, inquiring, experience and understanding of self-exciting vibrations' nature is very helpful.

5. CONCLUDING REMARKS

Commonly used limit cycle stability plot for straight track, e.g. [1], does not enable to conclude about vehicle stability in circular curves for the range of velocities from v_n to v_d. Our approach gives a kind of map of vehicle properties in this range. Such maps, in form of the shown figure pairs, permit to analyse lateral stability in full. The method can be used to compare stability of the same vehicle but with particular parameters variation (suspension ones, wheel/rail pair, etc.). Effects of accuracy of modelling (e.g. contact geometry) on the stability can also be studied. Stability properties of vehicles of entirely different structure can be compared, too.

REFERENCES

1. Chengrong , H., Feisheng, Z.: The numerical bifurcation method of non-linear lateral stability analysis of a locomotive, Proc. 13th IAVSD Symp., suppl. to Veh. Syst. Dyn. 23 (1993), pp.234-245.
2. True, H.: Railway vehicle chaos and asymmetric hunting, Proc. 12th IAVSD Symposium, suppl. to Veh. Syst. Dyn. 20 (1991), pp. 625-637.
3. True, H., Jensem, J.Ch.: Parameter study of hunting and chaos in railway vehicle dynamics, Proc. 13th IAVSD Symposium, suppl. to Veh. Syst. Dyn. 23 (1993), pp. 508-521.
4. Zboiński, K.: Dynamical investigation of railway vehicles on a curved track, Eur. J. Mech.- A Solids 17(6) (1998), pp. 1001-1020.
5. Zboiński, K.: Dusza, M.: Simulation investigation of railway vehicle dynamics in curved track, Proc. XV Scientific-Technological Conference Rail-Vehicles 2 (2002), pp. 343-352 (in Polish).
6. Zboiński, K.: Importance of imaginary forces and kinematic type non-linearities for description of railway vehicle dynamics, Proc. Inst. Mech. Eng.- J. Rail & Rapid Tr. 213(F3) (1999), pp. 199-210.
7. Zboiński, K.: Relative kinematics exploited in Kane's approach to describe multibody systems in relative motion, Acta Mech. 147(1-4) (2001), pp. 19-34.

Vehicle System Dynamics Supplement 41 (2004), p.232-241 © Taylor & Francis Ltd.

Importance of track modeling to the determination of the critical speed of wagons

YAN QUAN SUN AND MANICKA DHANASEKAR[1]

SUMMARY

This paper presents the application of a three-dimensional wagon-track system dynamics (3DWTSD) model to examine the hunting characteristics of wagons containing three-piece bogies running on tangent track. In the 3DWTSD model, two types of track subsystem were considered – one as a layered viscoelastic track subsystem and the other as a 'rigid' track subsystem. Both the critical speed of wagons and the hunting frequency obtained using the viscoelastic track subsystem are smaller than those using 'rigid' track subsystem. The influence of the wheel profiles to the hunting characteristics is also examined. The worn wheel profile was found to increase the critical speed of wagons in addition to changing the hunting characteristics.

1   INTRODUCTION

Lateral dynamics of the wagon – track system is crucial to wagon stability and safety. Hunting of the wagons containing three-piece bogies and coned wheels is a serious problem to the heavy haul rail industry. As would be observed sometimes in the field, a sustained lateral oscillation of the wagon body or the wagon bogies occurs when wagons travel at or over their critical speed even on excellent track under the influence of some minor disturbance from the straight-line motion. This lateral oscillation of the wagon components leads to the wheel flange contacting the gauge face of the railhead associated with derailment potential. Due to the risks involved in wagon hunting, field testing is not an option and the investigation of this aspect must, therefore, be limited to modelling and simulation.

With the demand for increase in operating speed being more pressing than ever before, the instability analysis of wagons and/or bogies has recently attracted wider attention of railway researchers. There have been many mathematical models [1 ~ 8] established to analyse the lateral dynamic instability of wagons. Unfortunately, most

[1] *Corresponding Author:* A/Prof. Manicka Dhanasekar, director, Centre for Railway Engineering, Central Queensland University, Rockhampton, QLD 4702 AUSTRALIA. Tel.: 61 (07) 49309677; Fax: 61 (07) 49306984; E-mail: m.dhanasekar@cqu.edu.au

of these models consider the track either as a rigid or as a uniformly supported platform. With a view to truly representing the track as a discretely supported rail beam resting on a layered subsystem, a three-dimensional wagon-track system dynamics (3DWTSD) model has been developed at the Centre for Railway Engineering and has been widely published in the literature [9, 10]. In this paper, the 3DWTSD model is used to examine the hunting characteristics of wagons containing three-piece bogies running on tangent track. The results obtained from the layered viscoelastic track subsystem model are compared to the ones obtained from the 'rigid' track subsystem. The influence of the wheel profiles on the hunting characteristics is also examined and briefly discussed.

2   3DWTSD MODEL

The 3DWTSD model includes three subsystems, namely, the wagon subsystem, the track subsystem and the wheel-rail interface subsystem [10]. Their dynamic characteristics are briefly described in this section. Detailed formulation of the basic governing equations is recently published [9, 10].

2.1   WAGON SUBSYSTEM

Wagon subsystems with three-piece bogies include one wagon car body containing two bolsters and two bogies. Each bogie consists of two secondary suspension elements, two sideframes, and two wheelsets as shown in Fig. 1 (a) and (b).

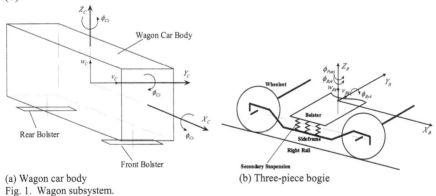

(a) Wagon car body   (b) Three-piece bogie
Fig. 1. Wagon subsystem.

Except for the secondary suspension elements, all the other components of the wagon are considered as rigid bodies with masses and mass inertia moments along the three Cartesian coordinate directions. The secondary suspension is represented by linear spring and damper elements that provide the lateral and vertical viscoelasticity. Due to the special connection between the wagon car body and the bogie (bolster, two sideframes and two wheelsets), the bogie exhibits relative rotation against the wagon car body known as the parallelogram rotation or lozenging of the bogie. Therefore, there are two rotations of the bogie about the vertical axis considered in the modeling – one the yaw rotation and the other the parallelogram rotation or lozenging. Except for the longitudinal dynamic motion, all other motions of the wagon subsystem are taken into account. The total degrees of freedom (DoFs) required to describe the lateral and the vertical displacements and rotations of the full wagon are 15 DoFs as listed in Table 1 in which u, v, w are the linear displacements and ϕ_x, ϕ_y, ϕ_z are the rotations about X, Y and Z axes respectively and ϕ_{Pa} is the parallelogram motion of the bogie about the vertical (Z) axis.

Table 1 DoFs of Wagon Subsystem Containing Three-piece Bogies

| Components | DoFs | | | | | | | No. of items | No. of DoFs |
	u	v	w	ϕ_x	ϕ_y	ϕ_z	ϕ_{Pa}		
Wagon Car Body	×	×	×	×	×			1	5
Bogie Structure	×	×			×	×	×	2	10
				Total DoFs					15

The equations of dynamic equilibrium can be written using multi-body mechanics method as shown below:

$$M_W \ddot{d}_W + C_W \dot{d}_W + K_W d_W = F_{WT} \qquad (1)$$

where M_W, C_W and K_W are the mass, damping and stiffness matrices of the wagon subsystem. These matrices are of size 15×15. d_W is the displacement vector of the wagon subsystem, and F_{WT} is the interface force vector between the wagon and the track subsystems consisting of the wheel-rail normal contact forces, tangent creep forces and creep moments about the normal direction in the wheel-rail contact plane.

2.2   Track Subsystem

The track subsystem contains three layers – rails, sleepers and ballast-subballast blocks based on the discretely supported distributed parameter approach. The schematic view of the model in the lateral and longitudinal direction is shown in Fig. 2 (a) and (b) respectively. In the model, all the track components used in the conventional ballasted heavy haul track structure are assembled in the same sequence as that of the actual structure. The track subsystem comprises two rails, n_s number of sleepers, $4 \times n_s$ fastener and pad assemblies, $2 \times n_s$ ballast-subballast blocks, and subgrade.

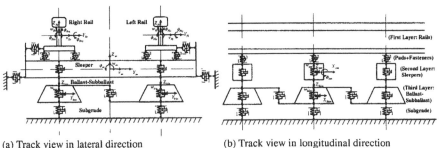

(a) Track view in lateral direction　　　　(b) Track view in longitudinal direction

Fig. 2. Track subsystem.

The lateral and the vertical bending and shear deformations of the rail beam are described using Timoshenko beam theory extended by considering the torque of the rail beam. Thus, five DoFs at any point along the longitudinal neutral axis of the rail beam, namely, lateral and vertical displacements and rotations about the lateral (Y) and vertical (Z) axes and the torsional rotation about the longitudinal (X) axis are used in the formulation of the rail beam.

The sleepers are considered as deformable short beams resting on elastic foundation and represented by their mass and viscoelastic properties at the rail seat location and represented by three DoFs per sleeper, namely, the lateral and vertical displacements and the rotation about the longitudinal (X) direction. The ballast-subballast layer is represented by its mass and viscoelastic properties with only one vertical displacement DoF per block. The equations of dynamic equilibrium of the rails, sleepers, the ballast-subballast blocks are assembled using multi-body mechanics methods. Finally, the governing equations of dynamic equilibrium for the track (rail and all other track components) are expressed in the following matrix form:

$$M_T \ddot{d}_T + C_T \dot{d}_T + K_T d_T = \tilde{F}_{WT} \qquad (2)$$

in which M_T, C_T and K_T (each of size $(10n_m + 5n_s) \times (10n_m + 5n_s)$ where n_m is the number of modes of the rail beam) are the mass, damping and stiffness matrices of the track subsystem. The vector d_T contains displacement of the track subsystem that includes the modal and physical displacements, and $\widetilde{F}_{WT}$ is the combined interface force vector between the wagon and the track subsystems.

2.3   WHEEL-RAIL INTERFACE SUBSYSTEM

Under rolling contact the wheel and the rail produce contact forces in the normal direction on the wheel-rail contact plane. In addition creep forces are generated in the longitudinal and the lateral directions tangential to the contacting plane, and creep moment about the normal direction. In this paper, the normal contact force due to the wheel-rail rolling contact is determined using the Hertzian static contact theory.

The creep forces and the creep moments are determined using Kalker's linear creep theory. The creep forces and moments are usually calculated without due consideration to the velocities of the rail [7], in other words, considering track as rigid. However in the 3DWTSD model the velocities of the rail in the lateral and spin directions are included in these calculations of creepages. Expressions for modified longitudinal, lateral and spin creepages at the wheel-rail contact plane are presented in Equation (3)

$$\begin{cases} \xi_{ix} = (1/V)[V(1 - r_i/r_0) \mp y_{wi}\dot{\phi}_{wz}] \\ \xi_{iy} = (1/V)[(\dot{v}_w + r_i\dot{\phi}_{wx} - V\phi_{wz}) - (\dot{v}_{Ri} - z_{Ri}\dot{\phi}_{Rix})] \\ \xi_{isp} = (1/V)[\dot{\phi}_{wz} \mp \delta_i V/r_0 - \dot{\phi}_{Riz}] \end{cases} \quad (3)$$

in which V is the wagon speed, the subscript i is either l for the left or r for the right wheel, r_i and y_{wi} are the wheel radius and the lateral coordinate of the wheelset at the centroid of the contact patch respectively, r_0 is the wheel nominal radius, z_{Ri} is the vertical coordinate of the rail at the centroid of the contact patch, δ_i is the wheel-rail contact angle at the centroid of the contact patch, and the symbol $\mp$ is negative - when $i = l$, and positive + when $i = r$.

As Kalker's linear theory best defines the creep forces only when the creepages are very small, Johnson-Vermeulen's approach presented in [7] is used to further modify the creep forces.

3   WAGON HUNTING ANALYSIS

An example was taken from a heavy haul wagon-track system in Queensland Rail [8] to carry out the wagon hunting analysis. It had been reported [8] that the empty wagon hunting was observed at an average operating speed of 75.5 km/h from the measured operational speed and lateral wheelset acceleration, and the hunting frequency was measured about 1 Hz, as shown in Fig. 3 (a) and (b).

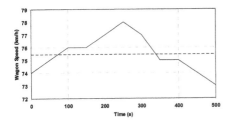 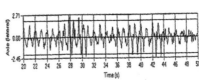

(a) Wagon speed during hunting   (b) Time histories of axle acceleration

Fig. 3. Experimental data [10].

In the wagon hunting simulations reported in this paper, two cases have been taken into account – new wheel profile and worn wheel profiles. Both cases were analysed using 3DWTSD model in layered viscoelastic track and rigid track.

3.1   HUNTING UNDER TAPERED (NEW) WHEEL PROFILE

The contact parameters of new wheel and new rail profiles were used. An initial condition of 0.2 mm lateral displacement was applied to the leading wheelset. Several simulations were carried out at varying operating speeds. The lateral displacement responses of the wagon components for these operating speeds (70 km/h, 74 km/h and 75 km/h) were simulated using 3DWTSD model in layered viscoelastic track. These results are presented in Figs 4 (a), (b) and (c) respectively for the three operating speeds.

Fig. 4 (a), (b) and (c) show the time histories of the lateral displacements of the front bogie, the rear bogie and the wagon car body. The onset of wagon lateral instability can be seen at 74 km/h (Fig. 4 (b)) and the hunting frequency is approximately 1.08 Hz.

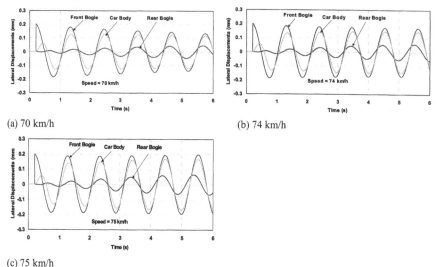

(a) 70 km/h   (b) 74 km/h

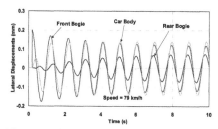

(c) 75 km/h

Fig. 4. Time histories of lateral displacements

Similarly, the lateral displacement responses of the wagon components at the critical speed of 79 km/h are simulated using 3D WTSD model in rigid track, and shown in Fig 5. The hunting frequency is about 1.11 Hz. Compared with those shown in Fig. 4 (b), both critical speed and hunting frequency are larger (although not significantly) using rigid track than using layered viscoelastic track.

Fig. 5. Time histories of lateral displacements at speed of 79 km/h

3.2 HUNTING UNDER HOLLOW (WORN) WHEEL PROFILES

The wheels used in heavy haul wagons are designed with a tapered tread. After being in service for a long period, the tapered tread becomes worn to the "hollow" profiles. In the simulation presented here, the hollow worn shape was modeled by an arc of certain pre-defined radius. Arcs with radii of 500 mm, 450 mm, 400 mm and 350 mm were selected to model the hollow worn wheel profiles.

The hunting characteristics of a wagon with all wheels having a hollow arc of radius 500 mm running on the layered viscoelastic track and the rigid track were shown in Fig. 6 (a) and (b) respectively.

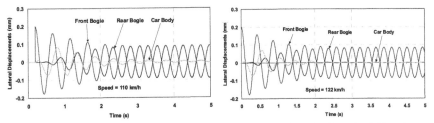

(a) Speed of 110 km/h on the layered viscoelastic track   (b) Speed of 122 km/h on the rigid track
Fig. 6. Time Histories of Lateral Displacements of Wagons with Hollow Wheels of Arc 500m

It can be seen from Fig. 6 that the hollow wheel profiles significantly increase the critical hunting speed and affect the hunting mode relative to the new wheel. The critical hunting speed determined from the layered viscoelastic track (110 km/h) is less than that determined from the rigid track (122 km/h). It was also seen that the hunting of the car body is significantly reduced as the wheel wears out, with a corresponding increase in the hunting of the rear bogie (in anti-phase with the front bogie).

Fig. 7 (a), (b) and (c) present the lateral displacement time series of the wagon components (front and rear bogies, and the car body) simulated using the layered viscoelastic track modelling at their respective speeds for the hollow arc radius of 450 mm, 400 mm and 350 mm respectively. It can be seen from Fig. 7 (a) that the hunting is limited to bogies, with the front and the rear bogies hunting in anti-phase with equal magnitude. Fig. 7 (b) shows a different type of bogie hunting in which the magnitude of the lateral displacements of the front and the rear bogies alternated. In other words, when the front bogie attained maximum lateral displacement, the rear bogie exhibited minimum lateral displacement and vice versa. In spite of this difference, the two bogies maintained anti-phase lateral displacements thereby preventing car body hunting. Fig. 7 (c) shows hunting that includes the wagon car body as well as the two bogies. In this case, the two bogies hunt in the same phase, allowing for the car body hunting. Simulations of hunting due to worn wheel tread were also carried out using the rigid track model. The comparison between the critical speeds and frequencies using the models with the layered viscoelastic track and the rigid track respectively is summarised in Table 2.

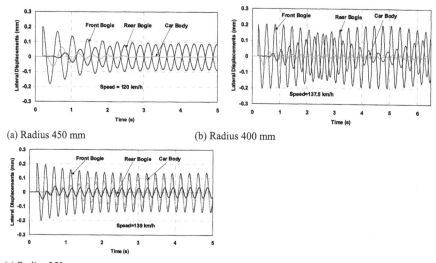

(a) Radius 450 mm   (b) Radius 400 mm

(c) Radius 350 mm
Fig. 7 Time Histories of Lateral Displacements for Various Hollow Arc Radii

Table 2 Comparisons between Critical Speeds and Frequencies

Radius of Wheel Tread (mm)	Critical Speeds (km/h)			Frequencies (Hz)		
	Layered Track	Rigid Track	% Increase	Layered Track	Rigid Track	% Increase
(New)	74	79	6.8	1.08	1.11	2.8
500	110	122	10.9	2.1	2.58	22.9
450	120	136	13.3	2.42	3.0	24.0
400	137.5	150	9.0	3.22	3.82	18.6
350	139	159	14.4	4.33	4.91	13.4

From Table 2, the maximum differences of the critical speed and hunting frequency between results from the rigid track and from the layered track are 20 km/h (or 14.4%) and 0.58 Hz (or 24.0%) respectively.

4   CONCLUSION

In conclusion, it could be said that the layered viscoelastic track modeling has provided conservative critical speed values. The lower critical speeds were associated with lower hunting frequencies. One of the potential reasons for the reduction in critical speed in the layered viscoelastic modelling is the effect of lateral velocity of rail to the lateral creepage.

The lateral velocity of the rail reduces the lateral creepage (in relation to the rigid track where the lateral velocity is zero), which subsequently leads to reduction in critical speed. This conclusion is worth taking note of the railway mechanical engineering practitioners, as they routinely model the track as rigid or continuously elastically supported. The knowledge on reduced critical speed and frequency could potentially improve the design and safety issues. The simulation also has proved that the severely worn wheels provide better stability to the wagon dynamics.

Another finding from the simulation is the mode of hunting. As there are cases where bogies hunt severely while keeping the car body stable. This highlights the need for full a understanding of the wagon dynamics mechanism as relying on driver instinct to avoid derailment would not be possible in such cases.

REFERENCES

1. Tuten, J.M., Law, E.H., and Cooperrider, N.K.: Lateral stability of freight cars with axles having different wheel profiles and asymmetric loading. Journal of Engineering for Industry, Transactions of the ASME 101(1979), pp. 1-16.
2. He, Q. and Cooperrider, N. K.: Optimum Design of a railroad freight car for lateral stability and curving performance. Proceeding of 9th IAVSD Symposium 1985, pp. 183-196.
3. No, M. and Hedrick, J. K.: High speed stability for rail vehicles considering varying conicity and creep coefficients. Vehicle System Dynamics, 13(1984), pp. 299-313.
4. Pater, A.D.: The lateral stability of a railway vehicle with two two-axle bogies. Proc. of 11th IAVSD Symposium, 1989, pp. 440-450.
5. Dukkipati, R.V.: Modelling and Simulation of the Hunting of A Three-Piece Railway Truck on NRC Curved Track Simulator. Proc. of 13th IAVSD Symposium, 1993, pp. 105-115.
6. Fujimoto, H., Tanifuji, K., and Miyamoto, M.: Influence of track gauge variation on rail vehicle dynamics (an examination based on comparison between data from a test train running on track with irregularity artificially set and numerical simulation. Proc Instn Mech Engrs, 214 (2000), Part F, pp. 223-230.
7. Garg, V.K. and Dukkipati, R.V.: Dynamics of railway vehicle systems. Academic Press Canada, ISBN 0-12-275950-8. 1984.
8. McClanachan, M.J.: Investigation of extreme wagon dynamics in central Queensland coal trains. Master Thesis, Central Queensland University, Australia, 1999.
9. Sun, Y.Q. and Dhanasekar, M.: A dynamic model for the vertical interaction of the rail track and wagon system. International Journal of Solids and Structures, 39 (2001), pp. 1337-1359.
10. Sun, Y.Q., Dhanasekar, M., and Roach, D.: A three-dimensional model for the lateral and vertical dynamics of wagon-track system. Proc. Instn Mech. Engrs Vol. (217) 2003 Part F, pp. 31-45.

Vehicle System Dynamics Supplement 41 (2004), p.242-251

Simulation of a Railway Vehicle's Running Behaviour: How Elastic Wheelsets Influence the Simulation Results

JÜRGEN ARNOLD*, INGO KAISER*,[1] AND GUNTER SCHUPP*

SUMMARY

The straight running behaviour of a passenger coach is investigated for the cases of the wheelsets considered as elastic and as rigid bodies. The model used for this investigation includes rotating elastic wheelsets modelled in an efficient way for the computation and a nonlinear wheel-rail contact.

The behaviour below and beyond the critical speed is studied for the configuration of rigid and elastic wheelsets. The comparison of the simulation results show a distinct influence of the structural elasticity of the wheelsets, especially a drop of the critical speed.

1. INTRODUCTION

For the simulation of the straight running of a railway vehicle, the wheelsets are usually considered as rigid bodies; this modelling is regarded to be sufficient for the low frequent dynamics located in the frequency range $f < 20$ Hz, see [1]. However, the wheel-rail contact can be very sensitive even to small displacements. Therefore the question arises, whether elastic deformations of the wheelsets can have an influence on the contact and, thereby, on the low-frequency running behaviour of the vehicle. For the low-frequency case of straight running of a railway vehicle, the critical speed $v_{0,crit}$ is of main interest. Below the critical speed, i.e. $v_0 < v_{0,crit}$, disturbances decrease; for $v_0 \geq v_{0,crit}$, disturbances can lead to a limit cycle motion of the vehicle, the so-called "hunting motion", which can lead to dangerous operational states of the vehicle and, therefore, should be avoided in regular operation.

Since $v_{0,crit}$ is calculated by simulations during the process of mechanical design of the vehicle, it is investigated, whether the consideration of the wheelset's elasticity leads to different simulation results than the consideration of the wheelsets as rigid bodies. To perform this investigation, a model of a railway vehicle considering

[1] Address correspondance to: Ingo Kaiser, Department Vehicle System Dynamics, Institute of Aeroelasticity, DLR – German Aerospace Center, D-82230 Wessling, Germany; E-mail: Ingo.Kaiser@DLR.de
*Department Vehicle System Dynamics, Institute of Aeroelasticity, DLR – German Aerospace Center, D-82230 Wessling, Germany.

wheelsets as elastic bodies is developed. The calculation results yielded by this model are compared with those obtained with rigid wheelsets; this comparison shows influences of the wheelset's elasticity.

2. STRUCTURE OF THE MODEL

The investigated vehicle is a passenger coach for long distance traffic. It is equipped with two bogies of the type MD (Minden-Deutz) 522; its maximum operational speed is 200 km/h. A characteristic of this bogie type is the yaw damping using dry friction. The parameters for the model are taken from [2] and [3]; they are slightly modified, especially for the yaw damping by dry friction. For the wheels, the profile S 1002 is used; the rails are of the type UIC 60 with an inclination of $1/40$.

The entire vehicle-track system is split up into three subsystems, namely "vehicle", "wheel-rail contact" and "track". Fig.1 shows these subsystems and their interactions with each other. The subsystem "wheel-rail contact" is connected to the subsystems

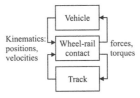

Fig. 1. Structure of the entire vehicle-track system.

"vehicle" and "track" at wheel nodes and rail nodes, respectively. At these nodes, the kinematics, i.e. the local translational and rotational positions and velocities, are taken from the vehicle and the track and are applied to the wheel-rail contact. The forces and torques, which are calculated from the kinematics, are applied to the vehicle and the track also at these nodes.

2.1. Vehicle

The subsystem "vehicle" is an elastic multi-body system (EMBS); Fig.2 shows the bodies which the EMBS consists of. The car body, the bogie frames and the bolsters are modelled as rigid bodies; for the wheelsets, the structural dynamics, i.e. their elasticity, are taken into account. For the car body, the bogie frames and the wheelsets, all six rigid body motions are possible; the bolsters can only perform relative yaw motions to the car body. The bodies are connected by linear springs and linear damping elements; between the car body and the bolster, a nonlinear force element consisting of a

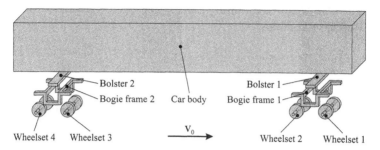

Bolster 2
Bogie frame 2   Car body   Bolster 1
Bogie frame 1

V_0

Wheelset 4   Wheelset 3   Wheelset 2   Wheelset 1

Fig. 2. Multi-body model of the car.

stiff spring and a friction contact according to Coulomb is used for the approximation of dry friction.

The kinematics of the wheelsets are described by a superposition of small elastic deformations to the rigid body motions of the wheelsets, i.e. a relative description is used. The description of the deformations is based on a modal synthesis; since the wheelsets are connected by force elements to the bogie frame and the track, no geometric boundary conditions exist, and, therefore, the eigenmodes of the free wheelset are used as shape functions.

Due to the rolling motions, the wheelsets perform large rotations related to the inertial system $\mathcal{I}$ and to the other bodies. This is avoided by introducing an intermediate coordinate system $\mathcal{A}$, which performs all rigid body motions of the wheelset except the rotation around the symmetry axis. Due to the rotational symmetry of the wheelset, the modal synthesis can be performed in the system $\mathcal{A}$; this leads to an Arbitrary Lagrangian-Eulerian (ALE) approach. Therefore, the position of a particle located at the point P is described as:

$$\mathbf{r}_{OP}^{\mathcal{I}} = \mathbf{r}_{OC}^{\mathcal{I}} + \mathbf{S}^{\mathcal{I}\mathcal{A}}\mathbf{r}_{CP}^{\mathcal{A}} = \mathbf{r}_{OC}^{\mathcal{I}} + \mathbf{S}^{\mathcal{I}\mathcal{A}}\left[\mathbf{r}_{CB}^{\mathcal{A}} + \mathbf{r}_{BP}^{\mathcal{A}}\left(\mathbf{r}_{CB}^{\mathcal{A}}, t\right)\right]$$

$$\text{with} \quad \mathbf{r}_{CB}^{\mathcal{A}} = \left[r\sin\theta \; y \; r\cos\theta\right]^{T}, \mathbf{r}_{BP}^{\mathcal{A}}\left(\mathbf{r}_{CB}^{\mathcal{A}}, t\right) = \sum_{i=1}^{n}\mathbf{W}_i(r, \theta, y)q_i(t) \quad (1)$$

Here, O and C signify the origin of the inertial system $\mathcal{I}$ and the center of mass of the wheelset, respectively. The point B denotes the position of the particle in the undeformed state. Due to the deformations, the particle is moved from B to the actual position P. Since the wheelset is a rotational symmetric body, the use of cylindrical coordinates r, θ, and y is obvious. Within the modal synthesis describing the elastic deformations, $\mathbf{W}_i(r, \theta, y)$ and $q_i(t)$ are the shape functions and the modal coordinates, respectively. Fig.3 shows the kinematics.

The cylindrical coordinate θ used in the intermediate system $\mathcal{A}$ is not a material

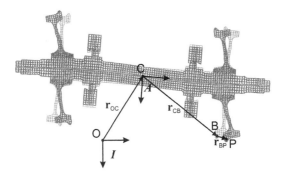

Fig. 3. Kinematics of a particle of the elastic wheelset.

coordinate, but a local one. Therefore, the velocity of the particle is given by:

$$\mathbf{v}_{OP}^{\mathcal{I}} = \mathbf{v}_{OC}^{\mathcal{I}} + \dot{\mathbf{S}}^{\mathcal{I}\mathcal{A}}\mathbf{r}_{CP}^{\mathcal{A}} + \mathbf{S}^{\mathcal{I}\mathcal{A}}\frac{d\mathbf{r}_{CP}^{\mathcal{A}}}{dt},$$

$$\frac{d\mathbf{r}_{CP}^{\mathcal{A}}}{dt} = \sum_{i=1}^{n}\mathbf{W}_i\dot{q}_i + \left[\left[r\cos\theta\ 0\ -r\sin\theta\right]^T + \sum_{i=1}^{n}\frac{\partial\mathbf{W}_i}{\partial\theta}q_i\right]\frac{d\theta}{dt} \qquad (2)$$

The intermediate system $\mathcal{A}$ performs only small rotations related to the inertial system, the large angle of the rolling motion does not appear itself, especially not as an argument of a trigonometric function, but only its time derivative. Therefore, after a linearization the subsystem vehicle can almost completely be described by linear equations with constant coefficients; this is a very simple mathematical structure which allows a very efficient calculation:

$$\mathbf{M}\ddot{\mathbf{y}}(t) + [\mathbf{D} + \Omega_0\mathbf{G}]\dot{\mathbf{y}}(t) + \left[\mathbf{K} + \Omega_0{}^2\mathbf{Z}\right]\mathbf{y}(t) = \mathbf{h}_{ext}(t) + \mathbf{h}_{nonlin}(\mathbf{y}(t)) \qquad (3)$$

Here, $\mathbf{M}$, $\mathbf{D}$ and $\mathbf{K}$ denote the mass matrix, the damping matrix and the stiffness matrix, respectively. Gyroscopic and centrifugal effects are described by the matrices $\Omega_0\mathbf{G}$ and $\Omega_0{}^2\mathbf{Z}$, respectively. The nominal angular velocity $\Omega_0 = v_0/r_0$ is the quotient of the travelling speed v_0 and the nominal rolling radius r_0. The wheel-rail forces and the gravitational forces are regarded as external forces; they are represented by the vector $\mathbf{h}_{ext}(t)$. The only nonlinear force elements within the subsystem "vehicle" are the friction elements acting between the car body and the bolsters; they are described by the vector $\mathbf{h}_{nonlin}(\mathbf{y}(t))$.

As mentioned before, the eigenmodes of the free wheelset are used as shape functions $\mathbf{W}_i$ in the modal synthesis. For the calculation of these eigenmodes, the FE-method is used. The calculation of the eigenmodes of a three-dimensional body can be reduced to a two-dimensional problem by taking advantage of the rotational symmetry of the wheelset, i.e. using trigonometric functions in circumferential direction

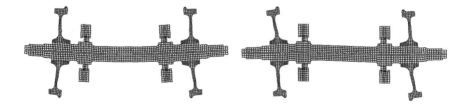

Fig. 4. Symmetric (left) and antimetric (right) eigenmodes of the wheelset, representing the lowest eigenfrequencies 84,2 Hz (symmetric) and 147,2 Hz (antimetric).

for the distribution of the deformations. Because of the rotational symmetry, the use
of cylindrical coordinates r, θ, and y is obvious. For the deformations R in radial, T in
tangential, and V in lateral direction, the following functions are an analytical solution
of Navier's equations:

$$R(r, \theta, y, t) = R_k(r, y) \cos(k\theta + \beta) \cos(\omega_{0,k}t + \phi) \tag{4}$$

$$T(r, \theta, y, t) = T_k(r, y) \sin(k\theta + \beta) \cos(\omega_{0,k}t + \phi) \tag{5}$$

$$V(r, \theta, y, t) = V_k(r, y) \cos(k\theta + \beta) \cos(\omega_{0,k}t + \phi) \tag{6}$$

For $k = 0$, rotational symmetric modes and torsional modes are described; bending
modes, i.e. modes which include a bending of the axle, are described by $k = 1$. For
$k \geq 2$, modes limited to the wheel discs or the brake discs occur. Further details concerning the eigenmodes of a wheelset can be found in [4]. Furthermore, the model can
be reduced by taking advantage of the symmetry due to the transverse middle plain of
the wheelset.

The elements used for the FE-models are ring elements with quadrangular cross
section, using trigonometric functions over the circumference and bilinear shape functions in the cross section. Two results of the FE-calculation are depicted in Fig.4, the
symmetric and antimetric bending modes of the wheelset related to the lowest eigenfrequencies. In the simulation model, all eigenmodes calculated with $0 \leq k \leq 4$ related to eigenfrequencies $f_0 \leq 5000$ Hz are considered.

2.2. Wheel-rail contact

The subsystem "wheel-rail contact" is a nonlinear force element acting between the
two nodes, where it is connected to the vehicle and the track, respectively. The input
for this force element are the kinematics of the nodes of the wheel and the rail, i.e. the
positions and the velocities with respect to translational and rotational motions.

Since the wheel rim is a massive part compared to the axle and the wheel disc,
it can be assumed, that it moves due to elastic deformations, but the profile remains

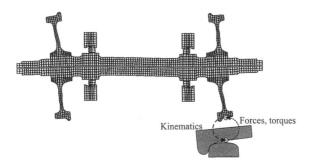

Fig. 5. Coupling of the elastic wheelset and the contact.

nearly unchanged. The same applies to the rail head. Therefore, the kinematics are applied at the wheel node and the rail node which do not necessarily have to coincide with the actual contact, so that the shift of the actual contact in performed internally in the subsystem "wheel-rail contact". This in shown in Fig.5.

In the first step, the deepest penetration of the undeformed surfaces of the wheel and the rail and its location are calculated. This location is assumed as the sole contact point. In the second step, the relative velocities in this contact point are calculated. In the third step, the forces acting in the contact are calculated by using the theory of Hertz (see [5]) for the normal force and the algorithm by Polach [6] based on the theory of Kalker for the tangential forces. In the fourth step, the forces and torques related to the nodes of the wheel and the rail are calculated; they are the output. Further details concerning the subsystem "wheel-rail contact" can be found in [7].

2.3. Track

For the track, a very simple model is used. Each wheelset is supported by a single track element, i.e. there is no interconnexion between the track elements. Fig.6 shows the mechanical model of one track element; it can perform lateral, vertical, and roll motions. This track model is the same as used in the multi-body simulation software *SIMPACK*; further details can be found in [8]. Although this model is very simple, it should be sufficient for the investigation of the wheelset elasticity's influence in the low-frequency range.

Fig. 6. Track element.

3. SIMULATION RESULTS

3.1. Transient behaviour

To investigate the transient behaviour below the critical speed, the front wheelset of the vehicle is excited by a lateral impact. Fig.7 shows the lateral motions performed by the center of mass of both wheelsets of the front bogie; the travelling speed is $v_0 = 270$ km/h. Here and in the following, the denominations of the wheelsets correspond to Fig.2. It can clearly be seen that the lateral motions decrease more slowly for the

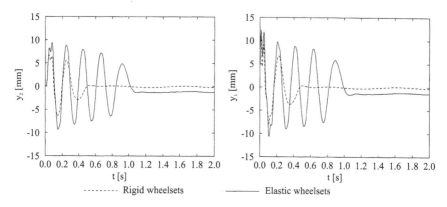

Fig. 7. Lateral motions of the wheelsets 2 (left) and 1 (right) due to a lateral impact on the wheelset 1.

elastic wheelsets than for the rigid ones. The offset displacement which occurs for the elastic wheelsets results from the dry friction damping between the bolster and the car body. Of course, this can also occur for rigid wheelsets; it is a pure coincidence that in this case the offset displacement for the rigid wheelsets is nearly zero.

To get an impression of the magnitude of the elastic deformations, the lateral deformations occuring at the left wheel nodes of both wheelsets are depicted in Fig.8.

3.2. Limit cycle behaviour

The limit cycle behaviour is calculated for the configurations of rigid and elastic wheelsets, respectively. Fig.9 shows the phase portraits of the lateral motion of the center of mass of the leading wheelset. For the rigid wheelsets, a sharp bend of the curves occurs at a lateral displacement of $y \approx \pm 6.6$ mm, which results from the flange hitting the rail head. On the contrary, the phase portraits for the elastic wheelsets are much smoother and also larger. This smoothing results from the structural elasticity; the wheelset acts as a spring cushioning the impact due to the flange contact.

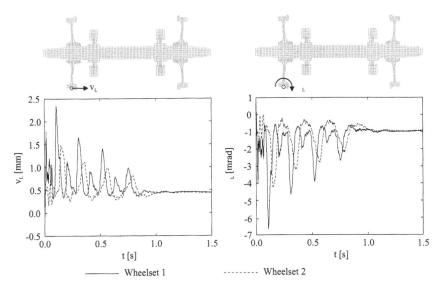

Fig. 8. Translational (left) and rotational (right) deformation at the left wheel nodes due to a lateral impact on the wheelset 1.

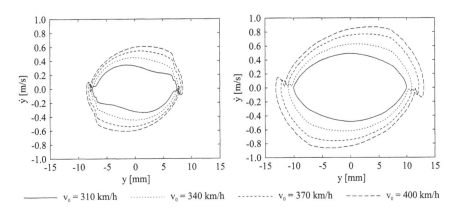

Fig. 9. Phase portraits of the lateral motions of the wheelset 1, modelled as a rigid body (left) and as an elastic body (right).

If the wheelsets are considered as rigid bodies, the limit cycle behaviour occurs at $v_{0,crit} = 304$ km/h. If only torsional deformations are considered, the critical speed is slightly shifted down to $v_{0,crit} = 298$ km/h; the amplitudes are hardly changed. The most drastic change of the limit cycle behaviour is caused by the bending elasticity: If only the bending elasticity of the wheelsets is taken into account, not only the critical

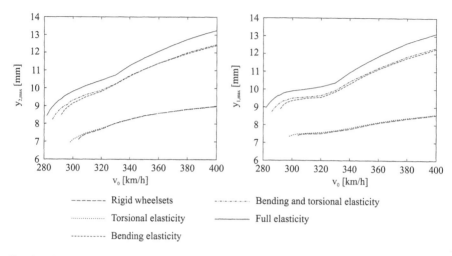

Fig. 10. Influence of the travelling speed on the lateral amplitude of the wheelset 2 (left) and 1 (right) for different types of elasticity.

speed drops down to $v_{0,crit} = 292$ km/h, but also the amplitudes increase dramatically. The amplitudes for the case of pure bending elasticity grow more strongly with increasing travelling speeds than for the case of rigid or torsionally elastic wheelsets; this is a result of the structural elasticity of the wheelset, which yields additional displacements due to the lateral deformations of the wheelsets. If the torsional elasticity is considered in addition to the bending elasticity, the same effect occurs as for the rigid wheelsets: The whole curve is shifted to lower travelling speeds; in this case, the critical speed drops from $v_{0,crit} = 292$ km/h down to $v_{0,crit} = 286$ km/h, while the amplitudes hardly increase. If full elasticity is considered, i.e. wheel deformations are taken into account in addition to the torsional and bending elasticity, the critical speed is shifted down again; it drops down to $v_{0,crit} = 282$ km/h. Furthermore, the lateral amplitudes increase again, since the wheelsets are getting "softer" due to the additional structural modes. The full elasticity causes a drop of the critical speed by $\Delta v_0 = 22$ km/h compared to the rigid wheelsets.

4. CONCLUSIONS

A model of a railway vehicle considering the structural elasticity of the wheelsets has been developed. The use of an Arbitrary Lagrangian-Eulerian (ALE) approach for the rotating elastic wheelsets provides a very efficient computation.

With this model, the straight running behaviour is investigated; the results are compared with those obtained for considering rigid wheelsets. Below the critical speed $v_{0,crit}$, lateral disturbances of the wheelsets decrease more slowly and with larger amplitudes in case of elastic wheelsets. Furthermore, the elasticity of the wheelsets leads to a distinct drop of $v_{0,crit}$ from 304 km/h down to 282 km/h and causes larger amplitudes beyond $v_{0,crit}$. Calculations taking into account different types of elasticity show that motions including the bending of the wheelset's axle make the major contribution to the difference between the behaviour of the elastic and the rigid wheelsets: The critical speed is shifted down, and also distinctly larger amplitudes occur.

Therefore, it can be concluded, that the structural elasticity of the wheelsets distinctly influences the simulation results of the low-frequent running behaviour. The drop of the critical speed $v_{0,crit}$ observed for elastic wheelsets may be of practical interest for the mechanical design of high-speed railway vehicles.

ACKNOWLEDGEMENTS

This work is a continuation of the project "Modeling and simulation of the mid-frequency behaviour of an elastic bogie" (project Po 136/19) finished in 2002; this project was part of the Priority Programme "System Dynamics and Long-Term Behaviour of Vehicle, Track and Subgrade" of the DFG (German Research Council). The financial support by the DFG is gratefully acknowledged.

REFERENCES

1. Vohla, G.K.W.: Werkzeuge zur realitätsnahen Simulation der Laufdynamik von Schienenfahrzeugen. VDI-Fortschrittbericht, Reihe 12, Nr. 270, VDI-Verlag Düsseldorf, 1996.
2. Diepen, P.: Horizontaldynamik von Drehgestellfahrzeugen – Berechnung und Optimierung des Laufverhaltens von schnellfahrenden Reisezugwagen mit konventionellen Laufwerken. Dissertation, TU Braunschweig, 1991.
3. Kim, K.H.: Verschleißgesetz des Rad-Schiene-Systems. Dissertation, RWTH Aachen, 1996.
4. Gasch, R. and Knothe, K.: Strukturdynamik, Band 2: Kontinua und ihre Diskretisierung. Springer-Verlag Berlin, Heidelberg, New York 1989.
5. Beitz, W. and Küttner, K.-H. (Eds.): Taschenbuch für den Maschinenbau / Dubbel. 16. Aufl. Springer-Verlag Berlin, Heidelberg, New York 1987.
6. Polach, O.: A Fast Wheel-Rail Forces Calculation Computer Code. *Vehicle System Dynamics Supplement* 33 (1999), pp. 728–739.
7. Kaiser, I. and Popp, K.: Modeling and Simulation of the Mid-frequency Behaviour of an Elastic Bogie. In: Popp, K. and Schiehlen, W. (Eds.): System dynamics and long-term behaviour of railway vehicles, track and subgrade, pp. 101–120, Springer-Verlag, Berlin Heidelberg 2003
8. Netter, H.: Rad-Schiene-Systeme in differential-algebraischer Darstellung. VDI-Fortschrittbericht, Reihe 12, Nr. 352, VDI-Verlag Düsseldorf, 1998.

Vehicle System Dynamics Supplement 41 (2004), p.252-261

Human-in-the-loop optimization of vehicle dynamics control with rollover prevention

BO-CHIUAN CHEN[1]

SUMMARY

Two Vehicle Dynamics Control (VDC) systems with rollover prevention function are proposed in this paper. Control gains and threshold values of both designs are optimized via the UMTRI preview driver model and Man-Off-The-Street course. The first design consists of yaw rate following and sideslip angle following for path tracking, and is optimized with the information of the lateral load transfer ratio (LTR) to improve its capability of rollover prevention. The second design is similar to the first one, but with additional rollover prevention control. Yaw rate and lateral acceleration are used to design the rollover prevention control. Open-loop and closed-loop maneuvers are used to evaluate the performance of both designs in Matlab/Simulink with a CarSim model.

1 INTRODUCTION

The average percentage of rollover occurrence in fatal crashes was significantly higher than that in other types of crashes [1]. The 2000 Fatality Analysis Reporting System (FARS) also shows that more than 50 percent of light vehicle occupant fatalities in single-vehicle crashes were involved in rollover event [2].

National Highway Traffic Safety Administration (NHTSA) proposed a Rollover Resistance Rating program in 2000 [3]. This program uses Static Stability Factor (SSF), which is the ratio of one half the track width to the C.G. height, to determine the rating. However, from automotive industries' comments, SSF was too simple without considering the dynamic effects of suspension deflection, tire traction, and Vehicle Dynamics Control (VDC) system. Therefore, NHTSA published an observation summary of rollover dynamic test maneuvers in 2002 [2].

It has been shown in [4,5] that VDC can utilize differential braking to generate a stabilizing yaw moment to (1) avoid directional instability and (2) improve the capability to follow the desired path. Since yaw-roll dynamics are coupled, it is possible to control the roll motion indirectly via controlling the yaw motion. Palkovics et al. [6] proposed a Roll-Over Prevention (ROP) system. If wheel lift-off is detected, full brake application will be activated by ROP. Knorr-Bremse developed a rollover prevention function in [7]. Whenever lateral acceleration is larger than a threshold value or wheel lift-off is detected, brake will be applied to reduce the vehicle speed. Thus, reduce the rollover danger. Wielenga [8] proposed an

[1] *Address correspondence to:* Department of Vehicle Engineering, National Taipei University of Technology, Taipei 106, TAIWAN. Tel.: 886-2-27712171 Ext. 3622; E-mail: bochen@ntut.edu.tw

Anti-Rollover Braking (ARB) by using differential braking instead of full brake application. Whenever wheel lift-off is detected or lateral acceleration is larger than the threshold value, ARB activates differential braking to make the vehicle understeer to reduce the rollover danger. Winkler et al. [9] proposed Rearward Amplification Suppression (RAMS) for articulated heavy-duty vehicles with multi-trailer combinations. RAMS uses differential braking to reduce the rearward amplification of lateral acceleration and thus prevent the last trailer from rolling over. The author and Peng proposed a rollover prevention system based on a Time-To-Rollover (TTR) metric [10,11] in [12]. Differential braking will be activated whenever the estimated TTR is less than the threshold value. Volvo introduced Roll Stability Control (RSC) on its 2003 XC90 Sport Utility Vehicle (SUV) [13]. RSC activates differential braking whenever an impending rollover is detected via a roll rate sensor.

Eisele and Peng proposed a VDC system with rollover prevention in [14]. Their VDC was able to limit roll angle and sideslip angle, and track the desired yaw rate under test maneuvers. However, the interaction between the driver and vehicle was not considered while designing the controller. The author and Peng proposed a VDC system with rollover prevention for SUVs in [15] and the control gains were optimized via the UMTRI driver model [16,17,18] and Man-Off-The-Street (MOTS) course [19]. Preliminary results showed improved rollover resistance while maintaining path tracking performance for evaluation maneuvers.

Two Vehicle Dynamics Control (VDC) systems with rollover prevention function are proposed in this paper. Instead of reducing the sideslip angle to below a certain threshold value, desired sideslip angle [20] is introduced for sideslip angle following. The first design consists of yaw rate following and sideslip angle following for path tracking, and is optimized with the information of the lateral load transfer ratio (LTR) to improve its capability of rollover prevention. The second design is similar to the first one, but with additional rollover prevention control. CarSim is used as a high-fidelity simulation platform to evaluate the performance of both designs.

The remainder of this paper is organized as follows: a brief introduction of the UMTRI driver model and CarSim SUV model is presented in Section 2. In Section 3, the proposed VDC control algorithm is defined and presented. The optimization of control gains and corresponding threshold values is presented in Section 4. Evaluation maneuvers and simulation results of the proposed deign are presented in Section 5. Finally, conclusions are made in Section 6.

2   MODELING

2.1 CARSIM VEHICLE MODEL

CarSim [21] is a software program developed by the Mechanical Simulation Corporation to simulate and analyze vehicle dynamic responses. The Big SUV model is selected to verify the proposed control in this paper. This model has front independent suspensions, rear solid axles, 14 multibody degrees of freedom, and 54 state variables.

2.2 UMTRI PREVIEW DRIVER MODEL

MacAdam developed the UMTRI preview driver model (see Figure 1) in 1980 [16,17]. The steering angle is selected such that the predicted future vehicle trajectory (dark solid line) will minimize the previewed path error (cross-shaded area).

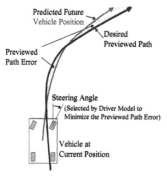

Fig. 1. UMTRI preview driver model.

CarSim has a built-in UMTRI preview driver model for steer control [21]. Two parameters, preview time and time delay, are used to specify the driver characteristics. Preview time describes the look ahead behavior.   Time delay simulates the neuromuscular time delay. Users can specify the desired path described in the global coordinates in CarSim and the driver model will generate steering angles to track the desired path. Preview time and time delay are set to be 1.5 sec and 0.25 sec [18], respectively, in this paper.

3   CONTROL ALGORITHM

3.1 DIRECT YAW MOMENT CONTROL

Koibuchi [22] showed that when the vehicle is cornering: (1) if inward moment is desired, differential braking should be applied on the rear-inner wheel; (2) if outward moment is desired, differential braking should be applied on the front-outer wheel. A direct yaw moment control (DYC) is designed to generate the desired yaw moment following the above guideline. Wheel brake distributions of the DYC are shown in Table 1. Counter clockwise rotation viewed from above are positive in this paper.

Table 1. Wheel brake distributions.

	Steering Angle > 0	Steering Angle < 0
Desired Yaw Moment > 0	Left Rear (LR) Wheel	Left Front (LF) Wheel
Desired Yaw Moment < 0	Right Front (RF) Wheel	Right Rear (RR) Wheel

After determining the brake distribution, DYC converts the desired yaw moment command M_d to the wheel brake pressure P_i by Equation (1).

$$P_i = \frac{2r_w}{K_i T} M_d \tag{1}$$

where the subscript i denotes for the selected wheel, r_w is the wheel radius, K_i is the brake gain of the selected wheel, and T is the track width.

Each wheel is equipped with an anti-lock brake system (ABS) to prevent wheel lock-up and retain a sufficient lateral tire force for directional control. The brake pressure is equal to the driver's brake input plus the DYC brake pressure from Equation (1), while modulated by the ABS.

3.2 VDC CONTROL ALGORITHM

The first VDC design consists of yaw rate following and sideslip angle following for path tracking, and is optimized with the information of the lateral load transfer ratio (LTR) to improve its capability of rollover prevention. The second VDC design is similar to the first one, but with additional rollover prevention control.

3.2.1 Yaw Rate Following

Yaw rate $\dot{\psi}$ is one of essential feedbacks for the driver to track the desired path. In order to reduce the deviation from normal vehicle behavior, $\dot{\psi}$ must be controlled to match the driver's desired yaw rate $\dot{\psi}_d$, which is the yaw rate response on a high friction road surface and is a nonlinear function of both vehicle speed and steering angle. $\dot{\psi}_d$ is represented by a 2-D lookup table in this paper (see Figure 2).

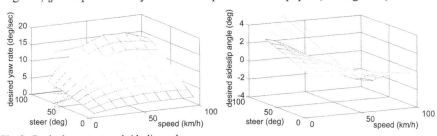

Fig. 2. Desired yaw rates and sideslip angles.

If the following condition is met, yaw rate following control will be activated.

$$\left| \dot{\psi} - \dot{\psi}_d \right| \ge Threshold_1 \tag{2}$$

This condition ensures that the control is not activated for normal driving when the yaw rate deviation is small. Proportional control is used for calculating the corresponding desired yaw moment M_1 as follows.

$$M_1 = K_{yaw}(\dot{\psi}_d - \dot{\psi}) \tag{3}$$

where K_{yaw} is the P-control gain.

3.2.2 Sideslip Angle Following

Instead of reducing the sideslip angle β to below a certain threshold value, desired sideslip angle β_d is introduced for sideslip angle following [20], which allows a more reasonable vehicle response under different vehicle operating conditions, and also prevents excessive β response on low friction road surface. β_d is a nonlinear function of both vehicle speed and steering angle and is represented by a 2-D lookup table in this paper (see Figure 2). If the following two conditions are met, sideslip angle following control will be activated.

$$|\beta - \beta_d| \geq Threshold_2 \tag{4}$$

$$\beta \cdot \dot{\beta} > 0 \tag{5}$$

Equation (4) ensures that the control is not triggered for normal driving when the sideslip angle deviation is small. Equation (5) indicates a possible directional instability. Proportional-derivative (PD) control is used for calculating the corresponding desired yaw moment command M_2 as follows.

$$M_2 = K_{\beta,p} \cdot (\beta_d - \beta) + K_{\beta,d} \cdot \dot{\beta} \tag{6}$$

where $K_{\beta,p}$ and $K_{\beta,d}$ are P- and D-control gains, respectively.

3.2.3 Rollover Prevention Control

Roll rate $\dot{\phi}$ and lateral acceleration a_y are used to detect the impending rollover. If the following two conditions are met, rollover prevention control will be activated.

$$|a_y| \geq Threshold_3 \tag{7}$$

$$a_y \cdot \dot{\phi} > 0 \tag{8}$$

Equation (7) ensures that the control is not triggered for normal driving when a_y is small. Large a_y only indicates possible large roll angle. If the roll angle is increasing in the same direction of a_y, possible rollover might happen for large roll angles. Thus, Equation (8) is used to indicate the roll instability.

Feeding back a_y [12] could make the vehicle understeer and make the roll dynamics less responsive to steering input. Proportional control is used for calculating the corresponding desired yaw moment command M_3 as follows. This part of VDC prevents the lateral acceleration from growing too large and causing rollover.

$$M_3 = -K_{ay} \cdot a_y \tag{9}$$

where K_{ay} is the P-control gain. The overall desired yaw moment commands for the first VDC design and the second VDC design are M_{d1} and M_{d2}, respectively.

$$M_{d1} = M_1 + M_2 \tag{10}$$

$$M_{d2} = M_1 + M_2 + M_3 \tag{11}$$

4   OPTIMIZATION

Four parts of the MOTS course [19] are selected for optimization (see Figure 3): (1) a large radius arc, (2) an obstacle avoidance maneuver, (3) a small radius arc, and (4) a fishhook maneuver. Entry speeds of each track are designed to be relatively challenging compared to the ordinary driving. Entry speeds are 85 km/h for the 1st track, 72 km/h for the 2nd track, 70 km/h for the 3rd track, and 65 km/h for the 4th track.

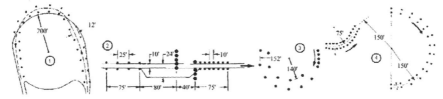

Fig. 3. MOTS test tracks [19]. (Units in ft.)

Three VDC designs are designed and evaluated in this paper. Control gains and threshold values of a traditional VDC without rollover prevention function are optimized with respect to the cost function J_1 using a human-in-the-loop (HIL) technique (i.e. UMTRI driver model in the loop between CarSim and MOTS tracks).

$$J_1 = \sum_{i=1}^{4} \left(\frac{PE_i}{PE_{off,i}} \right)^2 + \sum_{i=1}^{4} \left(\frac{\Delta V_i}{\Delta V_{off,i}} \right)^2 \tag{12}$$

PE is the path error defined as the area between the desired path and the actual traveled path. The subscript *off* denotes for the value obtained from the vehicle with VDC turned off. The smaller the ratio of PE to PE_{off}, the better the path tracking is. Penalties are imposed to prevent excessive speed reduction ΔV, which is defined as the initial speed minus the final speed within a certain simulation period. The larger the ratio of ΔV to ΔV_{off}, the greater the speed reduction is. The subscript i denotes for the ith test track.

The 1st VDC design has the same structure of the traditional VDC but is optimized by including penalties of rollover danger in the cost function. Lateral load transfer ratio (LTR) is selected to represent the rollover danger in the cost function J_2:

$$J_2 = J_1 + 4 \cdot \sum_{i=1}^{4} \left(\frac{\text{Max}(|LTR_i|)}{\text{Max}(|LTR_{off,i}|)} \right)^2 \tag{13}$$

where $\text{Max}(|LTR|)$ is the maximum absolute value of LTR. A weighting factor of 4 is imposed to add more penalties on LTR. If the driver is driving straight, LTR is 0. For extreme conditions, the absolute value of LTR can be 1 when either right or left side of tires lifts off the ground. The smaller the ratio of $\text{Max}(|LTR|)$ to $\text{Max}(|LTR_{off}|)$, the less the rollover danger is. The 2nd VDC design is designed based on the 1st VDC with additional rollover prevention control. HIL technique will only be applied to optimize K_{ay} and $Threshold_3$.

After optimizations, improvements of above three VDC systems are shown in Figure 4. The improvements of path error and LTR are defined as $(PE_{off}-PE)/PE_{off}$ and $[Max(|LTR_{off}|)-Max(|LTR|)]/Max(|LTR_{off}|)$, respectively. The larger the improvement of path error, the better the path tracking is. The larger the improvement of LTR, the better the rollover resistance is.

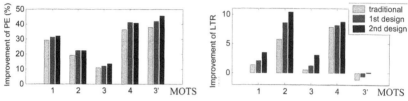

Fig. 4. Improvements of traditional VDC and two proposed VDC systems for the MOTS tracks.

In Figure 4, all three VDC systems have significant improvements of path error. The traditional VDC has significant LTR improvements for the 2nd and 4th MOTS tracks, but not for the 1st and 3rd MOTS tracks (cornering maneuvers). The 1st VDC design has slightly better LTR improvements than the traditional one, but is still worse than the 2nd VDC design, which has additional rollover prevention control. If the entry speed for the 3rd MOTS track is raised to 75 km/h (results are shown as 3' in Figure 4), only the 2nd VDC design can still has LTR improvement.

5   SIMULATION RESULTS

Performances of above three VDC systems are evaluated using both open-loop and closed-loop maneuvers. Because VDC can reduce the vehicle speed, automotive industries have different comments [2] regarding to whether the same rollover resistance credit should be given to vehicles with VDC. Evaluations with speed control to follow the speed of the vehicle without VDC are also studied in this section.

Figure 5 shows selected open-loop maneuvers in [2], NHTSA J-turn (entry speed at 80 km/h) and Roll Rate Feedback Fishhook (RRFF, entry speed at 65 km/h). The steering angle of RRFF is revered to the opposite direction when the first zero crossing of the roll rate between ±1.5 deg/sec is detected. Closed-loop maneuvers include the Path Specified Double Lane Change (PSDLC, entry speed at 72km/h) from Ford [23] and the closed-loop fishhook (entry speed at 70 km/h) proposed by the author and Peng in [15]. Both types of maneuvers are also shown in Figure 5.

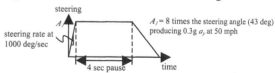

Open-loop maneuvers [2]   NHTSA J-Turn

steering rate at 1000 deg/sec

steering

4 sec pause   time

A_1 = 8 times the steering angle (43 deg) producing 0.3g a_y at 50 mph

crossing of the roll rate between ±1.5 deg/sec is detected. Closed-loop maneuvers include the Path Specified Double Lane Change (PSDLC, entry speed at 72km/h) from Ford [23] and the closed-loop fishhook (entry speed at 70 km/h) proposed by the author and Peng in [15]. Both types of maneuvers are also shown in Figure 5.

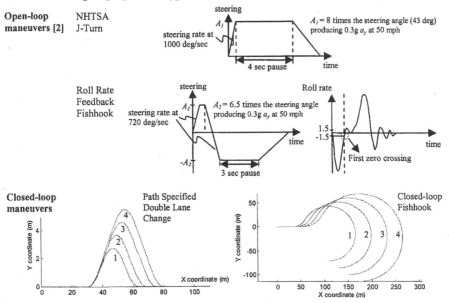

Fig. 5. Evaluation maneuvers.

Open-loop and closed-loop evaluation results are shown in Figure 6 and Figure 7, respectively. For NHTSA J-Turn, both proposed VDC designs have better rollover resistance than the traditional VDC. If the speed control is on, only the traditional VDC deteriorates the rollover resistance (i.e., worse than the vehicle with VDC turned off). For RRFF, both proposed VDC designs have better rollover resistance than the traditional VDC no matter the speed control is on or off. The performance of 1st VDC design and the 2nd one is about the same for the open-loop maneuvers selected by NHTSA, but better than the traditional VDC design.

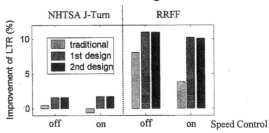

Fig. 6. Open-loop evaluation results.

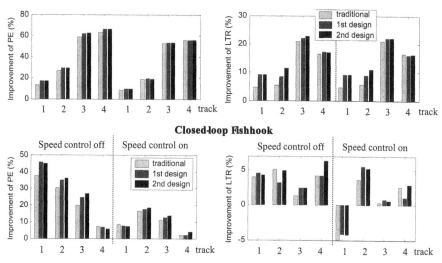

Fig. 7. Closed-loop evaluation results.

As can be seen from Figure 7, all three VDC designs can improve the path error and rollover resistance for PSDLC no matter the speed control is on or off. *PE* improvements of the 3rd and 4th tracks are much larger than that of the other two tracks, because the vehicle without VDC has lost its directional control for these two tracks. As for the closed-loop fishhook maneuvers, all three VDC designs can improve the path error and rollover resistance for speed control turned off. If the speed control is on, all three VDC designs deteriorate the rollover resistance for the 1st track.

6   CONCLUSIONS

Both proposed VDC designs have better rollover resistance than the traditional VDC for the open-loop maneuvers selected by NHTSA. It is possible to improve the rollover resistance of the traditional VDC if the system is optimized with LTR penalties in the cost function via HIL techniques. However, both designs have similar *PE* improvements and rollover resistance of the traditional design for closed-loop evaluations. For double lane change type maneuvers (the 2nd MOTS track, PSDLC, and the S section of the closed-loop fishhook), all three VDC designs might already have the capabilities to make the vehicle understeer at the right time to reduce the rollover danger. Additional rollover prevention control (the 2nd VDC design) might only be necessary to reduce the rollover danger for cornering maneuvers. If the speed control is on, VDC might deteriorate the rollover resistance. Therefore, speed reduction is a desired side effect of VDC to reduce the rollover danger.

ACKNOWLEDGEMENTS

This project was supported by the National Science Council in Taiwan under the contract NSC 91-2213-E-027-004.

REFERENCES

1. National Highway Traffic Safety Administration: Traffic Safety Facts 1993-2000: A Compilation of Motor Vehicle Crash Data from the fatality Analysis Reporting System and the General Estimates System, 1994-2001.
2. National Highway Traffic Safety Administration: Consumer Information Regulations; Federal Motor Vehicle Safety Standards; Rollover Resistance, Docket No. NHTSA-2001- 9663, Notice 2, 2002.
3. National Highway Traffic Safety Administration: DOT Announces Proposal to Add Rollover Ratings to Auto Safety Consumer Information Program, NHTSA Now, 6(7) (2000).
4. Zanten, A., Erhardt, R., and Pfaff, G.: VDC, the vehicle dynamics control system of Bosch, SAE Paper No. 950759.
5. Leffler, H., Auffhammer, R., Heyken, R, and Röth, H.: New Driving Stability Control System with Reduced Technical Effort for Compact and Medium Class Passenger Cars, SAE Paper No. 980234.
6. Palkovics, L., Semsey, A., and Gerum, E.: Roll-Over Prevention System for Commercial Vehicles – Additional Sensorless Function of the Electronic Brake System, Vehicle System Dynamics 32(4) (1999), pp. 285-297.
7. Hecker, F., Schramm, H., Beyer, C., Holler, G., and Bennett, M.: Heavy vehicle stability notification and assistance, SAE Paper No. 2000-01-3481.
8. Wielenga, T.J.: A Method for Reducing On-Road Rollovers – Anti-Rollover Braking, SAE Paper No. 1999-01-0123.
9. Winkler C., Fancher P., and Ervin R.: Intelligent Systems for Aiding the Truck Driver in Vehicle Control, SAE Paper No. 1999-01-1301.
10. Chen, B. and Peng, H.: A Real-time Rollover Threat Index for Sports Utility Vehicles, Automatic Control Conference, June 1999.
11. Chen, B. and Peng, H.: Rollover Warning For Articulated Vehicles Based on A Time-To-Rollover Metric, ASME International Mechanical Engineering Conference and Exhibition, November 1999.
12. Chen, B. and Peng, H.: Differential-Braking-Based Rollover Prevention for Sport Utility Vehicles with Human-in-the-loop Evaluations, Vehicle System Dynamics 36(4-5) (2001), pp. 359-389.
13. DeMeis, R.: This Volvo Won't Roll, Design News. Boston: 58(3) (2003), pp. 29.
14. Eisele, D.D. and Peng, H.: Vehicle Dynamics Control with Rollover Prevention for Articulated Heavy Trucks, AVEC 2000, August 2000.
15. Chen, B. and Peng, H.: Design of Vehicle Dynamics Control with Rollover Prevention via UMTRI Preview Driver Model, AVEC 2002, September 2002.
16. MacAdam, C.C.: An optimal preview control for linear systems, Journal of Dynamic Systems, Measurement and Control, Transactions of the ASME 102 (3) (1980), pp. 188-190.
17. MacAdam, C.C.: Application of an optimal preview control for simulation of closed-loop automobile driving, IEEE Transactions on Systems, Man and Cybernetics SMC-11(6) (1981), pp. 393-399.
18. MacAdam, C.C.: Mathematical Modeling of Driver Steering Control at UMTRI - An Overview, UMTRI Research Review 20 (1) (1989), pp. 1-13.
19. Rice, R.S., Dell'Amico, F., and Rasmussen, R.E.: Automobile Driver Characteristics – The Man-Off-The-Street, SAE Paper No. 760777.
20. van Zanten, A.T.: Bosch ESP System: 5 Years of Experience, SAE Paper No. 2000-01-1633.
21. Mechanical Simulation Corporation: CarSim User Manual Version 5.11, November 2001.
22. Koibuchi, K., Yamamoto, M., Fukada, Y., and Inagaki, S.: Vehicle Stability Control in Limit Corning by Active Brake, SAE Paper No. 960487.
23. Ford Motor Company: Comments to Consumer Information Regulations; Federal Motor Vehicle Safety Standards; Rollover Resistance, Docket No. NHTSA-2001-9663-24, 2001.

Vehicle System Dynamics Supplement 41 (2004), p.262-271   © Taylor & Francis Ltd.

Human-vehicle system modelling – focus on heuristic modelling of driver-operator reactions and mechatronic suspension

WLODZIMIERZ CHOROMANSKI AND JERZY KISILOWSKI[1]

SUMMARY

The paper presents issues dealing with modelling and simulation of human-vehicle-environment system with special attention on modelling of human behaviour in the filed of none mechanical phenomena. Heuristic techniques based mainly on the fuzzy set theory have been used, and the concept of active mechatronic suspension of driver's/machinist's seats which are adjustable to individual character traits of operators are also presented in it.

1. INTRODUCTION

The article presents issues related to the modelling of human-vehicle-environment system, with special attention on modelling of the human factor. In it, as seen in Fig 1 a human acts not only as a mechanical object but also as an essential control element. That controller-operator function is particularly important. An analysis of many accidents shows that the human factor is responsible for most of them. Yet, in the case of rail vehicles, automobiles or aircrafts the answer to what the reasonable compromise between automation of steering/navigation of a vehicle or a ship, and a human operator still remains open. So far, experience with the construction of the Airbus A320 and later analysis of, for example its accidents make it all the more difficult to give a definite answer to this question. And it is especially difficult to present a definite answer regarding automobiles and aircrafts. Perhaps it would be easier to do with rail vehicles such as for example metro/underground. In the case of these, today we better understand full remote controlled steering systems of vehicles without human operators. This particular dilemma also leads to a quest for a solution and a particular kind of engineering philosophy that would go beyond engineering, as we understands it today. One, which could help us find a satisfactory answer to these two questions:

---

[1] *Address correspondence to:* Warsaw University of Technology, Faculty of Transport, Koszykowa 75, 00-662 Warsaw, POLAND, e-mail: wch@it.pw.edu.pl

Can artificial intelligence replace human intelligence in operating a vehicle?; How much automation can be considered acceptable in steering a vehicle without limiting one's individual freedom? For example, the individual freedom to choice to drive one's own vehicle as one pleases.

There arises a need in the field of these studies for the construction of none mechanical human driver or machinist model, which would allow us to study man as a complex control system model. Most works, which deal with the human-vehicle-environment type of models, construct mechanical (bio-chemical) type of models. These types of androids as they are sometimes called are widely used in different types of simulation sessions, and are useful in conducting analysis of ex. collision related injuries. They are also used for analysis of the influence of different types of vibrations on man. We can mention a number of works here, such as for example [1]. This work describes particularly hazardous vibrations and their levels and scales. Generally, these vibrations are ascribed to acceleration. Griffin also introduced in his work a third term derived from transients after time called „jerk". According to Griffin, the value of its volume is the measure that has an impact on man. There are many standards, which control the influence of vibrations on man. Among those is the international standard ISO 2631-1. Many works on this subject deal with an anthropomorphic analysis of this problem. The source of data for those works is derived from statistics of a given population group. Their authors try to describe the optimal diameters of, for example the parameters of a rail vehicle's operator's cabin. These works are situated in the field of ergonomic science. It's also worthwhile to mention here the following works: [2] and more works on this subject by authors of this article [3,4].

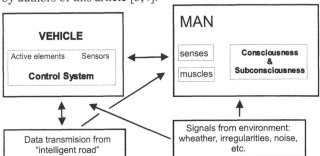

Fig. 1 Man as an element of a vehicle's control system.

However, works that study man not as just a geometric and biomechanical object but as a steering element in the human-vehicle-environment setting are very rare. In many of those works man is substituted by a simple automatic element i.e. an inertial element. This work, however, is an effort of taking a deeper approach to the issue i.e. an effort to model and simulate human behaviour, but not in the sphere of mechanical phenomena. The authors try to cover such objective facts as monotonous motion and the driver's poor psychophysical condition, which slows down the driver's reaction time and can cause an increased risk of accident. The second part of this work deals with problems of biomechanical nature. We will consider the concept of vehicle suspension synthesis, which would adjust to a driver's character. We know that the personal dispersal of biomechanical parameters of man is quite large. Therefore, a question arises: "Should not the construction of vehicles take under account the dispersal of human biomechanical parameters?"

The authors of this article conclude the following:

· It will still be long before a machine in steering vehicles; especially of automobiles and aircrafts will replace man. That is why in the modelling and simulation of the human-vehicle-environment layout, one should remember that the human model is a fundamental and most important steering element in a vehicle. And as such, it cannot be ignored in the engineering and construction of vehicles.

· The dispersed human biomechanical parameters are very sizable and should always be accounted for in the synthesis of a vehicle's suspension.

2. MODELLING AND SIMULATION OF HUMAN BEHAVIOUR IN THE FIELD OF NONE MECHANICAL PHENOMENA THE EXPERIMENTAL PLATFORM

The object of the analysis will be a system like the one shown in Fig. 1. The analysed system consists of a mechanical part i.e. the mechanical system of a vehicle, a part of the steering system i.e. ABS - automobile's antiskid system, and that, which is connected to man. The following general ideas used in this work will be understood as follows:

1. The term „machine" will indicate a vehicle. Further in the work it will become necessary to specify the kind of vehicle that is being discussed ex. a rail vehicle, an automobile etc.

2. The machine behaviour is defined by mechanics laws, which are typical for a given object with an account made for the steering system. Vehicle model should consider man's reaction to vehicle and vehicle's reaction to man.

3. A machine model should make possible modelling of a so-called „situation" i.e. various manoeuvres and actions made by the vehicle.

4. Model which describes interrelation between man and vehicle includes a number of parameters and vehicle feedback .

5. The overall man-vehicle model should enable the analyses of a vehicle's movement, the analysis of its impact on man, and a safety evaluation of the man-vehicle layout itself.

Let us analyse the mathematical construction of the human-vehicle-environment model now. It should be understood that a vehicle model is a group of models connected to the problem under analysis. In reality, we shell be dealing with dynamic models (ex. a critical analysis of the processes involved in stopping rail vehicles, or an analysis of processes engaged in passing vehicles by an automobile etc.), these will often be kinematical models (connected to ex. a simplified stopping model observed as a unvarying delayed movement) or logical models (ex. whether a vehicle stopped before or after a stop sign – probability). In the area of dynamic models, MBS packets should be taken under account; it is important for the model to take account of integrated steering systems, if available in a vehicle.

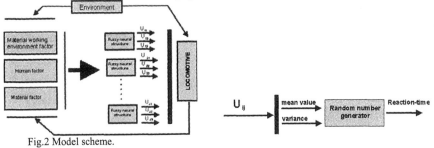

Fig.2 Model scheme.

The construction of such a model should also take account of signals generated by the man, who drives the vehicle. Man can have an influence over a vehicle through n-actions that are typical for the particular model of movement of a vehicle under analysis; for the given i-this action the influence on a vehicle is characterised by a three-elemental vector:

$$U_i = [u_{i1}, u_{i2}, u_{i3}]^T \qquad (1)$$

where:

u_{i1} - the time information was received; u_{i2} - the time a decision was made; u_{i3} - the time an action was taken; T – transposition.

We will limit the discussion to reaction time i.e. describes the operator's reaction time to a given situation (reaction time is a sum of u_{ij} where: $j=1,2,3$).

In earlier works of the authors [3] the three-elemental vector was defined as time of initiation of manoeuvre, the level of correctness of manoeuvre and its speed. We should notice that from the point of an analytic view of a definite situation or movement the material meaning of elements described in the relation (1) is very important, but omission of the material definition of the vector elements is admissible at some levels of a general discussion. The variables u_{ij} are in fact random variables at a specified average value and a given distribution of density of probability (see Fig.2). The authors establish that these variables have a bi-parameter distribution. The choice between the types of distribution must be based on a database of analytic experiments and a statistic system of procedures for testing of various theories on compatibility of empiric distribution with theoretic ones. Not all cases of action/movement will need to have all three components of the vector defined, of course. In the case of some simple actions, a component such as ex. u_{i2} could be omitted sometimes. It will also be vitally important to describe the initial conditions, such as the vehicle's speed etc. to illustrate its "behaviour pattern" for the purpose of conducting such simulation. The description of the components of vector poses a substantial difficulty. Therefore, we can assume that their values are the function of these three components:

$$u_{ij} = F(\overline{C}, \overline{M}, \overline{S}) \quad j = 1,2,3 \tag{2}$$

where:

$\overline{C}$ - vector i.e. human factor; $\overline{M}$-vector i.e. material factor; $\overline{S}$ - vector i.e. associated material working environment factor

The composite value of vector $\overline{C}$ is quite large and can have further subdivisions such as: psychophysical conditions, consumption of physical energy, volume of data transmitted to the mind, education, experience and employment related disorders etc.; however, we shell not list all components of vector here, because this is not the objective of this work. More information on can be found in [2,3,4], and in other works. Vectors $\overline{M}$ describe such components as: vehicle's technical condition, level of automation etc. Vector $\overline{S}$ is determined by such components as, for example: noise, mechanical vibrations. In the process of simulation some components of vectors can be assigned or considered arbitrarily as feedback from the simulation vehicle-environment layout model (see Fig. 2). The most important, and at the same time most difficult element in the construction of such a simulation model is the construction of its descriptive map (2). A lack of algorithms, which in the case of vehicles are determined by laws of mechanics and steering theory, turns us inevitably to heuristic methods. The authors suggest the utilization of neural networks [5] or models based on fuzzy sets [6,7]. In the proposed simulation model focus was placed on modelling man's influence on a machine (vehicle). The model structure of this influence is shown Fig.2. Signals, taken as elements of vectors $\overline{C}$, $\overline{M}$ and $\overline{S}$ will constitute data for the modelling of a given situation. And as mentioned earlier, they may also be treated as feedback data from a vehicle. But let us also notice that some vehicle feedback may have a shorter, a longer or a very long character ex. increasing exhaustion, development of job related disorders). Computer simulation of such processes is exceptionally complex and requires an analysis of many very complicated areas in the fields of physiology, medicine and biology. The authors only signalise the problem in this work. The most frequent difficulty is in identifying and verifying models described in a neural structure. In reality, neural networks are self-educating structures, but also ones, which require a great deal of practical experience. The authors propose using multi-layer non-linear neural networks and the most efficient backpropagation algorithm in the process of learning a network. The process of learning a network and collection of measurements should be an ongoing and essential procedure in the simulation. In the case of applying fuzzy sets to estimate vector elements and expert's knowledge, which was expressed in a "linguistic model" (collection of implications) can be used. The authors intend to employ this technique most frequently.

2.1   EXAMPLE RESULTS

The analysed example represents a fraction of the full documentation of results that are found in the author's work [3,4]. A vehicle's stopping rout is simulated. An assumption is made that the machinist suddenly notices a stop sign. The machinist acts upon the vehicle through: reaction time of stopping procedures from the moment a stop sign was noticed by the machinist, and the amount of stopping power used. It is assumed that the stopping power remains unchanging during the entire stopping procedure, which of course is a simplification of a real situation.

Human reaction time can be modelled on function of ex. factors such as:

monotony - mt -(mf1 -small, mf2 medium, mf3 - large); volume of information - wi - (mf1 - medium, mf2 - large); visibility - wd - (mf1 - bad, mf2 - good); noise - h - (mf1 - strenuous, mf2 - not strenuous); job adoptability - pp - (mf1 - poor, mf2 – very good); psychophysical condition -ps - (mf1 - bad, mf2 good)

Standardised levels of all factors used in range between <0,1> [3]. Reaction time –wyn - result is shown as the result of aggregation of defused subsets (mf1 small, mf2 medium small, mf3 normal, mf4 medium large, mf5 large, mf6 very large). In real values small reaction time is 278 ms and a large reaction time is 1500 ms (it is also shown in standard format in graphs). Mamdamiego model (Fig.3) was used to construct a model of reaction time that is dependent on the above factors. It was built on the funding blocks of knowledge of experts and expressed in the form of implications[2,4].

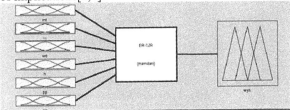

Fig. 3 Mamdani model to simulate reaction time.

Examples of membership functions in Mamdani scheme were shown in figures 4-5.

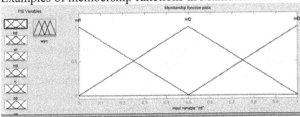

Fig. 4 Membership functions for inputs.

On figure 6 example graphs of output values that is the reaction time in the act of performing the parameters. And figure 7 depicts the result of the other simulation. One thousand simulations were conducted in accordance with the graph shown in Fig 7. The way was marked on the horizontal axis in normalised veriables. The value 0.9 represents an optimal situation where a stop was made 4 m from a sign. The first post in the graph represents the result from a fuzzy set model.

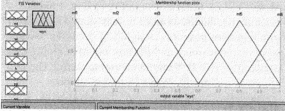

Fig. 5 Membership functions for outputs.

The secondrepresents the neuron model (description of the neural model was established in the works [2,3]. The example refers to this instance: little monotony, little noise, good visibility, and good adaptation to working conditions, good psychophysical condition, and good physical condition – Fig. 7 a). We can clearly see here a small chance of an accident (passing the stop sign). On the other hand, illustration shows analogical results the following reverse conditions: monotony, much noise, poor visibility, bad adaptation to working conditions, bad psychophysical condition and bad physical condition - Fig.7 b).

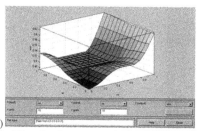 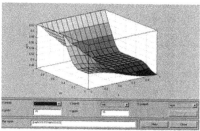

a)   b)

Fig.6 Reaction time versus: a) wi – visibility and mt – monotony b) mt monotony and h noise.

3. CONCEPT OF MECHATRONIC SUSPENSION SYNTHESIS THAT ACCOUNTS FOR DRIVER'S/MECHANIST'S BIOMECHANICS

In conducting syntheses of an active suspension, evaluating the comfort of vehicle etc. usually, standardised factors are employed to describe human biomechanics. Meanwhile, the dispersal of character traits of a given population (ex. body weight, parameters of the spinal bone, which is a fundamental supporting element of the body etc.) is subject to considerable variation depending on individual character traits. Therefore, it seems that in the synthesis of active driver's/mechanic's seats they should not be omitted. A brief, generalized concept of this system is illustrated in Fig. 8. Here, a new, important element is the effort to use feedback for steering of the adjustable active suspension and semi-active suspension through measurements of acceleration and of relative displacements on the driver's/machinist's

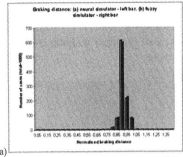 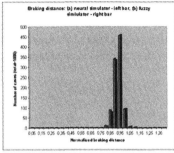

a)   b)

Fig.7 Braking distance; a) little monotony, little noise, good visibility, and good adaptation to working conditions, good psychophysical condition, and good physical condition b) monotony, much noise, poor visibility, bad adaptation to working conditions, bad psychophysical condition and bad physical condition.

body. This enables the suspension parameters to be adjustable to individual characteristic traits. The technical construction of the suspension will utilize electro hydraulic shock absorbers (of a viscidity determined by the value of an applied electric current) and electrohydraulic servomechanism.

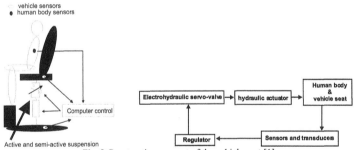

Fig. 8 Construction concept of the vehicle seat [1]

A schematic of such a serwomechanism is shown in Fig. 2. Its fundamental element is an electro hydraulic servo-valve. We can clearly see a limited frequency range of this element (up to 5-10 Hz). An essential element in the synthesis of the driver's/machinist's amortisation layout is a subsystem which describes the biomechanical human makeup. Some of the basic concepts of the system that have mainly dealt with the human spine are illustrated below. Mechanical transmission of vibrations/impulses to different parts of the body is conducted through the skeleton and muscles. From the point of view of effects of vibrations on the vehicle operator's organism (in a sitting posture), the element which is most engaged in the transmission of vibrations from the surface of the seat to higher parts of the corpus and the head is the spine. The spine is the most fundamental part of the human bone structure and is the centremost point of the entire skeleton. The difficulties in modelling man as a dynamic object arise mostly from: a) the diversity of the human population (his different anthropomorphic characters should be taken account of, his age and his individual character);b) the heterogeneity and variability of material that constitutes the physical makeup of each of his organs (the size and volume of many organs change through time) and as a result of that, there arises the problem of establishing the correct physical parameters of each type of physical makeup c) the complexity of structure of his skeletal and muscular system.

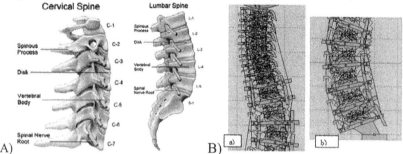

Fig.9 A) Anatomical model of the human spine B) Simulation model oh human spine: a)chest and cervical section, b) lumbar section.

The construction and the function of the human spine are quite complicated. That is why it is very difficult to build anything like it. In the proposed biochemical model the authors tried to incorporate in it all essential elements responsible for the proper functioning of the spine. The characteristic of this layout is presented below:

In the model, which the authors constructed (Fig, 5) there are three different sections: the cervical section, the chest section and the lumbar section. The neck section of the spine in our model consists of 7 neck vertebras, the cervical section of 6 chest vertebras, and the chest section of 5 lumbar vertebras. Because the chest section is the firmest and the least active of all, six chest vertebras in our model replaced 12 real chest vertebras. In other words, two (appropriately enlarged in size and mass) chest vertebras in our model replaced one real chest vertebra. The discussed model was created with the help of MBS Working Model 2D. Further in the work, the proposed biochemical human model will be used in human-seat-suspension (of vehicle) simulation studies. In the model, a seat without a back support was used. Therefore, a less favourable then a real instance was used. This was done for a purpose. The purpose was to create for the steering model the most difficult situation, which could generate the most data for a critical analysis

Now, lets to formulate the quality function Q. Lets describe it by relation:

$$Q = \sum_{j=1}^{4} \sum_{i=1}^{n} (f_{iw}\hat{x}^2_{ji} + f_{iw}\hat{y}^2_{ji}) \tag{3}$$

f_{iw} weight value of given frequency f_i; $\hat{x} / \hat{y}_{ij}$ -rms value effective in acceleration of head, loin area vertebras, chest area vertebras , neck area vertebras (j=1,2,3,4) for given f_i frequency in x/y direction

The steering synthesis (active part) of the principle of maximum Pantriagina (leading to the solution of the problem of Ricattiego, for linear layouts). Adequate parameters of firmness and suppression were chosen to minimise the factor Q in the half active part of the suspension. Rosenbrock's none gradient model was used here. In a simplified analysis the layouts (kinematics input at amplitudes of 4 mm and f=2-12 [Hz]) achieved a lesser quality index (more then two times in comparison to clasical suspension).

4. CONCLUSION

The described method and simulation results presented in the work for the analysis of the human factor in the man-vehicle-environment model seem to be very promising. The definite advantage of the method is its ability of taking account of „experts knowledge". Of course, the question as to how to verify the volume of the output data still remains open for discussion. The authors are currently conducting studies at the Institute of Aerial Medicine in Warsaw, Poland to verify the results. A separate problem, to which a general solution was drawn in the work, is the fitting of parameters of suspension (mostly the suspension of the driver's/mechanist's seat) to individual human character traits. It seems that the present level of technology would allow for the synthesis of such suspensions. Undoubtedly, their introduction into vehicles or aircrafts would give a rise to a new standard in the field. Besides the steering synthesis, which was mentioned earlier, it also requires a synthesis of measurement layout for measuring vibrations in particular parts of the human body.

ACKNOWLEDGEMENTS

This research was supported by the State Committee for Scientific Research (KBN) - grant no. 5T12C 061 22

REFERENCES

1. Griffin, M.J.: Handbook of Human Vibration, Academic Press, 1966.
2. Grabarek, I.: Ergonomic Diagnosis of the Operator_Railway Vehicle –Environment System, Warsaw University of Technology Press, 2003 /in polish/
3. Choromanski, W.: Simulation and Optimization in the Dynamics of Railway Vehicles, Warsaw University of Technology Press, 2000 /in polish/
4. Choromanski, W., Grabarek,I.:Fuzzy Sets in Modelling of Human Factor of Driver-locomotive-Environment System, Proceedings of the IAESTED International Conference, pp287-290 Pittsburg USA 2001.
5. Choromanski, W.: Application of Neural Network for Intelligent Wheelset and Railway Vehicles Suspension Designs , Proceedings of 14th IAVSD Symposium, Ann Arbor, USA,pp.87-99, 1995.
6. Mamdani, E.H. : Advances in the Linquistic Synthesis of Fuzzy Controllers, International Journal of Man-Machine Studies, Vol.8 pp.669-678, 1976.
7. Ronald, R., Yager D, P. Filev,P.,1994, Essentials of Fuzzy Modeling and Control, John Wiley & Sons,1994

Vehicle System Dynamics Supplement 41 (2004), p.272-281   © Taylor & Francis Ltd.

Driver perception sensitivity to changes in vehicle response

KATRIN STRANDEMAR[1] AND BORIS THORVALD

SUMMARY

This paper investigates driver sensitivity to vehicle ride response. The objective is to quantify driver perception limit for small changes in typical vehicle behavior. Cab motions are measured on a tractor semi trailer combination for different cab suspension settings. Measured motions are then realized in a motion simulator. Changes are made in small steps by combining measured signals from different settings. Driver sensitivity is finally quantified by time and frequency domain measures.

1. INTRODUCTION

The importance of improving vehicles dynamic properties is constantly growing in truck development. Various areas like active safety and driver environment are dependent on enhanced dynamic behavior. An essential tool in this process is the ability to quantify and grade these properties.

The dynamic response of a road vehicle may be divided into the sub-areas ride and handling. Where ride mainly refers to vehicle response upon road irregularities and handling refers to vehicle response to driver controlled inputs. Today the most common method to determine vehicle ride and handling qualities is by subjective rating tests [1]. Test drivers ratings will conclude which vehicle that has preferable dynamic behavior. However, the complex connection between ratings expressed by test drivers and explicit design parameters makes improvement work difficult. Lack of repeatability in subjective testing is furthermore a problem since test drivers change their acceptance level over time. A long time goal is therefore to develop objective measures for driving impressions [2].

Since decisions regarding vehicle dynamics e.g. chassis and cab suspension tuning, are based upon subjective evaluation, test driver perception sensitivity is one factor that will indicate how small changes that are meaningful to test. If the test driver's sensitivity to changes in vehicle response were well known effort could be spared in the iteration process for optimal parameter values.

In most existing objective evaluation methods for ride and handling, subjective ratings are one building block, e.g. linear regression models [3,4], this because it is crucial that the objective measures correspond to how most drivers apprehend the vehicle. A high correlation between vehicles with large dynamic differences and

---

[1] *Address correspondence to*: Scania CV AB, Department Vehicle Dynamics, SE-151 87 Södertälje, Sweden. Tel: +46 (0)8 553 819 45 Fax: +46 (0)8 553 856 04; E-mail: katrin.strandemar@scania.com

subjective ratings is easy to achieve but of little value in vehicle development where differences between design iterations often are small. Furthermore, if the subjective ratings do not originate from vehicle characteristics due to driver sensitivity being lower than the induced difference in response, problems with the objective measures will arise.

The objective of this work is to establish an effective test procedure to measure driver perception sensitivity to parameter changes affecting vehicle ride.

2. SIGNAL DETECTION THEORY

Signal detection theory [5], which is part of the science branch psychophysics, deals with human's ability to sense external stimuli in the form of e.g. audio or visual signals. Sensory thresholds are established by presenting a stimulus of different intensity to test persons while asking whether they can sense it or not. Two types of thresholds may be established; *absolute threshold*, which is the lowest intensity that can be detected and *difference threshold*, which is the intensity difference required for discrimination of two stimuli.

		STIMULUS present	STIMULUS absent
ANSWER	yes	hit	false alarm
	no	miss	correct rejection

Fig. 1. Possible outcomes from a trial in a detection test.

When presenting a stimulus of certain intensity at different occasions the test subject will typically sense it in some but not all cases. The perception threshold is therefore commonly defined as the intensity when the test subject can detect or discriminate the stimulus in 50 % of the cases. Each time the test subject indicates detection or discrimination without it being present must however also be considered. Otherwise a strategy to give positive answer in all cases would result in illusive low perception threshold. Depending on if stimulus is present or not and the test subject answer, there are four possible outcomes, fig. 1.

2.1 INTERNAL RESPONSE

In the process when the test subject takes decision to answer yes or no there are two factors contributing to uncertainty, internal and external noise. The surrounding environment defines external noise. Internal noise comes from the fact that the neural response in the brain (internal response) is noisy and variable. Identical stimulus intensity will thus result in different internal response at different occasions. Some internal response will also result when no stimulus is present. Assuming that internal response for noise alone and stimulus plus noise follow Gaussian distribution, fig. 2 may be used to illustrate the decision making process.

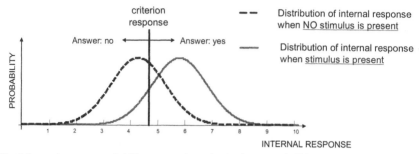

Fig. 2. Internal response probability curves when stimulus is present (solid) and no stimulus is present (dashed.).

When the internal response is above a certain level test subject will answer yes. Different test subjects however use different criteria for this decision. This is illustrated by the position of the vertical line *criterion response*. The ratio between hits and false alarms determines how separated the two distributions are, fig. 3.

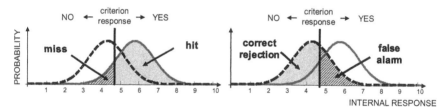

Fig. 3. Illustration of the probability for hits and false alarms given a distribution and criterion response.

If the distributions for noise alone and stimulus plus noise overlap each other no criteria choice can result in 100% hits and 0% false alarms. A stimulus with higher intensity will shift the stimulus plus noise distribution to the right, thus separating the two curves. Only when the two distributions are fully separated a 100% hit rate combined with 0% false alarm is possible.

By calculating *Receiver Operating Characteristic (ROC)*, possible choices of criterion for the test subject is illustrated in a single graph, fig. 4. The ROC curve is generated by shifting the criterion response line from right to left while looking at the ratio between hits and false alarms.

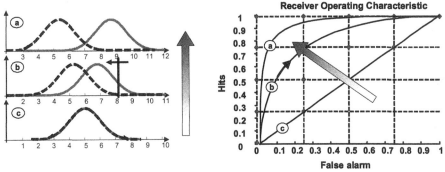

Fig. 4. Generation of receiver operating characteristic.

If the two distributions only have small overlap (fig. 4, case a) the ROC curve will have a steep beginning and a flat ending. It is thus possible to combine high percentage hits with low percentage false alarms. When the two distributions coincide (fig. 4, case c) the hit rate will be equal to the false alarm rate, independent of criterion choice.

2.2 PSYCHOMETRIC FUNCTION

A common way to illustrate sensitivity thresholds is by psychometric functions. These functions illustrate percent of cases for which a certain stimulus intensity is detected, fig. 5. In this work the *percent detections* is defined as the percent of hits when no false alarms occur, fig. 5. This means that the ability to detect differences is based on the distribution alone and not the choice of criterion response position. Comparison may thus be made between different test persons.

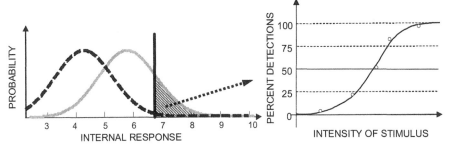

Fig.5. Psychometric function and definition of percent detections used in this work.

3   VEHICLE MEASUREMENTS

Human vibration sensitivity has, when it comes to fictitious signals, been investigated in earlier research [6]. This information is however difficult to apply in the case of complex vehicle motion. This work therefore focuses on sensitivity to changes in typical vehicle behavior.

Measures of cab accelerations are performed while driving a tractor semi trailer combination on a special track at Scania's test course. Each recording is 20 seconds long and consists of a number of transients with relatively high acceleration level during the whole sequence (*peak* $[m/s^2] = 8(x); 10(y); 7.5(z)$) and (*rms* $[m/s^2] = (1.9(x); 1.9(y); 1.5(z))$).

The sensors, three accelerometers and a gyroscope are positioned under the seat to measure the 6-degree of freedom motion of the cab. This measurement point is chosen since it corresponds to the motion generation point in the simulator and makes it possible to make corresponding measurements in the simulator, eliminating recalculation for comparison.

4   SIMULATOR TESTS

The measured cab motions are recreated in a motion simulator, fig. 6. Test drivers are seated in the same position as in the actual vehicle with the same driver seat. A steering wheel is also used to achieve the correct sitting position.

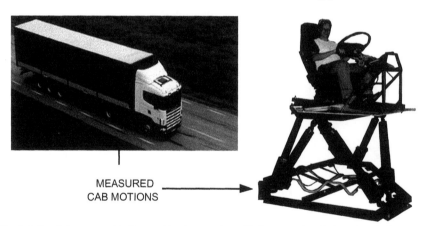

MEASURED
CAB MOTIONS

Fig. 6. A Scania tractor semi trailer combination together with the servo electrical motion platform used to generate motion for sensitivity measures.

4.1 COMBINING TIME SIGNALS TO GENERATE SMALL DIFFERENCES

To generate vehicle motion with small differences for the sensitivity tests, signals are varied using different combinations of two measured vehicle configurations, fig. 7. The utilized procedure enables continuous transition from the first to the second vehicle configuration and is possible to apply when time signals are synchronized.

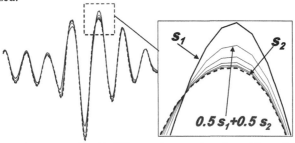

Fig. 7 Position signals from two trucks with different cab suspension, changed in small steps.

To simply scale the signal with a constant is not preferable since it does not correspond to how vehicle motion changes with different parameter settings. Design iterations rarely decrease amplitudes for all frequencies, but amplitude distribution change e.g. when reducing a single disturbing frequency. Sometimes it is just the transient's curve form or the phase lag between motions in different directions that make a vehicle more or less comfortable.

4.2 GENERATION OF PHASE EQUIVALENT SIGNALS (WITH DIFFERENT ENERGY)

A commonly used measure to quantify driver comfort is Power Spectral Density (PSD) of measured accelerations. In order to investigate to what extent PSD spectra capture differences in vehicle behavior, a virtual time signal is created. Starting with measured signals from two different vehicles, the virtual signal is created by combining the amplitude contents from signal 2, with the phase information from signal 1. This is achieved by scaling signal 1's amplitude spectrum with $H(f)$, eq. 1a. The virtual signal will thus have a PSD spectra close to signal 2, eq. 1b and be phase equivalent with signal 1.

$$H(f) = \sqrt{\frac{PSD_2(f)}{PSD_1(f)}} \tag{1a}$$

$$PSD_{new}(f) \approx PSD_2(f) \tag{1b}$$

4.3 TEST PROCEDURE

4 test persons; two trained test drivers, one inexperienced and one with big driving experience are used. Perception sensitivity of each person is investigated separately. All test persons start at the same difference intensity, depending on the result the difference intensity is either increased or decreased. All test drivers judge signals generated according to both 4.1 and 4.2

Table 1. Typical test sequence with random order of pairs used at one intensity level.

PAIR	ANSWER*	RESULT
2,2	no	corr. rej
1,1	no	corr. rej
2,1	yes	hit
2,1	no	miss
1,1	no	corr. rej
1,2	yes	hit
2,2	yes	false al.
1,2	yes	hit

*Did you feel a difference? (yes/no)

Test persons compare signals in eight pairs at each intensity level, table 1. Four out of eight pairs consists of two different signals and the remaining four consists of identical signals. The eight signal pairs are presented in random order.

5   DIFFERENCE MEASURES

In order to quantify differences in vehicle response for different configurations two measures are created. Since in this work all response differences are generated around a working point, Weber fractions [7] should have little influence on the result. In short Weber fractions may be illustrated as, if a person can barely detect the difference between something with intensity 10 and 10.1 the difference detection level around 20 will be 0.2. Both utilized measures are divided with the signal intensity to make them more applicable to other working points according to Weber fractions. This has however not yet been evaluated.

5.1 FREQUENCY DOMAIN DIFFERENCE MEASURE

In frequency domain the difference measure is calculated from PSD spectra of motion accelerations, fig. 8. and eq. 2. This measure will reflect changes in stationary vibration levels

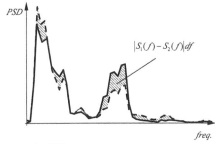

Fig. 8. Difference calculated from the PSD.

$$\Delta'_{12} = \frac{\int\limits_{0}^{F\,\text{max}} |S_1(f) - S_2(f)| df}{\int\limits_{0}^{F\,\text{max}} \frac{1}{2}(S_1(f) + S_2(f)) \cdot df} \tag{2}$$

5.2 TIME DOMAIN DIFFERENCE MEASURE

Human experience seems to put a lot of weight on transients [8]. The difference in time domain between two signals may therefore be calculated considering only the part of signals above a certain magnitude, fig. 9 and eq. 3.

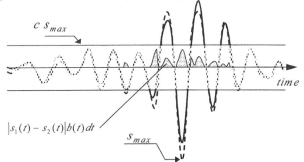

Fig. 9. Difference calculated from time domain capturing the transient differences and differences in curve form.

$$\Delta'_{12} = \frac{\int\limits_{0}^{T} |s_1(t) - s_2(t)| \cdot b(t) dt}{\int\limits_{0}^{T} \frac{1}{2} |s_1(t) + s_2(t)| \cdot b(t) \cdot dt} \tag{3a}$$

$$b(t) = 1 \quad c \cdot s_{\text{max}} < \frac{1}{2}(s_1(t) + s_2(t)) \tag{3b}$$
$$b(t) = 0 \quad \text{otherwise}$$

$$c = \frac{rms_1 + rms_2}{2 \cdot s_{\text{max}}} \tag{3c}$$

In fig. 9 it is illustrated how the difference between signal s_1 and s_2 is calculated. For both frequency and time domain measure one Δ is calculated in each direction and then combined, according to eq. 4.

$$\Delta = \sqrt{\Delta_x^2 + \Delta_y^2 + \Delta_z^2}$$

(4)

6 RESULTS

The rate of *false alarms* and *hits* for each stimulus intensity and each person will determine the detected differences in percent according to 2.1. This information is combined with either the frequency- or time difference measure and used to create psychometric functions describing each person's sensitivity threshold to changes in vehicle response. A good difference measure should be high if humans are sensitive to the applied difference and low if humans are insensitive to the motion difference.

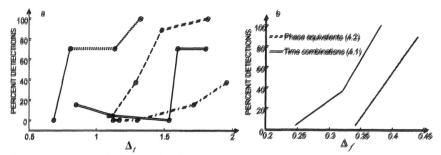

Figure 10 a) Psychometric functions based on time difference measure. b) Psychometric functions based on frequency difference measure.

The time difference measure capture all types of different signals, energy equivalents, time mixes and ordinary vehicle alterations. These tests resulted in four different psychometric functions, one for each test person, fig. 10a.

In fig. 10b, the dashed line represent signals generated with the phase equivalent method presented in 4.2. From this test it is possible to see that test persons have more difficulties separating signals with the same phase information, in spite of equally large differences in the PSD- spectra.

7 CONCLUSIONS

The presented methodology provides measures to investigate and quantify driver perception sensitivity to changes in vehicle response. In this work the smallest perceived difference starting from measured typical vehicle motion is examined.

Presented measures for driver perception sensitivity are suited for differences in real vehicles, first to be used in the development process when tuning vehicle parameters, secondly as information when objective evaluation methods are developed.

One advantage with the utilized methodology to recreate motion in a simulator is that vehicle measurement strategies may be validated against simulator subjective ratings. This will make subjective rating tests and vehicle measurements both more accurate and reliable, and also contribute in the search for objective measures of driving impressions.

The time difference measure corresponds better than the frequency difference measure to human's ability to detect differences. Perception threshold information needs to be used in vehicle development since its only meaningful to test design iteration steps larger than driver sensitivity. Results from tests with phase equivalent signals show that PSD spectras do not capture enough information to use as base for objective evaluation of vehicle response.

ACKNOWLEDGEMENTS

The authors would like to express gratitude to Bror Tingvall and Andi Wijaya at Luleå Technical University for their assistance during the simulator test and colleagues at Scania CV AB.

REFERENCES

1. Subjective Rating Scale for Vehicle Handling. SAE Recommended Practice mars. 98: SAE J1441
2. Ushijima, T. and Kumakawa, S.: Objective Harshness Evaluation. SAE 951374
3. Lee, S. K. and White, P. R.: Application of wavelet analysis to impact harshness of a vehicle. ImechE 2000 Proc Instn Mech Engrs Vol 214 Part C
4. Schröder, F. and Zhang T.: Objective and Subjective Evaluation of Suspension Harshness. SAE 97A2105
5. Heeger, D. : Course Notes for Introduction to Perception, Stanford University couse in psychology. ", D. Heeger, Stanford University 1999.
6. Griffin, M. J. Handbook of human Vibration. ISBN 0-12-303040-4
7. Paul Andrew Millman : Haptic Perception of Localized Features. Dissertation, Northwestern University, Field of Mechanical Engineering, Dec 1995
8. Giuliano, F. and Ugo, A.: Experimental Assessment of Ride Comfort: A Statistical Approach. SAE 92A280

A study of active steering strategies for railway bogie

SHUIWEN SHEN[1], T.X.MEI[1], R.M.GOODALL[2], J. PEARSON[2] AND
G.HIMMELSTEIN[3]

SUMMARY

This paper presents a fundamental study of active steering for a railway bogie and is focussed on the development of control strategies to provide a perfect curving performance. A number of active steering strategies are considered. One is to control the wheelset and the bogie to follow desired yaw movements on curves; the second is to control the lateral motion of the wheelset to take the pure rolling line; the third is to control the actuators such that low frequency zero force in the primary suspensions is achieved; and finally a control method using relative motions between the two wheelsets is also investigated to improvement of the performance robustness against parameter variations. A comprehensive assessment of the steering strategies is carried out.

1. INTRODUCTION

A trade-off between the high speed stability and low-speed steering for railway vehicles has been a difficult challenge for railway engineers for many years [1, 2, 3]. Active control [3, 4] may provide a solution for this problem as it offers much great design flexibility. There have been a number of studies on the active control/steering of railway vehicle, e.g. active yaw relaxation approach [5], active control of the lateral position and/or the yaw moment of the wheelset control [6], control of the yaw moment between the wheelsets and the bogie [7]. This study has taken an approach to develop the vehicle stabilisation and steering as two sub-systems and then integrate them together. As part of the study [8], an active controller is designed and implemented which increase the critical speed of a bogie vehicle from *100km/h* to over *300km/h* in test without the use of secondary yaw dampers. This paper presents the second part of the study, which investigates and assesses a broad range of active strategies for the bogie vehicle (i.e. pure rolling) without compromising the stability.

This paper is organised as follows. Four active steering strategies are presented in Section 2, and a description of a bogie vehicle is given in Section 3. After the strategies are evaluated by numerical experiments in Section 4, Section 5 gives conclusions and suggests the future work.

2. CONTROL STRATEGIES

[1] School of Electronic and Electrical Engineering, The University of Leeds, Leeds, LS2 9JT, UK.
  Tel: 44(0) 113 343 2066, Fax: 44 (0) 113 343 2032, E-mail: s.shen@ee.leeds.ac.uk
[2] Department of Electronic and Electrical Engineering, Loughborough University, Loughborough, LE11
  3TU, UK.
[3] Bombardier Transportation, Siegen, Germany.

The lateral forces on a railway vehicle on curves due to cant deficiency are balanced partly by the contact geometry (contact angle etc) and partly by the lateral creep forces at the wheel/rail contact points. In general, the presence of longitudinal creep forces is undesirable (except for traction) and it can only result in an inappropriate steering action. Ideally, perfect steering requires all wheelsets to have a desired angle of attack on curves just enough to balance the remaining centrifugal forces, and a desired lateral displacement to ensure the pure rolling of the wheelset. A number of steering strategies are possible as presented below.

Strategy 1: Control of the yaw movement of the wheelsets and the bogie. It can be readily shown that perfect steering can be achieved if the angle of attack for two wheelsets (in addition to the radial angular position) can be controlled to be equal and the bogie to be in line with the track on curves. This idea can be implemented by controlling the position of each actuator such that the wheelset forms an appropriate yaw angle with respect to the bogie [4]). This strategy may be implemented using a feedforward approach (refereed to as **Ff** in the study). The required yaw angle is determined by the track curvature, cant-deficiency and the creep coefficient, accurate knowledge of which obviously affects the effectiveness of any feedforward strategy. In addition, the longitudinal (yaw) stiffness of the primary suspension and the stiffness that exists in the steering linkages will have a significant influence on the achievable performance. On the other hand, a main advantage of the feedforward control approach is that it does not interfere with the vehicle stability and hence the control design can be decoupled. To enhance the steering performance, which may be degraded due to inaccuracy information of parameters such as stiffnesses, the strategy can also be implemented using a feedback control strategy. The feedback signals (yaw angles of the wheelsets and bogie) can be derived from a measurement of longitudinal and lateral movements of the primary suspension (**Fb1**). Information of the track curvature and cant-deficiency will still be required to calculate the yaw angle references of the feedback control.

Strategy 2: Control of lateral motions of wheelsets. This strategy is based on the idea that zero longitudinal creep force implies equal lateral creep forces for all wheelsets. Thus, it is possible to apply a torque in the yaw direction to each wheelset so as to achieve pure rolling (**Fb2**) [2]. The control strategy obviously relies on the assumption that the wheel conicity at the contact point is known for an accurate prediction of the pure rolling line, which may be difficult to achieve for profiled wheels. In addition to the track curvature, this strategy reduces the sensing requirements to the lateral displacement only, but this is a difficult parameter to measure in practice.

Strategy 3: Control of the primary suspensions. The longitudinal stiffness of the primary suspension forces the wheelsets away from the pure rolling and hence causes imperfect steering. It is therefore possible to apply active torque in a yaw torque such that it cancels out the effect of the stiffness on curves (at relative low frequencies). This can be realized by either measuring the relative yaw angle between the individual wheelset and the bogie and compensating for the primary forces **Fb3**) or by controlling the forces and/or moments [1,2] of the primary suspension. However cautions are necessary to ensure that it will not adversely affect the vehicle stability by restricting the action to low frequencies only, i.e. below the kinematic frequencies.

Strategy 4: Control of the relative movement between wheelsets. This is derived from the observation that zero difference in wheelset angle of attack and in lateral displacement leads to perfect steering. This strategy can be realised by controlling the two actuators such that the lateral suspension forces (or movement) at the two wheelsets are equalised and the ratio of the two control torques matches the that of the suspension yaw movements of the two wheelsets (referred as to **Fb4**). This approach requires the measurement of the movements (or forces) of primary suspension in the longitudinal and lateral directions, no knowledge of the track is necessary.

Extra benefits and/or improved performance can be gained by combining some of the strategies together. For example, an analysis indicates that combination of **Ff** and **Fb4 (Comb1)** reduces the undesirable high gains required for **Fb4**, and that combination of **Fb3** and **Fb1 (Comb2)** improves the dependency of **Fb3** on the knowledge of suspension stiffness.

3.   SYSTEM DESCRIPTION

This study concentrates on a bogie vehicle with two actuators. Figure.1 shows a half-vehicle plan view model, in which two actuators are mounted between wheelsets and the bogie frame in parallel with the normal primary suspension components. The actuation is applied via steering linkages, which have the effect of a series stiffness. The effects of body yaw are not thought as essential and will be taken into account for a later stage of a comprehensive evaluation of the performance.

The actuator, consisting of a DC motor and a gearbox, can be manipulated to apply a demanded torque at the wheelset and the bogie, or to realize a demanded relative position between the wheelset and the bogie. In view of integrating with the stability controller, only the former case is considered in this paper. The dynamics of the actuator are modelled and a torque control loop with an integral action is also designed for the implementation of active steering.

In summary, a simulation model, which is used to assess the active strategies in the next section, is of order twenty-two, and the key parameters are listed in Table.1.

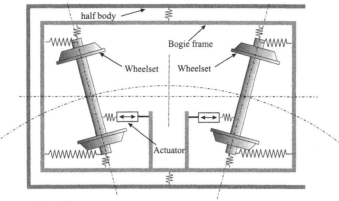

Fig.1 the vehicle configuration.

Table 1. Key parameters

Cant angle of the curve	104.7 *mrad*
Half gauge of wheelset	0.75 *m*
Half longitudinal spacing of the wheelsets	1.0 *m*
Half lateral spacing of primary suspension	1.225 *m*
Longitudinal primary stiffness	1.9 *MN/m* (per wheelset)
Lateral primary stiffness	9.4 *MN/m* (per wheelset)
Longitudinal primary damping	12 *kN s/m* (per wheelset)
Lateral primary damping	24 *kN s/m* (per wheelset)
Lateral secondary stiffness	490 *kN/m* (per bogie)
Lateral secondary damping	40 *kN s/m* (per bogie)
Lateral creep coefficient	10 *MN* (nominal)
Longitudinal creep coefficient	10 *MN* (nominal)
Travel speed	50 *m/s* (nominal)
Wheel conicity	0.2 (nominal)
Wheelset, bogie, half body mass	1376 *kg*; 3477 *kg*; 17230 *kg*
Wheelset, bogie inertia	766 *kg m²*; 3200 *kg m²*
Wheel radius	0.45 *m*
Actuator linkage stiffness	5.8 *MN/m*

4. SIMULATIONS AND ASSESSMENTS

The performances of the active different steering strategies are evaluated using a set of criteria [7]. Among them, the indices for the creep forces are considered particularly, representing the deviation from equalizing forces in lateral direction and the zero forces in longitudinal direction, i.e. deviations from the "perfect steering" condition. A larger lateral index would thus indicate that the lateral creep forces of leading and trailing wheelset are more unbalanced, while a larger longitudinal index represents undesirable longitudinal creep forces. On the other hand, the controllers of the active strategies are developed based on the accurate knowledge of the vehicle, which is not always possible. Thus, it is important to assess the steering performance under various uncertainties. The uncertainties assessed in this paper consists of a 50% decrease in creep coefficient, an 80% increase in the stiffness of the primary suspension and the actuator stiffness, and a nonlinear wheel profile.

Table 2. Steering performance (%) (V=50 m/s, Curve R= 1250 m.)

		Nominal concity=0.2	Creep coeff. Dec. 50%	Susp. stiffness Inc. 80%	Steering link Stiffness Inc 80%	Wheel profile
Ff	Long.	7.2%	15.1%	24.1%	7.2%	8.4%
	Lateral	4.0%	9.3%	10%	3.9%	5.2%
Fb1	Long.	6.2%	17.1%	9.1%	6.1%	11.6%
	Lateral	4.2%	4.5%	5.2%	4.2%	7.4%
Fb2	Long.	7.2%	6.8%	11.5%	17.1%	91%
	Lateral	3.8%	4.6%	5.1%	3.8%	56%
Fb3	Long.	7.9%	11.0%	25%	7.8%	8.6%
	Lateral	4.0%	6.3%	9.5%	4.0%	4.8%
Fb4	Long.	2.6%	1.9%	3.2%	2.5%	14.3%
	Lateral	3.3%	3.2%	3.4%	3.3%	9.8%
Comb 1	Long.	7.5%	8.9%	7.6%	7.5%	5.1%
	Lateral	4.2%	7.2%	4.5%	4.2%	4.1%
Comb 2	Long.	2.3%	2.5%	3.4%	2.3%	3.8%
	Lateral	3.1%	3.0%	3.5%	3.1%	4.1%

Table.2 gives a summary of the results from the selected control strategies. It compares the effectiveness of those approaches in the nominal condition, and with system uncertainties. Firstly, with the accurate knowledge of the vehicle, all the strategies are able to steer the vehicle correctly on a steady curve- an example time history is shown in Fig.2. The different vales of the longitudinal and lateral creep force indices are mainly contributed from the transition behaviours, which for the different strategies are quite different – this can be seen in the "Nominal" column of Table.2. Secondly, this table indicates that the strategies of **Ff** and **Fb3** are quite sensitive to the suspension stiffness (24.1% and 25% of the longitudinal indices respectively) while **Ff** and **Fb1** to the creep coefficient (15% and 17.1% of the longitudinal indices respectively). Also, **Fb2** is not a particular good strategy for the wheels with non-linear profiles due to the difficulty of finding the pure rolling line. Apart from not working well with nonlinear profile, **Fb4** is rather a robust strategy over parameter variations. Thirdly, the improvements from the combined strategies (**Comb1** and **Comb2**) are evident in the table. The application of the strategy **Comb1** reduces the longitudinal and lateral indices of **Fb3** with respect to the suspension stiffness variation from 25% and 9.5% to 7.6% and 4.5% respectively.

Similarly, the superiority of **Comb2** over **Fb4** is clearly shown with respect to the performance with a nonlinear profile.

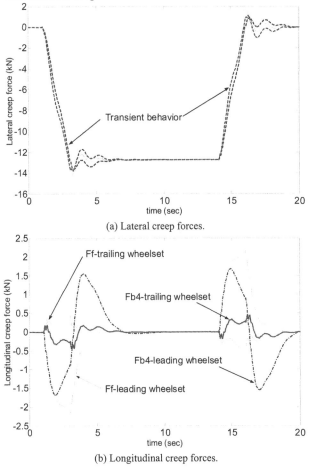

(a) Lateral creep forces.

(b) Longitudinal creep forces.

Fig.2. Creep forces for different strategies under nominal condition.

Time history simulation results are also examined in the study. Fig.3 shows the creep forces of the control strategies when the exact knowledge of creep coefficients is not known. It is clear that the strategy **Ff** and **Fb1** gives inappropriate steering on curves, which is consistent with Table.2 , but the reasons are different. **Fb1** is able to produce an equal lateral force, but it makes the longitudinal force even worse. On the other hand, though **Ff** is not able to steering the wheelset in either direction, it does not produce as high longitudinal forces as those of **Fb1**. This explains why the index shown in Table 2 for **Ff** is higher in lateral direction, whereas for **Fb1** it is higher in the longitudinal direction. Inappropriate steering of **Fb1** is because the desired relative yaw angle between the wheelsets and the bogie depends on the creep coefficient.

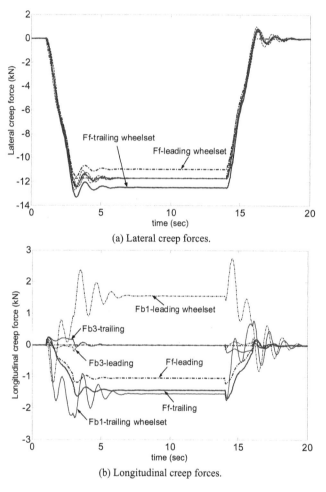

(a) Lateral creep forces.

(b) Longitudinal creep forces.

Fig.3. Creep forces for different strategies with creep coefficient variation.

Figure 4 gives a time historical assessment of the performance when the exact knowledge of suspension stiffness is not known. The results show that **Ff** and **Fb1** don't perform well on curves, in particular giving a much worse longitudinal force at the leading wheelset (more than 25% of the desirable lateral creep force). Consistent with what shown in Table 2, the steering behavior of **Ff** and **Fb1** are rather similar, both resulting in an uneven lateral force and worse longitudinal forces.

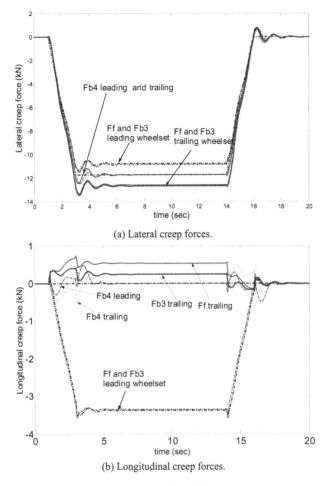

(a) Lateral creep forces.

(b) Longitudinal creep forces.

Fig.4. Creep forces for different strategies with suspension stiffness inaccuracy.

So far, the assessment is conducted at a higher vehicle speed, but the lower speed steering is more critical under the same cant deficiency. The study on a low speed tight curve (V=25m/s and R= 310m) has found that the actuators are working very near the pre-set torque limit (30Nm). The torque limit can be reached in some conditions, e.g. if there is a decrease in creep coefficients or an increase in suspension stiffness. Consequently there is a difference between the torque demand from the steering controller and the torque output from the actuator. Moreover, the use of an integral term within the actuator torque causes a so-called 'wind-up' effect. It is clear therefore that the full potential of the active steering can only be achieved if the limit is increased. For the bogie under investigation, either a more powerful actuator is required or the primary longitudinal stiffness must be reduced if it is to be used on very low speed curves. Based on this assumption of more powerful actuators are employed, the performances of the above mentioned steering strategies are assessed - see Table 3.

Table 3. Steering performance (%) (V=25 m/s, Curve R= 310 m.)

		Nominal concity=0.2	Creep coeff. Dec. 50%	Susp. stiffness Inc. 80%	Steering link Stiffness Inc 80%	Wheel profile
Ff	Long.	25.2%	28.1%	85.6%	24.2%	25.9%
	Lateral	4.5%	10.9%	19.8%	4.3%	6.8%
Fb1	Long.	38.2%	36.6%	58.1%	36.8%	48.0%
	Lateral	17.9%	18.3%	26.5%	17.3%	25.8%
Fb2	Long.	25.9%	20.8%	41.6%	24.9%	268.1%
	Lateral	5.2%	7.4%	9.0%	5.0%	163.8%
Fb3	Long.	27.5 %	28.7 %	86.7%	26.5 %	28.2 %
	Lateral	4.3 %	7.4 %	15.2 %	4.3 %	6. 8%
Fb4	Long.	11.5 %	8.5%	13.5%	11.0 %	28.3 %
	Lateral	6.3%	5.7%	7.3%	6.1%	18.7%

In the nominal condition, all active steering strategies are able to achieve perfect steering on a constant curve. However, there are noticeable increases in the performance indices when compared with those from the high speed assessments (given in Table 2). An examination of time history simulations shows that the increases are mainly caused by a less orderly behaviour on transitions. Moreover, the sensitivity of **Ff**, **Fb1** and **Fb3** to the knowledge of parameters, and the dependency of **Fb2** and **Fb4** on the wheelset profile, are the same as on the high speed curve.

5. CONCLUSIONS AND FURTHER WORK

This paper presents a study on the active steering strategies for the bogie vehicle. These strategies can be implemented with different control approaches, e.g. the feedforward and the feedback approaches are applied for steering strategy one. Thus, this paper is the general steering strategy orientated rather that the particular control approaches.

Four different strategies have been proposed and assessed. The high and low speed assessments indicate that, 1. All strategies achieve perfect steering if the parameters of the vehicle are exactly known; 2. The strategy one (feedback realization) requires accurate knowledge of the creep coefficient, while the strategy three depends on the suspension stiffness; 3. The strategy two is particularly bad with the absence of wheel profile, and it also requires the knowledge of the curves; 4. The strategy four has a rather robust performance, but it doesn't work particular well with nonlinear profile due to a worse transition behaviour; 5. The appropriate combination of different strategies can improve the robustness of the steering action significantly due to model uncertainties. However, lower speed steering is particularly difficult, and transition behaviour goes worse.

The assessment so far has only focussed on the sensitivity to parameter uncertainties, but the assessment of the performance under various sensor configurations and the effect of inaccuracy and noise are equally important. Furthermore, some strategies require some signals which are either difficult or even not possible to measure. Either an estimator (Kalman filter) or an H_{inf} controller has to be developed; both solutions introduce extra dynamics into the system, and further studies are essential.

REFERENCES

1. Wickens, A.H., "Stability criteria for artculated railway vehicles possessing perfecting steering", Journal of Vehicle System Dynamics 7 (1978), pp. 165--182.

2. Goodall, R.M., Li, H., "Solid axle and independently-rotating railway wheelsets – A control engineering assessment of stability", Journal of Vehicle System Dynamics, 33(2000), pp. 57-67.

3. Mei, T.X. and Goodall, R.M., "Modal controllers for active steering of railway vehicles with solid axle wheelsets", Journal of Vehicle System Dynamics, 34(2000), pp. 24-31.

4. Illingworth, R. and Pollard, M.G., The use of steering active suspension to reduce wheel and rail wear in curves", Proc. Instn. Mech. Engrsium, 196(1982), pp. 379-385.

5. Shen, G., Goodall, R.M. "Active Yaw Relaxation for Improve Bogie Performance", Journal of Vehicle System Dynamics, Vol.28, pp. 273-282, 1997.

6. Perez, J., Busturia, J.M., Goodall, R.M., "Control strategies for active steering of bogie-based railway vehicles" Control Engineering Practice, Vol.10, pp. 1005—1012, 2000.

7. Shen, S., Mei, T.X., Goodall, R.M., Pearson, J., Himmelstein, G., "Active Steering of Railway Vehicles: A Feedforward Strategy and Assessment of Parameter sensitivities", European Control Conference, Cambridge, UK, 2003.

8. Pearson, J., Goodall, R.M., Mei, T.X., Shen, S., Kossmann, C., Polach, O., Himmelstein, G., "Design and Experimental Implementation of an Active Stability System for a High Bogie", IAVSD'03, Kanagawa, Japan, 2003.

Vehicle System Dynamics Supplement 41 (2004), p.292-301   © Taylor & Francis Ltd.

Effects of bogie centre plate lubrication on vehicle curving and lateral stability

HUIMIN WU AND JAMES ROBEDA[1]

SUMMARY

A combination of track tests, NUCARS™ simulations, and bogie performance detector data analysis has indicated that lubricating the centre plate of a bogie could reduce wheel/rail wear on curves. Centre plate lubrication is more likely a secondary factor for wheel/rail lateral forces in curving as compared to wheel/rail profiles, track lubrication, and other dominant factors. Higher friction at the centre plate can improve vehicle lateral stability on tangent track. Constant-contact side bearings can significantly improve vehicle lateral stability by producing higher resistant moment to the bogie yaw motion.

1.   INTRODUCTION

High wheel lateral forces have been observed on some freight cars with dry centre plates in revenue service. Investigators have often reported dry centre plates as one of the observations in derailment reports. Applying liquid lubricants may only reduce lateral forces under certain circumstances. However, this reduction may not last long. According to observations, the reductions only lasted a few weeks before wheel lateral forces again exceeded the threshold. While some railways are emphasizing centre plate lubrication, others do not lubricate centre plates at all to reduce the risk of vehicle hunting.

Effects of centre plate lubrication on vehicle performance have been an area of concern for the railroad industry for the last decade. Limited studies have been conducted. Wolf et al.[1] examined the wheel lateral force due to centre plate lubrication on articulated double-stack cars and found, via the NUCARS™ simulation, curving lateral force could increase with increasing centre plate friction coefficient at a certain speed and track lubrication condition. No wheel/rail wear was investigated in this study. A performance study by Shust and Urban[2] using NUCARS™ simulations showed that curving wheel lateral force can increase considerably when the centre bowl friction coefficient was above 0.6. The friction coefficient above 0.6 is generally considered as the result of a defective centre bowl/centre plate.

---

[1] *Address correspondence to:* Transportation Technology Center, Inc, Pueblo, CO 81001 USA. Tel.: (719) 584-0533; Fax: (719) 584-0770; E-mail: huimin_wu@ttci.aar.com

NUCARS™ is a trademark of Transportation Technology Center, Inc., a wholly owned subsidiary of the Association of American Railroads.

In 2002, Transportation Technology Center, Inc. (TTCI) conducted a comprehensive study at the request of the AAR's Railway Technology Working Committee to gain a better understanding of the effect of centre plate lubrication and to provide guidelines for centre plate lubrication practice and maintenance. This paper discusses the levels and the circumstances in which centre plate lubrication could affect vehicle curving and lateral stability for conventional freight vehicles (with two bogies) in this study. Vehicle dynamics simulations were made in conjunction with on-track tests. Important parameters, such as wheel/rail profiles, rail lubrication, and constant-contact side bearings (CCSB), have been considered. This paper also includes a discussion on current centre plate lubrication practices and materials used.

2. BOGIE STEERING AND RESISTANT TURNING MOMENT

For proper curve negotiation, the bogies of a vehicle must rotate relative to the carbody, as Figure 1(a) illustrates. The rotation is resisted by the friction force generated at the contact surface of the centre bowl and centre plate. Friction from side bearings can also contribute to rotational resistance. On the other hand, the resistant moment at the centre plate (and side bearings) is an essential element for controlling the bogie yaw motion, which in turn controls vehicle hunting.

For bogies to turn relative to the carbody, a sufficient turning moment must be generated from the wheel/rail interface to overcome the friction resistance in the centre bowl (and side bearings). The bogie turning moment includes two components: 1) the steering moment; and 2) the warp moment. See Figure 1(b). The steering moment is the result of longitudinal forces at the wheel/rail interface. The warp moment is the result of lateral forces at the wheel/rail interface.

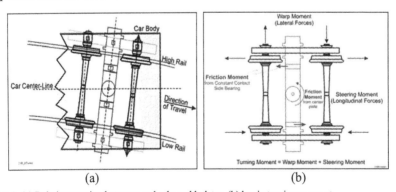

| (a) | (b) |

Figure 1. (a) Relative rotation between carbody and bolster, (b) bogie turning moment.

The distribution of vertical force at the centre plate may vary based on the contact situation between the centre plate and the centre bowl, which is influenced by contact surface conditions, loading conditions and bolster bending. Figure 2

293

displays the even distribution of the load (a), an approximate ring contact caused by the elastic banding of bolster (b), and the edge contact due to carbody roll (c).

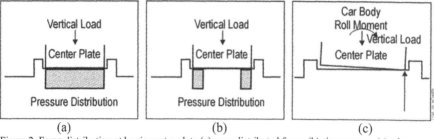

(a)   (b)   (c)

Figure 2. Force distribution at bogie centre plate (a) even distributed force, (b) ring contact, (c) edge contact.

The yaw-resistant moment, based on the Coulomb Friction Law, is a linear function of friction coefficient. For evenly distributed force on the centre bowl, it can be computed using Equation 1.

$$T = \int_{R_1}^{R_2} f * 2\pi r * r * dr$$

(1)

Where R_2 is the centre plate radius and R_1 is the centre hole radius for the evenly distributed force or the inner circle radius for the ring contact. The value f is the friction force of the unit area. f can be determined by the vertical load (P) and friction coefficient (μ) on the centre bowl (Equation 2).

$$f = \frac{P\mu}{\pi (R_2^2 - R_1^2)}$$

(2)

Uneven load distributions can occur as a result of carbody roll. In the extreme condition, the centre plate may only have an edge contact, as shown in Figure 2(c). When the entire vertical load is carried on the centre plate edge, the turning moment can be approximated as:

$$T = P\mu R_2$$

(3)

Due to the increase of the effective turning moment arm, the resistant moment under edge contact condition is higher than that of the evenly distributed cases, assuming the rotation is around the bogie centre. The increase ratio depends on the dimensions of the centre plate and the position of contact. For a 125-ton loaded hopper, the resistant moment produced by edge contact is about 1.5 times of that produced by the evenly distributed force (R_2=203 mm).

Further, if roller side bearings (RSB) contact one side of the carbody due to carbody roll, or if CCSBs are used, the resistant turning moment could be significantly higher due to the much larger moment arm. The total resistant moment depends on the load-sharing ratio between centre bowl and side bearings.

Figure 3 shows the comparison of resistant moment produced at one bogie of a 125-ton hopper (empty and loaded) under different friction coefficients at centre plate and side bearings. When the vehicle is equipped with CCSBs, each side of the CCSB has a pre-setting force of 20.5 kN with 635 mm of moment arm to the bogie centre. The resistant yaw moment produced only by the centre plate (without side bearing contact) is set as one. In the empty condition, about 40 percent of carbody load is shared by CCSB compared to 6 percent in the loaded condition. Therefore, the resistance moment of the empty car is more affected by the preloading of the CCSB and the friction level at the CCSB. As shown in Figure 3, compared to that measured without side bearing contact, the ratio of resistant moment with the CCSBs are generally around or below 1.5 under the loaded condition, and are much higher under the empty condition.

The centre plate contacting the vertical rim of the centre bowl can also contribute to the resistant moment. It likely occurs at the enter and exit spirals, during braking and under any conditions that cause the centre plate and centre bowl to have different velocities.

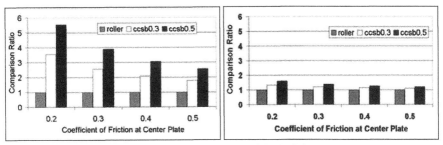

Figure 3. Comparison of resistant moment (left) empty car, (right) loaded car

3. EFFECT OF CENTRE PLATE LUBRICATION ON VEHICLE CURVING PERFORMANCE

3.1 WHEEL/RAIL LATERAL FORCE

A track test was conducted on a 125 ton loaded hopper equipped with roller side bearings. The relative yaw displacement between carbody and bolster was measured using string pots (Figure 4, left). The lateral forces were measured using a bogie performance detector installed at a curve with a 349-meter radius. The centre plate was dry (steel contact steel, $\mu \approx 0.5$) during the first day of testing and fully lubricated (liquid lubricant, $\mu \approx 0.1$) during the second day. The dry centre plate produced less relative displacement at the reverse curves (circled section, 349 m in radius) at the leading bogie (Figure 4, right). However, the average lateral forces of the leading axles measured at the middle of the circled section (Figure 5) showed small differences between dry and fully lubricated centre plates. In terms of percentage, the variations of lateral force on the trailing axles were larger from dry to fully lubricated centre plate. The actual values, however, were low.

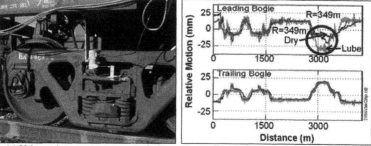

(a) Using string pots to measure relative yaw displacement between carbody and bolster   (b) Relative rotation measured

Figure 4. Relative yaw displacements measured at a reverse curve with 349-m radius for cars with fully lubricated and dry centre plates.

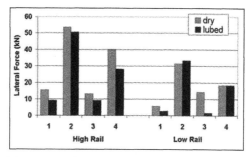

Figure 5. Average lateral force measured at the circled section (Axles 2 and 4 were the leading axles).

Figure 6 provides the simulation results for a 125-ton loaded hopper car with roller side bearings on a reverse curve 291 meters in radius. Similar to the track test results, varying friction at centre plate had little effect on the lateral force of the leading axle but a relatively large effect on the trailing axle on the reveres curve.

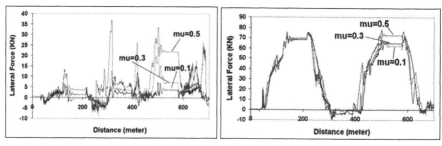

Figure 6. Wheel lateral forces under three levels of centre plate friction coefficient (μ=0.1, 0.3 and 0.5), left – leading wheels, right – trailing wheels

Figure 7 shows the wheelset angle of attack (AOA) measured in the test described in Figure 4. At the leading axle, the AOA varied from about 9 milliradians (mrad) from the lubricated centre plate to about 17 mrad for the dry centre plate. The AOA on the trailing axle varied from close to zero to about 13 mrad. Figure 8 may explain the variations of wheel lateral force and AOA observed in the test.[3] When the lateral creep forces were saturated (usually above 5 mrad of wheelset AOA), the lateral force would only result in a small change as the wheelset AOA increases. This was the situation at the leading axles. When the AOA increased from 0 to 13 mrad on the reverse curve at the trailing axle, a large increase in lateral forces resulted.

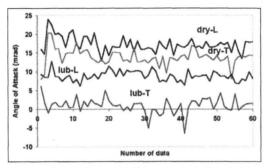

Figure 7. Wheelset angle-of-attack at leading and trailing axles; L=leading and T=trailing

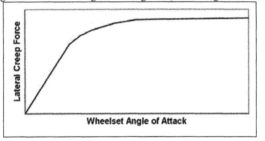

Figure 8. Lateral creep force versus wheelset angle-of-attack.

The wheel/rail profile is one of the dominant factors that has a more pronounced effect on lateral force than centre plate lubrication. Hollow-worn wheels tend to significantly reduce the bogie steering ability and result in a high AOA. Figure 9 shows the simulation results for the same vehicle and track conditions as described for Figure 6 with three types of wheel profiles: the AAR-1B wheel, tapered worn wheels, and 3-mm hollow worn wheels. The hollow-worn wheels produced higher wheelset AOA on both leading and trailing axles and produced much higher lateral forces on the trailing wheels only. Again, referring to Figure 8, the AOA at the leading axles varied from 7 mrad for the AAR-1B wheels to 15 mrad for the hollow-worn wheels. The AOA increased from 0 to above 10 mrad for the trailing axle, resulting in a high lateral force at the trailing wheels.

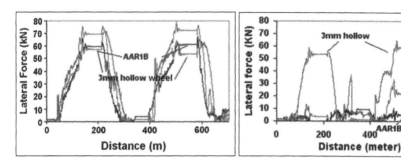

Figure 9. Wheel lateral force under three types of wheel profiles: AAR-1B, tapered worn, and 3-mm hollow-worn wheels, μ-centre plate = 0.5, (left) leading wheels, (right) trailing wheels.

Rail lubrication is another important factor that can significantly affect wheel lateral force when curving. The simulation results are described in *Reference 4* produced by the author. Track tests and observations resulting in a significant amount of wayside force data have validated the simulations.[5]

This study concluded that under normal vehicle/track conditions, the turning moment produced at the wheel/rail interface generally could make the bogie steer and the centre plate lubrication only has a small effect on wheel lateral forces. Dry centre plates require higher turning moment and may cause some increase in wheel lateral force at trailing wheels, especially on the reverse curves (Figure 9, left).

Hollow-worn wheels and differential track lubrication tend to significantly reduce the bogie turning moment at the wheel/rail interface, leading to increased lateral forces on both leading and trailing axles. Reducing the centre plate friction may reduce wheel lateral force under such conditions when they act separately (Figure 9, centre).

Combinations of multiple, undesired vehicle and track conditions—e.g., incompatible wheel/rail profiles, improper track lubrications, and very sharp curves— could cause high wheel forces on both leading and trailing axles. Lubricating the centre bowl showed little or no improvement to wheel lateral forces under those conditions because the other parameters dominated the situations (Figure 9, right).

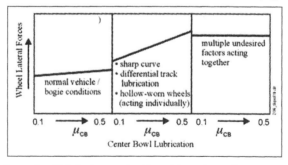

Figure 10. Effect of centre bowl lubrication on wheel lateral forces in curving.

3.2 ROLLING RESISTANCE

Equation 4 defines the vehicle rolling resistance, also known as the wear index. It is computed from the creepages (γ and ω) and creep forces (T and M) on all eight wheels of a vehicle.

$$Rolling\ Resistance = \sum_{n} T_x \gamma_x + T_y \gamma_y + M_z \omega_z \tag{4}$$

The previous section has shown that dry centre plates resulted in higher AOA. Equation 5 shows that the lateral creepage increases as wheelset AOA (ψ) increases under steady-state curving.[6] Also, dry centre plates resulted in higher longitudinal force and creepage because higher turning moment is required to make the bogie steer. Therefore, dry centre plates would produce higher rolling resistance. Figure 11 shows the results of rolling resistance for the same simulation illustrated in Figure 6.

$$\gamma_y = (\psi - \frac{\dot{y}}{V}) \sec(\delta) \tag{5}$$

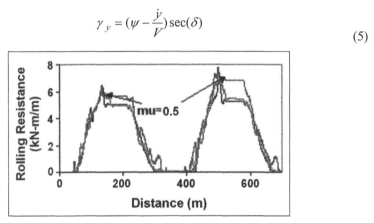

Figure 11. Rolling resistance for three level of centre bowl friction (μ = 0.1, 0.3, and 0.5) on reverse curves of 291-m radius.

In this case, the dry centre plates (μ=0.5) produced about 10 to 15 percent higher rolling resistance than lubricated centre plates (μ=0.1 and 0.3). Since rolling resistance directly relates to wheel and rail wear, dry centre plates can result in higher wheel/rail wear on curves than that of lubricated centre plates.

4. EFFECT OF CENTRE PLATE FRICTION ON VEHICLE LATERAL STABILITY

Based on the simulation results, the centre bowl friction level has considerable effect on vehicle lateral stability when vehicles are equipped with roller side bearings. When vehicles are equipped with CCSBs, the load shared at the CCSB has a bigger moment arm that produces higher resistance moment to bogie yaw motion. Therefore, CCSB can significantly improve vehicle lateral stability. When vehicles are equipped with CCSB, the centre bowl friction has only a small effect on the hunting speed. Compared to the vehicle with roller side bearings, the CCSBs increased the average hunting speed about 20 km/h for the simulated cases (Figure 12).

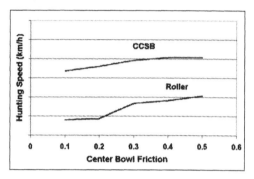

Figure 12. Centre bowl friction versus vehicle hunting speed.

5. CURRENT CENTRE PLATE LUBRICATION PRACTICES AND MATERIALS

Railways have been applying different practices to centre plate lubrication, ranging from no lubrication to continually full lubrication. Varying materials, including liquid, grease, molysulfide-based lubricants, and different types of polymers have been used. The dry practice is usually applied to cars operating on routes dominated by straight track to reduce the risk of vehicle hunting. Wheel/rail wear can be high when those cars run on curves. Consistent liquid lubrication is often applied on cars operating on very curvy routes to improve bogie steering. Frequent maintenance on centre plate lubrication for these cars could be costly. Also, the commonly used molysulfide-based lubricant in North America has been proved to have short service life of 30,000 km or less.

The introduction of polymer liners as centre bowl friction materials in past years has shown the possibility of having long-lasting materials with controlled friction levels for different operating conditions.

TTCI is currently conducting an evaluation of a number of existing polymer liners through railroad survey, track tests, and laboratory tests. The evaluation criteria include: coefficient of friction, endurance, geometry change in service, and cost. The friction coefficient of the polymer liner samples collected in this study ranged from 0.12 to 0.33. Three types of polymer liners tested by TTCI were in fair condition after 48,000 km of operation. The average friction coefficients were consistently around the designed values according to the air table turning moment tests conducted before and after 48,000 km of operation. The service life of these liners is expected to exceed 300,000 km.

The concern related to applying polymer liners is the reduction of thickness as they wear. It could increase the pre-load forces at the CCSBs. As more load shifts to the side bearings, the resistant moment to bogie turning could be significantly increased during vehicle curving. Therefore, selecting a material with high strength and good wear resistance is crucial for the centre plate liners.

Selecting materials with proper friction coefficient for the centre plate liners could improve vehicle performance and reduce wheel/rail wear. A low friction level benefits vehicle curving, while a higher friction level benefits vehicle lateral stability. For those operations in which both conditions must be considered, a medium level of friction should be chosen.

6. CONCLUSIONS

1. Lubricating centre plates can reduce wheel/rail wear on curves. Centre plate lubrication is more likely a secondary factor for the lateral force in curving compared to wheel/rail profiles, track lubrication, and other dominant factors.

2. High friction at the centre plate can improve vehicle lateral stability on tangent track. Constant contact side bearings can significantly improve vehicle stability by producing higher resistant moment to the bogie yaw motion.

3. Properly designed centre plate polymer liners have shown the possibility of having long-lasting materials with controlled friction levels for different operating conditions.

ACKNOWLEDGMENTS

This work was funded principally by the Association of American Railroads (AAR), as part of the Strategic Research Initiative program. The authors are grateful for the direction provided by the AAR's Vehicle/Track System Research Committee. The authors also wish to acknowledge many colleagues who made contributions to the project.

REFERENCES

1. Wolf, G.: "Final Report, Testing, Evaluation & Recommendations Curving Performance of 125 DS Cars," RSI – DDSF – 2, February 1993.

2. Shust, W., Urban, C., Lundberg, W. and Kramp, K.: "Truck Performance Research Consortium: Wayside Performance Detection Report," Report P-00-017, Transportation Technology Center, Inc., Pueblo, Colorado, February 2000.

3. Elkins, J. and Wu, H.: "Angle of Attack and Distance-Based Criteria for Flange Climb Derailment," The Dynamics of Vehicles on Roads and on Tracks, Vol. 33, 2000.

4. Wu, H.: "Effects of Centre Bowl Friction on Vehicle Curving and Lateral Stability," Report R-959, Association of American Railroads, Washington, D.C., December 2002.

5. Reiff, R., Robeda, J. and Wu, H.: "Friction Control", Proceedings, 8th Annual AAR Research Review, Pueblo, Colorado, March 2003.

6. Elkins, J.: "Prediction of Wheel/Rail Interaction: The State-of-Art", 12th IAVSD Symposium, Lyon, France, 1991.

Vehicle System Dynamics Supplement 41 (2004), p.302-310 © Taylor & Francis Ltd.

Nausea and comfort in tilting trains: Possible regression models for nausea

JOHAN FORSTBERG[1]

SUMMARY

Ride comfort, nausea and train speed in curves are some of the main issues of tilting trains. Higher speeds may reduce comfort due to higher lateral forces in curves. Tilting trains reduce perceived lateral acceleration by tilting the car body inwards in curves. However, it seems that strong tilt motions provoke nausea. In order to optimise the total performance of tilting trains, it is necessary to know the influence from speed, lateral and vertical accelerations and angular rotations on ride comfort and nausea as well as train performance. This paper describes a test in Norway with a new tilting train and proposes that a net motion dose of roll velocity, i.e. a motion dose with leakage (outflow of accumulated dose), is the main source for nausea provocation.

1. INTRODUCTION

Tilting trains are becoming quite popular, since they both shorten train journeys in time and enhance comfort by reducing lateral forces [1, 2]. Due to modern bogie technology, track forces and running performance are controlled and kept within safety limits [3, 4]. However, the risk of being motion sick is increased for some sensitive persons and this may give tilting trains a bad reputation [5]. The causes of tilt sickness are unclear but may be lateral, vertical or roll motions or a combination of these [6-8].

Tests in Sweden have shown that roll motions, not vertical or lateral motions, give the best correlation with degree of nausea [9, 10]. This corresponds well to findings in Japan, where a limitation of roll velocity to 0.1 rad/s (5°/s) and roll acceleration to 0.3 rad/s^2 (15°/s^2) is recommended [11]. To test the influence of roll motions in order to extract a possible regression model with accumulated motion dose with leakage (net motion dose), a test with NSB new tilting train BM73 was organised in Norway just before the start of commercial service [12].

1.1 MOTION SICKNESS HYPOTHESIS

The most common hypothesis of motion sickness is that of sensory conflict, i.e. a mismatch between the sensory information from the eyes and vestibular organs on

[1] *Address correspondence to*: Swedish National Road and Transport Research Institute (VTI),
SE - 581 95 Linkoping, SWEDEN, Tel.: +46 13 20 40 47; Fax: +46 13 14 14 36;
E-mail: johan.forstberg@vti.se

one hand, and the postural information from muscles and joints on the other, but also a mismatch to what is expected may provoke motion sickness [13, 14]. The sensory conflict has also been formulated in control theory terms, where the Central Nervous System (CNS) updates a copy of past and present expected sensory information from the sensory organs, and compares them with the actual sensory information [15, 16]. Another sharper version of the hypothesises is that the mismatch is between the *sensed* (by sensory organs) vertical g-vector in both amplitude and direction and the *subjective* (the CNS) copy of it [17]. However, for practical and measurement reasons, the conflict has been replaced with a weighting filter ($w_f \approx 0.08 - 0.35$ Hz) and an accumulation motion dose model (*MD*) [18-20]. A summary is also given in [21].

2. METHODS AND MATERIAL

The main hypothesis of the test was that tilting trains with their higher speed in curves are not degrading the passenger estimation of ride comfort and the ability to work during the journey compared with a conventional train. The design idea was to create conditions where the lateral accelerations and roll motions perceived by the test subjects could be separated. Normally these two quantities are bounded together with the selected tilt compensation. By selecting different speeds in curves with different tilt compensations, it is possible to find conditions where high tilt velocities are combined with low lateral acceleration perceived by the subjects and vice versa. Therefore, three different speed levels were selected with a corresponding lateral acceleration in track level of 1.0, 1.4 and 1.8 m/s^2, respectively. Table 1 gives an overview of the test conditions and the corresponding level of some specific track and tilt quantities. The conditions *I* and *II* used the 1.8 speed level with an effective compensation of 53% and 45% in the tilt system, respectively. The conditions *III* and *IV* used the 1.4 speed level with a compensation of 53% and 62%. The reference condition (*V*) was the 1.0 speed level with no compensation, i.e. non-tilting.

Table 1. Compilation of conditions I - V. For the different conditions the following quantities were noted: lateral acceleration in the track level and used tilt compensation ratio as well as typical maximum tilt angle, lateral acceleration in car body and roll velocity

Condition	I	II	III	IV	V (ref.)	Units
Speed level	1.8		1.4		1.0	
Lateral acc. in track plane	1.8		1.4		1.0	[m/s^2]
Cant deficiency	280		220		150	[mm]
Tilt compensation ratio	53 %	45 %	53 %	62 %	0 %	-
Tilt angle^1	5.6	4.7	4.7	5.4	0	[°]
Tilt angle velocity^1	2.8	2.35	2.1	2.5	0	[°/s]
Roll velocity^1	5.7	5.2	4.8	5.1	2.1	[°/s]
Lateral acc. in car body^1	0.85	1.0	0.7	0.6	1.0	[m/s^2]

Remark: ^{1} Typical maximum values.
Condition V was used as a reference condition without tilt motion.

2.1 TILTING TRAIN

The recently delivered tilting train, class BM73, belonging to NSB, was used in this test, see Figure 2. This train consists of four cars, see Figure 1. The train is allowed to have a lateral acceleration of 1.8 m/s^2 in the track plane, corresponding to a cant deficiency of 280 mm. The main focus was on the two middle cars: a bistro car and a second class car, since the end cars might have less ride comfort due to tilt motions. The middle cars might have different tilt compensation ratios in order to test two conditions during the same test run.

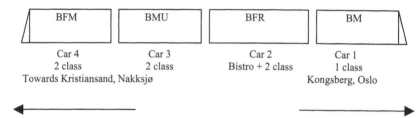

Fig. 1. Principal layout of the test train

Fig. 2. NSB test train of the class BM73 "Signatur".

2.2 TEST RUNS

Tests were conducted with five test runs during one week in October 1999, one run per day except for one occasion with two runs on one day. The test runs lasted altogether somewhat more than 2 hours plus stop time at the reversing station. During the test runs, the conditions *I* and *II* were combined with different compensation in car 2 and 3, conditions *III* and *IV* combined in the same way and condition *V* was tested alone.

2.3 TEST LINE

The chosen test line is a part of the south-west coast main line between Oslo and Kristiansand, a distance of about 94 km located between the stations Kongsberg and Nakksjoe, which is extremely curved. About 20% of the track length has curve radii less than 400 m, another 8% of the track length has curves with radii between 400 – 600 m and 30% of the track length consists of transition curves. About 18% of the distance is straight. Reverse curves are typical. Typical lengths of transition curves and curve elements are between 60 – 80 m corresponding to a time for passing the element of 2 – 3 s. The test line was divided into two sections of about the same length by the station Nordagutu, giving four test parts for evaluation, two on the outward run and two on the return run.

2.4 TEST SUBJECTS

The test subjects were recruited from a nearby college and each test run had 32 to 60 subjects. The mean age was 22.4 years and there was about equal distribution between females and males in each test run.

2.5 SCALES

The subjects were asked to estimate their comfort, ability to work and degree of nausea four times in the test run, after each test part. Ride comfort and ability to read and write was estimated on a five point scale from *very bad* (1) to *very good* (5). Motion sickness was estimated at the same time as nausea on a five point scale from *no symptoms* (0), *slight symptoms but no nausea* (1), *slight nausea* (2) *moderate nausea* (3) to *strong nausea* (4). The possible endpoint *vomiting* was not used (5).

2.6 MEASUREMENTS

The middle cars (#2 and #3) were each equipped with a box containing a lateral and a vertical accelerometer together with a roll velocity gyroscope. These signals, together with some signals from the train computer: tilt angles for these cars, lateral acceleration at bogie level and speed, were tape recorded for later analysis and evaluation. Motion Dose (*MD*) and Net Motion Dose (*NMD*) were calculated from the measurements, see definition below.

2.7 MOTION DOSE

MD is highly correlated with the degree of motion sickness in simulator tests and ship journeys of up to 8 hours [18, 22]. *MD* is defined as the square root over an integral of squared w_f filtered acceleration over time, defined for vertical acceleration as Motion Sickness Dose Value (*MSDV$_z$*) in ISO 2631-1 [20]. The w_f filter is principally a band-pass filter 0.08 – 0.35 Hz. *MD* can also be calculated as a summation over r.m.s.-values over a certain time interval Δt at time step t_n.

$$MD(t_n) = \left[\int_0^T a_{wf}^2(t)\,dt \right]^{0.5} = \left[\sum_{i=1}^n a_{wf,rms,\Delta t}^2(t_i) \cdot \Delta t \right]^{0.5} \qquad [\text{m/s}^{1.5}] \qquad (1)$$

The *NMD* [23] is defined as a recursion formula with an input corresponding to *MD* over the time step Δt at the time t_n and a leakage of *NMD* from the time step before.

$$NMD(t_n) = \left[(NMD(t_{n-1}) \cdot e^{c_l \cdot \Delta t})^2 + a_{wf,rms,\Delta t}^2(t_n) \cdot \Delta t \right]^{0.5} \qquad [\text{m/s}^{1.5}] \qquad (2)$$

where c_l is the time constant for leakage. If linear acceleration is changed to angular acceleration, the unit for *MD* and *NMD* is $[\text{rad/s}^{1.5}]$ and when changed to angular velocity, it is $[\text{rad/s}^{0.5}]$.

MD is mostly used for uniform conditions with steady state acceleration levels. A typical railway journey is a non-stationary condition, where you may have a large input of doses in one part and much lower input levels in other parts or at stations.

3. RESULTS

3.1 RIDE COMFORT AND WORK PERFORMANCE

Average ride comfort was estimated between 4.0 – 4.4 in the test conditions with a significantly higher level in the non-tilting condition for the test part 1, see Table 2. Other test parameters did not show any significant differences. Estimated work performance is between 3.6 and 4.0 with no significant differences, see also [24]. Ride comfort according to Wz was normally good or very good (below 2.0).

Table 2. Average estimated ride comfort for the test conditions on the test part 1.

Test conditions Speed level, Tilt compensation	Estimated ride comfort	Standard dev.	Standard error
I (1.8, 53%)	3.99	0.69	0.08
II (1.8, 45%)	4.19	0.52	0.09
III (1.4, 53%)	3.94	0.73	0.09
IV (1.4, 62%)	4.06	0.58	0.08
V (1.0, 0%, ref.)	4.36	0.50	0.09

Remark: Scale was 1 = very bad comfort to 5 = very good comfort.

3.2 MOTION SICKNESS AND NAUSEA

Degree of nausea (nausea ratings, *NR*) differs substantially between the conditions with tilting and the reference non-tilting condition, see Table 3. Higher speed levels and higher tilt compensation ratios seem to increase *NR*. Females normally have about twice as high *NR* as males.

Table 3 Nausea ratings for the different conditions for test part 2

| Test Conditions | Nausea | 95% Confidence Interval | |
Speed level, Tilt compensation	Rating (NR)	Lower bound	Upper bound
I (1.8, 53%)	0.64	0.47	0.81
II (1.8, 45%)	0.40	0.13	0.66
III (1.4, 53%)	0.50	0.31	0.69
IV (1.4, 62%)	0.48	0.27	0.69
V (1.0, 0%, ref.)	0.10	-0.19	0.39

Figure 3 shows a calculation of both lateral *NMD* and roll acceleration *NMD* for a part of a test run with the speed level 1.8. A condition with higher tilt compensation shows lower *NMD* for lateral acceleration but a higher *NMD* for roll acceleration. The calculation clearly shows that *NMD* increases in more curved sections and decreases at stations or in less curved sections because of the leakage of accumulated dose.

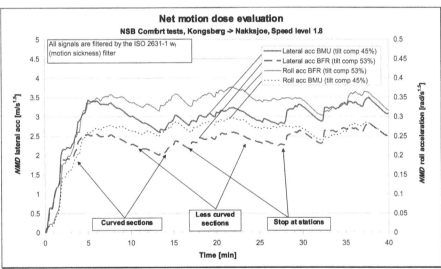

Fig. 3. Example of *NMD* calculation with a time constant of 15 min of leakage for both lateral and roll acceleration for tilt compensations of 45% and 53% at the speed level of 1.8.

3.3 REGRESSION MODEL FOR *NR*

The correlation between *NR* and the corresponding *NMD* values is shown in Table 4. A corresponding correlation between *NR* and *MD* will be low, since *MD* is growing over time but *NR* is not. The correlation factors are high for roll motions, less for lateral acceleration (but with a negative sign) and non-existent for vertical accelerations.

Table 4 Correlation between *NR* and *NMD*

Correlatbn NMD - NR	Vert. acc.	Lat. acc. car body	Roll vel.	Roll acc.
Pearson Correlation	-0.036	-0.524	0.749	0.743
Sig. (2-tailed)	0.831	0.001	0.000	0.000
N	38	38	38	38

The choice between roll acceleration or roll velocity for a regression model is mainly a practical question. Roll velocity is easier to measure and in most cases it can be defined by track geometry, as well. Possible regression models for *NR* are shown in Figure 4, one with linear and one with squared input of *NMD* of roll velocity. Both models are significant ($p < 0.001$) and have a reasonably high constant of explanation (r^2).

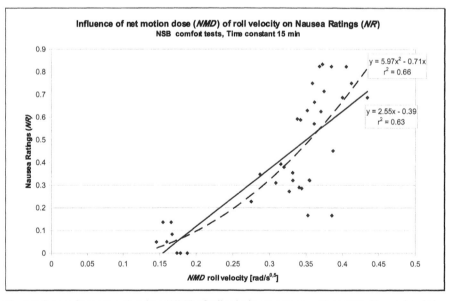

Fig. 4. Influence from net motion dose (*NMD*) of roll velocity on n nausea ratings (*NR*). The used scale is no symptoms (0), slight symptoms (1) to strong nausea (5).

4. DISCUSSIONS

These results are in line with earlier results from tests in Sweden [9] where roll motions are identified as the key provoker of nausea. It is also in accordance with tests in Japan and they imposed a limitation of both roll acceleration and roll velocity in their tilting trains [7, 8, 11]. Low frequency lateral accelerations may also provoke nausea [25, 26] and simulator tests with both lateral and roll motions have shown that a combination of lateral acceleration in the horizontal plane and roll velocity may be used as a predictor of nausea [9, 27]. However, because of the strong correlation between lateral (in the horizontal or car body plane) and roll motion, it is not feasible to use this combination in a regression model for nausea in tilting trains.

5. CONCLUSIONS AND FURTHER RESEARCH

Comfort and ability to work are little influenced by high curving speed in tilting trains whereas motion sickness is influenced by gender, tilt compensation and high curving speeds. A net motion dose model of roll acceleration/velocity seems to explain most of the degree of nausea. However future research has to prove the validity of this model by new tests, where *NR* is estimated at regular intervals and compared with motion data as well as doing tests on other routes. The concept of *NMD* has to be verified with this new data.

The European project FACT (Fast and comfortable trains) within the EU fifth research framework aims to both perform tests on two different lines in different countries and also simulator tests to verify the influence of roll and lateral motions on nausea.

ACKNOWLEDGMENTS

The Norwegian State Railway *NSB BA*, the Norwegian Track Administration *Jernbaneverket* and vehicle manufacturer *Bombardier Transportation* are greatly acknowledged for their helpful assistance.

REFERENCES

1. Andersson, E., von Bahr, H., and Nilstam, N.G.: Allowing higher speed on existing tracks - Design considerations of train X2000 for Swedish State Railways (SJ). Proceedings of the Institution of Mechanical Engineers. Part F: Journal of Rail and Rapid Transit, 209:2, (1995), p. 93-104.
2. Andersson, E. and Nilstam, N.: The development of advanced high speed vehicles in Sweden. Proceedings of the Institution of Mechanical Engineers, Part D, 198:15, (1984), p. 229-237.
3. Kufver, B. and Förstberg, J.: Research on specific aspects of tilting. in World Congress of Railway Research (WCRR '01). 2001. Köln: DB AG (Also pub as VTI Särtryck 347A, VTI).
4. Andersson, E., Nilstam, N.G., and Ohlsson, L.: Lateral track forces at high speed curving. Comparison between practical and theoretical results of the Swedish high speed train X2000, in 14th IAVSD

Symposium (Aug. 21-25 1995), L. Segel, Editor. Swetz & Zeitlinger: Ann Arbor (USA). 1996, p. 37-52.

5. Hughes, M.: Tilt nausea is bad business. Railway Gazette International, 153:4, (1997), p. 249.

6. Förstberg, J., Andersson, E., and Ledin, T.: Influence from lateral acceleration and roll motion on nausea: A simulator study on possible causes of nausea in tilting trains. in World Congress on Railway Research (WCRR '99). 1999. Tokyo: RTRI (Tokyo).

7. Ueno, M., et al.: Studies on motion sickness caused by high curve speed railway vehicles. Japanese Journal of Industrial Health, 28, (1986), p. 266-274.

8. Ohno, H.: What aspect is needed for a better understanding of tilt sickness? Quarterly report of RTRI, 37:1, (1996), p. 9-13.

9. Förstberg, J.: Ride comfort and motion sickness in tilting trains: Human responses to motion environments in train and simulator experiments, TRITA-FKT Report 2000:28 (also publ as VTI rapport 449A, Linkoping: Sweden), KTH Railway Technology: Stockholm. 2000.

10. Förstberg, J.: Regressions models for provoking motion sickness in tilting trains. in World Congress of Railway Research (WCRR '01). 2001. Köln: DB AG (Also pub as VTI Särtryck 347A, VTI).

11. Suzuki, H.: Research trends on riding comfort evaluation in Japan. Proceedings of the Institution of Mechanical Engineers: Part F Journal of Rail and Rapid Transit., 212:F1, (1998), p. 61-72.

12. Förstberg, J.: Influence from train motions on ride comfort and nausea in a train with active tilt system: Test performed with NSB BA tilting train BM73, autumn 1999. Final report, VTI rapport 482A (in preparation), VTI: Linköping. 2003.

13. Reason, J.T.: Motion sickness adaptation: a neural mismatch model. Journal of the Royal Society of Medicine, 71, (1978), p. 819-829.

14. Benson, A.J.: Motion sickness, in Aviation Medicine, K. Ernsting, Editor. Butterworths: London. 1988, p. 318-338.

15. Oman, C.M.: A heuristic mathematical model for the dynamics of sensory conflict and motion sickness. Acta Otolaryngol Suppl, 392, (1982), p. 1-44.

16. Oman, C.M.: Motion sickness: a synthesis and evaluation of the sensory conflict theory. Canadian Journal of Physiology and Pharmacology, 68:2, (1990), p. 294-303.

17. Bles, W., et al.: Motion sickness: only one provocative conflict? Brain Research Bulletin, 47:5, (1998), p. 481-7.

18. Griffin, M.J.: Handbook of Human Vibration. London: Academic Press. 1990.

19. Lawther, A. and Griffin, M.J.: Prediction of the incidence of motion sickness from the magnitude, frequency, and duration of vertical oscillation. Journal of Acoustical Society of America, 82:3, (1987), p. 957-966.

20. ISO: Mechanical vibration and shock - Evaluation of human exposure to whole body vibrations - Part 1: General requirements, ISO 2631-1.2:1997 (E), ISO: Geneva. 1997.

21. Förstberg, J. and Ledin, T.: Discomfort caused by low-frequency motions: A literature survey of hypotheses and possible causes of motion sickness, VTI Meddelande 802A. (Also published as TRITA-FKT report 1996:39, Stockholm: KTH), VTI: Linköping. 1996.

22. Griffin, M.J.: Physical characteristics of stimuli provoking motion sickness. AGARD: Neuilly sur Seine. 1991.

23. Kufver, B. and Förstberg, J.: A net dose model for development of nausea, VTI Särtryck 330-1999, VTI: Linköping (Sweden). 1999.

24. Förstberg, J.: Ride comfort and ability to work in tilting trains: Report from a train study carried out with NSB, Norway. in International Symposium on Speed-up and Service Technology for Railways and Maglev Systems 2003 (STECH '03). 2003. Tokyo.

25. Golding, J.F., Finch, M.I., and Stott, J.R.: Frequency effect of 0.35-1.0 Hz horizontal translational oscillation on motion sickness and the somatogravic illusion. Aviation, Space and Environmental Medicine, 68:5, (1997), p. 396-402.

26. Golding, J.F., Müller, A.G., and Gresty, M.A.: Maximum motion sickness is around 0.2 Hz across the 0.1 to 0.4 Hz range of low frequency horizontal translation oscillation. in 34th UK group meeting on human response to vibration. 1999. Ford Motor Comp. Dunton (England): Ford Motor Comp.

27. Förstberg, J.: Influence from lateral and/or roll motion on nausea and motion sickness: Experiments in a moving vehicle simulator, KTH-TRITA-FKT Report 2000:26, KTH: Stockholm. 2000.

Lateral Driving Assistance Using Robust Control and Embedded Driver-Vehicle-Road Model

SAID MAMMAR *, THIBAUT RAHARIJAONA †,
SEBASTIEN GLASER ‡ AND GILLES DUC†

SUMMARY

This paper addresses the problems of vehicle handling improvement and lane keeping support. The control method uses a coprime factors and linear fractional transformations (LFT) based feedback and feedforward H_∞ control. The control synthesis procedure uses a linear driver-vehicle-model model which includes the yaw motion and disturbance input with speed and road adhesion variations. The synthesis procedure allows the separate processing of reference signal and robust stabilization problem or disturbance rejection. The control action is performed as an additional steering angle and yaw moment generation by differential wheel braking, using a combination of the driver input, feedback of the yaw rate and vehicle positioning . The synthesized controller is tested for different speeds and road conditions on a nonlinear model in both disturbance rejection and driver imposed yaw reference tracking maneuvers and lane keeping.

1   INTRODUCTION

A large number of vehicle accidents results from inadequate yaw motion such as spin-out and lane departure. In addition, such type of accidents generally occurs on rural road, and about 30% of fatalities in France are due to accidents with the vehicle alone. In the framework of French ARCOS project, this paper is mainly concerned with the the development of an assistance which combines driver-vehicle-road interactions using individual wheel braking and active steering. Active safety technologies for vehicles have been mainly developed for powertrain control. Actually lateral dynamics improvement is addressed using independent wheel braking, but as pointed out in [1], vehicle active steering has attractive benefits with regard to vehicle handling improvement. In fact additional generated lateral forces can be used in order to reject yaw and roll torque disturbances even on

*Université d'Evry val-d'Essonne CNRS FRE-2494 LSC, 40 rue du Pelvoux CE1455, 91025 Evry Cedex France. Tel. 33-1 40 43 29 08, Fax. 33-1 40 43 29 30, E-mail: smam@iup.univ-evry.fr
†SUPELEC, Laboratoire d'automatique, 3 rue Joliot Curie, 91192 Gif-sur-yvette France.
‡LCPC-LIVIC, 13 route de la minière, 78000 Versailles, France.

poor road adhesion conditions. The controller output may consist in an additional steering command which is continuously added to the driver steering command. Active steering can be implementable using only on-board internal vehicle sensors, it may prevent spinning and improve the yaw dynamics. However, combination of video, INS and DGPS system makes possible today the definition of a robust axial guide to be followed by a vehicle [2].

The method presented here acts both as a feedback and feedforward controller. The feedback component uses available information from proprio and/or extero-ceptive sensors while the feedforward part processes the driver inputs and can be used for the compensation of sensed perturbations such as road curvature or for tracking reference signals. Using this configuration, it is possible to address both handling improvement and assistance according to the vehicle sensed environ-ment: lane, other vehicles, which defines the admissible trajectories. It is shown that this configuration allows achievement of robust model matching against pa-rameters variations and rejection of lateral forces and torque disturbances which may rise from wind forces. Lane keeping in enhanced. An H_∞ performance index is used for both criteria.

The remainder of the paper is organized as follows: section 2 introduces the model used for control synthesis. Control synthesis and simulation results are provided respectively in sections 3 and 4.

2   LOW COMPLEXITY VEHICLE AND DRIVER

The driver assistance approach which uses individual wheel braking and active steering is developed on the basis of a low complexity model version and then tested on the high level complexity [3]. This low complexity model is presented here.

2.1   Human driver model

Modelling of human driver is a difficult task. However, several components can be identified [4]. The first one is called structural model. It is constituted by a time delay representing inherent human processing time and neuromotor dynamics. This component represents the high frequency driver compensation component, modelled by a dead time $\tau_p = 0.15$ sec and a second order low pass filter with damping factor $\xi_n = 0.707$ and natural frequency $\omega_n = 10$ rad.s^{-1}. The second component corresponds to the driver lead and predictive actions. It is modelled by a first order lead filter, where the time constant τ_L is representative of the driver mental load. The third component is a simple gain representing the proportional action of the driver face to the perceived vehicle positioning relative to the driving environment which are the lateral displacement at some look-ahead distance and the relative yaw angle (Figure 1-a). Finally, the driver model presents a third

(a) : Driver model   (b) : Vehicle dynamics

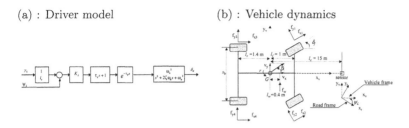

Fig. 1. Driver model and vehicle dynamics

input represented by the feedforward action taken from the driver estimation of the road curvature.

2.2   Vehicle dynamics for handling

The model used for control synthesis is derived from the high complexity model in which the longitudinal velocity v is assumed to be constant and not influenced by the yaw torque control input T_z used in differential braking. All the angles are also considered sufficiently small in order to allow linear approximations. The sideslip angle β is used as the first state variable, the second is the yaw rate r. It is also assumed equal cornering stiffness for the two front wheels $\frac{c_f}{2} = 25.2$ K.N/rad and the rear ones $\frac{c_r}{2} = 25.2$ K.N/rad. When the track width is neglected, the tire slip angles equal at front and rear wheels. The model takes the form

$$\dot{x} = \bar{A}x + \bar{B}_w w + \bar{B}_u u \qquad (1)$$

where $x = [\beta, r]^T$, $w = f_w$ is the disturbance wind force, $u = [\delta_f, T_z]^T$ is the control input where δ_f is the front tire steering angle.

$$\bar{A} = \begin{bmatrix} a_{11} & a_{12} \\ a_{21} & a_{22} \end{bmatrix}, \bar{B}_w = \begin{bmatrix} b_{w_1} \\ b_{w_2} \end{bmatrix}, \bar{B}_u = \begin{bmatrix} b_{u_1} & 0 \\ b_{u_2} & \frac{1}{J} \end{bmatrix} \qquad (2)$$

with

$$
\begin{array}{llll}
a_{11} = -\frac{c_r + c_f}{mv} & a_{12} = -1 + \frac{l_r c_r - l_f c_f}{mv^2} & b_{w_1} = \frac{1}{mv} & b_{u_1} = \frac{c_f}{mv} \\
a_{21} = \frac{l_r c_r - l_f c_f}{J} & a_{22} = -\frac{l_r^2 c_r + l_f^2 c_f}{Jv} & b_{w_2} = \frac{l_w}{J} & b_{u_2} = \frac{c_f l_f}{J}
\end{array} \qquad (3)
$$

Vehicle parameter variations, mainly the cornering stiffness and the speed and represented in an linear fractional transformation (LFT) form by defining extra input and output on the system and diagonal perturbation [6]. The mass $m = 1400$ Kg and the moment of inertia $J = 2750$ Kg.m^2 are constant. In the following, it is assumed that the additional steering angle is achieved by steer-by-wire and the yaw moment by differential wheel braking. The necessary control logic for the obtention of the desired actions are outside the scope of this paper and is not addressed in the remainder.

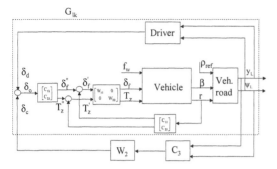

Fig. 2. Control architecture.

2.3 Additional dynamics for lane keeping assistance

The previous model has to be expanded with two supplementary equations for lane tracking. Let $\psi_L = \psi - \psi_d$ be the yaw angle error which is the angle between the vehicle heading and the tangent to the road. Differential equation for ψ_L is (figure 1-b)

$$\dot{\psi}_L = \dot{\psi} - \dot{\psi}_d = r - \dot{\psi}_d \qquad (4)$$

The road reference curvature ρ_{ref} is defined by $\dot{\psi}_d = v\rho_{ref}$. Denoting by l_s the look-ahead distance, the equations giving the evolution of the measurement of the lateral offset y_L from the centerline at sensor location is obtained by

$$\dot{y}_L = v\left(\beta + \psi_L\right) + l_s r \qquad (5)$$

The new state vector is $x = [\beta, r, \psi_L, y_s]^T$, $w = [f_w, \rho_{ref}]^T$ is the disturbance input, the control input remain the same. In addition to the yaw rate, it is assumed that y_L and ψ_L are measured using a video sensor and then available for feedback. The measurement vector is thus $y = [r, y_L, \psi_L]^T$.

3 CONTROL SYNTHESIS

The control philosophy processes an internal loop for handling improvement and an external loop for lane keeping assistance. Considering vehicle active steering, the tire steering angle is set in part by the driver steering angle δ_d through the vehicle classical steering mechanism while an additional steering angle is set by the controller. The yaw torque is directly set by the controller (Figure 2).

The internal loop uses dynamic feedback controller C_1 of the yaw rate to the yaw torque and a dynamic feedforward controllers C_{21} and C_{22} of the steering angle $\delta_o = \delta_d + \delta_c$, commanded by the outer loop to respectively the tire steering angle and the yaw moment. The inner loop control takes thus the form

$$
\begin{bmatrix} \delta_f' \\ T_z' \end{bmatrix} = \begin{bmatrix} C_1 & C_2 \end{bmatrix} \begin{bmatrix} r \\ \delta_d + \delta_c \end{bmatrix} = \begin{bmatrix} C_{11} & C_{12} \\ C_{21} & C_{22} \end{bmatrix} \begin{bmatrix} r \\ \delta_d + \delta_c \end{bmatrix} \tag{6}
$$

where $\begin{bmatrix} \delta_f', T_z' \end{bmatrix}^T$ is such that $[\delta_f, T_z]^T = W_1 \begin{bmatrix} \delta_f', T_z' \end{bmatrix}^T$, and W_1 is a shaping filter of control inputs which will be defined below.

As lane keeping is a disturbance rejection problem, the outer loop uses a feedback controller C_3 of the lateral displacement from lane centerline at a lookahead distance l_s and the relative yaw angle. This controller produces a steering angle $\delta_c = C_3 [y_L, \psi_L]^T$ which is added to that of the driver. The action of each control component is as follows

- The feedback controller C_1 ensures robust stability of the feedback loop with guaranteed damping enhancement on the yaw rate. It has also in charge fast disturbance rejection within driver reaction time. Controller C_2 acts as a prefilter of the reference signal by adding the feedforward action. This controller is synthesized such that to make the vehicle yaw rate response to robustly follow as close as possible the response of a reference model. This fact constitutes robust model matching [5].

- From the vehicle point of view, a lane keeping maneuver requires the controller to reject lateral acceleration and yaw rate disturbances caused by change in the radius of curvature. In fact, in this configuration, the reference curvature is an external input for the system. This is achieved by controller C_3.

In the following a two stages approach is adopted for the synthesis of feedback and feedforward components. At the first stage the feedback part C_1 of the controller is computed using the H$_\infty$ coprime based loop shaping method of [6]. Afterwards, the new vehicle model which incorporates the feedback controller is computed, thus the feedforward part C_2 is synthesized from a second H$_\infty$ optimization. The procedure used for the synthesis of C_1 is also used for C_3. The inner loop is processed first and then the outer one. This ensures that handling improvement is still optimal even when the lane keeping assistance is not connected because of insufficient accurate video detection. All the controllers are synthesized on a nominal linear system at speed of 20 m/s and full road adhesion.

3.1 Synthesis of $[C_{11}, C_{21}]^T$

We consider first the sub-system $G_{r[\delta_f, T_z]^T} = \begin{bmatrix} G_{r\delta_f}, G_{rT_z} \end{bmatrix}$ which maps the front wheels steering angle and the yaw moment T_z to the yaw rate r. In order to reject a constant step input perturbation on the yaw rate, a diagonal weighting compensator is added on the inputs of the system. The compensators are of the form of a PI filter for δ_f and a combination of two lead filter for T_z. This choice makes the yaw moment negligible after driver reaction time.

$$W_1(s) = diag\{W_{11}, W_{22}\} = diag\left\{0.3\frac{s+5}{s}, 375\frac{0.2s+1}{0.1s+1}\frac{100s+1}{10s+1}\right\} \quad (7)$$

Let now G_s be the shaped plant ($G_s = G_{r[\delta_f, T_z]^T} W_1$). According to the gap-metric, the stability of the vehicle for admissible parameters variations is guaranteed. The stabilizing H_∞ feedback controller $[C_{11}, C_{21}]^T$ is computed using the non-iterative procedure in [7] with a relaxed value of the maximal stability margin ($\gamma_1 = 1.5$). This controller C_1 provides the needed phase lead for close loop stabilization and ensures robust stability for all systems variations. The controller is implemented as shown on figure 2.

3.2   Synthesis of $[C_{12}, C_{22}]^T$

The weighting compensators are left at the system input, the shaped system described by $r = G_{r\delta_f} W_{22}\delta'_f + G_{rT_z} W_{22}T'_z$ is closed with the feedback controller C_1 such that the control input are respectively $\delta'_f = C_{11}r + \delta''_f$ and $T'_z = C_{21}r + T''_z$. The mapping from the two inputs $\left[\delta''_f, T''_z\right]^T$ to r is $r = G_{ff_1}\delta''_f + G_{ff_2}T''_z$. We seek now a single input, two output feedforward controller $[C_{22}, C_{22}]^T$ such that $\left[\delta''_f, T_z\right]^T = [C_{12}, C_{22}]^T \delta_o$. The controller C_{12} is set as a static speed depend controller such that the DC-gain from steering angle δ_0 to yaw rate is equal to that of the conventional vehicle without feedback control and when yaw moment input is zero. As the compensator W_{11} contains an integral action, one has to choose $C_{12}(v) = G_{r\delta_f}(0, v)C_{11}(0)$. The dynamic feedforward controller C_{22} will be designed with robust model matching purposes. Let T_0 be the desired transfer function between δ_o and r. In order to ensure at nominal speed, the same steady state value for the controlled and the conventional car, the reference model is chosen as a first order transfer function with the same steady state gain as the conventional car. It is of the form $T_0 = \frac{G_{r\delta_f}(0,v)}{0.15s+1}$. The settling time is about 0.5 sec. A first order model also avoids overshot on vehicle responses. The feedforward controller C_{22} has to keep the error signal z small in H_∞ sense for the class of perturbed systems according to vehicle parameter variations [8], [6]. The error signal z is computed from ($z = r - T_0\delta_o$). When including the controller C_{22}, the error signal is thus $z = (G_{ff_1}C_{12} + G_{ff_2}C_{22} - T_0)\delta_o$, it can be written in LFT form which is suitable for H_∞ optimization

$$z = lft\left(\begin{bmatrix} G_{ff_1}C_{12} - T_0 & G_{ff_2} \\ I & 0 \end{bmatrix}, C_{22}\right) \quad (8)$$

(a) : yaw rate r (b) : yaw moment T_z (c) : steering angle δ_f

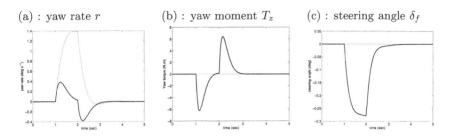

Fig. 3. Wind forces step input rejection for nominal system, solid : controlled, dotted : conventional.

3.3 Synthesis of C_3

After designing the controller C_{22}, the transfer function from δ_o to r is $G_h = G_{ff_1}C_{12} + G_{ff_2}C_{22}$. The control input δ_o is set in part by the driver and by the controller which performs lane keeping. The model is first completed with the two state equations for vehicle positioning relative to the lane The new measurement variables are the lateral displacement y_L at the look-ahead distance l_s and the relative yaw angle ψ_L. As shown on figure 2, the nominal system G_{lk} used for controller synthesis is obtained bay feeding back the model with the driver model. This model has the steering angle δ_c as the control input and two output $[y_L, \psi_L]^T$. As lane keeping is a disturbance rejection problem, this system is shaped at the input by the pre-compensator $W_2 = \frac{0.1}{0.1s+1}$. A stabilizing controller is finally synthesized for the shaped plant with ($\gamma_3 = 2.2$). It is finally implemented as shown on figure 2.

4   SIMULATION RESULTS

In all figures, solid lines correspond to the controlled car responses and dotted ones to the conventional car responses. A first set of simulations addresses yaw dynamics improvement, while the second set concerns lane keeping.

4.1   Handling improvement

Disturbance rejection
The vehicle is supposed to be at nominal speed and full road adhesion and is subject to a step disturbance wind force. The wind force appears at time $t_1 = 1$ sec and disappears at $t_2 = 2$ sec. It is assumed that the driver doesn't react to this disturbance. In this case, only controller C_1 is in action. One can note from Figure 3 that the yaw rate is greatly reduced and thus the controlled vehicle will

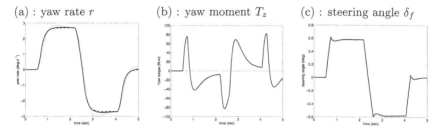

Fig. 4. Lane change maneuver, nominal system, solid : controlled, dotted : conventional, dashed : ref. model.

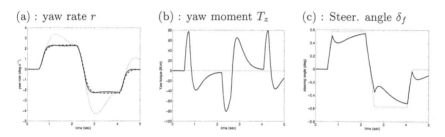

Fig. 5. Lane change maneuver, for nominal road adhesion and speed at 40 m/s (solid : controlled, dotted : conventional, dashed : reference model).

remain closer to road centerline. In addition, the maximum value of yaw rate during the transient phase is smaller than the one of the conventional car and the disturbance is practically rejected within driver reaction time. One can notice that yaw moment quickly vanishes.due to limiting effect of the shaping filter W_{22}. It was verified that the controller exhibits good stability and performance robustness.

Lane change maneuver

The handling improvement is now investigated in case of driver steering angle which corresponds to lane change maneuver (Figure 4-c, dotted line). In this case, both controllers C_1 and C_2 are in action. The dashed-dot line corresponds to the response of the reference model. Figure 4 shows results obtained at nominal speed with road adhesion equal to 1. Figure 5 shows results obtained for $v = 40$ m/s and nominal adhesion. Robust model matching occurs, and due to the speed scheduling of the gain parameter $\alpha(v)$, we ensure that the controlled vehicle and the conventional one present the same steady state behavior. Responses are degraded for high speed and low road adhesion values ($v = 40$ m/s and half adhesion). When the road adhesion is at its nominal value even when the speed varies, the control

318

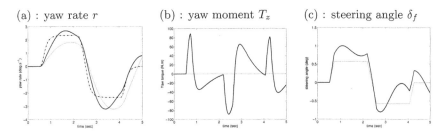

Fig. 6. Lane change maneuver for v=40 m/s and half road adhesion solid : controlled, dotted : conventional, dashed : reference model.

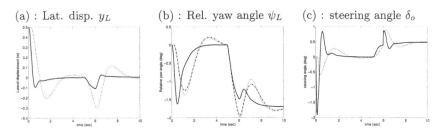

Fig. 7. Simple lane keeping maneuver, nominal system solid : controlled with LK controller, dotted : conventional, dashed : controlled with handling controller.

effort vanishes within driver reaction time which is assumed to be between 0.5 and 1 second. When the road adhesion is decreased, there is a remaining coontrol actions. Lane keeping capabilities are now investigated.

4.2 Lane keeping improvement

At the beginning of the simulation, the vehicle is on straight road section, at a lateral distance of 0.5 m from the lane centerline. The relative yaw angle is zero. As shown on Figure 7, in dotted line, without any control support, the driver gives a steering angle in order to make the vehicle close to the centerline. The overshot is about -0.2 m and the lateral displacement is near zero 3 sec later. At 5 sec, the vehicle enters a curved road section with $1/500$ m^{-1} of road curvature. The lateral displacement overshoots again to -0.3m but is less than 0.1 m, 2 sec later. On the same figure, dashed lines show the responses when handling improvement support is activated. Responses are rather the same but small reduction of peak values can be observed. Similarly, solid lines correspond to the vehicle with both handling improvement and lane keeping support. In this case, response time is

(a) : TLC nominal   (b) : TLC perturbed

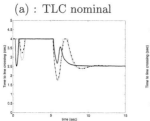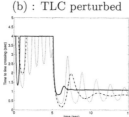

Fig. 8.   Time-to-line-crossing, nominal system solid : controlled with LK controller, dotted : conventional, dashed : controlled with handling controller.

less than 1 sec and overshoots form centerline are under 0.1 m. However a larger amount and faster steering angle is required. Finally, Figure 8-a and 8-b show the achieved time to line crossing (TLC) by each vehicle when they are first at nominal conditions and then at high speed respectively. The vehicle with both handling and lane keeping support presents the best TLC particularly when entering the curve.

5 CONCLUSION

In this paper, some aspects of combination of active steering and individual wheel braking have been explored. Both handling improvement and lane keeping support are addressed. On the basis of several simulated maneuvers, It has been shown that the controlled vehicle exhibits better yaw damping, faster and more accurate lane keeping. The control method uses a combination of feedforward and feedback H_∞ controllers.

REFERENCES

1. J. Ackermann, D. Odenthal and T. Bünte. Advantages of active steering for vehicle dynamics control., pages 263–270, 1999.
2. Bodor, et. al.: In-vehicle GPS-based lane sensing to prevent road departure, Proceedings of ITS World Congress 96, 1996.
3. S. Glaser, S. Mammar, J. Sainte-Marie, Lateral Driving Assistance using Embedded Driver-Vehicle-Road Model, Proceedings of ESDA2002, Istanbul, Turkey, July 2002.
4. A. Modjtahedzadeh and R. A. Hess, A model of driver steering control behavior for use in assessing vehicle handling qualities., 15:456–464, 1993.
5. E. M. Kasenally, D. J. N. Limebeer and D. Perkins, On the design of robust two degree of freedom controllers., 29.1, pp. 157–168, 1993.
6. S. Mammar and D. Koenig, Vehicle Handling improvement by Active Steering, Vehicle System Dynamics Journal, vol 38, No3, 211-242, 2002.
7. D. McFarlane and K. Glover. A loop shaping design procedure using H_∞ synthesis., 37, pp. 759–769, 1992.
8. A. Giusto, and F. Paganini , Robust synthesis of feedforward compensators, IEEE Transactions on Automatic Control, vol. 44, pp. 1578-1582, 1999.

Vehicle System Dynamics Supplement 41 (2004), p.321-331

A DRIVER MODEL FOR A TRUCK-SEMITRAILER COMBINATION

M. WEIGEL, P. LUGNER and M. PLÖCHL

Institute of Mechanics, Vienna University of Technology, Wiedner Hauptstr. 8-10, A-1040 Vienna, Austria

Abstract

To establish a driver model the equations of motion for a linearized truck-semitrailer were employed. Different decision layers were used to keep the vehicle at a given trajectory. Moreover the option was introduced to consider a truck based driver model in comparison to a driver model based on the truck-semitrailer combination. The driver models were tested using a complex Multi-Body-System-description of the vehicle when entering a curve manoeuvre.

1 Introduction

For the development of cars and trucks and the evaluation of their dynamic behaviour the simulation of the whole system including the driver becomes increasingly important. Quite a lot of driver models for passenger cars, e.g. [1, 2, 3], exist and are partially applied even with commercial software, e.g. [4]. For trucks and truck-trailer or semitrailer combinations, however, corresponding driver models need to be investigated in order to estimate the driver intentions correctly.

As most heavy traffic in Europe today is done with car-semitrailers, the focus for developing such a driver model and for showing the differences with respect to a single two-axle-unit is set on this type of road vehicles. An additional problem arises by the essential changes of the system properties due to the range of permissible loading conditions of the trailer. For example the moment of inertia for the vertical axis (CG_T of the trailer) can vary more than tenfold. A skilled driver will be aware of that and will adapt his steering input accordingly – which will lead to different parameters of the driver model.

2 System model

The structure of the system model is shown in Fig. 1, see [2, 3]. It is assumed that the driver can apply an anticipated feed forward control based on the knowledge or rather the estimation of the curvature of the vehicle trajectory ahead, see section 3.1. Defining a desired distance to the trajectory, he realizes a closed loop control (compensatory control loop) to keep the vehicle, respectively a reference point connected to the towing vehicle, on the desired path once again looking ahead of his momentary position.

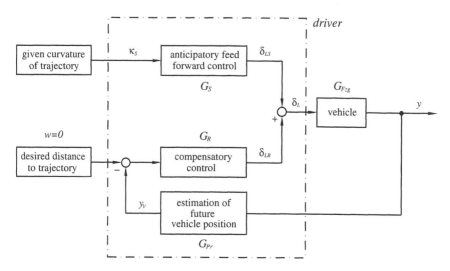

Fig. 1: Driver-vehicle system

For the simulation of the system behaviour the "vehicle" is modelled by a commercial Multi-Body-System (MBS) software [4], whereas for the design of the two control units a reduced linear model is applied. A flat, dry, horizontal road surface is assumed.

2.1 Vehicle model

A scheme of the complex model of the towing unit and semitrailer, based on the software ADAMS/Car [4], is shown in Fig. 2.

The two units comprise all details of the axle suspension systems, the braking system, the steering system of the towing truck and its drive train. For the fifth wheel the vertical rotational DOF is given free with small damping while the other possible motions are constrained by bushings of different stiffnesses and dampings. The parameters of the model are based on data given in [5, 6, 7].

The tyre model comprises the mutual influence of lateral and longitudinal slip for the horizontal tyre forces and takes into account the lateral tyre transient behaviour by the standard first order differential equation and the tyre transition length model. Model and parameters proposed in [8, 9] are used.

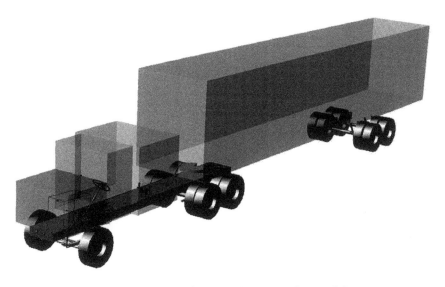

Fig. 2: ADAMS/Car truck-semitrailer model

2.2 Linear truck-trailer model

To establish a driver model the equations of motion for a linearized truck and semitrailer are employed. Fig. 3 shows the reduction of the system to a three-wheel-unit and its kinematic quantities and forces. The notation is generally self-explaining. The linearized accelerations $\underline{a}_{CG}$, $\underline{a}_T$ of the CGs of truck and semitrailer, described in coordinate frame 1 and 2, respectively, are with the curvature $\kappa = \frac{1}{\rho}$, $v_x \approx v$:

$$
\underline{a}_{CG/1} = \begin{bmatrix} -\dot{\psi}\, v_y \\ \dot{v}_y + \dot{\psi}\, v \\ 0 \end{bmatrix} = \begin{bmatrix} -v^2\, \kappa\, \beta \\ v^2\, \kappa \\ 0 \end{bmatrix} \tag{1}
$$

$$
\underline{a}_{T/2} = \begin{bmatrix} -v^2\, \kappa\, \beta - \psi_R\left(v^2\, \kappa - l_A\, \ddot{\psi}\right) \\ v^2\, \kappa - l_A\, \ddot{\psi} - a\, \ddot{\psi}_T \\ 0 \end{bmatrix} \tag{2}
$$

The linearized, substitutive forces corresponding to the properties of the axles and axle combinations with their suspensions and tyres can be written with

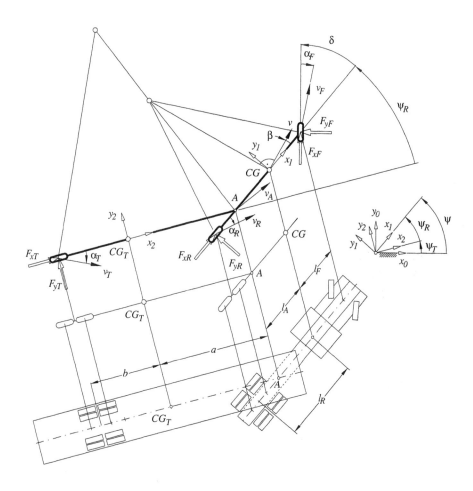

Fig. 3: Reduced truck-semitrailer model

$l = a + b$ in the form:

$$F_{yF} = c_F \alpha_F = c_F \left(\delta - \beta - \frac{l_F \dot{\psi}}{v} \right)$$

$$F_{yR} = c_R \alpha_R = c_R \left(-\beta + \frac{l_R \dot{\psi}}{v} \right) \qquad (3)$$

$$F_{yT} = c_T \alpha_T = c_T \left(-(\psi - \psi_T) - \beta + \frac{l_A \dot{\psi} + l \dot{\psi}_T}{v} \right)$$

Newton's law for the whole combination and Euler's equations for both parts, formulated for the hinge point A, are in linearized form (small jaw knife angle ψ_R):

$$m_T\, a_{T2} + m\, v^2\, \kappa = F_{yF} + F_{yR} + F_{yT} \tag{4}$$

$$I_{CG}\, \ddot{\psi} + m\, v^2\, \kappa\, l_A = F_{yF}\, (l_A + l_F) + F_{yR}\, (l_A - l_R) \tag{5}$$

$$I_T\, \ddot{\psi}_T - m_T\, a_{T2}\, a = -F_{yT}\, l \tag{6}$$

where a_{T2} is the second component of the acceleration vector (2). The masses of truck and semitrailer are m, m_T, and the moments of inertia with respect to CG and CG_T (vertical axes) are I_{CG} and I_T, respectively.

If the truck is considered only, in equation (4) m_T and F_{yT} are set equal to zero, and the so modified equation (4) together with (5) are then the equations of motion.

3 Driver model

A two-level driver model is introduced, e.g. [2, 3], based on the intention that the driver recognizes the trajectory and its (approximate) curvature in front of the vehicle and tries to follow it. If he anticipates that he may not keep the vehicle on the track with this feed forward steering only, he will have to initiate corrections. In this two-level model, shown in Fig. 1, the "knowledge" of the driver with respect to the behaviour of his vehicle is derived from the linear models, presented in section 2.2.

3.1 Feed forward steering

For designing this part of the driver model with the transfer function

$$G_S(s) = \frac{\delta_{LS}(s)}{\kappa_S(s)} \tag{7}$$

the properties of the vehicle and the capability of the driver to steer according to the desired curvature κ_S in front of him (prediction time T_P) have to be taken into account.

Based on experiments, [10], for the corresponding driver behaviour the following description was found:

$$T_{2s}^2\, \ddot{\delta}_{LS}(t) + T_{1s}\, \dot{\delta}_{LS}(t) + \delta_{LS}(t) = V_P\, \kappa_S(t + T_P) \tag{8}$$

or with a Taylor series expansion for $\kappa_S(t + T_P)$:

$$T_{2s}^2\, \ddot{\delta}_{LS}(t) + T_{1s}\, \dot{\delta}_{LS}(t) + \delta_{LS}(t) = V_P(\kappa_S(t) + T_P\dot{\kappa}_S(t) + \frac{T_P^2}{2!}\ddot{\kappa}_S(t) + \ldots) \tag{9}$$

with the time constants T_{1s}, T_{2s} and the gain factor V_P. If the driver behaviour (8) matches (7), he would be able to keep the vehicle ideally on the trajectory. Consequently the characteristic parameters of (8) should be determined with a determination of G_S.

To determine G_S, it is assumed that a point P in front of the truck, see Fig. 4, is kept on a path determined by using

$$\kappa_P = \frac{1}{v}\left(\dot{\beta}_P + \dot{\psi}\right), \qquad \beta_P \approx \beta + (l_F + l_P)\,\frac{\dot{\psi}}{v}$$

with

$$\kappa_P = \kappa + \frac{(l_F + l_P)}{v}\,\frac{\ddot{\psi}}{v} \tag{10}$$

Substituting (9), (10) into the equations of motion finally results in the 4th-order

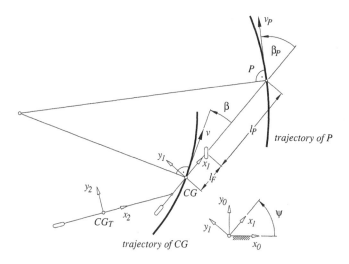

Fig. 4: Vehicle with leading point P on given trajectory

differential equation:

$$\frac{\delta_4}{\delta_0}\,\overset{(4)}{\delta}_{LS} + \frac{\delta_3}{\delta_0}\,\dddot{\delta}_{LS} + \frac{\delta_2}{\delta_0}\,\ddot{\delta}_{LS} + \frac{\delta_1}{\delta_0}\,\dot{\delta}_{LS} + \delta_{LS} =$$

$$= \frac{\kappa_1}{\delta_0}\left(\frac{\kappa_5}{\kappa_1}\,\overset{(4)}{\kappa}_P + \frac{\kappa_4}{\kappa_1}\,\dddot{\kappa}_P + \frac{\kappa_3}{\kappa_1}\,\ddot{\kappa}_P + \frac{\kappa_2}{\kappa_1}\,\dot{\kappa}_P + \kappa_P\right) \tag{11}$$

with the constants δ_i. Note that the constants κ_i do not include l_P but only parameters of the vehicle.

Comparing (11) with (9) and setting $\kappa_P = \kappa_S$ (point P is kept on the desired path) will determine an approximation for V_P and T_P by the right hand side of the equations. This leads to four different relations for T_P:

$$T_{P1} = \frac{\kappa_2}{\kappa_1}, \quad T_{P2} = \sqrt{\frac{2\,\kappa_3}{\kappa_1}}, \quad T_{P3} = \sqrt[3]{\frac{3!\,\kappa_4}{\kappa_1}}, \quad T_{P4} = \sqrt[4]{\frac{4!\,\kappa_5}{\kappa_1}} \tag{12}$$

and thereby to a possible range for the choice of $l_P = v\,T_P$. For the left hand side comparison of (11) with (9) the behaviour described by the 4th-order differential equation has to be approximated by a 2nd-order system behaviour. Thereby it is consequently assumed that also for the truck-semitrailer combination the driver can only steer according to the experimentally confirmed characteristics (8).

This approximation of the 4th-order system with PT2 characteristics was done in the frequency range using a nonlinear least squares method. The steering frequency range is selected according to the more smoother truck steering until 1 Hz. Moreover the approximation of the gain was dominant – the approximation of the phase shift is of course limited by the properties of the PT2 characteristics. As shown in Fig. 5, an acceptable agreement could be achieved. The velocity dependent free parameter l_P was chosen with $l_P = 2.5 \cdot l_F$ for $v = 20$ m/s, in the range given by (12).

For the sake of comparison the same procedure is also done for the towing truck only, starting with the equations of motion and setting $m_T = 0$, $F_{yT} = 0$ as told before. Compared to (11), in this case only expressions up to second order derivatives show up, both on the left and right hand side. So, V_P, T_{2s} and T_{1s} can be determined immediately while for T_{Pi} now two possible values are derived.

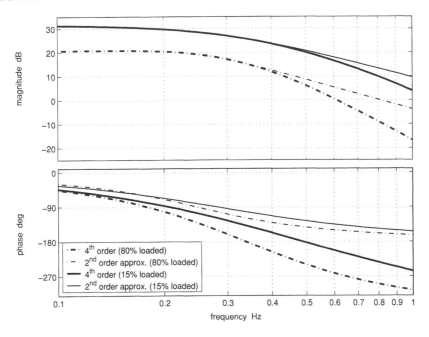

Fig. 5: Approximation of the linearized truck-trailer behaviour by a PT2 characteristic, $v = 20$ m/s, semitrailer 15% and 80% loaded, for data see Tab. 1

4 Compensatory Control

The human behaviour for correcting a predicted path deviation y_V by a steering input δ_{LR} is described by the transfer function [11]:

$$G_R(s) = \frac{\delta_{LR}(s)}{y_V(s)} = V_R \frac{1 + T_V s}{1 + T_N s} \cdot e^{-t_R \cdot s} \tag{13}$$

with the gain V_R and the time constants T_V, T_N. The human reaction time is chosen $t_R = 0.2$ s corresponding to [2]. For the predicted path deviation again a Taylor series expansion is applied,

$$y_V = y(t + T_P) \approx y(t) + T_P\,\dot{y}(t) + \frac{T_P^2}{2}\,\ddot{y}(t)$$

with the prediction time T_P according to l_P. The corresponding transfer function is

$$G_{Pr}(s) = \frac{y_V(s)}{y(s)} = 1 + T_P s + \frac{T_P^2}{2} s^2 \tag{14}$$

In the case of controller design for determining the vehicle transfer function the same equations of motion for the linearized system (4), (5), (6) and the corresponding descriptions for the tyre forces (3) are used. This finally results in the first equation of (15) where the constants a_i, b_i are lengthy expressions containing the vehicle parameters. If the towing truck is considered only, the transfer function reduces to the second equation of (15):

$$G_{Fzg}(s) = \frac{y(s)}{\delta_{LR}(s)} = \begin{cases} \dfrac{a_4\,s^4 + a_3\,s^3 + a_2\,s^2 + a_1\,s + a_0}{s^2(b_4\,s^4 + b_3\,s^3 + b_2\,s^2 + b_1\,s + b_0)} & truck - trailer \\[3mm] \dfrac{c_2\,s^2 + c_1\,s + c_0}{s^2(d_2\,s^2 + d_1\,s + d_0)} & truck \end{cases} \tag{15}$$

For the controller design (truck and truck-semitrailer combination) considering open loop properties in the interesting frequency range of $0.5 - 3$ Hz a gain decline of -20 dB/decade in the area of the cross over frequency, [11], and a phase margin of about $50°$ are demanded. The dominant pole assignment method was applied to derive the control constants of (13) (damping $D = \sqrt{2}/2$, fade-out time 2 s, final tolerance range of 0.05). The resulting parameters for the optimal control design are dependent on load and on driving velocity.

5 Results

The main data for the truck-semitrailer combination are listed in Tab.1. The indices 15 and 80 with the semitrailer indicate a loading condition of 15% and 80% of the maximum permissible load; the height h_{CG_T} of CG_T is not changed.

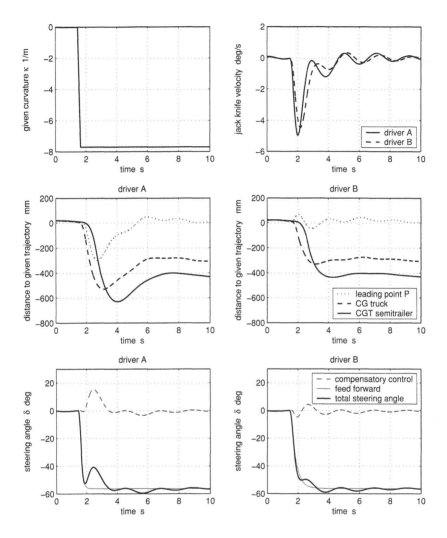

Fig. 6: Entering a curve with driver model A for the truck only and the driver model B for the truck-semitrailer combination, $v = 20$ m/s, trailer 15% loaded

Simulations of entering a curve with the 15% loaded truck-semitrailer combination show that for a standard transitional curve and small trailer load the control models of truck driver and truck-semitrailer driver do not differ essentially. But for a fast steering, represented in Fig. 6 by a very short transitional curve the differences become obvious even for 15% trailer loading. The smoother steering behaviour of the truck-semitrailer driver B needs less compensatory steer control and results in less vigorous vehicle response and a better decay of the initiated oscillations. Note, that the practically identical final deviations of

Parameter	Symbol	Unit	Value
Mass of the truck	m	kg	7150
Yaw moment of inertia of the truck	I_{CG}	$kg\,m^2$	93639
Front axle location of the truck	l_F	m	2.69
Rear axle location of the truck	l_R	m	3.42
Fifth wheel location	l_A	m	3.05
height of CG_T	h_{CG_T}	m	1.80
Mass of the trailer (15% loaded)	$m_{T,15}$	kg	11870
Yaw moment of inertia of trailer (15% loaded)	$I_{T,15}$	$kg\,m^2$	179067
Mass of the trailer (80% loaded)	$m_{T,80}$	kg	20260
Yaw moment of inertia of trailer (80% loaded)	$I_{T,80}$	$kg\,m^2$	444363
Distance between the king-pin and the CG_T	a	m	6.94
Distance between the CG_T and the trailer axle	b	m	4.28

Table 1: Nominal parameters of the model

the CGs are due to the adapted steering gains.

How essential it is for the driver model to take into account the existence of a trailer can be seen by the behaviour of an 80% loaded semitrailer, Fig. 7. For the same trajectory as in Fig. 6 now the dynamics is more critical. Especially if the driver model A for the truck only will be used for the loaded truck-semitrailer combination, a stabilization of the initiated semitrailer swing seems impossible.

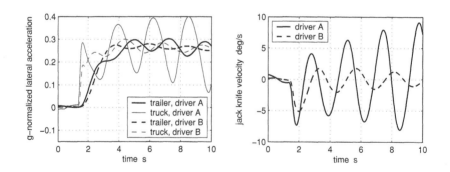

Fig. 7: Corresponding to Fig. 6 but with semitrailer 80% loaded, g-normalized lateral accelerations of truck and semitrailer and jack knife velocities $\dot\psi_R$

6 Conclusions

A comparison of a driver model design by the properties of the towing truck only and the driver model of the truck-semitrailer combination shows that especially for rapid changes in the trajectory curvature the smoother steering of the second approach is advantageous for the lane keeping of a truck-semitrailer combination. This is of particular importance if the semitrailer is heavily loaded. Consequently the driver and also the driver model need to take into account the semitrailer and its loading condition. The shown kind of driver model provides a first necessary step for investigations of closed loop truck-semitrailer manoeuvres.

References

[1] MacAdam, C.C.: *Understanding and Modeling the Human Driver.* Vehicle System Dynamics, Vol. 40, pp. 101-134, Swets & Zeitlinger, 2003.

[2] Mitschke, M.: *Dynamik des Kraftfahrzeuges, Band C, Fahrverhalten.* Springer Verlag Berlin Heidelberg, 1990.

[3] Plöchl, M., Lugner, P.: *A 3-level driver model and its application to driving simulations.* Vehicle System Dynamics, Supplement 33, pp. 71-80, Swets & Zeitlinger, 1999.

[4] *Using ADAMS/Solver.* Mechanical Dynamics, 2002.
Getting started using ADAMS/Car. Mechanical Dynamics, 2002.

[5] Gáspár, P., Palkovics, L., Bokor, J.: *Iterative Design of Vehicle Combinations for Stability Enhancement.* Vehicle System Dynamics Supplement 28, pp. 451-461, Swets & Zeitlinger, 1998.

[6] Valášec, M., Stejskal, V., Šika, Z., Vaculin, O., Kovanda, J.: *Dynamic Model of Truck for Suspension Control.* Vehicle System Dynamics Supplement 28, pp. 496-505, Swets & Zeitlinger, 1998.

[7] Fancher, P., Winkler, C., Ervin, R., Zhang, H.: *Using Braking to Control the Lateral Motions of Full Trailers.* Vehicle System Dynamics Supplement 28, pp. 462-478, Swets & Zeitlinger, 1998.

[8] Lugner, P., Mittermayr, P.: *A Measurement Based Tyre Characteristics Approximation.* Proc. of the 1st International Colloquium on Tyre Models for Vehicle Dynamics Analysis, Swets & Zeitlinger, 1993.

[9] Hirschberg, W., Rill, G., Weinfurter, H.: *User-Appropriate Tyre-Modeling for Vehicle Dynamics in Standard and Limit Situations.* Vehicle System Dynamics, Vol. 38 (2002), Nr. 2, pp. 103-125, Swets & Zeitlinger, 2002.

[10] Horn, A.: *Fahrer - Fahrzeug - Kurvenfahrt auf trockener Straße.* Dissertation TU Braunschweig, 1985.

[11] McRuer, D., Graham, B., Krendel, E.S., Reisner, W.: *Human pilot dynamics in compensatory systems.* AFFDL-TR-65-15, 1965.

Vehicle System Dynamics Supplement 41 (2004), p.332-340 © Taylor & Francis Ltd.

A Study on a Driver Model for Longitudinal Control on Heavy Duty Vehicle

RYO IWAKI[1] TETSUYA KANEKO[2] AND ICHIRO KAGEYAMA[3]

SUMMARY

This paper is concerned with a driver model of heavy-duty vehicles for longitudinal control using the multiple regression analysis. Firstly, as a longitudinal control, modeling of a fore-and-aft operation was constructed for each gear based on experiment data for heavy-duty vehicles. Next, the gear change model was constructed to select a longitudinal control model. Using the factor analysis for the model, the relationship between input and output was analyzed. Finally, the driver's control for longitudinal direction was described using this driver model.

Keywords: driver behavior, modeling, longitudinal control, heavy duty vehicle, regression analysis .

1. INTRODUCTION

These days, fuel consumption of vehicles must consider the viewpoint of economic and environmental problems. Normally, rate of fuel consumption of vehicles is evaluated using the 10- and 15-mode test. However, the rate in ordinary traffic flow is strongly affected by the control actions of the driver. Therefore, it is important to analyze the fuel consumption in the traffic flow from the viewpoint of human-machine-environment-system. In this study, a driver model is constructed to describe the control behavior for longitudinal direction.

2. DRIVER CONTROL ACTIONS

In general, it is considered that a driver goes through certain control actions by recognizing vehicle state values and the relative motion between the preceding vehicle and its own. That is to say, to analyze fuel consumption, it is necessary to cover not

[1] Address correspondance to: Graduate School of Nihon University 1-2-1 Izumi-cho, Narashino-shi, Chiba 275-8575, Japan.
[2] Osaka Sangyo University, Japan.
[3] College of Industrial Technology, Nihin University, Japan.

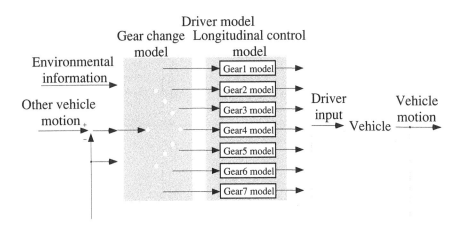

Fig. 1. Structure of a Model

only the vehicle itself but also the human and vehicle relationship. Firstly, the control actions of the driver can be classified into two main groups: lateral-control operation and longitudinal-control operation. The control actions of the driver, throttling, braking, steering, and gear-changing operation, are closely related. In this paper, we focus on the longitudinal control operation, such as throttling, braking, and gear-changing operation because of their effects on fuel consumption. Furthermore, throttling and braking operations are controlled separately for each gear. In the case of gear-changing operation, the gear shift is shifted up, down or in gear.

Several studies have been made on the analysis of control actions. In this study, to analyze differences in these control actions, the control actions of the driver are expressed mathematically, and the driver model was constructed to observe the information the driver processed. Regarding control operations of the driver, it is considered that each gear of the driver model was constructed appropriately. In other words, the driver model was divided into two parts: the gear-changing model and the longitudinal control model, as shown in fig.1.

The construction of the driver model of longitudinal control for heavy-duty vehicles and results of information processed are shown below.

Table 1. Vehicle Specifications.

Length	$[mm]$	11990
Width	$[mm]$	2490
Hight	$[mm]$	3775
Wheel Base	$[mm]$	7180
Weight	$[kg]$	12140
Displacement	$[cc]$	12882
Transmission		$7M/T$

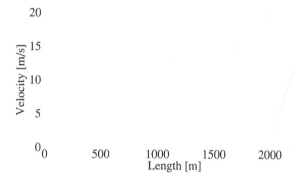

Fig. 2. Field Test Pattern

3. CONSTRUCTION OF THE DRIVER MODEL

3.1. Vehicle Test

To analyze the longitudinal control actions, the experiment used actual vehicles. The specifications of the vehicles for the experiment are as shown in Table 1. To compare two different drivers under the same condition, the experiments were carried out using two heavy-duty vehicles under the following conditions: a preceding vehicle controlled with the pattern-mode-running on a straight test course. Figure 2 shows pattern-mode-running with fixed velocity according to distance the vehicle progressed. In other words, The driving mileage and the running time of the experimental vehicles were the same during the whole test in principle. The experiment used two drivers: a good mileage driver and an aggressive driver. The twelve parameters of measurement were control actions of drivers: breaking force, throttling, and exhaust braking, and state values of vehicle: velocity, steering angle, right-and-left angular velocity of rolling tire, engine speed, and shift position, and environmental factor: distance between vehicles.

3.2. Longitudinal Control Model

The model, constructed using multiple regression analysis, had inputs of state variables, such as, relative distance between two vehicles, and output of control actions of the driver, such as, breaking force. The research of cross-correlation between inputs and output has indicated that coefficient of correlation with a time lag of zero has the highest figure. Thus, we have also determined that the model is static. The equation of the model using multiple regression analysis is as follows:

$$y = a_0 + a_1 \cdot x_1 + a_2 \cdot x_2 + \cdots + a_n \cdot x_n \qquad (1)$$

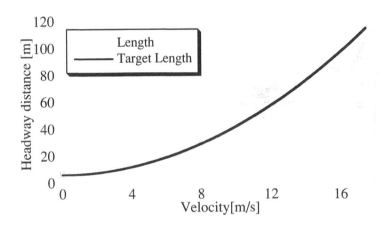

Fig. 3. Target length

The multiple regression analysis calculates the linear relational expression which is connected between dependent variable as the effect, and independent variable as the cause. And, the causal relationship between the dependent variable and the independent variable was analyzed by partial regression coefficient. The independent variables of this model were velocity, jerk, and distance between two vehicles; the dependent variables were throttling and braking force. For the dependent variables, the multi-input one-output structure of multiple regression analysis made the throttling and the braking operation into one variable. That is, non-dimension was carried out by dividing each data by the standard deviation, which assumed an average value of zero, respectively, to become one output. For the independent variables, vehicle state values were independent variables and relative deviation was information of distance between the two vehicles. The deviation is the difference between actual relative distance and desired relative distance determined by each driver. As fig.3 shows the results of pre-experiments, the desired relative distance is described by a quadratic equation to the vehicle speed. The deviations were the deviation of distance between two vehicles, integral of that deviation, and relative velocity. These make up the PID factors of relative distance. When the model was actually constructed, it was necessary to select one input using the step-wise-method because of the multicollinearity problem, which is the case of high correlation between inputs. The results of the step-wise method shows that inputs were velocity, acceleration, jerk, relative deviation, integral of relative deviation, and relative velocity. The results of the constructed model are displayed in Fig.4. These figures are the results of the longitudinal control model from each gear for 15 seconds. The vertical axis is "longitudinal control input" and the plus side shows normalized of throttling, and the minus side shows normalized of braking operation. In these fiugres, these models are accurate.

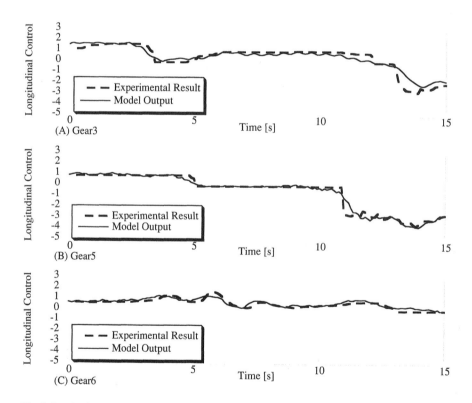

Fig. 4. Result of longitudinal control model

3.3. The gear-changing model

The gear-changing model was constructed by using multiple regression analysis. Output was shift position and inputs were velocity, acceleration, and engine speed. Therefore, when the driver changes gears on a M/T, we considered that the driver decides the shift position from not only the throttling operation, engine speed, and vehicle velocity but also environmental factors and acceleration. Since velocity of each gear is pretty much fixed, we judged that the gear-changing model could be constructed in linear form. And, the shift position of the model output was described with sufficient accuracy to calculate using threshold value, as shown in fig.5. However, the accuracy of the model fell when the gears were shifted up and down repeatedly. It is considered that modeling only velocity, acceleration, and engine speed is not sufficient to determine accurate results.

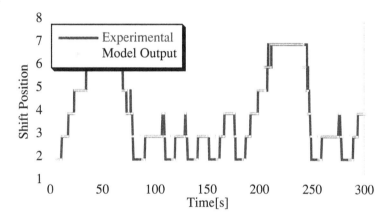

Fig. 5. Gear-changing model

These make up the complete driver model, which includes not only the fore-and-aft control model but also the gear-changing model, shown in fig.6.In the models of both drivers, coefficient of correlation between experimental results and the model output showed 0.95 or higher values.

4. THE FACTOR ANALYSIS

The factor analysis was performed by comparing the degrees of incidence for each independent variable. In this section, we compare drivers, and each gear. To compare the

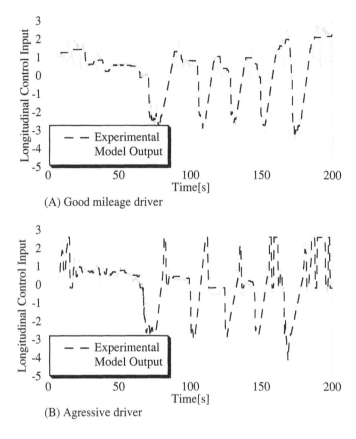

(A) Good mileage driver

(B) Agressive driver

Fig. 6. Comparison between field test and driver model

degree of influence, we used standard partial regression coefficient, which can compare relatively between the independent variable and the dependent variable. Standard partial regression coefficients were calculated by making the independent variable and dependent variable averages zero and the standard deviation one, so it would be non-dimensional. The results of the good-mileage driver are shown in fig.7(A) and the results of the aggressive driver are shown in fig.7(B). And, standard partial regression coefficient was given absolute value to cach parameter to make up the total, as shown in fig.7.

In comparison of degree of incidence for velocity between drivers, the good-mileage driver had a higher velocity at the lower gears, on the other hand, the aggressive driver had a lower velocity at the lower gears. That is, the good mileage driver in

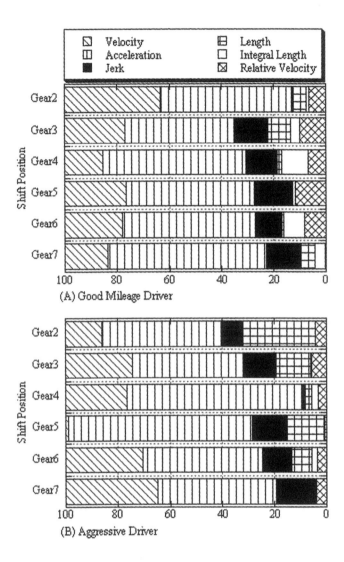

Fig. 7. Standardizede regression coefficient of longitudinal control input

low gear has more information about degree of incidence of engine speed because co-efficients of correlation between velocity and engine speed are 0.99 for each gear, so we can regard degree of incidence of velocity as engine speed. From the viewpoint of the low gear, the aggressive driver had higher parameters of relative variable between two vehicles than the good-mileage driver. In this analysis, It is thought that the aggressive driver, in second gear performs control actions while the driver is conscious of the preceding vehicle. And, the aggressive driver has more information about the degree of incidence of the vehicle, like acceleration and velocity. That is, the aggressive driver keeps smooth acceleration in mind at low speeds. From the viewpoint of the fourth and fifth gears, it is clear that the aggressive driver has wider parameters of acceleration than the good-mileage driver. The aggressive driver had a higher parameter of relative distance between the two vehicles than the good-mileage driver. And, the good-mileage driver had a higher parameter of relative velocity between the two vehicles than the aggressive driver. From this analysis, we concluded that the aggressive driver performs control actions while the driver is conscious of the distanse-between-two-vehicles. And the good-mileage driver performs control actions while predicting the future position of the preceeding vehicle.

Since these results model a driver's characteristics, we concluded that it describes driver behavior.

5. CONCLUSION

In this paper, when pursuing the preceding-vehicle, to describe the behavior of drivers who have different characteristics, we constructed the longitudinal control model by using multiple regression analysis. The longitudinal control model was constructed for each gear, and by analyzing the factors of the model, the information processed by each driver were clearly different. And, the gear-changed model was constructed by using multiple regression analysis too. As the output of the gear-changed model was a continuous value, the correct shift position was calculated using threshold value. Therefore, these make up the complete-driver model, which includes not only fore-and-aft control model but also the gear-changed model. That is, the model of longitudinal control directions of drivers for heavy-duty vehicle was constructed.

REFERENCES

1. Shintaro Ohnishi et al: A Study on Construction of Driver Model using Multivariate Analysis. *JSAE* 20005436, Japanese, 2000

Vehicle System Dynamics Supplement 41 (2004), p.341-350   © Taylor & Francis Ltd.

A semi-empirical, three-dimensional, tyre model for rolling over arbitrary road unevennesses

A.J.C. SCHMEITZ[1] AND H.B. PACEJKA

SUMMARY

In this paper a three-dimensional effective road surface description is introduced and a new three-dimensional enveloping tyre model is proposed. This new enveloping tyre model is used in combination with the well-known rigid ring dynamic tyre model to simulate both tyre in-plane and out-of-plane forces and moments. Simulation results of the tyre rolling over several oblique cleats have been compared with measurements for both the quasi-static rolling tyre and the tyre impacting cleats at high velocities. It is shown by comparing simulation results with measurements for both cases, that the proposed model works rather well.

1. INTRODUCTION

It is well known that dynamic tyre behaviour has a major impact on both the ride and handling properties of vehicles. Many significant ride events involve high frequency vibrations generated by short road irregularities that are not very long compared to the tyre contact patch. Therefore, tyre models are required that accurately describe wheel loads for high frequencies and that include tyre enveloping behaviour when rolling over arbitrary road profiles. Several models have been presented so far but many of these are not able to calculate the response on short irregularities such as (oblique) curbs, cleats and potholes accurately or are too complex to be used in vehicle dynamic simulations.

At the Delft University of Technology, research [1,2,3,4] was carried out during the last decade resulting in a rigid ring tyre model (SWIFT - Short Wavelength Intermediate Frequency Tyre - model) that can describe tyre behaviour under pure and combined slip conditions at relatively high frequency excitation. The model can deal with road irregularities of both short and long wavelengths. For modelling the tyre response on short irregularities such as curbs, cleats and potholes the highly non-linear tyre enveloping properties are taken into account. A suitable enveloping model for calculating these properties accurately for arbitrarily shaped two-dimensional road unevennesses was developed in the past four years and it was

---

[1] *Address correspondence to:* Faculty of Design, Engineering and Production, Delft University of Technology, Mekelweg 2, 2628 CD Delft, The Netherlands; Tel.: +31 15 278 6644, Fax: +31 15 278 2492; E-mail: a.j.c.schmeitz@wbmt.tudelft.nl

shown that the rigid ring model including this enveloping model is able to simulate tyre in-plane forces (longitudinal and vertical) and wheel spin velocity variations on these arbitrary road surfaces in a quite satisfactory way.

During the last year, research at Delft University has been focussed on extending the present model to account for three-dimensional road surfaces as well. An effective road surface description for three-dimensional road surfaces was proposed and a suitable enveloping model was developed. The rigid ring model including this new enveloping model is able to simulate both tyre in-plane and out-of-plane forces and moments.

The content of this paper is as follows. In the next section a brief overall model description is given. Subsequently it is described in detail how the tyre enveloping behaviour is modelled. After that, dynamic simulation results are compared with the results of oblique cleat experiments. Finally, the paper is summarised and conclusions are drawn.

2. MODEL DESCRIPTION

The dynamic tyre model consists of two main components: the well-known rigid ring model and an enveloping model. The enveloping model transforms the actual road surface into an effective road surface that serves as input to the rigid ring model. Besides the output forces and moments at the wheel axle the rigid ring model returns the contact patch dimensions (length and width) that serve as input to the enveloping model. See Figure 1.

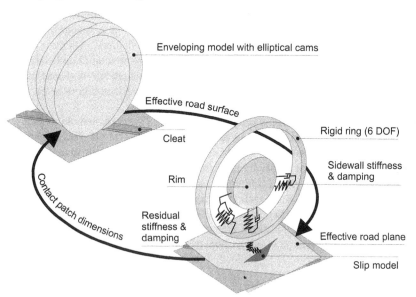

Fig. 1. Schematic representation of the tyre model consisting of the rigid ring and enveloping models.

The enveloping model consists of a number of parallel tandems. Each tandem consists of two elliptical cams that move over the actual road surface. From the vertical displacements of the cams the effective road surface position and orientation is derived.

The rigid ring model consists of the following components: The rim is modelled as a rigid body, the tyre sidewalls with pressurised air are modelled as springs and dampers, the tyre tread-band is modelled as a rigid ring and the contact model consists of residual stiffness and damping elements and a slip model (transient and Magic Formula models [2]). The rigid ring model is able to represent those motions of the tyre where the shape of the tread-band/ring remains circular, i.e. the flexible belt modes are neglected. The valid frequency range is therefore up to 50-80 Hz.

3. MODELLING THE TYRE ENVELOPING BEHAVIOUR

The rigid ring model has a single-point tyre-road interface. For relatively long wavelengths (much larger than the length of the contact patch), the geometry of the road surface can serve directly as input to the model. For short obstacles this method is not valid anymore and an effective road surface is used instead. The concept behind this effective road surface is that the quasi-static response of a tyre model with a single-point tyre-road interface on the effective road surface is similar to the quasi-static response of the (real) tyre on the actual road surface. In addition the assumption is used that the tyre contact zone dynamically deforms mainly in the same way as it does quasi-statically and that local dynamic effects can be neglected. Consequently, the effective road surface excitation can be assessed from the quasi-static enveloping properties of the tyre that comprehend the capability of the tyre to cushion (filter) small irregularities and it can be used to excite a dynamic tyre model. Comparison of numerous simulation results with measurements in [4] and [3] showed that these assumptions are valid for the in-plane, two-dimensional, case where the obstacles are positioned perpendicular to the rolling direction and cover the whole tyre width.

3.1 THE THREE-DIMENSIONAL EFFECTIVE ROAD SURFACE

The three-dimensional effective road surface is composed of four effective inputs. These inputs are the three two-dimensional effective inputs, that are the effective plane height (w) and the effective forward slope (β_y), that were introduced by Davis [5] and the effective rolling radius variation (r_e), that was introduced by Zegelaar [4], and in addition the here introduced effective road camber angle (β_x).

The two-dimensional effective road surface can be determined by conducting an experiment where the wheel is rolled at very low velocity over an uneven road surface at constant vertical load [3]. It is assumed that the resulting force (with the relatively small rolling resistance force omitted, that is: we assume that the tyre rolls on a frictionless road surface) is directed perpendicular to the effective road plane.

The effective plane height (w) is defined as the distance the wheel centre has to move vertically in order to keep the vertical load constant. From w and the vertical axle displacement (z_a) the vertical spindle force (K_z) or the radial deflection (ρ_z) can be obtained:

$$w - z_a = \frac{K_z}{C_z} = \rho_z \cos \beta_y \tag{1}$$

where C_z is the tyre vertical stiffness.

The effective forward slope can be obtained by dividing the longitudinal horizontal force (K_x) (after having subtracted the relatively small initial rolling resistance force ($F_r = -K_{x0}$)) by the vertical force:

$$\tan \beta_y = \frac{K_x - K_{x0}}{K_z} \tag{2}$$

Zegelaar [4] showed that the variation in effective rolling radius can be obtained approximately from the first two effective inputs:

$$\tilde{r}_e = r_{e0} \, \rho_z \frac{d\beta_y}{dx} \tag{3}$$

where r_{e0} is the effective rolling radius on a flat horizontal road surface and x the travelled distance. When the two-dimensional effective road surface is used in combination with a (dynamic) tyre model it is assumed that the slip model acts on the effective road plane.

In case of three-dimensional road surface unevennesses it is again assumed that, when the tyre moves quasi-statically over an uneven (assumedly frictionless ($\mu=0$)) road surface, the resulting force is directed perpendicular to the effective road plane, that now may also show a transverse slope $\tan\beta_x$. For small slopes the following equation holds approximately:

$$\frac{\tan \beta_x}{\cos \beta_y} \approx -\frac{K_y - F_y}{K_z} \tag{4}$$

The difficulty that now arises is: a tyre rolling at very low velocity over a three-dimensional road surface experiences camber and side slip. The consequence is that the unknown slip force (F_y) influences the resulting measured force on a friction surface. Since it is impossible to measure at $\mu=0$ conditions, camber and side slip forces must be taken into account when attempting to assess the height and slopes of the effective road plane. The unknown effective transverse slope may, in principle, be obtained by using Equation (4) in conjunction with a transient tyre model that accounts for the transient response of the tyre side force to camber variations ($\gamma = -\beta_x$). Through optimisation, the course of the effective transverse slope angle β_x may be established. To circumvent this rather tedious procedure, it is assumed that the existing single track enveloping model (that is: the tandem cam system, cf. [2]) extended to a double or multiple track system can be used to tackle the 3D unevenness.

3.2 THE THREE-DIMENSIONAL ENVELOPING MODEL

A finite element tyre model was used to study the deformation of a non-rolling tyre at zero friction on an oblique step. It was observed that in this specific case horizontal forces could only be transmitted along the sharp edge of the oblique step. Consequently, at zero friction the tangent of the cleat angle relates the slopes β_x and β_y. This observation did help in the development of the enveloping model.

The developed 3D enveloping model is an extension of the 2D enveloping model with elliptical cams that was developed by Schmeitz and is described in [2]. The assumption of the tandem model with elliptical cams is that the tyre enveloping behaviour can be modelled with two connected rigid elliptical cams moving over the actual road surface. Each ellipse contacts the road surface in at least one point. Simulations show that the distance between the connected elliptical cams (l_s), the so-called shift, is related to the contact length of the tyre. The ellipse shape parameters are assessed by fitting model responses to measurements of the tyre rolling at very low velocity over steps of different heights at a number of constant vertical loads. As may be expected, the elliptical cam shape turns out to be practically identical to the contour of the tyre in side view just in front of the contact zone (up to the height of the highest step considered in the fitting process). Once the ellipse parameters have been established, the cam dimensions are not affected anymore during the simulations. In Figure 2 the enveloping model moving over an oblique cleat is illustrated.

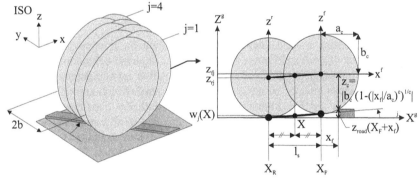

Fig. 2. Enveloping model with elliptical cams. On the left the 3D model consisting of several parallel tandems (in this case n=4) is shown. On the right one tandem is depicted.

The effective road surface is derived from the vertical displacements of the cams. The equations for obtaining the effective plane height and forward slope read:

$$w = \frac{1}{n}\sum_{j}^{n} w_j = \frac{1}{n}\sum_{j=1}^{n}\left(\frac{z_{fj}+z_{rj}}{2}-b_c\right) \tag{5}$$

$$\tan\beta_y = \frac{1}{n}\sum_{j=1}^{n}\left(\frac{z_{rj}-z_{fj}}{l_s}\right) \tag{6}$$

where n is the number of parallel tandems and z_{fj} and z_{rj} are the cam centre heights of the front and rear cams, respectively, of parallel tandem j. Notice that the effective forward slope is obtained by taking the average of the forward slopes of all parallel tandems. The forward slope of one tandem is calculated from the difference in front and rear cam heights (or basic profile heights) divided by the tandem length (l_s). The basic profile is defined as the profile that is obtained by moving with one cam over the actual road surface at a specific lateral position.

The effective road camber angle is obtained by calculating the average transverse slope of the connection lines of the extreme left and right basic profiles over the length of the tandem (l_s). Since the basic profile height at a certain longitudinal position can be obtained from the cam height at that position, the equation for calculating the effective road camber angle with m longitudinal cams reads:

$$\tan \beta_x = \frac{1}{m} \frac{1}{n-1} \sum_{i=1}^{m} \sum_{j=1}^{n-1} \left(\frac{z_{i,j+1} - z_{i,j}}{2b/(n-1)} \right) = \frac{1}{m} \sum_{i=1}^{m} \left(\frac{z_{i,n} - z_{i,1}}{2b} \right) \tag{7}$$

where $2b$ is the contact width of the tyre and $z_{i,j}$ the height of the ith cam centre of parallel tandem number j. Notice that the extreme right and left tandems have the numbers $j=1$ and $j=n$, respectively. Possibly, if one wants higher accuracy, weighting functions can be used that give more importance to cams at the front and rear edge of the contact patch.

Alternatively it can be shown that Equation (7) can be written as:

$$\tan \beta_x = \int_0^X \left(\frac{z_{fn} - z_{f1} - z_{rn} + z_{r1}}{2b \, l_s} \right) dX \tag{8}$$

where the effective road camber angle is obtained by integration over the travelled distance of the wheel centre (X). Notice that only the heights of the front and rear, extreme left and right, cams are required, which saves computational effort considerably. Disadvantages of using Equation (8) instead of Equation (7) are that one has to start on a flat horizontal road surface and that a (small) progressing integration error cannot be avoided. In general, the use of Equation (8) is very suitable for simulating oblique cleat impacts and the use of Equation (7) for the general case of relatively long arbitrarily uneven road surfaces. For a detailed discussion reference is made to [6]. All results shown in this paper were obtained with Equation (8).

The number of parallel tandems that has to be used depends on the contact width of the tyre and, in case of oblique cleats, the cleat angle (θ). In general one should aim to get smooth curves for the effective road surface quantities, which implies that if one parallel cam has traversed the obstacle the neighbouring cam (in lateral direction) should have started moving over it. For the tyre and obstacles used in this study at least 10 parallel tandems appeared to be necessary to obtain smooth curves at the highest vertical load (largest lateral distance between parallel tandems) and the largest cleat angle (60 degrees). If Equation (7) is used, at least 10 longitudinal cams are required to obtain a good approximation.

3.3 QUASI-STATIC MODEL VALIDATION

As discussed before, the three-dimensional enveloping model can only be validated in combination with a slip model. In this study the transient tyre model of Pacejka and Besselink described in [7,2] is used. This model can be regarded as a simplified version of the slip model [1,2] that is part of the SWIFT model. In this paper only contact patch side and camber slip as a result of (road) camber (β_x=-γ) occurs, since the wheel slip angle has been kept equal to zero. The (transient) slip forces and moments together with the normal force, which act from the effective road surface to the tyre, are transformed to obtain the spindle forces and moments.

Numerous cleat experiments at low velocity (2.3 cm/s) were carried out to assess the necessary parameters and to validate the enveloping model (with the transient slip model). In Figure 3 measurement results of spindle forces and moments are compared with simulations of the tyre rolling over oblique cleats of different angles (θ = 0°, -30°, -45° and -60°) at a constant vertical load of 4000 N. The cleat cross-section is 10x50 mm. It can be seen that for larger cleat angles the obstacle is filtered more and the response becomes longer. Furthermore, the amplitudes of the lateral forces (K_y), the overturning moments (T_x) and aligning torques (T_z) increase, while those of the longitudinal forces (K_x) and spin velocities (Ω) decrease. Notice that the amplitudes of the lateral and longitudinal forces are about equal at -60° and not as might have been expected at -45°. This is caused by side forces due to camber. In general the simulation results show a rather good agreement with the experimental results, certainly if one takes into account the simplicity of the model. The measured lateral forces show a somewhat larger relaxation effect than the simulations, i.e. the lateral force is still present after obstacle contact. This also causes the differences in spindle overturning moment. Furthermore, the model is not able to describe the first positive peak in the aligning torque responses.

4. VALIDATION OF THE DYNAMIC TYRE MODEL

To validate the theory of the three-dimensional effective road surface and to assess the performance of the developed enveloping model in combination with the SWIFT model, experiments with oblique cleats at various higher velocities (15, 25, 39, 59, 80 km/h) were carried out on the 2.5m diameter drum test stand of the Delft University at fixed axle position. As an example simulation results are compared with measurements for a 205/60 R15 tyre rolling over the drum with a velocity of 39 km/h and an initial static vertical load of 4000 N in Figure 4. The oblique cleat is 10 mm high, 50 mm long and mounted at an angle of 34 degrees with respect to the vertical plane through the wheel spin axis (slip angle zero). The responses of the vertical, longitudinal and lateral forces and those of the overturning moments and aligning torques and the spin velocity variations are indicated. In addition the power spectra of these quantities are shown.

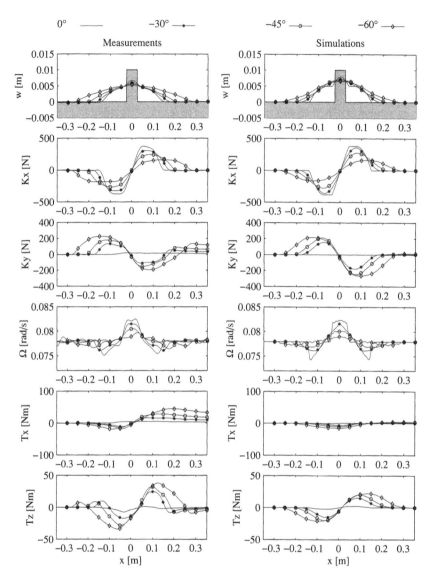

Fig. 3. Comparison of measurement results of spindle forces (K) and moments (T) (left) with simulations (right) of the tyre rolling at 2.3 cm/s over oblique cleats of different angles ($\theta = 0°$, -30°, -45° and -60°) at a constant vertical load of 4000 N. The cleat cross-section is 10x50 mm. ISO sign conventions are used.

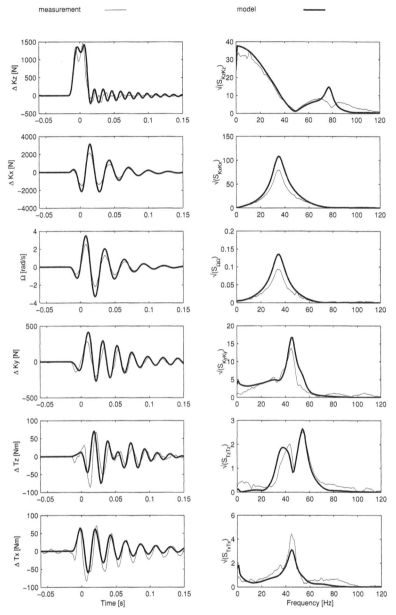

Fig. 4. Comparison of simulated spindle forces (K) and moments (T) with measurements for a 205/60 R15 tyre rolling over the drum with a velocity of 39 km/h at fixed axle height and an initial static vertical load of 4000 N. The oblique cleat is 10 mm high, 50 mm long and mounted at an angle of 34 degrees with respect to the vertical plane through the wheel spin axis (wheel slip angle is zero). ISO sign conventions are used.

Figure 4 shows that simulation results agree rather well with measurements. The amplitudes of the time responses are about right and the resonance peaks of the tyre are predicted quite well. The following mode shapes can be observed: rotational mode at 35 Hz (K_x and Ω), camber mode at 46 Hz (K_y and T_x) and yaw mode at 54 Hz (T_z). It is interesting to notice the amplitude differences between the longitudinal and lateral forces. The figure shows that the longitudinal forces are about 10 times larger than the lateral forces, while this factor is much less for the quasi-static case (Fig. 3). The reason for this difference is that due to effective rolling radius variations longitudinal slip is generated that results in large (additional) longitudinal forces.

Finally, it may be remarked that the present model can easily be extended for the case of moving over uneven roads at a (varying) wheel inclination angle (wheel camber).

5. SUMMARY AND CONCLUSION

In this paper a three-dimensional effective road surface description is introduced and a new three-dimensional enveloping tyre model is proposed. In combination with the well-known rigid ring dynamic tyre model, both tyre in-plane and out-of-plane forces and moments can be simulated. Simulation results of the tyre rolling over several oblique cleats have been compared with measurements for both the quasi-static rolling tyre and the tyre impacting cleats at high velocities. It is shown that simulation results agree quite well with measurements for both cases. An extensive description of the research will be published in [6].

REFERENCES

1. Maurice, J.P.: Short Wavelength and Dynamic Tyre Behaviour under Lateral and Combined Slip Conditions. PhD Thesis, Delft University of Technology, Delft, The Netherlands, 1999.
2. Pacejka, H.B.: Tyre and Vehicle Dynamics. Butterworth-Heinemann, An imprint of Elsevier Science, ISBN 0-7506-5141-5, www.bh.com, 2002.
3. Schmeitz A.J.C. and Pauwelussen J.P.: An Efficient Dynamic Ride and Handling Tyre Model for Arbitrary Road Unevennesses. VDI-Berichte, NR. 1632, pp. 173-199, VDI Verlag GmbH, Düsseldorf, Germany, 2001.
4. Zegelaar, P.W.A.: The Dynamic Response of Tyres to Brake Torque Variations and Road Unevennesses. PhD Thesis, Delft University of Technology, Delft, The Netherlands, 1998.
5. Davis, D.C.: A radial-spring terrain-enveloping tyre model. Vehicle System Dynamics, Vol. 3, 1974, pp. 55-69.
6. Schmeitz, A.J.C.: A semi-empirical three-dimensional model of the pneumatic tyre rolling over arbitrarily uneven road surfaces. PhD Thesis, Delft University of Technology, Delft, The Netherlands, to be published in 2004.
7. Pacejka, H.B. and Besselink, I.J.M.: Magic formula tyre model with transient properties. 2nd International Colloquium on Tyre Models for Vehicle Dynamic Analysis, Berlin, Germany, February 20-21, 1997, Vehicle System Dynamics, Vol. 27 supplement, 1996, pp. 234-249.

Vehicle System Dynamics Supplement 41 (2004), p.351-360 © Taylor & Francis Ltd.

Investigation of steady-state tyre force and moment generation under combined longitudinal and lateral slip conditions

GEORGE MAVROS, HOMER RAHNEJAT AND PAUL KING[1]

SUMMARY

The paper provides an insight into the contact mechanical behaviour of pneumatic tyres in a wide range of steady-state operating conditions. Tyre forces and self-aligning moment generation during steady-state manoeuvres are studied in some depth. For this purpose, two different versions of a dynamic model of a tyre are developed. The simplest version consists of a one-dimensional series of bristles distributed on the tyre periphery. The bristles incorporate anisotropic stiffness and damping in the lateral and longitudinal directions, while the distributed tread mass is also taken into account. The vertical pressure distribution along the contact patch is assumed to be parabolic and the length of the contact area is assumed to be known apriori. The friction forces developed on the contact patch follow a stick-slip friction law.

The second version of the tyre model improves the potential of the simple model by introducing radial and tangential stiffness and damping, as well as a Kelvin element for rubber behaviour in the simulation of the impact on the leading edge of the contact area. The Kelvin model closely conforms to the semi-infinite incompressible nature of rubber. The tyre models show effective reproduction of measured longitudinal and lateral forces, as well as the self-aligning moment, under pure side-slip, pure longitudinal slip and combined slip situations. The generated curves show qualitative concordance with the results obtained experimentally, or by semi-empirical models such as the Pacejka's Magic Formula. In addition, the tyre models seem to be capable of reproducing the generated contact pressure profiles and the shape of the observed variations in tyre forces between side-slipping, braking and traction diagrams. An investigation of these three situations reveals the different mechanisms that result in the different shapes of the diagrams.

Finally, a study is carried out for tyre behaviour at very high speeds, which indicates deviations from the results of traditional investigations.

1. INTRODUCTION

The primary aim in any tyre analysis is the prediction of tyre forces and moments applied on the wheel-hub, under various driving conditions. Due to the highly non linear and interdependent nature of these forces, it is often necessary to look into a number of different aspects of tyre behaviour.

Pacejka and Sharp [1] provide with a comprehensive review of modelling aspects depicting all the current trends in steady state tyre modelling. Increasingly complex points of view are presented, including anisotropic tyre behaviour, various shapes of contact pressure distribution, tyre belt deformation and camber angle effects. It is

---

[1] Wolfson School of Mechanical and Manufacturing Eng. Loughborough University LE11 3TU, UK

evident that the use of brush-type models has enhanced our understanding of tyre behaviour and has also contributed in the area of tyre simulation. Sakai [2-5] provides with a detailed analysis on tyre force generation, covering almost all aspects of it. The author's investigations start from basic definitions and the presentation of simple tyre models. The frictional behaviour of rubber, the shape of the contact pressure distribution and some experimental and simulation techniques are also discussed. Bernard et al [6] use a brush model accompanied by a simple friction law and a normal pressure distribution known beforehand in order to develop a tyre model appropriate for computer simulation. Sharp et al [7] use a multi-spoke model that is able to generate tyre shear forces as well as the normal pressure distribution, while Levin [8] uses an anisotropic brush model with distributed mass that emphasises on the shape of the pressure distribution and the generation of tyre forces in high speeds.

The models presented in this paper attempt to enhance understanding of tyre behaviour under various driving conditions. A critical view is maintained and the effects of different assumptions are discussed in detail. Some limitations of the brush models are presented and the effect of high rolling velocities is also investigated.

2. DESCRIPTION OF THE SIMPLE VERSION OF THE TYRE MODEL

2.1 MODELLING OF THE TYRE

The behaviour of tyre can be ascertained by discretisation of the continuum, as is the case for all contact mechanical treatments. The generation of shear forces is based on the brush concept. The tyre tread is modelled as a one-dimensional series of bristles distributed on the tyre periphery. The bristles incorporate anisotropic stiffness and damping in the lateral and longitudinal directions and the distributed tread mass on the tyre periphery is also taken into account by attaching an infinitesimal mass to the end of each bristle. Initially, for the sake of simplicity, the length of the contact patch is assumed constant and the vertical pressure distribution is regarded as parabolic.

All tyre properties are expressed per unit length of the tread. The mass distribution along the tyre tread is $C = dm/dx$, the stiffness coefficients per unit tread length are $K_x = dF/dxdx$, $K_y = dF/dydx$ for the longitudinal and lateral deformations of the bristles respectively, and the coefficients of damping are $D_x = dF/du_x dx$, $D_y = dF/du_y dx$, where u_x, u_y denote the rate of change of the longitudinal and lateral deformations of the bristles respectively.

Figure 1 shows a side view of the tyre model including the normal pressure distribution, and a top view of the model, depicting an arbitrary position of the infinitesimal mass dm, which corresponds to a length dx of the tyre tread. A bristle connecting the mass to the wheel periphery, is deformed laterally as well as longitudinally and, the mass may or may not be sliding on the ground, depending on the forces applied by the bristle (viscoelastic element), the vertical force at the specific position and the local coefficient of friction.

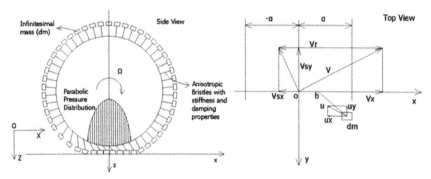

Fig. 1. Side and Top view of the tyre model, depicting an arbitrary position of infinitesimal mass.

The equation of the parabola, yielding the vertical force distribution per tyre tread length can be easily derived, given the length of the contact patch $l = 2 \cdot a$ and taking into account the fact that the integral of the force along the contact patch balances the normal load applied on the wheel hub by the suspension:

$$F_{vertical} = \frac{3 \cdot F_z}{4 \cdot \alpha} \cdot \left(1 - \frac{x}{a}\right)^2 \tag{1}$$

where F_z denotes the total vertical force applied on the wheel hub.

Two different sets of axes are used in order to describe the motion of the tyre and its components. The global frame of reference (O, X, Y, Z) is attached to the ground, while a second frame of reference (o, x, y, z) has its origin on a point in the contact patch, where the vertical line from the centre of the wheel plane meets the ground. Both frames are shown in Figure 1.

The longitudinal component of the velocity of the bristle base relative to the ground (i.e. in the global frame of reference) is V_{sx}, the lateral component being V_{sy}. When the vertical force results in the generation of a high enough frictional force, the infinitesimal mass dm sticks to the ground. In any other case, the mass moves with respect to the ground with a sliding velocity u, consisting of u_x and u_y in the longitudinal and lateral directions respectively.

For the purpose of the analysis, the motion of the infinitesimal mass is traced throughout the contact patch and beyond the end of it. The analysis, therefore, considers the physics of motion of a typical infinitesimal mass through the contact to be representative of all such discrete elements. Thus, the proposed model represents steady state contact conditions.

2.2 MODELLING OF FRICTION

Friction f between the infinitesimal mass and the road is assumed to follow a stick-slip friction law. Karnopp [9] provides with a way of modelling the stick-slip behaviour for use in computer simulations. This approach is valid for hard, almost elastic materials such as steel. Visco-elastic materials with low stiffness and considerable internal damping (such as rubber) show a more complicated frictional behaviour that strongly depends on the vertical load, apparent contact area, sliding velocity and temperature [2]. Nevertheless, experimental work [10] indicates that stick-slip behaviour is not only evident in rubber contacts, but in some cases the friction laws appear to be very similar to the laws describing hard materials. Thus, as a starting point for the present analysis, a typical stick-slip law is chosen with a transition between the static and a constant kinetic friction.

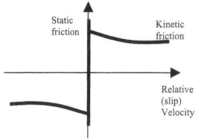

Fig. 2. Stick-Slip Friction Law.

2.3   EQUATIONS OF MOTION FOR THE INFINITESIMAL MASS

The motion of an infinitesimally small length dx of the tyre tread is considered. According to the definitions given previously for the mass, stiffness and damping distributions, the equations of motion of the infinitesimal mass dm which corresponds to length dx, are written as follows (in accordance with Figure 1):

$$\ddot{u}_x \cdot C \cdot dx = (x_s - x) \cdot K_x \cdot dx + (V_{sx} - u_x) \cdot D_x \cdot dx - f_x \tag{2}$$

$$\ddot{u}_y \cdot C \cdot dx = (y_s - y) \cdot K_y \cdot dx + (V_{sy} - u_y) \cdot D_y \cdot dx - f_y \tag{3}$$

$$\dot{x}_s = V_{sx} \tag{4}$$

$$\dot{y}_s = V_{sy} \tag{5}$$

where f_x, f_y denote the friction forces in the longitudinal and lateral directions respectively. Forces f_x and f_y depend on the normal force at the specified point of the contact patch, the friction coefficient and the direction of motion of the infinitesimal mass.

3. EXPANSION OF THE SIMPLE TYRE MODEL

A mechanism has to be incorporated in order to simulate the build-up of vertical pressure distribution on the tyre contact patch. For this purpose, a couple of modifications are made to the aforementioned simple model. First, radial and tangential stiffness and damping characteristics are introduced. Secondly, a Kelvin element is used to connect the infinitesimal mass to the ground. The Kelvin element is used mainly for the simulation of the impact of a viscoelastic solid, while the combined effect of tangential and radial compliance replaces the longitudinal stiffness and damping of the bristles. The modified tyre model is shown in Figure 3.

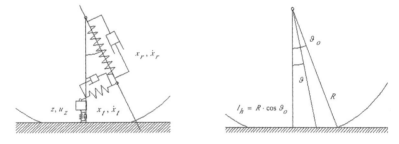

Fig. 3. The modified tyre model and some important dimensions

The derivation of the equations of motion for the infinitesimal mass is achieved by depicting a random position of the mass, after it has entered the contact patch. Referring to Figure 4, equations (2)-(5) can be re-written as follows:

$$\ddot{u}_x \cdot C \cdot dx = K_r \cdot x_r \cdot \sin \vartheta \cdot dx - K_t \cdot x_t \cdot \cos \vartheta \cdot dx$$
$$+ D_r \cdot \dot{x}_r \cdot \sin \vartheta \cdot dx - D_t \cdot \dot{x}_t \cdot \cos \vartheta \cdot dx - f_x \tag{6}$$

$$\ddot{u}_y \cdot C \cdot dx = (y_s - y) \cdot K_y \cdot dx + (V_{sy} - u_y) \cdot D_y \cdot dx - f_y \tag{7}$$

$$\ddot{u}_z \cdot C \cdot dx = C \cdot g \cdot dx + K_r \cdot x_r \cdot \cos \vartheta \cdot dx + K_t \cdot x_t \cdot \sin \vartheta \cdot dx$$
$$+ D_r \cdot \dot{x}_r \cdot \cos \vartheta \cdot dx + D_t \cdot \dot{x}_t \cdot \sin \vartheta \cdot dx - K_z \cdot x_z \cdot dx - D_z \cdot u_z \cdot dx \tag{8}$$

$$\dot{x} = u_x \tag{9}$$

$$\dot{y} = u_y \tag{10}$$

$$\dot{z} = u_z \tag{11}$$

$$\dot{y}_s = V_{sy} \tag{12}$$

If the state variables of the dynamic system are chosen as $x, y, z, u_x, u_y, u_z, y_s, x_r, x_t$, it becomes obvious that the set of differential equations is insufficient to arrive at an analytic solution. Furthermore, the absence of variable x_s disables the definition of longitudinal slip.

The problem is solved by introducing a set of kinematic constraints. Referring to Figure 3, the radius of the tyre just before entering the contact patch of the vertically loaded rolling tyre is R. This radius is assumed to remain constant in the area outside the contact patch and equal to the radius of the unloaded tyre. The angle between R and the vertical line, connecting the tyre centre to the ground is ϑ_o. Given R, the height of the tyre centre above the ground is $l_h = R \cdot \cos \vartheta_o$. This height is assumed to be equal to the radius of the vertically loaded tyre under pure rolling conditions, R_d.

At an arbitrary position of the mass inside the contact patch, the total deflection ΔR is given by the following scalar constraint function:

$$\Delta R = R - R \cdot \frac{\cos \vartheta_o}{\cos \vartheta} \tag{13}$$

The vertical projection of the total deflection ΔR is:

$$\Delta R_{vertical} = \Delta R \cdot \cos \vartheta = R \cdot (\cos \vartheta - \cos \vartheta_o) \tag{14}$$

At any instance of time, $\Delta R_{vertical}$ is equal to the vertical deflection of the Kelvin element (i.e. the "local" deflection) added to the vertical components of the deflections of the radial and tangential elements (i.e. the "global" deformation of the continuum), as described in the following equation:

$$x_r \cdot \cos \vartheta + x_t \cdot \sin \vartheta + x_z = R \cdot (\cos \vartheta - \cos \vartheta_o) \tag{15}$$

Analogously, the horizontal components of the deflections of the radial and tangential elements are equal to $x_s - x$, which represents the distances between the base and the tip of the bristles, as defined previously for the simple tyre model. Consequently, the following constraint function also holds true:

$$x_r \cdot \sin \vartheta - x_t \cdot \cos \vartheta = x_s - x \qquad (16)$$

Equations (15), (16), provide a way of expressing x_r, x_t with respect to x, x_s and ϑ, while their derivatives provide two more relations, which include velocities $\dot{x}_r, \dot{x}_t, \dot{\vartheta} = \omega$, as follows:

$$\dot{x}_r \cdot \cos \vartheta - x_r \cdot \omega \cdot \sin \vartheta + \dot{x}_t \cdot \sin \vartheta + x_t \cdot \omega \cdot \cos \vartheta + u_z = -R \cdot \omega \cdot \sin \vartheta \qquad (17)$$

$$\dot{x}_r \cdot \sin \vartheta + x_r \cdot \omega \cdot \cos \vartheta - \dot{x}_t \cdot \cos \vartheta + x_t \cdot \omega \cdot \sin \vartheta = V_{sx} - u_x \qquad (18)$$

Equations (15)-(18) result in the omission of variables $x_r, x_t, \dot{x}_r, \dot{x}_t$ and enable the formulation of the system of differential equations in a way that x_s is retained as a state variable, thus providing a means for the definition of longitudinal slip.

The rate of change of angle ϑ can be obtained by solving the equation of motion of the wheel for the rotational degree of freedom, while height l_h, or alternatively angle ϑ_o (for a given radius R) can be obtained by solving the equation of motion of the wheel for the vertical degree of freedom.

4. RESULTS AND DISCUSSION

A number of simulations were carried out using the 4th order fixed step Runge Kutta method to solve the non-linear system of differential equations. The various tyre parameters were chosen according to [2-5], [8] and [10].

Figure 4 shows a typical response of the tyre model in pure cornering conditions, while Figure 5 presents a typical braking force diagram and a case of combined braking and cornering. The force diagrams appear to be smooth and show qualitative concordance with experimental and empirical curves. A good quantitative agreement with experiments was achieved by modifying tyre parameters; nevertheless a consistent parameter identification technique is yet to be developed.

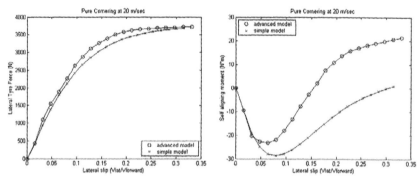

Fig. 4. Cornering force and self aligning moment

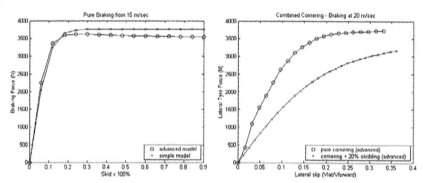

Fig. 5. Braking Force and lateral force during combined cornering - braking

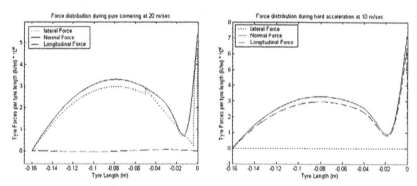

Fig. 6. Normal pressure distribution for purely cornering and traction conditions

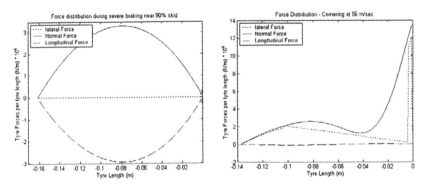

Fig. 7. Normal pressure distribution for hard braking and high speed cornering

Figures 6 and 7 deal with the shape of the normal pressure distribution in relation with the build up of tyre shear forces. According to the diagrams, the intensity of the impact in the beginning of the contact patch determines the shape of the Force-Slip relations. Pure cornering lies midway between traction and braking. The rolling velocity of the tyre corresponds to a forward velocity of 20 m/sec and the leading edge impact does not cause much distortion to the parabolic shape of the distribution. During hard acceleration, the increased rolling velocity causes deviations from the parabolic shape, while the opposite effect is observed during braking, when rolling velocity deteriorates. High velocity cornering, as shown in the second diagram of Figure 7, causes a higher distortion of the shape of the normal pressure distribution and this in turn effects the generation of the cornering force.

Simulation also indicates that the generation of shear forces depends largely on the initial conditions of the infinitesimal mass. Assuming a rigid tyre belt with the bristles connected to it, it is sensible to use V_{sx} and/or V_{sy} as initial velocities for the infinitesimal mass. By doing so, the mass is initially sliding and has to be decelerated by friction in order to enter the "stick" phase. This produces an area of increased friction force in the beginning of the contact patch, as shown in Figures 6 and 7.

In reality, the tyre belt deforms laterally as well as longitudinally. In case of cornering, the direction of the mass just before entering the contact patch is such that part of the circumferential velocity compensates for V_{sy}. Thus, the relative initial velocity of the mass can be less than V_{sy} and in cases of moderate cornering could even reduce to zero. Consequently, the overall friction forces decrease for the same normal pressure distribution. This finding points out a major shortcoming of brush type models and indicates the need of their integration with a belt-carcass deformation model.

While the peak at the leading edge of the contact patch is evident in experimental results [8], it is not as intense as predicted by the current analysis. This is explained by the fact that the impact at the leading edge of the contact patch is not the only mechanism for the vertical deceleration of the mass. In fact, most of the deceleration is achieved internally, by the local deformation of the tyre belt in the neighborhood of the leading edge. Again, the need of integration of the brush model with a belt deformation model is obvious.

5. CONCLUSION

A new formulation of the generation of normal and shear tyre forces is presented. With the aid of numerical integration, a number of different operating conditions are studied and some comments are made concerning the effects of different kinds of manoeuvres - namely cornering, braking and traction - on tyre behaviour. Finally, two basic shortcomings of the brush-type models are highlighted and the effect of considerably high operating speeds is addressed.

REFERENCES

1. Pacejka H.B. and Sharp R.S.: Shear Force Development by Pneumatic Tyres in Steady-State Condition: A Review of Modelling Aspects. Vehicle System Dynamics 20 (1991), pp. 121-176.
2. Sakai H.: Theoretical and Experimental Studies on the Dynamic Properties of Tyres, Part 1: Review of theories of Rubber Friction. Int. J. of Vehicle Design 2(1) (1981), pp. 78-110.
3. Sakai H.: Theoretical and Experimental Studies on the Dynamic Properties of Tyres, Part 2: Experimental Investigation of Rubber Friction and deformation of a tyre. Int. J. of Vehicle Design 2(2) (1981), pp. 182-226.
4. Sakai H.: Theoretical and Experimental Studies on the Dynamic Properties of Tyres, Part 3: Calculation of the six components of force and moment of a tyre. Int. J. of Vehicle Design 2(3) (1981), pp. 335-372.
5. Sakai H.: Theoretical and Experimental Studies on the Dynamic Properties of Tyres, Part 4: Investigations of the influences of running conditions by calculation and experiment. Int. J. of Vehicle Design 3(3) (1982), pp. 333-375.
6. Bernard J.E., Segel L. and Wild R.E.: Tyre Shear Force Generation During Combined steering and Braking Maneuvers. SAE paper 770852 (1977), pp. 2953-2969.
7. Sharp R.S. and El-Nashar M.A.: A Generally applicable Digital Computer Based Mathematical Model for the Generation of Shear Forces by Pneumatic Tyres. Vehicle System Dynamics 15 (1986), pp. 187-209.
8. Levin M.A.: Investigation of features of Tyre Rolling at Non-Small Velocities on the Basis of a Simple Tyre Model with Distributed Mass Periphery. Vehicle System Dynamics 23 (1994), pp. 441-466.
9. Karnopp D.: Computer Simulation of Stick-Slip Friction in Mechanical Dynamic Systems. Transactions of the ASME – Journal of Dynamic Systems, Measurement and Control 107, pp. 100-103.
10. Braghin F., Cheli F. and Resta F.: Friction Law Identification For Rubber Compounds on Rough Surfaces at Medium Sliding Speeds. 3rd AIMETA International Tribology conference, Salerno, Italy, 18-20 September 2002
11. Denti Eugenio and Daniele Fanteria.: Models of Wheel Contact Dynamics: An Analytical Study on the In-Plane Transient Responses of a Brush Model. Vehicle System Dynamics 34 (2000), pp. 199-225.
12. Yu Zeng-Xin, Tan Hui-Feng, Du Xing-Wen and Sun-Li: A Simple Analysis Method for Contact Deformation of Rolling Tire. Vehicle System Dynamics 36(6) (2001), pp. 435-443.

Vehicle System Dynamics Supplement 41 (2004), p.361-370 © Taylor & Francis Ltd.

The Development of The Tire Side Force Model Considering The Dependence of Surface Temperature of Tire

MASAHIKO MIZUNO[1], HIDEKI SAKAI[*,2], KOZO OYAMA[*] AND YOSHITAKA ISOMURA[**]

SUMMARY

The aim of this paper is to develop the tire side force model including the influence of tire surface temperature. The model consists of the part of thermodynamics and the part of side force change. The parameters of the tire model are identified using the test results of indoor test facility.

The calculation results of the change of tire side force and surface temperature by the developed model agreed very well with the measurements.

Keywords: Tire Model, Thermodynamics, Simulation.

1. INTRODUCTION

In vehicle dynamics studies, an accurate description of tire characteristics is essentially important for a wide range of vehicle behaviour. For example, in the development of the control systems to improve the vehicle stability and control such as an ABS and VSC (Vehicle Stability Control system), the accurate tire model to cover the wide range is required in the simulation tools for the evaluation of those systems.

In general, the measured results by indoor test facilities are used to develop the tire models with high accuracy. However there are some problems in the measurement method using indoor test facilities. The one of those problems is that the tire measuring input conditions do not fit to the actual driving conditions of a certain vehicle and tire combination.

[1] Address correspondance to: Research Domain 16, Toyota Central R&D Labs., Inc., Nagakute, Aichi, 4801192 JAPAN. Tel:(561)63-6747; Fax:(561)63-8346; E-mail:mizuno@vdlab.tytlabs.co.jp
[2] Address correspondence to: Chassis System Development Div., Vehicle Engineering Group, Toyota Motor Corporation, 1200, Mishuku, Susono, Shizuoka, 4101193 JAPAN. Tel:(55)997-7893; Fax:(55)997-7872; E-mail: hideki@sakai.tec.toyota.co.jp
[*] Chassis System Development Div., Vehicle Engineering Group, Toyota Motor Corporation, 1200, Mishuku, Susono, Shizuoka, 4101193 JAPAN.
[**] Chassis System Development Div., Vehicle Engineering Group, Toyota Motor Corporation, 1, Toyota-cho, Toyota, Aichi, 4718572 JAPAN.

From this point of view, the authors tried to make the tire model using the measured data of a vehicle running on actual roads[1]. On the other hand, the test procedure for indoor test facilities was developed[2] to measure the tire characteristics under the realistic driving conditions for modelling the tire force and moment properties.

Even if such modelling and measuring techniques are applied, other conditions like tire surface temperature cannot be fitted to all the cases of actual running conditions.

Therefore, the new tire side force model considering the change of tire surface temperature was developed and will be presented in this paper.

In the next chapter, the model structure will be explained. Then the deviation method of the model parameters will be shown.

Finally, the calculated results by the new model will be compared with the measured results by the indoor test facility.

2. THE MATHEMATICAL MODEL OF THE TIRE SIDE FORCE CONSIDERING THE DEPENDENCE OF TIRE SURFACE TEMPERATURE

The authors propose the two functions to build the new tire side force model considering the dependence of tire surface temperature. There are as follows:

- The thermodynamic model to describe the change of tire surface temperature
- The tire side force model considering the change of tire surface temperature

After here, these models are described.

2.1. The Thermodynamics Model of Tire Surface

The thermodynamic model of tire surface was defined with the assumption as follows:

1. The thermal input is occurred in the contact area.
2. The thermal output is also occurred in the contact area.
3. Those input and output are simultaneously occurred in the same area.

From those, the tire surface temperature T is described as,

$$W\frac{dT}{dt} = q - \lambda A(T - T_0) \tag{1}$$

where,
$\quad W$: heat capacity
$\quad q$: heat flux
$\quad \lambda$: thermal conductivity

A : contact area

T_0 : ambient temperature

Next, the heat flux q is described with the assumption that the work of tire to the road surface is changed to the thermal energy as follows:

$$q = F_y V \alpha \tag{2}$$

where,

F_y : tire side force

V : vehicle velocity

α : slip angle

The Eq. (1) is rewritten by using Eq. (2) as follows:

$$\frac{dT}{dt} = \frac{1}{W} [F_y V \alpha - \lambda A (T - T_0)] \tag{3}$$

The Eq. (3) is the differential equation that describes the relationship between the tire side force and the tire surface temperature, so the surface temperature can be calculated by using the Eq. (3)

In this equation, the parameters to be derived are two, one is W and another is λ.

2.2. The Tire Side Force Model Considering the Effect of Tire Surface Temperature

The tire side force model considering the effect of the tire surface temperature was made by using the force adjusting function, which describes the relationship between the change of the surface temperature and the change of the tire forces.

The side force adjusting model was defined with the assumption as follows:

1. The side force is changed linearly by the tire surface temperature.
2. The changing ratio of side force versus tire surface temperature is effected by the slip angle and the vertical load.

From those point of view, the force adjusting model is described as follows:

$$F_y (F_z, \alpha, \gamma, T) = F_y^* (F_z, \alpha, \gamma)$$
$$\cdot \left[1 + \left(1 + \frac{F_z - F_z^n}{F_z^n} \cdot \frac{\partial \mu}{\partial F_z} \right) \cdot \left(1 + \frac{\partial \mu}{\partial \alpha} |\alpha| \right) \cdot \frac{\partial \mu}{\partial T} (T - T_m) \right] \tag{4}$$

where,

F_y^* : steady state tire side force model like Magic Formula tire model

F_z^n : nominal load

F_z : vertical load

γ : camber angle

T_m : average tire surface temperature during measuring the steady state tire model data

$\partial\mu/\partial T$: variation of tire side force with the tire surface temperature

$\partial\mu/\partial F_z$: variation of the parameter $\partial\mu/\partial T$ with vertical load

$\partial\mu/\partial\alpha$: variation of the parameter $\partial\mu/\partial T$ with slip angle

In this equation, the parameters to be derived are three. The parameters of tire surface temperature effect are $\partial\mu/\partial\alpha$ and $\partial\mu/\partial T$, and the parameter of vertical load effect are $\partial\mu/\partial F_z$.

In next section, the derivation method of the parameters will be shown in Eq. (3) and (4).

3. THE DERIVATION OF THE MODEL PARAMETERS

3.1. The Method of Derivation of Parameters in the Function of Tire Surface Temperature

At first, the experimental methods of derivation of the parameters λ and W in Eq. (3) are described. The Eq. (3) can be described as,

$$\frac{dT}{dt} = \frac{F_y V\alpha}{W} - \kappa\left(T - T_0\right) \tag{5}$$

Where κ is the new parameter as follows:

$$\kappa = \frac{\lambda A}{W} \tag{6}$$

The parameter κ can be measured when the tire side force is zero and the tire surface temperature is changing. This is the situation of tire surface temperature cooling down. The sequence of the slip angle change in the measurement test is shown in figure 1.

When $\alpha = 0$, the parameter κ can be calculate using the measured result of tire surface temperature as follows:

$$\kappa = \frac{dT}{dt} \cdot \frac{1}{T - T_0} \tag{7}$$

The example of measured result of κ is shown in figure 2. The parameter κ is changing as time goes on. In this case, the parameter κ is the average value of the stable situation (between 10 and 20 seconds) to remove the thermal noise on measuring systems. The measured tire size is 205/65R16 for passenger vehicle. After here, the experimental result is the measured data of 205/65R16 PC Tire.

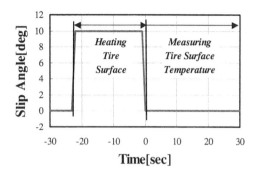

Fig. 1. Slip Angle to Measure the Parameter κ and W

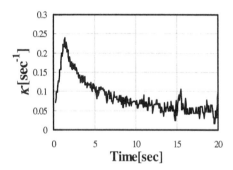

Fig. 2. Example of Measured Result of κ [Vertical load : Nominal Load, Camber Angle : 0]

Secondly, the experimental method of derivation of the parameter W is described. The Eq. (3) can be described as,

$$W = \frac{F_y V \alpha}{\left[\dfrac{dT}{dt} + \kappa\,(T - T_0)\right]} \tag{8}$$

The parameter W can be measured after κ is derived. The measurement sequence of tire indoor test facility is the sequence that the tire surface temperature is grown up. So an example of sequence is the warm up sequence of measuring the parameter κ between -22 and -2 seconds in figure 1.

Figure 3 is the example of measured results of the parameter W. The absolute values of the parameter W when slip angle is +10 degrees and -10 degrees is different, and these parameters are changing as time goes on. The parameter W was determined by averaging the results at plus slip angle and at minus slip angle and by the stable situation between -15 and -2 seconds because of the same reason of the parameter κ.

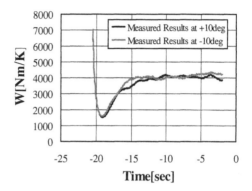

Fig. 3. Measured Results of Parameter W [Vertical Load : Nominal Load, Camber Angle : 0]

3.2. The Method of Deviation of Parameters in the Side Force Adjusting Function

The Eq.(4) can be described as follows:

$$
\frac{F_y(F_z, \alpha, \gamma, T) - F_y^*(F_z, \alpha.\gamma)}{F_y^*(F_z, \alpha, \gamma)}
$$
$$
= \left[\left(1 + \frac{F_z - F_z^n}{F_z^n} \cdot \frac{\partial \mu}{\partial F_z} \right) \cdot \left(1 + \frac{\partial \mu}{\partial \alpha} |\alpha| \right) \cdot \frac{\partial \mu}{\partial T} (T - T_m) \right] \quad (9)
$$

If the Eq. (9) is differentiated from T, the equation is replaced as follows:

$$
\frac{dF_y}{dT} / F_y^* = \frac{\partial \mu}{\partial T} \left[1 + \frac{F_z - F_z^n}{F_z^n} \cdot \frac{\partial \mu}{\partial F_z} \right] \left[1 + \frac{\partial \mu}{\partial \alpha} |\alpha| \right] \quad (10)
$$

First, the method of deviation about $\partial \mu/\partial T$ and $\partial \mu/\partial \alpha$ is described.

In Eq. (10), dF_y/dT can be derived by using the measured tire side force and tire surface temperature at certain slip angle, vertical load and camber angle as shown in Figure 4. The value of dF_y/dT can measured with the plus and minus slip angle. So the value of dF_y/dT at certain slip angle is average of the plus and minus measured value.

F_y^* can be calculated by using the steady state tire side force model. The left term of Eq. (10) can be measured here.

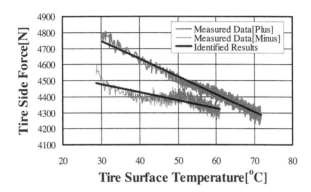

Fig. 4. Example of Measurement Result and Identified Result of dFy/dT [slip angle : ±10 degrees, vertical load : Nominal Load, camber angle : 0]

Especially if $F_z = F_z^n$, Eq. (10) is rewritten as follows:

$$\frac{dF_y}{dT}/F_y^* = \frac{\partial \mu}{\partial T}\left[1 + \frac{\partial \mu}{\partial \alpha}|\alpha|\right] \tag{11}$$

From this equation, the parameters $\partial\mu/\partial T$ and $\partial\mu/\partial\alpha$ can be calculated using the measurement results of $(dFy/dT)/F_y^*$ at different slip angles.

Figure 5 shows the measured results of $(dFy/dT)/F_y^*$ at different slip angles.

Next, the method of derivation about the parameter $\partial\mu/\partial F_z$ is described. The parameter $\partial\mu/\partial F_z$ is the gradient of the parameters $(dFy/dT)/F_y^*$ versus the vertical load as shown in figure 6. The parameter $\partial\mu/\partial F_z$ can be calculated using this results.

4. THE COMPARISON OF THE NEWLY DEVELOPED TIRE MODEL AND THE TEST RESULTS

The simulation result of the new tire model is compared with the measured data on indoor test facility. The test sequence is shown in figure 7. The ambient temperature and the road surface temperature were kept to the same level during measurement.To change the tire surface temperature, the slip angular velocity was changed.

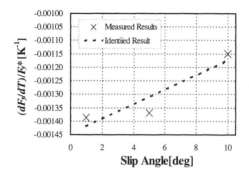

Fig. 5. Example of Measured Results and Identified Result of $\dfrac{dF_y}{dT}/F_y^*$ at different slip angles [Vertical Load : Nominal Load]

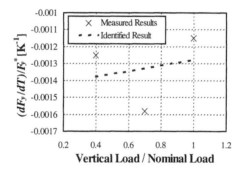

Fig. 6. Example of Measured Results and Identified Result of $\dfrac{dF_y}{dT}/F_y^*$ at different vertical load [Slip Angle : 10 degrees]

The test results and simulation results of the side force and the surface temperature are shown in figure 8 and 9. Left figures are the tire side force versus tire slip angle, and right figures are the tire surface temperature versus tire slip angle.

The simulation model is added the Eq. (12) that is the effect of relaxation length σ. The side force include the effect of σ, $\tilde{F}_y$ is drawn in figure 9.

$$\sigma\frac{d\tilde{F}_y}{dt} + V\tilde{F}_y = VF_y(F_z, \alpha, \gamma, T) \tag{12}$$

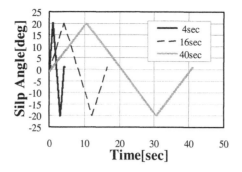

Fig. 7. Test Sequence to Compare the Test Result and Simulation Result

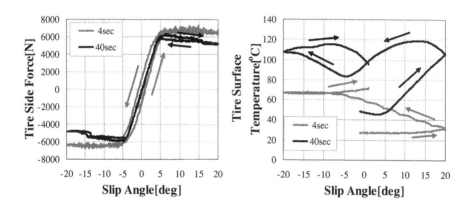

Fig. 8. The Measured Results of Tire Characteristics with Different Slip Angular Velocity

The measured data of steady state (based) tire side force model for the considering tire surface temperature is measured on indoor test facility and the test sequence of slip angle is shown in figure 7 by dashed line.

It is shown that a characteristics of the change is similar when a measured result is compared with a simulation result. These are the drop of side force during large slip angle and the hysteresis of side force during increase slip angle and decrease slip angle in 40 seconds sequence. In 4 seconds sequence, the curve shape of side force and surface temperature in simulation is similar as in test results, too.

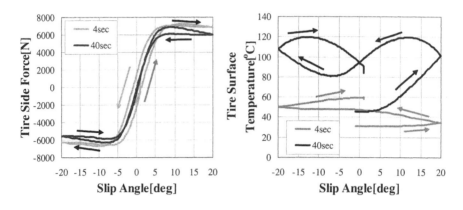

Fig. 9. The Simulation Results of Tire Characteristics with Different Slip Angular Velocity

5. CONCLUSIONS

The conclusions of this study will be summarised as follows:

1. The new tire side force model that is considering the tire surface temperature is developed.
2. The measurement method of tire parameters of new tire model is considered and the examples of test were executed.
3. The simulation result was compared with the measurement result. The calculation results of the change of tire side force and surface temperature by the developed model agreed very well with the measurements.

ACKNOWLEDGEMENTS

The authors are thankful to Mr. T. Takahashi in Toyota CRDL, Mr. T. Haraguchi and Mr. M. Yamamoto in Toyota Motor Corp. for suggestions about this study.

REFERENCES

1. Mizuno, M., Takahashi, T. and Hada, M.: Magic Formula Tire Model Using the Measured Data of a Vehicle Running on Actual Road. *AVEC '98*, September 1998, Nagoya, Japan.
2. Oosten, J.J.M. van e.o.: TIME, Tire Measurements, Forces and Moments, Final report, February 1999, *EC DG XII*, Standards, Measurements & Testing, TNO, Delft, Netherlands.

Vehicle System Dynamics Supplement 41 (2004), p.371-380 © Taylor & Francis Ltd.

Modelling and validation of steering system response to road and driver induced forces

JONAS JARLMARK[*]

SUMMARY

The driver serves both as a controller for the vehicle direction as well as a sensor for the road induced forces. A vehicle model is developed with both computer simulations as well as driving simulator use in mind. Therefore, the included steering system model was divided into one part below and one part above the steering gear. The presented steering system model focuses on force transmission from the wheels to the steering rack and the rack movements including a friction model. Above the steering gear, the model focuses on servo control through the twist of a torsion bar and includes steering wheel inertia and steering column stiffness and friction. The vehicle model has ten degrees of freedom, includes a Magic Formula 5.1 tyre model and is validated together with the steering system model as well as a geometric road model as a complete vehicle-environment model. The validation is made using data from the KTH experimental vehicle in a number of isolated driving cases performed at the Volvo Hällered test track. The simulations performed with the model are mainly focused on the steering wheel moment sensed by the driver.

1. INTRODUCTION

To make high speed driving of automobiles on motorways relaxed, the driver has to be given the total control over the vehicle. This feeling of control seems to be intimately linked to the vehicle reaction upon both driver- and road-induced forces.

The driver uses the steering wheel mainly as a tool to control and direct the vehicle but also as a sensor for road condition and vehicle behaviour. When designing a steering system, it is important to consider both the control and the sensor aspects in order to facilitate the vehicle with both "road feel" and the sensation of "control".

This control can be described as unbiased closed-loop feedback [1,2] with short time delay [3]. Because of this, it is very important how the road- and driver-induced forces are transmitted through the system. Furthermore, the alterations to the forces made on the way from the driver to the road surface and back again are important. Depending on the design of the steering system, it is possible to block forces or movements in the system that are essential for the closed-loop control.

There are many aspects to consider in the development of a good steering system [4-6]. To study the two steering system tasks, control and sensor, as decoupled and free to change without affecting the other should provide insight. To do this, a model of a real vehicle and steering system is needed so that the behaviour

* Division of Vehicle Dynamics, Royal Institute of Technology, SE-100 44, Stockholm, Sweden,
Tel: +46-8-790 7805, Fax: +46-8-790 9304, E-mail: jarlmark@kth.se

is in parity with current automobile standards and alterations can be made from this baseline. This paper describes the modelling, measurement and validation of a vehicle with its steering system and interactions between them. The results are intended to form the basis for a model to be implemented in a driving simulator.

To model both the control and sensor aspect of the steering system, a model capable of transmitting forces and movements in both directions is considered, figure 1. Normally this is modelled as a set of differential equations based on the mechanical layout of the system [7-9].

As the aim here is to investigate the possibilities of unrealised steering systems, the model cannot be dictated by the current layout of most systems. Therefore, a separation of tasks was made in the model such that the levels of the parameters affecting the behaviour are different depending on direction of the forces in the system.

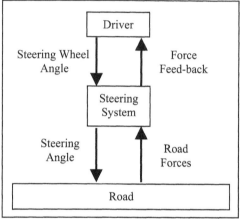

Figure 1. The steering system is functioning both as control device and sensor for the closed loop vehicle-driver-environment.

The first task is to transmit the driver-induced forces to the wheels and the second to send the road induced forces up to the driver. The model was developed to have full freedom to change the parameters affecting one task without altering the other, this to be able to find a desirable solution to the duality of its task, even if the physical incarnation would be impossible to achieve.

Many of the parameters involved will have an "optimum" value as they affect both tasks, for example the overall friction level: When the driver of an automobile directs the vehicle, a steering moment is applied to the steering wheel to gain a desired steering angle. The actual following of the intended path is mainly evaluated by visual means and is corrected by applying more or less moment to the steering wheel, until the steering angle change. If the road-induced forces vary and the friction in the steering system is low, it will be hard for the driver to keep the intended steering angle. On the other hand, if the friction is high, the variation in steering wheel moment during ordinary steering angle control will be great and a sensation of stick-slip behaviour will follow. The "optimum" in this case will be the trade between control and feel.

2.   STEERING MODEL BELOW THE STEERING GEAR

The model of the steering system is chosen to be split at the pinion and rack contact point, figure 2. The model has a geometry that can be found in most modern, front wheel drive passenger car. The force and angle equilibrium between the rack and pinion are solved so that the model as a whole is functioning as a single unit.

The road-induced forces are created in the contact between the road and the tyre. The forces involved are modelled through a vehicle model, a geometric road model and a Magic Formula 5.1 tyre model; they are described in chapter 4. The forces created by these models are the longitudinal (F_X), lateral (F_Y) and vertical (F_Z) tyre forces as well as the aligning torque (M_Z) of the tyre. In the lateral force, the camber force is included.

These forces have their own points of action as described by the Magic Formula and is shown in figure 3. Interesting from a steering system point of view are the moments around the left and right kingpin axes (M_{kpL} and M_{kpR}), figure 2.

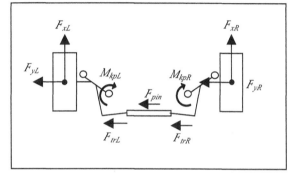

Figure 2. Definition of directions of the forces acting on the lower part of the steering system. The indicated moments are the kingpin axis moments.

To calculate a kingpin axis moment, the tyre forces need to be projected normal to the kingpin axis and multiplied with the shortest distance between the axis of the force and the kingpin axis. For the lateral and longitudinal forces the geometry is easily drawn in 2D. Here, the unconventional true force-to-axis distance is used rather than the more widely spread projection along the ground plane, i.e. caster- and kingpin-offsets. For small casters and kingpin inclinations, the errors are normally minimal with the conventional way, but as the angles grow and the road gets locally cambered, the errors can be significant.

The kingpin offset is only applicable to forces originating from outboard brakes or to other suspension reacting forces such as the traction forces of a solid axle. For a car using inboard brakes or chassis reacting drive shafts, the moment applied to the wheel can be seen as a free moment without reaction forces around the kingpin axis. This gives that the traction forces has to be transferred to the wheel centre giving the resulting traction offset, normally larger than the kingpin offset, figure 3.

In straight line driving, the vertical force creates a major input to the kingpin axis moment, as the vertical force is very large. When projecting the force using the sine of the total axis inclination, the projected force is still large even for normal inclination angles. In a straight line it gives typically ten times the moment of the lateral and longitudinal forces with ordinary toe and camber angles. If there is a small variation in moment arm due to tyre belt deformation over a cambered surface, this will influence the equilibrium of the rack.

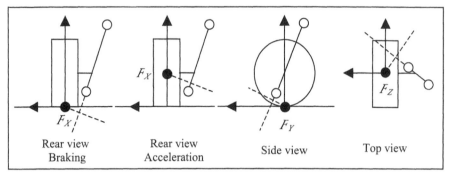

Figure 3. The front suspension seen in line with the applied force, note that acceleration and braking have two different vertical positions of action for the applied force.

Therefore, the Magic Formula 5.1 is expanded to cover the position of the forces within the tyre path, both laterally and longitudinally. The lateral change in position of the vertical force is modelled as a look up table where the input is the camber angle between road and tyre. The values can be measured statically on a simple rig and the lateral-progress characteristics can be compared between different tyres.

The kingpin axis moments are defined in equations 1 and 2. They include the moments due to the three tyre forces ($M_{kpFx, Fy, Fz}$) and the aligning moment (M_{kpMz}). The kingpin moments are transformed into the two tie rod forces (F_{trL} and F_{trR}) and added to the force from the steering gear pinion (F_{pin}) and steering servo (F_{servo}) to give the total resulting force on the steering rack (F_{rack}), equation 3. During normal motorway driving, the two tie rod forces are relatively large and opposed, so that the force from the pinion and servo can be small and balance the rack.

$$M_{kpL} = M_{kpFxL} + M_{kpFyL} + M_{kpFzL} + M_{kpMzL} \qquad (1)$$
$$M_{kpR} = M_{kpFxR} + M_{kpFyR} + M_{kpFzR} + M_{kpMzR} \qquad (2)$$
$$F_{rack} = F_{trL} + F_{trR} + F_{pin} + F_{servo} \pm F_{frict} \qquad (3)$$

Depending on the direction of rack movement, a friction force (F_{frict}) is added to the equation 3 that will dampen the movement of the rack. Two friction models for numeric simulation are a very stiff damper with saturation and a very stiff spring with sliding saturation, so that force falls back into the compliant area as soon as the direction of travel is changed, see figure 4. These models have a disadvantage in the compromise between initial stiffness and numerical oscillation. To make the model quick, the spring or damping stiffness has to be low, and to model stick-slip friction the stiffness has to be high. The use of these models often result in a compromise, where the model is prone to oscillate and yet does not show the stick-slip properties of natural friction.

To solve this, a friction model that has the possibility to constrain the related degree of freedom was developed. The logical functions controlling the model are shown in equation 4. When the rack force is large enough to overcome the friction force, the degree of freedom is opened and the total force is applied to the rack body generating rack acceleration. When the rack force is small enough and the predicted velocity in the next calculation step is either zero or of opposite sign, the degree of freedom is constrained and the velocity and total force set to zero. In other words, the force of the friction is exactly the size of the outer forces, but opposed, so there is no acceleration.

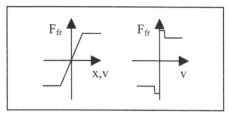

Figure 4. The classical friction model to the left, compared to the constraining friction model to the right.

This solution is considered to be valid as long as the outer forces are smaller than the maximum friction force. The limits for the friction forces of the stick and slip regions are possible to change independent of each other.

$$v_2 = v_1 + t_{step} * (F_{rack} - sgn(v_1) * F_{frslip}) / m \qquad (4)$$

$$\begin{aligned}
&\text{if } |F_{rack}| > F_{frstick} \\
&\quad \text{if } sgn(v_1) = 0 \\
&\qquad F_{tot} = F_{rack} - sgn(F_{rack}) * F_{frstick} \\
&\quad \text{else} \\
&\qquad F_{tot} = F_{rack} - sgn(v_1) * F_{frslip} \\
&\quad \text{end} \\
&\text{else} \\
&\quad \text{if } sgn(v_2) = sgn(v_1) \\
&\qquad F_{tot} = F_{rack} - sgn(v_1) * F_{frslip} \\
&\quad \text{else} \\
&\qquad v_2 = 0 \\
&\qquad F_{tot} = 0 \\
&\quad \text{end} \\
&\text{end}
\end{aligned}$$

F_{tot} is the total force accelerating the rack, including either of the friction levels, $F_{frstick}$, stick friction or F_{frslip}, slip friction. F_{rack} is all forces on the rack excluding the friction. v_1 is the current velocity and v_2 is the predicted velocity for the next time step. m is the mass that is accelerated by F_{tot}.

3.   STEERING MODEL ABOVE THE STEERING GEAR

The model of the steering system is selectable depending on the type of layout that is to be studied. For example, a normal pinion-column-wheel system with or without hydraulics can be used. In this model the additional force from the hydraulic servo is controlled by the valve angle, i.e. the twist of the servo torsion bar, α_{tb}, figure 5.

A hybrid system with a servo that has multiple controlling parameters, such as velocity, lateral acceleration, steering wheel velocity, damper movement, steering wheel torque, is also considered. The mechanical link between the steering wheel and the rack is still present, and the angular ratios are more or less fixed; only the force from the servo can be modulated. This could be realised with an electronically controlled servo connected to the CAN-bus of the vehicle.

Also a steer-by-wire system is modelled where the rack is totally controlled by a steering actuator and the steering wheel only serves as a sensor for desired driving path. With this model, the steering characteristics of the driver to vehicle part of the control loop are fully free to change with software. The feedback is possible to vary.

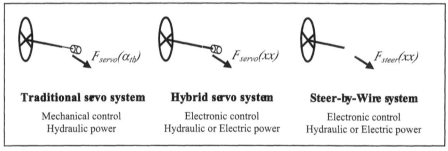

Traditional servo system	Hybrid servo system	Steer-by-Wire system
Mechanical control	Electronic control	Electronic control
Hydraulic power	Hydraulic or Electric power	Hydraulic or Electric power

Figure 5. Both the traditional servo system and the hybrid system still have the mechanical link between the steering wheel and the rack, but different controlling systems.

To be able to verify the complete model, the steering system of the Volvo V40 was modelled, including the force transfer efficiency of the rack and pinion, which is different depending on the direction of the force input. If the force is being excited from the tyre/road contact, driving the steering wheel, the efficiency is rather low and when the system is driven by the pinion, the efficiency is considerable higher.

It is found that the inertia of the steering wheel around the steering column is important to give the vehicle model the same free steer properties as the real vehicle. Servo characteristics are important to the driver in parking and limit handling situations. In motorway situations, the servo is normally within the dead-band of the servo valve and will therefore add no force to the rack. In situations where the road is locally cambered and the steering torque required by the driver increases, the servo valve can go outside its dead-band and set the servo into action.

4. VEHICLE MODEL

The vehicle model describes the KTH experimental vehicle, a Volvo V40, and the data for the model are measurements either from Volvo Car Corporation or from KTH. The model was tested in all development steps and was found to give reasonable results for the applicable level of detail. .In [10], the basis for the vehicle model and the validation of it is described in more detail

In this step, the vehicle model has six degrees of freedom for the body and one degree each for the vertical movement of the four wheels, generating a total of ten, see figure 6. Added to this are the varying degrees of freedom for the steering system.

A geometric road model was created to give input to the tyre and suspension models. The road is described by values of vertical position arranged in a matrix with a lateral division of 3.5 mm and a longitudinal division of 1.0 m. The high lateral resolution is needed to

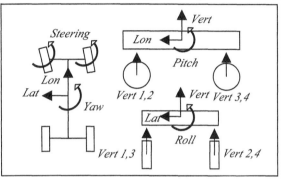

Figure 6. There are six degrees of freedom for the vehicle body, four for the vertical movement of the tyres.

calculate the lateral position of the vertical force within the tyre contact patch, giving fluctuations in top view moment arm. It is also needed for the camber force calculations. The longitudinal profile of the road provides the fluctuations in vertical force amplitude.

For the forces in the road plane, a Magic Formula 5.1 tyre model was used with parameters originating from the tyre manufacturer. As the main purpose is to model straight line behaviour in typical motorway situations, the toe and camber properties of the suspension was introduced by measured data of both camber and toe angle change as a function of lateral force, roll angle and vertical movement. This expansion of the model enables the tyre model to work under conditions similar to the behaviour of the real suspension when subjected to small variations in forces.

This vehicle model is combined with the steering system model and they constitute together the complete vehicle-environment model used for the simulations. The verification of this model was done using standardised and specialised manoeuvres on both the experimental vehicle and the simulation model.

5. MEASUREMENTS

To facilitate a validation of the vehicle model and its steering system models, measurements were made using the KTH experimental vehicle on a thoroughly prepared measurement track at the Volvo Hällered test facility. The vehicle was equipped with four VELOS rim force transducers measuring the wheel forces and moments in all directions. It was also equipped with accelerometers and gyros for the body movements as well as a steering angle sensor, a steering wheel torque sensor and two pressure transducers for the two sides of the servo piston.

The measurements were made on a flat, smooth, horizontal, high friction surface giving measurements of general vehicle behaviour and steering system dynamics. The experiments were designed to cover many aspects of the interaction between the steering system and the vehicle.

Furthermore, sensitivity to tyre forces were measured both for the vehicle itself with locked steering wheel, with normal driver corrections and for the combination of vehicle and steering system with free steering wheel. Acceleration, braking, free rolling and constant velocity was used during all experiments to define different combinations of lateral and longitudinal load cases to validate the simulation models against.

6. SIMULATION AND VALIDATION

The combined model of the vehicle and steering system was validated for lateral accelerations up to 8.5 m/s² and for frequencies ranging from 0.2 to 3.5 Hz using random input, one wave sinusoidal input, step steer input, constant steering wheel angle input, free steer input as well as normal driver corrections. Vehicle velocity range has been 50 to 180 km/h. During the course of validations, the vehicle model has been found to reproduce the experiments with good accuracy. For example the yaw rate gain is within 5%

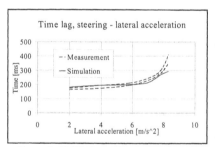

Figure 7. Validation of the model through the time lag between steering wheel angle and lateral acceleration.

and the yaw rate phase is within 4° for the entire frequency range. The time lags from steering wheel angle to vehicle reactions have also been validated over the lateral acceleration range 2 to 8.5 m/s², figure 7 shows the time lag to lateral acceleration.

To cover the normal range used during motorway driving, simulations have been made using sinusoidal steering wheel input with low amplitude, 0.2 rad and frequency, 0.5 Hz, , figure 8. Vehicle velocity has been kept at 100 km/h through constant traction forces, F_X. This generates a slow slalom motion with low lateral accelerations, typically 1.6 m/s² in peak, and controlled, slow body roll.

To find the amount of change in total kingpin moment, the moment arm of the vertical force around the left side kingpin axis is varied. As can be seen in figure 9, the level of the kingpin moment due to vertical force changes drastically with the three different offsets of the force. Also, the varying vertical force (due to roll) shows larger sensitivity at positive offset than for the nominal offset. The peak-to-peak vertical force variation is 1195 N due to roll with a static force of 4080 N.

At steady driving straight ahead, the two kingpin moments are balancing each other. If the car is subjected to either a one-wheel vertical force- or offset-change, the rack force will be non-zero. If the force is large enough to overcome the friction force, it will be felt as a movement and moment in the steering wheel. This could be the case when changing tyres from a narrow tyre with high aspect ratio to a low profile tyre with a wide and stiff carcass. The increased tyre stiffness and weight will generate higher vertical tyre forces by a certain one-wheel bump. In combination with that, the offset change due to local camber, as in a longitudinal rut, is higher for the stiffer tyre and will also generate higher kingpin moments. Simulations of a longitudinal rut with several one-wheel bumps should the best way to compare a soft, narrow and a stiff, wide tyre.

The stick-slip properties of the friction model are demonstrated in figure 10. The rack is stopping at its extreme position as the steering wheel is changing direction. The slip is initiated when the torsion bar has been wound up in the other direction to overcome the stick friction.

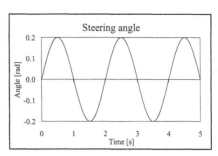

Figure 8. Steering wheel input. Low amplitude and frequency to generate levels realistic to motorway driving

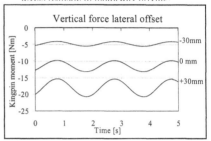

Figure 9. Variation in vertical force offset by +/- 30 mm from standard position.

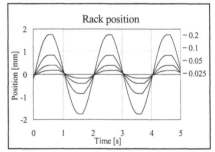

Figure 10. With varying amplitude of the steering wheel angle, the stick and slip of the rack due to friction shows clearly. Steering wheel angle peak values of 0.2, 0.1, 0.05 and 0.025 rad are shown

7. DISCUSSION AND CONCLUSIONS

This work describes modelling and validation of steering system response to road and driver induced forces. The steering system model has been divided in two tasks; control and sensor to enable studies of steer-by-wire systems. Validation using the KTH experimental vehicle show that the moment around the kingpin axis created by the vertical force can give a significant contribution to the total kingpin moment depending on the front wheel suspension, figures 2 and 3. Even a small variation in the vertical tyre force or the corresponding offset can severely change the balance of the steering rack and demand the driver to respond with a steering wheel moment to counteract the change in road induced force.

Naturally, more rack friction will partly mask the problem, but also introduce new problems for the driver control task; this is why a compromise between returned road force, rack friction and steering wheel feedback has to be made. This will also be partly driver dependant and therefore, the combination of simulations and a driving simulator will be a useful tool in steering system development.

This compromise will be more and more pronounced, as the current trend is to equip cars with larger rims using stiffer and wider low profile tyres. The higher stiffness and increased unsprung mass will give a larger variation in tyre vertical force running over a given bump. The width and stiffness of the tyre belt will also make the vertical force offset change more rapidly due to local road camber.

ACKNOWLEDGEMENT

The author is pleased to acknowledge the financial support of Volvo Car Corporation and VINNOVA (The Swedish Agency for Innovation Systems). The author also wish to thank professors Staffan Nordmark, VTI and Annika Stensson, KTH for academic guidance. The author gratefully acknowledges Johan Wedlin, Alf Söderström and Lars Lidner, Volvo Car Corporation and Stefan Edlund, Volvo Truck Corporation for their support during the course of this work.

REFERENCES

1. Hisaoka, Y; Yamamoto, M and Fujinami, H.: A study on desirable steering responses and steering torque for driver's feeling. AVEC'96, pp.295-305.
2. Adams, F. J.: Power steering "road-feel", SAE Technical Paper Series 830998.
3. Nishimura, S and Matsunaga; T.: Analysis of response lag in hydraulic power steering system.
4. Jerand, A.: Improvement, validation and multivariate analysis of a real time vehicle model, Licentiate Thesis, Department of Vehicle Engineering, Royal Institute for Technology, Stockholm, 1997.
5. Jaksch, F. O.: The steering characteristics of the Volvo concept car. 8th ESV Conference, Oct. 1980.
6. Springer, H.: Über den einfluss des lenkungssystems auf allgemeine ebene und dissipative bewegungen eines kraftfahrzeuges. Vehicle System Dynamics vol. 2 pp. 93-116, 1973.
7. Roos, G; Rollet, R; Kriens, R.F.C.: Numerical simulation of vehicle behaviour during straight line keeping on undulating road surfaces, Vehicle System Dynamics vol. 27 pp. 267-283, 1997.
8. Mitschke, M.: Dynamik der Kraftfahrzeuge 2nd Ed, Band C Fahrverhalten, Springer-Verlag 1990.
9. Ammon, D; Gisper, M; Rauh, J; Wimmer, J.: High performance system dynamics simulation of the entire system tire-suspension-steering-vehicle, Vehicle System Dynamics vol. 27 pp. 435-455, 1997.
10. Jarlmark, J.: Driver-Vehicle Interaction under Influence of Crosswind Gusts, Licentiate Thesis, Department of Vehicle Engineering, Royal Institute of Technology, Stockholm, Sweden, 2002.

BILATERAL DRIVER MODEL FOR STEER-BY-WIRE CONTROLLER DESIGN

-The Improvement of Lane-Change Stability Using

Optimal *SAT* Rate Feedback + Lead Steering Method-

THEERAWAT LIMPIBUNTERNG AND TAKEHIKO FUJIOKA[1]

SUMMARY

This paper proposes the bilateral driver model, which simultaneously takes the steering wheel angle and the driver torque into account. By analyzing this bilateral driver model and vehicle system, this paper shows a new design strategy for improving the vehicle stability, while preserving the handling feeling during lane changing. These improvements consist of the calculation of optimal steering gain and the compensation of system delay time, which is comprised of the vehicle delay and the driver delay. The optimal steering gain calculated by the bilateral driver model is slightly higher than those calculated by the conventional angle-control driver model. The vehicle's delay time can be easily compensated using only lead steering; however, to compensate the driver's delay time, the combination of lead steering together with the feedback of steering torque derivative is necessary.

1 INTRODUCTION

To increase the fuel consumption efficiency, Electric Power Steering (EPS) has been recently developed. In an EPS system, an electric motor is used to reduce the steering torque instead of the hydraulic system of the conventional power steering system. However, in the near future, the EPS system would be developed to Steer-By-Wire (SBW) system.

Because the SBW system conceptually doesn't need any mechanical linkage, the front-wheel steering angle and the steering torque can be freely designed and can improve the vehicle stability.

So far, in most researches the SBW controllers have been designed as an angle-control system with the input of steering wheel angle and the output of front-wheel angle; therefore, a driver model with the output of only steering angle such as 1st-order look-ahead driver model is enough for driver-vehicle system

---

[1] *Address correspondence to:* Department of Engineering Synthesis, The University of Tokyo, Tokyo, 113-8656 JAPAN, Tel./Fax: +81-3-5841-6386, E-mail: theerawa@vdl.t.u-tokyo.ac.jp and fujioka@ingram.t.u-tokyo.ac.jp

simulation to evaluate the controller. However, when steering torque that is another important characteristic of the steering system is also considered, the conventional angle-output driver model becomes incapable to simulate and evaluate the driver-vehicle system.

As a result, this paper introduces a new driver model called "bilateral driver model", proposed by the authors in [1] for designing SBW controller. Superior to the conventional steering angle-control driver model, the bilateral driver model can simultaneously deal with both steering angle controller and steering torque controller. After that, based on the closed-loop analysis using the bilateral driver model, this paper shows some design examples to improve the vehicle stability during lane changing. These design examples include how to find the optimal steering gain and how to compensation of system delay time. Finally, the designed controllers will be validated by the experiments using driving simulator.

2   BILATERAL DRIVER MODEL

The bilateral driver model introduced in [1] is defined as Equation (1).

$$\frac{\delta_h}{K_\delta} + \frac{T_d}{K_T} = \Delta Y_{future} e^{-t_d s} \tag{1}$$

when
$$\Delta Y_{future} = \{Y_{des} - (Y + L\psi)\} \tag{2}$$

Different to the conventional angle-output driver models, the output of bilateral driver model is not only the steering wheel angle δ_h, but also the driver torque T_d. Thus, it is more similar to the human driver.

The input of the model is the future lateral deviation to the target course ΔY_{future} with the driver's delay time of t_d. ΔY_{future} can be calculated using any algorithm; however, in this paper we use the well-known simple look-ahead algorithm expressed in Equation (2), where Y_{des} is the target lateral position, L is the look-ahead distance and ψ is the yaw angle of vehicle.

K_δ and K_T represent the steering wheel angle gain and the driver torque gain respectively. Actually, K_δ and K_T can be any function, but for simplicity K_δ and K_T are assumed to be constant in this paper. The values of K_δ and K_T show how the driver drives the vehicle. For example, when $K_T = \infty$ it means that the driver drives the vehicle by controlling only steering angle as the conventional driver models; on the other hand, when $K_\delta = \infty$ it means that the driver drives the vehicle by controlling only steering torque. When K_δ and K_T are between 0 and ∞, it means that the driver drives the vehicle as a hybrid control of angle and torque.

In [1], the bilateral driver model has been confirmed that it can efficiently represent the behavior of human driver in lane changing under various kinds of steering characteristic. It is also found that, in the normal driving conditions such as lane changing without disturbance, the driver operates the steering wheel relatively slow. Thus, the moment of inertia and the viscous friction of steering system become small and negligible, then

$$T_d = -T \tag{3}$$

when T is the steering torque of the vehicle.

As a result, the bilateral driver model in Equation (1) can be simplified as Equation (4), which can be explained that the driver's arms can be simply represented as a spring with the stiffness of K_T/K_δ.

$$\frac{\delta_h}{K_\delta} - \frac{T}{K_T} = \frac{\Delta Y_{future}}{1 + t_d s} \tag{4}$$

3   THE MODEL OF SBW SYSTEM

In this paper, we adopt the "dual-port system" presented in [2] for describing the steering system including SBW system. The dual-port system describes a steering system using two equations: the steering angle equation and the steering torque equation, and can be expressed as Equation (5).

$$\begin{bmatrix} \delta_f \\ T \end{bmatrix} = \begin{bmatrix} \alpha_{11} & \alpha_{12} \\ \alpha_{21} & \alpha_{22} \end{bmatrix} \begin{bmatrix} \delta_h \\ SAT \end{bmatrix} \tag{5}$$

The inputs of the system are the steering wheel angle δ_h and the self-aligning torque of tires SAT. The outputs are the front-wheel steering angle δ_f and the steering torque T. In SBW system, α_{11}, α_{12}, α_{21} and α_{22}, can be independently and freely assigned to one another. For simplicity, this paper assumes $\alpha_{12} = \alpha_{21} = 0$. That means the δ_f is the function of δ_h only, while T is the function of SAT only.

4   THE IMPROVEMENT OF LANE-CHANGE STABILITY

4.1 The Causes of Vehicle's Load during Lane Changing

To improve the vehicle stability during lane changing, we have to investigate what causes the vehicle's load during lane changing. In this paper, we investigate the causes of vehicle's load by performing the computer simulations under lane

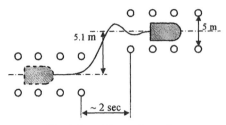

Fig. 1. Lane-Change Scenario.

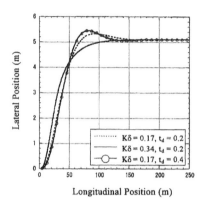

Fig. 2. Lateral Position at 60 km/h.

changing at 60 km/h and 150 km/h. The lane-change scenario used in this paper can be shown in Fig. 1.

Fig. 2 and Fig. 3 show the typical results of lateral position Y and driver's steering wheel angle δ_h. Fig. 4 shows the yaw rate-side slip angle $(\gamma$-$\beta)$ diagram, whose size shows the load of vehicle.

From the results in Fig. 4, we found that the largest load of vehicle during lane changing occurs in the initial stage of the operation. This largest load can be divided into two parts: the load of approaching phase L_a and the load of correction phase L_c, and their causes can be summarized in table 1.

The load of approaching phase L_a and the total load on the vehicle $L_a + L_c$ are affected by the steering angle gain of driver and vehicle. If the steering angle gain of driver or vehicle is high, L_a and $L_a + L_c$ will be high.

While, the load ratio of the correction phase to the approaching phase L_c/L_a is affected by the delay time of driver and vehicle. If the delay time is large, L_c/L_a will be large.

As a result, to increase the vehicle stability during lane changing, we may reduce either $L_a + L_c$ or L_c/L_a.

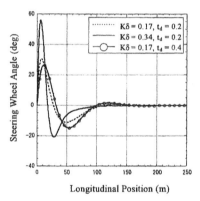

Fig. 3. Steering Wheel Angle at 60 km/h

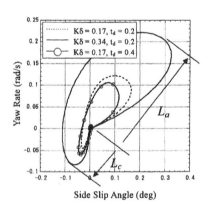

Fig. 4. γ-β Diagram at 60 km/h.

Table 1. Causes of Loads of the Vehicle During Lane Changing

Load	Driver's Cause	Vehicle's Cause
L_a and L_a+L_c	Driver's Steering Angle Gain	Vehicle's Steering Angle Gain
L_c/L_a	Driver's Delay Time	Vehicle's Delay Time

4.2 The Reduction of L_a and L_a+L_c: Optimal Steering Angle Gain

From the study of the parameter effects mention above, we can reduce L_a and L_a+L_c, by either informing the driver to reduce his/her steering gain or reducing the steering angle gain α_{11} of the SBW controller.

However, if we inform the driver to adjust his/her steering angle gain to reduce the vehicle's load, the driver's burden will increase. As a result, it is better to reduce the steering angle gain α_{11} of SBW instead. Unfortunately, if we reduce α_{11} too much, the closed-loop task performance of the lane changing will decrease.

As a result, to compromise between the vehicle stability and the closed-loop task performance, we apply the method of optimal control for finding the optimal α_{11}. The performance index expressed in Equation (6) is chosen for finding the optimal α_{11}. ΔY is the lateral deviation of vehicle's position from the target position, A_y is the lateral acceleration of the vehicle, ω_y is the weight of lateral deviation, and ω_a is the weight of lateral acceleration.

$$J = \int_0^\infty \left[\omega_y (\Delta Y)^2 + \omega_a (A_y)^2 \right] dt \tag{6}$$

The proportion of ω_a to ω_y represents the driving style of driver. If ω_a/ω_y is high, it means that the driver will give priority to the lateral acceleration. In contrast, if ω_a/ω_y is low, it means that the driver will focus on keeping the lateral deviation as small as possible.

To keep the driver's burden low, the optimal α_{11} must be calculated by assuming that the driver's steering angle gain K_δ is constant at all velocities. Therefore, before calculating the optimal α_{11}, we have to calculate this constant K_δ. In this paper, we assume that the driver always drives the vehicle with the constant K_δ that is optimal for the conventional vehicle ($\alpha_{11} = \alpha_{22} = 1/Steering\ Ratio$ and $\alpha_{12} = \alpha_{21} = 0$) at 60

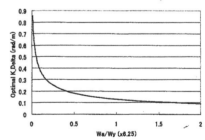

Fig. 5. The Optimal K_δ for Conventional Vehicle at 60 km/h.

km/h. By using the performance index in Equation (6), The optimal K_δ at 60 km/h of each driving style can be shown as Fig. 5. It shows that the more the driver pays attention to the lateral acceleration, the smaller K_δ is.

Unfortunately, to calculate the optimal α_{11}, the values of α_{22} must be also given. From [1], it is found that, if we reduce α_{11} too much without adjusting α_{22}, the handling feeling will be deteriorated. It is shown in [1] that, when $\alpha_{12} = \alpha_{21} = 0$, the handling feeling can be adjusted by the value of α_{11} multiplied by α_{22}. If the value of $\alpha_{11}\alpha_{22}$ is kept constant, the handling feeling will also unchanged regardless of the values of α_{11} and α_{22}. That means, while controlling α_{11} to improve the vehicle stability, we can preserve the handling feeling at any desired feeling by adjusting α_{22} such that

$$\alpha_{22} = \frac{(\alpha_{11}\alpha_{22})_{desired}}{\alpha_{11}} \tag{7}$$

In this paper, we choose to preserve the handling feeling of the SBW vehicle to be the same as the conventional vehicle. That is α_{22} is assigned such that

$$\alpha_{22} = \frac{1}{\alpha_{11}(SteeringRatio)^2} \tag{8}$$

By using the values of K_δ in Fig. 5 and the value of α_{22} in Equation (8), the optimal α_{11} for each driving style can be shown in Fig. 6. Fig. 6 shows that, at high velocity, the optimal α_{11} is lower than the steering gain of conventional vehicle. This can be explained that, for most of driver, the steering angle gain should be lowered at high velocity. In contrast, at low velocity, the steering angle gain should be increased.

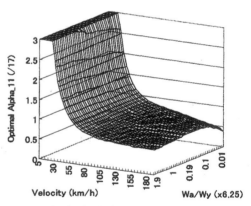

Fig. 6. The Optimal α_{11} Calculated by Bilateral Driver Model.

By comparing the first peak value of δ_h at 60 km/h of computer simulation results to those of experimental results, we found that the range of ω_d/ω_y of human drivers lies between 0.6-9.3. When we consider the optimal α_{11} only in this range, we found that the optimal α_{11} of each ω_d/ω_y is not much different as shown in Fig. 7. It is also found that, calculated by bilateral driver model, the optimal α_{11} at low velocity has pretty similar characteristic to the variable steering ratio with constant lateral acceleration gain. While, the optimal α_{11} at high velocity becomes very close to the 4WS-like variable steering ratio, which is about 60-70% of the conventional steering system.

Compared to the optimal α_{11} calculated by the conventional angle-control driver model, we can see from Fig. 7 that the optimal α_{11} at high velocity calculated by the conventional angle-control driver model is about 20% smaller than the value calculated by the bilateral driver model. That means the vehicle with the optimal α_{11} calculated by the conventional angle-control driver model will responds too slow for the human drivers.

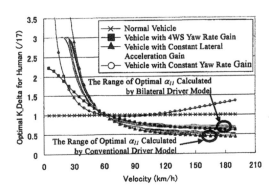

Fig. 7. The Optimal α_{11} in the Range of ω_d/ω_y of Human Drivers.

4.3 The Reduction of L_c/L_a

To further reduce the vehicle load, this section will show some examples to reduce the load ratio of correction phase to approaching phase L_c/L_a. Because the ratio of L_c/L_a is caused by the system's delay, we have to compensate the system's delay, which is comprised of the delay of the vehicle itself and the delay of driver.

4.3.1 The Compensation of Vehicle Delay: Lead Steering

Because the delay time of vehicle is the delay time between the vehicle responses and front-wheel steering angle, we can easily compensate by utilizing the "lead steering", which augments the additional steering angle as a function of angle velocity of steering wheel. In this paper, the gain of the lead steering τ_v is assigned equal to the time constant of characteristic equation of vehicle derived from the linear-bicycle model, which is a 2nd-order system. Thus the SBW controller to compensate the vehicle's delay can be written as Equation (9), with the lead steering gain of Equation (10). Equation (10) shows that the delay of vehicle is proportional to the vehicle's velocity, and this is the reason why L_c/L_a at high velocity is more significant than at low velocity.

$$\begin{bmatrix} \delta_f \\ T \end{bmatrix} = \begin{bmatrix} \alpha_{11}(1+\tau_v s) & 0 \\ 0 & \alpha_{22} \end{bmatrix} \begin{bmatrix} \delta_h \\ SAT \end{bmatrix} \tag{9}$$

$$\tau_v = \frac{V}{\left[\dfrac{\left(l_f^2 K_f + l_r^2 K_r\right)}{I} + \dfrac{\left(K_f + K_r\right)}{m} \right]} \tag{10}$$

4.3.2 The Compensation of Driver Delay: Lead Steering + Steering Torque Rate Feedback

Unlike the compensation of vehicle's delay, the compensation of the driver's delay is more complicated and cannot achieve by using only the "lead steering". This is because the driver's delay time also has effect on the steering torque.

From the analysis using the block diagram, we found that, in addition to the lead steering, if we include the "feedback of the *SAT* derivative" such in Fig. (8), the driver's delay time can be compensated.

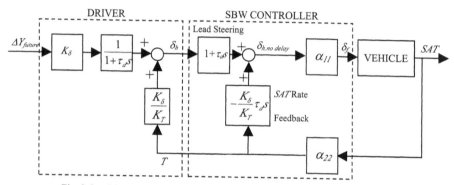

Fig. 8. Lead Steering + *SAT* Rate Feedback for Driver's Delay Compensation.

In this case, because the driver delay seems rather constant, the gain of lead steering and *SAT* rate feedback τ_d may be chosen to be a constant. Therefore, the SBW controller for driver's delay compensation can be written as Equation (11).

$$\begin{bmatrix} \delta_f \\ T \end{bmatrix} = \begin{bmatrix} \alpha_{11}(1+\tau_d s) & -\dfrac{K_\delta}{K_T}\alpha_{11}\alpha_{22}\tau_d s \\ 0 & \alpha_{22} \end{bmatrix} \begin{bmatrix} \delta_h \\ SAT \end{bmatrix} \qquad (11)$$

Unfortunately, from stability analysis, we found that *SAT* rate feedback will change the vehicle's characteristic equation and cause the vehicle unstable. However, as we know from [1] that during lane change we can assume $T_d = -T$, this problem can be solved by feedback the derivative of driver torque T_d instead of the derivative of *SAT*. As a result, to compensate the driver's delay, the SBW controller may be designed as in Equation (12).

$$\begin{bmatrix} \delta_f \\ T \end{bmatrix} = \begin{bmatrix} \alpha_{11}(1+\tau_d s) & 0 & \dfrac{K_\delta}{K_T}\alpha_{11}\tau_d s \\ 0 & \alpha_{22} & 0 \end{bmatrix} \begin{bmatrix} \delta_h \\ SAT \\ T_d \end{bmatrix} \qquad (12)$$

388

5   THE EVALUATION OF DESIGNED SBW CONTROLLERS

In this section, the optimal steering angle gain and the delay time compensator described in section 4 will be evaluated. The lane-change experiments using driving simulator were performed at constant velocities of 60 and 150 km/h under 4 conditions shown in table 2. The samples of experimental results at 150 km/h are shown in Fig. 9-12.

Table 2. Four Conditions for Evaluation of SBW Controllers

Case	Conditions
1	Conventional Vehicle ($\alpha_{11} = \alpha_{22} = 1/17$ and $\alpha_{12} = \alpha_{21} = 0$)
2	Vehicle with Optimal α_{11} (as shown in Fig. 7)
3	Vehicle with Optimal α_{11} + Vehicle Delay Compensation
4	Vehicle with Optimal α_{11} + Driver Delay Compensation

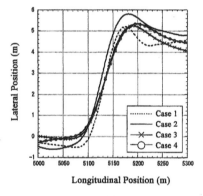

Fig. 9. Lateral Position at 150 km/h.

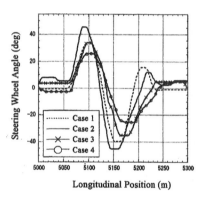

Fig. 10. Steering Wheel Angle at 150 km/h.

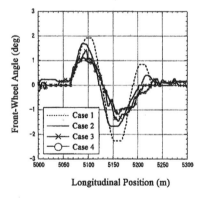

Fig. 11. Front-Wheel Angle at 150 km/h.

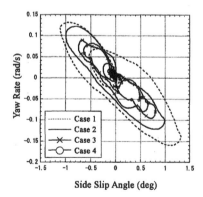

Fig. 12. γ-β Diagram at 150 km/h.

389

The results prove that, with the optimal $\alpha_{//}$ we can reduce the load of vehicle during lane changing without deteriorate the task performance too much. While either "lead steering" or "lead steering + driver torque derivative feedback" can effectively compensate the system delay and further reduce the load of vehicle.

6 CONCLUSIONS

6.1 The bilateral driver model, which considers both steering angle and steering torque simultaneously, has been introduced for SBW controller design.

6.2 During lane changing, the largest load on vehicle occurs in the initial stage of the lane changing, and can be divided into two parts: the load of approaching phase L_a and the load of correction phase L_c.

6.3 The load of approaching phase L_a and the total load L_a+L_c are affected by the system gain. These loads can be reduced by lowering the steering angle gain $\alpha_{//}$. To minimize these loads, while retaining the satisfied close-loop task performance, the optimal $\alpha_{//}$ has been calculated using bilateral driver model. The vehicle with this optimal $\alpha_{//}$ is similar to the vehicle with constant lateral acceleration gain at low velocity and has the same yaw rate gain as 4WS with zero steady-state side-slip angle at high velocity.

6.4 The load ratio of the correction phase to the approaching phase L_c/L_a is affected by the delay time of system, and becomes more significant at high velocity. The vehicle delay can be compensated using only lead steering, while the driver delay compensation needs not only the lead steering, but also the derivative feedback of driver torque.

6.5 The lane-change experiments using driving simulator confirmed that, the calculated optimal $\alpha_{//}$ helps to reduce the total load of vehicle without deteriorate the closed-loop performance and handling feeling. In addition, the delay compensation also helps to further reduce the vehicle's load.

REFERENCES

1. Limpibunterng T., and Fujioka T., "Bilateral Driver Model for Steer-By-Wire Controller Design", Proceedings of JSAE Autumn Conference, No. 117, 2002

2. Limpibunterng T., and Fujioka T., "A New Design Approach for Steer-By-Wire System by Dual-Port System", Proceedings of 17th IAVSD Symposium, pp. 197-208, 2001

3. Segawa M., et al., "A Study of Vehicle Stability Control by Steer By Wire System", Proceedings of AVEC 2000, CD-ROM, 2000.

4. Wolfgang K., et al., "Potential Function And Benefits of Electronic Steering Assistance", CSAT Czech Republic, 1996.

5. Karnopp D., "Active Steering Systems", Report Dept. of Mechanical, Aeronautical Materials Engineering, University of California, Davis, 1992.

On-Center Steer Feel Evaluation based on Non-linear Vibration Analytical Method

HIROYUKI TOKUNAGA, KAZUHITO MISAJI, YASUO SHIMIZU[1]

SUMMARY

This paper develops quantitative evaluation method for on-centre steer feel at high speeds. This evaluation method applied "Analytical Method of Equivalent Linear System using the Restoring Force Model of Power Function Type" to the analysis of the vehicles response characteristic. By this method, non-linear characteristics of vehicles response can be grasped with concise indices. Actual driving data of the vehicles with which steer feel evaluation differs were analyzed, and indices were computed. The relation between these indices and subjective evaluations by expert drivers was considered. Consequently, correlation nature was able to be found out among these. In order to verify the validity of this evaluation method, the target characteristics were set up by using these indices, and it analyzed using simulation tools. And the vehicle in which the analysis result was made to reflect was built. Expert drivers using the vehicle that was modified based on this method confirmed the effectiveness of the evaluation method.

1   INTRODUCTION

Drivers require safe and comfortable driving performance from their vehicles. Accurate steering is necessary to achieving this, particularly at high speeds. At this point, steer feel is an extremely important factor.

To begin the analysis and evaluation of steer feel, one of the important steps is to quantify the physical characteristics related to steer feel.

In real driving situations, steering input angle and frequency vary considerably, and the dynamic characteristics of the vehicles, especially the steering torque characteristics, show strong non-linear dependency on the input angle and frequency due to the effect of tire and power steering assist characteristics. To analyze steer feel, it is therefore necessary to understand input angle and frequency dependency of vehicle characteristics. However, most examples of a research report from such a viewpoint cannot be found.

[1] *Address correspondence to:* Honda R&D Co.,Ltd. Tochigi R&D Center, 4630 Shimotakanezawa, Haga-machi, Haga-gun, Tochigi, 321-3393 JAPAN. Tel.: (028) 677-7625; Fax: (028) 677-7510; E-mail: hiroyuki_tokunaga@n.t.rd.honda.co.jp

In this paper, first, subjective evaluation by expert drivers was carried out to investigate how steer feel varies according to steering conditions.

Three vehicles (vehicles A, B and C) were selected and tested to determine how subjective evaluations of steer feel vary in response to steering input conditions. The following test conditions were established to reproduce high-speed driving in the on-centre area:

Speed: 140km/h
Task: Slalom with fixed maximum steering wheel angle and steering frequency
Maximum steering wheel angle: 5-15 [deg]
Steering wheel turning frequency: 0.25-1.5 [Hz]

A five-stage subjective evaluation of vehicle steering torque response and yaw rate response was conducted. The main evaluation criteria were steering torque level, steering torque build-up performance, linearity of steering torque, linearity of yaw rate response, and responsiveness. Six expert drivers formed the test subjects, and their responses were averaged to provide data. Fig.1 shows results of evaluations of response characteristics (steering wheel angle vs. steering torque and steering wheel angle vs. yaw rate). The results demonstrate the trend of steer feel variation with steering input conditions.

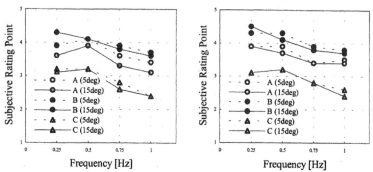

Fig. 1. Subjective Rating Results (Steering Angle – Steering Torque, Steering Angle – Yaw Rate)

2 ANALYTICAL METHOD

In this study, we examine a new evaluation method that has capability to evaluate steering torque and vehicle yaw rate characteristics against steering wheel angle or more.

For the purpose of this study, the following were established as conditions for analytical method of vehicle characteristics for steer feel evaluation.

1) Capable of treating non-linearity of vehicle characteristics.
2) Capable of expressing steering frequency and input angle dependency.

3) Capable of simple correlation with subjective evaluations (Capable of concise expression).

The "Analytical Method of Equivalent Linear System using the Restoring Force Model of Power Function Type", utilized in the analysis of non-linear systems demonstrating viscoelastic characteristics, was selected as a method fulfilling these conditions.

2.1 Analytical method of equivalent linear system using the restoring force model of power function type

The equation of motion for a spring particle system when $m_0 \cos \omega t$ is operating as the external force is:

$$I\ddot{\theta} + m(\theta,\dot{\theta}) = m_0 \cos \omega t \qquad (1)$$

where I is the moment of inertia, θ is the angle, $m(\theta, \dot{\theta})$ is the restoring moment, m_0 is the amplitude of steering torque, ω is the angular velocity of the steering wheel, t is time, m_S and θ_S are steering torque and steering angle at the linear limit respectively, ω_S is the natural frequency in the linear limit area, and θ_0 is the amplitude of steering angle.

It is also possible to express equation (1) as equation (2).

$$\frac{d^2\phi}{d\tau^2} + M(\phi,\phi') = M_0 \cos\eta\tau \qquad (2)$$

$$\left(\phi = \theta / \theta_S , \phi_0 = \theta_0 / \theta_S , \omega_S^2 = m_S / (\theta_S \cdot I), \eta = \omega / \omega_S , \tau = \omega_S t, M_0 = m_0 \right)$$

in which $M(\phi,\phi') =$ dimensionless parameter of restoring force.

The following equation is obtained by replacing the hysteresis vibration system of equation (2) with the equivalent linear system:

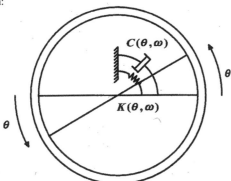

Fig. 2. Steering vibration model

$$\frac{d^2\phi}{d\tau^2} + 2H_{eq}\frac{d\phi}{d\tau} + K_{eq}\phi = M_0\cos\eta\tau \tag{3}$$

H_{eq}, K_{eq} are expressed by the following equations respectively.

$$H_{eq} = -\frac{1}{2\pi\eta}\left(\frac{1}{\phi_0}\right)^2 A(\phi_0,\eta) \tag{4}$$

$$K_{eq} = \frac{1}{\pi}\left(\frac{1}{\phi_0}\right)^2 \int_{-\phi_0}^{\phi_0} P(\phi)\cdot R(\phi,\eta)d\phi \tag{5}$$

in which $A(\phi_0,\eta)$ = the area of hysteresis loop expressed as a function of the amplitude of angle ϕ_0 at frequency η. $R(\phi,\eta)$ is added in ascending and descending branches of the loop

And the fundamental equations of power function type are expressed as follows:
The bone curve

$$M(\phi,\eta) = k\phi^\alpha$$

The ascending branches

$$M(\phi,\eta) = 2k\left[\frac{1}{2}(\phi_0+\phi)\right]^\alpha - k\phi_0^\alpha \tag{6}$$

The descending branches

$$M(\phi,\eta) = -2k\left[\frac{1}{2}(\phi_0-\phi)\right]^\alpha + k\phi_0^\alpha$$

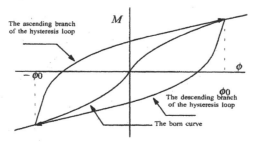

Fig. 3. Restoring Force Model of Power Function Type
In the following equations, H_{eq} and K_{eq} are obtained by using equations (4), (5) and (6).

$$H_{eq} = \frac{2k}{\pi}\frac{1}{\eta}\left(\frac{1-\alpha}{1+\alpha}\right)\phi_0^{\alpha-1} \tag{7}$$

$$K_{eq} = \frac{4k}{\sqrt{\pi}} \left(\frac{\alpha}{\alpha+1} \right) \frac{\Gamma(\alpha+0.5)}{\Gamma(\alpha+1)} \phi_0^{\alpha-1} \tag{8}$$

α, k are given by

$$\alpha(\phi_0,\eta) = \frac{4M_0(\phi_0,\eta) \cdot \phi_0 - G_0(\phi_0,\eta)}{4M_0(\phi_0,\eta) \cdot \phi_0 + G_0(\phi_0,\eta)} \tag{9}$$

$$k(\phi_0,\eta) = \frac{M_0(\phi_0,\eta)}{\phi_0^\alpha} \tag{10}$$

The spring constant K and the damping constant C dependent on steering wheel angle and steering speed (frequency) can be found using equations (11) and (12).

$$K(\theta_0,\omega_0) = k_s \cdot K_{eq} \tag{11}$$

$$C(\theta_0,\omega_0) = \frac{2m_s}{\omega_s \cdot \theta_s} \cdot H_{eq} \tag{12}$$

3   ANALYSIS OF VEHICLE CHARACTERISTICS

To investigate the steering wheel turning frequency dependency and the input angle dependency of steering torque and yaw rate characteristics of each vehicle, steering angle, steering torque and yaw rate were measured and analysed by using the "Analytical Method of Equivalent Linear System using the Restoring Force Model of Power Function Type".

3.1   Test Vehicle

The three vehicles (vehicles A, B and C) used in subjective evaluations

3.2   Test Method

Sine wave slalom at 140 [km/h]
Steering wheel turning frequency: 0.25 – 1.0 [Hz]
Maximum steering wheel angle: 5 – 25 [deg]

3.3   Test Results

3.3.1   Hysteresis Curves

Figure 4 shows examples of the hysteresis curves generated by measured data.
(steering wheel turning frequency: 0.5 [Hz], input steering wheel angle: 15 [deg])

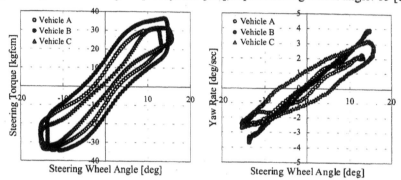

Fig. 4. Hysteresis curves

3.3.2   Analysis Results

Figures (5)-(6) show results from the application of equations (11)-(12) to these
hysteresis curves. Analysis results demonstrate that it is possible to concisely express
the steering wheel input frequency dependence of vehicle characteristics.

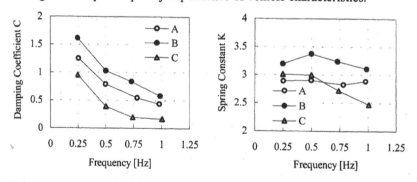

Fig. 5. Frequency Dependency of C, K / Steering Wheel Angle – Steering Torque (5deg)

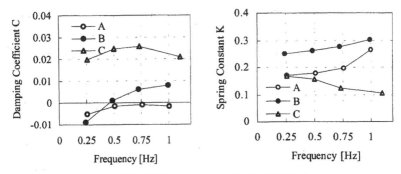

Fig. 6. Frequency Dependency of C, K / Steering Angle – Yaw Rate (15deg)

4   CONSIDERATIONS

4.1   Steering Wheel Angle - Steering Torque

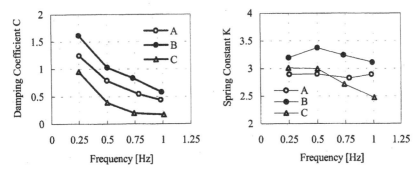

Fig. 7. Frequency Dependency of Damping Coefficient C and Dynamic Spring Constant K
Steering Wheel Angle – Steering Torque (5deg)

Fig. 7 shows the results of analysis of steering wheel turning frequency dependency at steering wheel input angle of 5deg.

Damping coefficient C in steering angle -steering torque characteristics means steering torque time lead level against input steering wheel angle. The values are considered to relate to the expression of damping feel during subjective evaluations. The damping coefficients for vehicle C are relatively lower than those for the other vehicles. This shows that Vehicle C has less damping feel.

Dynamic spring constant K in steering angle - steering torque characteristics means steering torque gain against input steering wheel angle. Fig.7 shows that dynamic spring constant of Vehicle C decreases when steering wheel turning frequency increases. Vehicle C shows large steering torque variation with its steering wheel turning frequency. Viscous elements in a turning control mechanism such as a steering wheel improve operation of vehicle. The relationship between spring and damping led to the hypothesis that their ratio has an effect on steer feeling.

Fig. 8 shows the ratio of damping coefficient, C, and spring constant, K, and its relationship to subjective evaluations. This relationship demonstrates that the ratio of C and K can be correlated with subjective evaluations. However, for steering wheel turning frequency at 0.25Hz, especially steering wheel angle at 5deg, this correlation becomes worse. It seems that static friction factor affects damping coefficient in lower steering wheel speed, so that the linearity of steering torque on the change of steering wheel turning speed become worse in these conditions.

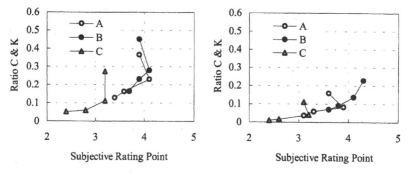

Fig. 8. Relation between Subjective Rating Point and Ratio C & K (5deg/Left, 15deg/Right)

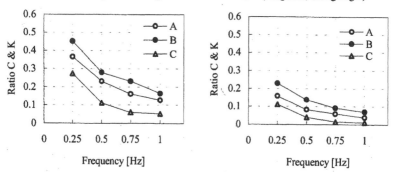

Fig. 9. Frequency Dependency of Ratio C & K (5deg/Left, 15deg/Right)

When the ratio of C to K is too high, there is too much damping, which will worsen driver's control. If the ratio is too low, sensitivity to both human factors (hand tremors, inattentive steering) and external factors is increased, and this results in a negative effect on controllability. Although it is necessary to take account of inertia term during precise analysis, since the vehicles used in this test are from the same category, the inertia terms of the vehicles are assumed to be identical and are not included in this study.

4.2 Steering Wheel Angle – Yaw Rate

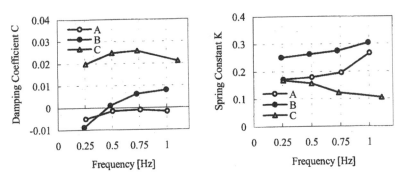

Fig. 10. Damping Coefficient C and Spring Constant K / Steering Angle – Yaw Rate (15deg)

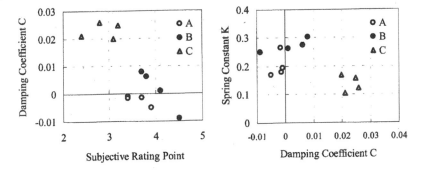

Fig. 11. Relation between Subjective Rating Point and C / Relation between C and K

Fig. 10 shows the results of analysis of the frequency dependence of yaw rate response at a maximum steering wheel input angle of 15deg. Damping coefficient C for Vehicle C becomes relatively large across the frequency domain in comparison with the other vehicles. Here, damping coefficient C in steering angle -yaw rate characteristics means yaw rate response lag level to input steering wheel angle. This shows that Vehicle C has larger lag in yaw rate response compared to other vehicles. Dynamic spring constant K means yaw rate gain against steering wheel angle. In vehicles A and B, dynamic spring constant, K, shows a tendency to increase with increased frequency, but in vehicle C, it shows a tendency to decrease. This shows that vehicle C has poor response to steering speed. Looking at correlation between damping constant, C, and subjective evaluations (Fig.11), it can be observed that the lower the value of C, the higher the evaluation. In other words, a lower lag in yaw rate response to steering input angle is desirable in terms of steer feel.

5   VERIFICATION

In this study, the development target of on-centre steer feel can be set by physical values which have correlation with subjective evaluations, so that the analysis using the simulation tools can be performed. Simulation tools were utilized to optimize the characteristic value of each component of suspension and steering systems. And based on the results, the test vehicle was modified to achieve the target. Fig. 12 shows the vehicle characteristics before and after the modification. As a result of subjective evaluation by expert drivers, the vehicle was highly evaluated in on-centre steer feel.

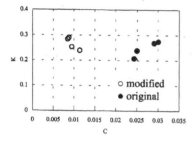
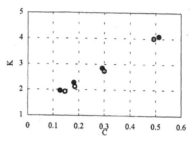

Fig. 12.   steering wheel angle – yaw rate

steering wheel angle – steering torque (0.5 Hz)

6   CONCLUSIONS

This paper has discussed on the application of the "Analytical Method of Equivalent Linear System using the Restoring Force Model of Power Function Type" to the analysis of on-centre steer feel at high speeds.
The following conclusions emerged from this study:
1) The results of subjective evaluation show clear difference on evaluations of individual vehicle, and each steer wheel turning frequency.
2) Correlation between these indices and subjective evaluations by expert drivers has been verified.
3) Effectiveness of this method has been verified by expert drivers using the modified test vehicle.

REFERENCES

1.   Shibata, K., Misaji, K. and Kato, H.,Vibration Characteristics of Rubber (Nonlinear Vibration Characteristics Depending on Frequency and Amplitude of Displacement),Trans. Jpn. Soc.Mech. Eng.,(in Japanese),Vol.59,No.564,C(1993),p.144.
2.   Misaji, K., Kato, H. and Shibata, K., Vibration Characteristics of Rubber Vibration Isolators of Vehicle (Analysis of Nonlinear Vibration Response), Trans. Jpn. Soc. Mech. Eng.,(in Japanese),Vol.60,No.578,C(1994),p.42.
3.   Misaji, K., Hirose, S. and Shibata, K., Vibration Analysis of Rubber Vibration Isolators of Vehicle Using the Restoring Force Model of Power Function Type (Analysis of Nonlinear Vibration Using Frequency Characteristics Determined by Hysteresis Loop),JSME International Journal Series C,Vol.38,No.4,pp.679-685,1995.

Vehicle System Dynamics Supplement 41 (2004), p.401-410

Analytical Method for Modeling Driver in Vehicle Directional Control

KONGHUI GUO[1,2], YING CHENG[1], HAITAO DING[1]

SUMMARY

Driver modeling is identified as an important process, which may aid the evaluation and optimization of the vehicle performance and the development of intelligent vehicles as well. This paper aims to establish a directional control driver model with efficiency, accuracy and driver preference. Error Elimination Algorithm (EEA), an analytical method in driver modeling based on the Preview-Follower theory and the error tracking analysis, is presented. Driver preference parameter is then discussed. Numerical simulations of the proposed driver model in typical maneuvers are studied and compared with human driver test results.

1 INTRODUCTION

With the improvement of the automobile performance, one of the major problems encountered by researchers and engineers is the active safety of the automobile, which calls for the in-depth understanding of the interaction between the dynamics of the automobile and the human driver behavior, as well as the reliable construction of the driver model. Many driver models based on the optimal control methods or artificial intelligence approaches have been proposed in literatures [1-8]. When used for on-board control or a closed-loop directional control simulation associated with a sophisticated vehicle model, the driver model has to be both effective and efficient.

Most of the existing directional control driver models are sufficiently accurate for automobile active safety evaluation if parameters of the driver model are properly chosen; however, techniques for an efficient driver model are less well developed until now. The crossover model proposed by Ashkens and McRuer [1] provides a compensation control to the vehicle, but it did not include the preview behavior of the driver when it was originally proposed. However Lin et al [2] added a preview effect on to the crossover model making it similar to existing preview tracking driver models. In late 1960s, Yoshimoto [3] presented a second order prediction driver model with

[1] State Key Lab of Automobile Dynamic Simulation, Jilin University, Changchun, CHINA.
[2] *Address correspondence to:* State Key Lab of Automobile Dynamic Simulation, Jilin University, No.142 Renmin Avenue, Changchun 130025, CHINA. Tel: (86) 431 5687676-6101; Fax: (86) 431 5687676-6209; E-mail:guo-kong-hui@sohu.com

torque input. Lately Sakai et al [4] adopted a second order prediction driver model with steering angle input to examine the performance of the electric power steering (EPS) system and employed an evaluating index to determine parameters in the driver model through optimization. Stefan et al [5] proposed a neurocontroller for lateral vehicle guidance. This neural network driver model is fed with measured human-driving data. MacAdam and Johnson [6] constructed a two-layer neural network to represent driver steering behavior. The network is trained using sampled data collected by sensors of an on-road car.

In the following sections, a Preview Optimal Artificial Neural Network (POANN) driver model, especially a simplest POANN driver model (POSANN) for vehicle directional control and its application issues are introduced. An analytical method in driver modeling, Error Elimination Algorithm (EEA), is presented. Numerical simulations in typical steering task including double lane change and slalom are then employed to compare with human driver test results. Finally, the parameter representing driver preference is analyzed.

2 POSANN MODEL AND APPLICATION ISSUES

For vehicle directional control, A Preview Optimal driver model based on the Artificial Neural Network (POANN) [7] can be structured as Figure 1, whose inputs are preview path $f(x(t))$ and the vehicle state feedback y, $\dot{y}$, $\ddot{y}$.

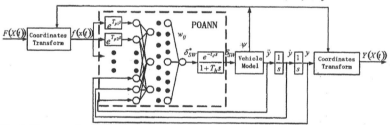

Fig. 1. The generalized architecture of the POANN driver model.

It has been proved by driving simulator test that a Preview Optimal Simplest Artificial Neural Network (POSANN) driver model [7], consisting just one single neuron with 4 inputs (Fig. 2), has shown satisfactory performance for vehicle directional control compared with multi-layer and more neurons ANN models (a neuron is defined as a joint point of multi-inputs with single output).

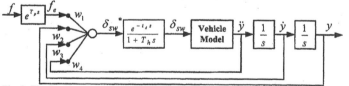

Fig. 2. The Preview Optimal Simplest Artificial Neural Network (POSANN) driver model.

Driver neural response delay time t_d and driver action delay time T_h are included in the modeling of POSANN. The weighting coefficients, $w_1 \sim w_4$, can be obtained through a training process, where the desired path of the vehicle is used as the ideal output and there is no need of the experimental input/output data. The training process (Fig. 3) is based on an optimization algorithm aiming at minimizing the cost function as,

$$J_T = \frac{1}{t_n} \int_0^{t_n} \left\{ [f(t) - y(t)]^2 / \hat{E}^2 + \left(\dot{\delta}_{SW} / \hat{\dot{\delta}}_{SW} \right)^2 + (\ddot{y}/\hat{\ddot{y}})^2 \right\} dt \tag{1}$$

where the thresholds $\hat{E}$, $\hat{\dot{\delta}}$ and $\hat{\ddot{y}}$ act as inverse weighting factors, representing driver's different personality in balancing the tracking accuracy, handling busyness and driving safety, respectively.

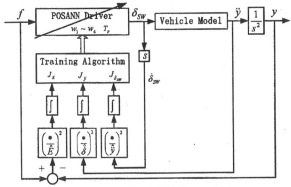

Fig. 3. The training process of POSANN driver model.

POSANN driver model is useful in most cases, but the training process is usually time-consuming, especially when it is used for on-board control, such as an automatic driving system, or for a closed-loop directional control simulation associated with a sophisticated vehicle model in a commercial code, like ADAMS®. As discussed in the next section, EEA has made it possible to determine the weighting coefficients in Figure 1 from a mathematical point of view.

3   ERROR ELIMINATION ALGORITHM IN DRIVER MODELLING

3.1   Preview-Follower theory and tracking error analysis

The Error Elimination Algorithm (EEA) is an analytical method based on the Preview-Follower theory and the tracking error analysis. The Preview-Follower theory proposed by Guo [8] considers the transfer function of the whole driver/vehicle system as two parts, the preview function $P(s)$, representing driver's preview process, and the follower function $F(s)$ including driver's steering behavior and vehicle response, see Figure 2.

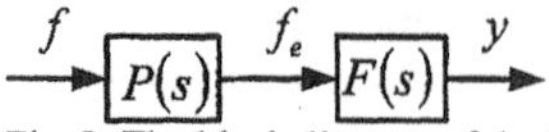

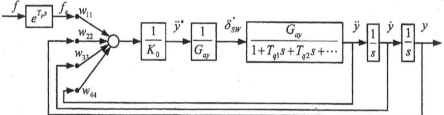

Fig. 2. The block diagram of the Preview-Follower theory

In Figure 2, f is the path input, f_e the effective preview input and y the lateral position of the vehicle. The tracking error of the driver/vehicle system is given by,

$$e(s) = f(s) - y(s) \tag{2}$$

$$\frac{e}{f}(s) = 1 - P(s) \cdot F(s) \tag{3}$$

In order to eliminate the tracking error function, the relation of $P(s)$ and $F(s)$, as well as that of their Taylor expansions, has to meet Equation (4),

$$P(s) = F^{-1}(s)$$
$$P_0 + P_1 \cdot s + P_2 \cdot s^2 + P_3 \cdot s^3 + \cdots = F_0 + F_1 \cdot s + F_2 \cdot s^2 + F_3 \cdot s^3 + \cdots \tag{4}$$

Further constructing Figure 1 as Figure 3 can help to conduct the error analysis.

Fig. 3. The block diagram of EEA in driver modeling.

As shown in Figure 3, T_p is the preview time, K_0, G_{ay} gain factors and T_{q1}, T_{q2} parameters of the equivalent system including driver delays and vehicle dynamics,

$$\frac{e^{-t_d \cdot s}}{1 + T_h \cdot s} \cdot G_{ay} \cdot \frac{T_{y2} \cdot s^2 + T_{y1} \cdot s + 1}{T_2 \cdot s^2 + T_1 \cdot s + 1} = G_{ay} \cdot \frac{1}{T_{q2} \cdot s^2 + T_{q1} \cdot s + 1} \tag{5}$$

$$\begin{cases} T_{q1} = T_h + t_d + T_1 - T_{y1} \\ T_{q2} = T_h \cdot T_1 + t_d^2/2 + (t_d - T_{y1}) \cdot (T_h + T_1 - T_{y1}) + T_2 - T_{y2} \end{cases} \tag{6}$$

where $G_{ay}(T_{y2} \cdot s^2 + T_{y1} \cdot s + 1)/(T_2 \cdot s^2 + T_1 \cdot s + 1)$ represents the a two degrees of freedom vehicle model (bicycle model). The block diagram of Figure 3 can be expressed as Equation (7),

$$(w_{11} \cdot f_e + w_{22} \cdot y + w_{33} \cdot \dot{y} + w_{44} \cdot \ddot{y}) \cdot \frac{1}{K_0} \cdot \frac{1}{G_{ay}} \cdot \frac{G_{ay}}{1 + T_{q1} \cdot s + T_{q2} \cdot s^2 + \cdots} = \ddot{y} \tag{7}$$

Notice that $w_{11} = -w_{22}$ provided by the non-drift condition, we have

$$F^{-1}(s) = \frac{f_e}{y}(s) = w_{11} - w_{33} \cdot s + (K_0 - w_{44}) \cdot s^2 + K_0 \cdot T_{q1} \cdot s^3 + K_0 \cdot T_{q2} \cdot s^4 \qquad (8)$$

Given the single point preview strategy where the Taylor expansion of the preview function is

$$P(s) = e^{T_p s} = 1 + T_p \cdot s + T_p^2 \cdot s^2 / 2! + T_p^3 \cdot s^3 / 3! + \cdots \qquad (9)$$

Equation (4) turns to

$$\begin{cases} P_0 - F_0 = 1 - w_{11} = 0 \\ P_1 - F_1 = T_p + w_{33} = 0 \\ P_2 - F_2 = T_p^2 / 2 - K_0 + w_{44} = 0 \\ P_3 - F_3 = T_p^3 / 3! - K_0 \cdot T_{q1} = 0 \\ P_4 - F_4 = T_p^4 / 4! - K_0 \cdot T_{q2} = 0 \end{cases} \qquad (10)$$

from which parameters of the driver model can be obtained as

$$\begin{cases} w_{11} = -w_{22} = 1, w_{33} = -T_p, w_{44} = K_0 - T_p^2 / 2 \\ K_0 = T_p^3 / (6T_{q1}) \\ T_p^* = 4T_{q2} / T_{q1} \end{cases} \qquad (11)$$

where T_p^* is the optimal preview time. When the preview time is extended larger than T_p^*, a more relaxed driving is realized, however some tracking error will be introduced, which represents driver different preferences as further discussed in section 4.

3.2   Complexity of the tracking path and empirical modification

Since EEA is eliminating the first few terms of the Taylor Expansion of the tracking error function (Equations [3], [4]), it is more suitable for tracking path with low order polynomial (such as slalom) than that with higher order, lane change for example. An empirical modification can be introduced to higher order tracking path, where $T_p = T_p^* + \Delta T_p$ (ΔT_p is usually chosen as 0.1s).

3.3   Performance of the driver model compared with human drivers

Numerical simulation is applied here for different speeds ranging from 80km/h to 140km/h and maneuvers involving a double lane change and slalom. Comparisons of EEA simulation results with the human driver experimental results carried out at JLU driving simulator are shown in Figures 4 and 5, where parameters of the driver model are set as t_d=0.4s and T_h=0.1s. As seen, a good match is achieved between the simulation results of the proposed driver model using EEA and those of human drivers.

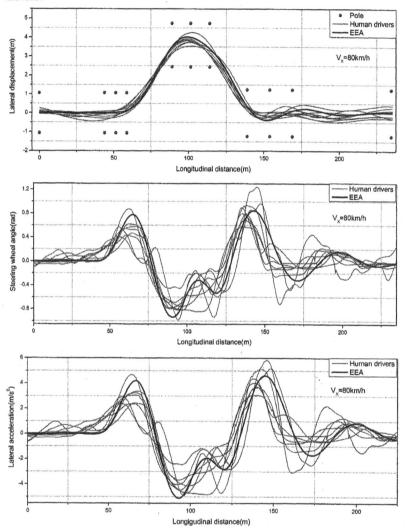

Fig. 4. Simulation results compared with experimental results (Vx=80km/h double lane change).

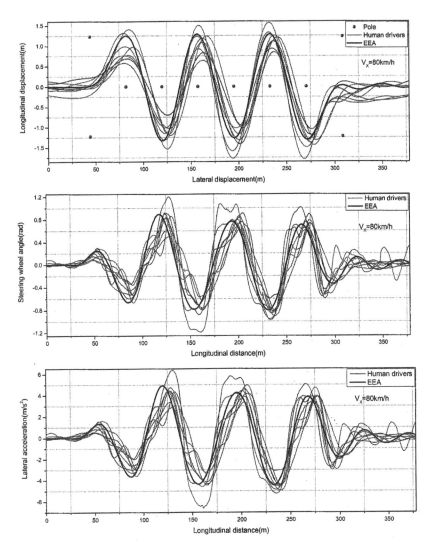

Fig. 5. Simulation results compared with experimental results (Vx=80km/h slalom).

4   DRIVER PREFERENCE

Different drivers may have different driving preferences, among which there are two basic rules that all drivers prefer: accurate as well as safe tracking, and relaxed handling. In other words, a driver always tries to follow the path as accurately as possible and to reduce the handling busyness. EEA proposed in this paper assures the maximal tracking accuracy in driving, with a recommended T_p^*. When the actual preview time is extended larger than T_p^* depending on different driver preferences, a more relaxed driving at the expense of losing some tracking accuracy is realized. Driver's steering behavior is investigated in Figure 6, which shows the relationship between the ideal steering wheel angle δ_{Sw}^* and the effective steering wheel angle δ_e.

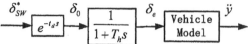

Fig. 6. The relationship between steering wheel angles.

The effective steering wheel rate can be obtained from Figure 6 as,

$$\dot{\delta}_e = (\delta_0 - \delta_e)/T_h \tag{12}$$

which reflects driver's handling busyness. The larger the driver's action delay time T_h, the less busy the handling. As given in Equations (11) and (6), driver's optimal preview time T_p^* is extended along with T_h, which is illustrated in Figure 7. Simulation results under different T_h are shown in Figure 8. Experimental tests of human drivers of the same manoeuvre are also shown in the same figure. It can be seen from Figure 8 that when the value of T_h is set at 0.3, the handling busyness is reduced while some tracking error is introduced at the same time. This can be used to model a driver preferring more relaxed handling. When T_h is set at 0.1, the tracking error is reduced while $\dot{\delta}_e$ is increased, which can be applied to the modeling of a driver preferring more tracking accuracy.

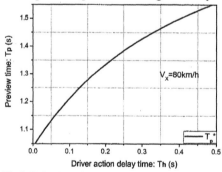

Fig. 7. Driver's optimal preview time vs. action delay time.

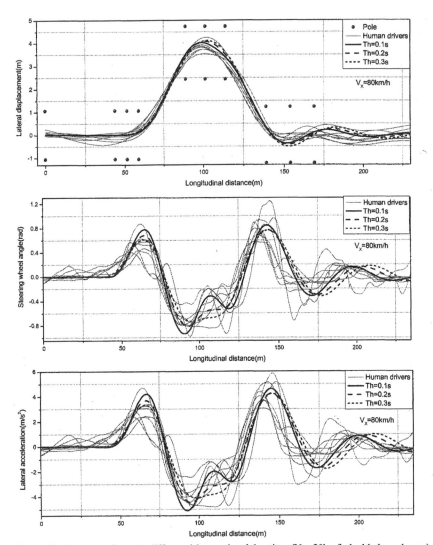

Fig. 8. Simulation results under different driver action delay time (Vx=80km/h double lane change).

5   CONCLUSIONS AND DISCUSSIONS

• An analytical method for driver modeling, Error Elimination Algorithm (EEA), is presented in this paper. By eliminating the tracking error function, EEA assures the maximal tracking accuracy.

409

- Driver preference is also analyzed. Driver's action delay time T_h can be chosen as a variable to describe different driver preferences.
- Simulations show that the new analytical modeling method is quite efficient and reliable in application to driver-vehicle closed-loop simulation and evaluation, as well as directional control of automatic driving.
- Although the vehicle model within linear range is used here, it is good enough for most cases in normal driving.
- When a non-linear vehicle model is applied to the driver/vehicle system, some kind of sigmoid transfer function may be used in the neural network in order to represent the driver steering behavior in controlling the non-linear vehicle, such as on the icy road surface.

REFERENCES

1. Ashkens, I.L. and McRuer, D.T.: A Theory of Handling Qualities Derived form Pilot/Vehicle System Consideration. *Aerospace Engineering*, No.2, 1962.
2. Lin, M., Popov, A.A. and McWilliam, S.: Handling Studies of Driver-Vehicle Systems. *Proceedings of the 6th International Symposium on Advanced Vehicle Control*, Hiroshima, September 2002.
3. Yoshimoto, K.: Simulation of Driver-Vehicle System Using Steering Behavior Model Including Prediction. *Journal of JSAE*, Vol.71, No.596, 1968, pp. 13-18.
4. Sakai, K., Yoneda, A. and Shimizu, Y.: Improvement in Control Performance of Driver-Vehicle System with EPS using Cables to Connect the Steering Wheel and Gearbox. *Proceedings of the 6th International Symposium on Advanced Vehicle Control*, Hiroshima, September 2002.
5. Stefan, N., et al: Neurocontrol for Lateral Vehicle Guidance. *IEEE Micro*, February 1993, pp. 57-65.
6. MacAdam, C.C., Gergory, E.J.: Application of Elementary Neural Networks and Preview Sensors for Representing Driver Steering Control Behavior. *Vehicle System Dynamics*, 25(1996), pp. 3-30.
7. MacAdam, C.C.: Application of an Optimal Preview Control for Simulation of closed-loop Automobile Driving. *IEEE Transactions on Systems, Man, and Cybernetics*, Vol.11, 1981, pp. 393-399.
8. Guo, K.H., Pan, F., Cheng, Y. and Ding, H.T.: Driver Model Based on the Preview Optimal Artificial Neural Network. *Proceedings of the 6th International Symposium on Advanced Vehicle Control*, Hiroshima, September 2002.
9. Guo, K.H., et al: Preview-Follower Method for Modeling Closed-loop Vehicle Directional Control. *Symposium of 19th Annual Conference on Manual Control*, Cambridge Massachusetts, May 1983.

Driver Independent Road Curve Characterisation

MAGNUS KARLSSON[1]

SUMMARY

A method to determine the statistical distribution of curvature of road curves is presented. The curvature is the inverse of the radius at any point. From road construction requirements it is reasonable to model the curvature of a road curve with a trapezoid. The algorithm presented detects curves from measurements and fits a trapezoid to each curve. The residual between the fitted trapezoid and the measured curvature can be seen as the driver influence. The method is verified against proving ground measurements and applied to a series of measurements in Brazil with different drivers on the same road. The distributions of the curvature for the different drivers are presented and the variance of the residuals in the curves is used to compare the effects of the different driving styles.

1. INTRODUCTION

A major problem when predicting fatigue life for vehicle components is to determine the service loading conditions that the vehicle will experience during its life. Due to insufficient knowledge about the loading conditions the strength of components are often exaggerated through high safety factors. Many of these unnecessarily strong constructions can be avoided with a better understanding of the fatigue load in the real operating environment. The natural way to acquire information about the service loading is to make field measurements of forces, strains etc. However, it is a troublesome and expensive task, especially since there exists a great variation between different customers' way of using their vehicles.

Since the field measurements have seldom been considered long enough they have often been performed in such a way that the driver has been driving at high speed over a bad road stretch, thus guaranteeing that the measurement is severe enough. In the same way the durability test tracks have been constructed to replicate the worst-case loading conditions found during field measurements. [1]

There are other suggestions for how to get the fatigue load design specifications more related to the actual customer environment, by using for instance questionnaire surveys, (see [1-3].), but these methods do not appear to be reliable enough.

---

[1] *Address correspondence to:* Volvo Truck Corporation, Department 26780 A2, SE-40508 Göteborg, Sweden. Tel.: +46 31 668 286; Fax: +46 31 508 764; E-mail: magnus.k.karlsson@volvo.com

In recent years, the ambition in the automotive industry has been to design and optimise components towards a finite fatigue life. This optimisation has been based on the almost opposing requirements of lightweight design on one hand and safety and durability requirements on the other hand. It has become obvious that a more thorough understanding of the complex fatigue loading conditions is necessary.

Most techniques today directly try to describe the loading conditions in terms of forces and moments. It appears to be practice of the date to separate different kind of load origins from each other, see [4,5]. The most important loads should be those coming from driving into potholes and bumps, road vibration, curvature, braking and accelerations. These can be seen as vertical, lateral and longitudinal load origins. The vertical loads have been considered in a series of articles by Volvo, see [6-8]. This particular work focuses on the lateral loads. These loads, caused by curvature, are arguably not as great as those from the vertical loads, but will have a huge influence on, for instance, steering components.

This article will in section 2 describe a new concept for describing loads based on separation of the effects of the operating environment, driver's influence, transport mission and the vehicle design. These ideas will then be applied to lateral loads. In section 3 a mathematical description of a curve is suggested, based on road construction requirements. In section 4 an algorithm for detecting curves from measurement is described and the mathematical description of a curve is used to find the curvature for each detected curve. These methods are verified against proving ground measurement. In section 5 the ideas are applied to measurements in Brazil. The distribution of curvature is found. In section 6 some conclusions about the method are drawn.

2.   A CONCEPT FOR DESCRIBING LOADS

An alternative to the idea of trying to describe the loads directly, is to ask the interesting question: "What causes the loadings?" Naturally there are several factors, but the most important should be the following:

-   The operating environment: What kind of roads is the vehicle driven on?
-   The vehicle utilisation: How is the driver driving the vehicle?
-   The transport mission: What is the vehicle used for?
-   The vehicle design: What kind of vehicle do we have?

An idea could therefore be to describe these factors separately, in order to actually understand from where the loads come, and how important the above listed sources are. The main purpose of this idea is to separate the description of the road from the influence of the driving style, the vehicle design and the effects of surrounding traffic. This means that the road description will be approximately the same irrespective of the driver and at what time the measurement is performed. The description of today will therefore hold in the future as well. The influence of driving style can then be described separately. The method therefore gives an opportunity to study to what extent the driver influence affects the fatigue life. This can then be compared to the influence of the market where the truck is used as well as the mission of the truck. The description of the operating environment can also be of great use when performing handling and comfort analyses.

3. A SIMPLE MATHEMATICAL MODEL OF A CURVE

When constructing a curve a crucial part is to make sure that vehicles driving through the curve will not experience too high lateral accelerations. In order to avoid this and to get a smooth ride through the curve it is therefore useful to have a constant curve radius over a longer time, say a couple of seconds. One way of modelling a curve would therefore be to think of it as having a constant radius over the entire curve. However if the curve is narrow, i.e. has a small radius, it is probably better to have an area at the beginning (and the end) of the curve where the radius is adapted to the part with a small constant radius. Thus it would be easier for a driver to adjust to the smaller radius.

The Swedish National Road Administration [6] constructs curves according to the following concepts: If the radius is big (at least 900 m on a major road), the curve should follow the arc of a circle, i.e. has a constant radius. If the curve is narrow it should first have a linear change of the curvature, then the radius stays constant for a certain time and finally increases again until the road is straight.

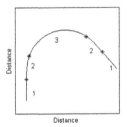 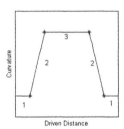

Fig. 1. To the right: A straight part before entering the curve (1), an adaptation area where the radius decreases (2), and an area of constant radius (3). To the left: The curvature for the same curve

According to these ideas it is therefore reasonable to see the curvature, here defined as the inverse of the radius at any point of the road, as following a trapezoid. See Fig. 1 above.

An equation for the trapezoid model used in this paper is as follows:

$$C(x) = \begin{cases} k \cdot (x - (x_0 - l)), & x_0 - l \le x < x_0 - l + h/k \\ h, & x_0 - l + h/k \le x < x_0 + l - h/k \\ k \cdot (x_0 + l - x), & x_0 + l - h/k \le x \le x_0 + l \end{cases} \qquad (1)$$

where k is the slope of the side of the trapezoid, x_0 is the distance from the starting point of the road, l is the distance from the middle of the curve to the endpoints, and h is the height of the trapezoid. (See Fig. 2.) The trapezoid is assumed to have the same slope on both sides.

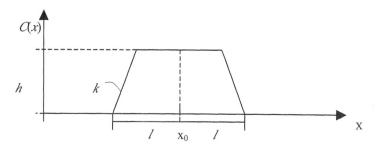

Fig. 2   A trapezoid with the parameters in the model.

4.   DETECTING AND MEASURING CURVES

4.1 DIFFERENT DRIVING MODES

Before it is possible to use real measurements to analyse the curvature of a road with the above-mentioned trapezoid method, two problems have to be solved. First we have to know where there is a curve. That is, we have to be able to automatically detect a curve simply by running the measurements through a programme. Second, we have to use the measurements to somehow estimate the curvature, where we have detected a curve.

The problem of detecting the curves can be solved in the following way. Each point of the measurements is sorted into any of the following three categories: *manoeuvres*, *curves* and *straights*. The term manoeuvres is in this case for instance parking manoeuvres or reversing for loading or unloading. The differentiation can be made in the following way. Manoeuvres are characterised by high wheel angle and low speed. Thus the measurement is searched for points fulfilling these demands. Secondly the measurement is searched for characteristics for curves such as high lateral acceleration under a certain time, and high speed relative to the manoeuvres. What remains after these two searches is considered to be straight parts.

4.2 AN ESTIMATE OF THE CURVATURE

The curvature of the road can be found using measurements in which the yaw rate and the velocity have been recorded. The curvature can be estimated by the ratio of the yaw rate and the velocity at any point. Naturally these measurements will not exactly follow a trapezoid for each curve and then stay zero for the straights. The idea is instead that the curvature can be described as a sum of a deterministic function and a zero mean random process. The deterministic function corresponds to the trapezoid and the random process to the residual between the actual measurement and the trapezoid fit.

In practise, the trapezoid fit is made in the following way. For every curve detected, the signal corresponding to the curvature is picked out, and the distance driven is found by integrating the velocity for this particular part of the road. A trapezoid as a function of the driven distance is fitted to the measured curvature.

Under the assumption that we get the true curvature from the measurement above, how do we then explain the residual? One part can be explained by a simple measurement error. This can be seen as the high-frequency component. The low-frequency component on the other hand can be regarded as the driver's influence, such as over-compensating when taking the curve. Other possible factors causing errors are disturbances from potholes and road roughness, the dynamics of the vehicle, and the velocity. It should be possible to model the curvature (C) as a deterministic function corresponding to the trapezoids ($C_{trapezoid}$) and a stochastic process corresponding to the residual (δ), where the residual is a combination of the factors mentioned above:

$$C = C_{trapezoid} + \delta = C_{trapezoid} + \delta_{driver} + \delta_{measure} + \delta_{road} + \delta_{vehicle} + \delta_{velocity} \qquad (2)$$

4.3 REFERENCE MEASUREMENTS AND VERIFICATION OF THE ALGORITHM

The way of detecting curves from the measurement is naturally based on a great deal of subjective judgement. Questions like: "What is low speed?" can of course be raised. In order to verify that the method actually works, the separation of the measurement was compared to the actual situation on the road by studying video films, which were recorded simultaneously with the measurements, as well as comparing the result with known curves at a proving ground.

The result was that it is possible to use this sort of decomposition. On the proving ground measurements, two very small curves that did not exist according to the map were detected out of a total amount of thirty-five curves. On the other hand all existing curves were detected. Since the decomposition is based on threshold values there is of course a small number of questionable cases. The most sensitive part is whether or not over-turns and such things as turning in order to avoid potholes should be considered as curves or not. In this study most of them are not. The key is how long time it should take to drive through a curve. Making this parameter smaller places the above two examples in the class of curves.

In order to verify that the method of fitting a trapezoid to each detected curve works, measurements from a proving ground with known curvature were used. The error from the measurements were found in the following way:

$$\varepsilon = \frac{r_{fitted} - r_{true}}{r_{true}} \tag{3}$$

where r_{fitted} is the smallest radius found in each curve, and r_{true} is the radius according to a map of the proving ground. The program was run over a series of such measurements, containing altogether thirty-five curves. The mean error was very close to zero (-0.0033) and thus it seems like r_{fitted} is an unbiased estimate of r_{true}. The standard deviation of ε was found to be 0.054. To better understand these figures estimating a $r_{fitted} = 90m$ when $r_{true} = 100m$ gives an error $\varepsilon = 0.10$. Thus it seems like the estimate is good, especially if we consider the fact that the drivers do not necessarily follow the track of the road exactly, but may very well try to take a somewhat shorter way through a curve, or may have to take a longer way due to the surrounding traffic.

Another way of verifying the algorithm is to use the measurement of the curvature to find the road driven and compare that to the road that is found by using the fitted trapezoids. Although this is not the purpose of this model it should give a good hint about the quality of the model. The results come from a measurement in Brazil. As can be seen in Fig. 3, there is a rather small difference between the two estimated roads. The difference appears about halfway through the road, where the trapezoid method gives a less emphasised curve to the left than the original measurement, which is then propagated through the rest of the measurement. At the very end of the road there is also a clear mistake. Nevertheless the results seem to verify the quality of the trapezoid model. It would be of interest to compare the resulting roads curve by curve, but since the original measurement contains no information on exactly where the curves are located such a direct comparison is not possible. It is therefore difficult to draw any further conclusions more than that it looks as if the maps are very similar.

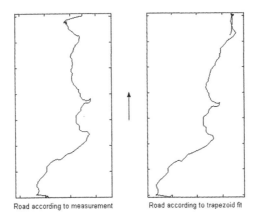

Road according to measurement   Road according to trapezoid fit

Fig. 3. The road according to the measurement and according to the fitted trapezoids. The arrow indicates the direction in which the truck was driven.

4.4 THE RESIDUAL

As a first idea for a measure of the drivers' unsteadiness, the variance of the residuals in the curves can be used, thus giving us a possibility to compare the different drivers to each other. An unsteady driver gives a high variance, which will cause more damage.

5. METHOD APPLIED TO MEASUREMENTS IN BRAZIL

5.1 DESCRIPTION OF MEASUREMENTS

A series of measurements performed in Brazil was used to try out the suggested method. During these measurements five different drivers were driving the same route consisting of approximately 25 km highway, 25 km secondary road, 7 km unpaved road and 2.5 km city driving. One of the drivers was driving the stretch of roads four times, in order to see how great the variation for a single driver is.

According to the trapezoid model it should be possible to estimate the curvature of the road at every point by fitting a trapezoid to the measured curvature where a curve has been detected. Thus when letting different drivers driving over the same route the results of the estimated curvature should, at least approximately, be the same.

The algorithm described above was used to detect curves and estimate the curvature of each detected curve for all eight measurements. Since the route was not exactly the same (some drivers skipped the city driving or parts of it), some of the most obvious curves were not included for all measurements. Nevertheless, it is still possible to study the empirical distribution of, for instance, the curvature or the curve length, which should be the same for all drivers.

5.2 RESULTS FROM MEASUREMENTS

In Fig. 4 the empirical distributions for the curvature for the five different drivers are plotted. As a reference the empirical distribution from another measurement performed in Brazil is also plotted. This measurement was 57 km long and consisted mainly of secondary and unpaved road. In Fig. 4 it is referred to as "reference road".

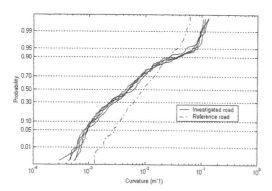

Fig. 4. Empirical distribution for the curvature for different measurements on the same road as well as a reference measurement on another road. (Plotted on lognormal paper.)

In the upper right corner of the figure an area can be seen where the distribution for the investigated road makes a bend. This part corresponds to the city driving part where the curves are narrow and similar in terms of curvature. On the other hand, in the left part of the figure the investigated road gets lower values on the curvature, than the reference road. This part comes from the highway section where the curves are designed to make it possible to drive fast. The reference road has a more homogeneous distribution of curves.

Since not all drivers took the same route the city driving part was excluded when comparing the residuals. The variance of the residuals in the remaining curves was recorded. In Fig. 5 the variance of the residuals for the different drivers are shown normalized by the highest value. Driver No. 5 drove the route four times.

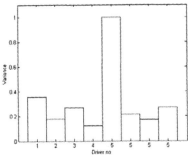

Fig. 5. The variance of the residuals for the different drivers normalized by the highest variance.

5.3 DISCUSSION ON THE RESULTS

The results from the Brazilian measurements indicate that it is possible to use the detection algorithm and the trapezoid method to measure the curvature on the roads with a fairly high precision. The empirical distributions for different drivers appear to be very similar from one driver to another, and can thus be seen as driver independent.

A co-driver from Volvo made a subjective judgement on the different drivers, which makes it possible to check the result from the variance of the residuals.

Table 1. The subjective judgement of a co-driver (a Volvo employee).

Driver no:	1	2	3	4	5
Judged as:	Fast and somewhat reckless	Fast and experienced	Careful	Careful	Fast
Nationality:	Brazilian	Brazilian	Brazilian	Brazilian	Swedish

A comparison shows that the reckless driver has the second highest variance, although he was acquainted with the course. The Swedish driver was not familiar with the Brazilian conditions during his first measurement, which is shown clearly. In the following measurements he is on the same levels as the Brazilian drivers. Notable is that one of the "careful" drivers (No. 3) did not get a very low value.

6. CONCLUSION

The results indicate that it is possible to use the above-suggested ideas to detect the curves and measure the curvature on the roads with a fairly high precision. This gives us a useful tool to detect the road conditions which is easy to use in, for instance, complete vehicle models.

The residual gives us information about to what extent a driver is driving unsteady. Thus we can find a bad driver by looking at the residual. However it should be pointed out that the most severe effect on fatigue life is the vehicle speed.

7. ACKNOWLEDGEMENTS

This work has been a joint work between Volvo Truck Corporation, Fraunhofer-Chalmers Research Centre for Industrial Mathematics (FCC), and the Department of Mathematical Statistics at Chalmers University of Technology, where the author is a graduate student. The author wishes to express his gratitude towards his supervisors Dr. Thomas Svensson, FCC, Prof. Jacques de Maré, Chalmers and Dr. Bengt Johannesson, Volvo. The project is funded by the Swedish Agency for Innovation Systems.

REFERENCES

1. Olofsson, M.: Evaluation of Estimates of Extreme Fatigue Load – Enhanced by Data from Questionnaires. Licentiate of Engineering Thesis, Chalmers University of Technology, Göteborg, Sweden, 2000.
2. Niemand, L.: Probabilistic Establishment of Vehicle Fatigue Test Requirements. Master of Engineering Thesis, University of Pretoria, Pretoria, 1996.
3. Thomas, J.J., Perroud, G., Bignonnet, A., Monnet, D.: Fatigue Design and Reliability in the Automotive Industry. In Marquis, G. Solin, J. (editors), Fatigue Design and Reliability, ESIS Publication 23. Oxford, UK: Elsevier Science, pp.1-12, 1999.
4. Grubisic, V., Fischer, G.: Methodology for Effective Design Evaluation and Durability Approval of Car Suspension Components. SAE Technical Paper 970094, 1997.
5. Fischer, G.: Design and Quality Assurance of Vehicle Components based on the Usage Related Stress Analysis. XXIX Convegno Nazionale dell'Associazione Italiana per l'Analisi delle Solleciatazioni, Lucca, 6-9 Sept. 2000.
6. Aurell J., Edlund S.: Operating Severity Distribution a Base for Vehicle Optimization, 11th IAVSD-Symposium on the Dynamics of Vehicles on Roads and Tracks, Kingston, Canada, 1989.
7. Aurell J., Edlund S.: Prediction of Vibration Environment in Real Operations Described by Q-distributions, 14th IAVSD - Symposium on the Dynamics of Vehicles on Roads and Tracks, Ann Arbor, USA, 1995.
8. Öijer, F., Edlund S.: Complete Vehicle Durability Assessments Using Discrete Sets of Random Roads and Transient Obstacles Based on Q-distributions, 17th IAVSD - Symposium on the Dynamics of Vehicles on Roads and Tracks, Lyngby, Denmark, 2001.
9. Vägverket: Vägutforming-94 version S-1, del 6 Linjeföring. Vägverket Publikation 1994-52, Borlänge, 1994. (In Swedish).

Vehicle System Dynamics Supplement 41 (2004), p.421-430 © Taylor & Francis Ltd.

Estimation of Tire Grip Margin Using Electric Power Steering System

EIICHI ONO[1], KENJI ASANO AND KEN KOIBUCHI[2]

SUMMARY

This paper develops an estimation method of tire grip margin from sensor signals of an electric power steering system and vehicle stability control system. Tire grip margin is defined as a ratio of tire force margin to maximum tire force that corresponds to friction characteristics between tire and road. In this paper, SAT (Self Aligning Torque) can be described as a function of the lateral and longitudinal forces of the tire and the tire grip margin, and an estimation method of the tire grip margin is proposed. The estimation method is based on a brush model, so that the robustness of the estimation is verified by applying the method to a magic formula tire model. Furthermore, on line identification algorithm, which adapts to change of tire characteristics such as inflation pressure and types of tires, is applied to estimation method of tire grip margin for practical use. The estimation method is evaluated by experiment on a proving ground that has a constant friction coefficient between tire and road.

1. INTRODUCTION

SAT (Self Aligning Torque) has characteristics that saturate more in lower lateral acceleration turning motions than lateral force, so that an estimation algorithm of tire characteristics using the relation between SAT and lateral force (or slip angle) was developed [1-3]. However, these methods can be applied to pure lateral slip motions. In this study, the relation between lateral grip margin and SAT with combined lateral and longitudinal forces is clarified by using brush models [4, 5], and an estimation method of grip margin from an EPS (Electric Power Steering) sensor signal is proposed. It is expected that the estimated tire grip margin can be applied to control systems enhancing vehicle dynamics performance [3].

[1] *Address correspondence to:* Vehicle Control Lab., Toyota Central R&D Labs., Inc., Nagakute, Aichi, 4801192 Japan. Tel.: (561) 63-6197; Fax: (561) 63-6119; E-mail: e-ono@mosk.tytlabs.co.jp

[2] Chassis System Development Div., Component & System Development Center, Toyota Motor Corporation, 1200, Mishuku, Susono, Sizuoka, 4101193, Japan. Tel.: (55) 997-7930; Fax: (55) 997-7872; E-mail: koibuchi@mail.tec.toyota.co.jp

2. THEORETICAL ANALYSIS OF TIRE GRIP MARGIN BASED ON BRUSH MODEL

2.1 SAT CHARACTERISTICS DESCRIBED BY BRUSH MODEL

By using the brush model [5], SAT T_{sat} can be described as the function of slip angle, longitudinal slip and friction coefficient between tire and road.

SAT T_{sat} can be described as

(*Braking*)

$$T_{sat} = \frac{lK_\beta \tan \beta}{2(1-s)} \xi_s^2 \left(1 - \frac{4}{3}\xi_s\right) - \frac{3}{2} l\mu F_z \sin \theta \cdot \xi_s^2 (1-\xi_s)^2$$

$$+ \frac{2lK_s s \tan \beta}{3(1-s)^2} \xi_s^3 + \frac{3l(\mu F_z)^2 \sin \theta \cdot \cos \theta}{5K_\beta} \left(1 - 10\xi_s^3 + 15\xi_s^4 - 6\xi_s^5\right), \quad (1)$$

(*Traction*)

$$T_{sat} = \frac{l}{2} K_\beta (1+s) \tan \beta \cdot \xi_s^2 \left(1 - \frac{4}{3}\xi_s\right) - \frac{3}{2} l\mu F_z \sin \theta \cdot \xi_s^2 (1-\xi_s)^2$$

$$+ \frac{2}{3} lK_s (1+s)s \tan \beta \cdot \xi_s^3 + \frac{3l(\mu F_z)^2 \sin \theta \cdot \cos \theta}{5K_\beta} \left(1 - 10\xi_s^3 + 15\xi_s^4 - 6\xi_s^5\right),$$

$$\quad (2)$$

where, β: slip angle, s: longitudinal slip, l: ground contact length, μ: friction coefficient between tire and road, F_z: normal force, θ: force direction,

$$K_s \equiv \frac{\partial F_x}{\partial s}\bigg|_{s=0, \beta=0}, \quad K_\beta \equiv \frac{\partial F_y}{\partial \beta}\bigg|_{s=0, \beta=0}, \quad \text{and}$$

$$\xi_s = \max\left\{1 - \frac{K_s}{3\mu F_z}\kappa, \ 0\right\}, \quad (3)$$

(*Braking*)

$$\kappa = \sqrt{s^2 + \left(\frac{K_\beta}{K_s}\right)^2 \tan^2 \beta}, \quad (4)$$

(*Traction*)

$$\kappa = \sqrt{s^2 + \left(\frac{K_\beta}{K_s}\right)^2 (1+s)^2 \tan^2 \beta}. \quad (5)$$

By assuming that force direction θ coincides with slip direction as

(*Braking*)

$$\tan\theta = \frac{K_\beta \tan\beta}{K_s s}, \tag{6}$$

(*Traction*)

$$\tan\theta = \frac{K_\beta \tan\beta(1+s)}{K_s s}, \tag{7}$$

and constant friction coefficient $\mu=1.0$, SAT characteristics can be shown as Fig.1. This figure shows that longitudinal slip complicates SAT characteristics, so that it seems hard to estimate friction characteristics by using SAT with combined lateral and longitudinal slip.

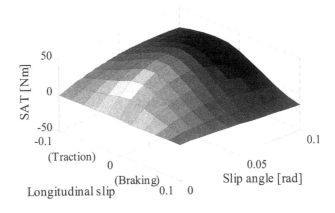

Fig.1. SAT characteristics of brush model ($\mu=1.0$).

2.2 TIRE GRIP MARGIN

In order to coincide in description of brush models of braking and traction states, the lateral and longitudinal slip are defined as follows.
(*Braking*)

$$\kappa_x = \frac{s}{1-s} \tag{8}$$

$$\kappa_y = \frac{K_\beta}{K_s} \frac{\tan\beta}{1-s} \tag{9}$$

(*Traction*)

$$\kappa_x = s \tag{10}$$

$$\kappa_y = \frac{K_\beta}{K_s}(1+s)\tan\beta \tag{11}$$

Here, κ_x : longitudinal slip and κ_y : lateral slip. By using (10) and (11), SAT T_{sat} can be rewritten as

$$T_{sat} = \frac{1}{2}K_s\kappa_y \cdot \xi_s^2\left(1-\frac{4}{3}\xi_s\right) - \frac{3}{2}l\mu F_z \sin\theta \cdot \xi_s^2\left(1-\xi_s\right)^2$$

$$+\frac{2lK_s^2}{3K_\beta}\kappa_x\kappa_y \cdot \xi_s^3 + \frac{3l(\mu F_z)^2 \sin\theta \cdot \cos\theta}{5K_\beta}\left(1-10\xi_s^3+15\xi_s^4-6\xi_s^5\right), \tag{12}$$

and the assumption that force direction θ coincides with slip direction can be rewritten as

$$\kappa_x = \kappa\cos\theta, \tag{13}$$
$$\kappa_y = \kappa\sin\theta, \tag{14}$$

and,

$$\kappa = \sqrt{\kappa_x^2 + \kappa_y^2}. \tag{15}$$

Further, the friction force can be described as

$$F = \mu F_z\left(1-\xi_s^3\right), \tag{16}$$
$$F_x = -F\cos\theta, \tag{17}$$
$$F_y = -F\sin\theta. \tag{18}$$

From (16), the tire grip margin can be described as

$$\varepsilon \equiv 1 - \frac{F}{\mu F_z} = \xi_s^{1/3}. \tag{19}$$

The following relations can be derived from (16)-(18).

$$\mu F_z \cos\theta = -\frac{F_x}{1-\xi_s^3} \tag{20}$$

$$\mu F_z \sin\theta = -\frac{F_y}{1-\xi_s^3} \tag{21}$$

Further, longitudinal slip κ_x and lateral slip κ_y can be rewritten as

$$\kappa_x = \frac{3\mu F_z}{K_s}\left(1-\xi_s\right)\cos\theta = -\frac{3F_x}{K_s\left(1+\xi_s+\xi_s^2\right)} \tag{22}$$

and

$$\kappa_y = \frac{3\mu F_z}{K_s}\left(1-\xi_s\right)\sin\theta = -\frac{3F_y}{K_s\left(1+\xi_s+\xi_s^2\right)} \tag{23}$$

from (3), (13), (14), (20) and (21). By substituting (19)-(23) in (12), SAT T_{sat} can be rewritten as

$$T_{sat} = \left\{ \frac{l}{2} \frac{\varepsilon}{1+\varepsilon^{1/3}+\varepsilon^{2/3}} + \frac{3l}{5} \frac{F_x}{K_\beta} \cdot \frac{1+2\varepsilon^{1/3}+3\varepsilon^{2/3}+4\varepsilon}{\left(1+\varepsilon^{1/3}+\varepsilon^{2/3}\right)^2} \right\} F_y. \tag{24}$$

This equation shows the relation among tire grip margin, SAT, and longitudinal and lateral forces.

This paper defines a nominal model of SAT T_{sat0} that is derived from (24) by substituting $\varepsilon=1$ as follows.

$$T_{sat0} = \left\{ \frac{l}{6} + \frac{2l}{3} \cdot \frac{F_x}{K_\beta} \right\} F_y \tag{25}$$

The nominal model (25) indicates SAT under the condition that friction coefficient between tire and road $\mu=\infty$. Then, SAT model rate γ is defined as the ratio of SAT to the nominal model of SAT, i. e.,

$$\gamma = \frac{T_{SAT}}{T_{SAT0}}. \tag{26}$$

From (24)-(26), the relation among SAT model rate, longitudinal force and tire grip margin can be derived as

$$\left(\frac{1}{6} + \frac{2}{3} \frac{F_x}{K_\beta} \right) \gamma \left(1+\varepsilon^{1/3}+\varepsilon^{2/3}\right)^2$$
$$= \frac{1}{2}\varepsilon\left(1+\varepsilon^{1/3}+\varepsilon^{2/3}\right) + \frac{3}{5} \cdot \frac{F_x}{K_\beta} \cdot \left(1+2\varepsilon^{1/3}+3\varepsilon^{2/3}+4\varepsilon\right). \tag{27}$$

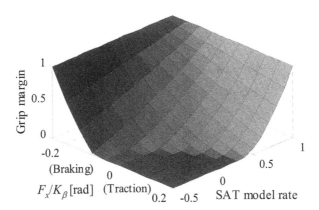

Fig. 2. Relation between γ, F_x/K_β and ε.

Figure 2 shows tire grip margin ε as a function of SAT model rate γ and normalized longitudinal force F_x/K_β, which is calculated by (27). This figure shows that the tire grip margin can be described by a monotonous function of the SAT model rate and normalized longitudinal force. Then, tire grip margin can be estimated from γ and F_x/K_β by using the three-dimensional map shown in Fig. 2. Further, (27) does not include vehicle parameters, such as contact length, so the map shown in Fig. 2 does not change according to vehicle parameters.

3. ROBUSTNESS OF THE ESTIMATION METHOD

The estimation method of the tire grip margin, which is proposed in the previous section, is based on a brush model. In this section, the robustness of the estimation is evaluated using magic formula tire models [6, 7].

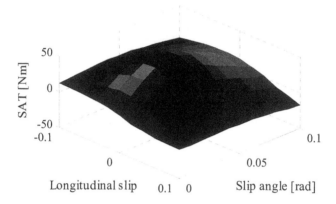

Fig. 3. SAT characteristics of magic formula model.

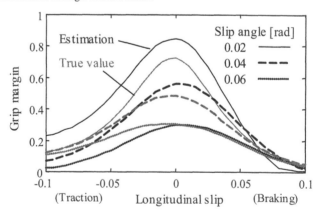

Fig. 4. Estimation of grip margin for magic formula model.

Figure 3 shows SAT characteristics of the magic formula model with the condition that normal force = 4000 N and camber angle = 0 rad. Figure 4 shows the estimation results of grip margin for the magic formula model. The estimation value indicates the calculation from the proposed method with the three-dimensional map in Fig. 2, and the true value ε_{true} indicates the calculation as follows,

$$\varepsilon_{true} = 1 - \frac{F}{F_{max}} = 1 - \frac{\sqrt{F_x^2 + F_y^2}}{F_{max}} \tag{28}$$

where, F_x, F_y: longitudinal and lateral forces of the magic formula model, and F_{max}: maximum value of friction force F. The parameters K_β and l for estimation of tire grip margin are designed by using the relation among slip angle, lateral force and SAT of the magic formula model around an origin. This figure shows that the estimation method of grip margin has good robustness against tire model perturbations between the brush model and magic formula model.

4. ESTIMATION USING EPS SYSTEM

SAT can be estimated from steering torque (= driver's torque + EPS assist torque). Since steering torque includes coulomb friction of the steering system, it is necessary to remove coulomb friction from steering torque in order to estimate SAT.

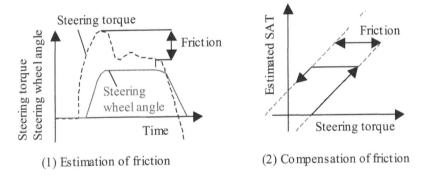

(1) Estimation of friction   (2) Compensation of friction

Fig. 5. Estimation of SAT.

Coulomb friction can be estimated from a difference between maximum steering torque during an increasing steering wheel angle and the steering torque when the driver begins to decrease the steering wheel angle (see Fig. 5 (1)). Figure 5 (2) shows an algorithm of compensation for coulomb friction. The area between the two gray dashed lines indicates the deadband for steering torque as an input signal. The width of the deadband indicates the value of estimated coulomb friction. When the steering torque increases from a neutral point, SAT is 0 in the deadband and increases according to steering torque on the edge of the deadband. When the steering torque decreases, SAT does not change in the deadband, and decreases according to steering torque on the edge of the deadband.

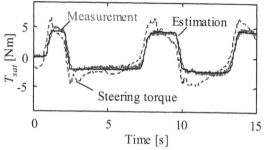

Fig. 6. Experimental result of SAT estimation.

Figure 6 shows the experimental result of SAT estimation. The measurement value of SAT is measured by the wheel dynamometer [8], and estimated SAT is calculated from steering torque by removing friction. This figure shows that estimated SAT coincides with measured SAT.

The nominal model of SAT (25) includes ground contact length. Further, ground contact length changes according to tire inflation pressure, vertical load of tires, types of tires, etc. This implies that adaptation to changes of tire characteristics is necessary for practical use. Therefore, in this study, the nominal pneumatic trail

$$K_{SAT} \equiv \frac{\partial T_{SAT}}{\partial \hat{F}_y}\bigg|_{F_y=0, F_x=0} \tag{29}$$

is estimated by on line identification for adaptation to tire inflation pressure and types of tires, and the nominal SAT model is calculated as

$$T_{SAT0} = \left\{ \left(\frac{\hat{F}_z}{F_{z0}} \right)^{1/2} + 4 \cdot \frac{\hat{F}_x}{K_{\beta 0}} \left(\frac{\hat{F}_z}{F_{z0}} \right)^{-1/2} \right\} K_{SAT} \hat{F}_y, \tag{30}$$

where, ^: estimated value and subscript 0 indicates a nominal value. In this study, the nominal pneumatic trail of a studless snow tire is used as a initial value of estimation K_{SAT0}, and an increase of the nominal pneumatic trail, which is occasioned by under-inflation of tire, an increase in superimposed load, changing tire types from studless to summer, etc., are adapted by applying on line the least squares method [9]. Figure 7 shows the estimation result of K_{SAT} / K_{SAT0} under the conditions of low inflation of tire (100 kPa) and change of tire type (summer tire). The friction coefficient of the proving ground changes from μ=1.0 to 0.35 at t=60 [s] in the experiment, and estimated ρ decreases. By using the maximum value of ρ for calculating grip margin, an accurate estimation of lateral grip margin is expected after adaptation ($t \cong 40$ [s]), however, ρ decreases according to a decrease in lateral grip margin.

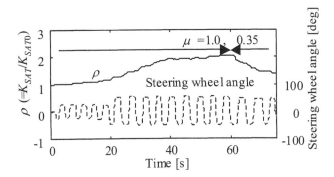

Fig. 7. Estimation of K_{SAT} / K_{SAT0} (v=11m/s).

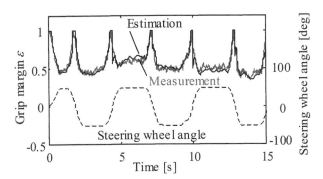

Fig. 8. Grip margin without longitudinal slip.

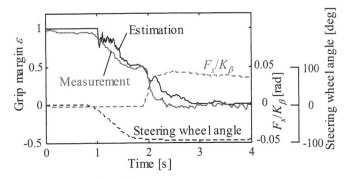

Fig. 9. Grip margin with longitudinal slip.

Figure 8 shows the estimation result of the grip margin on a low friction road (μ=0.35) after adaptation of ρ. The measurement value is the calculated value from acceleration at the front axis g_{front} and μg as follows

$$\varepsilon = 1 - \frac{g_{front}}{\mu g}, \tag{31}$$

where g is the gravitational constant. Figure 9 shows the estimation result of the grip margin on a low friction road with longitudinal slip (traction force). These figures show that the estimation value of the grip margin coincides with the measurement value.

5. CONCLUSIONS

This paper develops an estimation method of tire grip margin from sensor signals of an electric power steering system and vehicle stability control system. The estimation method of the tire grip margin is based on a brush model, and the robustness of the estimation is verified by applying the method to a magic formula tire model. Furthermore, the estimation method is evaluated by experiment on proving ground that has a constant friction coefficient between tire and road.

ACKNOWLEDGMENT

The authors thank Mr. Mizuno at Toyota Central R&D Labs. for designing the magic formula model.

REFERENCES

1. Pasterkamp, W. R. and Pacejka, H. B.: On Line Estimation Of Tyre Characteristics For Vehicle Control, Proceedings of the 2nd International Symposium on Advanced Vehicle Control 1994, (1994), pp. 521-526
2. Fukada: Japanese Patent 1995-137647, (1995)
3. Muragishi, Y., Ono, E., Yasui, Y., Tanaka, W., Momiyama, M., Asano, K. and Imoto, T. : Estimation of Grip State Based on Self Aligning Torque and Its Application to Enhance Vehicle Stability, JSAE Paper No. 20035105, in Japanese, (2003)
4. Bernard, J. E., Segel, L. and Wild, R. E.: Tire Shear Force Generation During Combined Steering And Braking Maneuvers, SAE paper, 770852, (1977)
5. Abe, M.: Vehicle Dynamics and Control, in Japanese pp.30-39, Sankai-do (1992)
6. Pacejka, H. B.: The Tyre as a Vehicle Component, Proc. XXVI FISITA Congress (1996)
7. Bakker, E., Pacejka, H. B. and Linder, L.: A New Tyre Model with an Application in Vehicle Dynamics Studies, 4th Autotechnologies Conference, Monte Carlo, SAE 890087, pp. 83-95 (1989)
8. Burkard, H. and Clame, C.: Rotating Wheel Dynamometer With High Frequency Response, Tire Technology International 1998, (1998), pp.154-158
9. Niedźwiecki, M.: Identification of Time-varying Processes, p.105, John Wiley & Sons, LTD (2000)

Measurement and analysis of interaction between track and road
-Analysis of forward slip phenomenon-

AKIHIKO SHIMURA AND KEISUKE UEMURA[1]

SUMMARY

This paper describes the forward slip phenomenon of tracks. The mechanism of the well-known phenomenon is discussed with real measurement and analytical modeling and simulation.

The motion of track shoes of a real heavy tracked vehicle is measured by video motion tracker. As a result, quantitative data of the forward slip is obtained. Displacement of the forward slip is approximately 100mm at no traction condition. And it decreases with increase in traction. The simulation on multi-body dynamics software is carried out at almost same condition of the real measurement. The results of simulation show very similar to the real measurement. By consideration of the both results, it is estimated that the forward slip phenomenon is caused by pitch motion of track shoes at approach to ground on leading edge and at departure from ground on trailing edge.

1 INTRODUCTION

When building simulators for combat vehicle, consideration of track - road force is very important. There are many available case examples of tire - road force models, whereas there aren't so many case examples of track - road force models. Therefore understanding interaction between track and road is required to build tracked vehicle model on the simulator. In order to measure the track - road force, experiments by using a real heavy tracked vehicle and towed testing rig are carried out. Analysis by a 1/4 vehicle model on multi-body dynamics (MBD) software is carried out also. As a result, mechanism of the well-known forward slip phenomenon of track was unraveled [1]. Nevertheless, there are some unclear points remained. In this paper, further discussion about these points is attempted by re-building a 1/2 vehicle model on MBD software.

2 MEASUREMENT OF MOTION OF TRACK SHOES

In this section, measurement of motion of track shoes is summarized. A heavy tracked vehicle that is attached some color makers on chassis, track shoes and sprocket is measured with video motion tracker as shown in Fig. 1. The heavy tracked vehicle is

[1] *Address correspondense to*: Vehicle System research Section, 2nd Division, 4th Research Center, Technical Research and Development Institute, Japan Defense Agency, 2-9-54 Fuchinobe, Sagamihara, Kanagawa 229-0006 Japan. Tel.: (81)42 752 2941; Fax.: (81)42 752 5462; E-mail: poko@jda-trdi.go.jp

Fig. 1. Measured heavy tracked vehicle.

pulled to rearward by testing rig. The rig can make pulling force by using its brake. Mass of heavy tracked vehicle and the rig are 50,000 kg and 9,000 - 15,000 kg. The tractive (being pull by the testing rig) force is measured with load cell on the towing bar between the measured vehicle and the testing rig. The running speed is measured with non-contact vehicle velocity meter on the testing rig. The testing rig can change its mass by adding or removing of weights. The contact length of tracks and ground of the measured vehicle is 4.55m. The measurement is carried out at tractive force from 5 kN to 40 kN and running speed 10, 20 and 30 km/h. The measurement is carried out on paved road and on soil. However, in this paper, the data on paved road is discussed only, to compare with analytical model result. As a result of measurement, the motion of the chassis, shoes and sprocket that is shown in Fig. 2 is obtained. It shows that the track shoes moves forward during contact with ground at most condition. The existence of this forward slip phenomenon is well-known [2]. By measurement at many conditions, quantitative data of the forward slip phenomenon that is shown in Fig. 3 is obtained. Displacement of the forward slip is approximately 100mm at no tractive force condition. And it decreases with increase in tractive force.

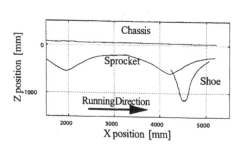
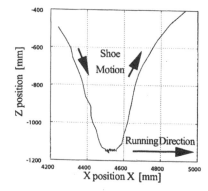

Fig. 2. Measurement results of the video tracker.

432

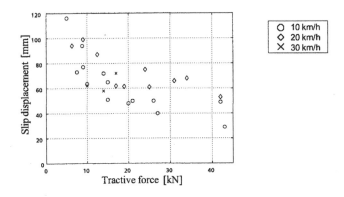

Fig. 3. Measurement results of tractive force and slip displacement of track shoe during contact to ground.

3 ANALYSYS BY MBD MODEL OF TRACKED VEHICLE

The former 1/4 model requires some transformation on comparing result of the measurement and the analysis. Because there is difference of contact length of tracks and ground between the model and the measured vehicle. Therefore, a 1/2 analytical model that has same contact length and the number of track shoes and road wheels to the measured vehicle is built on MBD software DADS [3] as shown in Fig. 4. The model consists of 77 shoes and connectors, 6 road wheels, 3 support rollers, idler, sprocket, chassis and some parts. The mass of chassis is half to the measured vehicle. The properties of the model except mass of chassis are same as the measured vehicle. All parts are constrained on XZ plane, so all parts are allowed only surge, bounce and pitch motion. There are contact definitions between the track shoe pads and the ground and between the inner surface of track shoes and the road wheels, support rollers, idler and sprocket as shown as in Fig. 5. By using point - segment contact force element of DADS, the contact force is based on the depth of penetration and the relative velocity

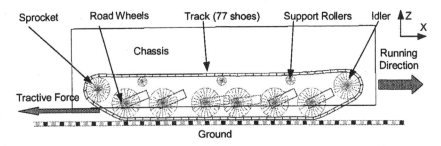

Fig. 4. Analitycal model of tracked vehicle (1/2 MBD model).

433

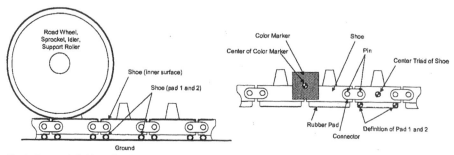

Fig. 5. Contact definition of analytical model and position of triad.

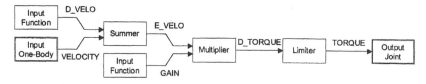

Fig. 6. Flow of control elements.

normal to the contact surface, while friction forces are calculated based upon the relative velocities tangential to the surface. The track shoe pads, road wheels, support rollers, idler and sprocket are treaded as points of the contact force element that have specific radius. The model is driven with torque on sprocket that is controlled to sustain the running speed. The flow of control elements is shown in Fig. 6. The desired velocity is defined with taking the running resistance into consideration. Simulations are carried out on similar condition of the measurement by real heavy tracked vehicle.

4 COMPARISON MEASUREMENT AND ANALYSYS

The results of the multi-body simulations are shown in Fig. 7. The trajectory of track shoe by analysis is very close to measured result as shown in Fig. 7 (a). Fig. 7 (b)~(d) show the time history of longitudinal motion of track shoe at some tractive conditions. These figures show the forward slip of track shoe and its trend of decreasing with increase on the tractive force. The displacement of track slip and the tractive force are plotted on Fig. 7 (e). From view point of the relation between tractive force and slip displacement, result of the analysis accords with result of the measurement well. In both results, displacement of the forward slip is approximately 100mm at no tractive force. And it decreases with increase in tractive force. The 1st order approximated line of the analysis results at 10 km/h matches it of the real measurement well. On 20 km/h condition, the analysis line doesn't match with the real line so well. But, the trend that slip displacement at 20 km/h is larger than it at 10 km/h especially at large traction force is same. On 30 km/h condition, there isn't the line of the real measurement be-

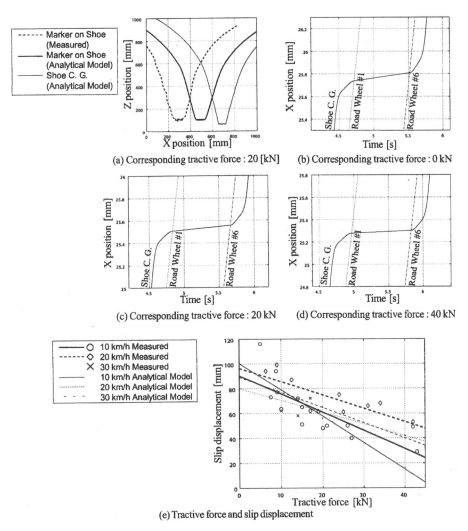

(a) Corresponding tractive force : 20 [kN]

(b) Corresponding tractive force : 0 kN

(c) Corresponding tractive force : 20 kN

(d) Corresponding tractive force : 40 kN

(e) Tractive force and slip displacement

Fig. 7. Multibody simulation results.

cause of limited sample number on this condition. But, the line of analysis is very close to the real measurement plot points.

The analysis results of longitudinal and vertical contact force on the track shoe pads that is shown in Fig. 8 have same characteristics with the result of former research work [4] that is shown in Fig. 9 by measurement of the other real tracked vehicle that has 5 road wheels. From mentioned above, appropriateness of the analysis model is validated. The three charts of Fig. 9 show that longitudinal force on track shoe pads occurs when the track shoe pad moves rearward simultaneously with vertical force. Through the discussion of above, the mechanism of the forward slip phenomenon is estimated as

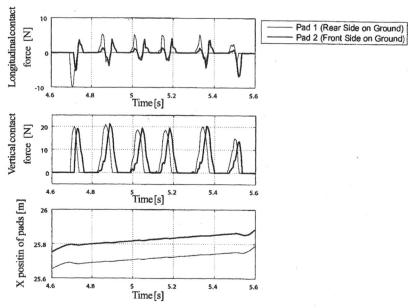

Fig. 8. The contact force on pads and longitudinal motion of pads by multibody simulation.

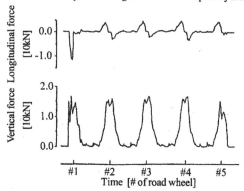

Fig. 9. Measured contact force on pad of 5 wheeled vehicle[4].

follow section.

5 MECHANISM OF FORWARD SLIP PHENOMENON

When the track shoe approaches to the ground and departs from the ground, the track shoes make pitch motion as shown in Fig. 10. By this pitch motion, the track shoe pad has angular velocity in addition to translation velocity. Accordingly, the track shoe pad at the point has higher velocity to rearward and the track shoe pads at the other point

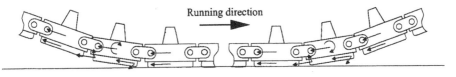

Running direction

Departure at trailing edge.   Approach at leading edge.

Fig. 10. Motion of pads that causes forward slip.

have relatively lower velocity. From view of vehicle coordinate, velocity of the ground becomes between the higher velocity and lower velocity because the ground comes in contact with shoe pads that have the higher velocity and lower velocity. Consequently, the track shoes at the other point have relatively lower velocity than the ground and make slip to forward against to the ground.

The existence of the forward slip is undesirable because it increases the running resistance. The discussion above shows that thinner shoe pad will reduce the forward slip by reducing the longitudinal component of angular velocity on shoe pad. In addition, lower vertical load of the road wheel at both ends will reduce the forward slip also.

6 CONCLUSIONS

In this paper, the forward slip phenomenon of track shoes is discussed through the measurement of real heavy tracked vehicle and analysis by a 1/2 tracked vehicle model on MBD software. The results of the analysis are quite appropriate for the result of the measurement. By discussion on both results, the mechanism of the well-known forward slip phenomenon has been unraveled. The phenomenon is caused with behavior of track shoe pads on approaching to ground and departing from ground.

ACKNOWLEDGMENT

Our special thanks are due to Northern Army of JGSDF for operating vehicles on the measurement.

REFERENCES

1. Uemura, K., Shimura, A., Inoue, Y., Kamozaki, H. and Tanaka, N.: Measurement of Track - Road Force - Analysis of Forward Slip of Track -. *Proceedings of the 2000 JSAE Spring Convention* No.4-00 (2000), pp5-8. (in Japanese)
2. Itoh, N., *et al.*, *Design Engineering for Tracks* (2000), Japanese society for terramechanics. (in Japanese)
3. Haug, E. J., *Computer Aided Kinematics and Dynamimcs of Mechanical Systems* (1989), Allyn and Bacon.
4. Gotoh, K., Sawagashira, K. and Kinoh, J.: Research on controllablity and stability (3rd report). *Technical Report of TRDI* No. 572 (1974). (in Japanese)

Modelling Freight Wagon Dynamics

M. MCCLANACHAN, Y. HANDOKO, M. DHANASEKAR[†],
D. SKERMAN AND J. DAVEY

SUMMARY

This paper presents some dynamic responses recorded during normal operation of a QR freight container wagon and the associated computer simulation. For the purpose of computer simulation, the track was modelled using track geometry car data and the freight wagon containing three-piece bogies was modelled using the known characteristics of the wagon components and the loading data. VAMPIRE® wagon modelling software was used for the simulation. The simulated and the measured dynamic response of the freight wagons enable more complete understanding and deeper insight into the causes and response of track induced wagon dynamics. The improved understanding and the validated computer simulation models potentially extend the capability of the theoretical framework in solving complex practical problems.

1 INTRODUCTION

The dynamics of railway wagons are affected by a number of variables including track irregularities, track curvature, track grade, bogie characteristics, wheel conicity and longitudinal train dynamics. As part of ongoing research on wagon dynamics, extensive data were recorded from a QR (one of the major railway operators and network owners of Australia) wagon running on their network. The field-testing was performed as joint research between the Centre for Railway Engineering (CRE) and QR. One of the objectives of the field-testing was to evaluate the wagon dynamics induced by the track geometry characteristics and to use the data as a source of identifying and rectifying tracks with severe geometry defects. This paper describes the principles and methods of data recording and the associated computer simulation.

[†] Address correspondence to: Dr. M. Dhanasekar, Director, Centre for Railway Engineering, Central Queensland University, Rockhampton 4702, Australia. Ph. +61 7 4930 9677; Fax. +61 7 4930 6984; email: m.dhanasekar@cqu.edu.au

2   FIELD TESTING

2.1   Test Equipment

The field data collection was carried out using an instrumented freight container wagon with a tare of 17.2 tonnes and a maximum carry load of 62.8 tonnes. The wagon uses three-piece bogies with only secondary suspension containing constant friction dampers. The wagon was tested with various loads ranging from 10 tonnes to 50 tonnes. A typical loading configuration is shown in Figure 1. Parameters measured included vertical wheel bearing force, lateral bogie force, coupler force and coupler movement. Track position was determined by a GPS receiver, speed transducer and bogie yaw measurement.

Figure 1. Freight Container Wagon

Vertical wheel bearing forces and lateral bogie forces were measured using strain gauges placed on the bogies, as shown in Figure 2. The strain gauges were positioned on the bogie such to achieve maximum signal output of the force being measured with minimal noise from other forces. Although the bogie instrumentation provides only an indirect measure of the wheel-rail interaction forces, it is less expensive and better suited for use in an operational railway environment than the alternative of using sophisticated instrumented wheelsets.

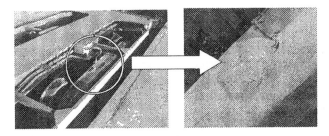

Figure 2. Strain gauge measuring vertical wheel bearing force.

2.2   Test Procedure

Field testing was carried out in commercially operating freight trains on the main freight routes within the state of Queensland that cover almost 3000 kilometres of track. Data were recorded only when the wagon was moving and data storage allowed over 10 days of continuous operation. Data collected represented 5000 km in distance travelled in 11 days. Track geometry of the route was recorded separately prior to the wagon tests by the QR track recording car.

2.3   Data Searching

The field data collected was searched for extreme dynamics, and various dynamic events such as the pitch, the bounce, the roll and the body yaw were identified [1]. These identified events with the exception of body yaw, are compared to the predictions of the computer simulation.

3   COMPUTER SIMULATION

3.1   Computer Model

The wagon model consisted of 11 masses defined by 62 DOF. Suspension elements with the capability of modelling the non-linear characteristics caused by friction and clearances in the suspension system were included in the model. Full non-linear representation of the wheel-rail contact was used such that the transition response and the kinematic motion of the bogie were modelled. The base model data were supplied by QR. Some data are similar to the values reported by Sun et al [2,3]. VAMPIRE® software was used in the simulation of the model.

The damping provided by the friction wedge between the bolster and the sideframe was modelled using a nonlinear two-dimensional coulomb friction element. The same type of friction element was also used to model the surface friction in the centre plate, side bearer and friction between axle box and side frame. The coefficient of friction is very sensitive and changes during normal haulage operation. Values between 0.30 and 0.40 were found to be appropriate in modelling, particularly for the bolster friction wedge.

Lift-off elements are used in modelling the connections of the centre plate, side bearer and between axlebox and sideframe. The mass of the wagon load was determined by averaging the vertical wheel force data from the field tests. The centre of gravity position of the load was calculated by assuming the 2.5 m high shipping containers were 100% full.

3.2   Track Model

The geometrical track input data was constructed from field data measured by the QR track recording car. The QR track recording car uses a chord based measurement system, therefore the output requires processing to produce the actual track geometry. In this chord based measurement two reference points on the rail, a set distance apart, form a baseline. At a third point, generally mid way on the baseline the rail geometry is measured with respect to the baseline. The reconstruction process accounts for the changes in the reference points due to the changing track geometry, and is reported in [4].

3.3   Simulation Output Filtering

The simulation output channels were chosen to provide results that could be compared to the measurement data from the field tests. Low pass filtering at 5 Hz was applied to the simulation output to match the 5 Hz low pass filtering that was applied to the experimental data.

4   MODEL VALIDATION

The wheel forces from the simulation and the field data were compared, to validate the model. Comparison was made for dynamic events of the wagon of pitch, bounce and roll.

4.1   Roll Event

It was found that the roll events are affected by the torsional stiffness of the wagon body. To represent the effect of the body torsional stiffness, the wagon body was modelled as a flexible component that has a torsional mode.

Figure 3 shows a roll event predicted by the rigid and flexible wagon models along with the corresponding data from field measurements. The data are presented as the percentage of static wheel force. The flexible model provides better correlation with the measured field data in that it reduces the force offset between the left and right wheels present in the rigid model output.

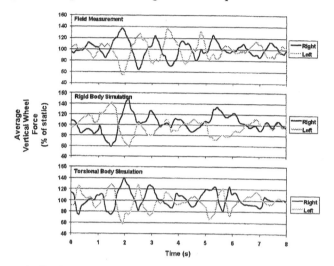

Figure 3. Roll event leading bogie forces, 50 tonne load at 54 km/hr.

4.1.1   Track Geometry

The track geometry for the roll event had an irregularity with a wavelength of 15 m, [1]. This is close to the distance between bogies of 13.6 m and was expected to produce a roll motion.

4.1.2   Resonance

The computer model was simulated at various speeds over the actual track profile. The simulated output is represented as a power spectral density (PSD) in Figure 4. The PSD shows that the model roll resonance occurs at 0.6 Hz and 30.6 km/hr.

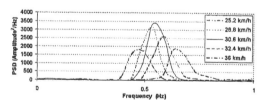

Figure 4. PSD of Roll event, 50 tonne load at 54 km/hr.

4.2   Bounce Event

Wagon bounce occurs when the vertical force on all wheels are in phase. Figure 5 compares a bounce event from the experimental data [1] with the simulated result. The experimental bounce response has a frequency of 2.1 Hz for a 50 tonne load at 57 km/hr. The model's friction wedge had a friction coefficient of 0.4, and normal force of 8.5kN. The normal force on rear bogie was increased by 40% to achieve a simulated result close to that of the field data. This can be confirmed somewhat, as during the static calibration of the bogies the wedges on the rear bogie produced a larger noise when the vertical force was increased, indicating a larger friction force.

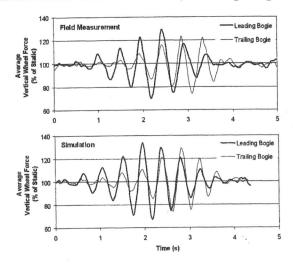

Figure 5. Bounce Event – 50tonne load at 57km/hr .

4.2.1   Track Geometry

The track geometry used in the simulation was derived from the actual QR track recording car data at the location of this event as shown in Figure 6.   The wavelength of the track irregularity was 7.2 m which is approximately half the bogie spacing of the vehicle of 13.6 m and therefore was expected to cause a bounce motion.

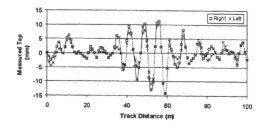

Figure 6. Bounce event – track geometry data as recorded by the QR track recording car[1].

The QR track recording car uses a chord based system to measure the vertical movement of the rail and the output produced is not the true geometrical representation of the track. The chord length is 6.5 m with vertical displacement measured 3 m from the front reference point. Because the wavelength of the irregularity is close to the chord length, the amplitude of the track was estimated to be half that recorded by the QR track recording car. The corrected track data used in the simulation is shown in Figure 7. Both left and right data were averaged for ease of simulation.

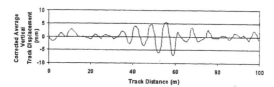

Figure 7. Bounce event – corrected average track geometry used in simulation.

4.2.2 Resonance

The computer model was simulated at various speeds. The simulated output is represented as a power spectral density plot in Figure 8 which shows at a speed of 57 km/hr the model produces a frequency of 2.2 Hz which corresponds to the experimental frequency of 2.1 Hz.

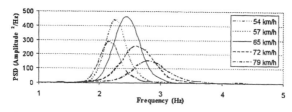

Figure 8. Simulated bounce power spectral density – 54 to 79kph.

From elementary theory the natural bounce frequency is calculated as 2.3 Hz which would occur at a slightly faster speed of 60km/hr. The PSD peaks in the speed range from 57 km/hr to 65 km/hr with a frequency of about 2.3 Hz. The largest PSD amplitude occurs at 65 km/hr with a frequency of 2.44Hz. The motion at 65 km/hr was confirmed to have a component of pitch motion by restricting pitch by applying a larger pitch inertia to the wagon body of 3.2×10^6 kg.m^2, as used in the pitch simulation. The pitch dynamic component was produced as two track wavelengths are slightly larger than the bogie spacing and also because the track irregularity amplitude increases with distance.

4.3   Pitch Event

Wagon pitch occurs when the vertical force on the leading and trailing bogies are out of phase. Figure 9 compares a pitch event from the experimental data [1] with the simulated result. To obtain a close match between the experimental and simulation the pitch inertia was increased to 3.2×10^6 kg.m^2, reducing the force amplitude Also the smaller force amplitude of the trailing bogie was simulated by a larger friction wedge normal force value on the trailing bogie of 21kN. This friction value exceeds the expected range and it is proposed that the wagon used in the field tests has an uneven weight distribution which was not accurately represented in the model. Future track tests are planned that will be more controlled in terms of loading from these tests it should be possible to fine tune the model's spring stiffness and friction properties.

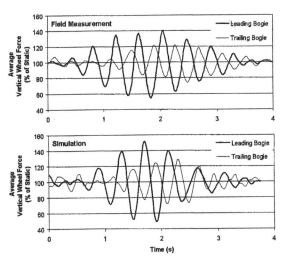

Figure 9. Pitch Event – 50tonne load at 78km/hr.

The field data has a pitch frequency of about 2.4 Hz which correlates to the simulated frequency of 2.5 Hz.

4.3.1   Track Geometry

The track geometry measured by the QR track recording car indicated an irregularity on both rails with a wavelength of 8.2 m. Similar to the bounce event the measured track data was corrected by a reduction in the amplitude to remove the effect of the chord measuring system on the QR track recording car.   As the irregularity's wavelength was 8.2 m the amplitude of the irregularity was multiplied by 0.56, shown in Figure 10. Pitch motion was expected because the distance between bogies of 13.6 m is approximately 1.5 times the irregularity's wavelength.

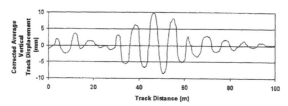

Figure 10. Pitch event – corrected average track geometry used in simulation.

4.3.2   Resonance Speeds

The computer model was simulated at various speeds to determine the resonance frequency. The simulated output is represented as a power spectral density in Figure 811, showing at a speed of 79 km/hr the model produces a frequency of 2.4 Hz which corresponds to the experimental frequency. The PSD predicts resonance at 65km/hr, but due to the unconfirmed loading distribution this result would need verification in the future field tests.

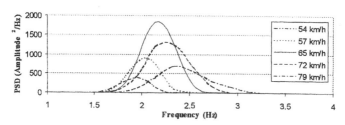

Figure 11. Pitch event – corrected average track geometry used in simulation.

The bogie spring stiffness of 6.3 MN/m and bogie spacing of 13.6 m gives an equivalent wagon pitch spring rate of 580×10^6 N.m/rad. The estimated moment of inertia was 3.2×10^6 kg.m^2, based on a loading of two 6.1 m containers each 25 tonnes situated on the wagon ends. Using these values the wagon's theoretical natural pitch frequency was calculated to be 2.1 Hz. This theoretical pitch frequency is only an estimate, as the exact loading configuration and centre of mass are not known. The theoretical pitch frequency increases to 2.6 Hz for a loading of three 6.1 m containers at 17 tonnes each and 100% full.

5 CONCLUSIONS

The comparison of field data and simulation results shows it is possible to adequately model freight wagons containing three-piece bogies using VAMPIRE®. The use of field data enables the refinement of the computer model resulting in a more accurate model. The wagon roll response is affected by both the wagon body's torsional stiffness and track input. A flexible body model is required to predict the roll event. Bounce and pitch dynamics are predicted well by the rigid body model. This indicates track geometry affects bounce and pitch more than the wagon body characteristics.

From field data it is apparent that resonant responses occur. Simulation will aid the understanding of these dynamics and allow the formulation of additional standards for track geometry to reduce the possibility of derailments. These may include variable limits on the amplitude to track irregularities based on their wavelength and the number of cycles. Suitable temporary operational constraints can also be investigated which may include speed restrictions or vehicle loading limits. Simulation is also being used to reduce the amount of physical testing necessary to certify that new rollingstock meets dynamic performance standards.

REFERENCES

1. McClanachan, M., Dhanasekar, M., Skerman, D., Davey, J.: Monitoring the dynamics of freight wagons. *Proceedings CORE2002, Wollongong*, pp. 213-224.

2. Sun, Y. Q., Dhanasekar, M. and Roach, D.: A three-dimensional model for the lateral and vertical dynamics of wagon-track systems. *Proc. Instn Mech. Engrs Vol 217 Part F: Journal of Rail and Rapid Transit.* IMechE 2003, pp. 31-45.

3. Sun, Y. Q. and Dhanasekar, M.: A dynamic model for the vertical interaction of the rail track and wagon system. *International Journal of Solids and Structures* 39 (2002), pp. 1337-1359.

4. Processing versine measurements in VAMPIRE®. *LR/VDU/92/071*, AEA Technology Rail, 1992.

Vehicle System Dynamics Supplement 41 (2004), p.448-457 © Taylor & Francis Ltd.

Innovative System Dynamics Analyses of Rail Vehicles

ANTON STRIBERSKY[1], FRANZ MOSER[1], JEAN-LUC PETERS[2], AND GERHARD SCHANDL[3]

SUMMARY

The use of simulation technologies has become increasingly important for the rail vehicle development process. The paper describes three key components of a recently developed virtual product development (VPD) system, which supports the development of innovative rail vehicle systems. Aerodynamics has a major impact on the cross wind sensitivity of a car. Using flexible car body structures embedded in the vehicle dynamics model, the ride comfort can be predicted. Finally, the structure gauge assessment can be supported by the use of an active digital mock-up.

1 INTRODUCTION

Due to growing quality demands for innovative designs, shorter times to market, and the continuous effort to reduce the development cost, the use of simulation technologies has become increasingly important for the rail vehicle development process. Still, the combination of numerical simulation and experimental testing is a compelling requirement to ensure the quality of a new product. Nevertheless, virtual product development systems can simulate the entire life cycle of rail vehicles and therefore are involved in the overall product development process.

In the past the major modeling and simulation disciplines like computer aided design, finite and boundary element methods, computer aided control engineering, and multibody system dynamics have followed their separate paths. Recently, with the use of networks of such tools the analysis of even very complex rail vehicle systems becomes possible, as can be seen in Fig. 1. Three-dimensional geometry, volume and material data is generated using a computer aided design (CAD) tool, which holds the basic design data, the design history, and the model parameterization. For the numerical simulation a certain design status is extracted and used to model the elastic structures and aerodynamic shapes using the finite element method (FEM) and to model the overall vehicle as a multibody system (MBS). The success of the multibody system approach in combination with efficient numerical methods generated powerful software tools, which are now extensively

---

[1] Siemens SGP Verkehrstechnik GmbH, Vienna, Austria. Tel.: +43 51707 41686; E-mail: anton.stribersky@siemens.com
[2] Siemens AG, Transportation Systems, Munich, Germany.
[3] Vienna University of Technology, Institute for Mechanics, Austria.

used by the development engineers. To perform a dynamic simulation on the basis of geometrical information, an active digital mock-up (DMU) simulation tool can be used.

To meet the demands for the dynamic analysis and simulation of innovative conceptual designs, different engineering disciplines have been applied, e.g. modeling of flexible bodies like light-weight car body structures [1], analysis of the aerodynamics of high speed trains [2], the simulation of coupled mechanical and electronical systems like traction or braking systems, and finally the approach of collaborative dynamic simulation of full vehicles to support the systems integration task. The paper describes three key components of a recently developed virtual product development (VPD) system [3], which supports the development of innovative rail vehicle systems.

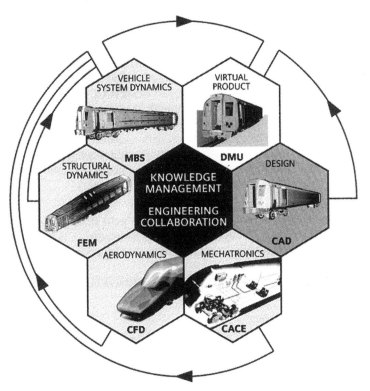

Fig. 1. Virtual product development system. MBS (multibody system), DMU (digital mock-up), CAD (computer aided design), CACE (computer aided control engineering), CFD (computational fluid dynamics), FEM (finite element method).

2   AERODYNAMICS OF A HIGH SPEED POWER CAR

Aerodynamics has a major impact on rail vehicles and their environment. In this paper, only the effects on the vehicles themselves will be considered. These effects may be classified under two major aspects: cost-effectiveness and safety.

2.1   COST-EFFECTIVENESS

At the operational speeds of modern high speed trains, more than 80% of the total resistance to motion is caused by aerodynamic drag [2]. The optimization of the drag will then clearly be one of the major ways of reducing the operational costs of high speed trains.

Many of the newly built high speed lines feature a large proportion of tunnels. When entering these tunnels, trains trigger a pressure wave that propagates through the tunnel at the speed of sound [4]. This wave is reflected back and forth at the tunnel extremities, changing each time its sign. This results in quickly alternating pressures on the trains in the tunnel. For aural comfort, the trains have to be sealed, which creates additional costs. The walls of sealed trains are then subjected to constantly changing loads, affecting the Life Cycle Costs (LCC) of the vehicles. So, a train that generates a lower tunnel entry pressure has a positive effect on its LCC.

Finally the head pressure perturbation of a train [5] may affect the comfort of passengers in passing trains, or even, under certain circumstances, damage older passing rolling stock. An aerodynamically well shaped leading car nose will significantly reduce this head pressure pulse, eliminating possible additional costs.

2.2   SAFETY

The loads generated by strong cross winds on a train depend greatly on its aerodynamic shape [6]. At worst, they can lead to the derailment of the train and are always one of the major inputs in the running dynamics calculations.

2.3   METHODS OF OPTIMIZING THE AERODYNAMICS OF RAIL VEHICLES

2.3.1   Experimental methods

Wind tunnel tests have been and still are one of the best and most convenient research and development tools for train aerodynamicists [2]. With a few precautions with regard to measuring equipment, wind tunnel choice, Reynolds-numbers and model mounting, they can provide very reliable results. Other experimental methods are usually much more costly, and, in the case of full scale tests, are not possible during the development phase of a new train anyway.

2.3.2   Computational Fluid Dynamics

Particularly in the last decade, Computational Fluid Dynamics (CFD) has established itself as an essential complement to the experimental methods.

Computational fluid dynamics is essentially based on the governing equations of fluid dynamics: the continuity, momentum and energy equations. Computational fluid dynamics is, in part, the art of replacing the governing partial differential equations of fluid flow with numbers, and advancing these numbers in space and/or time to obtain a final numerical description of the complete flowfield of interest. It should be noted that the results of CFD are only as valid as the physical models incorporated in the governing equations and boundary conditions. In particular, as most CFD solutions of turbulent flows contain turbulence models which are just approximations of the real physics, and which depend on empirical data for various constants that go into the turbulence models, they are subject to inaccuracy.

This having been said, excellent results can be obtained by calculating the head pressure pulse of locomotives by means of a panel method [5], one of the oldest and most simple CFD solutions. More elaborate CFD codes are however necessary to calculate the forces produced by cross wind on a rail vehicle.

For the reasons mentioned above, calculation results of aerodynamic phenomena must be validated by means of experimental data. That is why CFD calculations were systematically coupled to a wind tunnel optimization of high speed train nose shapes with regard to cross wind sensitivity.

2.4   COMPARISON OF EXPERIMENTAL AND NUMERICAL RESULTS

The object of the comparison was the power car of the AVE S102, scheduled to run from the year 2004 on between Madrid and Barcelona in Spain. The aerodynamics of this power car has been developed at the Locomotives Division of Siemens Transportation Systems (TS) in Munich, Germany, in an attempt not only to obtain optimal cost-effectiveness and safety, but also to have a minimal environmental impact [6].

There is a wide consensus in the CFD community that even the best computational resources available today are not yet adequate to reliably calculate the drag of a train. So the drag was optimized by using the large amount of experimental data available at Siemens TS [2].

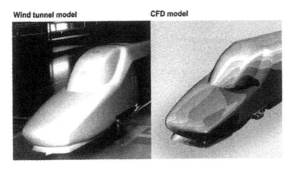

Fig. 2. "Duck bill" nose shape. Left = 10% wind tunnel model. Right = CFD (FLUENT) model showing pressure distribution at yaw (darker areas = positive pressures).

The head pressure pulse could however be optimized by means of a panel method, checking at the same time that the conditions for a minimum tunnel entry pressure wave gradient were met. These conditions were calculated by the Japanese RTRI [4] by means of an extensively validated one-dimensional CFD code.

The cross wind behavior of the power car was first tested in the wind tunnel with the experimental results currently being used for the validation of the CFD code.

2.4.1   Wind tunnel tests

A major contribution of these tests to the success of the research program was a significant improvement in the quality of the measurements carried out in several wind tunnels with different balances. Using an improved 6-component internal balance, reliable results could be obtained for all the 6 aerodynamic coefficients [2], and in particular for the most relevant rolling moment coefficient (Cmx), up to yaw angles of at least 50°.

A combination of total power car 6-component measurements, nose 6-component measurements, pressure distribution measurements and flow visualizations showed that the so called "duck bill" shape of the AVE S102 power car nose (Fig. 2) improves the cross wind behavior of the power car not only due to its particular shape, but also to the result of its influence on the aerodynamic behavior of the rest of the car body. The behavior with yaw angle of the rolling moment coefficient Cmx, the side force coefficient Cy, the lift coefficient Cz and the pitching moment coefficient Cmy of a power car with the "duck bill" nose is shown on Fig. 3.

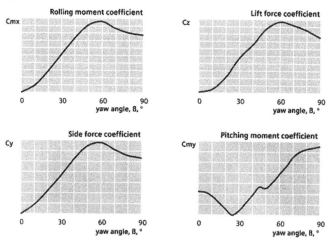

Fig. 3.   Rolling moment coefficient Cmx, side force coefficient Cy, lift coefficient Cz and pitching moment coefficient Cmy as a function of yaw angle ß of a high speed power car.

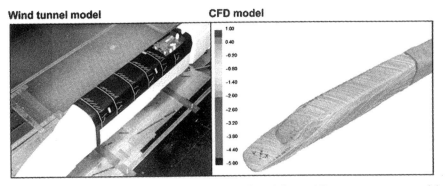

Wind tunnel model   **CFD model**

Fig. 4. Comparison of calculated (right) with visualized (left) path lines at 45° yaw on a power car model.

2.4.2   CFD validation

The original results of the numerical solution of the governing equation of fluid mechanics by the CFD program are the pressure distribution on the vehicle (as shown on Fig. 2 right) and the line paths. Fig. 4 shows a comparison of the calculated line paths around the power car at 45° yaw angle and an air speed of 50 m/s with the line paths made visible on the wind tunnel model by means of wool tufts. An excellent agreement may be observed.

In the second stage, the CFD program integrates the pressure distribution, providing the forces and moments on the vehicle. First comparisons between calculated and measured results are very encouraging, with a good reproduction of the general tendency with yaw angle and the correct order of magnitude of the values, at least for yaw angles up to 45°.

The calculations were carried out for a reproduction of the wind tunnel conditions, i.e. 10% scale model and a wrong simulation of the airflow over the ground. Should CFD correctly reproduce the wind tunnel results, then the real scale and, more importantly, the real ground conditions could be simulated, which might produce physically more authentic results than the wind tunnel.

3   RIDE COMFORT ANALYSIS OF A MAINLINE CAR

Ride comfort simulations for an intermediate car of a mainline train have been conducted using the VPD system. Results from the FEM analysis for the flexible car body, as shown in Fig. 1, have been introduced in the MBS simulation by using a modal approach [1, 3]. The MBS-model includes 32 bodies, 68 joint position states, with 20 of these states being restricted by constraints, and 92 force elements. The flexible full-vehicle simulation model predicts in the medium frequency range eigenmodes with dominant elastic deformations for the diagonal distortion (1, jelly mode), vertical bending (2), torsion (3), body shell breathing (4), and lateral bending (5) of the car body. In Fig. 5 these mode shapes for the fully equipped car body structure without bogies are shown.

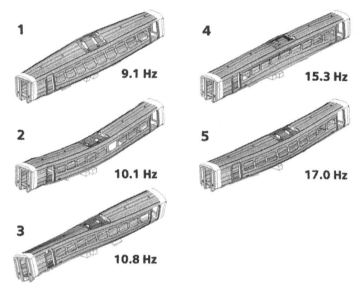

Fig. 5. Example mode shapes and eigenfrequencies for the fully equipped mainline car body structure.

To improve the identification task for a total of 28 eigenmodes up to 25 Hz, the recently developed active digital mock-up for the rail vehicle has been used, which also allows the visualization of the elastic deformations of the car body structure. In Fig. 6 the cross section through the DMU combined with the visualization of the diagonal distortion of the car body can be seen. The car body is made out of large aluminum extrusions, which are welded together. The original finite element model, also shown in Fig. 6, consists of approximately 99800 finite elements, 73000 nodes, and 434700 degrees of freedom. For the Guyan reduction process, 593 master nodes and 1851 master degrees of freedom have been used.

Digital mock-up   **Finite element model**

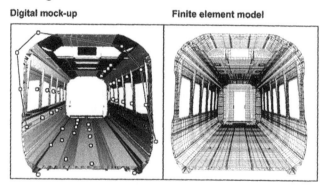

Fig. 6. Visualization of the structural jelly mode for the virtual vehicle model using a digital mock-up (left) together with the original finite element model (right).

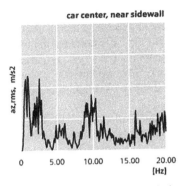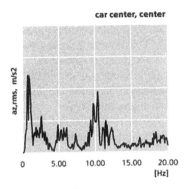

Fig. 7. Simulated frequency spectra of the vertical acceleration on the floor

In Fig. 7 simulated frequency spectra for the vertical acceleration (rms-values) measured on the floor of the mainline car can be seen. The simulation has been done for the vehicle running with a speed of 160 km/h over a track with irregularities (British track 160). The influence of the eigenfrequencies related to the elastic car body structure is clearly visible.

4   DYNAMIC COLLISION ANALYSIS FOR RAIL VEHICLES

To ensure that a railway vehicle runs safely along a track, a structure gauge assessment has to be done. This assessment has to prove that the vehicle does not come unacceptably close to lineside structures, equipment, or to vehicles running on the adjacent track. A number of different approaches are used for vehicle gauging. It should be mentioned that in the UK procedures like shadow gauging and absolute gauging are applied.

Using the multibody system approach, the dynamics for even very large rail vehicle models can be calculated. Utilizing the motion data, which are calculated by multibody simulation, an active digital mock-up (DMU) tool (CENTRIC) can be used for dynamic collision detection of railway vehicles. The DMU can be applied to perform dynamic gauging analysis and clearance studies for vehicle parts under different running conditions. The DMU model for the rail vehicle investigated together with the platform from a train station can be seen in Fig. 1.

4.1   BASIC APPROACH

At first the complex motion of a railway vehicle has to be evaluated. This is done by a multibody system (MBS) simulation using the multibody simulation tool SIMPACK. The geometry of each moving body is contained in an "ensemble". For rigid bodies the movement of each ensemble is described by the body fixed reference frame of the body, to which the ensemble is attached. Simple flexible geometries, e. g. coil springs, are described by additional scaling factors, which shorten or lengthen the part in x-, y- and z-direction.

Via a recently developed interface between the MBS tool and the DMU tool, the results of the multibody simulation can be injected into the component attributes of the DMU [7]. This interface creates two files. The first file contains the names of the ensembles, their initial positions (x-, y-, z-translation and x-, y-, z-rotation) and their initial scaling factors (x-, y- and z-scaling). The second file contains the position information (translation, rotation and scaling in absolute coordinates) for each ensemble at any output step.

The active DMU reads this file and creates a simulation file. This file consists of a configuration folder, which includes the information on the ensembles, their initial positions, and a result file containing the position information for any ensemble at any output step. Then the real geometry must be connected with the MBS-ensembles. For that the 3D-geometries of the car body and bogies of the rail vehicle and additional infrastructure data, e. g. station platforms, which are available in a 3D-CAD-format, are transformed into a DMU-representation with a triangulated surface description. This reduces the amount of data and allows the handling of even large structures. The geometry is connected with the ensembles in the configuration folder by a simple drag and drop operation, in which each geometry part is assigned to the corresponding MBS-ensemble.

Afterwards, it is necessary to make the coordinate systems of the multibody simulation results and the geometry data consistent. For this purpose a global offset, describing the translational and rotational differences between the initial system of the MBS and the initial system of the CAD system is fed into the simulation, allowing the transformation of coordinate information from one system to the other.

Then the simulation scenario is ready to start. Step by step each part of the vehicle is moved from one discrete output point to the next. At every step a collision detection is performed, using a polygon-to-polygon algorithm to test each part of the vehicle on interference with infrastructure or with other parts of the vehicle. The polygon-to-polygon algorithm allows the study of the interaction of rigid mechanical bodies. If a collision occurs, the interfering parts change their color. A cross section of the DMU can easily be produced to analyze the reason for the interference.

4.2 NUMERICAL RESULTS FROM THE DYNAMIC COLLISION ANALYSIS

To perform a dynamic collision detection, a full-vehicle car model for a mainline train has been investigated. First, a curve-entry into a curve with a radius of 160 m at a speed of 55 km/h has been simulated. The anti-roll bar on the leading bogie has been deactivated, to investigate a failure of this part which increases the body roll of the vehicle. The result of this simulation is shown in Fig. 8 (left). It can clearly be seen that the running board of the vehicle collides with the virtual platform, which includes a safety margin of 50 mm.

In a second scenario, the vehicle with the anti-roll bar failure moves along the same track, but at a speed of only 5 km/h. Due to the cant excess the car body rolls inwards. A collision occurs on the car body structure of the vehicle, as shown in Fig. 8 (right).

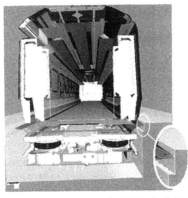 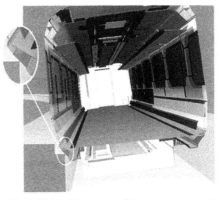

Fig. 8. Curve entry at a speed of 55 km/h (left) and 5 km/h (right). Shown are collisions due to an anti-roll bar failure at the leading bogie.

5 CONCLUSIONS

In the aerodynamic design of new rail vehicles, numerical approaches (CFD) are gaining momentum, and are sometimes able to advantageously replace traditional experimental methods. Still, as they are often based on inaccurate physical models, they need to be constantly checked by experiments. This new reference role further increases the requirements with regards to the accuracy of measured results, for the benefit of the quality of the final product.

Complex structural dynamics models have been developed to predict the ride comfort of rail vehicles. Further, an active digital mock-up has been used to support a structure gauge assessment for a mainline vehicle. Using a VPD system saves time and cost. All of the example models discussed in the paper have been developed to support the product development process of innovative rail vehicle systems.

REFERENCES

1. *Stribersky, A., F. Moser and W. Rulka: Structural Dynamics and Ride Comfort of a Rail Vehicle System. Advances in Engineering Software 33 (2002), pp. 541 - 552.*
2. *Peters, J.L.: Aerodynamics of high-speed trains and maglev vehicles: State of the art and future potential. Int. J. of Vehicle Design, SP3 (1983), pp. 308 – 341.*
3. *Stribersky, A., F. Moser, W. Rulka and W. Trautenberg: Advances in Combined Structural Dynamics and System Dynamics Analyses of Rail Vehicles. 17th IAVSD Symposium, Copenhagen, Swets & Zeitlinger B.V., Lisse (2003), pp. 465 – 477.*
4. *Maeda, T., Matsumara, T., Iida, M., Nakatani, K. and Uchida, K.: Effect of train nose on compression wave generated by train entering tunnel. Int. Conf. on Speedup Technology for Railway and Maglev Vehicles, JSME, Yokohama, Japan, 22-26 Nov. 1993.*
5. *Peters, J.L.: Computational design and aerodynamic optimization of locomotive nose shapes. Computers in Railways VII, WIT Press, (2000), pp. 757 – 765.*
6. *Peters, J.L.: Tunnel optimized train nose shape, Proc. of 10th Int. Symp. on Aerodynamics and Ventilation of Vehicle Tunnels, Boston, USA, October 2000, pp. 1015 – 1021.*
7. *Schandl, G.: Werkzeuge zur dynamischen Geometriesimulation von Schienenfahrzeugen. Diploma Thesis, Vienna University of Technology, (2002).*

Vehicle System Dynamics Supplement 41 (2004), p.458-467 © Taylor & Francis Ltd.

Computational Bifurcation Analysis of Mechanical Systems with Applications to Railway Vehicles

GUNTER SCHUPP*

SUMMARY

For identifying the long time behaviour of nonlinear dynamical systems with respect to the influence of one or more system parameters, numerical bifurcation analysis is an ideal method. The objective of the paper is to describe a software environment for such an analysis basing on the principles of path–following or continuation under the specific viewpoint of an application on mechanical systems or, more specifically, on railway vehicles being modelled as multibody system. Their stationary as well as their periodic behaviour is considered. Three major topics are of primary interest: The integration of the bifurcation software into a software package for the simulation of arbitrary mechanical systems; the direct calculation of periodic solutions (limit cycles); and the handling of differential algebraic equations (DAE). The algorithms are applied finally on the 'realistic' simulation model of a railway vehicle running on a straight track.

1 INTRODUCTION

Numerical bifurcation analysis is a computer aided method for the examination of nonlinear dynamical systems with respect to the parameter dependency of their long time behaviour ($t \rightarrow \infty$). The results of such an analysis are usually given as bifurcation diagrams. Herein, the system's steady state is characterised as stationary, periodic, quasi–periodic or chaotic, being either stable or unstable and with the varied system parameter as abscissa. Of particular significance are distinct parameter values for which the system's dynamical behaviour changes abruptly and radically – the bifurcation points; for a more detailed description see e.g. [1, 2].

For analysing the running behaviour of railway vehicles by computer simulations within an industrial design process, commercially available software is applied, [3]. Being usually based on a *MultiBody System* (MBS) approach, the suitable software has to provide specific additional features like an efficient modelling/computation of the highly nonlinear contact geometry and mechanics between wheel and rail.

The critical velocity v_{crit} constitutes an important design parameter characteristic for each type of railway vehicle. The background for this parameter is build by the vehicle running with constant velocity v along an ideally straight track. The critical

*Address correspondance to: Department Vehicle System Dynamics, Institute of Aeroelasticity, DLR – German Aerospace Center, D-82230 Wessling, Germany; E-mail: Gunter.Schupp@DLR.de

velocity separates the parameter range for which an (initial) disturbance decays down to a stationary behaviour from the one for which a disturbance initiates a limit cycle or hunting behaviour (Fig. 1). The latter one is characterised as a periodic, mainly lateral and yawing motion of the vehicle and has to be avoided in regular operations. The methods of linear system analysis yield as critical velocity the velocity at the Hopf bifurcation, $v_{crit} := v_{lin}$ (Fig. 1). In contrast to this a bifurcation analysis yields as critical velocity the lower velocity at the saddle–node, $v_{crit} := v_{nlin} < v_{lin}$, see e.g. [4]. Therefore, for the design of a vehicle, the lower velocity v_{nlin} has to be taken.

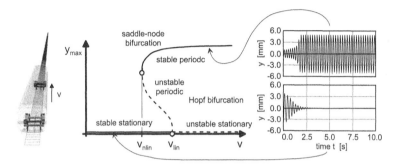

Fig. 1. Typical bifurcation diagram of a wheel–rail system. The bifurcation parameter is the velocity v along the track, y is the lateral displacement of a wheelset with respect to the track centre line.

The paper presents a software environment for a continuation based bifurcation analysis of arbitrary mechanical systems; as application such an analysis is performed for a railway vehicle's simulation model. Algorithms and analysis are restricted to stationary and periodic behaviour, i.e. to the range of technical–industrial relevance within railway vehicle dynamics.

2 BIFURCATION ANALYSIS OF MULTIBODY SYSTEMS

The integration of bifurcation analysis as an additional analysis tool into an MBS–software enables its efficient application on sophisticated simulation models of arbitrary mechanical systems. On behalf of this, PATH[1] as a software tool for the continuation based bifurcation analysis of general dynamical systems, see [5], and SIMPACK[2] as a software tool for the computer–aided generation and analysis of the equations of motion of general technical mechanical systems (and especially of wheel–rail systems, too), see e.g. [6], have been interconnected.

---

[1] PATH has been developed at the Technical University of Denmark (DTU), Copenhagen, Denmark.
[2] SIMPACK is commercially distributed, maintained etc. by Intec GmbH, Wessling, Germany.

2.1 Simulation of Railway Vehicles by an MBS Approach

For the computational analysis of a railway vehicle's running behaviour usually a multibody system approach is applied. In doing so, the vehicle is abstracted as a system of rigid and elastic bodies interconnected via massless force elements and joints, [7]. Due to the relative motion of the system's bodies, force elements generate applied forces and torques; a typical application is the modelling of a vehicle's suspension system consisting of springs and dampers. Contrarily, joints give rise to constraint forces by constraining the relative motion of the system's bodies.

Interpreting the interaction between wheel and rail as ideally rigid, i.e. as a *kinematic constraint*, the equations of motion of a wheel–rail system can be formulated in the position coordinates $\mathbf{p}(t)$ and the velocity coordinates $\mathbf{v}(t)$ as Lagrange equations of the first kind, yielding a system of *Differential–Algebraic Equations* (DAE) of differential index three:

$$\dot{\mathbf{p}} = \mathbf{v} \tag{1}$$

$$\mathbf{M}(\mathbf{p}, t)\,\dot{\mathbf{v}} = \mathbf{f}(\mathbf{p}, \mathbf{v}, \boldsymbol{\lambda}, t) - \mathbf{G}^T(\mathbf{p}, t)\,\boldsymbol{\lambda} \tag{2}$$

$$0 = \mathbf{g}(\mathbf{p}, t) \ . \tag{3}$$

Herein, the algebraic constraints (3) describing the contact conditions between wheel and rail are coupled to the dynamic equations (2) by the Lagrangian multipliers $\boldsymbol{\lambda}(t)$. The constraint forces $-\mathbf{G}^T(\mathbf{p}, t)\,\boldsymbol{\lambda}$ with $\mathbf{G}(\mathbf{p}) := \partial \mathbf{g}(\mathbf{p})/\partial \mathbf{p}$ ensure the kinematic constraints to be always fulfilled. If the algebraic constraints are formulated in an appropriate manner, the Lagrangian multipliers $\boldsymbol{\lambda}$ represent directly the normal forces acting in the point of contact. The force vector $\mathbf{f}(\mathbf{p}, \mathbf{v}, \boldsymbol{\lambda}, t)$ comprises all applied forces (including coriolis forces etc.), thus including the friction forces between wheel and rail which depend directly on the constraint forces $\boldsymbol{\lambda}(t)$, too. The symmetric mass matrix is denoted $\mathbf{M}(\mathbf{p})$.

2.2 Continuation Based Bifurcation Analysis

The purpose of a comprehensive computer aided bifurcation analysis of a nonlinear dynamical system is to find all its steady state solutions (i.e. $t \to \infty$) with respect to a predefined range of a selected system parameter. Even if the value of this parameter remains unchanged, depending on the current initial states qualitatively different steady state solutions can occur – a phenomenon called *coexistence* of steady state solutions, emerging in Figure 1, too. As a result of such an analysis, all of the system's potential behaviour patterns (like stationary or periodic motions) are given for a defined parameter range within a bifurcation diagram. Of substantial interest are bifurcation points, identifying parameter values for which the system's dynamical behaviour changes abruptly and radically; examples are saddle–node and Hopf bifurcations.

Within the scope of this paper, for bifurcation analysis only the principle of path–following is of interest. It combines methods for the direct computation of a system's steady state with the continuation of a one–dimensional curve in a higher dimensional space. Regarding bifurcation analysis, such a spatial curve is built discretely by a sequence of single solution points of a nonlinear system in its state–parameter space. This *path* of a system usually is visualised with the help of a bifurcation diagram. Often, path–following algorithms are implemented as predictor/corrector schemes. The task of the corrector is to find directly a steady state solution from an initial predictor estimation. Especially in the case of periodic solutions (limit cycles), this is the by far most challenging part of a bifurcation analysis and therefore will be described in more detail in the following.

3 DIRECT COMPUTATION OF PERIODIC MOTIONS

3.1 Formulation as Boundary Value Problem

Starting from an initial estimation, the task of a bifurcation software's corrector module is to generate a steady state solution of the given system equations in a direct manner. Preliminarily, the equations of motion are assumed to be defined for simplicity reasons as a system of n autonomous Ordinary Differential Equations (ODE) with p being the varied system parameter ($\mathbf{y} := (\mathbf{p}^T, \mathbf{v}^T)^T$, see Equations (1), (2)):

$$\dot{\mathbf{y}} = \mathbf{f}(\mathbf{y}, p), \qquad \mathbf{y}(t{=}t_0) = \mathbf{y}_0, \qquad \mathbf{y} \in \mathbb{R}^n . \qquad (4)$$

Then, in the stationary case, the corrector only has to solve iteratively the nonlinear system $\mathbf{f}(\mathbf{y}, p) = \mathbf{0}$ – a usually trouble–free computation even if high accuracy is demanded. Whereas finding a periodic solution of the system equations with period T_P, characterised by $\mathbf{y}(t_0 + T_P) = \mathbf{y}(t_0)$, is significantly more complicated. Here, the task is formulated as a Boundary Value Problem (BVP) with the initial time $t_0 = 0$, the unknown initial states $\mathbf{s} := \mathbf{y}(t_0)$ and the also unknown period T_P:

$$\dot{\mathbf{y}} = \mathbf{f}(\mathbf{y}, p) \qquad (5)$$

$$\mathbf{0} = \mathbf{y}(T_P, \mathbf{s}) - \mathbf{s} . \qquad (6)$$

Discussed below is the *Poincaré Map Method*, a special kind of a single shooting method implemented in the software tool PATH to solve such BVPs, see also [2, 5].

3.2 The Poincaré Map Method

A Poincaré plane of the system (4) is a $(n{-}1)$–dimensional hyperplane in the n–dimensional state space being transversal to the flow $\varphi(t, \mathbf{s})$ of the system, see Figure 2. Starting from a point s near a periodic solution and being located on a suitable

Poincaré plane Σ of (4), $s \in \Sigma$, the flow $\varphi(t, s)$ will hit Σ for the first time in the same direction again after the return time T_R, $\varphi(T_R, s) \in \Sigma$. Then, via the time discrete Poincaré map $\mathbf{P}: s \to \varphi(T_R, s)$, the *residual map* $\mathbf{Q}(s): s \to q$ with

$$\mathbf{Q}(s) := \mathbf{P}(s) - s = \varphi(T_R, s) - s \qquad (\mathbf{Q}(s): \Sigma \to \Sigma) \qquad (7)$$

can be defined (Fig. 2). A fixed point s_p of the Poincaré map $\mathbf{P}(s)$ representing a periodic solution of the system (4) results as zero of the residual map:

$$\mathbf{Q}(s_p) = \varphi(T_P, s_p) - s_p = 0 . \qquad (8)$$

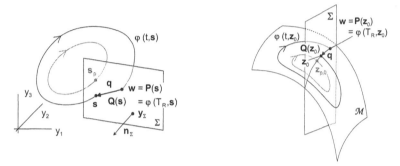

Fig. 2. Poincaré plane, Poincaré map, and residual map of a dynamical system's flow $\varphi(t, s)$ in $\mathbb{R}^3$: *Left:* Equations of motion given in ODE–form; *Right:* ... given in DAE–form (refer to section 4).

Thus, to compute a periodic solution, the system of nonlinear equations (8) has to be solved. This is done usually by a Newton iteration of the type $\mathbf{Q}_s^j \cdot \Delta s^j = -\mathbf{Q}^j$ with the increment $\Delta s^j = s^{j+1} - s^j$ and the Jacobian matrix $\mathbf{Q}_s = \partial \mathbf{Q}/\partial s$. Within the software tool PATH, the columns of the Jacobian $\mathbf{Q}_s^j$ have been approximated up to now by finite differences, see [5]. This approach implies the evaluation of the residual map (7) and therefore the time integration of the system equations (4) up to the return time T_R once per column. Since these integrations are – as usual – performed independently, according to [8] an unavoidable total error of the size $\varepsilon_g = \mathcal{O}(\sqrt{TOL})$ per column has to be expected; TOL is the user–specified error tolerance of the integration scheme. A pragmatic (but, generally, rather uncertain) solution is now to reduce the error tolerance of the integrations extremely (down to $TOL = 10^{-15}$ might be necessary), but this leads to a sharp increase of the computation time, hence is simply not applicable in case of comprehensive simulation models.

3.3 Utilisation of the Variational Differential Equations

A more reliable and more efficient way for the computation of the Jacobian $\mathbf{Q}_s$ is to combine the outlined shooting method with an integrated sensitivity analysis of the system equations. The basic procedure is to decompose the Jacobian $\mathbf{Q_s}$ analytically into fundamental terms that can be generated by the solution of appropriate differential equations. Differentiation of the residual map (7) with respect to the unknown initial states s yields

$$\mathbf{Q_s} := \frac{\partial \mathbf{Q(s)}}{\partial \mathbf{s}} = \frac{\partial \varphi(T_R(\mathbf{s}), \mathbf{s})}{\partial \mathbf{s}} + \frac{\partial \varphi(T_R(\mathbf{s}), \mathbf{s})}{\partial T_R} \cdot \frac{\partial T_R(\mathbf{s})}{\partial \mathbf{s}} - \frac{\partial \mathbf{s}}{\partial \mathbf{s}} . \tag{9}$$

Introducing the sensitivity matrix $\mathbf{S}(t, \mathbf{s}) := \partial \mathbf{y}(t, \mathbf{s})/\partial \mathbf{s}$, the analogue differentiation of the system equations (4) leads to their *Variational (Differential) Equations*,

$$\dot{\mathbf{S}} = \frac{\partial \mathbf{f}}{\partial \mathbf{y}}(\mathbf{y}(t, \mathbf{s})) \cdot \mathbf{S} , \quad \mathbf{S}(t_0 = 0) = \mathbf{I} , \tag{10}$$

a set of altogether $n \cdot n$ differential equations in the unknown sensitivities $S_{i,j} := \partial y_i / \partial s_j, \ i, j = 1, \ldots, n$. Taking into account $\varphi(t, \mathbf{s}) := \mathbf{y}(t, \mathbf{s})$, it becomes obvious that the sensitivity matrix evaluated for the return time T_R represents the first term of (9), $\mathbf{S}(T_R, \mathbf{s}) \equiv \partial \varphi(T_R, \mathbf{s})/\partial \mathbf{s}$. Further, for the second term of (9) $\partial \varphi(T_R, \mathbf{s})/\partial T_R = \mathbf{f}(\mathbf{y}(T_R, \mathbf{s}))$ is valid and $\partial T_R/\partial \mathbf{s}$ is a function of the sensitivity matrix $\mathbf{S}(T_R, \mathbf{s})$. In the case of a periodic solution $\mathbf{y}_p(t, \mathbf{s}_p)$, according to [1] $\mathbf{S}(T_P, \mathbf{s}_p)$ is the appropriate *monodromy matrix* enabling the stability analysis of $\mathbf{y}_p$.

Consequently, besides solving the system equations (4) (the nominal system) for the flow $\varphi(T_R, \mathbf{s})$, this method requires the n^2 variational equations (10) (the sensitivity system) to be solved for the sensitivity matrix $\mathbf{S}(T_R, \mathbf{s})$ additionally. The basic principle now is the synchronous integration of the nominal and the sensitivity system by means of the code DAGSL, a derivative of the famous DAE–solver DASSL, see [9] for the latter. Following roughly the algorithm described in [10], in every single time step t_m DAGSL first calculates the discrete solution $\mathbf{y}_m$ of the nominal system (4) by a BDF–approach (BDF – *Backward Differentiation Formula*) as usually. Then, the n independent, discrete sensitivity vectors $S_{i,m} = \partial \mathbf{y}_m/\partial s_i, i = 1, \ldots, n$ follow from an analogous BDF–discretisation of the variational equations (10) in a subsequent, sequential loop. Due to the linearity of (10) this means merely the additional solution of n systems of linear equations. Since these n systems are set up only internally on the base of the data from the iteration of the nominal system's discrete solution $\mathbf{y}_m$, it is not necessary to define the variational equations explicitly. Compared to the former finite differences approach, the approximation error of the Jacobian $\mathbf{Q_s}$ is reduced to the size $\varepsilon_g = \mathcal{O}(TOL)$ by this algorithm.

4 FROM ODE TO DAE BY A LOCAL STATE SPACE FORM

One way to handle so–called kinematically closed loops within multibody systems is to introduce additional algebraic constraint equations into its equations of motion.

Then, as already discussed in section 2.1, the Lagrange equations of the first kind can be found as differential–algebraic equations (1)–(3) of differential index three.

The fundamental principle in order to enable the application of the outlined path–following algorithms on such DAE–systems is their reduction to a local state space form, an equivalent system of differential equations in a minimum number of $2\,n_f$ reduced or minimum coordinates; $2\,n_f$ is also the number of degrees of freedom of the system. This reduction corresponds to a transformation to a local parametrisation of the constraint manifold $\mathcal{M}$ (with $\mathbf{x} := (\mathbf{p}^T, \mathbf{v}^T)^T$),

$$\mathcal{M} := \big\{(\mathbf{p}, \mathbf{v}) \in \mathbb{R}^{n_x} \mid \mathbf{g}(\mathbf{p}) = \mathbf{0}, \; \mathbf{G}(\mathbf{p})\,\mathbf{v} = \mathbf{0}\big\} \in \mathbb{R}^{n_x} , \tag{11}$$

being defined by the algebraic constraint equation (3) and its first time derivative.

With $\mathbf{z}$ being the $(2\,n_f \times 1)$–vector of reduced coordinates, the local state space form of the DAE–system (1)–(3) can be found according to [11] by the reduction $\mathbf{z} = \widetilde{\mathbf{V}}^T(\mathbf{x}_c) \cdot (\mathbf{x} - \mathbf{x}_c)$ as

$$\dot{\mathbf{z}} = \widetilde{\mathbf{V}}^T(\mathbf{x}_c) \cdot \mathbf{f}\big(\boldsymbol{\phi}_{\mathbf{x}_c}(\mathbf{z})\big) . \tag{12}$$

Herein, $\mathbf{x}_c$ defines a point on the manifold $\mathcal{M}$; the inverse mapping $\boldsymbol{\phi}_{\mathbf{x}_c} : \mathbf{z} \to \mathbf{x}$ with $\mathbf{x} = \boldsymbol{\phi}_{\mathbf{x}_c}(\mathbf{z}) \in \mathcal{M}$ can be performed under certain conditions by solving the system of nonlinear equations built by the constraint equation (3) and its first time derivative.

The state space form is *not* unique but depends on the parametrisation matrix $\widetilde{\mathbf{V}}(\mathbf{x}_c)$. Regarding only multibody systems and using the DAE– as well as the state space form of the equations of motion in a combined manner as described below, usually a pretty simple variant of the coordinate partitioning method (see [12]) is sufficient. The positional coordinates are partitioned into independent coordinates $\mathbf{p}_i$ and into dependent coordinates $\mathbf{p}_d$ a priori as part of the pre–processing such that $\mathbf{p}_i$, $\mathbf{p}_d$ are alternative subsets of $\mathbf{p}$, $\mathbf{p}_i \subset \mathbf{p}$, $\mathbf{p}_d \subset \mathbf{p}$ with $\mathbf{p}_i \cap \mathbf{p}_d = \{\}$; $\mathbf{v}$ is partitioned analogously. This implicates that the parametrisation matrix $\widetilde{\mathbf{V}}$ is constant throughout the complete continuation process; for the reduced coordinates $\mathbf{z} = (\mathbf{p}_i^T, \mathbf{v}_i^T)^T$ is valid.

The flow $\boldsymbol{\varphi}(t, \mathbf{y}_0)$ ($\mathbf{y} := (\mathbf{p}^T, \mathbf{v}^T, \boldsymbol{\lambda}^T)^T$) of the DAE–form (1)–(3) is restricted for all t to the constraint manifold $\mathcal{M}$ defined in (11), refer to Figure 2, right. Therefore, the concept of a reduced Poincaré map $\widetilde{\mathbf{P}} = \widetilde{\mathbf{V}}^T \mathbf{P}(\boldsymbol{\phi}(\mathbf{z}_0))$ allows the direct computation of periodic solutions with nearly the same algorithm as the one described in section 3. Now, a zero of the *reduced residual map* $\widetilde{\mathbf{Q}}(\mathbf{z}_0) : \Sigma \cap \mathcal{M} \to \Sigma \cap \mathcal{M}$ with

$$\widetilde{\mathbf{Q}}(\mathbf{z}_0) := \widetilde{\mathbf{P}}(\mathbf{z}_0) - \mathbf{z}_0 = \widetilde{\mathbf{V}}^T \, \boldsymbol{\varphi}\big(T_R, \boldsymbol{\phi}(\mathbf{z}_0)\big) - \mathbf{z}_0 . \tag{13}$$

has to be found.

From an implementational point of view, by applying the DAE–form (1)–(3) together with an equivalent state space form (12) in a consecutive manner, the path–following algorithm can be subdivided into two operational blocs: Within an inner bloc the residual of the defining nonlinear system of equations (e.g. Equation (8)) is evaluated on the basis of the DAE–form. Within an outer bloc the defining system is iterated and the path is continued on the basis of the corresponding state space (ODE) form. This operational subdivision has two major advantages: the state space form can be based on the described simple constant coordinate partitioning and only the sensitivities with respect to the initial values z_0 have to be computed in the periodic case. In this way, a considerable amount of computation time can be saved.

5 BIFURCATION ANALYSIS OF A PASSENGER CAR'S MODEL

In this section, the results of applying the path–following algorithms outlined above on a 'complete' simulation model of a passenger car are illustrated; 'complete' should mean that the simulation model could also be used for a comprehensive analysis of the vehicle's running behaviour with nearly no modifications necessary.

5.1 The Simulatiom Model

Fig. 3.   Simulation model of an *Avmz*–passenger car with Fiat 0270 bogies.

As vehicle a 1st class *Avmz* coach is chosen. The simulation model is given in Figure 3, the modelling is according to the benchmark description [13].

The mechanical model of the passenger car consists of 15 rigid bodies: one car body, two bogie frames with two wheelsets each, as well as a right– and a left–hand trailing arm per wheelset. The primary suspensions consist of 2 flexicoil springs per wheelset with nearly parallel dampers. The secondary suspension again is built by 2x2 flexicoil springs with nearly parallel dampers, completed by 2 yaw and 2 lateral dampers as well as one stiff lateral bump stop between each bogie and the car body. All the springs are modelled with constant stiffnesses while the dampers show a only piecewise linear force–velocity–characteristic with a serial stiffness superposed;

hence, the eigendynamics of the latter have to be considered. The trailing arms are connected to the wheelsets via rotational joints, their bogie–sided bushings are represented by three–dimensional force elements with once again linear characteristics.

The state space of the vehicle model is described by altogether 114 position, velocity, and algebraic coordinates. Hence, a total of 9576 equations has to be integrated for limit cycle calculations (nominal system and variational equations).

5.2 Bifurcation Analysis

Applying the outlined algorithms for a bifurcation analysis of the *Avmz* car's simulation model yields the results displayed in extracts in Figure 4. The dynamic behaviour of the vehicle depending on the velocity v as varied system parameter is represented in the bifurcation diagram to the left by the maximum lateral deviation y of the leading wheelset with respect to the track. Therein, the two Hopf bifurcations A, B are marked with ○, the saddle–node bifurcation C is marked with ▽. The stability of the periodic solutions can be evaluated by the evolution of the complex Floquet–multipliers (the eigenvalues of the monodromy matrix, see section 3.3) given to the right of Figure 4: a solution is stable if the moduli of all of these multipliers are less than one.

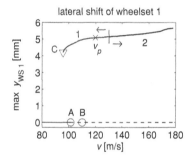 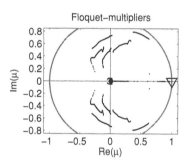

Fig. 4. Bifurcation analysis of the *Avmz* passenger car. *Left:* Numerically computed bifurcation diagram. *Right:* Floquet–multipliers of the periodic solutions.

The stationary solutions are continued from nearly zero velocity up to the first Hopf bifurcation A, characterised by the velocity $v_{lin} = v_A = 101.62\,\text{m/s}$, and beyond, where a second Hopf bifurcation B is detected. Up to now, it is not possible to continue the unstable periodic solutions branching off from Hopf bifurcation A. Therefore, the first step to continue periodic attractors is to generate an initial estimation with the help of a conventional, external time integration. Here, this was done for a velocity $v_p = 130.0\,\text{m/s}$. For decreasing velocities, the path of stable periodic solutions ends in the saddle–node bifurcation C, $v_{nlin} = v_C = 95.62\,\text{m/s}$. This bifurcation is indicated by the Floquet–multipliers with one of them crossing the stability limit, i.e. the unit circle, along the real axis. For increasing velocities, the amplitude of the limit cycle oscillations grows very slowly due to the nonlinear geometry of the wheel profiles and the increasing intensity of the contact between the flanges of the wheels and the rails.

6 CONCLUSIONS

The paper describes a software environment for the continuation based bifurcation analysis of complex technical mechanical systems. As a result of the project forming the basis of this paper, such a method is now available as an additional analysis tool of a software package for the simulation of mechanical systems – including railway vehicles. The bifurcation analysis of a passenger car's 'complete' simulation model proves the applicability of the developed software on detailed, realistic and therefore necessarily complex models being of industrial relevance. Though throughout the paper a particular emphasis is laid on wheel–rail systems, the outlined algorithms' potential range of application is of course not restricted to this specific case.

REFERENCES

1. Seydel, R. *Practical Bifurcation and Stability Analysis: From Equilibrium to Chaos.* Springer–Verlag, New York, 1994.
2. Nayfeh, A.H. and Balachandran, B. *Applied Nonlinear Dynamics: Analytical, Computational, and Experimental Methods.* John Wiley and Sons, Chichester, New York, 1995.
3. Iwnicki, S., editor. *The Manchester Benchmarks for Rail Vehicle Simulation*, volume 31 of *Supplement to Vehicle System Dynamics.* Swets & Zeitlinger, Lisse, 1999.
4. True, H. and Jensen, J.C. Parameter Study of Hunting and Chaos in Railway Vehicle Dynamics. In *The Dynamics of Vehicles on Roads and Tracks*, 13th *IAVSD–Symposium, Chengdu, China, 1993*, pages 508–521. Swets & Zeitlinger B.V., Amsterdam and Lisse, 23–27August 1993.
5. Kaas–Petersen, C. Chaos in a Railway Bogie. *Acta Mechanica*, 61:89–107, 1986.
6. Rulka, W. *Effiziente Simulation der Dynamik mechatronischer Systeme für industrielle Anwendungen.* PhD thesis, Technical University of Vienna, Austria, 1998.
7. Popp, K. and Schiehlen, W. *Fahrzeugdynamik: Eine Einführung in die Dynamik des Systems Fahrzeug – Fahrweg.* Teubner, Stuttgart, 1993.
8. Buchauer, O., Hiltmann, P., and Kiehl, M. Sensitivity Analysis of Initial–Value Problems with Application to Shooting Techniques. *Numerische Mathematik*, 67:151–159, 1994.
9. Brenan, K.E., Campbell, S.L., and Petzold, L.R. *Numerical Solution of Initial–Value Problems in Differential–Algebraic Equations.* SIAM, Philadelphia, USA, 1996.
10. Feery, W.F., Tolsma, J.E., and Barton, P.I. Efficient Sensitivity Analysis of Large–Scale Differential–Algebraic Systems. *Applied Numerical Mathematics*, 25:41–54, 1997.
11. Eich–Soellner, E. and Führer, C. *Numerical Methods in Multibody Dynamics.* Teubner, Stuttgart, 1998. (European Consortium for Mathematics in Industry).
12. Wehage, R.A. and Haug, A.J. Generalized Coordinate Partitioning for Dimension Reduction in Analysis of Constrained Dynamic Systems. *J. of Mech. Design*, 104:247–255, 1982.
13. Vehicle Data for 1st Class Avmz Coach with Fiat 0270 Bogies (Appendix 1). In: ERRI B 176/DT 290: B176/3 *Benchmark Problem Results and Assessment.* Utrecht, Holland, 1993.

Vehicle System Dynamics Supplement 41 (2004), p.468-476 © Taylor & Francis Ltd.

Fault-detection-and-isolation system for a railway vehicle bogie

KENJIRO GODA[1] AND ROGER GOODALL[2]

SUMMARY

To detect suspension faults and the wheel-rail condition, a fault-detection-and-isolation system for a railway vehicle bogie -- with a model-based estimation method using a Kalman-Bucy filter and isolation scheme -- was developed. The advantage of this system is that many faults can be detected by using a small number of sensors. Time-history simulations were carried out to investigate the performance of the new fault-detection and isolation system. The simulation results show that faults in the suspensions and changes in the wheel-rail interface can be detected from the residual given by the Kalman-Bucy filter. They also show that each fault can be identified by the fault-isolation scheme. It is thus concluded that the model-based estimation method using a Kalman-Bucy filter and isolation scheme provides an effective basis for a fault-detection-and-isolation system for a railway vehicle bogie.

1 INTRODUCTION

As the maximum speed of railway vehicles continues to increase, to achieve good ride comfort, good curving performance and high stability, active suspension systems for high-speed trains are being demanded [1]. Safety information on, for example, bogie-suspension faults and wheel defects, needs to be detected early to ensure safety and reliability for high speed vehicles. This information is important especially for a vehicle with an active suspension. On the other hand, there has been increasing requirements for condition-based maintenance, which will help to reduce the maintenance costs of railway transportation systems radically. Early detection of incipient component faults can help to make an effective maintenance scheme for reducing costs and to prevent the occurrence of dangerous situations. This condition-

[1] *Address correspondence to:* Mechanical Engineering Research Laboratory, Hitachi, Ltd. 502, Kandatsu, Tsuchiura, Ibaraki 300-0013, JAPAN. Tel.: +81(298) 832 8245; Fax: +81(298) 832 2807; E-mail: goda@merl.hitachi.co.jp
[2] *Address correspondence to:* Department of Electronic and Electrical Engineering, Loughborough University, Leicestershire LE11 3TU UK. Tel.: (+44) 1509 227009; Fax: (+44) 1509 227008; E-mail: R.M.Goodall@lboro.ac.uk

based approach needs a variety of maintenance information, such as the conditions of the vehicle suspension and the wheel-rail interface.

For both these reasons, (i.e., safety information and maintenance information), the need for safety information and maintenance information is therefore increasing. To provide this information, a fault-detection-and-isolation system, which will help keep a high level of safety and reliability and reduce maintenance costs, should be developed.

2   FAULT-DETECTION-AND-ISOLATION SCHEME

2.1   Vehicle and sensor modelling

A model-based estimation method using a Kalman–Bucy filter was developed to detect suspension faults and evaluate the wheel-rail interface. To simplify the estimator design, a linearised plan-view vehicle model with four-axles, was developed. The model is shown in Figure 1. It consists of seven rigid bodies (one carbody, two bogies, and four wheelsets) and has 14 degrees of freedom (each rigid body has lateral and yaw degrees of freedom).

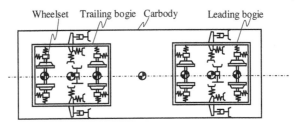

Fig. 1. Vehicle Model.

For the purpose of the Kalman-Bucy filter design, the state equation for this vehicle model is given as Equation 1.

$$x = A \cdot x + G \cdot w \qquad (1)$$

where x is a state vector, which consists of vehicle displacements and velocities, and w is track irregularity, which is taken into account as the system noise for the Kalman–Bucy filter.

To measure lateral acceleration and yaw velocity, seven accelerometers and seven gyros were assumed. In this simulation, a full set of sensors, is used. The output equation for the measurement is given as Equation 2.

$$y = C \cdot x + v \qquad (2)$$

where y is the output vector from the sensors, v is the sensor noise, which is taken into account as the measurement noise of the sensor.

2.2 Fault-detection-and-isolation scheme

The Kalman-Bucy filter shown in Equation 3 is used to estimate the state vector and output vector in Equation 1 and Equation 2:

$$\hat{x} = A \cdot \hat{x} + K \cdot (y - \hat{y}) \tag{3}$$

where $\hat{x}$ is the state estimate, $\hat{y}$ is the output estimate, and K is the Kalman-Bucy filter [2].

The Kalman-Bucy filter is normally expressed in continuous form, although it is straightforward to convert into discrete-time form for processor implementation; when done properly there will only be marginal differences in performance. Note also that the gain for the Kalman-Bucy filter is calculated off-line, unlike the full Kalman filter which is inherently discrete-time and its gain is re-calculated at every time step. Previous experience with model-based estimation applied to rail vehicles has shown that the linear time-invariant form of the Kalman-Bucy filter is effective, and its relative simplicity is an obvious advantage.

A block diagram of the model-based estimation method using a Kalman-Bucy filter is shown in Figure 2. The residual, i.e., the difference between the real sensor signal and the estimated sensor signal, is focused on here. Without a fault it should theoretically be Gaussian white noise with a small magnitude, but a change in vehicle-model parameters increases the size of the residual significantly, and also chabges its frequency distribution. This residual change provides a good indicator of the occurrence of a fault in the bogie system. To avoid false alarms, the residual change is precisely detected by using a frequency-weighting filter and a moving RMS analysis. The frequency-weighting filter is then designed to extract the frequency range in which the largest residual change occured. The RMS analysis was performed to decrease the deviation of the residual. The time history of the residuals after the filtering and RMS analysis was processed by a threshold detector to detect faults. A fault-isolation scheme was devised to separate the possible fault types and their locations.

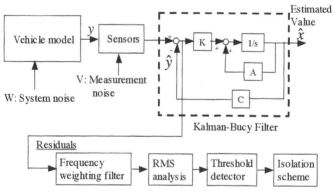

Fig. 2. Fault-detection and isolation system for railway-vehicle bogie with Kalman-Bucy filter.

3   CALCULATION RESULTS

3.1   Calculation conditions

To estimate the performance of the developed fault-detection-and-isolation system, time-histories were calculated. Vehicle parameters for a 381-series test-vehicle, a typical Japanese limited express train, were used. Running speed is 130 km/h, which is the maximum operating speed of a Japanese limited express.

The power spectrum density (PSD) of lateral irregularity data (Equation 4) for the Joban line, one of the suburban lines around Tokyo, was used as the track-irregularity input:

$$PSD = \frac{1.2 \times 10^{-7}}{F^2} \qquad (\text{m}^2/(\text{cycle/m})) \qquad (4)$$

where F is the spatial frequency of the track.

The time-history of lateral irregularity shown in Figure 3 is derived from the PSD for use in the time-history simulation. The measurement noise, namely sensor noise, was assumed to be white noise. It was set to 4 % of the RMS value of the sensor output signal in response to the track irregularity without sensor noise. This is a generalized technique which provides sensible values for measurement noise without having to choose a particular sensor. The value can be decreased to present a better sensor, for example. In any particular real design specific noise values provided by the sensor manufacturer would be used instead.

Three types of model-parameter changes were assumed to occur as faults during train operation. Yaw damper and lateral damper faults, which affect the stability and the ride comfort at high speed, were assumed to occur as suspension faults. Conicity change, which affects stability and the wheel-rail interface, was assumed to occur as the change in the wheel-rail condition.

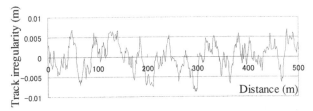

Fig. 3. Time history of track irregularity.

3.2   Tuning filter parameter

A 2nd-order low pass filter is used as the frequency-weighting filter. The parameters for the filter are defined from the measured frequency spectrum of the residual. Figure 4 shows the power spectrum density (PSD) of the residual of the leading-bogie yaw velocity. The difference between the residual with a yaw-damper fault and the residual without yaw-damper fault can be clearly seen. The power-spectrum analysis shows that the yaw-damper fault affects the frequency-response of the yaw velocity residual at less than 10 Hz. In the same way, the yaw-damper fault mainly affects the frequency response of the other residuals at less than 10 Hz. The frequency response less than 10 Hz should therefore be extracted to detect the faults precisely. The power-spectrum analysis for the other two faults, the lateral damper fault and conicity changes, showed the similar results; namely the data with a fault have different frequency responses below 10 Hz. Accordingly, the cut-off frequency of the low pass filter was set to 10 Hz.

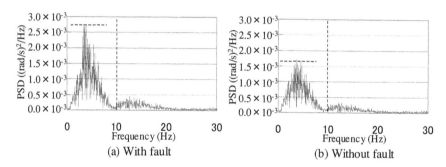

(a) With fault   (b) Without fault

Fig. 4. PSD of yaw velocity of leading bogie with yaw-damper fault and without fault.

In RMS analysis, the length of an RMS moving window produces a trade-off between the detection speed and number of false alarms. The effect of the length of the window on the residual was investigated. Figure 5 compares the leading bogie yaw velocity in the case of different window lengths; namely 1s, 2s, or 3s under the condition that a yaw-damper fault occurs. If the length of window is short, the deviation of the residual becomes large and the probability of a false alarm will increase. If the length of window is long, the deviation of the residual become small, but the detection speed becomes slower. The window length of 2 s was chosen from the view-point of the reduction of false alarms and the speed of fault detection.

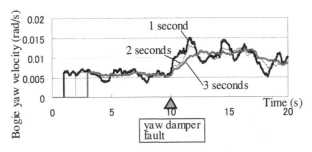

Fig. 5. Effect of window length of RMS moving window.

3.3   Detection performance

3.3.1   Yaw-damper fault

The detection performance for the yaw damper fault was evaluated first. The damping force was assumed to drop to 5% of the normal force after 10 s. The front-left yaw damper, one of the four yaw dampers, was assumed to fail. The total yaw damping on the leading bogie was reduced by nearly 50% in this case. Time-history simulation showed that only the residual of leading-bogie yaw velocity changes significantly after the yaw-damper fault. Figure 6 shows the time history of the residual (left side) and the sensor signal (right side) of leading bogie yaw velocity. The residual changes significantly after the yaw-damper faults occur after 10 s, whereas the sensor signal changes only slightly. This means that the appropriate threshold value can be set for residual change easily, but it is difficult to set the threshold value for the sensor signal. This indicates the faults of the yaw damper can be detected from the residual changes when the sensor signal changes slightly. The model-based estimation can thus be said to be effective for detecting the yaw-damper faults.

3.3.2   Lateral-damper fault

The detection performance for a lateral-damper fault was evaluated next. The damping force was assumed to drop to 5% of normal force after 10 s. The lateral damper at the leading bogie, one of the two lateral dampers, was assumed to fail in this case. Time-history simulation showed that the following residual values change significantly after the lateral damper fault occurs.

 - lateral acceleration of carbody and leading-bogie
 - yaw velocity of carbody

Figure 7 shows the time history of the residual and the sensor signal of carbody lateral acceleration. The residual changes significantly after a lateral-damper fault occurs, whereas the sensor signal changes only slightly. The occurrence of a lateral-damper fault can only be detected from the residual. The model-based estimation method can also be said to be effective for detecting lateral-damper faults.

3.3.3   Conicity change

Next the detection performance for conicity change was evaluated. The conicity was assumed to change from 0.15 to 0.20 suddenly after 10 seconds. In an actual wheel-rail interface the conicity changes gradually due to the wheel-rail wear. To compare the levels of residuals or sensor signals with different conicity in one time-history, the conicity changes were assumed to occur at one point in this simulation. Time-history simulation showed that the following residuals change after the change of conicity.
  - Lateral acceleration of leading and trailing bogie and wheelset 1-4
  - Yaw velocity of leading and trailing bogie and wheelset 1-4
   Figure 8 shows the time history of the residual and the sensor signal of bogie lateral acceleration. The occurrence of conicity change can only be detected from the residual. The effect of conicity change on vehicle dynamic motion (sensor signal) is smaller compared to the effect of conicity change on residual. This result shows that the model-based estimation method is also effective for detecting the conicity change and can thus provide maintenance information.

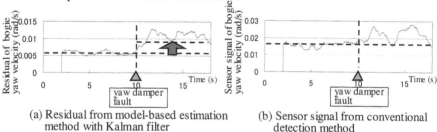

(a) Residual from model-based estimation   (b) Sensor signal from conventional
    method with Kalman filter   detection method

Fig. 6. Time history of bogie yaw velocity in the case of a yaw-damper fault.

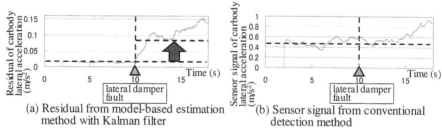

(a) Residual from model-based estimation   (b) Sensor signal from conventional
    method with Kalman filter   detection method

Fig. 7. Time history of carbody lateral acceleration in the case of a lateral-damper fault.

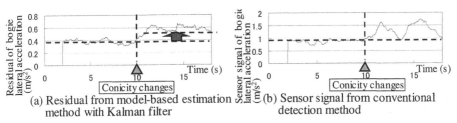

(a) Residual from model-based estimation method with Kalman filter

(b) Sensor signal from conventional detection method

Fig. 8. Time history of bogie lateral acceleration in the case of conicity changes.

3.4 Isolation performance

A fault-isolation scheme for separating the three kinds of faults (a yaw damper fault, a lateral damper fault and conicity changes) was devised. In this study all three types of fault, which were evaluated in the previous section, were assumed to occur. Table 1 lists the logic sequence which indicates how each fault affects the change of the residual. It is clear that the pattern of residual changes (i.e., tick marks) is different for each fault. In the case of a leading-yaw damper fault, only the residual of the leading-bogie yaw velocity changes. In the case of a leading-lateral-damper fault, the residuals of the carbody lateral acceleration, the leading-bogie lateral acceleration and the carbody-yaw velocity change. The pattern of residual change is thus different and this sequence indicates that the three faults can be identified by comparing the changes of the residuals.

Table 1. Logic sequence for detection and isolation

	Lateral acceleration							Yaw velocity						
	C	Bl	Bt	w1	w2	w3	w4	C	Bl	Bt	w1	w2	w3	w4
Yaw-damper fault									✓					
Lateral-damper fault	✓	✓						✓						
Conicity change		✓	✓	✓	✓	✓	✓		✓	✓	✓	✓	✓	✓

C:carbody; Bl, Bt : leading bogie, trailing bogie; w*: wheelset 1～4

4   CONCLUSION

A fault-detection and isolation system for railway vehicle bogie was developed. This system consists of a model-based estimation method using a Kalman-Bucy filter and an isolation scheme. Time-history simulations were carried out to investigate the detection and isolation performance of the developed system. The simulation results show that the model-based estimation method using a Kalman-Bucy filter and an isolation scheme provide an effective basis for a fault-detection and isolation system for railway vehicle bogies.

REFERENCES

1.   Goodall, R.: Active railway suspensions: implementation status and technological trends. *Vehicle System Dynamics*, Vol. 28 (1997), pp. 87-117.
2.   Kalman, R.E., Bucy, R.S.: New Results in Linear Filtering and Prediction Theory. *Transactions ASME, Journal Basic Engineering*, Series 83D, Mar (1961), pp. 95-108.
3.   Mei, T.X., Goodall, R.M., Li, H.: Kalman Filter for the State Estimation of a 2-Axle Railway Vehicle. *5th European Control Conference 99*, CA-10-F812, Aug-Sep (1999).
4.   Li, H., Goodall, R.M.,: Fault-Tolerant Sensing for Rail Vehicles. *Proc.14th int. Conf. on Systems Engineeering*,ICSE 2000, Sep (1999), pp. 381-385

Stability and Performance Studies of Driver-Vehicle Systems with Electronic Chassis Control

M. LIN, A.A. POPOV[1] AND S. MCWILLIAM

SUMMARY

A classical two-level preview type driver model with a 6-DOF vehicle model is presented in this paper within the framework of driver-vehicle system approach. The model provides a firm basis for analysing vehicle handling dynamics. The paper aims to provide performance analysis of driver-vehicle system with ABS control and to explore the effects of active chassis intervention systems on the driver-vehicle interaction. Generic mathematical models of vehicle and driver are implemented. The time domain simulations in combined cornering and hard braking manoeuvre are studied through numerical simulations. The essential driver model parameters, namely driver preview time constant and driver delay time constant, are identified. Their effects are compared for systems with and without ABS control.

1. INTRODUCTION

Mathematical models for driver-vehicle system simulations have been developed to a great extent and they are now accurate enough for a number of studies, for example, vehicle dynamic analysis, interactive driving simulation, vehicle testing, etc. The model complexity and solution procedures are defined according to a given application. Donges [1], McRuer *et al.* [2] and Allen [3,4] made significant contributions to the understanding of driver's control strategies and the interaction between driver and vehicle. Their models can be used to assess the drivability, manoeuvrability and safety of vehicles under various driving conditions.

At the same time, due to the desire for personal mobility, automotive chassis enhancement systems are introduced into vehicles. They are targeting on providing safety, stability and comfort, and minimising the environmental impacts. However, it is argued that in some cases these chassis enhancement systems can cause more harm than good. In [5], Sharp pointed out that the assessment of driver-vehicle dynamics qualities in the context of electronic enhanced vehicles contains many separate quality issues and many design conflicts. This involves driver-vehicle speed control and its relationship with directional/steering control, which has only recently received attention. A detailed review on automotive chassis enhancement systems in heavy vehicles, by Palkovics and Fries [6], includes systems such as anti-lock braking system (ABS), traction control system (TCS), rear axle steering system and dynamic stability control system. It is suggested that the driver is kept in the loop, as driver's intention is necessary to activate the systems. By making a vehicle easier to control, drivers may be encouraged to drive closer to the vehicle limits, therefore

---

[1] *Address correspondence to:* School of Mechanical, Materials, Manufacturing Engineering and Management, University of Nottingham, Nottingham NG7 2RD, U.K. Tel: +44 (0)115 9513783; Fax: +44 (0)115 9513800; E-mail: atanas.popov@nottingham.ac.uk

affecting the intended safety benefits. In the researches of Allen *et al.* [3,4], it is argued that under combined cornering and hard braking manoeuvres, vehicles exercise the most severe oversteer tendencies, which could aggravate the vehicle responses and cause a spin-out. It is interesting to see whether electronic chassis controls, for example ABS control, can act effectively in this case and whether the driver model influences in any way systems incorporating ABS control.

In the following sections, a brief description of the driver-vehicle system will be firstly given to outline the system structure and the control strategies of the driver. Secondly, the simulations under the combined cornering and hard braking manoeuvre are analysed and the effects of essential driver parameters are discussed.

2. NONLINEAR VEHICLE MODEL

Vehicle dynamics modelling starts with the consideration of forces and moments equations, where the dominant physical forcing is from tyres. In order to achieve model validity throughout the vehicle manoeuvring range, this requires a tyre model that reacts correctly to the full range of the operating conditions (slips, cambers and vertical loads), and a steering/suspension model that correctly provides the input information. The nonlinear vehicle model used here has six degrees of freedom: the longitudinal, lateral, yaw and roll motions for the vehicle body, and front and rear axle spin modes for the wheels, as shown in Fig.1. The suspensions are represented by a simplified description of body roll assuming a fixed roll axis defined by the heights of the roll centres of the front and rear axles of the vehicle. Values for the vehicle model parameters are reported in the Appendix.

2.1 EQUATIONS OF MOTION

The equations of motion using axes fixed to the vehicle body are given by Ellis [7],

$$m(\dot{u} + rv) = F_{xr} + F_{xf}\cos\delta - F_{yf}\sin\delta - F_{ax}$$

$$m(\dot{v} + ru) = F_{yr} + F_{yf}\cos\delta + F_{xf}\sin\delta$$

$$I_z \cdot \dot{r} - I_{xz} \cdot \dot{p} = a \cdot (F_{yf}\cos\delta + F_{xf}\sin\delta) - b \cdot F_{yr} - h\sin\phi(F_{xr} + F_{xf}\cos\delta - F_{yf}\sin\delta) \qquad (1)$$

$$I_x \cdot \dot{p} - I_{xz} \cdot \dot{r} + (k_{\phi f} + k_{\phi r})\phi + (c_{\phi f} + c_{\phi r})p = h\sin\phi(F_{zf} + F_{zr}) + h\cos\phi(F_{yr} + F_{yf}\cos\delta - F_{xf}\sin\delta)$$

$$(I_f + I_e G_r^2)\ddot{\theta}_f = T_f - F_{xf}R_r$$

$$I_r\ddot{\theta}_r = T_r - F_{xr}R_r$$

where u, v, r, and p are longitudinal velocity, lateral velocity, yaw rate and roll rate, respectively; θ_f, θ_r are front and rear axle angular positions; F_{xf}, F_{xr}, F_{yf}, F_{yr}, and F_{zf}, F_{zr} are front and rear axle longitudinal, lateral and vertical forces.

F_{ax} is the aerodynamic drag. When the vehicle is running at constant speed, the longitudinal motion can be uncoupled from the equations of motion. Under braking conditions, the model accounts for brake torque proportioned between the front and rear axles. A fixed proportioning ratio (60/40) is taken in the following simulations.

478

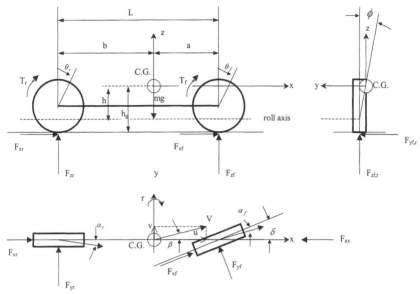

Fig. 1. 6-DOF nonlinear vehicle model.

For a complete description of tyre behaviour, there are three main variables: sideslip angles α_f and α_r defined in terms of vehicle motion variables plus steering components; longitudinal slips σ_f and σ_r linked with wheel spin modes to account for different wheel and vehicle speeds; camber angles γ_f and γ_r can be simply accounted for as a function of vehicle roll.

$$\alpha_f = a\tan\left(\frac{v + a \cdot r + hp\cos\phi}{u - rh\sin\phi}\right) - \varepsilon_f\phi - \delta, \quad \sigma_f = 1 - \dot{\theta}_f R_r / v_s, \quad \gamma_f = \xi_f\phi$$

$$\alpha_r = a\tan\left(\frac{v - b \cdot r + hp\cos\phi}{u - rh\sin\phi}\right) - \varepsilon_r\phi, \qquad \sigma_r = 1 - \dot{\theta}_r R_r / v_s, \quad \gamma_r = \xi_r\phi$$

(2)

where v_s is the vehicle forward speed with $v_s = \sqrt{u^2 + v^2}$. The dynamics of the nonlinear vehicle model includes the influence of the nonlinear tyre characteristics, which are modelled by the 'magic formula' [8] taking into account the combined longitudinal and lateral slips. Tyre force lags are expressed as first order lags that are a function of the tyre relaxation lengths and vehicle forward speed. The effects of lateral and longitudinal load transfers for braking have been incorporated through a steady state approximation.

$$\Delta F_{fz\_lat} = \frac{mru}{t_f}\left\{\frac{b(h_g - h)}{L} + d_{\phi f} h\right\}, \qquad \Delta F_{rz\_lat} = \frac{mru}{t_r}\left\{\frac{a(h_g - h)}{L} + d_{\phi r} h\right\}$$

$$\Delta F_{fz\_long} = -(F_{fx} + F_{fr})\cdot h_g / L, \qquad \Delta F_{rz\_long} = +(F_{fx} + F_{fr})\cdot h_g / L \qquad (3)$$

2.2 GENERIC ABS MODELLING

An antilock brake system (ABS) controls the longitudinal slip of wheels to prevent them from locking, so that a high friction is achieved and steerability is maintained. Referring to the wheel spin modes in Equation (1), a generic ABS system is showed in Fig.2. ABSs are installed on both axles of the vehicle. The control objective is to follow a reference tyre longitudinal slip value, improving the production of longitudinal forces during braking. Under combined cornering and braking manoeuvres, the influence of lateral tyre slips on longitudinal tyre slips is also taken into account.

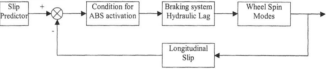

Fig. 2. Generic antilock brake system (ABS).

3. DRIVER-VEHICLE INTERACTION

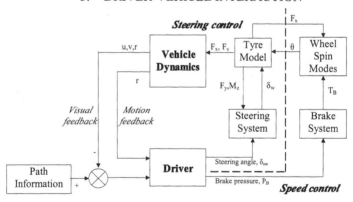

Fig. 3. Driver-vehicle system model.

As driver and vehicle form a closely coupled man-machine system, driver-vehicle system modelling provides a basis for vehicle handling design in a virtual environment. Driver's effort is concerned with steering movements and throttle/brake pedal actions. The driver has visual and motion feedbacks for developing steering movements. Throttle/brake pedal actions are generated according to acceleration/deceleration commands. An illustration of the operational blocks of the whole system is shown in Fig.3. For steering control, drivers can use preview behaviour to follow curved paths. A vehicle will follow a curved path for a given steering angle, so the driver can match horizontal road curvature with appropriate steer angle, and the remaining lane displacement can be handled with compensatory control actions. For speed control, the driver tries to match road grade with a throttle angle, although the correct perception of road grade is much more difficult and imprecise than the perception for horizontal curvature [4].

3.1 DRIVER BEHAVIOUR THROUGH PATH PREVIEW

For the driver modelling, a classical two-level model [1] (preview tracking level and compensatory tracking level) governs the steering control. Driver's steering movements, enabling the preview point to follow the prescribed preview point trajectory, must be found in the closed-loop mode. Driver behaviour through path preview involves actions based on perception of steering commands and path information. The driver exerts steering control to maintain lane position through preview control, and to manoeuvre the vehicle during curve negotiation, lane change or obstacle avoidance. Unpredictable road disturbances can randomly move the vehicle within the lane, and the driver must counteract these disturbances with compensatory control.

Weir and McRuer [9] argued that systems structured to control vehicle heading/yaw angle and lateral position, or path angle and lateral position, offered good closed-loop characteristics. Therefore, following Allen's approach [4], it is assumed that the driver develops steering corrections based on perceived heading/yaw and lane position errors. By setting a preview distance L_P on the vehicle-fixed x-axis with a single preview point, a predictive behaviour is incorporated into the system. As Kondo and Ajimine [10] suggested, the preview distance L_P can be approximated as the product of vehicle forward speed and a certain margin of time, which gives $L_P = T_P \cdot v_s$. Here, T_P is the driver preview time constant. The control action can be defined using the following equations

$$\frac{\delta(s)}{\psi_c(s)} = K_d e^{-\tau s}\left\{1 + \frac{K_\psi}{s}\right\}, \qquad \psi_c = y_e / L_P + (\psi - \psi_P), \qquad (4)$$

where y_e is the lane position error, and ψ_c, ψ and ψ_P are the composite heading angle, the actual heading angle and the initial heading angle input. The transfer function contains a gain K_d that sets the magnitude of road steering angle corrections for given heading error ψ_c, a time delay $e^{-\tau s}$ approximating driver's reaction time delay and the neuromuscular delay, and a lateral deviation trim function gain K_ψ that eliminates vehicle attitude errors. Generally, the K_d-value is mainly determined by the vehicle design parameters and varies with the vehicle forward speed. It also contains information about the gear ratio of the steering system k_s. Instead of separately perceiving both heading angle and lane position errors, the driver needs only to perceive the angular error ψ_c to the preview point down the road. Please refer to [11] for more details about driver behaviour through path preview.

3.2 DRIVER STRATEGY UNDER COMBINED CORNERING AND BRAKING CONDITION

Under combined cornering and braking condition, the variation of the speed plays an important role in the behaviour of the system. As stated above, three parameters in the driver model, namely driver preview distance L_P, driver delay time constant τ, and the steering angle gain K_d, are key parameters in the driver model. L_P will not only be determined by T_P, but also by the vehicle forward speed. This is consistent with everyday experience, which shows the driver sees nearer distance at lower speeds and further distance at higher speeds. Unlike K_d, which is defined by the vehicle design parameters, the choice of T_P and τ values depends more on driver's physical abilities. Driver's physical abilities are limited in respect to the steering wheel angle and acceleration. Their maximum possible values are strongly interdependent. For the case here, since the vehicle is under hard braking (0.6g's), the acceleration capability is not concerned. It is supposed that in normal conditions the maximum driver steering-wheel angle should not exceed 180 deg. The preview distance is governed in real-life conditions by the visibility (lighting, fog, obstructions) and manoeuvre difficulty (road curvature, tyre adhesion, vehicle speed): the parameter values range between 10m to 300m. Extensive experiments in [10] showed preview time variation for each individual tested driver, and it ranged from 2-4s for constant radius turns, with the radius of the turn varied from 15m to 100m. The values for driver delay time also spans for individuals [12], from 0.08s for a professional sports car driver to more than 0.25s for an occasional driver.

4. SIMULATION OF HARD BRAKING IN TURN

Now, consider the combined cornering and braking manoeuvre in a constant radius turn. Fig.4 illustrates a comparison between the characteristic responses of the vehicle open-loop model (Fig.4(a,b,c)) and the driver-vehicle closed-loop model (Fig.4(d,e,f)). The driver enters a 300m-radius turn with a speed of 100km/h and starts to brake. The braking deceleration command is set at 0.6g's. This manoeuvre results in excessive lateral acceleration over 0.3g's (Fig.4(b,e)). Statistically [3], a cautious driver aims at low to moderate lateral acceleration below 0.26g's while driving. In this case, lateral acceleration is reduced to an appropriate percentage below 0.26g's, corresponding with a reduced speed of 50km/h. The vehicle will then negotiate the curve with a reduced constant speed. Therefore, a combined hard braking in a moderate turn scenario is set up.

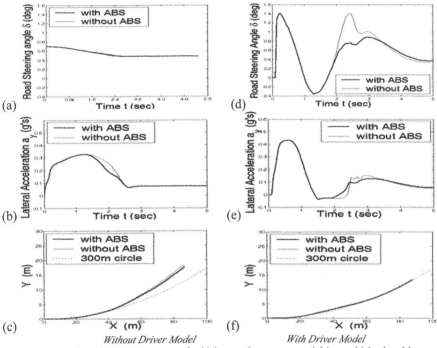

(a) (b) (c) *Without Driver Model*   (d) (e) (f) *With Driver Model*

Fig. 4. Comparisons between responses of vehicle open-loop system and driver-vehicle closed-loop system in braking in turn.

In Fig.4 (c), it is noted that without a driver in the loop, the vehicle cannot follow the desired path successfully either with the help of ABS or without it. When incorporating a driver, the task is completed (Fig.4 (f)). When the driver model begins braking, the lateral acceleration is subsequently brought up to the maximum level around 0.4g's (Fig.4 (e)). It is obvious that under this severe manoeuvre the understeer vehicle shows oversteer tendency for about 1 second, featured by the sign change of the steering input (Fig.4 (d)) and the lateral acceleration (Fig.4 (e)). This is the result of a reduction in the cornering force coefficient for the rear axle. The resulting vehicle dynamics contains instability, which creates a very difficult steering control task for the driver. In this simulation, the driver is able to reduce speed and bring the vehicle back under stable control in understeer condition. For a system with ABS, it seems that the driver's steering workload is reduced (Fig.4 (d)). However, significant differences between systems with ABS and without ABS cannot be found. With the application of ABS control, the aim is to prevent the wheel-locking and tyre force saturation. The characteristic axle angular speed and longitudinal slip plots in Fig.5 show the performance of the rear axle, where ABS is activated first. ABS manages to maintain the longitudinal slip oscillating around the reference value of 10% during braking. But it does not have much contribution to preventing the reduction in the cornering force coefficient for the rear axle within such a short activation period.

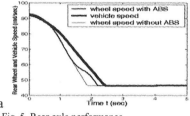 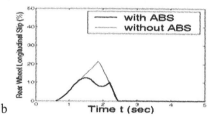

a   b

Fig. 5. Rear axle performance.

5.   PARAMETER STUDY OF THE DRIVER MODEL

Driver model parameters discussed in the description of the driver model are examined in this section. With the variation of these parameters, it is interesting to see how the system reacts. It is also important to identify sensitivities of the system with ABS to the changes in driver parameters. Simulations with varied driver model parameters are conducted under the combined cornering and hard braking manoeuvre, which is described in the previous section. Two parameters are considered here: driver preview time constant T_p and driver delay time constant τ.

Fig.6 presents road steering responses with varied T_p-values (0.5s, 1s, 2s and 3s) and a fixed τ-value of 0.1s. Fig.6(a) and (b) are for systems with and without ABS control respectively. Fig.7 presents road steering responses with varied τ-values (0s, 0.1s and 0.2s) and a T_p-value of 1s. Fig.7(a) and (b) are for systems with and without ABS control respectively.

In Fig.6, with the increasing T_p-value, the oscillation of the responses increases with a high overshoot. The system with ABS control shows that driver's steering loads are generally lower than the system without ABS control. In Fig.6(a), the road steering angle reaches 3.4 deg and about 50 deg for driver steering-wheel angle while T_p is set to 3s, which is about three times lower when compared with Fig.6(b). For other values of the preview time constant, the differences are relatively smaller. In other words, the system with ABS control is less sensitive to the parameter T_p. The same amount of change in T_p causes less influence on the system with ABS control, which shows higher robustness in this case.

Fig.7 presents the effect of driver delay time τ on the driver-vehicle system with ABS control. The driver starts to react after the delay time. In Fig.7(a), the responses become more oscillatory as τ increases. For τ =0.1s, the driver has slightly heavier movements than τ =0s, which is an ideal status for the driver. For both τ =0s and τ =0.1s, the system shows reasonable responses without overburdening the driver in either the transient period or the steady state. However, when the value increases up to 0.2s, the responses start to diverge, which indicates the instability. For system responses without ABS, (Fig.7(b)) shows similar values for 0s and 0.1s. For the response with 0.2s, the divergence is insensibly significant, which is not plotted here. Therefore, with this kind of severe manoeuvre, the effect of the driver delay time comes to a critical position. Ideally, the shorter the delay time is, the better the task is accomplished. However, not every driver is able to perform the task. It requires a prompt reaction from the driver under vehicle's limiting performance. For this reason, the driver reaction-time test is one of the standard tests for selecting Formula One drivers.

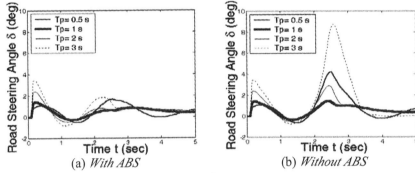

(a) *With ABS*   (b) *Without ABS*

Fig. 6. Road steering angle responses with varied T_P values.

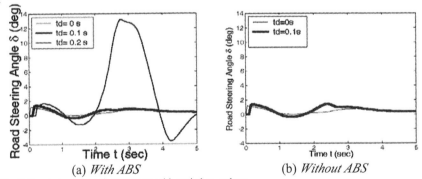

(a) *With ABS*   (b) *Without ABS*

Fig. 7. Road steering angle responses with varied τ values.

6. CONCLUSIONS

This paper has endeavoured to summarise the effects of driver model parameters on the driver-vehicle system with ABS control and without ABS control under the severe manoeuvre of combined cornering and hard braking. With the implemented models and by means of extensive computer simulations, the following results are obtained:

1. Rational vehicle and driver models are implemented by taking into account the vehicle characteristics and vehicle/road kinematics.

2. An attempt at an objective assessment of the consequences of introducing means of ABS controls is made. Under the designative manoeuvre, it is found that ABS is activated to maintain the longitudinal slip at the predicted value, preventing the rear axle from locking. However, it does not have much contribution on preventing the reduction in the rear axle cornering force coefficient, which causes the oversteer tendency.

3. The effect of driver parameters on the driver-vehicle system handling is evaluated theoretically. The system with ABS shows lower sensitivity to driver preview constant than the system without ABS control, which gives higher robustness to the change of the preview constant. The driver delay time constant is identified as an essential parameter for the system with ABS control. Larger delay time may cause instability to the system.

REFERENCES

1. Donges E.: *A Two-Level Model of Driver Steering Behaviour, Human factors 20(6), 1978, pp.691-707.*

2. McRuer D.T., Allen R.W., Weir D.H. and Klein R.H.: *New Results in Driver Steering Control Models, Human Factors 19(4), 1977, pp.381-397.*

3. Allen R.W., Rosenthal T.J. and Szostak H.T.: *Steady State and Transient Analysis of Ground Vehicle Handling, SAE paper 870495, 1987.*

4. Allen R.W.: *Analysis and Computer Simulation of Driver/Vehicle Interaction, SAE paper 871086, 1987.*

5. Sharp R.S.: *Some Contemporary Problems in Road Vehicle Dynamics, Proc. of Instn. Mech. Engrs., J. Mech. Eng. Sci, Vol.214, 1999, pp. 137-148.*

6. Palkovics L. and Fries A.: *Intelligent Electronic Systems in Commercial Vehicles for Enhanced Traffic Safety, Vehicle System Dynamics, 35(2001), No.4-5, pp.227-289*

7. Ellis J.R.: *Vehicle Handling Dynamics, Mechanical Engineering Publication Limited, London, 1994.*

8. Pacejka H.B. and Bakker E.: *The Magic Formula Tyre Model, Vehicle System Dynamics, vol.20(1991), pp.1-17.*

9. Weir D.H. and McRuer D.T.: *Dynamics of Driver Vehicle Steering Control, Automatica, Vol.6, 1970, pp.87-98.*

10. Kondo M. and A. Ajimine: *Driver's Sight Point and Dynamics of Driver/Vehicle System Related to It, SAE paper 680104, 1968.*

11. Lin M., Popov A.A. and McWilliam S.: *Handling Studies of Driver-Vehicle Systems, Proceedings of AVEC2002, 6th Int'l Symposium on Advanced Vehicle Control, Hiroshima, Japan, 2002.*

12. Genta G.: *Motor Vehicle Dynamics: Modeling and Simulation, Singapore: World Scientific, 1997.*

APPENDIX: Vehicle Model Parameters [12]

Vehicle mass m=830 kg, Vehicle yaw inertia I_z=1210 kgm^2, Vehicle roll inertia I_x=290 kgm^2, Cross-product inertia I_{xz}=-84 kgm^2, Front axle polar moment of inertia I_f=45.4 kgm^2, Rear axle polar moment of inertia I_r=30.9 kgm^2, Engine moment of inertia I_e=0.1 kgm^2, C.G. height h_g=0.53 m, Distance of C.G. from the roll axis h=0.4m, Wheel base L=2.16 m, Distance of C.G. from the front axle a=0.87 m, Distance of C.G. from the rear axle b=1.29 m, Front track t_f=1.284 m, Rear track t=1.277 m, Wheel Radius R_r=0.3m, Gear ratio of the steering system k_s= 16.3, Front axle roll stiffness $k_{\phi f}$=11252 Nm/rad, Rear axle roll stiffness $k_{\phi r}$=11130 Nm/rad, Front axle roll damping coefficient $c_{\phi f}$=824 Nms/rad, Rear axle roll damping coefficient $c_{\phi r}$=815 Nms/rad, Front axle roll steer coefficient ε_f=0.04, Rear axle roll steer coefficient ε_r=-0.06, Front axle roll camber coefficient ξ_f =0.60, Rear axle roll camber coefficient ξ_r =0.90, Front axle roll stiffness proportion $d_{\phi f}$=50.27%, Rear axle roll stiffness proportion $d_{\phi r}$=49.73%.

Side Slip Control of Small-Scale Electric Vehicle by DYC

MOTOKI SHINO[1], PONGSATHORN RAKSINCHAROENSAK[2],

MINORU KAMATA[1] , and MASAO NAGAI[2]

SUMMARY

This paper proposes a chassis control system utilizing DYC by using model matching control method to make the steering response of body side slip angle be constantly zero. For realizing the proposed system in an actual electric vehicle, it is necessary to estimate the side slip angle instead of measuring it directly. A speed-dependent observer is proposed in this paper. With the goal of enhancing stability, the control objective is to obtain desirable steering response by controlling the body side slip angle of the vehicle. Experiments using actual small-scale electric vehicle are carried out to verify the effectiveness of the proposed chassis control system on cornering performance in a J-turn and Lane-change tests.

1   INTRODUCTION

Nowadays, development of battery, performance of drive system, etc., for supporting the widespread of electric vehicle (EV) has grown rapidly. EV is expected to play a key role in the future private and public transportation systems[1]. It is also expected that small-scale electric vehicles can expand the mobility of pedestrians and senior citizen dramatically. However, active safety of such vehicle is not conducted extensively[2][3]. From the viewpoint of vehicle dynamics, EV can be designed to have different types of drive systems. For the design of electric vehicle, the sophisticated configuration called "In-Wheel-Motor" has been developed by integrating the motor into each drive wheel so that these motors are possible to be controlled independently of each other. Typically, the electric motor has comparatively faster torque response, and easy measurability of torque compared with the conventional engine and hydraulic machine[4].

Lack of vehicle stability due to its excessive side slip is recognized as the main cause of traffic accidents in Japan[5]. A concept to enhance vehicle stability during

[1] *Address correspondence to:* Department of Engineering Synthesis, The University of Tokyo, 7-3-1 Hongo, Bunkyo-ku, Tokyo, 113-8656 Japan, Tel.:(+81)-3-5841-6402, Fax.:(+81)-3-3818-0835, E-mail: motoki@sl.t.u-tokyo.ac.jp

[2] Department of Mechanical Systems Engineering, Tokyo University of Agriculture and Technology

cornering is to generate additional yaw moment by differential longitudinal forces. As EV has in-wheel-motor structure which their motors can be controlled to generate wheel torques independently of each other, differential longitudinal forces (traction or braking) can be precisely achieved. This control strategy is academically named "Direct Yaw Moment control (DYC)". Utilizing these merits in structure and characteristic, this study examines the feasibility of DYC system for enhancing vehicle stability by computer simulations and also experiments on small-scale EV "NOVEL". This paper proposes a chassis control system utilizing DYC by using model matching control method to make the steering response of body side slip angle be constantly zero. The experiments by using actual small-scale EV with In-wheel-motor equipped at rear wheel, are carried out to verify the effectiveness of the proposed chassis control system on cornering performance at J-turn and lane-change tests.

2 DESCRIPTION OF SMALL-SCALE ELECTRIC VEHICLE "NOVEL"

A small-scale EV named "NOVEL" is one-person-car with in-wheel-motor equipped at each rear wheel as real-wheel-drive type shown in Fig. 1. The vehicle is 400 kilograms in weight (driver included), has 50km/h in maximum speed. It is equipped with vehicle motion sensing system and digital signal processing system for implementation of yaw moment controller, which is going to be explained in section 4. It can generate different wheel torques between the left and right sides, which are independent on driver pedal operation on line. This can be said that NOVEL is equipped with "Drive-by-Wire" system. Practically, it measures the driver's steering maneuver, the vehicle speed, the yaw rate at the center of gravity, all wheel velocity, and inputs all of them into host computer via digital signal processing, and then calculated the additional yaw moment inside the computer as shown in Fig.1. Additional yaw moment is then realized by increment and decrement of each wheel torque simultaneously via digital signal processing again.

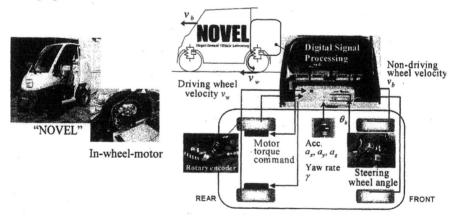

Fig. 1. On-board measurement and control system of small-scele electric vehicle "NOVEL".

3 IDENTIFICATION OF VEHICLE MODEL

For designing the control system, it is necessary to know the cornering characteristics with respect to steering input. The cornering characteristics are determined by identification method from equivalent linear two-wheel model of the vehicle in planar motion to avoid complexity as shown in Fig. 2. The vehicle model has two degree-of-freedom, i.e. the yaw, and the lateral motions. The coordinate system is fixed on the vehicle when setting up the equation of motions. Using linear model, tire lateral force is assumed to be proportional to tire side slip angle. Under the assumptions, the governing equations of lateral and yaw motions of the vehicle can be expressed as follows:

Lateral motion:

$$mV(\dot{\beta}+\gamma) = -2C_f(\beta+\gamma l_f/V-\delta_f)-2C_r(\beta-\gamma l_r/V) \tag{1}$$

Yaw motion:

$$I_z\dot{\gamma} = -2C_f l_f(\beta+\gamma l_f/V-\delta_f)+2C_r l_r(\beta-\gamma l_r/V)+M \tag{2}$$

where, m denotes mass of the body, β the side slip angle, γ the yaw rate, $l_f(l_r)$ the distance from C.G. to front (rear) axle, V the chassis velocity, I_z the yaw moment of inertia. $C_f(C_r)$ equivalent front (rear) cornering stiffness, δ_f the front steering angle input from driver, M the direct yaw moment control input.

From the equations (1), (2), the equivalent two-wheel model needs the following parameters: m, I_z, l_f, l_r, C_f, C_r. In these parameters, the tire cornering stiffness is very difficult to measure directly because their values include the dynamics characteristics of suspension, tire etc. In this paper, Based on an experiment on a small-scale electric vehicle and simulation from an analytical model, the cornering stiffnesses is determined by trial and error. When the same steering maneuver is input in both the experiment and the simulation at a constant velocity of 35km/h, the output of yaw rate are shown in Fig. 3. Fig. 3 (a) shows the time response of yaw rate when operating J-turn maneuver on dry asphalt road. Fig. 3 (b) shows the frequency response of yaw rate with respect to the front steering input. As a result, as compared the experiment with the simulation, the identification result is quite satisfactory. The measured and identified parameters of NOVEL are shown in Table 1.

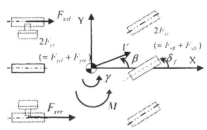

Fig. 2. Two-wheel vehicle model.

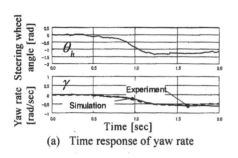

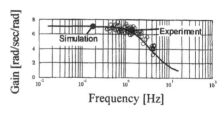

(a) Time response of yaw rate

(b) Frequency response of yaw rate with respect to the front steering input

Fig. 3. Identification results of NOVEL obtained from identification (V=35km/h).

Table 1. Parameters of experiment vehicle "NOVEL".

DEFINITION	SYMBOL	UNIT	VALUE
Vehicle mass (Included driver)	m	kg	400
Yaw moment of inertia	I_z	kgm^2	160
Wheel base	l	m	1.28
Axle tread	d	m	0.82
Distance from C.G. to front axle	l_f	m	0.75
Distance from C.G. to rear axle	l_r	m	0.53
Height of C.G.	h	m	0.4
Front cornering stiffness	C_f	N/rad	10000
Rear cornering stiffness	C_r	N/rad	16000
Steer gear ratio	n	$-$	18.7

From the equations (1), (2), let the derivative term be zero, the side slip angle of the vehicle in steady state cornering can be calculated as follows:

$$\beta_{st} = \frac{1 - \dfrac{ml_f}{2ll_rC_r}V^2}{1 + \dfrac{m}{2l^2}\dfrac{l_rC_r - l_fC_f}{C_fC_r}}\frac{l_r}{l}\delta_{fst} \tag{3}$$

Substituting the parameters of NOVEL and normal passenger cars (1500kg) into equation (3), the relationship between vehicle velocity and body side slip angle can be compared as shown in Fig.4. It was found that although NOVEL has maximum speed of 50km/h, its body side slip angle during steady state cornering is in the same level the high speed region of passenger car. According to statistic analysis of accidents, it was

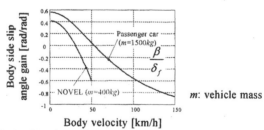

m: vehicle mass

Fig. 4. Comparison of body side slip angle between NOVEL and passenger car.

reported that must of traffic accidents occur due to excessive side slip which causes difficultly in steering vehicle as driver intends[5]. Therefore, based on this theoretical analysis, this paper proposes the DYC system which aims to suppress the body side slip angle.

4　DIRECT YAW-MOMENT CONTROL SYSTEM DESIGN

4.1　Control objective

For enhancing handling and stability of vehicle, the side slip angle and yaw rate of the vehicle are controlled the desired values by using direct yaw-moment system. The direct yaw-moment generated by the traction force at rear axle is employed as the control input to make actual response trace the desired values. With the application of model matching control technique, the control system consists of a feedforward compensator with respect to the steering angle, and a feedback compensator depending on state deviations of side slip angle and yaw rate as shown in Fig.5.

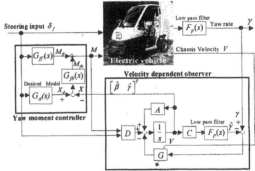

Fig. 5.　Direct yaw-moment control system.

4.2　Design of DYC for suppressing side slip angle

To make the steering response of side slip angle be constantly zero with respect to the steering input, this paper applies the model matching theory to design the system.

First, from equation (1) and (2), the state equation of vehicle model for designing the controller can be obtained as follows:

$$\begin{bmatrix} \dot{\beta} \\ \dot{\gamma} \end{bmatrix} = \begin{bmatrix} a_{11} & a_{12} \\ a_{21} & a_{22} \end{bmatrix} \begin{bmatrix} \beta \\ \gamma \end{bmatrix} + \begin{bmatrix} b_1 \\ b_2 \end{bmatrix} M + \begin{bmatrix} h_1 \\ h_2 \end{bmatrix} \delta_f \tag{4}$$

$$\dot{X} = AX + BM + H\delta_f,$$

where, $X = \begin{bmatrix} \beta \\ \gamma \end{bmatrix}$, $\begin{bmatrix} b_1 \\ b_2 \end{bmatrix} = \begin{bmatrix} 0 \\ \dfrac{1}{I_z} \end{bmatrix}$, $\begin{bmatrix} h_1 \\ h_2 \end{bmatrix} = \begin{bmatrix} \dfrac{2C_f}{mV} \\ \dfrac{2C_f l_f}{I_z} \end{bmatrix}$, $\begin{bmatrix} a_{11} & a_{12} \\ a_{21} & a_{22} \end{bmatrix} = \begin{bmatrix} \dfrac{-2(C_f + C_r)}{mV} & \dfrac{-2(C_f l_f - C_r l_r)}{mV^2} - 1 \\ \dfrac{-2(C_f l_f - C_r l_r)}{I_z} & \dfrac{-2(C_f l_f^2 + C_r l_r^2)}{I_z V} \end{bmatrix}$.

Moreover, the transfer functions are derived as follows:

$$\beta(s) = \frac{(h_1 s - h_1 a_{22} + h_2 a_{12})\delta_f(s) + b_2 a_{12} M(s)}{(s - a_{11})(s - a_{22}) - a_{21} a_{12}}.$$

(5)

$$\gamma(s) = \frac{(h_2 s + h_1 a_{21} - h_2 a_{11})\delta_f(s) + b_2(s - a_{11})M(s)}{(s - a_{11})(s - a_{22}) - a_{12} a_{21}}.$$

(6)

The feedforward compensation is designed to regulate the side slip angle at steady state cornering. The mathematical expression of feedforward compensator is as follows:

$$\text{Feedforward compensator: } M_{ff} = \frac{4 l l_r C_f C_r - 2 l_f C_f m V^2}{m V^2 + 2(l_f C_f - l_r C_r)} \delta_f.$$

(7)

As a result, the transfer function from the front steering angle to the yaw rate can be written by substituting equation (7) into (6).

$$\frac{\gamma(s)}{\delta_f(s)} = \frac{\left\{\dfrac{4C_f}{I_z}\dfrac{l_f^2 C_f + l_r^2 C_r}{mV^2 + 2(l_f C_f - l_r C_r)}\right\} s + \dfrac{2C_f V}{mV^2 + 2(l_f C_f - l_r C_r)}\left\{\dfrac{4l^2 C_f C_r}{ml_z V^2} - \dfrac{2(l_f C_f - l_r C_r)}{I_z}\right\}}{s^2 + \left\{\dfrac{2(C_f + C_r)}{mV} + \dfrac{2(l_f^2 C_f + l_r^2 C_r)}{I_z V}\right\} s + \left\{\dfrac{4l^2 C_f C_r}{ml_z V^2} - \dfrac{2(l_f C_f - l_r C_r)}{I_z}\right\}}.$$

(8)

Moreover, the desired model is determined such that the side slip angle response $\beta_d(s)$ is constantly zero, the yaw rate response $\gamma_d(s)$ is set to be the first order delay with respect to the front steering angle as follows:

$$\dot{X}_d = A_d X_d + H_d \delta_f$$

(9)

where, $X_d = \begin{bmatrix} \beta_d \\ \gamma_d \end{bmatrix}$, $A_d = \begin{bmatrix} 0 & 0 \\ 0 & -\dfrac{1}{\tau_{yd}} \end{bmatrix}$, $H_d = \begin{bmatrix} 0 \\ \dfrac{k_{yd}}{\tau_{yd}} \end{bmatrix}$.

k_{yd} and τ_{yd} are steady state gain and time constant of yaw rate response, respectively. Moreover, it is important to consider what is an ideal response of the desired model in this controller. By comparing the transfer function of the desired yaw rate with the transfer function in equation (8), steady state gain and time constant of the yaw rate response can be obtained as follows:

$$k_{yd} = \frac{2C_f V}{mV^2 + 2(l_f C_f - l_r C_r)}, \quad \tau_{yd} = \frac{I_z V}{2(l_f^2 C_f + l_r^2 C_r)}.$$

(10)

To compensate side slip angle and yaw rate during transient steering maneuver, it is necessary to combine the feedback compensator with feedforward compensator. Here, the state feedback of side slip angle and yaw rate is presented.

First, the state deviation between the desired value X_d and actual value X is assumed to be as follows:

$$E = X - X_d.$$

(11)

The differentiated value of the above equation can be obtained as follows:

$$\dot{E} = \dot{X} - \dot{X}_d$$
$$= A(X - X_d) + BM_{fb} + (A - A_d)X_d + (H - H_d)\delta_f$$
$$= AE + BM_{fb} + W. \tag{12}$$

Let the disturbance terms be zero, and the optimal control input can be calculated by state feedback of deviations of side slip angle and yaw rate as follows:

$$M_{fb} = -g_{fb1}(\beta - \beta_d) - g_{fb2}(\gamma - \gamma_d) \tag{13}$$

where, the feedback gains g_{fb1}, g_{fb2} are determined by the application of optimal control theory to minimize the following performance index.

$$J = \int_0^\infty \left\{ \left(\frac{\beta - \beta_d}{q_\beta} \right)^2 + \left(\frac{\gamma - \gamma_d}{q_\gamma} \right)^2 + \left(\frac{M_{fb}}{r_{fb}} \right)^2 \right\} dt \tag{14}$$

where, the weighting coefficients q_β, q_γ indicate the maximum allowable values of the state variables and the coefficients r_{fb} indicates the limit of control input.

To realize the total yaw-moment control input in the actual electric vehicle, the differential traction forces must be produced. With the merit of in-wheel-motor structure, the traction forces of the left and right wheels can be distributed easily.

If the longitudinal acceleration is given as a_x, the longitudinal motion can be expressed as:

$$ma_x = F_{xl} + F_{xr}. \tag{15}$$

The yaw-moment control input which is determined from the controller is given as:

$$M = \frac{d}{2}(-F_{xl} + F_{xr}). \tag{16}$$

By rearranging equations (15) and (16), the command traction force of each tire can be derived as follows:

$$\text{Left side: } F_{xl} = \frac{ma_x}{2} - \frac{M}{d}, \quad \text{Right side: } F_{xr} = \frac{ma_x}{2} + \frac{M}{d}. \tag{17}$$

4.3 Design of speed-dependent side slip angle observer

For realizing the proposed DYC system in the actual vehicle "NOVEL", it is necessary to estimate the side slip angle instead of measuring it directly, since the side slip angle sensor is expensive and cannot be installed in the vehicle. According to the observer theory, the side slip angle can be estimated by measuring the steering wheel angle, the yaw rate, and the chassis velocity. As the body side slip angle has speed-dependent characteristic, it is necessary to design the side slip angle observer with variable gain depending on chassis velocity.

When the state variables of observer are assumed to be $\hat{X} = \begin{bmatrix} \hat{\beta} & \hat{\gamma} \end{bmatrix}^T$, the state equation of observer can be expressed as follows:

$$\dot{\hat{X}} = F\hat{X} + GY + DU. \tag{18}$$

The observer is designed in the way that the error between the actual state variables and their estimated values must be zero. So, let the state error variables $e_{ob} = X - \hat{X}$ so that the following equation can be obtained from equation (4) and (18).

$$\dot{e}_{ob} = (A - GC)X - F\hat{X} + (B - D)U$$
$$= (A - GC)e_{ob}.$$

(19)

Here, the observer gains, G, are determined in such a way that the eigenvalues of matrix. F exist in the left side of the complex plane without the imaginary part in their values. Then, when the eigenvalues of $A-GC$ are assumed to be λ_1, λ_2, the observer gains can be calculated as follows:

$$G = \begin{bmatrix} g_{\beta\beta}(V) \\ G_{\beta\gamma}(V) \end{bmatrix} = \begin{bmatrix} -\{a_{11}(-a_{11} + \lambda_1 + \lambda_2) - \lambda_1\lambda_2 - a_{21}a_{12}\}/a_{21} \\ a_{11} + a_{22} - (\lambda_1 + \lambda_2) \end{bmatrix}.$$

(20)

It can be seen from the above equations that the observer gains include the speed-dependent coefficients a_{11}, a_{12}, a_{21}, a_{22}. Thus the observer gains are also speed-dependent. This observer is also built-in the low-pass filter to decrease the influence of sensor noise. The cut-off frequency of low-pass filter is chosen from the frequency analysis of experiment data. Finally, the direct yaw-moment system, as shown in Fig.5, consists of the yaw-moment controller and speed-dependent side slip angle observer.

Fig.6 shows the experiment results with respect to the given front steering angle that is equivalent to slalom running test with random speed. This study assumes that the body side slip angle is correctly estimated when the estimated yaw rate matches the measured yaw rate. As a result, Fig.6 shows that the estimated yaw rate can effectively trace the actual one by the use of the proposed observer against the change in velocity. From repeated experiments, it was confirmed that the proposed observer is effective against any arbitrary steering inputs.

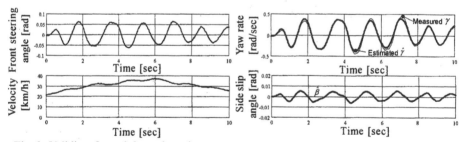

Fig. 6. Validity of speed-dependent observer

5   EFFECT OF YAW-MOMENT CONTROL SYSTEM

This section will discuss the validity of the proposed chassis control system on cornering performance as described in the previous section by experiments using small-scale electric vehicle NOVEL. All experiments are conducted under dry asphalt road. The air pressure of each tire is set as 150kPa at front tire, and 175kPa at rear tire. The front steering angle is calculated from the measured steering wheel angle divided by overall steering gear ratio, while neglecting the steering dynamics. The J-turn and lane-change tests were conducted respectively.

5.1 Control effect in J-turn test

In J-turn test shown in Fig.7, after the vehicle runs straightly at constant speed of 35km/h, the front steering angle pattern, as shown in the upper graph of Fig.7, is executed by the driver. This test is used to investigate the vehicle behavior for transient and steady-state steering maneuvers. Fig.7 shows the vehicle behavior with respect to the given steering maneuver input. It was found that, by using DYC, the body side slip angle was significantly suppressed during the steady-state cornering. Fig.7 also shows that the body side slip angle during transient state can be effectively suppressed by using feedback compensator. Moreover, it was confirmed that the DYC control input is realized by controlling each wheel torque independently to achieve differential traction in real time.

5.2 Control effect during Lane-change test

In lane-change test shown in Fig.8, the front steering angle as shown in the upper graph of Fig.8, is executed at vehicle speed of 35km/h. The course layout is set up equivalently to emergency obstacle avoidance. Fig.8 shows the vehicle behavior against lane-change steering input considered as continuous steering maneuver input. Fig.8 shows that, by using only feedforwad controller, the body side slip angle can be suppressed as it becomes closer to zero, as compared with the uncontrolled vehicle. It is also confirmed that the body side slip angle can be more suppressed by the combination with feedback compensator. Moreover, in the case of without control, large body side slip angle is generated during lane change maneuver which causes difficultly in tracing the given course. Collisions with pylon sometimes occurred during experiment. With regard to the subjective evaluation of the driver, DYC system significantly enhances handling performance and cornering stability so that driver can steer the vehicle much more easily.

6 CONCLUSIONS

Based on the concept of "Drive by wire" strategy, this paper investigates the effectiveness of the proposed direct yaw-moment control system. The major conclusions to be drawn from experiments studies by using a small-scale electric vehicle, "NOVEL" are:

1) The small-scale electric vehicle, "NOVEL" was developed to be able to generate the additional yaw moment by individual wheel torque control.
2) The feasibility of using speed-dependent observer for side slip angle estimation was clarified.
3) The effectiveness of the yaw-moment control system including yaw-moment controller and speed-dependent side slip angle observer was clarified.
4) The effect in enhancing the handling and stability in J-turn , as well as lane-change tests was proved by using NOVEL.

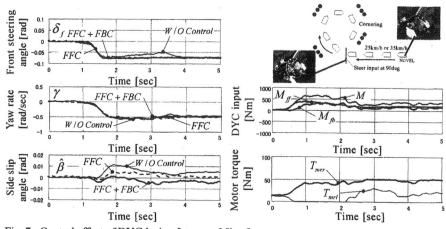

Fig. 7. Control effect of DYC during J-turn at 35km/h
(FFC:Feedforward controller, FBC: Feedback controller)

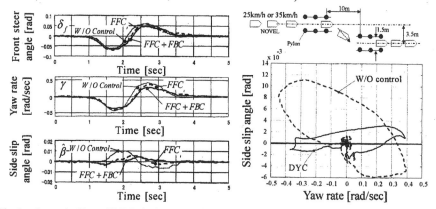

Fig. 8. Control effect of DYC during single lane-change at 35km/h

REFERENCES

1. Kamata, M., et al.: Study on Human Friendly Vehicle Usable for Elderly People, Journal of JSME C, Vol.68 No.669, pp.144-152, 2002. (Abstract in English)
2. Hattori, Y., et al.: Force and Moment Control with Nonlinear Optimum Distribution for Vehicle Dynamics, Proc. of AVEC2002., pp.595-600, 2002.
3. Sakai, K., et al.: Improvement in Control Performance of Driver-Vehicle System with EPS using Cables to Connect the Steering Wheel and Gearbox, Proc. of AVEC2002, pp.641-646. 2002.
4. Shino, M., et al. : Yaw-moment control of electric vehicle for Improving handling and stability, JSAE Review, Vol. 22, pp.473-480, 2001.
5. Inoue, H., et al. :Development of Vehicle Stability Control System for Active Safety, Journal of JSAE, pp.45-51, 1995. (Abstract in English)

Vehicle System Dynamics Supplement 41 (2004), p.497-506 © Taylor & Francis Ltd.

DRIVING DYNAMICS FOR HYBRID ELECTRIC VEHICLES CONSIDERING HANDLING AND CONTROL ARCHITECTURE

Johan Andreasson* and Leo Laine[†]

SUMMARY

The use of hybrid techniques together with the increasing demands on vehicle performance require an improved vehicle architecture to be feasible in the long run. In this paper, a generic control architecture is suggested and especially the information flow between driver's intentions to vehicle motion is discussed.

The idea is that the driver's intentions are transformed to a global force equivalent. Then, a practical approach is utilised to solve the control allocation problem of distributing the global forces to local wheel unit forces. A strategy to find wheel angles and wheel spin from desired wheel forces has been suggested and implemented.

1   INTRODUCTION

Although the research on Hybrid Electric Vehicles (HEV) is driven by environmental reasons, it is relevant to study the new technology from a driving dynamics point-of-view to be able to design competitive vehicles in the future. Except for the environmental advantages, two other aspects can be identified; the potential of improved handling as well as a need for a more structured control architecture.

Additional electric propulsion improves controllability of the vehicle behaviour compared to a conventional power-train for several reasons [1]. The torque applied to the wheels axes can be faster and more precise than with an Internal Combustion Engine (ICE) and hydraulic brakes. For an Anti lock Braking System (ABS), a more precise brake torque makes it possible to reduce vibrations and quicker response can be used to improve performance. Also, it is reasonable to have more than one electric motor for propulsion and thus it is possible to achieve active wheel torque distribution without advanced differentials. Even bidirectional torque distribution is possible with electric machines

*Div. of Vehicle Dynamics, KTH, SE-100 44 Stockholm, Sweden, e-mail: johan@fkt.kth.se, phone: +46 8 790 77 14, fax: +46 8 790 93 04

[†]Div. of Mechatronics, Chalmers University of Technology, SE-412 96, Gothenburg, Sweden, e-mail: laine@mvs.chalmers.se, phone: +46 31 772 58 52, fax: +46 31 772 13 80

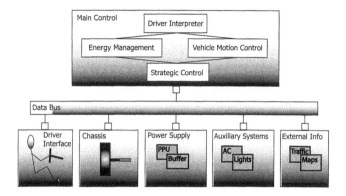

Figure 1: Schematic sketch of the functional architecture for a vehicle.

which adds possibilities to enhance the functionality of today's Vehicle Stability Controllers (VSC).

The other aspect relates to the fact that different subsystems are needed to be integrated in today's vehicles for better drivability and handling [2]. This increases the complexity of the vehicle system. Also, there are a variety of HEV configurations and most of them add more components to the vehicle. To handle this in an efficient way a hierarchical control architecture is suggested with generic interface signals. This opens up for reusability for different hardware configurations in a modular fashion [3].

2   GENERIC CONTROL ARCHITECTURE

Functional decomposition was used to identify functions within a HEV and place them in an hierarchical structure. The purpose was to make a generic control architecture[1] for HEVs [4]. Generic interface control signals was utilised between the identified functions to make it easy to change configuration [5]. Different components may perform similar tasks, e.g. electric wheel motors can perform as braking discs and in this work, all components within a wheel unit is seen as a function for applying force to the ground. This to allow tight integration of the different actuators for applying longitudinal, lateral, and vertical forces within the wheel unit.

The suggested functional architecture is shown in Figure 1. It is based on generic assumptions such that a vehicle must have a driver interface, an interaction with the ground (Chassis), power source(s) (Power Supply) and possibly also external information functions and auxiliary systems that are not involved in the vehicle motion. The system needs to communicate with the driver (Driver

---

[1]Generic Control architecture: A reusable control architecture that is not hardware dependent or configuration dependent.

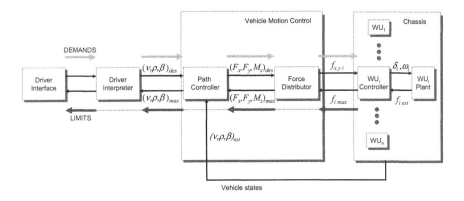

Figure 2: Signal flow between driver interface and chassis wheel units (WU). Note that only one WU is shown.

Interpreter), to control the vehicle's motion (Vehicle Motion Control), and to manage the energy flow (Energy Management). The chassis is considered as number of bodies, each with number of wheel units that can generate forces. A more thorough description is found in [3].

In [6], a structured and hierarchical way to handle the integration of different wheel controls is shown. However, all systems are based on a traditional car in the sense that there is a division into power train, chassis, brakes etc. that makes them less suitable for HEVs in general. Here, a Vehicle Motion Control that considers the desired motion and distributes forces to the wheel units is presented.

3 STRUCTURE FOR DRIVER'S INTENTIONS TO VEHICLE MOTION

Figure 2 shows the signal flow between Driver Interface and Chassis in more detail. Four isolated functions are used to transform the Driver Interface signals into vehicle motion and respond with suitable feedback; driver interpretation, path control, force distribution and wheel unit control. Each function has to set demands to the next one and send limits to the previous in order to guarantee that it can fulfill its own demands.

3.1 Driver interpretation

The task is to interpret the signals from the driver interface to a suitable path that is achievable according to the limitation set up by the path control. Also feedback signals are calculated and sent to the driver. The communication signals exchanged with the driver interface are all in percentage of their maximum values, respectively. Thus, the driver interface hardware can be exchanged from steering wheel and pedals and to e.g. a joystick. This step is left out of the further discussion within this work and instead a predefined path is given.

3.2 Path control

The task of the path controller is to follow the path set up by the driver interpreter by giving force and torque demands to the force distribution. The path is described by the velocity v, the vehicle's slip angle β and the curvature ρ. These values are chosen to give the opportunity to keep the path well defined even at low speeds and standing still. Both the current and desired path are treated as public information within the vehicle since they are considered to be generic signals. Together with the force distribution, this is the Vehicle Motion Controller. Within this work a P-controller is used but the structure opens up for more advanced solutions as well.

3.3 Force distribution

The distribution of forces depends on the controllability and the number of wheel units. This can be considered as a linear control allocation problem

$$\mathbf{B}u(t) = v(t) \tag{1}$$

where $v(t)$ is the desired global forces F_x, F_y, M_z, $u(t)$ is the desired wheel unit forces $f_{x,i}, f_{y,i}$ and $\mathbf{B}$ is a 3×8 transformation matrix. This is similar to control allocation problems for flight control, see e.g. [7] for a good overview.

However, while aircrafts normally have to deal with componentwise rudder deflection limitations, vehicles equipped with tyres instead have nonlinear, coupled constraints due to tyre friction ellipses. In [8], circular constraints are replaced by polygons, allowing standard solvers to be used. Possibly, the problem can also be rewritten into a second order cone program [9] that can be solved by e.g. interior point method.

For ground vehicles, a nonlinear optimisation algorithm is suggested for the case where individual torque control can be applied [6], but a combination of individual steering and drive is not found by the authors. Within this paper, finding an optimal solution is not the main focus and thus a practical approach was chosen for the control allocation problem, that is carried out in a few steps.

Consider Figure 3, first the division of yaw torque M_z between lateral and longitudinal forces are done with a weighting function $k(F_x, F_y)$ such that $M_z = k\Delta F_x + (1-k)\Delta F_y$. The weighting function k is here realised by summation of two second order polynomials, shown in Figure 4.

$$k = \frac{1}{2}(a_x + b_x|F_x| + c_x F_x^2 + a_y + b_y|F_y| + c_y F_y^2)$$

$$a_x = 0.5, \ b_x = (1+\sqrt{2})/F_{xmax}, \ c_x = -0.5b_x^2 \tag{2}$$

$$a_y = 0.5, \ b_y = -(1+\sqrt{2})/F_{ymax}, \ c_y = 0.5b_y^2$$

The idea is to avoid using forces that are near saturation so k should be small for high F_x. Also very low F_x should generate small k since it otherwise requires

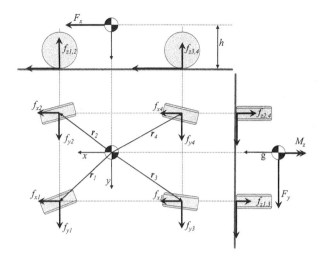

Figure 3: Vehicle model used for force distribution.

reversed forces left-right or front-rear. For mid range F_x, k is instead maximised, see Figure 4, left. For F_y, $(1-k)$ is instead considered, making k antisymmetric around $F_x/F_{xmax} = F_y/F_{ymax}$. The polynomials are used to give a smooth behaviour but of course, any other appropriate mapping could be chosen for this purpose.

Then, left and right longitudinal forces as well as the front and rear lateral forces are calculated according to:

$$\frac{(r_1 + r_3)}{2} \times (f_{x1} + f_{x3}) + \frac{(r_2 + 4)}{2} \times (f_{x2} + f_{x4}) = \Delta F_x$$
$$(f_{x2} + f_{x4}) + (f_{x1} + f_{x3}) = F_x$$
$$\frac{(r_1 + r_2)}{2} \times (f_{y1} + f_{y2}) + \frac{(r_3 + r_4)}{2} \times (f_{y3} + f_{y4}) = \Delta F_y$$
$$(f_{y1} + f_{y2}) + (f_{y3} + f_{y4}) = F_y$$

$$(3)$$

The last step is to decide the internal distribution to each wheel and it is done to distribute the force reserve at each wheel evenly. The maximum contact force for each wheel unit is assumed to be a function of ground conditions and the normal force. Thus, the distribution of the longitudinal forces between front and rear are defined by the lateral force reserve at the front and rear respectively. The lateral forces are distributed correspondingly between left and right.

3.4 Wheel Unit control

At each wheel unit, the desired forces have to be transformed into proper steering angles δ_i and wheel spin velocities ω_i. This is not done in the vehicle motion

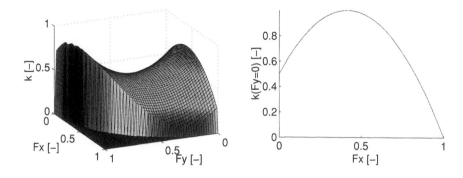

Figure 4: Normalised weighting surface $k(F_x, F_y)$, left, and $k(F_x, F_y = 0)$, right.

control since it requires that the vehicle has knowledge about each tyre's properties and thus $(\delta, \omega)_i$ cannot be generic.

Assuming that the tyre properties can be found by the wheel unit itself, there are two main problems to overcome when calculating $(\delta, \omega)_i$ from $(f_x, f_y)_i$. These are the nonlinearities of the tyre and the nonlinear transformation from chassis frame to wheel frame $\mathbf{T}(\delta_i)$.

The tyre nonlinearities are handled by approximating the tyre characteristics with a polynomial

$$f(s) = bs + cs^2 + ds^3 \tag{4}$$

where s is the magnitude of the slip and f the force respectively. The coefficients can be calculated from the Magic Formula [10] parameters according to

$$b = BCD$$
$$c = -1/2(BCD + 3D(BCs_{max} - 2)/s_{max})/s_{max} \tag{5}$$
$$d = D(BCs_{max} - 2)/s_{max}^3$$

In Figure 5, a comparison is shown between a variety of tyre properties generated by the Magic Formula and the corresponding polynomials.

The advantage with this representation is that $f(s)$ is easily invertible using Cardanus' formula. Since $f(s) \leq D$ for all relevant s there are always three real solutions to s of which the following is the proper one:

$$s = -\frac{1}{12d}Q_1^{1/3} + \frac{-c^2 + 3db}{3dQ_1^{1/3}} - \frac{1c}{3d} + \frac{1}{2}\sqrt{-3}\left(\frac{1}{12d}Q_1^{1/3} + \frac{-c^2 + 3db}{3dQ_1^{1/3}}\right)$$
$$Q_1 = 36bcd + 108f(s)d^2 - 8c^3 + 12\sqrt{3}Q_2d \tag{6}$$
$$Q_2 = \sqrt{-c^2b^2 + 4db^3 + 18bcdf(s) + 27f(s)^2d^2 - f(s)c^3}$$

Once the slip magnitude s is found, the components are calculated as $(s_x, s_y) = \frac{s}{f}(f_x, f_y)$. However, due to the steering angle the following nonlinear relation

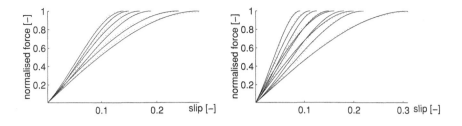

Figure 5: Comparison of generated polynomial (solid) with corresponding Magic Formula characteristics (dotted), these are generated with B between 3 and 6, C between 1.4 and 1.8 while D and E are kept constant at 1 and -20, respectively.

between the slips expressed in chassis frame $\bar{s}_i = (s_x, s_y)_i^T$ and wheel frame $\bar{s}_{iW} = (\kappa, \alpha)_i^T$

$$\bar{s}_i = T(\delta_i)\bar{s}_{iW} = \begin{pmatrix} \cos\delta_i & -\sin\delta_i \\ \sin\delta_i & \cos\delta_i \end{pmatrix} \bar{s}_{iW} \qquad (7)$$

Since $\alpha_i = \beta_i - \delta_i$, it is difficult to find δ_i from equation 7. Instead $\mathbf{T}(\delta_i)$ is linearised around $\delta_i = \beta_i$ which corresponds to zero tyre side slip, giving

$$T(\delta_i)|_{\delta_i=\beta_i} = \begin{pmatrix} \cos\beta_i & -\sin\beta_i \\ \sin\beta_i & \cos\beta_i \end{pmatrix} + \begin{pmatrix} \sin\beta_i & \cos\beta_i \\ -\cos\beta_i & \sin\beta_i \end{pmatrix}\alpha_i \qquad (8)$$

This is relevant as long as the maximum side force is generated at small angles. To improve accuracy for larger slip angles, δ_i from previous time step can be used.

Equation 8 is now used to solve equation 7 giving a system of polynomial equations with the solution

$$\alpha_i = \frac{1}{6}\left(108Q_4 + 12Q_3\right)^{1/3} - 2\frac{(-1 + Q_5)}{Q_5\left(-108Q_4 + 12Q_3\right)^{1/3}}$$

$$\kappa_i = Q_5 - \alpha_i^2$$

$$Q_3 = \sqrt{-12Q_5^3 + 36Q_5^2 - 36Q_5 + 12 + 81Q_4^2} \qquad (9)$$

$$Q_4 = -\sin\beta_i s_{xi} + \cos\beta_i s_{yi}$$

$$Q_5 = -\cos\beta_i s_{xi} - \sin\beta_i s_{yi}$$

Finally, the desired steering angle and wheel speed are calculated as

$$\omega_i = \frac{v_{xiW}}{R_e}(1 + \kappa_i)$$

$$\delta_i = \beta_i - \alpha_i \qquad (10)$$

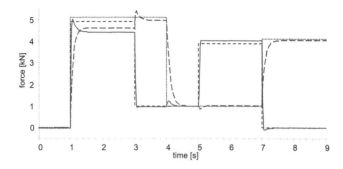

Figure 6: Desired and achieved forces for one wheel unit mounted on a test rig with predefined motion. Desired f_x (dotted), desired f_y (dashed), actual f_x (long-dashed) and actual f_y (solid).

4 SAMPLE SIMULATIONS

As a test bench for the control architecture for driving dynamics, a vehicle model in Modelica [11] is built according to Figure 1. Driver Interface, Driver Interpreter, Energy Management and Auxiliary Systems are made simple to facilitate evaluation of the results. The path control and the force distribution described above are implemented in the Vehicle Motion Control. The wheel units are realised with the controller suggested above together with a mechanical linkage suggested for autonomous corner modules presented in [12], steering actuators and wheel motors are modelled as first order filters with rate limits. To illustrate the ability of the implemented structure, two example simulations are shown, more information about the model can be found in [13].

In the first simulation, one WU is forced along a pre-defined path and is commanded to generate forces in series of steps(f_x, f_y). An available tyre model [14] with dynamics was used and the tyre characteristics was estimated separately. The simulation result is shown in Figure 6. During time=1-3s, the WU is commanded to generate more force than possible and the actual force is thus downscaled. Due to the linearisation in equation 8, actual f_y is sligthly too high for large slips as seen when time=5-7s.

The second simulation is of a full vehicle model following a lane change path, Figure 7. At this stage, only a simple P-controller is used to follow an intended path. In figure 8 a screen shot from the animated result is shown.

5 CONCLUSIONS AND FUTURE WORK

- A generic control architecture for driver's intentions to vehicle motion as described in Figures 1 and 2 has been implemented.

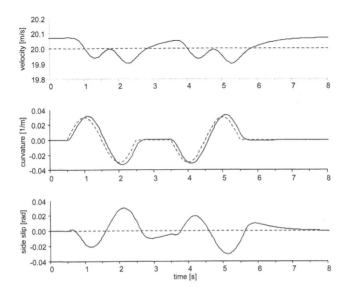

Figure 7: Desired (dashed) and achieved (solid) path for a vehicle with four wheel units.

- Forces are used as control signals between Vehicle Motion Control and Chassis which have been proven to be generic.

- A practical approach is utilised to solve the control allocation problem of distributing global forces to local wheel unit forces.

- A strategy to find wheel angles and wheel spin from desired wheel forces has been suggested and implemented.

The work is intended to continue with the following aspects in mind: 1) As suggested in section 3.3, the control allocation is not optimised. It will be further examined whether available control allocation theory can be used to improve performance of the force distributor. Especially when taking into account the dynamic limitations of tyres and actuators 2) The tyre characteristics has to be estimated continuously onboard the vehicle. Possibly, the friction circle assumption and the polynomial approximation must be refined to handle all types of tyre characteristics. 3) Today's cars have constraints between the wheels such as rack steering, differentials etc. These restrict the wheels' motion and thereby the force generation. This must be handled by the force distribution in a proper way. 4) Currently, the sensor information used is always accurate. It has to be examined how inaccuracies affects the performance.

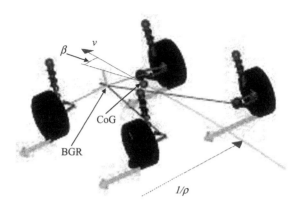

Figure 8: The animated vehicle with wheel unit linkages according to [12]. Vectors at the wheels indicate the generated tyre forces. Side slip angle (β), velocity (v), path curvature (ρ), Body Geometric Reference (BGR) and Centre of Gravity (CoG) are also indicated.

REFERENCES

1. S. Sakai and Y. Hori. Advanced vehicle motion control of electric vehicle based on the fast motor torque response. *Proceedings of the 5th International Symposium on Advanced Vehicle Control, Ann Arbor, Michigan, August 22-24*, 2000.
2. M. Yamamoto. Active control strategy for improved handling and stability. *SAE transactions, paper 911902*, pages 1638–1648, 1991.
3. L. Laine and J. Andreasson. Generic control architecture applied to a hybrid electric sports utility vehicle. *To be presented at the 20th International Electric Vehicle Symposium, Long Beach, CA, november 15-19*, 2003.
4. A. Phillips. Functional decomposition in a vehicle control system. *Proceedings of the American Control Conference, Anchorage, AK May 8-10*, 2002.
5. E. Coelingh and et al. Open-interface definitions for automotive systems. *SAE transactions, paper 2002-01-0267*, 2002.
6. Y. Hattori and et al. Force and moment control with nonlinear optimum distribution for vehcile dynamics. *Proceedings of the 6th International Symposium on Advanced Vehicle Control, Hiroshima, Japan, September 9-13*, 2002.
7. M. Bodson. Evaluation of optimization methods for control allocation. *AIAA Guidance, Navigation, and Control Conference and Exhibit, Montreal, Canada, August*, 2001.
8. L. Lindfors. Thrust allocation method for the dynamic positioning system. *10th International Ship Control Systems Symposium (SCSS'93), Ottawa, Canada*, 1993.
9. M. Akerblad and A. Hansson. Efficient solution of second order cone program for model predictive control. *IFAC 2002, July* 2002.
10. E. Bakker, H.B. Pacejka, and L. Lidner. A new tire model with application in vehicle dynamics studies. *SAE transactions, paper 890087*, pages 83–93, 1989.
11. Modelica Association. http://www.modelica.org.
12. S. Zetterstrom. Electromechanical steering, suspension, drive and brake modules. *VTC 2002-Fall, Vancouver, Canada, September 24-28*, 2002.
13. J. Andreasson and L.Laine. Implementation of a generic hybrid electric vehicle library in Modelica. *To be submitted*, 2004.
14. J. Andreasson and J. Jarlmark. Modularised tyre modelling in Modelica. *Proceedings of Modelica'2002 Conference*, 2002.

Experiment and analysis for improvement of curving performance with friction control between wheel and rail

YOSHIHIRO SUDA[1], TAKESHI FUJII[2], TAKASHI IWASA[2],

HISANAO KOMINE[2], MASAO TOMEOKA[3], KOSUKE MATSUMOTO[3],

NOBUYUKI UBUKATA[3], MASUHISA TANIMOTO[4], MACHI NAKATA[5],

TAKUJI NAKAI[5]

SUMMARY

This paper presents experimental and analytical results for improvement of curving performance with friction control between wheel and rail. Controlling the friction between wheel and rail is direct and very effective measure to improve the curving performances of railway trucks, because the curving performances depend much on friction characteristics. Authors have proposed a method, "friction control", which utilizes friction modifier with onboard spraying system. With the method, not only friction coefficient, but also friction characteristics can be controlled as required. In this paper, results of simulation and experiment with two roller machine, 1/10 scaled model vehicle and actual vehicle are reported which play an important role to realize the new method.

1. INTRODUCTION

In Japan, compatibility between high-speed stability and curving performance is a severe task ahead of urban railway that has high-speed section in suburbs and so many tight curves in urban area. On the view point of curving performance, kinds of problems related to wheel / rail contact are observed; increasing lateral force acting between wheel and rail, occurrence of squeal noise, progress of rail / wheel wear, corrugation on the top of rail and so on.

Most of the train operator companies have taken a measure, lubrication, that applying grease on the gage corner of high rail to prevent the wear of the rail/wheel or loud flange noise, and on the top of the low rail to decrease the lateral force or prevent the occurrence of corrugation. Although lubrication between wheel and rail is a traditional method to resolve these problems, it has a restriction of place to apply because the friction coefficient of grease is excessively low. That means, in

[1] *Address correspondence to:* Center for Collaborative Research & Center of Institute of Industrial Science, The University of Tokyo, Tel: +81-3-5452-6193, Fax:+81-3-5452-6194 suda@iis.u-tokyo.ac.jp

[2] Institute of Industrial Science, The University of Tokyo

[3] Teito Rapid Transit Authority

[4] Sumitomo Metal Technology Inc.

[5] Sumitomo Metal Industries, Ltd.

the case that grease is applied to the track just before a station, it contains a fear of skid in braking that gives severe damages to the wheel and rail[1-4]).

In this study, authors have proposed a method "friction control" between wheel and rail with friction modifier (KELTRACK™ HPF [5]) to improve curving performances in tight curves. Friction modifier can maintain appropriate friction coefficient between wheel and rail, and has appropriate friction coefficient and positive creep characteristics as described in section 2. To realize the concept of friction control, authors selected a system with onboard type device of train, in which liquid with friction modifier is sprayed onto the top of low rail as mist from the nozzle behind the last axle of train, and the following trains receive the benefit of it (Fig. 1).

The advantages of this onboard friction control system are as follows: the friction modifier can be supplied onto rail uniformly through the curve by control of its amount in proportional to the vehicle's velocity, the train with friction control device can supply the friction modifier immediately to any curve of the service line where wheel/rail contact condition should be improved unexpectedly, and the supply of friction modifier or maintenance of the device could be performed in equipped place of a car depot not at track.

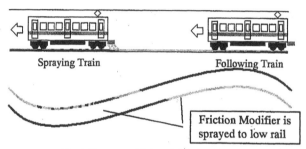

Fig. 1 Concept of Onboard Friction Control

Fig. 2 Field Running Test with Commercial Train

Figure 2 shows the photo of commercial train equipped onboard friction control system running through a test curve section with spraying friction modifier.
In this report, results of experiment and numerical analysis are reported. First, to evaluate the friction characteristics of friction modifier, authors carried out fundamental experiments with two-roller-rig machine focused on the creep characteristics between wheel and rail. Then, for the purpose of recognizing the effect of friction control, authors demonstrated numerical analyses regarding the characteristics of friction modifier, which is obtained by the experiment with two-roller-rig machine, utilizing the multibody simulation software modified from A'GEM[6]. Finally, in order to evaluate the effect of friction control, experiments with 1/10-scaled model vehicle were carried out, in which one-bogie model was selected for the purpose of simplifying the system of model.

2. FUNDAMENTAL TEST

Because the characteristics of friction modifier are very important in the method of friction control that is proposed by authors, experiments with two-roller-rig machine was carried out to recognize the fundamental characteristics of friction modifier.

2.1 Two-Roller-Rig Testing Machine
The two-roller-rig testing machine (shown in Fig. 3) was developed to evaluate the creep characteristics under various contact conditions between wheel and rail, such as dry, wet and lubricated by conventional grease. Especially, the creep force characteristics of friction modifier applied to the contact region were investigated.

Fig. 3 Appearance of Two-Roller-Rig Testing Machine

The testing machine can produce small slip between two rollers as required, and creep force generated by the slip can be measured very accurately. The diameters of both two rollers are 172 mm. Profile of wheel-roller is cylindrical and that of rail-roller has 100 mm convex single arc. These profiles of two rollers are designed considering the contact face ellipse and the contact pressure to model the actual wheel/rail contact.

The following procedures to analyze the relationship between creep force and slip-rate were made. First, two rollers are driven with given rolling speed and contact pressure. Then, controlling the differential gear mechanically produces required slip. As the result, the torque force of the rail-wheel can be measured in real time, and creep characteristics between wheel/rail roller are acquired with these measured data.

2.2 Test Procedure

To evaluate the creep force characteristics of friction modifier, the following test procedures were conducted; 1) The surface of each rollers are ground by sandpaper. 2) Two rollers are rotated with a certain creep rate without friction modifier until the coefficient of traction becomes approximately constant that means the conditions of two rollers surface becomes stable. 3) Friction modifier is applied on the surface of the rail-roller to evaluate creep force characteristics.

The spray nozzle, which was used in tests with commercial train in practical, was used to spray friction modifier to rail roller as a mist, and principal parameters to spray are described in Table 1.

Table 1 Parameters to Spray Friction Modifier

Friction Modifier	KELTRACK™ HPF SPRAY
Air Pressure	200 kPa
Height of Nozzle	0.3 m
Spray Time	2.72 sec
Diameter of Rail / Wheel Roller	0.172 m
Radius of the Top of the Rail Roller	0.1 m
Tangential Velocity of Contact Position between Rail / Wheel Roller	3.15 m/s

2.3 Creep Characteristics of Friction Modifier

To evaluate the creep characteristics of friction modifier, fundamental experiments with two-roller-rig machine were carried out. In the experiments, slip rate was continuously changed from 0 to 2.0 % in a test, even though contact force and rolling velocity were kept 200 N and 350 rpm constant. The test result of dry condition and friction control are shown in Fig. 4. With the results, creep coefficient of friction modifier is appropriate for curving compared to the one of dry condition, and creep characteristics of friction modifier is positive against creepage.

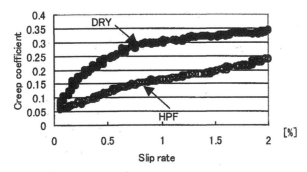

Fig. 4 Creep Force Characteristics

2.4 Effective Duration of Friction Modifier

To evaluate the effective duration of friction modifier for different slip rate, long time tests are conducted. The test result is shown in Fig. 5.

The time to reach the peak value of coefficient of traction is shorter, as the given slip rate becomes larger. It means that consumption (or required amount) of friction modifier depends on slip rate (creepage). That is to say, various elements related to slip rate, i.e., radius of curvature, velocity of the vehicle etc., must be considered to decide the amount of friction modifier to be sprayed.

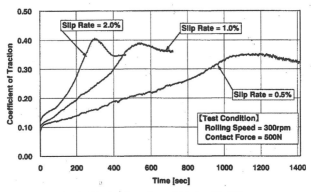

Fig. 5 Effective Duration of Friction Modifier

3. Simulation Results

To confirm experimental results with friction control, numerical analyses were carried out with the multibody simulation software modified from A'GEM to evaluate sufficiently the effect of friction-control. Simulation parameters and creep force characteristics were set to match with the experimental ones performed in chapter 2.3. The results of the analyses are shown in Table 2 and Fig. 6. With the result of simulations, lateral force at outside wheel of the leading axle is decreased drastically with the effect of friction control.

Table 2   Results of numerical analyses

Frictional condition		DRY	Friction Control
Q/P value observed in inner rail		0.377	0.318
Lateral force at outside wheel of leading axle [N]		71.3	58.0
Attack angle [deg]	Leading	0.912	1.008
	Trailing	0.042	0.057
Lateral displacement [mm]	Leading	9.57	9.58
	Trailing	5.66	4.71
Revolution of axles [rev]	Leading	19.2	19.0
	Trailing	19.2	19.3

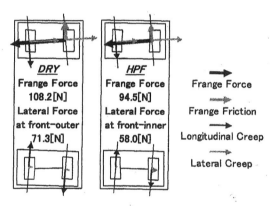

Fig. 6   Simulation results with illustration about creep force for a bogie

3. 1/10-SCALED MODEL VEHICLE TESTS AND SIMULATION

3.1 1/10-Scaled Model Vehicle Experimental Platform

In order to evaluate the effect of the proposed method, running test was performed on experimental equipment as shown in Fig. 7 and 8. This equipment was manufactured with scaled down from the actual railway system for 1/10 size and has a tight curve (3.3m radius). Model vehicle imitates a one bogie model without any motive power itself, i.e., the vehicle run through the curve by the force of inertia. Some sensors are put on the vehicle and rails so that various dynamical properties can be measured when the train passes there.

Conditions of the series of tests are described in table 3. The amount of friction modifier to be sprayed was set to match with the one for two-roller-rig tests by controlling the air pressure for spraying, height of spraying nozzle and spraying time.

Fig. 7 1/10-scaled model vehicle experimental platform

Fig. 8 Model Vehicle

Table 3   Condition of Experiments

Friction Modifier	KELTRACK™ HPF SPRAY
Mass of the Experimental Equipment	80 kg
Air Pressure	200 kPa
Height of Nozzle	0.4 m
Spray Time per rail length	2.5 s/m
Diameter of Model Vehicle's Wheel	0.086 m
Profile of the Top of the Rails	10 1/m Circular
Velocity of the Model Vehicle	Approximately 1.0 m/s
Temperature	23 to 26 deg
Humidity	54 to 76 %RH

3.2 Result of the Model Vehicle Tests

Some series of running test with the experimental platform were carried out. Experimental results are shown in Fig. 9 to 13. Before the running test 21st and 51st, the friction modifier was applied to the top of low rail through the curve. As the effect, the car behavior had changed drastically as shown in Fig. 9.

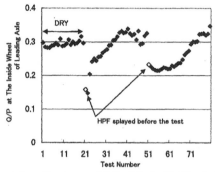

Fig. 9 Experimental results: change of Q/P at The Inside Wheel of Leading Axle

Figure 9 shows the changes of Q/P (lateral force divided by wheel load) at inside wheel of leading axle through the series of tests measured by strain-gage put on the rails. As usual, Q/P at the inside wheel shows the friction coefficient, because flange of inside wheel doesn't contact on rail. With this result, Q/P, i.e., friction coefficient is obviously decreased just after the friction modifier was sprayed. However, as the tests going on after the 21st test, Q/P had been gradually increased over the level of DRY condition, and with the effect of the spray before the 51st test, the Q/P had been decreased again. This result corresponds to the results of fundamental experiments by two-roller-rig machine. This result points out the importance of the timing to spray in order to obtain the effect of friction modifier.

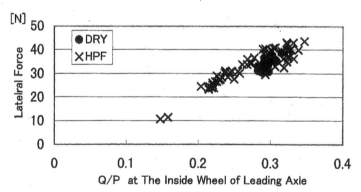

Fig. 10 Relationship between Lateral Force and Friction Coefficient

Figure 10 shows the relationship between Q/P at inside wheel and lateral force, which is dominated flange contact force. With the effect of decrease of friction coefficient, the lateral force of outside wheel at the leading axle was reduced drastically.

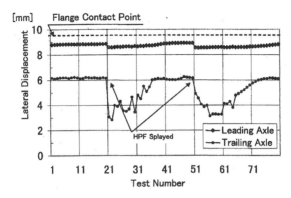

Fig. 11 Change of Lateral Displacement of Each Axle

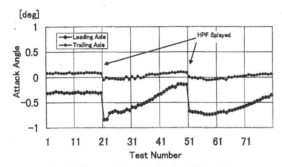

Fig. 12 Change of Attack Angle of Each Axle

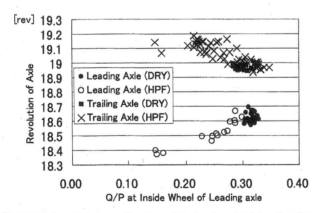

Fig 13 Relationship between Rotations of Each axle and Friction Coefficient

Figure 11 and 12 show the change of displacement and attack angle of each axle. With the effect of friction modifier, the lateral displacement of trailing axle was decreased and the attack angle of leading axle was increased as shown in figures. This means that friction control effects not only on lateral force but also on posture of bogie. It indicates that the friction control changes the balance of moment on bogie.

Figure 13 shows the relationship between rotations of each axle and Q/P at the inside wheel. This is very curious phenomenon that the friction control enlarges the differences of rotations between leading axle and trailing axle. This result indicates the possibility that differences of rotations could be a parameter of the friction control.

4. CONCLUSION

The friction control between wheel and rail was proposed to improve curving performance and experiments were made. The following results were obtained:

(1) It was found that the friction modifier has appropriate creep coefficient for curving and positive creep characteristics against the creepage from the results of fundamental experiment with two-roller-rig machine.

(2) The consumption of friction modifier depends on creepage.

(3) The effectiveness on curving performance of friction modifier sprayed to the top of low rail was confirmed by the numerical analyses and 1/10-scaled model vehicle tests.

(4) The phenomenon was recognized in model tests with 1/10-scaled model vehicle that the effect of friction modifier disappears with the number of wheel/rail contacts, which are seen in field tests with commercial train.

REFERENCES

[1] Tomeoka, M., Kabe, N., Tanimoto, M., Miyauchi, E., Nakata, M.: Friction Control between Wheel and Rail by means of on-board lubrication, *Wear 253 (2002)* pp. 124-129

[2] Suda, Y., Komine, H., Iwasa, T., Tomeoka, M., Matsumoto, K., Ubukata, N., Tanimoto, M., Nakata, M., Nakai, T. : Imrovement of Curving Performance with Friction Control between Wheel and Rail, *the 17th IAVSD Symposium Poster Session*, Lyngby, Denmark, August 2001.

[3] Suda, Y., Iwasa, T., Komine, H., Tomeoka, M., Nakazawa, H., Matsumoto, K., Nakai, T., Tanimoto, M., Kishimoto, Y. : Development of Onboard Friction Control System, *the 6th International Conference of Contact Mechanics and Wear of Rail/Wheel Systems (CM2003)*, Gothenburg, Sweden, June 2003. pp.321-326

[4] Suda, Y., Iwasa, T., Komine, H., Fujii, T., Matsumoto, K., Ubukata, N., Nakai, T., Tanimoto, M., Kishimoto, Y. : The Basic Study on Friction Control between Wheel and Rail (Experiments by Test Machine and Scale Model Vehicle), *the 6th International Conference of Contact Mechanics and Wear of Rail/Wheel Systems (CM2003)*, Gothenburg, Sweden, June 2003. pp.343-348

[5] Eadie, D., Kalousek, J., Chiddick K.: The Role of High Positive Friction (HPF) Modifier in the Control of Short Pitch Corrugations and Related Phenomena, *Proc. the 5th International Conference of Contact Mechanics and Wear of Rail/Wheel Systems (CM2000)*, Tokyo, Japan, July 2000. pp.42-49.

[6] Anderson, R. : A'GEM Rail Vehicle Dynamics Software Package User's Manual,1994.

Vehicle System Dynamics Supplement 41 (2004), p.517-526 © Taylor & Francis Ltd.

Optimisation of railway wheel profiles using a genetic algorithm

PERSSON I.[1] AND IWNICKI S.D.[2]

SUMMARY

This paper presents the procedures and preliminary results of a novel method for designing wheel profiles for railway vehicles using a genetic algorithm. Two existing wheel profiles are chosen as parents and genes are formed to represent these profiles. These genes are mated to produce offspring genes and then reconstructed into profiles that have random combinations of the properties of the parents. Each of the offspring profiles are evaluated by running a computer simulation of the behaviour of a vehicle fitted with these wheel profiles and calculating a penalty index. The inverted penalty index is used as the fitness value in the genetic algorithm.

The method has been used to produce optimised wheel profiles for two variants of a typical vehicle, one with a relatively soft primary suspension and the other with a relatively stiff primary suspension.

1. BACKGROUND

The selection of railway wheel and rail profiles is a challenge that has faced engineers since the dawn of the railway age. From the first cylindrical wheels running on flat plates, wheels were made conical to give better guidance and flanges were added for safety. Modern wheels often have complex profiles based on the shape of worn wheels in an attempt to improve their life. Rails also now have a complex profile with different radii on the rail head where the wheel tread contacts and on the corner where the flange contacts.

A high level of conicity will allow good curving behaviour even in the tightest curve without flange contact. This can however, lead to a relatively low critical speed and possibly dangerous hunting instability. A low level of conicity on the other hand will allow stable operation at high speeds but the flangeway clearance will quickly be used up in curves, resulting in flange contact and possible flange climb derailment. Flange angle and root radius are also variables that can have a significant effect on the possibility of derailment. In addition to the vehicle behaviour engineers must consider the stresses on the wheel and on the rail. These have a major influence on the development of rolling contact fatigue which can have expensive and sometimes dangerous consequences.

The methods used to design these profiles have been mainly based on the experience of railway operators using various rules of thumb regarding conicity, flange angle, flange root radius etc [1],[2].

---

[1]DESolver, Optand 2876, S-831 Ostersund, Sweden
[2]Manchester Metropolitan University, Chester Street, Manchester M1 5GD, UK

In nature these complex design problems are often faced and solutions found, sometimes of astonishing complexity, effectiveness and efficiency. In this work the same evolutionary principles are used to obtain profiles optimised for performance on typical vehicles and track. Genetic algorithms have been used in other engineering applications including optimisation of suspension components and have given useful results [3].

2.   THE GENES

The cross sectional profile of the wheel is initially described with a series of x,y coordinates and these are converted into a binary sequence. Several methods have been tried for producing the binary sequence including direct conversion of the cartesian coordinates of the profile into a series of binary numbers and a fourier transform of the profile description into a series of harmonic components but the best results so far are from a series of consecutive derivatives of the coordinates. From the nominal running circle to the inside of the wheel the 4th order derivative of the profile, calculated every 0.5mm, is used. On the flange side of the nominal running circle the 1st order derivatives are used. These numbers are converted into binary and joined together to form the gene. The profiles are digitised at 0.5mm spacing but in reforming the profiles spline interpolation is used to increase the number of data points to the 0.1mm used in Gensys.

The genes for the two parent profiles are mated by taking random sections from each to make a child. The children will represent different profiles to the parent but will share similar characteristics. Mutations are also made by randomly changing the genes to introduce occasional larger variations and are used to avoid local minima in the optimisation process.

3.   THE SIMULATION

The selected wheel profile is incorporated into a GENSYS [4] simulation model of a simple motored bogie vehicle with an axle load of 20 tonnes. The vehicle bodies are assumed to be rigid and the main primary and secondary suspension stiffness is linear. The vehicle has vertical primary dampers as well as secondary lateral, vertical and yaw dampers. Traction rods and anti-roll bars are included in the model and the yaw dampers have blow off valves and include series stiffness. The nominal wheel diameter is 1m.

For these tests two versions of the vehicle were set up, one with soft primary suspension and the other with a stiffer primary suspension and no yaw damper. The main vehicle parameters are shown in table 1.

The track selected for the tests was a section of Swedish main line (class K1). The vehicle was run at 145 km/h on straight track for 275m then into a 120m linear transition onto a curve of 800m radius and cant of 150mm (130mm cant deficiency). The rails are inclined at 1:40 and the track standard according to CEN/TC 256WG at 145 km/h is QN2. Measured track irregularities are included and the average gauge is 1430.76. After running for 227m around the curve the simulation was stopped and the results used to calculate the penalty index.

4.   THE PENALTY INDEX

In order to evaluate the effectiveness of each profile a penalty index has been introduced and is calculated after each simulation run. The aim of the penalty index is to provide an assessment of the vehicle behaviour in a single value.

In the current work factors have been included in the penalty index to reflect the maximum contact stress; the maximum lateral force on the track; the maximum derailment quotient (lateral force divided by vertical force at the contact point); total wear and the total ride index. The penalty index is the sum of the individual penalty factors.

Each of the factors that make up the penalty index can be weighted to reflect their importance to the designer or the particular service type. Other factors could also be included in the penalty function if they were felt to be important.

4.1.   RIDE COMFORT PENALTY FACTOR

The ride comfort penalty factor consists of the sum of the filtered RMS accelerations weighted according to ERRI Question B153 at the floor of the vehicle over both bogies. A lateral and a vertical value of the ride comfort penalty factor is calculated.

4.2.   LATERAL TRACK-SHIFTING FORCE PENALTY FACTOR

The lateral track force at each wheel is first filtered with a sliding 2m window and then the maximum value for all axles is taken. A non linear weighting is applied to this force to establish the penalty factor. If the maximum track-shift force is less than 40% of the permissible value (the Prud'Homme limit) the penalty factor will be 0. If the maximum track-shift force is between 40-90% of the permissible value, the penalty factor will increase linearly to 2.0, if the maximum track-shift force is over 90% of the permissible value the penalty factor increases to 100 at 100%.

4.3.   MAXIMUM DERAILMENT QUOTIENT PENALTY FACTOR

The likelihood of derailment is linked to the widely used derailment quotient and this has been selected as one of the factors in the penalty index. The lateral force is divided by the vertical force at each contact point and this is filtered in a sliding 50ms window. The largest maximum value for the simulation is stored and is weighted in exactly the same way as the maximum track shifting force with the maximum permissible limit being taken as 0.8.

4.4.   WEAR PENALTY FACTOR

The level of wear at the wheel and rail surface is indicated by the energy dissipated in the contact patch and this is calculated by taking the product of creepage and creep force (including spin). A high value of energy here is seen as indicating a high rate of wear of the wheel profile and to be avoided. Some wear is however seen as beneficial as it will probably remove small cracks which develop through rolling contact. A very low level of wear is also undesirable as it probably indicates that wear is taking place over a very limited section of the profile and the shape of the profile will not then be stable. For these reasons the weighting on the wear energy dissipation at values between 80 and 120 Nm/m is set to 0 and not included in the wear penalty factor. Outside this range the value of the wear penalty factor increases linearly to 0.5 at 0. and at 200.

4.5.   MAXIMUM CONTACT STRESS PENALTY FACTOR

High levels of contact stress are likely to result in rolling contact fatigue or other rail or wheel damage. The evaluation of the maximum contact stress penalty factor takes the 99.85[th] percentile of the contact stress at all wheels (taken as the normal force divided by the contact area) and retains the maximum value. This value is divided by 1e9 N/m^2 to produce the normalised penalty value.

5.   SELECTION OF THE FITTEST PROFILES

Once the penalty index has been calculated for all the profiles the 'parents' for the next generation are selected. A random selection weighted by the penalty index is used to select the parents for the next generation and mating as described above is carried out. Four profiles are also retained into the next generation (using the same weighted random selection).

Simulations are carried out for the 'child' profiles and the process is repeated. In this way a chance mating which results in poorer performance than the parents will not damage the population excessively.

6.   RESULTS

6.1.   PROFILES

The genetic algorithm and assessment simulation was run until no further improvement was seen (110 generations for the vehicle with the soft bogie and 121 generations for the stiff bogie vehicle). The profiles produced are shown in figures 1 and 2.

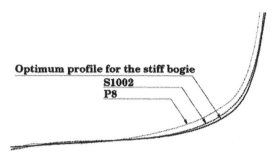

Figure 1. The profile produced by the genetic algorithm for the stiff bogie

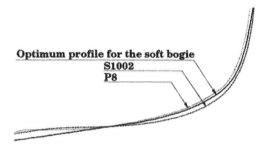

Figure 2. The profile produced by the genetic algorithm for the soft bogie

Figure 3. shows the conicity produced by each of the profiles. The P8 profile has been developed for rails with a 1:20 inclination and is steeper in the tread area than the S1002 profile, which is usually used on 1:40 or 1:30 rails. For the stiff bogie, the genetic algorithm has produced a wheel profile with a lower conicity than either the P8 or the S1002 profiles. The soft bogie has, in contrast, resulted in a wheel profile with a high conicity.

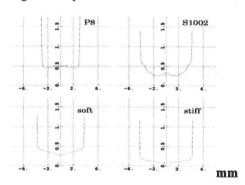

Figure 3. Conicity of the original and new profiles against lateral displacement of the wheelset

6.2. WEAR

Figure 4. shows the calculated wear index from the Gensys simulation as the vehicle runs over the virtual test track. The upper two plots show the results for the vehicle with the stiff primary suspension and the lower two are for the vehicle with the soft primary suspension. The first and third plots show the results with the S1002 wheel profiles and the second and fourth profiles are with the best profile from the genetic algorithm for the specific vehicle as presented in section 5.4. above. The solid line shows the results for the outer leading wheel on bogie 1 and the dashed line shows the results for the outer leading wheel on bogie 2. Table 2 summarises the wear penalty factors for the different cases.

Table 2. Wear penalty factor

Wear penalty factor	Stiff bogie	Soft bogie
Wheel profile S1002	3.68	0.509
Best wheel profile from Genetic Algorithm	3.24	0.306

It can be seen that the profiles produced by the genetic algorithm give reduced wear for both vehicles. The soft bogie already has better wear performance but one significant peak has been eliminated and the reduction in the wear penalty index is still significant.

6.3. CONTACT STRESS

Figure 5. shows the simulated contact stress for the s1002 wheel profile and the best profile from the Genetic Algorithm for both vehicles and table 3. shows the penalty factors. The solid line on the plot is the right wheel and the dashed line is the left wheel, both on the leading axle. For the vehicle with the stiff bogie there is an overall reduction in the contact stress in the curved section of the track with a corresponding 10% reduction in the penalty index. For the vehicle with the soft bogie some clear peak stress values have also been eliminated.

Table 3. Contact stress penalty factor

Contact stress penalty factor	Stiff bogie	Soft bogie
Wheel profile S1002	1.28	1.3
Best wheel profile from Genetic Algorithm	1.16	1.03

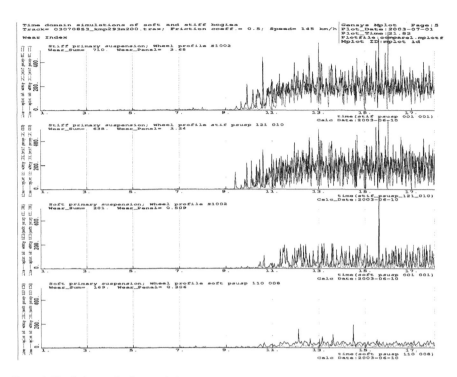

Figure 4. Simulation results for wear index

6.4.   DERAILMENT QUOTIENT

The derailment quotient penalty factor values are also shown in table 4. For the stiff vehicle with the S1002 profiles the derailment quotient is always below this limit and with the soft bogie there is only one small excursion above this lower limit. The Genetic Algorithm has eliminated the 1 excursion for the soft vehicle but has actually increased the penalty factor for the stiff vehicle to bring the overall level of the derailment quotient up towards the lower limit.

Table 4. Derailment quotient penalty factor

Derailment quotient penalty fator	Stiff bogie	Soft bogie
Wheel profile S1002	0	0.23
Best wheel profile from Genetic Algorithm	0.1	0.

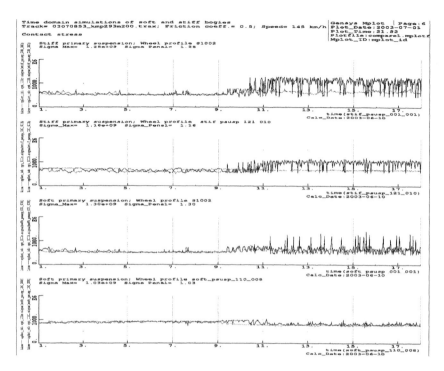

Figure 5. Simulation results for contact stress

6.5.   TRACK SHIFT FORCES

The results for the Lateral Track-Shifting force Penalty are shown in table 5.

Table 5. Lateral track-shifting force penalty factor

Lateral track-shifting force penalty factor	Stiff bogie	Soft bogie
Wheel profile S1002	0.989	0.
Best wheel profile from Genetic Algorithm	0.730	0.

For the stiff bogie the lateral forces have been reduced by using the new wheel profile and the maximum lateral track-shifting force is lower. The soft bogie does not generate lateral forces higher than 40% of the Prud'Homme limit.

6.6. PASSENGER COMFORT

The results for the ride comfort penalty are shown in table 6.

Table 6. Ride comfort penalty

Ride Comfort Penalty	Stiff bogie	Soft bogie
Wheel profile S1002	3.66	3.62
Best wheel profile from Genetic Algorithm	3.18	3.29

It can be seen that the best profile from the Genetic Algorithm gives an improvement in the vehicle ride in both cases. The improvement is greater for the stiff bogie.

7. CONCLUSIONS

This work has shown that a genetic algorithm can be used to optimise the wheel profiles for a typical railway vehicle running on main line track and that different profiles are produced for different vehicles. The method can be tuned to reflect the importance of various factors such as contact stress or ride comfort etc. The method could also be used to optimise rail profiles and the wheel/rail profile combination.

Further investigation is required to test the profiles produced by the genetic algorithm method. In particular the effect of wear of the wheel and the change this will cause to the profile must be understood. A possible initial alternative to practical testing may be to use the methods developed by Jendel and Berg [5] where 'load collectives' are used to model typical vehicle traffic including traction braking, curving etc on wear.

REFERENCES

1. *Esveld C. 'One procedure for optimal design of wheel profile' Proc. 1st Wheel-Rail Interface management conference, Amsterdam 2002, IQPC London*
2. *Shen G., Ayasse J.B., Chollet H. and Pratt I. 'A unique design method for wheel profiles by considering the contact angle function' Proc. Instn. Mech. Engrs. Vol.217 Part F. 2003*
3. *Mei T.X. and Goodall R. M. 'use of multiobjective genetic algorithms to optimise inter-vehicle active suspensions', Proc. Instn. Mech. Engrs. Part F 2002, 216, pp53-63*
4. *Persson I. 'Using GENSYS 0203' ISBN 91-631-3110-2 DEsolver 2002*
5. *Jendel T. and Berg M. 'Prediction of wheel profile wear methodology and verification' Proc. 17th IAVSD Symposium, Copenhagen 2001, ISBN 0042-3114*

Figure 6. Simulation results for derailment quotient

Parameter	Stiff bogie	Soft bogie	Units
PRIMARY SUSPENSION			
Lateral semi-spacing to primary suspension	1	1	m
Semi-wheelbase	1.5	1.5	m
Longitudinal primary stiffness	20e6	4e6	N/m
Vertical primary stiffness	1200e3	1200e3	N/m
Viscous damping in primary stiffness	2e3	2e3	Ns/m
Vertical damping	30e3	30e3	Ns/m
Vertical bump-stop clearance	0.05*	0.05*	m
Vertical bump-stop stiffness	20e6*	20e6*	N/m
SECONDARY SUSPENSION			
Lateral spacing to secondary suspension	1	1	m
Bogie centre spacing	16	16	m
Vertical secondary stiffness	1000e3	1000e3	N/m
Lateral secondary stiffness	600e3	600e3	N/m
Longitudinal secondary stiffness	600e3	600e3	N/m
Lateral bump stop clearance	0.09*	0.09*	m
Lateral bump stop stiffness	22e6*	22e6*	N/m
Lateral damping	40e3	40e3	Ns/m
Vertical damping	40e3	40e3	Ns/m
Yaw damping	0	500e3*	Ns/m
INERTIA PROPERTIES			
Bogie mass	10000	10000	kg
Centre of gravity above track	0.7	0.7	m
Bogie roll moment of inertia	3000	3000	kgm^2
Bogie pitch moment of inertia	10000	10000	kgm^2
Bogie yaw moment of inertia	15000	15000	kgm^2
Wheelset mass	2000	2000	kg
Centre of gravity above track	0.5	0.5	m
Wheelset roll moment of inertia	1200	1200	kgm^2
Wheelset pitch moment of inertia	200	200	kgm^2
Wheelset yaw moment of inertia	1200	1200	kgm^2
Body mass	52e3	52e3	kg
Centre of gravity above track	2	2	M
Body roll moment of inertia	80e3	80e3	kgm^2
Body pitch moment of inertia	1800e3	1800e3	kgm^2
Body yaw moment of inertia	1800e3	1800e3	kgm^2

(* non-linear cmponent, simplified linear properties given)

Investigation of vehicle dynamic influence on rolling contact fatigue on UK railways

J.R.EVANS[1] AND M.A.DEMBOSKY[2]

SUMMARY

This paper describes research undertaken to understand the influence of vehicle dynamic response in the formation of Rolling Contact Fatigue (RCF) in rails. The use of both quasi-static curving simulations and detailed dynamic simulations are described. Site-specific Vampire® dynamic simulations for a range of vehicles and wheel profiles are compared with the actual incidence of RCF at the sites. Good agreement was found between variations in the wheel/rail contact conditions and the presence of RCF through the site. This not only allows the underlying causes of the RCF at the specific sites to be identified with confidence, but also sheds valuable light on the mechanisms causing RCF on the UK rail network.

1. INTRODUCTION

Rolling Contact Fatigue (RCF) of rails has been an increasing problem on the UK rail network for a number of years. The aftermath of the Hatfield derailment of October 2000 crippled the UK rail network as many hundreds of miles of cracked rails were replaced, and RCF continues to develop at a high rate, in some cases even in the newly replaced rails.

The RCF occurs most commonly in turnouts and on the high rail in curves. In curves RCF is most severe in high-speed curves of moderate radius, and less common in tighter curves. RCF is also sometimes found on plain straight track.

This paper describes studies using vehicle dynamic simulation to understand the influence of vehicle dynamic behaviour on the occurrence of RCF and to establish the wheel/rail contact conditions which lead to RCF.

2. PARAMETERS OF ROLLING CONTACT FATIGUE

Two of the key parameters that are believed to drive the incidence of RCF are the contact stress and the tangential force in the contact patch. These parameters can be obtained most readily from a vehicle dynamic simulation; direct measurement is much more difficult and costly.

---

[1] AEA Technology Rail, 4 St Christopher's Way, Pride Park, Derby DE24 8LY, United Kingdom.
e-mail Jerry.Evans@aeat.co.uk

[2] Network Rail, 40 Melton Street, London NW1 2EE, United Kingdom.
e-mail Mark.Dembosky@networkrail.co.uk

RCF is considered to be a consequence of ratchetting in the rail surface, which is described by shakedown theory [1]. In this theory, tangential forces are normalised by dividing by the normal load to give the traction coefficient. The theory gives threshold levels of contact stress and traction coefficient above which RCF is predicted to occur. However, field evidence suggests that at high levels of creepage, tangential forces lead to wear rather than cracking, which is ultimately a less damaging condition.

3. QUASI-STATIC SIMULATIONS

In the first stage of the investigation, quasi-static curving studies were undertaken using VAMPIRE® to predict the contact stress, tangential forces and wear rates in idealised curves. This work was funded by Network Rail in the UK.

Almost 2500 simulations were undertaken to investigate the effect of curve radius, vehicle type, cant deficiency, friction, wheel profiles, rail profiles, track gauge and applied traction.

An example of a shakedown diagram for a selection of vehicles on a 1500m radius curve at 150mm cant deficiency, with 0.45 friction, is shown in Figure 1. Illustrative shakedown limit lines are shown are based on the work of Bower and Johnson [1] for two grades of rail steel. Up to four combinations of new and worn wheels on measured or design-case new rail profiles are plotted for each vehicle.

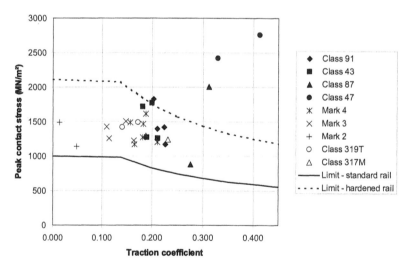

Fig. 1. Quasi-static shakedown diagram - 1500m curve.

Subsequent dynamic simulations have shown that track irregularities, which are ignored in the idealised quasi-static simulations, have an important effect on the relative ranking of different states of wheel wear and different vehicle types.

Nevertheless, valuable conclusions can be drawn from the quasi-static work:

- All the vehicles exceed the theoretical shakedown limit for the standard rail, while only a minority of curves of this radius suffer from RCF, suggesting that the real limit may be substantially higher, possibly due to work hardening.
- Traction coefficients and contact stresses are higher on tighter radius curves, but the incidence of RCF tends to be lower. This supports the view that high levels of traction coefficient and contact stress lead to wear rather than RCF.
- Older passenger vehicles such as the Mark 2 coach, with soft primary yaw suspension, are less damaging than modern vehicles.
- Although older locomotives such as classes 47 and 87 shown in Figure 1 can generate very high levels of contact stress and traction coefficient, this may lead to high wear rates rather than RCF.

4.   DYNAMIC SIMULATIONS

RCF is often found in clusters, separated by sections of crack-free rail. Sometimes the clusters are initiated by a weld or other discrete trigger in the track, but mostly they appear to be a result of longer wavelength dynamic response of the trains to the track geometry. The dynamic response of vehicles to changes in track geometry is also important in the formation of RCF in turnouts.

Study of adjacent sections of undamaged and damaged rail on the same line subject to the same traffic is very valuable, as it allows threshold levels of conditions at the wheel/rail interface to be investigated by dynamic simulation.

Detailed VAMPIRE® dynamic simulations were undertaken for a selection of vehicles running over two test sites. The results were then compared with the incidence of clusters of RCF. This work was funded by the UK Rail Safety and Standards Board, in conjunction with the First Great Western Pilot RCF project funded by the UK Wheel Rail Interface System Authority and Network Rail.

4.1 REPRESENTATIVE VEHICLES AND WHEEL PROFILES

Three vehicle types represent over 90% of the traffic over the chosen sites. These are the High Speed Train (HST) power car and trailer car, and an electric multiple unit (EMU) vehicle. The HST power car was modelled with traction applied.

All three vehicle types use the P8 wheel profile. However, when investigating RCF, the state of wear of the wheel profiles is a vital parameter.

Representative HST wheel profiles were selected from measurements of 191 trailer car and 56 power car profile pairs from vehicles operating on the route. Four profiles for each vehicle type were chosen: newly turned, average worn, most treadworn and most flangeworn.

The EMU vehicles had a consistent pattern of wheel wear with mileage. The wear was asymmetric, with the left wheels being significantly smaller than the right.

84 profile pairs were measured, from which four were chosen to be representative of trains with newly turned profiles, and after 80,000, 140,000 and 210,000 miles.

4.2 ACTON TEST SITE

The Acton test site included part of a 3220m radius right hand curve with 65mm cant, followed by part of the exit transition. Clusters of RCF were present well into the exit transition. Train speeds were typically 90 mile/h for HST and 100 mile/h for the EMU, giving cant deficiencies of 13mm and 31mm respectively.

63 pairs of Miniprof rail profile measurements and cant and gauge measurements were made through the site. Track alignment geometry was obtained from a recent track recording car run, and carefully synchronised with the site measurements by comparison with the manual cant and gauge measurements.

The variation of RCF severity was recorded through the site. The lateral position and angle relative to the rail of the cracks were also recorded at intervals.

Finally, paint band records were taken of the running bands of a number of trains of each type passing over the site. A typical example is shown in Figure 2, on which examples of the cracks can also be seen. The HST running bands were consistently closer to the crown of the rail than the RCF, while some, but not all, EMU running bands were on the shoulder of the rail where the RCF was found, as in Figure 2.

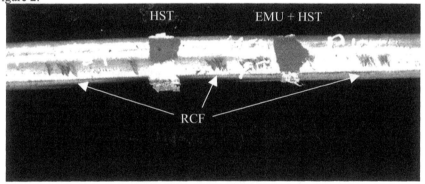

Fig. 2. Paint band rolling line records

4.2.1. Simulation Results from Acton

Twelve base simulations were undertaken, for the three vehicle types, each with four measured wheel profiles. A high coefficient of friction of 0.45 was used.

The simulation results were validated by comparison with the measured running bands, crack angles and variation of rail wear through the site.

Figure 3 compares the predicted contact positions (relative to the gauge face) for the HST power car with average worn wheels and the EMU with 140,000-mile worn wheels. Also shown are the lateral crack positions across the rail at various locations through the site. The contact position for the HST vehicle consistently misses the positions of the cracks. However, the very cyclic response of the EMU vehicle leads to contact in the cracked area, which is consistent with the measured running bands.

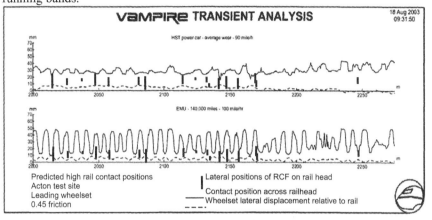

Fig. 3. Predicted contact positions at Acton

Figure 3 also shows the lateral motion of the wheelsets relative to the rail. The lateral motion of the HST power car wheel gives only a slightly larger change in contact position, while relatively small motion of the EMU wheel gives very large changes in contact position. This is a sign of unusually close conformality between the wheel and rail profiles. The jump in the contact position gives a step change in rolling radius, high longitudinal forces and hence RCF. The motion is self-exciting, as the effective conicity at the step in the rolling radius is too high for the wheelset to remain stable.

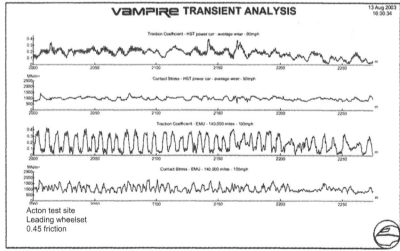

Fig. 4. Shakedown parameters at Acton test site

Figure 4 shows the traction coefficient and contact stress developed by the two vehicles. When the EMU contacts over the RCF, the traction coefficient and contact stress both exhibit peaks that are significantly higher than the HST power car.

Figure 5 shows the same data in the form of a dynamic shakedown diagram for the first 100m of the site. There is a clear difference between the conditions generated by the EMU which does cause cracking and the HST power car which does not. A notional boundary line has been drawn to separate the two cases. This is only an illustrative threshold above which RCF is likely to occur, and work continues to quantify the threshold more precisely.

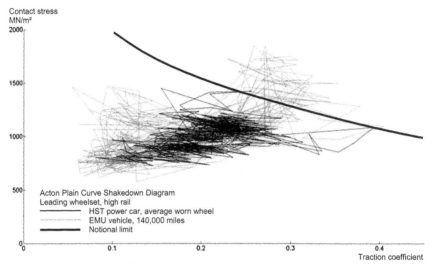

Fig. 5. Dynamic shakedown diagram at Acton

4.2.2 Causes and remedies of RCF at Acton

Having established good agreement between the base simulations and conditions on the ground, it was possible to run what-if scenarios to explore causes and remedies.

The RCF was found to result from closely-conformal worn wheel and rail profiles giving a limit-cycle instability. The behaviour was strongly influenced by the diameter differences between the wheels of the EMU vehicles forcing the wheels to run closer to the flange root – without this the wheels did not run in the conformal zone and forces and stresses were much lower. The particular worn shape of the left wheel, and the asymmetric treadwear were the consequence of wear on the low rail on a single sharp curve (of the opposite hand) further along the route of the EMUs.

The most promising solutions for this site would be to grind to maintain a new rail profile (but not an anti-RCF profile which is similar to the worn rail in this case), or to lubricate the low rail at the sharp curve to reduce the asymmetric wear. More frequent re-profiling of the wheels was ruled out as the problem started to appear after quite a low mileage. Turning the trains around was also not practical.

4.3 RUSCOMBE JUNCTION TEST SITE

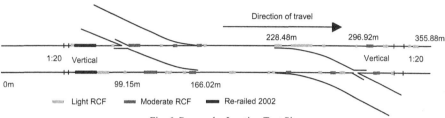

Fig. 6. Ruscombe Junction Test Site

The site covers a trailing and a facing turnout on straight track, as shown in Figure 7. The HST is the main traffic, typically travelling at 110 mile/h.

Detailed records were taken of RCF crack frequency, length, lateral position and angle relative to the rail. Clusters of RCF were present on both rails throughout the site, and are shown by grey blocks in Figure 6. The rail was approximately five years old apart from one section that had been re-railed in 2002 due to RCF damage (shown with a black block in Figure 6).

83 pairs of Miniprof rail profile measurements and cant and gauge measurements were made through the site. A theodolite survey was made of the track alignment through the first turnout, to supplement the information from the track recording car.

The measured track geometry from the track recording car is shown in Figure 7. The quality of the plain line approaching the site is good, but alignment through and immediately before the turnouts at the test site is very poor.

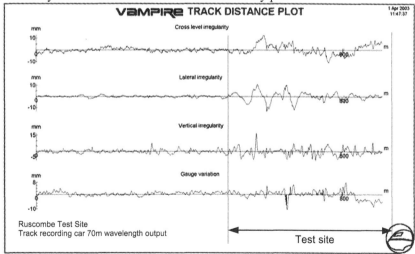

Fig. 7. Ruscombe Measured Track Geometry

4.3.1 Ruscombe Simulation Results

Simulation results for the HST power car with average worn wheels are shown in Figure 8. The locations of RCF and the re-railing are also shown by bars below the plots. The traction coefficient is only shown in the positive (traction) direction.

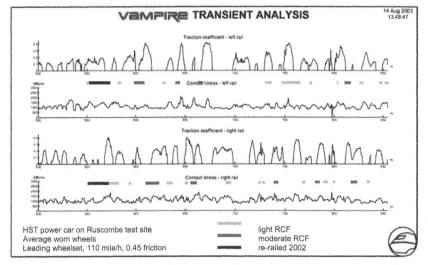

Fig. 8. Dynamic response of HST power car with average worn wheels at Ruscombe

The response at the left rail is also shown in the form of a dynamic shakedown diagram in Figure 9, with the notional limit line derived from Acton.

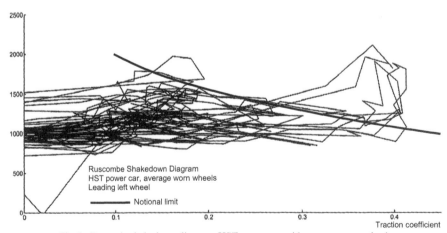

Fig. 9. Dynamic shakedown diagram, HST power car with average worn wheels

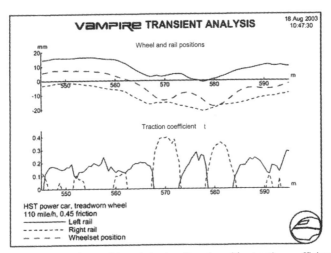

Fig. 10. Wheelset position relative to rails and resulting traction coefficient

Figure 10 shows the position of the wheelset relative to the two rails, 70m into the site. Initially, the wheelset tracks the rails well, but as the wavelength of the irregularity becomes less than the kinematic wavelength, there is convergent motion between the wheels and the rails, giving large peaks of traction coefficient. This shows that the dynamic response is due mainly to lateral track alignment, although the varying rail profiles through the switches and crossings also have an effect. The response becomes more severe and the wavelength becomes shorter with more worn wheels. Conversely, the difference between the HST power car and trailer is small.

4.2.3 Evaluating alternative RCF damage parameters

To test different damage parameters, a weighted sum of the damage from all the simulations is created, taking into account the varying age of the rails, and can be plotted as a map of the predicted damage across and along the rail head. Although shakedown limit exceedence remains a candidate, best results so far have been found with the wear energy $T\gamma$ (creep force * creepage). Figure 11 shows the very encouraging agreement with the actual RCF positions (outlined in black).

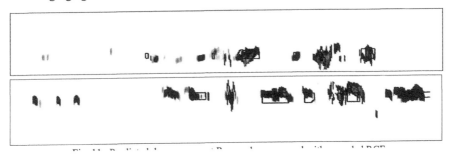

Fig. 11. Predicted damage ... Ty compared ... with actual RCF

4.3.3 Causes and remedies of RCF at Ruscombe

The RCF at Ruscombe is primarily due to vehicle dynamic response to the lateral alignment irregularities through the site, although the varying rail profile through the switches and crossings is also a significant factor. The severity of the response increases with wear of the wheel profiles, due to the increasing conicity.

The key to reducing RCF at this site is to improve the lateral alignment of the track, although even with perfect alignment, the varying switch and crossing rail profiles would give some risk of RCF. Grinding to improve the rail profiles and reduce the conicity through site would also be beneficial.

5. CONCLUSIONS

The study described in this paper has led to a major advance in the understanding of how RCF is generated on a high-speed inter-city railway, and indicates the importance of vehicle dynamic behaviour in this process. The authors believe that the ability to predict and model RCF in these situations is of immense value.

High creep forces (in the traction rather than braking direction) are the key factor in RCF development in rails, and can arise from five mechanisms:
1. Applied traction in locomotives and powered bogies
2. Quasi-static curving forces on the high rail
3. Dynamic response to lateral track irregularities
4. Dynamic response to rapid changes in rail profile
5. Self-excited instability at the wheel/rail interface

RCF is most likely when high creep forces are accompanied by high contact stress. Applied traction on its own is not normally accompanied by high contact stress, whereas high creep forces due to the remaining mechanisms all involve contact approaching the gauge corner, where contact stress is usually high. This may explain why traction alone does not appear to be a major cause of RCF in the UK.

The studies reported here have demonstrated RCF due to lateral track irregularities and profile changes at Ruscombe, and self-excited instability at Acton. At both sites RCF results from rapid changes in rolling radius, due either to closely-conformal wheels and rails giving a bi-stable motion, or convergent motion of the wheel and rail due to track irregularities shorter than the kinematic wavelength.

Evidence from the distribution of RCF in the UK suggests that quasi-static curving forces alone are generally not sufficient to cause RCF, otherwise every curve of a sufficient radius would be cracked. We also hypothesise that on tighter curves with hard flange contact, the high forces result in wear rather than RCF formation. Detailed dynamic modelling of sites with tighter radii is currently being undertaken to investigate these aspects.

REFERENCES

1. Bower, A.F. and Johnson, K.L.: Plastic flow and shakedown of the rail surface in repeated wheel-rail contact. Wear, 144 (1991) 1-18.

Vehicle System Dynamics Supplement 41 (2004), p.537-546 © Taylor & Francis Ltd.

Multi-Objective Robust Design of the Suspension System of Road Vehicles

Massimiliano GOBBI   Francesco LEVI   Giampiero MASTINU

SUMMARY

A new approach for the design of automotive suspension systems is addressed in the paper. The new approach is based not only on the theory of multi-objective optimisation but also on robust design. A simple two degree of freedom linear model has been used to derive a number of analytical formulae describing the dynamic behaviour of vehicles running on randomly profiled roads. Discomfort, road holding and working space are the performance indices to which reference will be made in the paper. The design variables are the suspension stiffness and damping (passively suspended system) considered as stochastic variables and the controller gains (actively suspended system). The vehicle body mass and the tyre radial stiffness have been considered as stochastic parameters. The optimal trade-off solutions (Pareto-optimal solutions) in a stochastic framework have been derived.

1. INTRODUCTION

A new approach for the design of automotive suspension systems is addressed in the paper. The new approach is based not only on the theory of multi-objective optimisation but also on robust design [1, 2]. The problem is dealt with in an academic perspective, actually a simple two degree of freedom linear model has been used to derive a number of analytical formulae describing the dynamic behaviour of vehicles running on randomly profiled roads [4, 5]. The analytical formulae can estimate with reasonable accuracy the dynamic behaviour of an actual road vehicle running on rough road.

With reference to wheel and body vibrations, a number of Authors have dealt with the problem of deriving basic concepts useful for road vehicle suspension tuning [15, 14, 10, 4, 5]. Many important relationships have been highlighted among vehicle suspension parameters and suspension performance indices. However a comprehensive Multi-objective and Robust approach seems to have never been addressed.

A new optimisation method - based on Multi-objective Programming and Robust Design Theory (Stochastic Multi-objective optimisation) is introduced and applied for the computation of the best compromise among conflicting performance indices pertaining to the vehicle suspension system. Discomfort, road holding and working space are the performance indices to which reference will be made in the paper. The de-

*Address correspondance to: Department of Mechanical Engineering, Politecnico di Milano (Technical University). Via La Masa, 34, 20158 Milan, Italy. E-mail: massimiliano.gobbi@polimi.it

sign variables to be optimised are the stiffness and the damping of the suspension for the passively suspended vehicle (PS), the controller gains for the actively suspended one (AS). Suspension stiffness and damping have been considered as stochastic design variables. Actually the spring stiffness and damping rate may vary with respect to their nominal value due to production tolerances and/or wear, ageing, ... The vehicle body mass and the tyre radial stiffness have been considered as stochastic parameters due to the variety of possible vehicle loading conditions and to the uncertainty of the tyre pressure of poorly maintained vehicles. The Stochastic Multi-objective optimisation method involves the simultaneous optimisation of the mean and variance of the performance indices.

The results that have been obtained are surely useful for academic purposes but also designers may take advantage from this simplified but very general theory.

2. SYSTEM MODEL: EQUATIONS OF MOTION AND RESPONSE TO STOCHASTIC EXCITATION

The adopted quarter-car system model for passively (PS) and actively (AS) suspended road vehicles is shown in Fig.1. The mass m_1 represents approximately the mass of the wheel plus part of the mass of the suspension arms, m_2 represents approximately $1/4$ of the body mass [15] and k_1 is the tyre radial stiffness. k_2 and r_2 are respectively the stiffness and the damping of the suspension and, for the AS model, g_1 and g_2 are the controller gains. The excitation comes from the road irregularity ξ. The model is generally reputed sufficiently accurate for capturing the essential features related to discomfort, road holding and working space (see [15]). The linear equations of motions of the system are

$$m_1\ddot{x}_1 - r_2\,(\dot{x}_2 - \dot{x}_1) - k_2\,(x_2 - x_1) + k_1\,(x_1 - \xi) - F_{act} = 0$$
$$m_2\ddot{x}_2 + r_2\,(\dot{x}_2 - \dot{x}_1) + k_2\,(x_2 - x_1) + F_{act} = 0 \tag{1}$$

The actuator force (AS system) is assumed to be function of the relative displacement wheel - vehicle body and of the absolute velocity of the vehicle body mass m_2 (sky-hook)

$$F_{act} = g_1(x_2 - x_1) + g_2\dot{x}_2 \tag{2}$$

The responses of the vehicle model are respectively the vertical vehicle body acceleration ($\ddot{x}_2$), the force applied between road and wheel (F_z), the relative displacement between wheel and vehicle body ($x_2 - x_1$).

Discomfort ($\sigma_{\ddot{x}_2}$, standard deviation of $\ddot{x}_2$), road holding (σ_{F_z}, standard deviation of F_z) and working space ($\sigma_{x_2-x_1}$, standard deviation of $x_2 - x_1$) are the *performance indices* to which reference will be made in the paper.

The displacement ξ (road irregularity) may be represented by a random variable defined by a stationary and ergodic stochastic process with zero mean value [3]. The power spectral density (PSD) of the process may be determined on the basis of experimental measurements and in the literature there are many different formulations for it (e.g. see [14] and [3]).

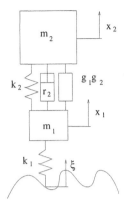

Figure. 1. Simplified quarter car vehicle model.

In the present paper for sake of simplicity the following spectrum has been considered

$$S_\xi(\omega) = \frac{A_b \, v}{\omega^2} \tag{3}$$

2.1. Formulae referring to the Passively Suspended (PS) model

The analytical formulae giving the discomfort, road holding and working space of a PS vehicle ($g_1=g_2=0$) have been already derived and presented in [4, 9], those formulae can be simplified by introducing the following non-dimensional variables

$$q = \frac{m_1}{m_2} \qquad K_x = k_2 \frac{(1+q)^2}{k_1 q} \qquad R_x = r_2 \sqrt{\frac{(1+q)^3}{k_1 m_2 q}} \tag{4}$$

So we obtain:
- Standard deviation of the relative displacement between wheel and vehicle body $x_2 - x_1$ (working space)

$$\sigma_{x_2-x_1} = \sqrt{1/2 \, A_b \, v} \, f_1 \qquad f_1^2 = \sqrt{\frac{m_2(1+q)^5}{k_1 q}} \left(\frac{1}{R_x}\right) \tag{5}$$

- Standard deviation of the vehicle body acceleration $\ddot{x}_2$ (discomfort)

$$\sigma_{\ddot{x}_2} = \sqrt{1/2 \, A_b v} \, f_2 \qquad f_2^2 = \sqrt{\frac{k_1^3 q^3}{m_2^3(1+q)^3}} \left(\frac{K_x^2}{R_x} + \frac{R_x}{q}\right) \tag{6}$$

- Standard deviation of the force acting between road and wheel F_z (road holding)

$$\sigma_{F_z} = \sqrt{1/2 \, A_b \, v} \, f_3 \qquad f_3^2 = \sqrt{k_1^3 m_2 q^3 (1+q)} \left(\frac{(K_x-1)^2}{R_x} + \frac{1}{q}(R_x + \frac{1}{R_x})\right) \tag{7}$$

2.2. Formulae referring to the Actively Suspended (AS) model

The analytical formulae giving the discomfort, road holding and working space are given in [5]. They have been obtained by setting $k_2 = 0$ and $r_2 = 0$, and they can be written more simply by introducing the following non-dimensional variables

$$q = \frac{m_1}{m_2} \qquad G_{1x} = g_1 \frac{(1+q)}{k_1} \qquad G_{2x} = g_2 \sqrt{\frac{q}{k_1 m_2}} \tag{8}$$

So we obtain:

- Standard deviation of the relative displacement between wheel and vehicle body $x_2 - x_1$ (working space)

$$\sigma_{x_2-x_1} = \sqrt{1/2 \, A_b \, v} \, h_1 \qquad h_1^2 = (\frac{m_2 q(1+q)^2}{k_1})^{1/2} \frac{1}{G_{1x}} (\frac{1}{G_{2x}} + G_{2x}(1+\frac{1}{q})) \tag{9}$$

- Standard deviation of the vehicle body acceleration $\ddot{x}_2$ (discomfort)

$$\sigma_{\ddot{x}_2} = \sqrt{1/2 \, A_b v} \, h_2 \qquad h_2^2 = (\frac{k_1^3 q}{m_2^3(1+q)^2})^{1/2} \frac{G_{1x}}{G_{2x}} \tag{10}$$

- Standard deviation of the force acting between road and wheel F_z (road holding)

$$\sigma_{F_z} = \sqrt{\frac{1}{2} \, A_b \, v} \, h_3 \qquad h_3^2 = (k_1^3 m_2(1+q)^2 q)^{\frac{1}{2}} (\frac{(G_{1x}-1)^2}{G_{1x}G_{2x}} + \frac{G_{2x}}{G_{1x}} + \frac{1}{G_{2x}(1+q)}) \tag{11}$$

3. MULTI-OBJECTIVE ROBUST OPTIMISATION

3.1. Deterministic problem

The theory of Multi-Objective Programming (MOP) can be found in [13]. This theory refers to the minimisation of a vector of (conflicting) objective functions.

In general, a MOP problem can be stated as

$$\min f(z) = \min (f_1(z), f_2(z), ..., f_k(z)) \tag{12}$$

where $z = [z_1, z_2, ..., z_n]$ are the n model design variables that may vary within a pre-defined domain Z (feasible domain) and $f = [f_1, f_2, ..., f_k]$ are the k objective functions.

The point z^* is a Pareto-optimal solution for a MOP problem if doesn't exist another solution z such that

$$\begin{cases} f_r(z) \le f_r(z^*) & r = 1, 2, 3, ..., k \\ \exists l : f_l(z) < f_l(z^*) \end{cases} \tag{13}$$

Pareto-solutions are in general not unique and constitute a set. Methods to find the whole set of efficient solutions are reported in [13].

3.2. Stochastic problem

A stochastic system is described by a mathematical model in which there are some quantities subject to uncertainty. These quantities can be a set of parameters c or a subset of the design variables vector z. The uncertainties on parameters or design variables are transmitted to the objective functions. The standard deviation of the objective function $f_i(z, c)$ due to the stochastic variation of uncertain design variables and/or parameters can be computed by means of a Taylor series expansion around the mean values of the uncertain quantities [11].

The variance of the objective functions (First order Taylor Expansion)

$$\sigma_{f_i}^2 = \sum_{i=1}^{n} (\frac{\partial f_i}{\partial z_i})^2 \sigma_{z_i}^2 + \sum_{i=1}^{m} (\frac{\partial f_i}{\partial c_i})^2 \sigma_{c_i}^2 \tag{14}$$

When considering a stochastic system, the solution can be obtained by transforming the original stochastic problem into an *equivalent deterministic problem*. The formulation of the equivalent deterministic problem adopted in the present paper is [1]

$$Given\ the\ probabilities \quad \beta_1, \beta_2, ..., \beta_k$$
$$\min(u_1, u_2, ..., u_k) \tag{15}$$
$$such\ that \quad Prob(f_i(z, c) \le u_i) \ge \beta_i \quad with \quad i = 1, 2, ..., k$$

The probability β_i describes the tolerable level of risk and it is fixed by the designer (failure probability F.P.=1-β_i). If we assume that the objectives have normal distribution, we can write

$$Given\ the\ probabilities \quad \beta_1, \beta_2, ..., \beta_k$$
$$\min(f_1(z, c) + \alpha_1 \sigma_{f_1}(z, c), f_2(z, c) + \alpha_2 \sigma_{f_2}(z, c), ..., f_k(z, c) + \alpha_k \sigma_{f_k}(z, c))$$
$$with \quad \alpha_i = \Phi^{-1}(\beta_i) \quad for \quad i = 1, 2, ..., k \tag{16}$$

where the function $\Phi^{-1}(x)$ is the inverse of the standard normal distribution and the percentile α_i corresponds to the objective functions f_i. So the Robust Design approach involves the simultaneous optimisation of the mean and variance of the objective functions.

3.3. Analytical method to find Pareto-optimal solutions

The analytical expression of the Pareto-optimal solutions can be derived by means of the *Fritz John necessary condition for Pareto optimality* [13]

Let the objective functions of the problem (12) be continuously differentiable at a decision vector $z^* \in Z$. A necessary condition for z^* to be Pareto optimal is that there exist a vector $0 < \lambda \in R^k$ such that

$$\sum_{i=1}^{k} \lambda_i \nabla f_i(z^*) = 0 \tag{17}$$

By writing the matrix of the gradients of the objective functions as

$$\nabla F = [\nabla f_1 \nabla f_2 ... \nabla f_k] = \begin{bmatrix} \frac{\partial f_1}{\partial z_1} & \cdots & \frac{\partial f_k}{\partial z_1} \\ \vdots & \ddots & \vdots \\ \frac{\partial f_1}{\partial z_n} & \cdots & \frac{\partial f_k}{\partial z_n} \end{bmatrix} \tag{18}$$

the Fritz-John condition can be written as

$$\nabla F \lambda = \mathbf{0} \Rightarrow (\nabla F^T \nabla F) \lambda = \mathbf{0} \Rightarrow det(\nabla F^T \nabla F) = 0 \qquad (19)$$

If $n = k = 2$ Eq. 19 leads to

$$\frac{\partial f_1}{\partial z_1} \cdot \frac{\partial f_2}{\partial z_2} = \frac{\partial f_1}{\partial z_2} \cdot \frac{\partial f_2}{\partial z_1} \qquad (20)$$

The Pareto-optimal set is therefore the curve described by Eq. 20 limited by the points of minimum of the two objective functions.

4. STOCHASTIC OBJECTIVE FUNCTIONS

The vehicle body mass m_2 and the tyre radial stiffness k_1 are considered as stochastic parameters due to the variety of possible loading conditions and to the variability of the tyre pressure. For the PS model the variables to be optimised are the stiffness k_2 and the damping r_2 of the suspension considered as stochastic design variables, while, for the AS model, the design variables are the gains g_1 and g_2. The variability of the parameters and design variables is described by the coefficient of variation $CV_i = \sigma_i / \mu_i$ (σ standard deviation, μ mean value). The reference values, variation ranges and CV_i for the design variables and parameters are reported in Tab. 1. The values of the coefficients of variation CV_i have been selected according to the data available in the literature [16], 3% for the spring stiffness, 10% for the damping coefficient, 10% for the tyre radial stiffness and 10% for the sprung mass.

The mean values of the objective functions are f_i (Eqs. 5-7,9-11). The standard deviations σ_{f_i} are computed in analytical form by means of Eq. 14. The generic objective function for the stochastic problem $\overline{f}_i$ for the PS model is

$$\overline{f}_i = f_i + \alpha_i \sigma_{f_i} = C_i \cdot F_{ix}(R_x, K_x, q, \alpha_i) \qquad (21)$$

Since C_i is a dimensional constant and the functions F_{ix} do not depend on k_1, m_1 and m_2, Eq. 20 states that the pareto-optimal set, in the plane of the non-dimensional variables R_x and K_x, depends only on the quantities α_i and q.

The same considerations hold for the AS model. The pareto-optimal set, in the plane of the non-dimensional variables G_{1x} and G_{2x}, will therefore depend only on the quantities α_i and q.

For the PS model, Fig. 2 shows that the increase in robustness (higher α) implies a shift of the optimal solutions to higher values of the design variable K_x (related to suspension stiffness k_2). Fig. 6 shows that the Pareto-optimal region for $\alpha = 6$ is an extension of the deterministic one ($\alpha = 0$) while the solutions for the robust design problem ($\alpha \rightarrow \infty$) are extremely different from the deterministic ones. For this reason the robust design solutions lead to poor mean performance while the mean values of the objective functions for $\alpha = 6$ are very close to the deterministic ones, see Fig. 3.

For the AS model, by increasing the robustness, the Pareto-optimal set for road holding-discomfort and road holding-working space moves to higher value of G_{1x}

Table 1. Design variables and parameters. Reference value, upper - lower bounds, coeff. of variation *(CV)*.

	unit	reference value	lower - upper bounds	*CV*
k_2	[N/m]	-	0-80000	.03
r_2	[Ns/m]	-	0-5000	.10
g_1	[N/m]	-	0-700000	-
g_2	[Ns/m]	-	0-50000	-
k_1	[N/m]	120000	-	.10
m_2	[kg]	229	-	.10
m_1	[kg]	31	-	-

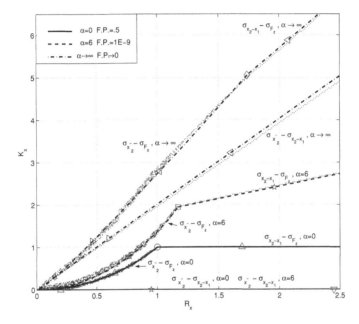

Figure. 2. Optimal design variables for the PS model in non-dimensional coordinates (Eq. 4). A numerical validation (points) is reported. For each curve some optimal points are highlighted with reference to Fig. 3.

(related to gain g_1), see Fig. 4. Instead the Pareto-optimal set for discomfort-working space problem moves from infinity to lower value of the variable G_{2x}. This causes a change in the optimal region for the robust design approach that is completely different from the deterministic one, see Fig. 6.

For both the models (PS and AS) the increase in robustness is shown, in Fig. 3 and 5, by the reduction of the length of the bars that represent the standard deviation of the objective functions. The advantages of the robust design approach ($\alpha \to \infty$), in terms of failure probability, is, in this case, negligible in comparison with the loss of mean performance. The optimisation of the mean and standard deviation together

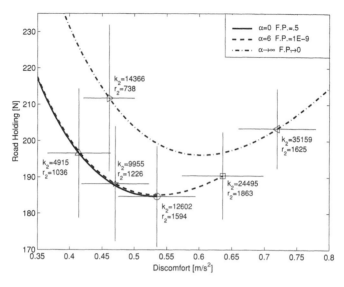

Figure. 3. Optimal objective functions for the PS model, $\sigma_{\ddot{x}_2} - \sigma_{F_z}$. For the optimal points with marker (see Fig. 2), the intervals of $\pm\sigma$ around the mean value and the values of the design variables are shown. Road roughness parameter A_b=6.9E-6, vehicle speed v=20m/s.

leads instead to a good compromise between robustness and efficiency.

5. CONCLUSION

A multi-objective stochastic optimisation method has been applied in the paper. A simple two degree of freedom linear model has been used to describe the dynamic behaviour of vehicles running on randomly profiled roads. Discomfort, road holding and working space are the objective functions which have been minimised. The design variables are the suspension stiffness and damping (PS model) and the controller gains (AS model). The vehicle body mass and the tyre radial stiffness have been considered as stochastic parameters due to the variety of possible vehicle loading conditions and to the uncertainty of the tyre pressure. Not only the uncertain parameters, but also the design variables have been considered as stochastic quantities which can deterio-rate the "robustness" of the system in terms of performance degradation. The optimal trade-off solutions (Pareto-optimal) in a stochastic framework have been found. In particular it was evident that the best compromise solutions computed without ac-counting for robustness were rather prone to become less efficient due to stochastic parameters variations. On the contrary solutions computed by means of a "robust de-sign" approach are not very efficient in terms of mean performance. Only the proposed multi-objective stochastic approach shows minor performance reduction with respect to a deterministic approach with relevant improvement in robustness. The efficiency of the procedure is clear and the adopted design method can be used for further studies.

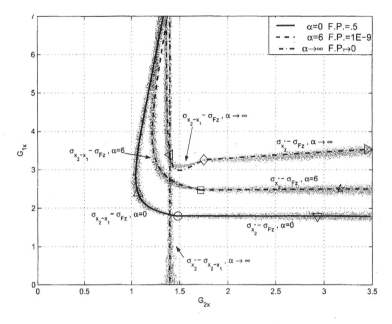

Figure. 4. Optimal design variables for the AS model in non-dimensional coordinates (Eq. 8). A numerical validation (points) is reported. For each curve some optimal points are highlighted with reference to Fig. 5.

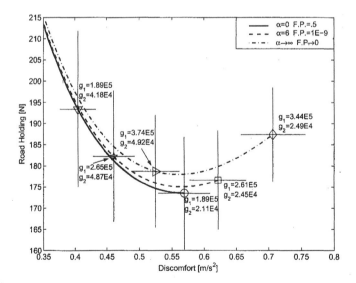

Figure. 5. Optimal objective functions for the AS model, $\sigma_{\ddot{x}_2} - \sigma_{F_z}$. For the optimal points with marker (see Fig. 4), the intervals of $\pm\sigma$ around the mean value and the values of the design variables are shown. Road roughness A_b=6.9E-6, vehicle speed v =20m/s.

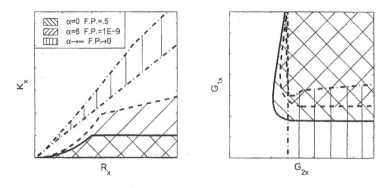

Figure. 6. Comparison between the Pareto-optimal region for three objective functions for PS (left) and AS (right) vehicle.

REFERENCES

1. R. Caballero et Al. 2001 *Journal of Optimization Theory and Applications 1*, 110, 53–74. Efficient Solution Concepts and their Relations in Stochastic Multiobjective Programming.
2. M. Gobbi and G. Mastinu 2003 *Proc. of the WCSMO5 Int. Conference, Lido di Jesolo.* Stochastic Multi-Objective Optimisation for the Design of Vehicle Systems.
3. C. Dodds and J. Robson 1973 *Journal of Sound and Vibration 2*, 31, 175–183. The description of road surface roughness.
4. M. Gobbi and G. Mastinu 2001 *Journal of Sound and Vibration 3*, 245, 457–481. Analytical Description and Optimization of the dynamic behaviour of passively suspended road vehicles.
5. M. Gobbi and G. Mastinu 2001 *Int. Journal of Vehicle Design*,28, n. 1/2/3. Symbolical multi-objective optimisation of the dynamic behaviour of actively suspended road vehicles.
6. M. Gobbi and Al. 1999 *Vehicle System Dynamics, Supplement 33.* Optimal & robust design of a road vehicle suspension system.
7. M. Gobbi, G. Mastinu and C. Doniselli 1999 *Vehicle System Dynamics Vol. 32 No.2-3.* Optimising a car chassis.
8. D. Hrovat 1993 *Transactions of the ASME, Journal of Dynamic Systems, Measurement, and Control 115* (June). Applications of optimal control to advanced automotive suspension design.
9. X. Lu and Al. 1984 *Int. Journal of Vehicle Design 5.* A design procedure for optimization of vehicle suspensions.
10. G. Mastinu 1988 *Proc. of the IMechE Conference - Advanced Suspensions* (London), Institution of Mechanical Engineers. Passive automobile suspension parameter adaptation.
11. A. Papoulis 1991 *Probability Random Variables, and Stochastic Processes.* McGraw-Hill
12. G. Mastinu 1995 *Smart Vehicles*, J. Pauwelussen and H. Pacejka, Eds. Swets & Zeitlinger Publishers. Integrated controls and interactive multi-objective programming for the improvement of ride and handling of road vehicles.
13. K. Miettinen 1999 *Nonlinear Multiobjective Optimization.* Kluwer Academic Publishers, Boston.
14. M. Mitschke 1990 *Dynamik der Kraftfahrzeuge.* Springer Verlag, Berlin.
15. R. Sharp and D. Crolla 1987 *Vehicle System Dynamics, vol.16, no.3.* Road vehicle suspension system design - a review.
16. J. Reimpell, H. Stoll 1989 *Fahrwerktechnik: Stoß-und Schwingungsdämpfer.* Buchverlag Würzburg.

Vehicle System Dynamics Supplement 41 (2004), p.547-556   © Taylor & Francis Ltd.

Handling and ride comfort optimisation of an intercity bus

DANIEL ANDERSSON[1] AND PETER ERIKSSON

SUMMARY

In this work the handling and ride comfort characteristics of an inter-city bus have been optimised. A rigid body simulation model of the bus has been used to evaluate the handling in terms of gain and time-lag of the yaw rate response due to a sinusoidal steering wheel input. A similar model but with a flexible bus body structure has been used to calculate the transient response due to obstacles on the road. The resulting discomfort within the bus has been evaluated in terms of vibration dose values according to the ride comfort standard ISO 2631:1997 at three locations within the bus. Both models have been validated against measurements. The stiffnesses of the bellows, the diameters of the anti-roll bars and the characteristics of the shock absorbers have been used as design variables in the optimisations. The results show that it is possible to improve the current ride comfort and handling characteristics of the bus by optimising the wheel suspension. It was, furthermore, shown that the handling and ride comfort criteria used in this study are in conflict with each other and a multi-criteria approach was therefore adopted to optimise both properties at the same time. The results show that it is possible to improve the ride comfort of the bus without affecting the handling properties considerably.

1   INTRODUCTION

As buses of today are getting more advanced, the passengers are becoming more fastidious and the demand for a more comfortable travelling environment is constantly increasing. At the same time, the handling performance of buses is getting more attention in the development process today than earlier. The wheel suspension in a bus is a critical component that not only affects the ride comfort but also the handling characteristics, the road holding ability etc. [1]. Hence, the designer is often forced to compromise between different requirements in order to design the wheel suspension with as good characteristics as possible. For example, a test driver might prefer a roll stiff wheel suspension from a handling perspective (the driver has a feeling of 'controlling' the vehicle) but if the ride comfort is taken into account, the same configuration might be regarded as too stiff (high accelerations from single-sided excitations on the road, e.g. potholes). It is, therefore, of major importance for bus manufacturers in the future to have reliable and robust tools to

---

[1] *Address correspondence to:* SCANIA CV AB, RBHS Dynamics and Simulations, Bus Chassis Development, SE-151 87 Södertälje, SWEDEN. Tel.: +46 8 553 50072; Fax: +46 8 553 50090; E-mail: daniel_p.andersson@scania.com

find the best design of the wheel suspension to meet the growing demand for good handling and ride comfort characteristics.

The purpose of this work is to investigate how numerical optimisation can be used in combination with simulation models in order to find a bus design with as good ride comfort and handling characteristics as possible.

2   BUS MODELS

In this work the Scania OmniLink bus has been studied, see Figure 1. This bus is intended for city/intercity traffic. It has a total length of approximately 12 meters with a wheel-base of 6 meters and a low floor with a single interior step after the middle door. The bus is equipped with a 9-litre engine which is installed longitudinally behind the rear axle and inclined 60 degrees. The total kerb weight of the bus is 11000 kg.

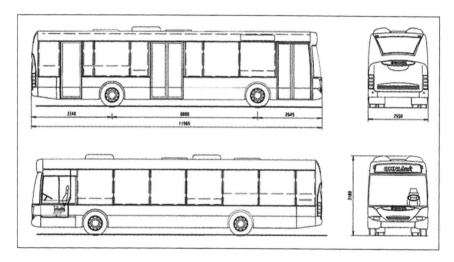

Fig. 1. Main dimensions of bus.

2.1   SIMULATION MODELS

Two vehicle models are used in the optimisation procedure for the handling and ride comfort evaluation. The multi-body system software MSC.ADAMS [2] has been used to create the models and solve the equations of motions. The first model, which is used for handling performance evaluation, consists of in total 37 rigid bodies including a bus body, representing the chassis frame and the bus body structure, two axles and a power unit. The axles are connected to the bus body via reaction rods, air bellows, shock absorbers and anti-roll bars. The reaction rods are modelled using rigid bars with non-linear springs and linear dampers, describing the behaviour of the bushings at each end of the bar. The air bellows are represented by linear springs and the shock absorbers by non-linear dashpots. The air bellow spring stiffness is simply changed by a single multiplier m_b and the non-linear shock absorber force-velocity curve is parameterised in such a way that the nominal damping force, for the current velocity, according to the nominal curve is multiplied by m_{sc} if the velocity is negative (compression) and m_{sr} if the velocity is positive (rebound). The anti-roll bar, with a diameter d_a, is modelled using rigid bodies, flexible links, revolute joints and a torsional spring with a stiffness calculated from the dimensions of the bar. The motions of the axles in the vertical direction are constrained by a number of bump-stops. These are modelled by non-linear springs that become very stiff when the bump-stops get in contact with the axle. The power unit is connected to the bus body by rubber mounts that have been modelled by linear springs and linear dashpots. The tires are modelled using an in-house tire model that is based on the Magic Formula model. Tire parameters have been provided by the tire manufacturer.

The second model, which is used for ride comfort evaluation, is basically the same as the handling model except the modelling of the bus body and the tires. In this model, the rigid bus body structure is replaced by a flexible structure using modal data (under 100 Hz) from a FE model of the complete bus body structure. The FE model, with about 40000 degrees of freedom, has originally been created in the FE-software I-DEAS v.9 and exported to ADAMS as a modal neutral file (MNF). The constraint nodes in the condensed FE model are coupled to the rigid bodies of the axles and the engine via fixed joints and mass-less links. The tires in the ride comfort evaluation model are modelled using a non-linear in-plane physically based tire model.

3   VALIDATION OF MODELS

Measurements of a corresponding test bus have been carried out to validate the handling and ride comfort simulation models. In Figure 2 the response of the bus, with a velocity of 40 km/h, due to a sinusoidal steering input is shown where the solid lines correspond to three measurements and the dashed lines to calculated results from the handling model. To validate the ride comfort model the accelerations at a number of positions in the bus were measured on the test track at SCANIA. A corresponding simulation was carried out and in Figure 3 the measured (solid) and calculated (dashed) vertical acceleration at the rear sofa in the back of the bus are shown in the time and frequency domain. The results from both models show a good conformity with the measured results.

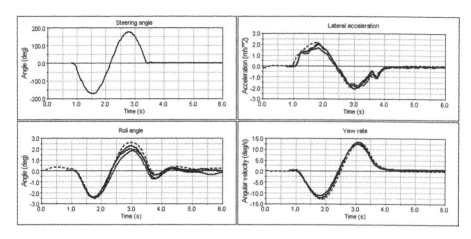

Fig. 2. Measured (—) and calculated (---) results for lane change manoeuvre at 40 km/h.

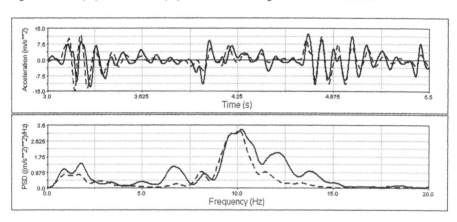

Fig. 3 Measured (—) and calculated (---) vertical accelerations at the rear sofa.

4   OPTIMISATION

A general non-linear optimisation problem can be defined as follows [3]:
Find an n-vector $\mathbf{x} = (x_1, x_2, ..., x_n)$ of design variables to minimise an objective function

$$f(\mathbf{x}) = f(x_1, x_2, ..., x_n) \tag{1}$$

subjected to the constraints

$$h_j(\mathbf{x}) = 0; \qquad j = 1 \text{ to } p \tag{2}$$

$$g_i(\mathbf{x}) \leq 0; \qquad i = 1 \text{ to } m \tag{3}$$

and simple bounds on the design variables

$$x_{kl} \leq x_k \leq x_{ku}; \qquad k = 1 \text{ to } n \tag{4}$$

The idea in this work is to optimise the ride comfort and handling characteristics at the same time and an approach how to deal with several objective functions must therefore be adopted. A common way to define the objective function $f(\mathbf{x})$, when several objective functions $f_k(\mathbf{x})$, $k = 1$ to r, are to be minimised is to use the definition given in Equation (5) which generates an Edgeworth-Pareto optimal solution [4].

$$f(\mathbf{x}) = \sum_{k=1}^{r} \left(w_k \left(\frac{f_k(\mathbf{x}) - f_k^*}{f_k^*} \right)^2 \right) \tag{5}$$

where f_k^* is the minimum corresponding to the minimisation of the single objective function $f(\mathbf{x}) = f_k(\mathbf{x})$ and w_k is a weighting coefficient, corresponding to $f_k(\mathbf{x})$, that can be used to control the importance of one objective function in the relation to the others in a multi-objective optimisation. In total, five objective functions ($r = 5$), described in the subsequent sections, will be used in this work.

4.1 OPTIMISATION SET-UP

The diameter d_a of the anti-roll bar, the stiffness m_b of the air bellows and the multipliers m_{sc} and m_{sr} associated with the compression/rebound forces of the shock absorbers of the front (index: f) and rear (index: r) wheel suspension have been used as design variables in the optimisation procedure. The lower and upper bounds of the normalised (with respect to the nominal value) design variables are given in Table 1. The basic lay-out of the bus, e.g. wheel base, mass, inertia and centre of gravity of the bus body, have substantial impact on the handling and ride comfort characteristics [5,6] but these variables have not been considered in this study since they are, in general, determined by other requirements. Main focus in this work is to study how components in the wheel suspension that are relatively easy to modify, such as shock absorbers, can be optimised to improve the handling and ride comfort of the bus.

Table 1. Lower and upper bounds on the normalised design variables.

Variable	Description	Lower	Upper
x_1	d_a^f : diameter of anti-roll bar front	0.75	1.50
x_2	m_b^f : bellow stiffness front	0.25	1.89
x_3	m_{sc}^f : shock absorber compression forces front	0.75	1.25
x_4	m_{sr}^f : shock absorber rebound forces front	0.75	1.25
x_5	d_a^r : diameter of anti-roll bar rear	0.6	1.20
x_6	m_b^r : bellow stiffness rear	0.44	3.30
x_7	m_{sc}^r : shock absorber compression forces rear	0.75	1.25
x_8	m_{sr}^r : shock absorber rebound forces rear	0.75	1.25

The sequential quadratic programming optimisation (SQP) algorithm in MSC.ADAMS has been used to solve the optimisation problem. Gradients have been calculated numerically using finite differences.

4.1.1   Ride comfort

In ride comfort evaluation routine, the bus traverses three rectangular obstacles on the road (introducing both single and double sided excitations of the left and right wheels) at a speed of 30 km/h and the response is evaluated at the floor at three locations within the bus: the driver seat position, the standing area in the middle of the bus and the sofa in the back of the bus. A similar type of ride comfort analysis of a bus has been carried out by Kuti [7]. The vertical accelerations at the three positions are thereafter frequency-weighted, using a comfort filter called Wk in the ride comfort standard ISO 2631:1997 [8] and the discomfort at these positions are evaluated by calculating the vibration dose values (VDVs) of the filtered signals according to Equation (6) below. The ride comfort performance index of the whole bus $f_c(\mathbf{x})$ is defined by the sum of the vibration dose values at the three positions.

$$f_{vdv} = \sqrt[4]{\int_{t_1}^{t_2} [a_w(t)]^4 \, dt}$$

(6)

where $T = t_2 - t_1$ is the considered time interval and $a_w(t)$ is the frequency weighted acceleration.

4.1.2   Handling

In the optimisation procedure, the handling performance is evaluated from the response of the bus during single lane change manoeuvres according to the guidelines given in ISO 14793 [9]. Similar handling evaluation methods of buses have been used by Cheng et al. [5] and Ahn [10]. The time lag and the gain of the yaw rate response with respect to the steering wheel input are used to quantify the handling performance. The objective in the optimisation is to minimise the time lag, denoted by $f_t(\mathbf{x})$, and to maximise the gain, denoted by $f_g(\mathbf{x})$. The handling properties are evaluated at two conditions in order to cover both normal and severe manoeuvres; at 40 km/h with a low steering wheel input (low lateral acceleration) and at 80 km/h with a large steering wheel input (high lateral acceleration). Thus, four handling performance indices are used in the optimisation, see Table 2 below. In addition to the handling objective functions mentioned above, the roll angle of the bus body has been controlled in the optimisation procedure via a roll angle constraint according to Equation (7).

$$g(\mathbf{x}) = |\phi(\mathbf{x})|_{max} - \phi_{ref} \qquad (7)$$

where $|\phi(\mathbf{x})|_{max}$ is the maximum roll angle of the bus body during the lane change manoeuvre in 40 km/h and $\phi_{ref} = 1.3°$ is permissible reference roll angle.

Table 2. Summary of objective functions.

Objective function	Target	Description
$f_1 = f_{g,40}$	Maximise[2]	Yaw rate gain at 40 km/h
$f_2 = f_{t,40}$	Minimise	Yaw rate time lag at 40 km/h
$f_3 = f_{g,80}$	Maximise[2]	Yaw rate gain at 80 km/h
$f_4 = f_{t,80}$	Minimise	Yaw rate time lag at 80 km/h
$f_5 = f_c$	Minimise	Sum of VDV values of vertical accelerations

---

[2] Maximising $f(\mathbf{x})$ is the same as minimising $-f(\mathbf{x})$. Therefore, in the combined objective function according to Equation (7) in which all objective functions $f_k(\mathbf{x})$ are to be minimised, $f_1 = -f_{g,40}$ and $f_3 = -f_{g,80}$ are used.

5   RESULTS

5.1   INDIVIDUAL OPTIMISATION

The performance indices were normalised with respect to their initial values corresponding to the nominal design $\mathbf{x}^{\text{nom}}$ and optimisations were thereafter executed with each of the five performance indices used as the objective function. The nominal design was used as the starting design. The results from the five optimisations are shown in Table 3 below.

Table 3. Individual normalised optimisation results.

Objective function	Target	Start	Optimal
$f_{g,40}$	Maximise	1.0000	1.0365
$f_{t,40}$	Minimise	1.0000	0.8694
$f_{g,80}$	Maximise	1.0000	1.06315
$f_{t,80}$	Minimise	1.0000	0.7666
f_c	Minimise	1.0000	0.8240

The results show that it is possible to improve all performance indices by optimising the wheel suspension of the bus. The results show furthermore that the yaw rate time-lags and the ride comfort are more affected by the tuning of the wheel suspension (the improvements are about 13-23 %) than yaw rate gains (improvement 3-6 %). The conflict between the performance indices can also be identified from the results. In the yaw rate gain optimisations, the rear wheel suspension parameters reached their upper bounds while the yaw rate time-lag optimisations resulted in the opposite. The ride comfort optimisation shows that the wheel suspension should be as soft as possible but still hard enough to prevent the axles to get in contact with the bump stops. The understeering characteristics were examined for buses with the optimal designs according to Table 3 and the results show that all buses are understeered in their operation range which, according to Whitehead [11], is usually considered as good handling behaviour of buses.

5.2   COMBINED OPTIMISATION

The handling and ride comfort performance indices were combined into a single objective function $f^{\text{com}}(\mathbf{x})$ according to Equation (5) with $w_k = 1$, $k = 1$ to 5. An optimisation was executed with the nominal design as starting design and an optimal design $\mathbf{x}^*$, which reduced the combined objective function by 26%, was found after 14 iterations (145 objective function evaluations). When the performance indices are evaluated separately at this design, it can be noticed that the improvement of the combined objective function is mainly caused by an improvement of the ride comfort; the handling performance indices are almost unchanged. The design histories, shown in Figure 4, show that the rear wheel suspension should have softer bellows and softer shock absorbers and an anti-roll bar with larger diameter. The corresponding front wheel suspension should have a stiffer anti-roll bar, stiffer air bellows and shock absorbers that are softer for compression and stiffer for rebound velocities.

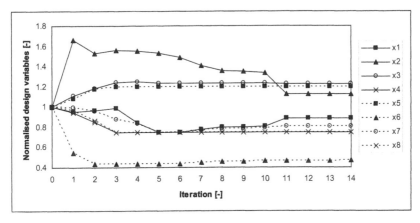

Fig 4. Histories of the design variables in the combined optimisation ($w_k = 1$, $k = 1$ to 5).

It can be noticed from the definition of the combined objective function and the results in Table 3 that the contributions of the time-lags and the ride comfort performance indices are much greater than contributions of the gain performance indices at the nominal design, i.e. the differences between $f_k(\mathbf{x}^{nom})$ and f_k^* are much greater for $f_{l,40}$, $f_{l,80}$ and f_c than $f_{g,40}$ and $f_{g,40}$. Thus, if the nominal design is used as starting design, the former indices will have more importance initially in the optimisation and the optimiser will probably try to reduce these contributions until the contributions from the gain performance indices become more significant. An optimisation was therefore executed with a modified combined objective function $f_{mod}^{com}(\mathbf{x})$ where the weighting coefficients w_k have been used to make the contributions from the five performance indices equally important initially, i.e.

$$w_k \left(\frac{f_k(\mathbf{x}^{nom}) - f_k^*}{f_k^*} \right)^2 = c \qquad k = 1 \text{ to } 5 \qquad (8)$$

where $c = 0.2$ so that $f_{mod}^{com}(\mathbf{x}^{nom}) = 1.0$. An optimal design $\mathbf{x}_{mod}^*$, which reduced $f_{mod}^{com}(\mathbf{x})$ by 30%, was found after 9 iterations (93 objective function evaluations). The results show that the gains and the ride comfort performance indices have been substantially improved but the time-lag performance indices have been worsened compared to the nominal design.

6 SUMMARY AND CONCLUSIONS

In this work, a handling and ride comfort optimisation of an inter-city bus have been presented. Two simulation models have been used in an optimisation procedure to evaluate the handling and ride comfort characteristics of the bus and the properties of the bellows, the anti-roll bars and the shock absorber in the front and rear wheel suspension have been used as design variables in the optimisation. The handling and ride comfort of the bus were first optimised separately and the results from these individual optimisations show that it is possible to improve the handling and ride comfort by optimising the wheel suspension. It was, furthermore, noticed that the performance criteria were in conflict with each other and a multi-criteria optimisation approach was therefore adopted. The results, using this approach with $f^{com}(\mathbf{x})$ as the objective function, show that it is possible to improve the ride comfort of the bus without affecting the handling properties considerably by optimising the wheel suspension. Depending on the choice of weighting coefficients in the multi-criteria approach the importance of one performance index in relation to the others can be controlled and the results show that, unsurprisingly, different optimal design can be found.

REFERENCES

1. Gobbi, M., Mastinu, G.: Analytical description and optimization of the dynamic behavior of passively suspended road vehicles. Journal of Sound and Vibration. 245, (2001), pp. 457-481.
2. Mechanical Dynamics Inc.: ADAMS version 12.0 Users Manual, Ann Arbor, Michigan, USA, 2002.
3. Arora., J.S.: Introduction to Optimum Design, McGraw-Hill, New York, USA, 1989.
4. Haftka, R.T., Gürdal, Z.: Elements of Structural Optimization, Kluwer, Dordrecht, Netherlands, 1992.
5. Cheng, B., Yanagisawa, M., Shimoyama, A.: Analysis and optimization on bus handling and stability by numerical simulation, SAE 958492, The Eighth International Pacific Conference on Automotive Engineering, November 4-9, Yokohama, Japan, 1995.
6. Lindqvist, Ö., Andersson, K., Lummer, J.P.: A study of the influence of some parameters on the stability and handling of inter-city coaches, IMechE Conference Publications 1986-8, International Conference on The Bus '86, September 9-10, London, U.K, 1986.
7. Kuti, I.: A computational procedure for non-linear dynamic analysis of vehicles, Vehicle System Dynamics, 30, (1998), pp. 37-54.
8. ISO: Mechanical vibration and shock – Evaluation of human exposure to whole-body vibration – Part 1: General requirements, ISO 2631-1:1997(E), Genève, Schwitzerland, 1997.
9. ISO: Road vehicles – Heavy commercial vehicles and buses – Lateral transient response test methods, ISO 14793:2003, Genève, Schwitzerland, 2003.
10. Ahn, K-W.: Analysis of the handling and ride comfort characteristics of a medium bus in a single lane change manoeuvre, SAE 982773, SAE International Truck & Bus Meeting and Exposition, November 16-18, Indianapolis, Indiana, USA, 1998.
11. Whitehead, J.P.: The handling characteristics of European intercity buses, SAE 912678, SAE International Truck & Bus Meeting and Exposition, November 18-21, Chicago, Illinois, USA, 1991.

A Robust Design Approach for Nonlinear Vehicle Dynamics Controller Based on Quality Engineering

SHINYA NOHTOMI[†], KAZUYUKI OKADA[†] AND SHINICHIRO HORIUCHI[*]

SUMMARY

This paper proposes a new design approach for achieving high robustness in the vehicle dynamics controller based on quality engineering. The focus of this paper is on studying the applicability of the robust design principle in quality engineering to simulation based design of nonlinear vehicle control systems. Vehicle parameters are taken as noise factors and the patterns of steering and braking inputs are taken as the signal factors. Controller parameters are treated as control factor. A metric of robust design is treated as response factors. A stochastic optimization algorithm is used to tune the control parameters. This paper illustrates the effectiveness of our approach through the design of integrated vehicle control system which coordinates four wheel braking control and four-wheel steering control.

Keywords: integrated vehicle control, quality engineering, robust design, nonlinear.

1. INTRODUCTION

In recent years, there are many researches to achieve higher vehicle dynamic stability and controllability by various kinds of sophisticated chassis controllers such as Vehicle Dynamics Control, Direct Yaw moment Control and Active Yaw Control. The authors have also proposed an integrated controller that can coordinate four-wheel braking control and four-wheel steering control based on extended Nonlinear Predictive Control Theory[1]. For the design of such vehicle control systems, standard methods rely on accurate model of the vehicle to be controlled. However, dynamics of vehicles vary significantly with such factors as their traveling speed, vehicle weight, location of the center of gravity, road surface condition, and so on. Therefore, robustness analysis for control systems is necessary to determine the possibility of instability or inadequate performance in the face of uncertainty. By using Stochastic Robustness Synthesis approach, the authors proposed a design method of a nonlinear controller that is robust to the changes in vehicle parameters, tire properties, and road surface conditions[2].

As an extension of our previous studies, this paper proposes a systematic design approach of a nonlinear vehicle control system based on quality engineering. On the

---
[*] *Address correspondance to*: Department of Mechanical Engineering, College of Science and Technology, Nihon University, 1-8 Kanda Surugadai, Chiyoda-ku, Tokyo, 101-8308, JAPAN, Fax +81-3-3293-8254; E-mail:horiuchi@mech.cst.nihon-u.ac.jp
[†] HONDA R&D Co. Ltd., Tochigi R&D Center

basis of the principle of quality engineering[3], parameters to be considered in the control system design are classified as control factors, noise factors, signal factors, and response factors. A stochastic metric of robust design that evaluates system stability, tracking performance, and control usage is treated as response factors. This robust design metric is calculated as probabilities of system instability and violation of certain performance requirements. The robust design metric has to be evaluated under extreme steering and braking conditions. To meet with this requirement, worst-case steering and braking inputs are used as a variation of control factors. This worst-case inputs are obtained as a solution of an optimal control problem that finds the inputs that maximize sideslip angle within the specified time. The theory for direct optimization[4] is applied to the computation of the worst-case inputs.

This paper is organized as follows: In Section 2, the robust design methodology for simulation-based nonlinear control system design is described. In Section 3, numerical method to determine the worst-case steering and brake inputs is described. Section 4 describes a structure and a design procedure of the integrated control system used in this paper. In Section 5, the effectiveness of proposed design procedure is demonstrated through a series of computer simulations.

2. ROBUST DESIGN APPROACH

2.1. Basic Design Procedure

The fundamental principle of quality engineering is to improve the quality of a product by minimizing the effects of variation in uncontrolled parameters.

Our design procedure consists of following steps.

STEP 1   Based on the principle of quality engineering, parameters to be considered in the control system design are classified as control factors, noise factors, signal factors, and response factors as shown in Figure1. In this study, parameters of the nonlinear vehicle controller are chosen as control factors which can be specified freely by the designer. Vehicle parameters such as vehicle mass, tire properties, mass moment of inertia, C.G. position are taken as noise factors that are not under the designer. These vehicle parameters are assumed as random variables that have given probability distribution. The patterns of steering and braking inputs are taken as the signal factors. A metric of robust design that evaluates stability, tracking performance, and control usage of the control system is treated as response factors.

STEP 2   The stochastic robust design metric J is defined as a weighted sum of the probabilities P_i that describe probabilities of unacceptable vehicle behavior. The probabilities have to be evaluated under extreme maneuvering conditions. Then the worst-case steering and braking inputs are calculated and used for the evaluation of the robust design metric.

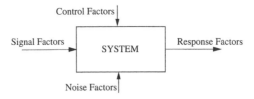

Fig. 1. Factors considered in robust design approach.

STEP 3   After the design parameters of the control system are selected, the synthesis of the controller resolves into stochastic optimization problem. The design variables are tuned to minimize the robust design metric. In this paper, Genetic Algorithm (GA) is used as a minimization method.

2.2. Metric of System Robustness

The stochastic robust design metric characterizes the robustness of a controller by calculating the probability that the closed-loop system will have unacceptable performance in the presence of possible parameter and input variations. In general, the probability depends on the following factors:

$H(v)$: the plant under consideration

$v \in R^n$: the plant parameters that may vary (noise factors)

$pr(v)$: the probabilioty density function of v

$C(d)$: the controller

$d \in R^r$: the design parameters of the controller (control factors)

u : control input (signal factors)

With H, v, C, d, u fixed, a binary decision can be made as to whether or not the matric is violated. The probability P_i that the controller does not satisfy the i-th design requirement can be defined as the integral of the binary indicator function I over the space of expected parameter variations.

$$P_i = \int_{-\infty}^{\infty} I_i[H(v), C(d), u]pr(v)dv \tag{1}$$

where $I_i = 1$ if the i-th design requirement is satisfied and $I_i = 0$ otherwise. Many different aspects of performance can be examined using this stochastic metric.

Generally Equation (1) cannot be evaluated analytically. The most practical and flexible altanative is to use Monte Carlo simulation. The probability is estimated by replacing the integral of Equation (1) with the summation:

$$\hat{P}_i = \frac{1}{N} \sum_{j=1}^{N} I_i[H(\mathbf{v}), C(\mathbf{d}), \mathbf{u}] \tag{2}$$

Using these probabilities, response factor would be

$$J = \sum_{i=1}^{N} w_i \hat{P}_i \tag{3}$$

where the w_i are positive scalar weights. The robustness is then maximized by finding $\mathbf{d}^*$ where:

$$\mathbf{d}^* : J(\mathbf{d}^*) = \min_{\mathbf{d}} J \tag{4}$$

In the practical search for $\mathbf{d}^*$, Monte Carlo estimates of the probabilities $\hat{P}_i$ must be used rather than the true probabilities P_i. The discrepancy between $\hat{P}_i$ and P_i adds apparent noise to the estimate of J and makes search more difficult. In this paper, Genetic Algorithm (GA) is adopted as an optimization method for its efficiency and flexibility.

3. WORST-CASE INPUTS

The one-player worst-case vehicle input problem can be defined as follows[5]:

Given a nonlinear vehicle dynamics model

$$\dot{\mathbf{x}}(t) = \mathbf{f}\{\mathbf{x}(t), \mathbf{u}(t)\}, \tag{5}$$

where $\mathbf{x}$ is an $n \times 1$ vector of states and $\mathbf{u}$ is an $m \times 1$ vector of controls which includes front wheel steering angle δ_f [deg] and brake command δ_b [G]. The prescribed initial conditions for the states are

$$\mathbf{x}(0) = \mathbf{x}_0, \tag{6}$$

Find, within saturation bound,

$$-\delta_{fmax} \leq \delta_f \leq \delta_{fmax} \tag{7}$$

$$-\delta_{bmax} \leq \delta_b \leq 0 \tag{8}$$

the time history of the control vector $\mathbf{u}(t)$ that will maximize a scalar performance function J_{WC}

$$J_{WC} = \phi(\mathbf{u}) = \int_0^{t_f} L\{\mathbf{x}(t), \mathbf{u}(t)\}dt \tag{9}$$

The function L is selected such that the vehicle sideslip angle is maximized.

In this paper, *direct method* is used to solve above stated worst-case problem. The basic idea behind the direct method involves *discretizing* the continuous control vector history. This technique allows the optimal control problem which has infinite number of variables to be transcribed into Nonlinear Programming Problem (NLP) with finite number of variables .

The conversion process begins by dividing the time interval of the optimal control problem into a N subintervals.

$$0 = t_0 < t_1 < t_2 < \ldots < t_{N-1} < t_N = t_f \qquad (10)$$

The individual time points are called nodes. Then, the function of time $u(t)$ is replaced by their values at the nodes u_k, $k = 0, 1, \ldots, N$ with some form of interpolation. The unknowns of the parameter optimization problem are the controls at the nodes u_k. Assume that values for the control parameters are given and placed in the unknown parameter vector U,

$$U = [\, u_0^T \quad u_1^T \quad \cdots \quad u_N^T \,] \qquad (11)$$

which contains $m(N + 1)$ elements. Because the control time history $u(t)$ are known, the state equation Equation(5) can be integrated numerically from $t = 0$ to $t = t_f$ to obtain the states at the nodes x_i, $i = 1, 2, \cdots, N$. For the integration of the state equation, it is assumed that the controls change linearly between the nodes. As a result of numerical integration, x_i become functions only of the unknown parameter vector U. Using controls and states at the nodes, the performance function (Equation(9)) can be rewritten as follows

$$J_{WC} = \Phi(U) \qquad (12)$$

Thus the optimal control problem can be reduced to the following parameter optimization problem.

$$\text{Find } U \text{ which maximize} \quad J_{WC} = \Phi(U) \qquad (13)$$

$$\text{Subject to}: \quad U_{min} \leq U \leq U_{max} \qquad (14)$$

This parameter optimization problem can be solved by mathematical programming technique such as Sequential Quadratic Programming (SQP) method [6].

4. NONLINEAR PREDICTIVE CONTROL

4.1. Vehicle Dynamics Model

The seven degrees-of-freedom model is used in the controller design. It is assumed that the vertical load on each wheel is a function of both the vehicle's weight and the dynamic weight transfer associated with longitudinal and lateral acceleration. Although the pitch and roll motions are not included in the vehicle model, their effect on the normal tire forces is accounted for. The dynamics of the tire-road interaction are dependent on the lateral and longitudinal wheel slips. The steering angles δ_f and δ_r have steering actuator dynamics which are represented as first order lag systems. The equations of motion for wheel rotation are also considered. Combining the above stated equations, the complete nonlinear vehicle model, including the vehicle planer motion, the wheel dynamics and the steering actuator dynamics, can be written as:

$$\dot{x}_1 = f_1(x) \tag{15}$$

$$\dot{x}_2 = f_2(x) + B_1 u_1 + B_2 u_2 \tag{16}$$

where $x = [\,x_1^T \quad x_2^T\,]^T$ represents vehicle state and

$$x_1 = [\,u \quad v \quad r\,]^T \tag{17}$$

$$x_2 = [\,\omega_1 \quad \omega_2 \quad \omega_3 \quad \omega_4 \quad \delta_f \quad \delta_r\,]^T \tag{18}$$

$$u_1 = [\,T_1 \quad T_2 \quad T_3 \quad T_4\,]^T \tag{19}$$

$$u_2 = [\,u_f \quad 0\,]^T \tag{20}$$

u_1 represents a input that is determined by the feedback controller and u_2 is a input that is determined in feed forward manner as a function of driver's steering command. u_f is a front wheel steering angle command.

A tire model derived from simplifying the Dugoff's tire model is used for controller design.

4.2. Nonlinear Predictive Controller

The design objective of the controller is to realize the desired forward velocity u, lateral velocity v, and yaw rate r to the driver's command with minimal control effort. The desired state trajectory $s = [\,u_m \ v_m \ r_m\,]^T$ is given by following reference model that represents ideal vehicle responses with respect to the driver's steering wheel angle command δ_{sw} and braking command δ_b.

$$\dot{s} = A_m s + B_m r \tag{21}$$

$$r = [\,\delta_{sw} \ \delta_b\,]^T \tag{22}$$

To attain the design objective, consider a performance index that penalizes the tracking error at time $t + \tau$ and current control input.

$$J_{NLP} = \frac{1}{2} e^T(t + \tau) Q e(t + \tau) + \frac{1}{2} u_1^T(t) R u_1(t) \tag{23}$$

The tracking error e is defined as:

$$e(t) = x_1(t) - s(t) \tag{24}$$

For given $x(t)$ and $s(t)$, the minimization of J_{NLP} by setting $\partial J_{NLP}/\partial u_1 = 0$ yields an optimal one-step-ahead predictive control law:

$$u_1(t) = -W^{-1}(t)\left(\frac{\tau^2}{2}G^T(t)QP(t)\right) \qquad (25)$$

where W, G, P are matrices calculated from vehicle states and inputs.

5. AN EXAMPLE OF INTEGRATED VEHICLE CONTROLLER DESIGN

5.1. Worst-Case Input

The worst-case steering and braking inputs are obtained by using the proposed method. A representative worst-case input and associated vehicle response are shown in Figure 2 The worst-case steering and braking switch between upper and lower boundary. The sideslip angle reaches 45 deg within 2 seconds.

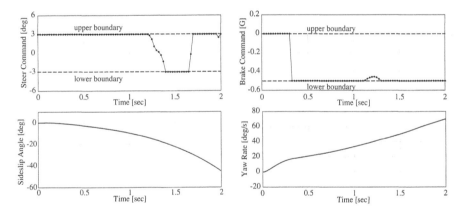

Fig. 2. Vehicle response under worst-case input.

5.2. Control Performance

The weight matrices in Equation (23) are assumed as diagonal

$$Q = \begin{bmatrix} q_1 & 0 & 0 \\ 0 & q_2 & 0 \\ 0 & 0 & q_3 \end{bmatrix}, \qquad R = \begin{bmatrix} r_1 & 0 & 0 & 0 \\ 0 & r_1 & 0 & 0 \\ 0 & 0 & r_2 & 0 \\ 0 & 0 & 0 & r_2 \end{bmatrix} \qquad (26)$$

With q_1 is fixed at 1.0, remaining 4 weighting parameters are tuned by using proposed robust design procedure. Therefore design parameter d is as follows:

$$d = \begin{bmatrix} q_2 & q_3 & r_1 & r_2 \end{bmatrix}^T \qquad (27)$$

Three patterns of inputs, the worst-case input, a step steering with 0.4G braking and a sinusoidal steering with 0.4G braking, are used as signal factor and 10 vehicle parameters are assumed to be uncertain. The optimal design parameter d^* is searched so that the performance function Equation (3) is minimized.

Figure 3 shows step response of the vehicle with optimized robust controller. As indicated in this figure, vehicle yaw rate and longitudinal velocity follow well the desired responses. Figure 4 shows the controlled vehicle response to the worst-case input. Yaw rate tracking performance is deteriorated by the worst-case input, whereas the longitudinal velocity follow well the desired velocity. Figure 5 shows the worst-case response of the vehicle with baseline controller which is tuned by designer with manual trial and error bases. Tracking performances of yaw rate and longitudinal velocity are deteriorated compared with the optimal robust design.

In Figure 6, results of 1000 Monte Carlo simulations for optimal robust design and baseline design are compared in terms of the probabilities of violation of 39 robust design metrics. As indicated in the metrics for yaw rate and longitudinal velocity, robust design shows lower probability of violation of the performance metric.

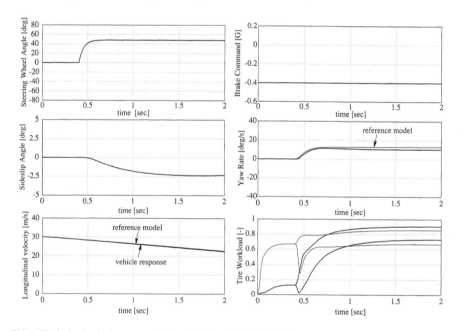

Fig. 3. Optimized vehicle response to step steering with 0.4G braking.

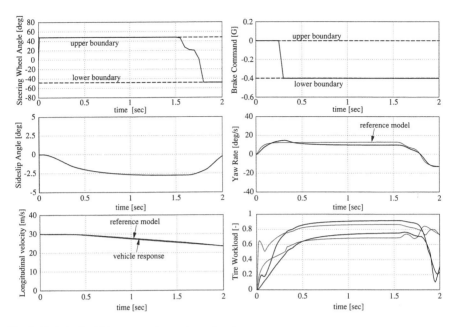

Fig. 4. Optimized vehicle response to worst-case input.

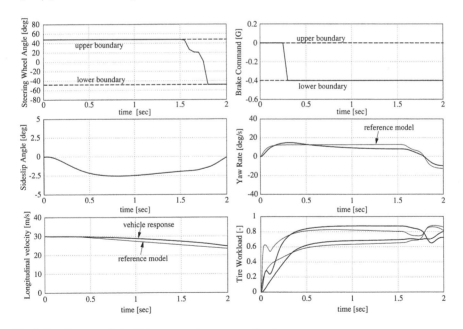

Fig. 5. Baseline controlled vehicle response to worst-case input.

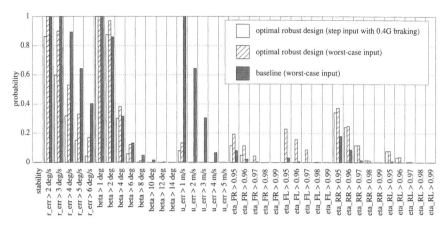

Fig. 6. Comparison of system robustness.

6. CONCLUSIONS

This paper proposes a stochastic design approach for the robust nonlinear vehicle controller based on quality engineering. It is illustrated through an example problem that this approach is useful for designing a nonlinear control system for improved stability and control even if the variations of vehicle parameters are exist. For future improvement, approximation technique such as response surface method and Kriging method could be used to replace the Monte Carlo simulation which requires significant amount of computations.

REFERENCES

1. Horiuchi, S., Okada, K. and Nohtomi, S.: Effects of Integrated Control of Active Four Wheel steering and Individual Wheel Torque on Vehicle Handling and Safety –A Comparison of Alternative Control Strategies–. *Suppl. to Vehicle System Dynamics* 33 (2000), pp. 680 – 691.
2. Horiuchi, S., Okada, K. and Nohtomi, S.: Stochastic Robustness Analysis and Synthesis of Nonlinear Vehicle Dynamics Controller. *Suppl. to Vehicle System Dynamics* 37 (2003), pp. 183 – 196.
3. Chen, W., Allen, J. K., Tsui, K. L. and Mistree, F.: A Procedure for Robust Design: Minimizing Variations Caused by Noise Factors and Control Factors. *Trans. of ASME, Jour. of Mechanical Design*, Vol. 118, (1996), pp. 478 – 485.
4. Hull, D. G.: Conversion of Optimal Control Problems into Parameter Optimization Problems, *Jour. of Guidance, Control, and Dynamics*, Vol. 20, (1997), pp. 57 – 60.
5. Ma, W. H. and Peng, H.: Worst-Case Vehicle Evaluation Methodology – Example on Truck Rollover/Jacknifing and Active Yaw Control Systems. *Vehicle System Dynamics*, Vol. 32, (1999), pp. 389 – 408.
6. Rao, S. S.: Engineering Optimization, Theory and Practice, Wiley, New York, (1996)

Vehicle System Dynamics Supplement 41 (2004), pp. 567-576

Application of Non-Linear Dynamics
to Instability Phenomena in Motorcycles

J. P. MEIJAARD[1] AND A. A. POPOV

SUMMARY

A procedure for generating vehicle models by a multibody system program and for analysing these models by numerical continuation methods is presented. This approach is applied to the dynamics of a motorcycle running straight ahead or negotiating a curve of constant radius.

1   INTRODUCTION

In this paper, the aim is to show how modern numerical techniques for investigating non-linear dynamical systems can be used to study the dynamical behaviour of motorcycles. Whereas techniques from the theory of non-linear vibrations have been used to analyse the behaviour of railway vehicles for almost half a century [1, 2], it appears that applications to the analysis of road vehicles have been rather limited. An early exception can be found in the study of the shimmy phenomenon by Pacejka [3]. The need to perform such a non-linear analysis is greater for motorcycles than for cars because of the former's inherent weaker stability, the large roll angles that occur in cornering and the more serious consequences of unexpected behaviour.

An extensive survey of past research in the field of motorcycle dynamics can be found in [4–6]. The first study that included the major contributions that affect the lateral stability of a motorcycle running straight ahead at a constant speed was made by Sharp [7]. It showed three basic ways in which the motion could become unstable: the capsize instability like in an inverted pendulum, the wobble in which the front wheel behaves as a shimmying caster wheel, and the weave in which the complete motorcycle makes a yawing and rolling motion. A more detailed model, which could also be used for the case of negotiating a curve with constant radius at a constant speed, was developed by Koenen [8]. Still, only the linearized motion about a stationary trim position was considered. The model was developed further in [9], where simulations with the non-linear model could be made.

As our main interest here is in showing how the analysis procedure can be used, the motorcycle model in this paper is taken from [7], but with a modification of the

[1]*Address correspondence to:* School of Mechanical, Materials, Manufacturing Engineering and Management, The University of Nottingham, University Park, Nottingham NG7 2RD, United Kingdom. Tel.: +44 115 8467682; Fax: +44 115 9513800; E-mail: jaap.meijaard@nottingham.ac.uk

tyre force model, so large deviations from the nominal position and the saturation of the tyre forces is allowed for. The behaviour of the rider is not included in the model; it is felt that a human control strategy should rather be based on the results of the calculations and the limitations of the rider at a later stage.

2   ANALYSIS METHODS

In imitation of the example in [9], the building of the mechanical model of the motorcycle and the derivation of the equations of motion is performed with the aid of a symbolic multibody dynamics program, AutoSim [10, 11]. This program can handle systems of rigid bodies that are interconnected by translational and revolute joints and combinations of these, and are arranged in a tree topology. Additional constraints can be imposed on the system for taking into account closed kinematic loops, special types of joints and non-holonomic constraints. Moreover, it is possible to describe a wide range of force elements. For our present needs, these rather limited modelling capabilities are sufficient.

The methods used for deriving the equations of motion are based on Kane's approach [12] with some modifications. The program is written in the language Lisp [13] and consists of a set of definitions of data structures, functions and macros. These definitions give procedures for manipulating symbolic expressions, for modelling components of multibody systems such as rigid bodies, points, joints and force elements, for formulating the equations of motion and for generating output. The input file for a specific system consists of a Lisp program and all standard functions and symbolic routines are available to the user. In the present investigation, the extensibility of the software has been exploited in the writing of procedures for generating the variational equations and derivatives of the equations with respect to parameters. These are needed for the continuation software that is used. The output of the program consists of a Fortran-77 main program which contains subroutines that calculate the right-hand sides of the dynamical equations, subroutines for the initialization, for reading parameter values and for writing output. The call to the numerical integration routine has been replaced by a call to the continuation subprogram.

For the analysis of the behaviour of the solutions of the equations of motion, the numerical continuation program AUTO [14, 15] is used, which is freely available. This program can perform a limited bifurcation analysis for systems of autonomous first-order differential equations of the form

$$\dot{x} = f(x, \lambda), \tag{1}$$

where the vector x contains the state variables and the vector λ the parameters. In particular, if one parameter is made variable, the program can determine branches of stationary solutions, in which the state variables have constant values, and branches of

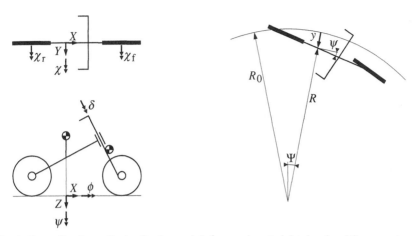

Fig. 1. Generalized coordinates for the models for running straight ahead and for cornering.

periodic solutions. The stability of the stationary solutions is investigated by examining the eigenvalues for the linearized equations. In general, a solution can change its stability at a saddle-node bifurcation, where the variable parameter reaches an extreme value, or at a Hopf bifurcation, where periodic solutions branch off. If the system and the stationary solution have a symmetry, pitchfork bifurcations can occur. Periodic solutions are found by piecewise approximating the solution by polynomials and making them satisfy the differential equations in a number of collocation points. Their stability is found by calculating their characteristic multipliers according to the Floquet theory. These multipliers measure the factors by which perturbations are multiplied after one period, so if the absolute values of these multipliers are smaller than one, perturbations decay and the solution is stable, while if the magnitude of some multiplier is larger than one, the solution is unstable. The bifurcations that can occur are saddle-node bifurcations and secondary Hopf bifurcations and also period-doubling bifurcations. More detailed information on bifurcations with many examples can be found in [16].

3   MOTORCYCLE MODEL

The model used in the present calculations is based on [7], but with an extension to produce a model valid for larger motions. The modifications mainly concern the contact of the tyre with the ground and the corresponding forces. Two different models are used: one in which the motorcycle runs on a straight level road and another in which the motorcycle negotiates a curve with constant radius. Only the generalized coordinates are different.

The mechanical model of the motorcycle consists of four rigid bodies: the rear frame with the rider rigidly attached to it, the front fork and the two wheels. These bodies are interconnected by revolute joints at the steering head between the rear frame and the front fork and at the two wheel hubs. No suspension system is included in the model. This model has nine degrees of freedom. In the reference position, the origin of the rear frame is directly below the centre of mass of the rear frame assembly at the road surface. For riding on a straight road, the generalized coordinates to describe the configuration of the system are the global Cartesian coordinates of the origin, X, Y and Z, the yaw angle, ψ, the pitch angle, χ, and the roll angle, ϕ, of the rear frame, the steering angle, δ, and the two relative rotation angles of the wheels, χ_r and χ_f. For curving, the Cartesian coordinates X and Y are replaced by polar coordinates with respect to the centre of the curve, namely the angle Ψ and the radius $R = R_0$ y, where the radius is split in a nominal curve radius R_0 and a lateral displacement y with respect to this radius. The yaw angle is split in two parts as $\Psi + \psi$. Figure 1 shows the directions of the axes.

The dimensions and mechanical properties are the same as those of [7], with the addition of the moments of inertia with respect to the pitch axis, which are taken as $I_{ry} = 42$ kgm^2 for the rear frame and $I_{fy} = 1.6$ kgm^2 for the front fork. The rear wheel is driven by the engine, which is assumed to have a reduced engine constant of $K_m = 10$ Nms. This means that the driving torque T_r on the rear wheel is given by

$$T_r = K_m(\dot{\chi}_r + V_0/R_e), \tag{2}$$

where V_0 is the velocity of the freely running motorcycle as determined by the throttle position and R_e is the nominal effective rolling radius of the rear wheel. Note that $\dot{\chi}_r$ is negative for forward motion.

The shape of the outer contour of the tyre tread is approximated by a toroidal surface. This surface is determined by the distance between the centre of the wheel and the centre of curvature of a meridional section of the tyre, c_r, and a radius of curvature of this section, ρ. A typical section is shown in the left part of Fig. 2.

Several unit vectors can be associated with the wheel, which is generically indicated with the letter 'w' (see Fig. 2, right). Firstly, the unit vectors in the directions of the global coordinate axes are n_X, n_Y and n_Z. The absolute position vector of the wheel centre is denoted by w_0. The body-fixed unit direction vectors attached to the centre of the wheel are w_x, w_y and w_z, where w_y points in the direction of the wheel hub axis. As the absolute rotation angle of the rotatory symmetric wheel is unimportant, it is convenient to introduce rotated unit vectors in the plane of the wheel, viz. w_{long} that is parallel to the ground surface and w_{rad} that points in a downward direction such that the triple (w_{long}, w_y, w_{rad}) forms a right-handed orthogonal system. In the ground surface a longitudinal direction vector parallel to w_{long} and a lateral unit direction vector w_{lat} are defined such that the triple (w_{long}, w_{lat}, n_Z) forms a right-handed orthogonal system. The inclination angle, or camber angle, is the angle between the unit vectors w_{rad} and n_Z, positive for an inclination to the right.

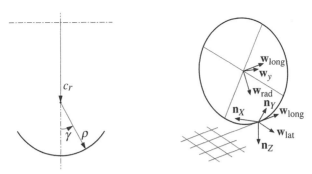

Fig. 2. Meridional section of the tread of a tyre (left) and unit direction vectors of a wheel (right).

Instead of the nominal contact point as in the recommended practice by the S.A.E. [17], which may be outside the actual contact patch of a round-section motorcycle tyre if the inclination angle is large, a point that is near the centre of the actual contact patch is used as a reference for the contact forces. We define the undeformed contact point as the point on the undeformed tread surface that would have the largest penetration in the ground surface if the tyre were rigid and the ground would deform (or, if contact is lost, the point with the smallest distance to the ground surface). The absolute position vector of this point, $\mathbf{w}_{cu}$, is given by the expression

$$\mathbf{w}_{cu} = \mathbf{w}_0 + c_r \mathbf{w}_{rad} + \rho \mathbf{n}_Z. \tag{3}$$

On paved roads the deformation of the road surface can usually be neglected in comparison with the deformation of the tyre. Therefore, we define the deformed contact point as the point that is directly above the undeformed contact point in the plane of the road surface. With the normal indentation

$$\epsilon_n = \mathbf{w}_{cu} \cdot \mathbf{n}_Z, \tag{4}$$

the absolute position vector of the deformed contact point, $\mathbf{w}_{cd}$, becomes

$$\mathbf{w}_{cd} = \mathbf{w}_{cu} \quad \epsilon_n \mathbf{n}_Z. \tag{5}$$

Three slips are defined in the contact point: the longitudinal slip velocity, the lateral slip velocity and the normal spin velocity. The longitudinal and lateral slip velocities are defined as

$$\dot{\epsilon}_{long} = \left[\dot{\mathbf{w}}_0 + \boldsymbol{\omega}_0 \times \left(\xi_{long} \mathbf{w}_{cu} + (1 \quad \xi_{long}) \mathbf{w}_{cd} \quad \mathbf{w}_0 \right) \right] \cdot \mathbf{w}_{long},$$
$$\dot{\epsilon}_{lat} = \left[\dot{\mathbf{w}}_0 + \boldsymbol{\omega}_0 \times \left(\xi_{lat} \mathbf{w}_{cu} + (1 \quad \xi_{lat}) \mathbf{w}_{cd} \quad \mathbf{w}_0 \right) \right] \cdot \mathbf{w}_{lat}, \tag{6}$$

where $\boldsymbol{\omega}_0$ is the angular velocity of the wheel and the coefficients ξ_{long} and ξ_{lat} take account of the effective rolling radius in longitudinal and lateral direction. The normal spin velocity is the component of the angular velocity vector in the normal direction,

$$\omega_n = \omega_0 \cdot \mathbf{n}_Z. \tag{7}$$

The dimensionless slip quantities are obtained by dividing the slip velocities by the rolling velocity $V_w = \dot{\mathbf{w}}_0 \cdot \mathbf{w}_{long}$ as

$$s_{long} = \dot{\epsilon}_{long}/V_w, \qquad s_{lat} = \dot{\epsilon}_{lat}/V_w, \qquad s_n = \omega_n R_w/V_w. \tag{8}$$

The nominal wheel radius R_w is introduced to make the normal spin velocity dimensionless. For zero camber angles, s_{long} is the negative of the usual longitudinal slip and s_{lat} is the tangent of the side slip angle.

The contact forces are defined as the energetic duals of the slip velocities and the normal indentation. These are the compressive normal force, F_n, the longitudinal force, F_{long}, acting in the direction of $\mathbf{w}_{long}$, the lateral force, F_{lat}, acting in the direction of $\mathbf{w}_{lat}$, and the aligning moment, M_n, in the direction of $\mathbf{n}_Z$. The normal force is assumed to be a linear function of the indentation and the rate of indentation, unless the contact with the road surface is lost, in which case the normal force becomes zero. The non-zero normal force is given by

$$F_n = K_n \epsilon_n + C_n \dot{\epsilon}_n. \tag{9}$$

The normal stiffness K_n is determined from the deflection in the reference position, which is estimated as 0.01 m, and the damping constant is chosen as $C_n = 500$ Nm/s, so the relative damping of the heave and pitch modes is 5–10 %. The evolution of the other three generalized contact forces are related to their steady-state values through a first-order low-pass filter as

$$\dot{F}_{long} = (V_w/\sigma_{long})[F_{long,ss}(s_{long}, s_{lat}, s_n, F_n, \gamma) \quad F_{long}],$$
$$\dot{F}_{lat} = (V_w/\sigma_{lat})[F_{lat,ss}(s_{long}, s_{lat}, s_n, F_n, \gamma) \quad F_{lat}], \tag{10}$$
$$\dot{M}_n = (V_w/\sigma_n)[M_{n,ss}(s_{long}, s_{lat}, s_n, F_n, \gamma) \quad M_n].$$

In these expressions, σ_{long}, σ_{lat} and σ_n are the relaxation lengths that depend on the normal load. These are chosen such that $\sigma_{long} = \sigma_{lat}$ and $\sigma_n = \frac{1}{2}\sigma_{lat}$. Two auxiliary quantities, which may be called the equivalent lateral slip and the total slip, are introduced as

$$s_{lat,eq} = s_{lat} \quad \xi_{eq} s_n, \qquad s_{tot}^2 = s_{long}^2 + s_{lat,eq}^2. \tag{11}$$

The factor ξ_{eq} takes into account the contribution of the normal spin to the lateral force. Since this is mainly due to the camber, the factor is chosen as the ratio of the camber and the cornering coefficients. With a magic-formula [18] type of expression, the steady state forces are given by

$$F_{long,ss} = \frac{C_{w1} s_{long}}{\sqrt{1 + B^2 s_{tot}^2}}, \quad F_{lat,ss} = \frac{C_{w1} s_{lat,eq}}{\sqrt{1 + B^2 s_{tot}^2}}, \quad M_{n,ss} = \frac{R_w C_{w2} s_{lat}}{1 + B^2 s_{tot}^2}. \tag{12}$$

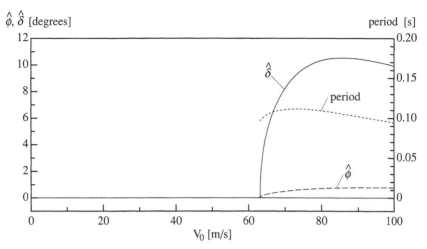

Fig. 3. Response diagram for running straight ahead. The nominal motion undergoes a Hopf bifurcation at the speed $V = 63.16$ m/s, at which a branch of periodic motions emerges. The amplitudes of the lean angle, $\hat{\phi}$, and of the steering angle, $\hat{\delta}$, as well as the period of vibration are shown.

The linearized coefficients are assumed to be proportional to the normal force and the coefficient B is given by $B = C_{w1}/(\mu_w F_n)$. Normally, the value $\mu_w = 1$ for the friction coefficient is chosen.

For riding straight ahead, no model for the rider input is needed, but for negotiating a curve, the rider is modelled as a weak PI-controller with the lateral deviations from the intended course as input and a steering torque as output.

4   RESULTS

Firstly, the stability when nominally running straight ahead under varying forward speed is investigated. The main parameter is V_0, the velocity of the freely running motorcycle. Without rider control, this nominal solution appears to be mildly unstable for all velocities. We call the solution mildly unstable if the unstable eigenvalues have an absolute value that is smaller than 2 s^{-1}, in which case it is fairly easy to stabilize these motions. No interval of stable solutions is found, as distinct from [7], which can be attributed to the difference in the contact model.

At $V_0 = 63.16$ m/s, a Hopf bifurcation occurs, which initiates a wobble motion. This bifurcation is found at a slightly higher value of the velocity in comparison with [7], which can be attributed to the inclusion of the pneumatic trail. Figure 3 shows the amplitudes of the steering angle, $\hat{\delta}$, and the roll angle, $\hat{\phi}$, and the period of the wobble

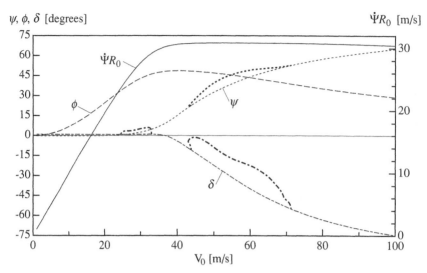

Fig. 4. Response diagram for negotiating a curve. The stationary values for the trajectory speed $\dot{\Psi} R_0$ (fully drawn), the relative yaw angle ψ (dotted), the roll angle ϕ (dashed) and the steering angle δ (dashed-dotted) are show. Corresponding maximal values of angles for periodic solutions are drawn with thick lines.

motion. It is seen that the bifurcation is of the soft type: the amplitudes increase continuously if the critical velocity is surpassed.

Increasing the steering damper by a factor of two removes the wobble instability, although the weave mode becomes less damped. Indeed, if the steering damping is increased by a factor of three, the weave mode becomes unstable between $V_0 = 43.09$ m/s and $V_0 = 59.62$ m/s. Lowering the mass centre of the rear frame assembly to half its value stabilizes the capsize mode between $V_0 = 5.30$ m/s and $V_0 = 15.90$ m/s and increases the critical velocity to 69.85 m/s. Lowering the lateral tyre stiffness with 20% increases the critical velocity to 68.27 m/s.

Next, the stationary cornering in a curve with radius $R_0 = 100$ m is examined. Figure 4 shows the relative yaw, lean and steering angle as functions of V_0. Also the speed along the desired trajectory, $\dot{\Psi} R_0$, is shown. Up to 25 m/s (90 km/h) the lean angle increases with speed to about 35°, while the yaw and steer angle remain small. At $V_0 = 24.51$ m/s, a Hopf bifurcation occurs, which can be characterized as a cornering wobble. A branch of periodic solutions, initially with frequency 5.67 Hz, connects this point to a second Hopf bifurcation at $V_0 = 32.94$ m/s. The maximal steering angle for this branch is shown in Fig. 4. Above this speeds, the stationary solution is only mildly unstable until the next Hopf bifurcation at $V_0 = 43.77$ m/s, after which the solution becomes more strongly unstable. Another branch of periodic solutions connects the third Hopf bifurcation to a fourth one at $V_0 = 71.23$ m/s.

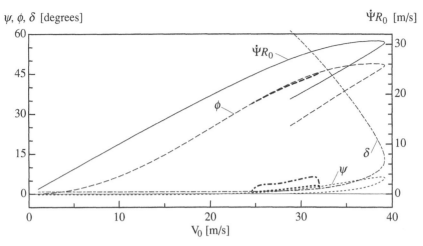

Fig. 5. Response diagram for negotiating a curve, as in Fig. 4, but with $\mu_r = 1.1$.

Initially, this is mainly a wobble instability, but the front wheel loses contact with the ground in most of this branch, where the transition points can be seen as the kinks in the amplitudes for the steering angle. The trajectory speed does not change much over a wide range of V_0, because the maximal lateral acceleration is limited by the friction coefficient, $\mu_r = \mu_f = 1$, while the relative yaw angle increases and the steering angle decreases. The limited grip is felt at first at the rear tyre. This behaviour reminds us of the oversteer behaviour for cars.

For the case in which the coefficient of friction at the rear wheel is higher, $\mu_r = 1.1$, the results are shown in Fig. 5. For the lower velocities there is not much difference with the previous case. Even the branch of periodic solutions connecting two Hopf bifurcation points are found at almost the same place. However, two other Hopf bifurcations are found at $V_0 = 28.95$ m/s and $V_0 = 36.12$ m/s. The ultimate behaviour of the curves is completely different. A turning point with a maximum of V_0 is found, and the steering angle increases up to $90°$ (only results for δ up to $60°$ are shown), while the relative yaw angle remains small. The limitation of the grip is now felt at first at the front tyre. This reminds us of the understeer behaviour for cars.

5 CONCLUSIONS

It has been shown how a multibody program and continuation software can be used to investigate the non-linear behaviour of a motorcycle. Realistically looking results were found that are difficult to obtain by other means. Future work will deal with the refinements of the motorcycle model, such as the inclusion of the suspension system and frame flexibility, and the inclusion of the rider input.

ACKNOWLEDGEMENT

This work is supported by a research grant from the Engineering and Physical Sciences Research Council (EPSRC) of the U.K.

REFERENCES

1. de Pater, A.D.: Étude du mouvement de lacet d'un véhicule de chemin de fer. *Appl. Sci. Res.* A 6 (1957), pp. 263–316.
2. Knothe, K. and Böhm, F.: History of stability of railway and road vehicles. *Vehicle System Dynamics* 31 (1999), pp. 283–323.
3. Pacejka, H.B.: *The Wheel Shimmy Phenomenon, a Theoretical and Experimental Investigation with Particular Reference to the Non-Linear Problem.* (Doctoral dissertation, Technische Hogeschool Delft) Kleine, Groningen, 1966.
4. Sharp, R.S.: A review of motorcycle steering behavior and straight line stability characteristics. SAE paper 780303; also in Anon.: *Motorcycle Dynamics and Rider Control*, SP-428, Society of Automotive Engineers, Warrendale, PA, 1978, pp. 1–6.
5. Sharp, R.S.: The lateral dynamics of motorcycles and bicycles. *Vehicle System Dynamics* 14 (1985), pp. 265–283.
6. Sharp, R.S.: Stability, control and steering responses of motorcycles. *Vehicle System Dynamics* 35 (2001), pp. 291–318.
7. Sharp, R.S.: The stability and control of motorcycles. *Proc. Inst. Mech. Engs. C, J. Mech. Engrg. Sci.* 13 (1971), pp. 316–329.
8. Koenen, C.: *The Dynamic Behaviour of a Motorcycle When Running Straight Ahead and When Cornering.* (Doctoral dissertation) Delft University Press, Delft, 1983.
9. Sharp, R.S. and Limebeer, D.J.N.: A motorcycle model for stability and control analysis. *Multibody System Dynamics* 6 (2001), pp. 123–142.
10. Sayers, M.W.: *Symbolic Computer Methods to Automatically Formulate Vehicle Simulation Codes.* (Doctoral dissertation) The University of Michigan, Ann Arbor, MI, 1990.
11. Anon.: *AutoSim*^{TM} *Reference Manual Version 2.5+.* Mechanical Simulation Corporation, Ann Arbor, MI, 1997.
12. Kane, T.R.: *Dynamics.* Holt, Rinehart and Winston, New York, 1968.
13. Steele, G.L., Jr: *Common Lisp, The Language.* (2nd ed.) Digital Press, U.S.A., 1990.
14. Doedel, E., Keller, H.B. and Kernevez, J.P.: Numerical Analysis and Control of Bifurcation Problems, (I) Bifurcation in Finite Dimensions, (II) Bifurcation in Infinite Dimensions. *Internat. J. Bifur. Chaos Appl. Sci. Engrg.* 1 (1991), pp. 493–520, 745–772.
15. Doedel, E.J., Champneys, A.R., Fairgrieve, T.F., Kuznetsov, Y.A., Sandstede, B. and Wang, X.: AUTO 97: Continuation and Bifurcation Software for Ordinary Differential Equations (with HomCont). Technical Report, Department of Computer Science, Concordia University, Montréal, 1997.
16. Thompson, J.M.T., and Stewart, H.B.: *Nonlinear Dynamics and Chaos, Geometrical Methods for Engineers and Scientists.* Wiley, Chichester, 1986.
17. Anon.: *Vehicle Dynamics Terminology*, SAE Recommended Practice J670e. Society of Automotive Engineers, Warrendale, PA, 1978.
18. Pacejka, H.B.: *Tyre and Vehicle Dynamics.* Butterworth and Heinemann, Oxford, 2002.

Vehicle System Dynamics Supplement 41 (2004), p.577-586   © Taylor & Francis Ltd.

Practical String Stability for Longitudinal Control of Automated Vehicles

XIAO-YUN LU *   and   J. KARL HEDRICK *

SUMMARY

This paper considers string stability for vehicle longitudinal control from a practical implementation viewpoint. A different parameterization is used compared to previous work. Two following strategies are considered. i.e. Adaptive Cruise Control (ACC) and vehicle platooning with inter-vehicle communication. A simplified model and transfer function are used for analysis. It shows that ACC cannot achieve string stability while platooning with communication can achieve it, which agrees with practical test.

1 INTRODUCTION

String stability describes the dynamic longitudinal inter-reaction between vehicles in short inter-vehicle distance following.

Previous work in [8] defines the *string stability* as an *asymptotic stability* of the overall system which is composed of a finite number of inter-connected subsystems with the same or similar dynamics. Necessarily, each closed-loop subsystem must be asymptotically stable. This is the ideal case for the dynamic behavior of the series of sub-system connected in a special fashion like a string.

For practical implementation, it is necessary to analyze overall system stability under nearly real circumstances. In practice, the following factors will affect the overall stability and performance:

(a) Time lag in sensor and actuators
(b) Pure time delays in sensor measurement and signal processing
(c) Model mismatch
(d) Measurement noises

The time delays will naturally cause measurement and actuation discrepancies. Due to such discrepancies in addition to model mismatch, measurement noise and external disturbances , each sub-system (a single vehicle) can only achieve ultimate boundedness in stability [2, 4], which coincides with experimental work. Thus to require strict attenuation of tracking error down stream (direction from the first vehicle to the last vehicle) along the platoon is too restrictive. Practically, *string stability* in vehicle following can only require that distance and speed tracking error will not propagate or has limited propagate

---
[1]PATH, U. C. Berkeley, Richmond Field Station Building 452, 1357 S. 46th Street, Richmond, CA 94804-4648, Email: xylu@nt.path.berkeley.edu khedrick@me.berkeley.edu

down stream in a platoon. However, for theoretical analysis, the definition for *string stability* in [8] is reasonable.

In ([9]), string stability for many vehicle following strategies has been considered. Time lag is also considered but not pure time delays. ([3]) considered both time lag caused by actuators and pure time delay caused by inter-vehicle communication. Due to the complication of the problem formulated, it is impossible to consider arbitrary design parameters. i.e. It only show that the system is string stable/unstable when some particular control parameters are chosen. This work uses a different parameterization approach compared to the work of [3, 9]. In this paper, two time delays are taken into consideration while model mismatch and measurement noise are ignored for simplicity. It greatly simplifies the problem from both theoretical analysis and practical implementation viewpoints. The results of this paper can be used for stability analysis of longitudinal control of any vehicles following, not just for ground vehicles.

String stability mainly depends on following strategy as discussed in [5] while the latter depends on availability of information from the preceding vehicles. If each vehicle follows its immediate preceding vehicle, it is the *Adaptive Cruise Control* (ACC) mode (no inter-vehicle communication). It will be shown mathematically that ACC mode cannot achieve string stability. Intuitively, a vicious circle exists in ACC which makes it string unstable. However, if certain amount of information from leader vehicle passed over by inter-vehicle communication is used, string stability can be achieved irrespective of the time delays. In fact, the key to achieve string stability in vehicle following is to appropriately incorporate leader vehicle information passed over by communication.[1] simplified the mathematical model as a finite number of linear spring-damper systems connected in a string fashion. The problem is that the spring and damping effects is expected to be produced from the closed controller instead of being assumed *a priori*.

The rest of the paper is organized as follows. Section 2 introduces string stability in vehicle following and its mathematical criteria. Section 3 is for string stability analysis using transfer functions for two typical following strategies in practice. Section 4 presents test results using 4 full size automated passenger cars with brief discussion. Section 5 present some concluding remarks.

Basic Notations

$x_i(t)$ or simply (x_i) − position of vehicle i in longitudinal direction. All the vehicles are with respect to a inertia frame.

$v_i(t), a_i(t)$− speed and acceleration of vehicle i

h_{p1}− time delay for obtaining front range

h_{p2}− time delay for obtaining preceding vehicle's speed and acceleration

h_l− time delay for on-car sensor measuring and for communication system to pass the leader vehicle's distance, speed and acceleration to other vehicles

L_i is the desired inter-vehicle distance with vehicle length accounted for

l− subscript for the leader vehicle

2 STRING STABILITY FOR VEHICLE FOLLOWING

This section will provide mathematical criteria for string stability in vehicle following.

2.1 String Stability of Vehicle Following

Let

$$\begin{aligned}
\varepsilon_i(t) &= x_i(t) - x_{i-1}(t) + L_i \\
\dot{\varepsilon}_i(t) &= v_i(t) - v_{i-1}(t) \\
\ddot{\varepsilon}_i(t) &= a_i(t) - a_{i-1}(t)
\end{aligned}$$

$E_i(s)$ is the Laplace transformation of $\varepsilon_i(t)$. $G(s)$ is the transfer function of the closed-loop dynamics $g(t)$ of sub-system i, which is the same for each vehicle. Then

$$G(s) = \frac{E_i(s)}{E_{i-1}(s)} \tag{1}$$

The string stability for a platoon of n vehicles requires that

$$\|\varepsilon_1\|_\infty \leq \|\varepsilon_2\|_\infty \leq \cdots \leq \|\varepsilon_n\|_\infty$$

From linear system theory

$$\begin{aligned}
\|\varepsilon_i\|_\infty &\leq \|g(t)\|_1 = \int_0^\infty |g(\tau)|\, d\tau \\
\|g * \varepsilon_i\|_\infty &\leq \|g(t)\|_1 \|\varepsilon_i\|_\infty \\
\|G(s)\|_\infty &\leq \|g(t)\|_1
\end{aligned} \tag{2}$$

Thus the inter-connected system is string stable if $\|g(t)\|_1 < 1$ and string unstable if $\|G(s)\|_\infty > 1$. To practically check it, one needs to evaluate $\|g(t)\|_1$.

2.2 Following Strategy and String Stability

There are two typical vehicle following strategies: (a) short distance following with inter-vehicle communication in Automated Highway Systems [7]; (b) *Adaptive Cruise Control (ACC)* without inter-vehicle communication in normal highway system. However, in practice, the string stability of automated vehicle platooning can be achieved for 8 vehicles as tested in PATH, while string stability cannot be achieved for *ACC* with more than 2 vehicles. The reason for this is that the most important factor, *time delay* which is due to sensor estimation and actuators, has been largely ignored in vehicle following.

Consider a simplified generic vehicle longitudinal dynamic model:

$$\begin{aligned}
\dot{x} &= v \\
\dot{v} &= \frac{1}{M}\left(U - C_a v^2 - F_r\right)
\end{aligned} \tag{3}$$

F_r— rolling resistance, C_a— aerodynamic drag coefficient, and U— force (control variable).

For string stability analysis, the above model can be simplified further for vehicle i as

$$\begin{aligned}
\dot{x}_i &= v_i \\
\dot{v}_i &= u_i
\end{aligned} \tag{4}$$

where u_i is the synthetic force. It can be considered as the upper level control. Such a simplification ignored all the actuator dynamics which may aggravate the performance in practice.

Let the sliding surface be defined as

$$S_i = \alpha \, \dot{\varepsilon}_i + \alpha q \varepsilon_i + (1 - \alpha)(v_i - v_l) + (1 - \alpha) q \left(x_i - x_l + \sum_{j=2}^{i} L_j \right)$$

where $\alpha \in [0, 1]$ is the interpolation parameter. Two extreme cases are: $\alpha = 1$ which means that each vehicle follows the preceding vehicle only and no lead vehicle information is used; $\alpha = 0$ which implies that each vehicle follows the leader vehicle only. However, the most interesting cases correspond to $0 < \alpha \leq 1$.

If the following sliding reachability condition

$$\dot{S}_i = -\lambda S_i$$

is used for vehicle i, with $\lambda > 0$, the controller (synthetic force) is solved out as

$$u_i^{(d)} = \alpha \, \ddot{x}_{i-1} + (1 - \alpha) \, \ddot{x}_l - \alpha(q + \lambda) \, \dot{\varepsilon}_i - \alpha \lambda q \varepsilon_i \\ - (1 - \alpha)(q + \lambda)(v_i - v_l) - \lambda q (1 - \alpha) \left(x_i - x_l + \sum_{j=2}^{i} L_j \right) \tag{5}$$

The design parameters (q, λ, α) are to be chosen such that
(a) The closed loop controller for each vehicle is stable;
(b) The overall system which is composed of finite number of inter-connected similar sub-systems is string stable.

2.3 Time Delays

In practice, there are two types of time delays: time lag and pure time delay. A first order filter is inserted to represent the effect of time lag as

$$\tau \, \dot{u}_i + u_i = u_{id}$$

which links the controller (5) and upper level vehicle model (4) with $\tau = 0.15[s]$.
There are two fundamentally different cases for Pure time delays.

Case 1: *With inter-vehicle communication*
Relative distance $(x_i - x_{i-1})$ is estimated from distance sensor reading, which causes pure time delay of $h_{p1} \approx 0.25[s]$ where $0.1[s]$ comes from radar/Lidar sensor delay (physical and radar internal signal processing) and $0.15[s]$ is due to signal processing in feed-forward control. (v_{pre}, a_{pre}) is passed over by communication which causes time delay about $h_{p2} \approx 0.1[s]$ in which $0.02[s]$ is the communication cycling period and $0.08[s]$ is due to the speed and acceleration sensor delays.

Assumption 1: The communication system passes information from the leader vehicle to each vehicle and from each vehicle to its follower simultaneously. The common time delays for each vehicle is $0.02[s]$ (time step used for control.).

Assumption 2: Pure sensor time delay on vehicle i with respect to the preceding vehicle is the same for all the vehicles. All the sensor measurement discrepancies can be ignored.

Case 2: *Without inter-vehicle communication*
All the three elements in $[(v_i - v_{i-1}), (a_i - a_{i-1}), (x_i - x_{i-1})]$ are estimated from measurement by Doppler radar, Lidar and video camera. In this case:

$h_{p1} \approx 0.25[s]$, $h_{p2} = 0.35[s]$. Note that leader vehicle information is unavailable directly.

2.4 Transfer Function Expression

To use frequency analysis approach to calculate the H_∞ gain, the transfer function for the closed-loop system of each vehicle is calculated as follows.

$$\tau \frac{d^3 \varepsilon_i}{dt^3} + \dddot{\varepsilon}_i = u_i^{(d)} - u_{i-1}^{(d)}$$
$$= -\alpha(q + \lambda)\dot{\varepsilon}_i - \alpha\lambda q \varepsilon_i - \lambda q(1 - \alpha)\varepsilon_i$$
$$- (1 - \alpha)(q + \lambda)(v_i(t) - v_{i-1}(t)) + \alpha \ddot{x}_{i-1}(t - h_{p2})$$
$$-\alpha \ddot{x}_{i-2}(t - h_{p2}) + \alpha(q + \lambda)(\dot{x}_{i-1}(t - h_{p2}) - \dot{x}_{i-2}(t - h_{p2}))$$
$$+\lambda\alpha q(x_{i-1}(t - h_{p1}) - x_{i-2}(t - h_{p1}))$$

Using Laplace transformation on both side to get

$$G(s) = \frac{E_i(s)}{E_{i-1}(s)} = \frac{\lambda\alpha q e^{-h_{p1}s} + \alpha s e^{-h_{p2}s}(s + (q + \lambda))}{\tau s^3 + s^2 + (q + \lambda)s + \lambda q} \tag{6}$$

3 STRING STABILITY ANALYSIS

This section analyzes the string stability with respect to the two typical following strategies above.

3.1 Vehicle Following without Communication (ACC)

In (6), set $\alpha = 1$ which is equivalent to using preceding vehicle information only the control law (5) becomes

$$u_i^{(d)} = \ddot{x}_i - (q + \lambda)\dot{\varepsilon}_i - q\lambda\varepsilon_i \tag{7}$$

It is obtained that

$$G(s) = \frac{\lambda q e^{-h_{p1}s} + s e^{-h_{p2}s}(s + (\lambda + q))}{\tau s^3 + s^2 + (\lambda + q)s + \lambda q}$$
$$= \frac{\lambda q + s e^{-(h_{p2} - h_{p1})s}(s + (\lambda + q))}{\tau s^3 + s^2 + (\lambda + q)s + \lambda q}e^{-h_{p1}s}$$

Additionally, for the stability of the feedback dynamics, it is necessary and sufficient that

$$D(s) = \tau s^3 + s^2 + (\lambda + q)s + \lambda q$$

be Hurwitz, which is equivalent to the parameter constraints:

$$\begin{aligned} \lambda q &> 0 \\ (\lambda + q) - \tau\lambda q &> 0 \end{aligned} \tag{8}$$

Because $e^{-h_{p1}s}$ does not effect the value of $|G(j\omega)|$ and thus it is ignored. Let $h_p = h_{p2} - h_{p1} > 0$ for simplicity. Thus

$$G(j\omega) = \frac{\lambda q + j\omega \left(\cos(\omega h_p) - j \sin(\omega h_p) \right) (j\omega + (\lambda + q))}{-\tau j\omega^3 - \omega^2 + (\lambda + q)j\omega + \lambda q}$$

Now it is necessary to evaluate $\|G\|_\infty = \max_\omega |G(j\omega)|$.

For considering only very small $\omega > 0$, the $3rd$ order terms are ignored to obtain

$$G(j\omega) \approx \frac{\lambda q + j\omega \left(1 - j\omega h_p \right) (j\omega + (\lambda + q))}{-\omega^2 + (\lambda + q)j\omega + \lambda q} = 1 + \frac{\omega^2 h_p(\lambda + q)}{(\lambda q - \omega^2) + (\lambda + q)j\omega}$$

$$= 1 + \frac{\omega^2 h_p(\lambda + q)(\lambda q - \omega^2)}{(\lambda q - \omega^2)^2 + (\lambda + q)^2 \omega^2} - \frac{(\lambda + q)j\omega}{(\lambda q - \omega^2)^2 + (\lambda + q)^2 \omega^2}$$

One can observe that for any feedback stabilization law (7) with constraints (8), if $\omega > 0$ is chosen very small such that $\omega^2 < \lambda q$ arbitrarily small $h_p > 0$ will lead to

$$\frac{\omega^2 h_p(\lambda + q)(\lambda q - \omega^2)}{(\lambda q - \omega^2)^2 + (\lambda + q)^2 \omega^2} > 0$$

Which means that the real part is positive and greater than 1. This implies from (2) that

$$\|G\|_\infty = \max_\omega |G(j\omega)| > 1$$

$$\|g(t)\|_1 > 1$$

It is thus concluded that the system is string unstable under any feedback control law in (7) although each sub-system is stablizable. This is summarized in the following theorem.

Theorem 1. For vehicle following in ACC mode, inter-vehicle range and range-rate are measured by radar, the system is string unstable for any stabilizing controller of the type (7).

3.2 Vehicle Following with Inter-vehicle Communication

Now (6) is directly considered with some information from the leader vehicle passed over by communication. $G(s)$ has the form

$$G(s) = \frac{\alpha \left[\lambda q e^{-h_{p1}s} + s e^{-h_{p2}s} \left(s + (q + \lambda) \right) \right]}{\tau s^3 + s^2 + (q + \lambda) s + \lambda q}, \quad 0 < \alpha < 1$$

To consider all the stabilizable feedback law, it is necessary and sufficient that the denominator

$$D = \tau s^3 + s^2 + (\lambda + q) s + \lambda q$$

be Hurwitz, which leads to the same constraint (8). Now $G(s) = \alpha G_0(s)$ with

$$G_0(s) = \frac{\lambda q e^{-h_{p1}s} + s e^{-h_{p2}s} \left(s + (q + \lambda) \right)}{\tau s^3 + s^2 + (q + \lambda) s + \lambda q}$$

Although by Initial Value Theorem and Final Value Theorem

$$\lim_{t \to 0} g_0(t) = \lim_{s \to \infty} sG_0(s) = 0$$

$$\lim_{t \to \infty} g_0(t) = \lim_{s \to 0} sG_0(s) = 0$$

where $g_0(t)$ is the inverse Laplace transformation of $G_0(s)$, one can expect that $g_0(t)$ is bounded, a rigorous proof is provided which can be used to calculate the bound in control design.

Lemma If the design parameters (λ, q) are chosen so that poles $s_k, (k = 1, 2, 3)$ are all simple. Then there exists a positive number $M > 0$ such that $\|g_0(t)\|_1 < M$.

Proof: By the definition of inverse Laplace transformation

$$
\begin{aligned}
g_0(t) &= \mathcal{L}^{-1}\left(G_0(s)\right) = \frac{1}{2\pi j} \int_{\sigma-j\infty}^{\sigma+j\infty} e^{st} G_0(s) ds \\
&= \frac{1}{2\pi j} \lim_{R \to \infty} \int_{\overline{ABCDEA}} e^{st} G_0(s) ds \\
&= \frac{1}{2\pi j} \lim_{R \to \infty} \int_{\overline{ABCDEA}} \left(\frac{\lambda q e^{s(t-h_{p1})}}{\tau s^3 + s^2 + (q+\lambda)s + \lambda q} \right) ds \\
&+ \frac{1}{2\pi j} \lim_{R \to \infty} \int_{\overline{ABCDEA}} \left(\frac{s(s+(q+\lambda))e^{s(t-h_{p2})}}{\tau s^3 + s^2 + (q+\lambda)s + \lambda q} \right) ds
\end{aligned}
\tag{9}
$$

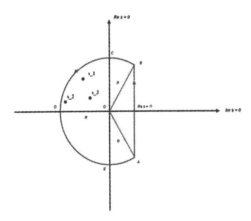

Figure 1: Integration along the contour

The integration can be implemented along the contour $\overline{ABCDEA}$ and take the limit for $R \to \infty$ (Fig. 1). By the *Residue Theorem* (p. 967 of [6])

$$g_0(t) = \frac{1}{2\pi j} \int_{\sigma-j\infty}^{\sigma+j\infty} e^{st} G_0(s) ds = \sum_{k=1}^{3} Res_{s_k}\left(G_0(s)\right)$$

For a simple pole s_k, $Res_{s_k}(G_0(s))$ is the coefficient of $\frac{1}{s-s_k}$ in the Laurent expansion of $G_0(s)$ at s_k, $(k = 1,2,3)$. Let the first integrand in (9) be partitioned into

$$\frac{\lambda q e^{s(t-h_{p1})}}{\tau s^3 + s^2 + (q+\lambda)s + \lambda q} = \frac{\lambda q e^{s(t-h_{p1})}}{\tau}\left(\frac{a^{(1)}}{s-s_1} + \frac{a^{(2)}}{s-s_2} + \frac{a^{(3)}}{s-s_3}\right)$$

where $a^{(i)} (i = 1,2,3)$ are constants. Expanding $e^{s(t-h_{p1})}$ as Taylor series at $s = s_i$ to produce

$$\frac{\lambda q e^{s(t-h_{p1})}}{\tau}\left(\frac{a^{(1)}}{s-s_1} + \frac{a^{(2)}}{s-s_2} + \frac{a^{(3)}}{s-s_3}\right) = \frac{\lambda q}{\tau}\left[\frac{a^{(1)}e^{s_1(t-h_{p1})}}{s-s_1}\sum_{n=1}^{\infty}\frac{(t-h_{p1})^n(s-s_1)^n}{n!} + \right.$$
$$\left. \frac{a^{(2)}e^{s_2(t-h_{p1})}}{s-s_2}\sum_{n=1}^{\infty}\frac{(t-h_{p1})^n(s-s_2)^n}{n!} + \frac{a^{(3)}e^{s_3(t-h_{p1})}}{s-s_3}\sum_{n=1}^{\infty}\frac{(t-h_{p1})^n(s-s_3)^n}{n!}\right]$$

Thus

$$\int_{ABCDEA}\left(\frac{\lambda q e^{s(t-h_{p1})}}{\tau s^3 + s^2 + (q+\lambda)s + \lambda q}\right)ds$$
$$= 2\pi j\left[a^{(1)}e^{s_1(t-h_{p1})} + a^{(2)}e^{s_2(t-h_{p1})} + a^{(3)}e^{s_3(t-h_{p1})}\right]$$

Similarly, if

$$\frac{s(s+(q+\lambda))}{\tau s^3 + s^2 + (q+\lambda)s + \lambda q} = \frac{b^{(1)}}{s-s_1} + \frac{b^{(2)}}{s-s_2} + \frac{b^{(3)}}{s-s_3}$$

where $b^{(i)} (i = 1,2,3)$ are constants, then

$$\int_{ABCDEA}\left(\frac{s(s+(q+\lambda))e^{s(t-h_{p2})}}{\tau s^3 + s^2 + (q+\lambda)s + \lambda q}\right)ds$$
$$= 2\pi j\left[b^{(1)}e^{s_1(t-h_{p2})} + b^{(2)}e^{s_2(t-h_{p2})} + b^{(3)}e^{s_3(t-h_{p2})}\right]$$

Thus

$$g_0(t) = a^{(1)}e^{s_1(t-h_{p1})} + a^{(2)}e^{s_2(t-h_{p1})} + a^{(3)}e^{s_3(t-h_{p1})} +$$
$$b^{(1)}e^{s_1(t-h_{p2})} + b^{(2)}e^{s_2(t-h_{p2})} + b^{(3)}e^{s_3(t-h_{p2})}$$

Notice that $s_i(i = 1,2,3)$ have negative real parts. Thus each term in $g_0(t)$ is bounded for $t \geq 0$.

This completes the proof. $\diamond$

Theorem 2. In vehicle following, if the control law (5) is used and if parameter λ and q are chosen such that the transfer function $G(s)$ has simple and stable poles. Then there exists α, with $0 < \alpha < 1$ such that (5) makes the overall system string stable irrespective the time delay caused by sensor measurement and signal processing.

Proof. By the lemma, one can choose α such that $0 < \alpha < \frac{1}{M}$. Then $\|g(t)\|_1 = \alpha\|g_0(t)\|_1 < 1$. This completes the proof. $\diamond$

Corollary Suppose that $s_i = -\sigma_i + j\eta_i$ with $\sigma_i > 0 (i = 1, 2, 3)$. Then M can be estimated as

$$|g_0(t)| \leq M = \left|a^{(1)}\right| e^{\sigma_1 h_{p1}} + \left|a^{(2)}\right| e^{\sigma_2 h_{p1}} + \left|a^{(3)}\right| e^{\sigma_3 h_{p1}}$$
$$+ \left|b^{(1)}\right| e^{\sigma_1 h_{p2}} + \left|b^{(2)}\right| e^{\sigma_2 h_{p2}} + \left|b^{(3)}\right| e^{\sigma_3 h_{p3}}$$

which provides an upper bound for $\alpha < \frac{1}{M}$.

Proof. Directly from the proof of the Lemma. $\diamond$

Remark 3.1 In theory, when every vehicle follows the leader vehicle only, the overall system is always string stable. This following strategy sounds ideal but impractical. This is due to the following reasons:

(a) Real-time distance estimation $(x_l - x_i)$ of each vehicle with respect to the leader vehicle is difficult to obtain. If radar distance with respect to the preceding vehicle is used for the estimation of $(x_l - x_i)$, error accumulation may increase down stream in the platoon even if inter-vehicle communication is available.

(b) For safety, each vehicle must avoid conflict with its immediate front vehicle, which requires α to be as large as possible. This suggests to choose $\alpha = \frac{1}{M} - \varepsilon$ with $\varepsilon (> 0)$ sufficiently small.

4 TEST RESULTS

For experimental work, 4 automated full size vehicle Buick Le Sabre have been used for experiment. The desired inter-vehicle distance is $6[m]$. Delco radar is used for distance measurement with internal delay about $50[ms]$. The above following strategy is used with $\alpha = 0.65$. Test results show that reasonably good string stability has been achieved (Fig. 2). However, both distance tracking error and speed tracking error did not monotonically decrease down stream in the platoon, which is not exactly as predicted by the analysis. This is due to measurement error, external disturbances and differences between vehicles which are ignored here.

5 CONCLUDING REMARKS

Automated vehicle short distance following is a hot research topic for both academic researchers and $R\&D$ of vehicle manufacturers. There are typically two following modes: Adaptive Cruise Control (ACC) and platooning with inter-vehicle communication. This paper uses a different parameterization approach to consider the string stability for those two vehicle following modes. Particularly, the main obstacle for vehicle following, i.e. the time lag and pure time delay are taken into consideration. It is shown that ACC cannot achieve string stability. Platooning, on the other hand, can achieve string stability regardless the time delay in the range sensing.

References

[1] Eyre, J., Yanakiev, D. and Kanallakopoulos, I., A simplified framework for string stability analysis of automated vehicles, *Vehicle System Dynamics*, **Vol. 30**, pp375-405, 1998

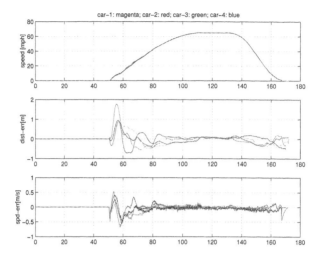

Figure 2: String stability test of 4 cars

[2] Khalil, H. K. , *Nonlinear Systems*, 2nd Ed., Prentice Hall, New Jersy, 1996

[3] Liu, X., Goldsmith, A., Mahal S. S. and Hedrick, J. K., Effects of communication delay on string stability in vehicle platoons, *Proc. of IEEE Conf. on Int. Trans. Syst.*- Oakland (CA), USA-August 25-29, p. 627-632, 2001

[4] Lu, X. Y. and Spurgeon, S. K., Robust sliding mode control of uncertain nonlinear systems, *Systems & Control Letters*, **Vol. 32, no. 2,** p75-90, 1997

[5] Lu, X. Y. and Hedrick, J. K., A panoramic view of fault management for longitudinal control of automated vehicle platooning, *Proc. of 2002 ASME IMECE, Dynamic Systems and Control Division, Advanced Automotive Technologies Symposium*, **IMECE2002-32106**, Nov. 17-22, New Orleans, 2002

[6] O'Neil, P. V., *Advanced Engineering Mathematics*, Wadsworth Publish Company, Belmont, California, 1983

[7] Rajamani, R., Tan, H.-S., Law B. and Zhang, W. B., Demonstration of integrated longitudinal and lateral control for the operation of automated vehicles in platoons, *IEEE Trans. on Control Systems Technology*, Vol. 8, No. 4, p. 695-708, 2000

[8] Swaroop, D. and Hedrick, J. K., String stability of interconnected systems, *IEEE Trans. Auto. Contr.*, **40, no.3**, p349-357, 1996

[9] Swaroop, D., String stability of interconnected systems: An application to platooning in Automated Highway Systems, *Ph. D. Thesis, Univ. of California, Berkeley*, 1994

The Dynamics of a Railway Freight Wagon Wheelset with Dry Friction Damping in the Suspension

HANS TRUE and LARS TRZEPACZ*

SUMMARY

We investigate a simplified model of a single-axle bogie that supports one end of a freight wagon. The suspension elements consist of linear springs and dry friction dampers, and various spring configurations are considered. The wagon runs on a straight track with constant speed V. The speed is the control parameter in the problem, and for the different spring configurations the resulting dynamics is calculated numerically. The dynamics is rather violent for speeds above 15 m/s. Although the suspension is only a crude model of a UIC standard leaf spring suspension certain dynamical features compare well with the dynamics of the UIC suspension.

Keywords: nonlinear nonsmooth dynamics, railway vehicle dynamics.

1. INTRODUCTION

Damping by dry friction has been used in vehicle constructions for centuries, and it is still used in railway vehicles today. The reason is that the dry friction dampers are much cheaper and more robust than the hydraulic dampers are. The drawbacks are that their characteristics are sensitive to weather conditions, pollution (dirt, oil) and to a high degree to wear. The designers use rules learned by experience for the applications of dry friction dampers in railway vehicles in part because there exist no generally accepted physical laws for their function.

True and Asmund [3] investigated the dynamics of a simplified model of a wheelset of a freight wagon. The wheels have a conical profile, no flanges and run on a rail profile, which is shaped as an arc of a circle. The wagon is supported horizontally by linear springs and dry friction dampers. It turned out that the wheelset derailed in large speed intervals. In the articles by True and Trzepacz [4] and [5] the dynamical model was improved in steps, until the wheelset stopped derailing. First a realistic wheel-rail geometry was introduced, then the play between the axlebox and the guidances was introduced in two steps - first laterally and then in addition longitudinally - and finally

*Address correspondance to: The Technical University of Denmark, Informatics and Mathematical Modelling, Bldg. 321, DK-2800 Lyngby, Denmark.

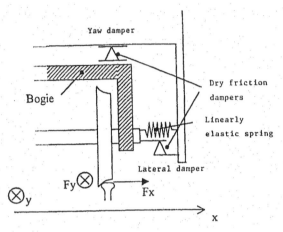

Figure 1: A diagram of the single-axle bogie model.

a linear spring was introduced as a model of the longitudinal restoring force created by the link suspension. In this way we gained insight into the functions of each of the single components in the suspension and their dynamical interactions. In this article we shall summarize our results so far and then discuss the dynamics of the complete model depending on the speed of the vehicle and investigate the effect of a variation of the stiffness of the longitudinal stabilizing spring. It corresponds to a change of the pendulum length in the suspension. The effect of a variation of the longitudinal and lateral dry friction damping forces was already discussed in the paper by True and Asmund [3].

2. THE VEHICLE MODEL

The original dynamical multibody system was formulated by True and Asmund [3] see figure 1. It is a mathematical model of the motion of a single-axle bogie running on a straight, horizontal and ideal track. It is assumed that all elements with exception of the suspension elements are ideally rigid and that the springs have linear characteristics. The dampers are dry friction dampers and a new model for the dry friction was formulated in the article by True and Asmund [3]. The bogie frame can turn horizontally in a frictionless pivot in the bottom of the car body.

In the articles by True and Trzepacz [4] and [5] and in this article we use UIC60 rails with a cant of 1/40 wheels and DSB 97-1 wear profiles. The DSB profile is a modified S-1002 profile, where the flanges are thinner in order to improve the dynamics of the vehicle on track with slightly narrower gauge. The contact is assumed to be Hertzian. Since we only investigate the horizontal dynamics it is assumed that the normal forces are constant and equally distributed and vertical springs and dampers are neglected. The equations of motion are formulated in a coordinate system, that moves along the center axis of the track with the constant speed of the wagon. The speed is the primary control parameter in our problem. We assume that the wheels and the rails remain in contact. We use the approximation by Shen, Hedrick and Elkins [2] for the calculation of the creep forces. The creep forces depend nonlinearly on the creepage and for high creepage they reach a saturation value, which corresponds to pure Coulomb friction. The wheel-rail contact points are calculated with W. Kik's routine RSGEO.

We have included stick-slip and hysteresis in our model of the dry friction and assume that Coulomb's friction law holds during the slip phase. For details please see the article by True and Asmund [2]. We thus have three important nonlinearities in our problem: The nonlinear creep-creep force relation, the nonlinear kinematic contact condition and the hysteresis with stick-slip of the dry friction dampers.

The model system has three degrees of freedom: Lateral (x-) translation of the car body and the wheelset, and yaw of the bogie around the frictionless pivot in the bottom of the car body.

The nonlinear dynamical system becomes a system of six first order differential equations with time t as the independent variable. The dependent variables in our problems are:

x_1 and x_2 The lateral translation and speed of the wheelset
x_3 and x_4 The yaw angle and angular speed of the wheelset
x_5 and x_6 The lateral translation and speed of the carbody

We use the parameter values:

k_F The lateral spring constant 10^6 m/N
$\nu_{lat}N = 923$ N, where ν_{lat} is the static adhesion coefficient of the lateral damper and N the normal force
$\mu_{lat}N = 773$ N where μ_{lat} is the dynamical coefficient of friction of the lateral damper
$\nu_{long}N = 414$ N where ν_{long} is the static adhesion coefficient of the longitudinal damper
$\mu_{long}N = 345$ where μ_{long} is the dynamical coefficient of friction of the longitudinal damper
m_{wheel} The mass of the wheelset 1022 kg
I_{bogie} The moment of inertia of the bogie 678 kgm^2
m_b The active mass of the carbody 10000 kg.

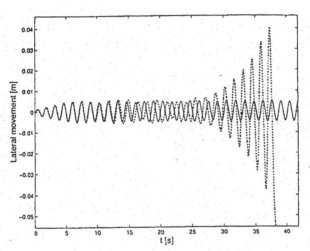

Figure 2: The sensitivity to changes in the vehicle speed. The lateral oscillations versus time for V = 10.00 m/s (full line) and at V = 10.05 m/s (broken line).

The dynamical system is solved numerically with appropriate initial conditions. We use a Runge/Kutta/Cash/Karp 5/6 order solver with adaptive steplength and error control. The program is implemented in C++. The speed of the calculations is then 1000 times higher than a MATLAB program. However MATLAB was used for the post processing.

3. THE RESULTS

In the work by True and Asmund [3] we found that the wheelset derails in large speed intervals and the dynamics is very sensitive to infinitesimal disturbances in the initial conditions or the parameters - even in the speed intervals where the model runs stably. As an example we show on figure 2 two curves of the lateral oscillations with the same initial conditions but with an infinitesimal speed difference . At the speed 10.00 m/s the wheelset stays in the track but at 10.05 m/s the wheelset derails within 38 seconds.

In the work by True and Asmund the wheel profile is conical without a flange, but True and Trzepacz [4] and [5] substituted the profiles with the realistic DSB 97-1 wheel profile in connection with the UIC60 rail profile with a rail cant of 1/40. *It is not sufficient to prevent derailment*. On figure 3 we plot the time until derailment versus the vehicle speed. If the time is longer than 200 s we assume that the wheelset will not derail. We have plotted the results for the realistic wheel-rail geometry together with the results from True and Asmund [3] in the same diagram. The black curve and

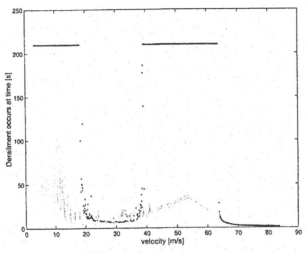

Figure 3: Comparison of the times until derailment versus the vehicle speed for realistic wheel-rail geometry (black) and conical profile without flanges (gray).

points correspond to the realistic wheel-rail geometry and the gray points and curves illustrate for comparison the results from True and Asmund. The realistic wheel-rail geometry does stabilize the motion in two large speed intervals, but between 20 m/s (72 km/h) and 40 m/s (144 km/h) the wheelset will derail within a few seconds.

True and Trzepacz [4] and [5] examined the influence on the motion of as well the stick phase as the slip phase of the dry friction force. They found that the dynamics is dominated by stick at low speed. At V = 15 m/s the motion changes in such a way that there is either full stick or full slip. When the speed increases further the slip becomes more dominant.

The wheelset still derails in the speed intervals 18 m/s < V < 38 m/s and above V = 64 m/s. The resulting lateral displacement is shown on figure 4.

The values for the lateral and longitudinal plays in the axlebox guidance were found in the standards for the UIC suspension or in the book by Hanneforth and Fischer p.36 [1]. Through the impact of the axlebox the guidance will deform. The lateral restoring force is modelled as a linearly elastic spring with a spring constant of 1500 kN/m. True and Trzepacz [4] and [5] showed that the lateral guidance *has no effect* on the lateral oscillations. True and Trzepacz then fixed the bogie frame in the car body and gave the axlebox a longitudinal play of 22.5 mm. The restoring force is very big in this case and the assumption of an elastic impact with $E = 2.1 \cdot 10^{11}$ made the dynamical system so stiff that the calculation time became unacceptably high. The impact was therefore approximated by an ideally elastic one, where the yaw speed of the wheelset is the

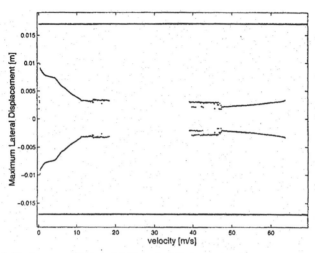

Figure 4: Maximum amplitudes of the lateral oscillation versus the speed of the vehicle.

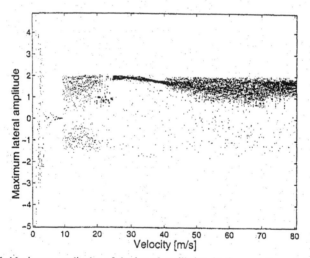

Figure 5: Maximum amplitudes of the lateral oscillation [cm] versus the speed of the vehicle when the motion is limited by the guidances.

same before and after impact but the direction of the motion changes. They compared some results of computations under each of the two assumptions and they agree very well.

As seen on figure 5 we have now finally achieved what we want: *The wheelset derails no more.*

Figure 5 shows the maximum amplitudes of the lateral oscillations versus the speed of the vehicle when as well lateral as longitudinal plays are present. The limit on the yaw motion stabilizes the motion of the wheelset, but due to the impacts with the guidances the motion is apparently chaotic in the entire speed range. That alone, however, is not harmful.

On figure 5 the different forms of motion are clearly seen. After the big oscillations at 0 m/s $<$ V $<$ 5 m/s follow small oscillations at 5 m/s $<$ V $<$ 10 m/s. Then a range 10 m/s $<$ V $<$ 19 m/s follows with big chaotic oscillations followed by an interval 19 m/s $<$ V $<$ 25 m/s where we must investigate the motion more carefully. The motion is not clearly chaotic. In the speed interval 25 m/s $<$ V $<$ 40 m/s the scatter of the chaotic amplitudes is small but the amplitudes are all close to the maximum value. Obviously the limitation of the yaw motion is very important in this speed interval. In the interval 40 m/s $<$ V $<$ 64 m/s the motion is strongly chaotic, but above V $=$ 64 m/s the amplitudes of the chaos decrease.

The results demonstrate that alone the longitudinal motion limiter will prevent a derailment in the large speed range 18 m/s $<$ V $<$ 38 m/s and above V $=$ 64 m/s. The impacts between the axle box and the guidances give rise to strongly chaotic dynamics, which is known also from other dynamical problems. Some of the kinds of chaotic motion we have found are, however, not known to us from earlier investigations so we shall continue our work with a more detailed analysis of these chaotic motions.

Finally a linear spring acting longitudinally on the axle box is introduced. We have calculated its stiffnesses on the basis of the known data from the wellknown UIC standard link suspension (see Hanneforth and Fischer p.37 [1]). We calculate the spring constant k from $k = mg/L$, where m is the mass carried and L are the lengths 180 mm or 35 mm of the pendulum, depending on the number of links that participate in the oscillation. Due to a motion limiter in the UIC suspension the restoring force will increase stepwise with the displacement. It turns out that the longitudinal spring with these characteristics has a very small stabilizing effect at low speeds. At higher speeds it has no effects at all.

We have found four basically different kinds of dynamical behaviour A-D in our theoretical model. They are shown on figure 6.

We now illustrate our main results on figure 7, where we show the lateral displacement versus time in discrete intervals. Each interval is of twenty minutes in order to secure that the transients die out. The results are summarized in the following table.

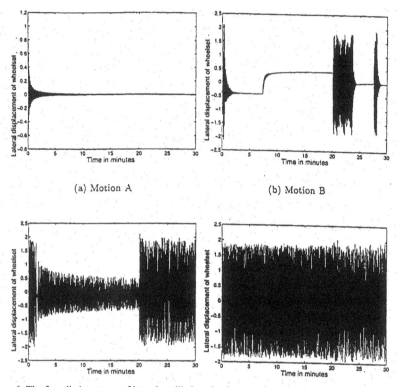

Figure 6: The four distinct types of lateral oscillations [cm] versus time.

Time interval [min]	Speed [m/s]	Motion
0-20	5	A
20-40	9.5	B
40-60	15	D1
60-80	22	C
80-100	30	D2
100-120	55	D3
120-140	80	D4

The Ds all indicate chaotic dynamics, but there are variations. The main variation is the difference in the number of impacts per 'cycle', whereby one must keep in mind that the motion is aperiodic, so a cycle in the strict meaning of the word does not exist.

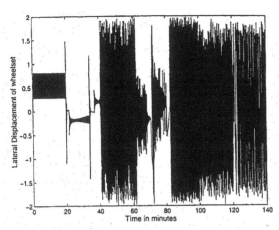

Figure 7: The lateral displacement [cm] versus time with jumps in the speed every twenty minutes

We finally want to investigate the possibility to improve the dynamics by changing the stiffness of the longitudinal spring. As a unit spring force we have chosen the restoring force yielded by a link with an effective pendulum length of 35 mm. On figure 8 we have outlined the domains A-D4 that characterize the dynamic behaviour in dependence on the speed and a multiple of the unit spring force. The influence of the dry friction damper was already investigated in the article by True and Asmund [3]. The 'A' type motion is the preferable one, and it is seen that it is necessary to increase the restoring force by a factor of forty in order to achieve an undisturbed ride at 35 m/s ~ 126 km/h keeping all other vehicle parameters constant. Such a strong restoring force may prevent the wheelset from turning into a radial position in the curves, and it is therefore unacceptable.

4. CONCLUSION

This work is a contribution to the analysis of the dynamics that may arise in railway vehicles due to dry friction in the suspensions. In the earlier paper by True and Asmund [3] the danger of derailment existed at almost any speed, and it was concluded that this was due to the strong simplifications introduced in that model.

In this article we demonstrate that only a limitation of the yaw angle of the wheelset will guarantee the safe run of the vehicle.

The dynamics is chaotic in large speed intervals, and speed intervals exist where it is connected with strong oscillations and violent impacts between the axle boxes and the guidances. The oscillations may create high stresses and large track forces that lead

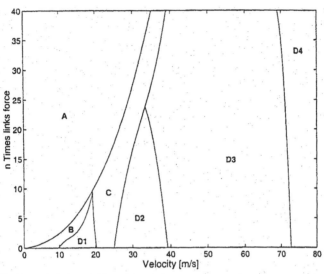

Figure 8: The dynamics of the single-axle bogie in dependence on the speed and the horizontal spring force

to fatigue and high wear of the rails. The dynamics may be improved at speeds up to 126 km/h by an increase of the longitudinal restoring force on the wheelset. The force must, however, be so large that the wheelset may not turn into a radial position, when the wagon enters a curve. Such an increase of the restoring force therefore cannot be recommended, and other solutions must be sought.

The research continues with a model of a Hbbills 311 freight wagon and J. Piotrowski's model of the UIC suspension. The first results are ready for publication.

REFERENCES

1. Hanneforth, W. and Fischer, W.: Laufwerke, 1986, Transpress, Berlin, DDR.
2. Shen, Z.Y., Hedrick, J.K. and Elkins, J.A.: A Comparison of alternative Creep-Force models for Rail Vehicle Dynamical Analysis, 1984, Proc. 8th IAVSD Symposium on Vehicle System Dynamics, The Dynamics of Vehicles on Roads and Tracks, pp. 591–605, Swets & Zeitlinger, Lisse, Holland.
3. True, H. and Asmund, R.: The Dynamics of a Railway Freight Wagon Wheelset with Dry Friction Damping, 2002, Vehicle System Dynamics, Vol.38, pp. 149-163.
4. True, H. and Trzepacz, L.: The Dynamics of a Railway Freight Wagon Wheelset with Dry Friction Damping in the Suspension, in print, Proc. 8th Miniconf. on Vehicle System Dynamics, Identification and Anomalies, Budapest, Nov. 11-13, 2002, Technical University of Budapest, Budapest, Hungary.
5. True, H. and Trzepacz, L.: Die Dynamik eines Güterwagenradsatzes mit Trockenreibungsdämpfung, 2003, Der Eisenbahningenieur, Vol.54, 7, pp. 37–42.

Vehicle System Dynamics Supplement 41 (2004), p.597-606   © Taylor & Francis Ltd.

Experimental and numerical modal analyses of a loco wheelset

NIZAR CHAAR AND MATS BERG[1]

SUMMARY

Wheelset structural flexibility is believed to have a major influence on the vehicle-track dynamics. Several studies have related the structural flexibility of the wheelset to fluctuations of wheel-rail forces, rail and wheel corrugation, etc. This paper reports part of an ongoing project that studies the effects of wheelset structural flexibility on the vehicle-track dynamics. The paper focuses on experimental and numerical modal analyses of a loco wheelset in the frequency range of 0-500 Hz. Major issues related to modal analyses and modelling of wheelset are presented along with respective results. The results from numerical modal analysis were in good agreement with those obtained from the experiment. In addition, the wheelset had fairly low eigenfrequencies. Reduced versions of the generated wheelset model will be used in coming work in on-track numerical simulations in order to determine the effects of wheelset structural flexibility on the vehicle-track dynamics. Results from these simulations will be validated against existing experimental on-track results.

1   INTRODUCTION

The wheelset is a fundamental component of rail vehicles. Its functions are to support the load of the vehicle during rolling, guide the vehicle on the track and transfer longitudinal forces at braking traction. A wheelset consists of two wheels that usually are rigidly connected to a common wheel axle. Axle boxes are often situated outside the wheels and support the weight of the carbody and the bogie frames via the primary suspension. Passenger vehicles and locomotives usually have axle or wheel mounted brake discs. For powered vehicles, traction gear is also introduced on the wheel axle.

The wheelset structural flexibility, which refers to the deformation of the wheelset, is believed to have a major influence on the vehicle-track dynamics. Different studies have investigated the effects of wheelset structural flexibility on various phenomena such as: lateral stability of railway vehicles, variation of wheel-rail forces, rail and wheel corrugation, wheel out-of-roundness and noise propagation. Popp et al. [1], mentioned that in the frequency range of 0-500 Hz, the

---

[1] *Address correspondence to:* Division of Railway Technology, Department of Aeronautical and Vehicle Engineering, Royal Institute of Technology (KTH), SE-100 44, Stockholm, Sweden. Tel.: +46 8 790 76 52; Fax: +46 8 790 76 29; E-mail: nizo@kth.se

elastic deformation of the wheelset and its components plays a major role in the vehicle-track interaction. In [2], the influence of wheelset structural flexibility was reviewed together with the various modelling methods adopted to represent this flexibility. Finite element models, lumped models and continuous models are the common methods used to represent wheelset structural flexibility. In addition, laboratory testing together with on-track measurements are necessary to evaluate the effects of wheelset structural flexibility, [2].

This paper presents part of an ongoing project at the Royal Institute of Technology (KTH); the aim of the project is to study the effects of the wheelset structural flexibility on vehicle-track dynamics. The present paper covers the experimental and numerical modal analyses of a wheelset. A finite element model of the investigated wheelset is developed and validated against experiment. This model will be used in further studies to study the overall effects of wheelset structural flexibility in the frequency range of 0-100 Hz.

2   WHEELSET UNDER INVESTIGATION

The wheelset under investigation is a powered wheelset pertaining to a Swedish Rc locomotive. The wheelset (Figure 1) consists of two solid coned wheels of 1.3 m in diameter that are rigidly connected to a common 2.11 m long axle. Axle boxes and traction gear are also attached on the wheel axle. The total mass of the wheelset amounts to 2960 kg, see Table 1 for a summation.

Due to the presence of the traction gear, the wheelset is neither symmetric (left-right side) nor axisymmetric anymore. The presence of the traction gear will cause a shift of the wheelset centre of gravity by 156 mm along the axle towards the gear side (Figure 1). The rigid body properties of the wheelset will be detailed in Section 4.

It is important to mention that the investigated wheelset was used during on-track measurements to measure the wheel-rail forces through strain gauges attached to wheel plates. In contrast to this instrumented wheelset, the ordinary loco wheelset has wheel mounted brake discs (two on each wheel) in addition to some other minor differences that will be discussed later in Section 4.

Fig. 1. Loco wheelset under investigation and lateral shift of its centre of gravity.

Table 1. Mass of the wheelset and its components.

Wheelset components	Mass (kg)
Wheel axle	516
Wheels	647*2
Axle boxes	121*2
Axle gear wheel	228
Gear box and its components (rotor axle, rotor gear wheel, coupler)	680
	Σ 2960 kg

3 EXPERIMENTAL MODAL ANALYSIS (EMA)

3.1 EXPERIMENTAL SET-UP

EMA of the loco wheelset was carried out at the Marcus Wallenberg Laboratory (MWL), KTH. The frequency range of interest is 0-500 Hz. The purpose of this analysis was to determine the eigenfrequencies, mode shapes and damping.

The wheelset was softly suspended in the ceiling through four ropes (Figure 2a). The aim behind this installation was to reproduce free boundary conditions. By using such a set-up, the rigid body motions were kept at frequencies far below the first bending mode and the effects of these motions on the flexible modes were small. The gear box end was suspended in a way to keep its orientation similar to the real orientation in the bogie (Figure 2b and 2c) and the axle boxes were turned upside down for convenience (Figure 2d).

A total of twelve accelerometers manufactured by Brüel and Kjaer (Type 4507 B 005) were used to measure the wheelset response. These accelerometers are highly sensitive and were calibrated by the manufacturer just before the test. Accelerations in the vertical, lateral and longitudinal directions were determined at 105 measurements points distributed on different components of the wheelset as follows: 14 points along the axle, 32 points on each wheel, 5 points on each axle box and 17 points on the gear box (Figure 2e).

The wheelset was excited by means of an electromagnetic shaker that induces excitation in the structure. The shaker was connected to the wheelset through a push rod that has a force transducer mounted at its end. The force transducer was glued to the wheel rim. The angular position of the excitation point was at 135 degrees from the top position (Figure 2f). In this way, bending of the wheelset can be excited in both the vertical and horizontal planes. Torsional excitation was not performed.

The signals from the twelve accelerometers and from the force transducer went through two amplifiers, which in turn were connected to a signal control unit. Two signal sources were used to excite the wheelset: random noise and sine excitation. Sine excitation results in accurate outcome since all excitation and response energy is concentrated to a single frequency range.

In order to reduce noise contribution 100 Time domain averagings were applied for each measurement point during random noise excitation. The coherence function, at each point, offered an indication on the accuracy and consistency of the results.

Fig. 2. Experimental set-up: (a) Free boundary conditions; (b) and (c) Attachment of the gear box; (d) Attachment of an axle box; (e) Some of the measurement points; (f) Excitation point.

3.2   ANALYSIS TECHNIQUES AND RESULTS

Extractions of the modal parameters (eigenfrequencies, mode shapes and damping) were performed using the software I-DEAS, [3]. Measurement points were represented by a simple wheelset model (Figure 4a). Data from the measurements were loaded into this model. Modal analysis was performed using the circle-fit method which assumes that at frequencies close to the natural frequencies, the accelerance can be approximated to that of a single-degree-of-freedom system, [4].

The eigenfrequencies in the range of 0-500 Hz are shown in Figure 3. The figure describes the average accelerance in the vertical direction along the wheel axle. The first five eigenfrequencies and the respective measured modal damping are given in Figure 3 and the respective mode shapes are shown in Figures 4b to 4f. The first bending mode (Figure 4b) occurs at only 55 Hz and the second bending mode (Figure 4c) at 82 Hz. For the following two modes the wheels deform in an umbrella-like fashion. The first (symmetric) umbrella mode and the second (anti-symmetric) umbrella occur at around 121 and 163 Hz respectively (Figures 4d and

4e). The wheel deformation mode with two nodal diameters occurs at 204 Hz (Figure 4f).

Sine and random noise excitations gave similar results and the rigid body motions were kept at low frequencies (Figure 3). The investigated wheelset has relatively low eigenfrequencies as compared to most wheelsets.

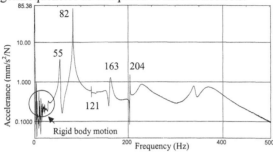

Fig. 3. Average accelerance in the vertical direction along the wheel axle (random noise excitation).

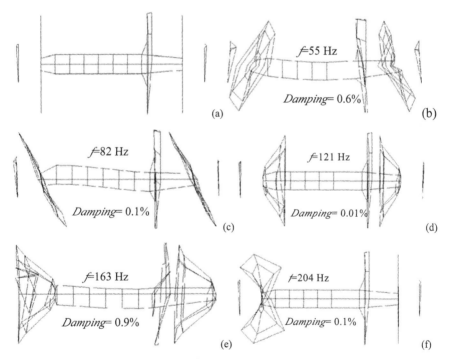

Fig. 4. EMA: (a) Undeformed wheelset; (b) First bending mode; (c) Second bending mode; (d) First umbrella mode; (e) Second umbrella mode; (f) Wheel deformation with 2 nodal diameters.

4   NUMERICAL MODAL ANALYSIS (NMA)

A finite element (FE) model of the investigated wheelset was developed and NMA was performed using the finite element software I-DEAS. Once validated, the model will be used in further studies for vehicle-track numerical simulations.
The main assumptions and simplifications adopted in the model are:
- The 1.3 m wheels and the axle gear wheel are rigidly connected to the axle.
- The axle boxes are rigidly connected to the axle despite that the axle boxes can perform some roll and yaw motion through the rolling bearings (for simulated torsional modes the axle boxes were not included).
- The rotor gear wheel and the motor coupler are rigidly connected to the rotor axle, which in turn is assumed to be rigidly connected to the gear box.
- Simplifications are introduced while modelling the gear box due its complicated geometry (Figures 5a and 5b).
- The interface between the gear box and the wheel axle is ensured through rolling bearings. In the simulation model, springs in lateral (k_l) and radial directions (k_{rv} and k_{rh}) are used to model these bearings (Figure 5c). For the simulated torsion modes the gear box was not included.

Solid 10-noded tetrahedral elements were used to model the wheelset. Each node has 3 degrees of freedom. The geometrical model used for the finite element modelling is shown in Figure 5d. A convergence study was performed and a typical mesh size of 50 mm was found appropriate. The complete wheelset model comprised around 180 000 degrees of freedom (dofs) distributed as follow: 90000 dofs representing both wheels, 22500 dofs for the axle and axle boxes and 67500 dofs for the gear box and its components. Free boundary conditions are assumed and the Lanczos solver is adopted to compute the eigenvalues and eigenvectors, [3].
For the bending motion of the wheelset, the following cases are studied:
- Case A: Wheelset with rigid wheels and flexible wheel axle.
- Case B: Wheelset with flexible wheels and wheel axle.
- Case C: As case B but axle boxes and axle gear wheel are added.
- Case D: Complete wheelset (including the gear box and its components).

At the wheel axle-gear box interface, a spring stiffness of 500 MN/m was assumed in both lateral (k_l) and radial vertical (k_{rv}) directions while a spring stiffness of 1 MN/m is assumed in the radial horizontal direction (k_{rh}), to represent some clearance of the bearings. The material properties used in the solution are as follows: modulus of elasticity 210 GPa, Poisson's ratio 0.3 and mass density 7820 kg/m^3.
The rigid body properties for each of the studied case are listed in Table 2. The wheelset centre of gravity (COG) in case A is already shifted by 8 mm towards the gear seat. In Table 2, moments of inertia are calculated with respect to the wheelset COG which in turn varies between the cases. Frequency results obtained for each case are listed in Table 3.

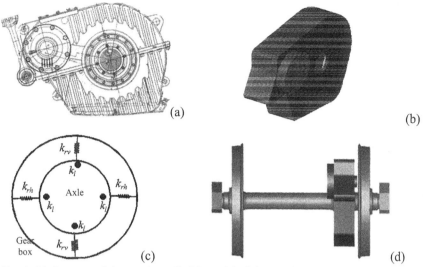

Fig. 5. (a) Geometry of the gear box; (b) 3-D model of the gear box; (c) Radial and lateral springs connecting the gear box to the wheel axle; (d) Wheelset geometry for FEM.

Table 2: Rigid body properties of the instrumented wheelset.

Rigid body properties	Cases A and B	Case C	Case D
Mass (kg)	1810	2280	2960
Centre of gravity lateral shift (mm)	7.9	56.5	156
Roll moment of inertia (kgm^2)	1094	1405	1636
Pitch moment of inertia (kgm^2)	354	384	545
Yaw moment of inertia (kgm^2)	1094	1403	1552

Table 3: Simulated eigenfrequencies (Hz) for cases defined above as well as EMA results.

Mode shapes	Case A	Case B	Case C	Case D	EMA
First bending (2 modes)	71.8	56.2	54.9	52.7, 56.8	55
Second bending (2 modes)	160.6	80.8	80.3	79.4, 80.4	82
First umbrella	-----	115.7	115.7	115.7	121
Second umbrella	-----	176.1	156.2	162.9	163
Wheel deform. with 2 nodal diam. (2 modes)	-----	206.1	206.1	206.1	204

Although many studies, [2], suggest that the wheels essentially remains undeformed in the frequency range of 0-200 Hz, the structural flexibility of the 1.3 m wheels in the present study showed significant effects on the overall dynamics, even at lower frequencies (case A and B). Table 3 also shows that the introduction of the axle boxes and the gear components affects the first bending mode while the second bending mode shows small variation (cases B, C and D). The first umbrella mode and the wheel deformation mode with two nodal diameters are not affected by the addition of any of these components. Finally, the second umbrella mode is highly sensitive to the introduction of the components. In case D, the first bending mode is separated into two perpendicular mode shapes at 52.7 and 56.8 Hz respectively. A separation also occurs for the second bending mode. Separation of the first and second bending modes is negligible in cases A, B and C due to the absence of the (non-axisymmetric) axle box.

The discrepancies in the simulation results (Case D vs. EMA) are due to the modelling of the rolling bearing interface between the wheel axle and the gear box, inaccuracies related to mass distribution and flexibility of the gear box and, to some extent, uncertainties in the material properties of the wheelset.

Figure 6a shows the undeformed FE wheelset model whereas eigenfrequencies and the respective mode shapes are shown in Figures 6b to 6f. Mode shapes derived from simulations showed good agreement with the experiment cf. Figure 4. Additional eigenfrequencies and mode shapes are obtained for cases C and D. These modes are dominated by gear box motion and are not detected in the experiment.

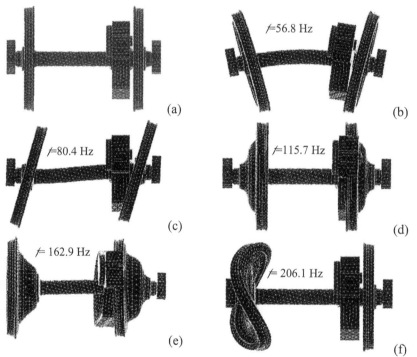

Fig. 6. NMA: (a) Undeformed wheelset model; (b) First bending mode; (c) Second bending mode; (d) First umbrella mode; (e) Second umbrella mode; (f) Wheel deformation with 2 nodal diameters as obtained from simulations.

The simulated torsion model consists of the two 1.3 m wheels and the axle gear wheel rigidly connected to the axle (case B + axle gear wheel). This subsystem has a mass of 2038 kg. The first simulated antisymmetric torsion mode where the wheels rotate 180 degrees out of phase occurs at only 48.1 Hz. The second symmetric torsional mode, where both wheels rotate in phase but out of phase with respect to the axle gear wheel, occurs at 362 Hz.

4.1   NUMERICAL MODAL ANALYSIS OF AN ORDINARY WHEELSET

As mentioned above, the ordinary wheelset has four wheel mounted brake discs together with some minor differences such as: larger axle diameter, minor changes in the wheel rim geometry and modified gear ratio. The total mass of the ordinary wheelset amounts to 3420 kg. For the bending motion, different cases are studied again. In the first case (case B) a plain wheelset is considered. Four brake discs are then rigidly added to the wheels (case B + brake discs). Each brake discs weighs around 88 kg. In case C, also the axle gear wheel and the axle boxes are added to the wheelset and in case D a complete wheelset is studied. The same assumptions adopted for the instrumented wheelset are assumed for the ordinary wheelset. The rigid body properties for each case are listed in Table 4. The frequency results obtained from the modal analysis are summarized in Table 5.

Table 4: Rigid body properties of the ordinary wheelset.

Rigid body properties	Case B	Case B +brake discs	Case C +brake discs	Case D
Mass (kg)	1856	2207	2732	3420
COG lateral shift (mm)	12	12	56	156
Roll moment of inertia (kgm^2)	1102	1355	1673	1895
Pitch moment of inertia (kgm^2)	350	457	493	660
Yaw moment of inertia (kgm^2)	1102	1355	1670	1805

Table 5: Simulated eigenfrequencies (Hz) for different cases.

Mode shapes	Case B	Case B +brake discs	Case C +brake discs	Case D
First bending (2 modes)	62.1	55.6	54.2	54.2, 58.4
Second bending (2 modes)	90.9	83.1	82.4	81.2, 82.3
First umbrella	128.2	122.9	121.8	121.4
Second umbrella	191	200.9	174.8	168.9
Wheel deform. with 2 nod. diam. (2 modes)	198.2	240.8	240.8	240.8

For case B, the ordinary wheelset gives higher frequencies (Table 5) as compared to the instrumented one (Table 4). From Table 6 we also see that the introduction of the brake discs lower the first three frequencies whereas the stiffening effect of the braked discs lead to an increase in the other two (Case B vs. Case B + brake discs). The introduction of the axle gear wheel and the axle boxes (Case C + brake discs) gives a trend similar to that obtained with the instrumented wheelset except for a slight frequency change for the first umbrella mode. Finally the introduction of the gear box (Case D) affects slightly the first two bending modes and to higher extent the second umbrella mode.

The simulated torsion model consisted of the two 1.3 m wheels, the axle gear wheel and the four brake discs. This subsystem has a mass of 2485 kg. The first simulated anti-symmetric torsional mode occurred at 46.1 Hz. The second symmetric torsional mode occurs at 356 Hz.

5   CONCLUSIONS AND FUTURE WORKS

A finite element model of a loco wheelset was developed and numerical modal analysis was performed. The model was validated against results obtained from experimental modal analysis. Numerical results showed good agreement with experimental ones. For the lowest five eigenfrequencies the relative error was less than 3%, except for the first umbrella mode. The modelling of the rolling bearing interface between the wheel axle and gear box is the primary cause of discrepancies in the simulation results. The eigenfrequencies of the investigated wheelset occurred at relatively low frequencies as compared to most wheelsets. Modal damping was also determined from the experimental modal analysis.

The upcoming work will focus on studying the effects of wheelset structural flexibility on vehicle-track dynamics in the frequency range of 0-100 Hz. Reduced versions of the FE models will be introduced into a full vehicle model. On-track multibody dynamics simulations will be performed validated against existing on-track experimental results.

ACKNOWLEDGEMENTS

We would like to express our gratitude for the financial support and the assistance we received from Banverket (Swedish National Rail Administration), TrainTech Engineering, Bombardier Transportation (Sweden), Vinnova (The Swedish Agency for Innovation Systems), SL (Stockholm Transport), SJ (Swedish State Railways) and Green Cargo. We would also like to thank the personnel at the MWL (The Marcus Wallenberg Laboratory for Sound and Vibration Research) for their help.

REFERENCES

1. Popp, K., Kruse, H. and Kaiser. I.: Vehicle-Track Dynamics in the Mid-Frequency Range. Vehicle Systems Dynamics, 31(1999), pp. 423-463.
2. Chaar, N.: Structural Flexibility Models of Wheelsets for Rail Vehicle Dynamics Analysis - A Pilot Study. Report TRITA-FKT, ISSN 1103-470X, Division of Railway Technology, Department of Vehicle Engineering, Royal Institute of Technology (KTH), Stockholm, Sweden, 2002.
3. I-DEAS Master Series 9, student guide, Structural Dynamics Research Corporation, 2000 Eastman Drive, Milford, OH 45150.
4. Ewins, D.J.: Modal Testing- Theory and Practice. Research Studies Press LTD, Taunton, Somerset, England, 2000.

Results of investigations of railway vehicles properties with the new design of wheelsets using the refined theoretical models and field tests

LEONID VINNIK AND GUENRIKH BOURTCHAK [1]
DMITRY POGORELOV [2]

SUMMARY

Results of linking research on analysis of the railway vehicle properties with the new design of wheelsets allowing various angular velocities of the tread rotations relative to the wheel centres are considered. On the basis of an analysis performed using the computer model the advantages of the given design connected with the decrease in wear of the wheels and the rails and improvement of the dynamic characteristics are found. The field tests on tram and metro car verified the results received using the models.

1   INTRODUCTION

In a number of papers of the authors the properties of the railway vehicle, i.e. passenger and freight car, tram, metro car, locomotive and so on, with the new design of wheelsets with differential rotation of the wheels (WSDR) are described. This design provides various angular velocities of rotation of separate wheels without employing a special system to run this process. Such a phenomenon is connected with the presence of clearance δ between the treads and the wheel centres of the wheels, which provides a connection between the tread rotations of separate wheel of gravitational – frictional torsion coupling type.

The advantages of the considered design cover first of all the decrease in intensity of wear of the wheels and rails, the improvement of dynamic properties and decrease of energy consumption to create the motion. Besides, the transition of lubrication inside the wheel improves the environment.

The use of the wheels of differential rotation (WDR) allowed the tram in the complicated conditions of running over the track with very many irregularities and curves with small radii in Moscow (Russia) to make the distance of more than 110,000 km without turning of the wheels. The change in flange thickness amounted to 0,2mm, and by the trams with traditional wheels running on the same track the analogous value was equal to 2mm. Further, after making the distance of 90,000km

[1] Address correspondence to: Electric Rolling Stock Repair Plant (ZREPS JSC), 125171, Moscow, Russia, Leningradskoe shosse 2a. Tel.: 7(095) 158-76-86; Fax: 7(095) 158-65-16; E-mail: leonid_vinnik@hotmail.com; Web-site: http://www.zreps.ru
[2] Address correspondence to: Bryansky State Technical University, 241035, Bryansk, Russia, b. 50 let Octyabrya. Tel.: 7(083) 256-86-37; Fax:7(083) 256-24-08; E-mail: pogorelov@bitmcnit.bryansk.su

and several turnings a large number of treads of the traditional wheelsets are usually replaced with the new ones.

The necessity of explanation of some new experimental phenomena, received in the recent time required the refinement and expansion of mathematical model of the system [1].

2 MODELLING OF WSDR AND INNER CONTACT CHARACTERISTICS

A model of the wheelset (WS) consisting of three absolutely rigid bodies, i.e. a central part, incorporating an axle of the wheelset and two wheel centres rigidly connected with it, and two treads) and having ten degrees of freedom was created to investigate the dynamic properties of the vehicle equipped with WSDR. Six degrees of freedom define the motion of the central part relative the inertial frame of reference. Each tread has two degrees of freedom relative to the wheel centre (Fig. 1). Thus, the model assumes the lack of free play of the treads in the direction of the axle of the wheelset. More than that, the detachment of the tread from the wheel centre is not acceptable, i.e. wheel centre is in the constant contact with the corresponding tread.

The given model is used to describe the variation in time of the macrokinematic characteristics of the WS. Further, by investigation of the interaction between the wheel centre and the tread the compliance in this connection is considered. In contradistinction to paper [1], the characteristics of this interaction are defined not on basis of phenomenological and experimental considerations, but in the context of theoretical solutions to the contact problems of the elasticity theory.

Taking into account, that the inner contact due to smallness of the clearance δ (Fig.1a) cannot be considered as non-compliant from Hertz's standpoint, the investigation was carried out using both 2D and 3D FEM models. However, use of these models in the simulation process is very time consuming.

Therefore, the transitional compromised solution to formally use the non-linear creep theory was taken. The consideration of the compliance of the contact is "included" in the non-linear function of the creep coefficient.

Thus, the interaction between the tread and the wheel centre is performed under acting of normal force N_{oc} directed to the wheel centre and tangential creep force F_{oc}. So long as the angle φ, which defines the location of the point inside the contact is a small value, the ratio $N_{oc} = F_{oz}$ has a great degree of accuracy, where F_{oz} is the vertical component of the summarised force acting in tread-rail interaction. The torque of interaction forces between the tread and the wheel centre in the inner contact relative to the axle of the wheelset M_{oc} consists of torque of the creep force and the torque of the forces of viscous resistance $\alpha_{oc}\dot{\varphi}$, specially introduced to damp the parasite high frequency oscillations of tread relative to the wheel centre, i.e.

$$M_{oc} = \mu_{oc} k_{oc} \left(\xi_{\omega oc}\right) N_{oc} r_c + \alpha_{oc} \dot{\varphi},$$

where μ_{oc} is a coefficient of friction in the inner contact, $\xi_{\omega oc}$ is an "angular creep", which dependence on coordinates and their derivatives in time has the following form:

$$\xi_{\omega oc} = \frac{\Delta \omega}{V_c} = \frac{\xi_{oc}}{r_c} = \left(\left(r_c + \delta\right)\dot{\varphi}_o - r_c\dot{\varphi} - r_c\dot{\varphi}_c\right)/\left(r_c V_c\right)$$

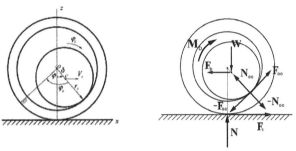

Fig. 1. Definition of the creep in the inner contact.

The non-linear function $k_{oc}\left(\xi_{\omega oc}\right)$ was found theoretically through finding a solution to the contact problem using FEM (Fig. 2).

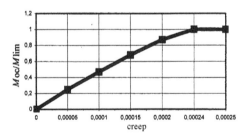

Fig. 2. Dependence of the relative torque of the creep force on the angular creep.

As for the compliance in the tread-rail contact, so, as the FEM calculations showed, all the ratios received are close to the analogous ones, based on the Hertz's theory, excluding the worn tread contact.

3 COMPUTER MODELS OF THE RAILWAY VEHICLES AND PROCEDURE OF MODELING

The numerical modelling of dynamics of the given vehicles (Fig.3) was carried out using the programme package "Universal mechanism" developed at Bryansk

State Technical University [2], www.umlab.ru. The multibody car models have 46 and 64 degrees of freedom for the vehicles with traditional wheels (TWS) and WSDR, respectively; the models of tram have 48 and 66 degrees of freedom.

Fig. 3. Models of metro and tramcar.

To calculate the creep forces in the wheel-rail contact in case of profiles with the possibility of two-point contact the Kalker's algorithm FASTSIM [3], and in case of one-point contact profiles the modified algorithm of Kik and Piotrowski [4,5] were used.

Hertz's contact theory cannot be used in studies of rail-worn wheel contact, as the radii of curvature of the wheel-rail profiles are subject to considerable change inside the contact region. To solve these kind of problems the special model of contact forces [5] based on the simplified Kalker's theory was developed. The peculiarities of this model include the application of the model of Winkler's elastic foundation to solve the normal contact problem and a modified algorithm FASTSIM for definition of the tangential contact forces. The work of creep forces acting within the contact patch was calculated to assess the wear of the profile:

$$ A = V \int_{0}^{t_M} \int\int_{C} \left[q_x \left(\xi_x - \phi y \right) + q_y \left(\xi_y + \phi x \right) \right] dx dy dt , $$

where V is the velocity of vehicle motion , C is the contact region, t_M is the simulation time, ξ_x, ξ_y are creeps, φ is spin, q_x, q_y are the components of the tangential stress.

Modelling of dynamics of the vehicle was performed through the integration of equations of motion generated automatically using the programme. Due to small mass of the WDR tread, the equations of motion of the separate WSDR are stiff, that is why for their numerical integration the implicit scheme of the Park's method [6] was used. The method is A-stable, but, however, like the other implicit methods, it requires the calculations of Jacobian matrices of the equations of motion and leads to a considerable deceleration of the simulation process. The procedure for calculation of the approximate Jacobian matrices was elaborated to make this process faster.

The Jacobian matrices are calculated only for the stiffest forces, which include the creep forces in the outer and inner contacts. According to the calculating data, the friction coefficient in the inner contact has a much greater value than in the outer contact. Therefore, the effect of creep forces in the inner contact on Jacobian matrix is determinative. In the simulation process the Jacobian matrix of the equations of motion is formed of the separate and pre-calculated local matrices, which correspond to the various WSDR and their motion modes. This procedure allowed to make the numerical integration to 30 times faster, comparing to the use of explicit scheme of Park's method.

The calculations were performed using the profiles of the wheel and rails with the possibility of both two-point and one-point contact. Computer-aided modelling allowed to assess the decrease in both the flange and wheel running surface wear of the vehicle with WSDR comparing to the vehicles with the conventional wheelsets by curve negotiation. Comparative analysis of the distribution of the creep forces over the wheelsets and also of angle of attack values for these models allowed to clarify the mechanism of the wear decrease for the car equipped with WDR.

4 RESULTS OF MODELLING

At the first stage of the studies of the metro car models the comparison of the dynamic characteristics of the models assuming that the friction coefficient in the inner contact of WDR is considerably high and does not allow the accelerated revolution, when the wheel centre rolls over the tread with sliding, is carried out. As a result, it was found, that in the mode of reducer revolution, when the velocities of contact points of the wheel centre and tread are equal for all the WDR, i.e. in the mode of the blocking of tread-wheel centre contact without allowing the friction force to achieve the limiting value, the force values in the wheel-rail contact coincide with the corresponding values of the model with TWS. The natural presumption is that the positive effects connected with WDR, are possible only in the state of accelerated revolution. Therewith, an analysis of the conditions of the transition into the mode of accelerated revolution is possible on the base of an analysis of the car model with TWS. To this purpose the term of boundary value of the friction coefficient μ_{oc}^* in the tread-wheel centre contact is introduced, which allows the accelerated revolution, i.e. saturation of the creep force in the tread-wheel centre contact to its limiting value occurs. The given parameter is defined though the ratio:

$$\mu_{oc}^*(t) = \frac{F_{oc}}{N_{oc}} \approx \frac{F_{oc}}{F_{oz}} \approx \frac{M_{oc}}{F_{oz}r_c} ,$$

where F_{oc} is the current value of creep force in the inner contact, and F_{oz} is the summarised vertical component of the interaction forces between WDR tread and rail. However, by neglecting the WDR tread mass relative to the whole vehicle, we may write $F_{oz} \approx N_{oc}$.

Coefficient μ_{oc}^* is calculated while ideal curve negotiation, i.e. without irregularities by the car model with TWS, running at the constant velocity in the trailer mode. The results of calculation allowed us to conclude, that the accelerated revolution occurs more likely at the rear wheelsets of the bogies. At small velocity, i.e. less than the balanced one, when the uncompensated acceleration is directed inside the curve, the accelerated revolution in the right curve is observed by the outer wheel relative to the curve, but at great velocity, i.e. more than balanced one, by the inner wheel. The conditions of slippage for WDR front and rear bogie are about the same. Lastly, with decrease of the curve radius the value of the boundary friction coefficient at the rear wheelsets is increased. For example, in 500m curve $\mu_{oc}^* \in [0.14, 0.16]$ is received at various velocities, in 400m curve we have $\mu_{oc}^* \in [0.18, 0.2]$, and in 200m curve $\mu_{oc}^* \in [0.27, 0.3]$.

The effect of the accelerated revolution, accompanied by the rolling with sliding on wear characteristics of the profiles by the variation of the friction coefficient μ_{oc} in the tread-wheel centre contact in the values range less than boundary one while negotiating the curves with various radii and running at different velocities, is investigated. By the decrease of friction coefficient of the rear wheelsets with WDR, which have an accelerated revolution, the friction force in the inner contact and, consequently, the summarised longitudinal force of wheel-rail interaction and wear of the wheel running surfaces of the rear wheelsets are decreased. The changes in the wheelsets positioning occur simultaneously and it produces the decrease in wear on the climbing wheel of the first wheelset. For example, by the friction coefficient equal to 0.06 in 300m curve the decrease in work of friction forces at flange of the climbing wheels in two-point contact amounts to about 15% (Fig.4).

In this figure and also below the following symbols are used: A is the work of friction forces; index f corresponds to the flange contact with the side surface of the rail in case of two-point contact, index O is a contact within the running surface; left wheel in the direction of the motion has an index L and right wheel R; the number defines the wheelset in the direction of the motion; the superscript $+$ corresponds to the climbing wheel. Thus, A_{fL1}^+ is a work of friction at flange contact of the left climbing wheel of the first wheelset running in the right curve.

In the traction mode together with the limiting coefficient μ_{oc}^* introduced above and defining the upper boundary of the friction coefficient in the inner contact of the separate WDR, it is necessary to introduce the lower boundary μ_{oc}' to provide the necessary adhesion level. This parameter is introduced for each separate WSDR, however, by $\mu_{oc} < \mu_{oc}'$ the both wheels of the given WSDR would be in the mode of accelerated revolution and by $\mu_{oc} > \mu'_{oc}$ the blocking of one of the wheels occurs. Due to the fact that the decrease of μ_{oc} gives rise to increase in effectiveness of WSDR, its recommended value must be located in the interval close to the lower boundary.

The μ'_{oc} value in each moment of time is calculated by the motion of the vehicle with TWS in the traction mode in accordance with the approximate formula:

$$\mu_{oc}' = \frac{M_D}{(F_{ozl} + F_{ozr})r_c},$$

where M_D is the current value of the traction torque at wheelset axle; F_{ozL} and F_{ozR} are summarised vertical components of interaction forces between the WSDR left and right tread and a rail, however, as it was specified above, $F_{oz} \approx N_{oc}$ for both wheels.

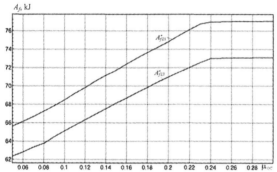

Fig.4. Work of friction forces at flange depending on the friction coefficient in the inner contact. Curve of 300m radius, steady curve 150m, transitions 50m each, $V = 8$m/s.

As result of simulation it was found, that for the metro car the lower boundary of the friction coefficient in the inner contact in the traction mode is $\mu'_{oc} = 0.17$. It is worthy to note that this μ'_{oc} value provides the realisation of the traction characteristic completely, i.e. WSDR introduction did not decrease the traction capabilities of the vehicle. The decrease in flange wear in the traction mode by decrease of friction coefficient to $\mu_{oc} = 0.17$ achieves the values from 10% to 15% by the motion of the metro car in curve of 300m radius at various velocities.

In process of investigation of dynamics of the tram models is was found, that, as by the metro car the effect of flange wear decrease of the climbing WSDR wheels increases with decrease of curve radius. This leads to more considerable wear decrease in curves of small radii (down to 20m). For example, in the numerical experiments the accelerated WDR revolution of the tram in curves in the traction motion was observed already at $\mu_{oc} < 0.32$, and the decrease in flange wear in two-point contact and at $\mu_{oc} = 0.1$ had achieved 21% by motion at small speed in 30m curve (Fig.5). By decrease of speed the effect of flange wear reduction increases.

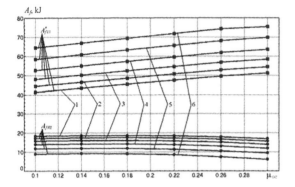

Fig.5. Work of friction forces at WDR flange of the tram depending on friction coefficient in the inner contact. Graphs one to six correspond to the velocities from three to eight m/s.

It is interesting to note, that the maximum relative decrease of flange wear is received by the motion of the tram in 80-100m curves and it reaches 25%, though in absolute values this characteristic increases with radius decrease (Fig.6).

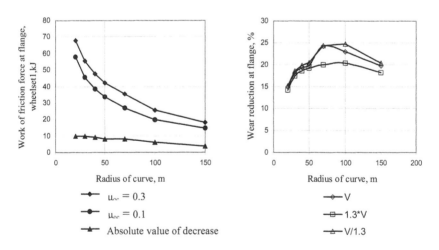

Fig.6. Wear values at flange by running in curves with various radii.

In the given research the dependencies of the wear rates on the friction coefficient in the inner contact using the profiles with one-point contact are investigated as well. The outlines of these profiles are close to the natural worn treads. It was found that in a number of cases the climbing wheels of the rear wheelsets of the each bogie are in the state of rolling with sliding in the inner contact. The boundary friction coefficient for these wheels lies in the range [0.26, 0.3]. The work of friction forces at the left WDR of the rear wheelsets, which operate in the mode of accelerated revolution, is significantly decreased by the μ_{oc} diminution to about 0.22.

Together with the wear decrease, the use of WSDR also allows to improve some values characterising the horizontal dynamics of the tram by running on the straight track and in curves. Table 1 contains the quantities defined through the oscillations in the horizontal plane by the motion in the trailer mode in the straight track, considering the irregularities, which correspond to the satisfactory conditions of the track. The following values are provided: 4 Max- an average of four maximum values, RMS – root mean square deviation of the process. By calculation of the first quantity the four maximum values on the modulus of the variable value are chosen, the highest one is discarded and for the three left an average is defined. The results show that both the maximum values and RMS for all the calculated quantities are decreased. At the same time, the use of WSDR in fact does not affect the rates connected with vertical oscillations.

As for dynamic rates received in curves, it is worthy to note that the use of WSDR leads to the balancing of the frame forces both at front and rear wheelsets of the bogies.

Table 1. The dynamic properties of tram models running on straight track, $V = 17$m/s.

Measurement	4 Max		RMS	
Vehicle	TWS	WSDR	TWS	WSDR
Lateral acceleratin, m/s²	1,72	1,52	0,43	0,41
Frame force at WS 1, kN	12,3	10,1	3,2	3,0
Frame force at WS 2, kN	7,6	7,2	2,5	2,3
Side force at left wheel of WS 1, kN	20,4	18,6	3,8	3,5
Side force at left wheel of WS 2, kN	8,5	6,3	1,5	1,3

By the tram running in the traction mode, the lower limit of the friction coefficient in the inner contact is $\mu'_{oc} = 0.13$. Let us remind that this coefficient defines the conditions, at which at least one of the WDR of the wheelset is in the mode of reducer revolution, and by the less values of the coefficient the accelerated simultaneous revolution in both wheels of one of the WSDR is possible. It is also worthy to note, that by the tram the μ'_{oc} value does not practically depend on the macrogeometry of the track and is mainly defined by the motion velocity and the chosen stage of rheostat steering of the motors. At μ'_{oc} = 0,13 the wear decrease rates are close to the corresponding values in the trailer mode.

5 CONCLUSIONS

The developed computer model based on the certified programme package "Universal mechanism" allowed to assess the advantages of the proposed design of the railway vehicles, which include WSDR as a component.

In particular, at a curtain range of tribological characteristics in the tread-wheel centre contact of the vehicles equipped with all leading axles it is possible, without diminution of the traction level required, to achieve 10-25% decrease in flange and tread wear of the wheels.

The performed analysis of the metro and tram car computer models also showed, that the use of WSDR with decreased friction coefficient in the inner contact allows to receive positive effects connected not only with the decrease in wheel and rail wear by the motion at constant velocity in both trailer and the traction mode, but also to achieve the more improved rates of dynamic properties of the vehicle, which are connected with the characteristics of yaw motion.

Besides, the advantages of the proposed design with the new WSDR wheelsets are validated by the tests carried out both at full scale rigs and at the physical 1:5 scale models. More than that, the accuracy of the numerical tests has correlated with the results received on the models made of optically active materials. As for field tests, they are described in the introduction above.

In general, according to the authors, the proposed system can give rise to the considerable effect by the improvement of some constructive component.

REFERENCIES

1. Vinnik, L. V. and Bourtchak, G..P.: Railway vehicles with wheelsets of differential rotation (Analysis of the non-linear dynamic properties). True, H.(Ed.): The Dynamics of Vehicles on Roads and on Tracks, Proc. 17th IAVSD Symposium, Lyngby, Denmark, August 2001. Swets & Zeitlinger, Lisse, 2003.

2. Pogorelov, D.Y.: On numerical methods of modelling large multibody systems. Mechanism and Machine Theory 34 (1999), pp. 791-800.

3. Kalker, J.J.: Three-dimensional Elastic Bodies in Rolling Contact. Kluwer Academic Publishers, Dordrecht, 1990.

4. Kik, W. and Piotrowski, J. A.: Fast approximate method to calculate normal load at contact between wheel and rail and creep forces during rolling. Zobory, I. (Ed.): Contact Mechanics and Wear of Rail/Wheel Systems, Proc.2nd Mini Conference, Budapest, Hungary, July 1996. Technical University of Budapest, Budapest, 1996.

5. Yazykov, V.N.: Some Results of Wheel-Rail Contact Modelling, Volume 1, Preprints of the NATO Advanced Study Institute on Virtual Non-linear Multibody Systems, Czech Technical University, Prague, 2002, pp. 236-241.

6. Park, K.C.: An improved stiffly stable method for direct integration of non-linear structural dynamic equations. J. Appl. Mech., ASME, (1975), pp. 464-470.

Vehicle System Dynamics Supplement 41 (2004), p.617-626 © Taylor & Francis Ltd.

Mechanisms of high-frequency force generation in hydraulic automotive dampers

DAVID J. COLE[1] and VICTOR Y. B. YUNG

SUMMARY

The paper describes progress of research into the mechanisms of high-frequency (up to 500 Hz) force generation in hydraulic automotive dampers. A test facility for making measurements of high-frequency behaviour is described. The force response of a monotube damper to double-sinusoid displacement inputs is presented. Wavelet analysis is used to aid identification of the force generation mechanisms. Friction is found to make a significant contribution, particularly at changes in sign of damper velocity and at high frequencies. An existing mathematical model of a damper is modified to improve the representation of friction, and it is found that prediction of force response to a random displacement input is significantly improved at frequencies above 50 Hz.

1. INTRODUCTION

The damper in a passenger car suspension strongly influences vehicle handling, ride comfort, and the level of noise, vibration, and harshness (NVH). Reducing NVH is an increasingly important objective for vehicle manufacturers as consumer expectations increase. Existing mathematical models of automotive dampers focus on predicting behaviour at frequencies below 30 Hz. While such models are suitable for determining vehicle handling characteristics and low frequency ride comfort, design for NVH requires attention to higher frequencies.

There are many published reports on damper performance and modelling; however, there is little information about behaviour at frequencies above 30 Hz, see [1] for a brief review. The mathematical prediction of high frequency behaviour is not well developed and thus the design and development of NVH tends to be experimentally based. Such empirical methods are often time-consuming and costly, and generally do not result in designs that are optimal or robust.

This paper describes progress made in understanding damper behaviour at high frequencies and in evaluating and improving existing mathematical damper models. An earlier paper [1] reported the results from experiments using excitations up to 64 Hz. In section 2 of the present paper, a new test rig is described that provides excitations up to about 200 Hz. Measurements on a monotube damper are presented in section 3 and wavelet analysis is used to examine the time variation in damping force at different frequencies. The contribution of friction is examined and in

---

[1] *Address correspondence to:* Department of Engineering, University of Cambridge, Trumpington Street, Cambridge, CB2 1PZ, UK. Tel: +44 1223 332600; Fax: +44 1223 332662; E-mail: djc13@eng.cam.ac.uk

section 4, the friction model of an existing damper model is improved. Conclusions are given in the final section of the paper.

2. TEST FACILITY

Figure 1 shows the monotube damper mounted on the test rig. The damper body was connected to the ram of a hydraulic actuator, and the damper rod was connected to a grounded force transducer. No axially flexible mounts were included, although steel flexure couplings were used at each end of the damper to minimise side loads. Pressure transducers tapped into the side of the damper body measured hydraulic pressure on each side of the damper piston. A linear potentiometer and an accelerometer measured the motion of the damper body.

The transducer signals were filtered with low-pass first-order Butterworth analogue filters set to a cut-off frequency of 1 kHz. Data acquisition was performed with a multifunction analogue input and output card installed in a computer. The card was used to sample the transducer signals at a frequency of 5 kHz, and to generate an actuator displacement demand signal that was updated at a rate of 5 kHz. An analogue circuit controlled the movement of the actuator via a high-speed servovalve. An iterative control algorithm was used to update the demand signal so that desired damper motion at frequencies up to 200 Hz (above the closed-loop bandwidth of the system) could be obtained.

The temperature of the damper body at the start of each test run was between 30°C and 31°C.

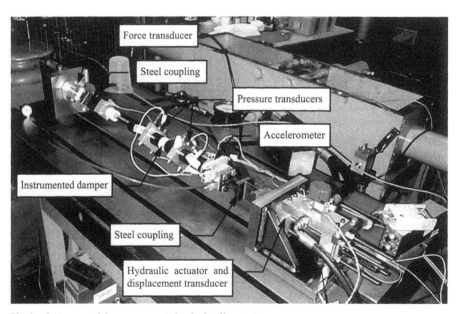

Fig. 1.   Instrumented damper connected to hydraulic actuator.

3. DAMPER BEHAVIOUR

Some insight into the behaviour of a hydraulic damper can be obtained by exciting the damper with a double sinusoid velocity input, consisting of two sine waves of different frequencies and amplitudes. Figure 2 shows the force response of the damper to excitation consisting of 1 Hz, 100 mm/s and 128 Hz, 10 mm/s sinusoids.

Figure 3 shows the normalized scalogram of the force signal over the same time period. The scalogram is calculated using a wavelet transform, and shows the changing frequency content of the signal as a function of time. Thus, the scalogram is able to identify if the generation of certain frequencies is localized in time. The normalized scalograms were calculated using the discrete wavelet transform algorithm and the Gabor wavelet [2, 3]. The force responses at 1 Hz and 128 Hz are clearly visible as ridges running in the direction of the time axis. However, there are also significant responses at other frequencies, which arise largely from the nonlinearity of the damper, see [1] for further examples of this behaviour. It is noticeable that the response at 128 Hz varies with time. In particular, the response in the region of negative damper velocities (bump) is less than elsewhere, due to the smaller damping rate traditionally provided in bump for automotive dampers.

Further insight into the mechanisms of force generation can be obtained by examining the normalized scalogram of the hydraulic component of the total damper force, as determined from the pressure transducers. Differences between the scalograms of the total force and the hydraulic force indicate the effect of friction on the damper response. Cross-sections of the scalograms are plotted to show clearly the time-variation of specific frequencies (Fig. 4), and to show the spectra at specific times (Fig. 5).

Figure 4 shows the time-variation of two different frequencies from cross-sections of the total force and pressure force scalograms. The difference between the total and pressure forces indicates that friction increases the damping force response at each frequency. However, the effect is more localized in time (near regions of zero damper velocity) at higher frequencies.

Figure 5a shows the total and pressure force spectra at 6.55 s from cross-sections of their respective scalograms. At this time, the damper velocity is near zero and the flow of oil through the valve assembly is low. The pressure force spectrum shows a decrease in the response to the 128 Hz, 10 mm/s excitation. However, the total force response remains relatively high due to friction, and friction adds generally to the response at frequencies above 11 Hz. Friction has little effect on the force spectra at 6.80 s (Figure 5b) when the damper is at its peak bump velocity.

Although side forces in the test damper were minimised, friction would be expected to play a more significant part in the response of the damper when installed in a vehicle, particularly a strut suspension intended to carry significant side load or bending moment. Further tests are desirable to examine the damper behaviour under conditions with side load.

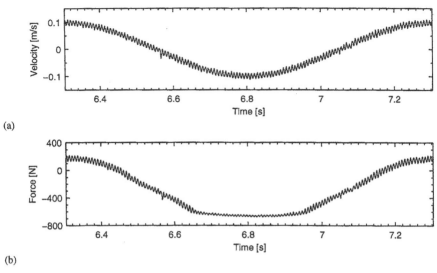

(a)

(b)

Fig. 2. (a) Velocity input and (b) force response for 1 Hz, 100 mm/s sinusoid plus 128 Hz, 10 mm/s sinusoid excitation.

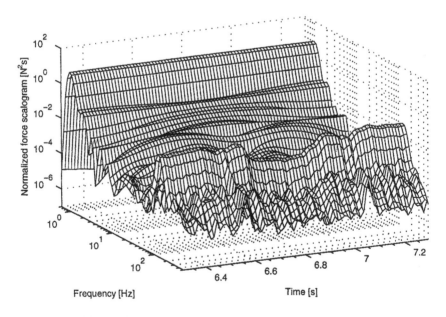

Fig. 3. Normalized force scalogram for 1 Hz, 100 mm/s plus 128 Hz, 10 mm/s.

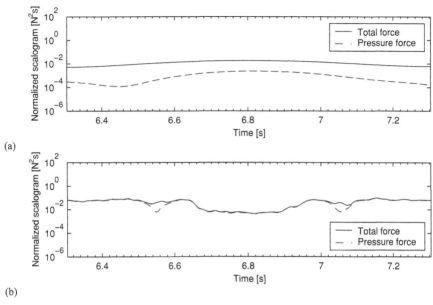

(a)

(b)

Fig. 4. Normalized total and pressure force scalogram cross-sections at (a) 3 Hz and (b) 124 Hz for 1 Hz, 100 mm/s sinusoid plus 128 Hz, 10 mm/s sinusoid excitation.

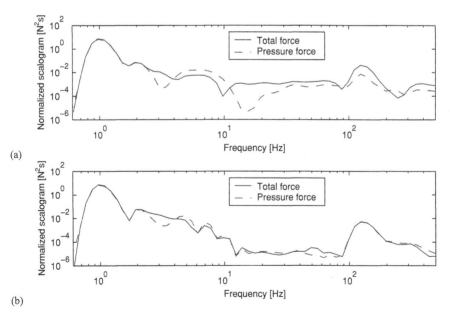

(a)

(b)

Fig. 5. Normalized total and pressure force scalogram cross-sections at (a) 6.55 s and (b) 6.80 s, for 1 Hz, 100 mm/s sinusoid plus 128 Hz, 10 mm/s sinusoid excitation.

4. MATHEMATICAL MODEL

The mathematical damper model used for this study has been described previously [1], and is based on work in [4]. The model is attractive to use in vehicle vibration studies because: it has relatively few parameters; the parameters relate to physical properties of the damper; and the parameter values can be obtained largely from straightforward tests on the complete damper.

The piston valve assembly is modelled as a combination of orifice type restrictions and an ideal blow-off valve. The flow through the valves is assumed to be turbulent and the valve parameters are independent of flow rate. The oil is assumed to be compressible. A Coulomb model represents the sliding friction force on the piston and rod. The parameters of the model were determined from two tests: a quasi-static compression and extension of the damper, and a sine sweep. The results of the parameter fitting are given in tables 1 to 4.

Table 1. Geometric damper model parameters.

Parameter		Value	Units
rod diameter	d_{rod}	14	mm
pressure tube diameter	d_{pt}	36	mm
pressure tube length	L_{pt}	380	mm
piston static position	x_0	151	mm

Table 2. User-defined damper model parameters.

Parameter		Value	Units
polytropic compression exponent	k	1.4	
smoothing factor for rebound valve	G_{reb}	8	
smoothing factor for compression valve	G_{com}	8	
effective compressibility	α	1×10^{-9}	Pa^{-1}
oil viscosity	ν	21×10^{-6}	m^2/s
velocity scaling (for friction)	$v_{friction}$	0.001	m/s

Table 3. Damper model parameters determined from quasi-static test.

Parameter		Value	Units
initial compression chamber pressure	$p_{com,0}$	2.3186×10^6	Pa
initial compression chamber gas volume	$V_{com,gas,0}$	1.3384×10^{-4}	m^3
friction force	$F_{friction}$	29.3648	N

Table 4. Damper model parameters determined from dynamic sine sweep test.

Parameter		Rebound Valve	Compression Valve	Units
blow-off valve opening pressure	Δp_0	3.2103×10^5	2.8626×10^5	Pa
blow-off valve coefficient	K_{spring}	6.1308×10^{12}	9.0561×10^{11}	$Pa^{3/2}/(m^3/s)$
leak coefficient	K_{leak}	1.3270×10^{14}	1.3989×10^{14}	$Pa/(m^{23/4}/s^2)$
port coefficient	K_{port}	2.8230×10^{12}	-6.6903×10^{11}	$Pa/(m^{23/4}/s^2)$

The validity of the model was examined by comparing the measured and simulated response to a random displacement input with a spectral density typical of that generated by the vehicle operating on a random road surface [5]. Comparison of the RMS forces generated by the damper and the model give some indication of the validity of the model. The relative RMS error values of about 11% indicate reasonable overall agreement between the model and the measurements, and are similar to values given in the literature [4]. However, the amplitudes of high frequency force components are small, so relatively large errors in the prediction of high frequency forces may only result in a small relative RMS force error.

Wavelet analysis can provide information to identify the contribution that different parts of the model make to high frequency errors. A wavelet-based transfer function (WTF) between measured and simulated forces (Fig. 8) shows the magnitude and location of modelling errors in the time-frequency plane. In the theoretical case of a perfect model and no measurement noise, the measured force to predicted force WTF has unity value everywhere in the time-frequency plane.

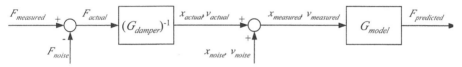

Fig. 8. Block diagram of equivalent signal path for assessment of simulation error.

Figure 9 shows the results for a 0.2 s period of the random input test. The arrows at the top of the graphs mark the times when the velocity crosses the zero point. The force time history shows significant errors in the simulation predictions at these points. Specifically, the simulation shows a sudden step change in force at each zero velocity crossing (as a result of the Coulomb model of friction), whereas the measured force data shows a smaller or more gradual change. This difference is most evident around 6.69 s. The dark regions in the WTF at the zero crossing points indicate that the simulation over predicts the intensity of the spectrum above 50 Hz. The light regions adjacent to the zero crossing points indicate that the simulation under predicts the intensity of high frequency components at low damper speeds.

Friction arises at three locations within the damper: between the rod seal and the rod; between the piston valve assembly seal and the tube; and between the floating piston seal and the tube. When the damper is stroked, it is expected that some deformation will occur at each seal before the onset of sliding. This suggests that the friction model should have a dependency on displacement. Since the seals have different geometry and may also have different material properties, each seal can be expected to have a different stiffness characteristic. Also, the amount of displacement that occurs before sliding begins is likely to be different for each seal.

While the behaviour of each seal could be modelled individually, this level of complexity may not be warranted if a simple overall model provides sufficient accuracy. A friction model was chosen based on the work of Berg [6] in developing a non-linear model for rubber springs used in rail vehicles. The friction force is predicted using the displacement and two parameters, F_{fmax} and x_2. The model also keeps track of the displacement and friction force each time the displacement changes direction.

After incorporating the Berg friction model into the damper model, the maximum friction force parameter was set to $F_{fmax} = 29.3648$ N (equal to the maximum force in the original Coulomb model). The value for the displacement parameter, $x_2 = 2$ μm, was selected using an iterative process to produce a good match of force time histories with the sine sweep excitation. The response of the friction model to a sinusoidal displacement excitation is shown in Figure 10. The friction force has an exponential-like response each time the motion changes direction. The damper force time histories and WTF corresponding to the random excitation are shown in Figure 11. They clearly show the improved predictions achieved with the new friction model. The dark regions that were previously present in the WTF at the zero velocity crossings are greatly reduced. This corresponds to the improved matching between the simulated and measured force time histories at the zero velocity crossings. The improvement is also apparent from the time histories in Figures 9 and 11, particularly in the region of 6.69 s.

5. CONCLUSIONS

(i)   A test rig for measuring the high frequency behaviour of automotive dampers was developed.

(ii)   Measurements on a monotube damper using double sinusoid inputs revealed nonlinear behaviour. Wavelet analysis aided identification of physical mechanisms. Friction was found to contribute across a broad frequency range, particularly in the region of velocity sign changes.

(iii)   An existing mathematical damper model was found to perform well at frequencies up to about 50 Hz, but significant discrepancies existed above this frequency. An inadequate model of friction was thought be a significant contributor to the discrepancy.

(iv)   The Coulomb friction model in the existing damper model was replaced with an improved friction model. Significant improvement in prediction accuracy was achieved at frequencies between 50 Hz and 500 Hz.

(v)   Work is underway to improve other aspects of the model to further improve simulation accuracy at high frequencies, and to examine the vibration behaviour of the damper in series with a flexible mounting bush.

ACKNOWLEDGEMENT

The authors would like to acknowledge the support of The Rt. Hon. Sir Winston Spencer Churchill Society of Edmonton; the Committee of Vice-Chancellors and Principals of UK Universities; and Jaguar Cars Limited.

REFERENCES

1. *Yung, V.Y.B. and Cole, D.J.: Analysis of high frequency forces generated by hydraulic automotive dampers. Proc. 17ᵗʰ IAVSD Symposium, Lyngby, Denmark, Aug. 20-24, 2001. Supplement to Vehicle System Dynamics 37 (2003), pp. 441-452.*
2. *Mallat, S.: A Wavelet Tour of Signal Processing, Second Edition. Academic Press, London, UK, 1999.*
3. *WaveLab Toolbox software and documentation available for download on the Internet at http://www-stat.stanford.edu/~wavelab/.*
4. *Duym, S.W.R.: Simulation tools, modelling and identification, for an automotive shock absorber in the context of vehicle dynamics. Vehicle System Dynamics 33 (2000), pp.261-285.*
5. *Kritayakirana, K. and Cole, D.J.: Evaluation of nonlinear models of a damper mounting bush. submitted to Vehicle System Dynamics.*
6. *Berg, M.: A nonlinear rubber spring model for rail vehicle dynamics analysis. Vehicle System Dynamics 30 (1998), pp.197-212.*

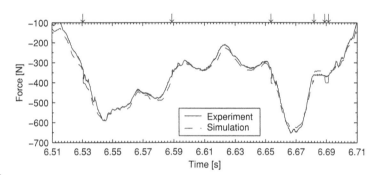

(a)

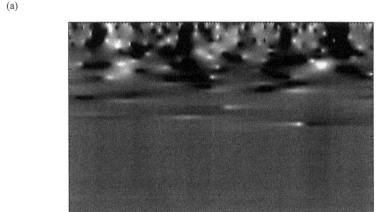

(b)

Fig. 9. (a) Force time history and (b) magnitude of (experiment input to simulation output) WTF for random excitation near zero velocity - baseline model.

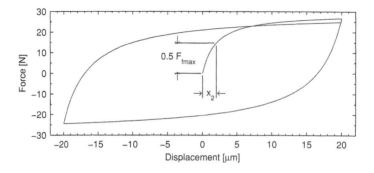

Fig. 10. Force-displacement characteristic for Berg friction model subject to sinusoidal excitation.

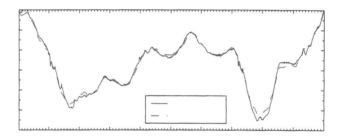

(a)

(b)

Fig. 11. (a) Force time history and (b) magnitude of (experiment input to simulation output) WTF for random excitation near zero velocity - with improved friction model.

Vehicle System Dynamics Supplement 41 (2004), p.627-636 © Taylor & Francis Ltd.

Controllable Vibration of the Car-Body Using Magnetorheological Fluid Damper

YANQING LIU, HIROSHI MATSUHISA, HIDEO UTSUNO, JEONG GYU PARK[1]

SUMMARY

The magnetorheological (MR) fluid can be changed from free-flowing and linear viscous liquid into semi-solid only in milliseconds by the applying magnetic field. The damper is one of the important components in a car suspension. Therefore, the MR fluid damper has been a very attractive subject for the controllable suspension of a car. In this paper, based on experimental data, a parametric model of MR damper consisted of hysteretic property and variable viscosity which was proportional to the applied voltage was proposed. MR damper as passive and semi-active types was taken as an electronic control unit in 4 degree-of-freedom half car model. The vertical and pitching responses of body and the suspension travel responses were analyzed, which indicated the advantages of MR damper.

1 INTRODUCTION

The magnetorheological (MR) fluid consists of micron-sized magnetically polarizable particles dispersed in a medium carrier such as mineral or silicone oil. When a magnetic field is applied, the MR fluid can become a semi-solid in milliseconds, which is a good controllable property [1]. Recently, because of this controllable property, the MR fluid damper has been widely studied in semi-active suspensions of the vehicle [1, 2]. Semi-active controllable suspension was proposed in the early 1970's [3], which is nearly as effective as full active suspension in improving ride comfort and handling stability. When the control system with MR damper fails, the suspension can still work as a passive device. Compared with active and passive suspension systems, the semi-active suspension combines the advantages of both [4, 5]. The characteristic of force vs. velocity of MR damper in theoretical suspension model was always taken as linear or piecewise which did not precisely describe the properties of the MR damper [6]. The applied voltage on MR damper in semi-active on-off suspension was always considered as constant, which caused shock acceleration at the switching moment [7].

In this paper, according to both the experimental results of the behavior of MR fluid and the mechanism of MR damper, suitable parameters of the theoretical model of MR damper were determined and the semi-active suspension of half car model and the dynamic responses of the body were analyzed. Comparisons of the responses of the semi-active and passive suspensions showed the advantages of the semi-active suspension with variable voltage, which could be useful to design and analyze the responses of the whole car system.

---

[1] Address correspondence to: Department of Precision Engineering, Kyoto University, Kyoto 606-8501, Japan, E-mail: yqliu@eng.mbox.media.kyoto-u.ac.jp

2   EXPERIMENTAL SETUP AND RESULTS

In order to build the theoretical model of the MR fluid damper, it is necessary to accurately understand the behavior of MR damper. The experimental rig shown in Fig.1 was designed for the purpose of obtaining basic characteristics of MR damper.

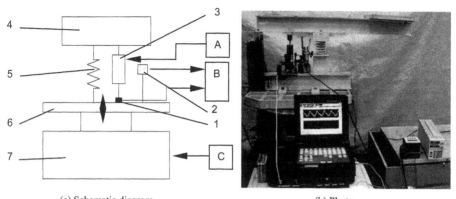

(a) Schematic diagram   (b) Photo
1: Force sensor   2: Laser displacement sensor   3: MR damper   4: Mass   5: Spring
6: Vibration platform   7: Vibrator   A: DC power   B: FFT analyzer   C: Signal generator
Fig. 1 Schematic diagram and photo of the test rig.

In this test rig, cantilever type 1-DOF (Degree of Freedom) vibration system is composed of a MR damper, a spring and mass. The MR damper is RD1005 produced by Lord Company, USA. The laser displacement sensor was installed between the mass and the vibrator platform. By the laser sensor, the relative displacement and velocity of the mass are measured. The force sensor was installed between the vibration platform and the MR damper to measure the output force of the MR damper. To change the strength of magnetic field inside the MR damper, a DC power supplier (A) was employed. The data acquisition system (B) consisted of a multi-channel FFT analyzer and a computer. The sinusoidal vibration waves from a signal generator (C) were input into the vibrator. The dynamic responses of the MR damper can be measured in a wide range of amplitude and frequency.

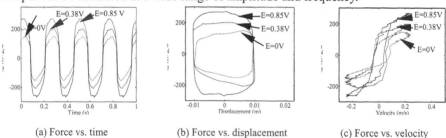

(a) Force vs. time   (b) Force vs. displacement   (c) Force vs. velocity
Fig. 2 The experimental results of MR damper.

The responses of damping force vs. time generated by sinusoidal vibration exciter with the frequency of 4.0 Hz and amplitude is 0.01m, are shown in Fig. 2(a). The applied voltage to the MR damper is three values, 0V, 0.38V and 0.85V. The relations of damping force vs. displacement and force vs. velocity are shown in Fig. 2(b) and (c). Fig. 2(c) shows that the force varies linearly with velocity in a large velocity region and the force has a hysteresis loop in a small velocity region

3   MECHANICAL MODEL OF MR DAMPER

Based on the characteristics of MR fluid and the structural mechanism of MR damper shown in Fig. 3, the theoretical model of MR damper can be proposed as showed in Fig. 4, in which the Coulomb friction and the hysteretic property are considered by Boc-Wen function [8].

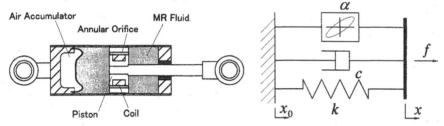

Fig.3 The structure of MR damper.   Fig.4 The model of the MR damper.

The total force f generated by this model is given as follows

$$f = -c(\dot{x} - \dot{x}_0) - \alpha z - k(x - x_0),$$ (1)

where k is the stiffness factor due to the air accumulator and the MR fluid, x and x_0 are input signal and output response, c is the damping coefficient of MR fluid and α is the parameter for the hysteresis loop. The differential equation governing the evolutionary variable z is given by

$$\dot{z} = -\gamma|\dot{x} - \dot{x}_0|z|z|^{n-1} - \beta(\dot{x} - \dot{x}_0)|z|^n + A(\dot{x} - \dot{x}_0),$$ (2)

where γ, β and A are parameters considering hysteretic loop and Coulomb friction, and n is a constant of the system. The parameters k, c and α are all approximately linear to the applied voltage E as follows

$$c = c_0 + c_E E, \quad \alpha = \alpha_0 + \alpha_E E, \quad k = k_0 + k_E E,$$ (3)

where c_0, α_0 and k_0 are constants when there is no voltage on MR damper, c_E, α_E and k_E are proportional coefficients. Based on experimental data, the parameters of the theoretical model can be obtained as shown in Table 1.

Fig. 5 shows the comparisons between experimental responses and calculated responses of three voltage states. It is clear that the calculated result of the model fits well to the experimental data both in hysteretic area and in larger velocity area. The proposed model can express the performance of MR damper.

Table 1. Parameters of the theoretical model of MR damper

Parameter	Value	Parameter	Value
c_0 (Ns/mm)	1.3175	c_E (Ns/Vmm)	0.3542
α_0 (N/mm)	0.742	α_E (N/Vmm)	1.58
k_0 (N/mm)	310	k_E (N/Vmm)	17.6
γ (mm^{-3})	200	A	96
β (mm^{-3})	60	n	3

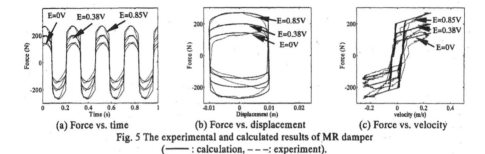

| (a) Force vs. time | (b) Force vs. displacement | (c) Force vs. velocity |

Fig. 5 The experimental and calculated results of MR damper
(———: calculation, – – –: experiment).

4   HALF CAR MODEL WITH CONTROLLABLE SUSPENSIONS

Based on the vehicle theory and structure, 4-DOF model of half car can be built as shown in Fig. 6. The MR damper as an ECU (Electronic Control Unit) is applied in the vehicle suspension between the body and axles to improve the ride comfort and handling stability. The nomenclatures of the half car model are shown in Table 2.

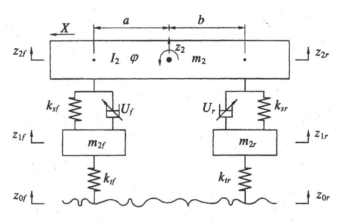

Fig. 6 The half-car model.

Talbel 2. Nomenclatures of the half car model.

f,r	Suffix of front and rear	z_{0f},z_{0r}	Displacement of road
k_{tf},k_{tr}	Stiffness of tires	z_{1f},z_{1r}	displacement of unsprung mass
k_{sf},k_{sr}	Stiffness of suspensions	z_{2f},z_{2r}	Displacement of sprung mass
m_{1f},m_{1r}	Unsprung masses	U_f,U_r	Damping forces of the MR damper
a,b	Distance of axles from centroid	z_2	Displacement of the body
I_2,m_2	Inertia moment and mass of body	φ	Pitching angle of the body

The equations of motion of the half car model is formulated as follows,

$$m_{1f}\ddot{z}_{1f} = U_f + k_{sf}(z_{2f}-z_{1f}) - k_{tf}(z_{1f}-z_{0f}), \tag{4}$$

$$m_{1r}\ddot{z}_{1r} = U_r + k_{sr}(z_{2r}-z_{1r}) - k_{tr}(z_{1r}-z_{0r}), \tag{5}$$

$$m_2\ddot{z}_2 = -[k_{sf}(z_{2f}-z_{1f})+U_f] - [k_{sr}(z_{2r}-z_{1r})+U_r], \tag{6}$$

$$I_2\ddot{\varphi}_2 = [k_{sf}(z_{2f}-z_{1f})+U_f]a - [k_{sr}(z_{2r}-z_{1r})+U_r]b \cdot \tag{7}$$

When the pitching angle φ is small, the relationship of displacements can be described as

$$z_{2f}=z_2-a\varphi, \quad z_{2r}=z_2+b\varphi. \tag{8, 9}$$

Substitution of Eqs. (8) and (9) into Eqs. (4)~(7) gives the state-space form as

$$\dot{Z}=AZ(t)+BU(t)+FW(t), \tag{10}$$

where

$$A = \begin{bmatrix} 0 & 1 & 0 & 0 & 0 & 0 & 0 & 0 \\ -\dfrac{k_{sf}+k_{tf}}{m_{1f}} & 0 & 0 & 0 & \dfrac{k_{sf}}{m_{1f}} & 0 & -\dfrac{k_{sf}a}{m_{1f}} & 0 \\ 0 & 0 & 0 & 1 & 0 & 0 & 0 & 0 \\ 0 & 0 & -\dfrac{k_{sr}+k_{tr}}{m_{1r}} & 0 & \dfrac{k_{sr}}{m_{1r}} & 0 & \dfrac{k_{sr}b}{m_{1r}} & 0 \\ 0 & 0 & 0 & 0 & 0 & 1 & 0 & 0 \\ \dfrac{k_{sf}}{m_2} & 0 & \dfrac{k_{sr}}{m_2} & 0 & -\dfrac{k_{sf}+k_{sr}}{m_2} & 0 & \dfrac{k_{sf}a-k_{sr}b}{m_2} & 0 \\ 0 & 0 & 0 & 0 & 0 & 0 & 0 & 1 \\ -\dfrac{k_{sf}a}{I_2} & 0 & \dfrac{k_{sr}b}{I_2} & 0 & \dfrac{k_{sf}a-k_{sr}b}{I_2} & 0 & -\dfrac{k_{sf}a^2+k_{sr}b^2}{I_2} & 0 \end{bmatrix}, \quad U=\begin{pmatrix}U_f\\U_r\end{pmatrix}, \quad W=\begin{pmatrix}z_{0f}\\z_{0r}\end{pmatrix},$$

$$B=\begin{bmatrix}0 & 1/m_{1f} & 0 & 0 & 0 & -1/m_2 & 0 & a/I_2 \\ 0 & 0 & 0 & 1/m_{1r} & 0 & -1/m_2 & 0 & -b/I_2\end{bmatrix}^T, \quad F=\begin{bmatrix}0 & k_{tf}/m_{1f} & 0 & 0 & 0 & 0 & 0 & 0 \\ 0 & 0 & 0 & k_{tr}/m_{1r} & 0 & 0 & 0 & 0\end{bmatrix}^T,$$

$$Z=\begin{pmatrix}z_{1f} & \dot{z}_{1f} & z_{1r} & \dot{z}_{1r} & z_2 & \dot{z}_2 & \varphi & \dot{\varphi}\end{pmatrix}^T \cdot$$

In this paper, relative control method is adopted, in which the accelerations of body in vertical and pitching direction are minimized. The control forces cancel the forces exerted by springs such that the mass does not receive force as,

$$U_f=-k_{sf}(z_{2f}-z_{1f}), \quad U_r=-k_{sr}(z_{2r}-z_{1r}). \tag{11, 12}$$

These control forces are designed on the basis of active control system, but MR damper is a passive system and the direction of force is always opposite to its velocity.

Therefore, when the force exerted by the MR damper (named damping force) has the opposite direction to the theoretical control force, the damper must be cut off, which is named on-off control method [3, 9]. The damping force is caused by the relative velocity of suspension, so the on-off control method can be described as

$$U_s = \begin{cases} U & if \ U(\dot{z}_2 - \dot{z}_1) \geq 0, \\ U_0 & if \ U(\dot{z}_2 - \dot{z}_1) \leq 0 \end{cases} \tag{13}$$

where U_s is the damping force produced by MR damper, $(\dot{z}_2 - \dot{z}_1)$ is the relative velocity of front and rear suspension, and U_0 is the minimum forces of MR damper. Because the MR damper is controlled by voltage, the force should be changed into voltage. According to the theoretical model of MR damper, the voltage is given by

$$E = \frac{U_s - U_0}{c_E(\dot{z}_2 - \dot{z}_1) + \alpha_E z + k_E(z_2 - z_1)}, \tag{14}$$

where E is the voltage applied on MR damper during "on" state, and U_0 is the force duing $E=0V$.

5 SIMULATION AND ANALYSIS

To investigate the dynamic response of the body with MR damper, the values of parameters for one certain car are chosen as follows: $m_2=650kg$, $m_{1f}=45kg$, $m_{1r}=40kg$, $I_2=1222 kg \cdot m^2$, $a=1.3 m$, $b=1.5 m$, $k_{sf}=19kN/m$, $k_{sr}=19kN/m$, $k_{tf}=200kN/m$, $k_{tr}=200kN/m$. The following three types of MR damper are considered in half car model: (1) the voltage of MR damper is 0V, named passive damper, (2) the voltage of MR damper is 2V and 0V for on and off, named on-off damper, (3) the voltage of MR damper is a variable with on-off control method, named variable damper. The dynamic responses of system with bump and random road conditions are analyzed.

5.1 Simulation of bump response

The input signal on the front and rear wheel is shock signal like road bump with width 0.10m and height 0.10m. The delayed time of the rear wheel was calculated according to the distance 2.80m between front and rear axis of the wheel and the car velocity, 20 m/s. Fig. 7 and Fig. 8 show the time responses of body accelerations in vertical and pitching direction and Fig. 9 and Fig. 10 show the time responses of front and rear suspension travels due to the bump.

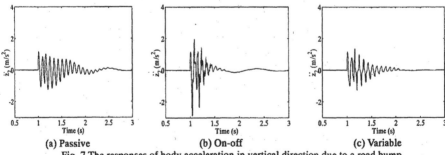

(a) Passive   (b) On-off   (c) Variable

Fig. 7 The responses of body acceleration in vertical direction due to a road bump

632

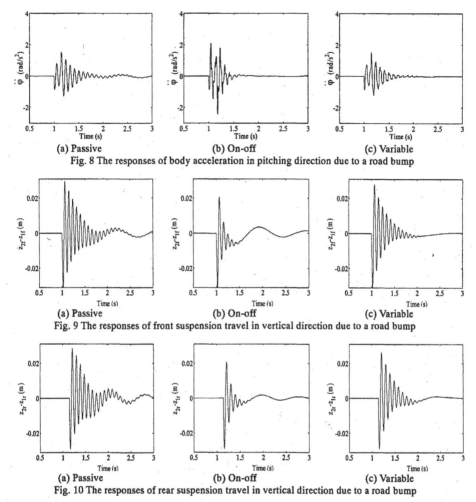

(a) Passive   (b) On-off   (c) Variable

Fig. 8 The responses of body acceleration in pitching direction due to a road bump

(a) Passive   (b) On-off   (c) Variable

Fig. 9 The responses of front suspension travel in vertical direction due to a road bump

(a) Passive   (b) On-off   (c) Variable

Fig. 10 The responses of rear suspension travel in vertical direction due to a road bump

According to the simulated results, the suspension travels of the semi-active dampers attenuate much faster than those with the passive damper. The suspension travels are attenuated a little quickly by the on-off damper. The acceleration of on-off damper is larger than that of variable damper. These results indicate that the semi-active dampers have better dynamic characteristics than that of passive damper, but comparison between the semi-active dampers should be studied further in the random road condition.

5.2 Simulation of random response

Based on the road roughness classification proposed by the International Standardization Organization (ISO), the random signal for the simulation was generated by Matlab Simulink tools. The signal corresponds to the road class C and the velocity of 20 m/s. According to Eq. (14), the voltages for MR dampers can be calculated as shown in Fig. 11.

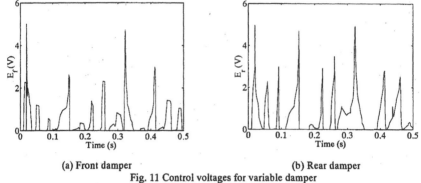

(a) Front damper   (b) Rear damper

Fig. 11 Control voltages for variable damper

Fig. 12 and Fig. 13 show the power spectral density (PSD) responses of acceleration for the body in vertical and pitching directions. Fig. 14 and Fig. 15 show the PSD responses of front and rear suspension travels. Table 3 lists the root mean square (RMS) values of body accelerations and suspension travels.

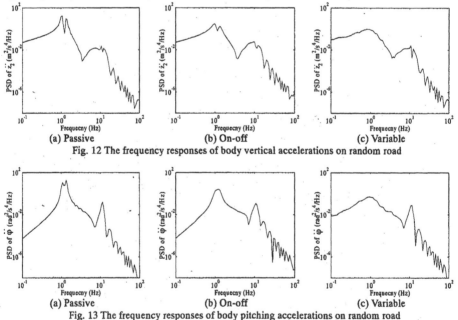

(a) Passive   (b) On-off   (c) Variable

Fig. 12 The frequency responses of body vertical accelerations on random road

(a) Passive   (b) On-off   (c) Variable

Fig. 13 The frequency responses of body pitching accelerations on random road

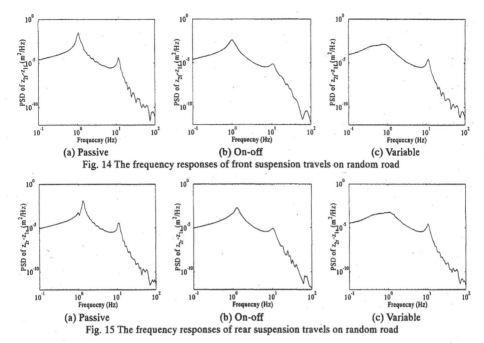

(a) Passive   (b) On-off   (c) Variable

Fig. 14 The frequency responses of front suspension travels on random road

(a) Passive   (b) On-off   (c) Variable

Fig. 15 The frequency responses of rear suspension travels on random road

From the frequency responses, it is easily found that the acceleration of the body and suspension travel of the passive damper at the resonance frequencies (1Hz and 11Hz) are larger than those of the semi-actives. The acceleration of variable damper is smaller than that of on-off damper at 1Hz, but they are almost same at 11Hz. The suspension travel of variable damper is smaller than that of on-off damper at 1Hz, but it is a little larger than that of on-off damper at 11Hz. So the suspension with variable damper can decrease the vibration of sprung mass trading off a little vibration of unsprung mass.

Regarding the RMS values of suspension travel and body acceleration, the values of passive damper are all larger than those of the semi-active dampers. The acceleration values of on-off damper are larger than those of variable damper by 30.8% in vertical and 31.2% in pitching directions. The suspension travels are a little smaller than those of variable damper. It can be concluded that the car with variable damper has good dynamic characteristics in random road condition.

Table 3 RMS values of suspension travel and body acceleration

	Suspension travel (mm)		Body acceleration	
	Front	Rear	Vertical (m/s^2)	Pitching (rad/s^2)
Passive	15.9	11.6	0.5777	0.4786
On-off	12.0	8.9	0.4686	0.3509
Variable	12.7	9.6	0.3583	0.2674

6 Conclusions

A prototype MR damper has been tested and its theoretical model is proposed, which can effectively describe the hysteretic nonlinear property of MR damper in a wide range of controllable damping force.

In the 4 DOF half car model, the MR damper can be an ECU to control the vibration of suspension and body to improve the car ride comfort and handling stability. By controlling the voltage applied on MR damper, MR damper can be the passive and semi-active dampers. Comparisons of the simulated results among the passive and two kinds of semi-active controls with bump and random road conditions, the car with variable voltage of MR damper has best dynamic characteristics.

REFERNCES

1. Pan, G., Matsuhisa, H., Honda, Y.: Analytical model of a magnetorhological damper and its application to the vibration control. 2000 IEEE International Conference on Industrial Electronics, Nagoya, Control and Instrumentation, October, 2000, pp22-28.
2. Nader, J.: A comparative study and analysis of semi-active vibration control systems. The ASME Journal of Vibration and Acoustics, 2002, 124(10), pp593-605.
3. Karnopp, D. C., Crosby, M. J., Harwood, R. A.: Vibration control using semi-active force generators. ASME Journal of Engineering for Industry, 1974, 96(2), pp619-626.
4. Kawabe, T., Isobe, O., Watanabe, Y.: New semi-active suspension controller designs using quasi-linearization and frequency shaping. Control Engineering Practice, 1998, 70(6), pp1183-1191.
5. Yoshimura, T., Nakaminami, K., Kurimoto, M.: Active suspension of passenger cars using linear and fuzzy logic controls. Control Engineering Practice, 1999, 145(7), pp41-47.
6. Liu, Y., Zhang, J.: Nonlinear dynamic responses of twin-tube hydraulic shock absorber. Mechanics Research Communications, 2002, 29(5), pp359-365.
7. Pan, G.: Investigation to the semi-active control method with MR damper, Ph.D Dissertation. Kyoto University, Japan, 2001.
8. Spencer, B.F., Dyke, S.J., Sain, M.K.: Phenomenological model of a magnetorheological damper. ASCE Journal of Engineering Mechanics, 1997, 123(3), pp230-238.
9. Rakheja, S., Sankar, S.: Vibration and shock isolation performance of a semi-active on-off damper. ASME Journal of Vibration, Acoustics, Stress and Reliability in Design, 1985, 107(4), pp398-403.

Vehicle System Dynamics Supplement 41 (2004), p.637-646 © Taylor & Francis Ltd.

Study on Electromagnetic Damper for Automobiles with Nonlinear Damping Force Characteristics
(Road Test and Theoretical Analysis)

YOSHIHIRO SUDA[1], TAICHI SHIIBA[1], KOJI HIO[1], YASUHIRO KAWAMOTO[1], TAKUHIRO KONDO[2], HIDEKI YAMAGATA[2]

SUMMARY

This paper presents study on the electromagnetic damper (EMD) with nonlinear damping force characteristics for automobiles and its road tests. The authors proposed EMD for automobiles for improved ride comfort and compatibility between stability and isolation. General dampers for automobiles have nonlinear damping force characteristics to satisfy this compatibility. The power electronic circuits for the EMD were developed to achieve any nonlinear damping force characteristics. These circuits have the following characteristics: 1) No external power source is required for circuits because the EMD regenerate power, 2) High controllability and tuning of nonlinear damping characteristics were made. The experiments with shaker and road tests proved that the proposed EMD system achieves the nonlinear damping force characteristics for automobiles, and has suitable ability for automobiles dampers.

1   INTRODUCTION

The improvement of automobile suspension is important to get compatibility between comfort, stability and steering ability of vehicles. Recently a lot of studies of active suspension system using hydraulic actuator were made, and these innovative systems are now using for commercial use. However, the energy consumption of active systems is one of the weak points for social demands. So, the semi-active suspension is practically used to control the damping coefficient without energy consumption. The major system of semi-active control is switching of the valve in the oil damper.

In the meantime, recent automobile is being mounted on many electrical devises. The electric device is very convenient to carry out various controls. For example, electro-motion power steering replaces the hydraulic power steering. Moreover, electric device has several possibilities to get new functions for suspensions. One is

---

[1] *Address correspondence to:* Institute of Industrial Science, The University of Tokyo, 4-6-1, Komaba, Meguro-ku, Tokyo, 153-8505, JAPAN.
[2] System Research Sect. Basic Technology R&D Center, Kayaba Industry Co., Ltd., 1-12-1 Asamizodai, Sagamihara-shi, Kanagawa Pref., 228-0828, JAPAN

the function of energy regeneration from vibration energy. Some of authors proposed self-powered active control system using regenerated vibration energy[1,2] and examined to apply to truck cab suspension[3] and anti-rolling system for ships[4]. These systems are successfully developed and expected for practical use in future.

From these backgrounds, in this study, the electromagnetic suspension using electric motor/generator is proposed for the objective to realize the ideal suspension for passenger vehicles. To apply passenger car, the improvement of ride comfort is quit important. Then, to improve ride comfort and satisfy compatibility between driving stability and isolation, general dampers of suspension for automobiles have nonlinear damping force characteristics as shown in Fig. 1. The other requirement for recent passenger car is reduction of fuel consumption.

The authors have proposed and developed the electromagnetic damper ("EMD") for the purpose of highly efficient and electrical damper [5,6]. From the view point of energy consumption, the developed EMD is passive system with electric motor as generator in closed circuit. The EMD uses electromagnetic force for damping force, so it can electrically control damping force by regulating induced current in the closed circuit. The proposed damper has the following three characteristics: 1) Tuning of damping force, 2) Energy regeneration, and 3) High controllability.

The prototype actual dampers were made and tested for basic characteristic. From results of oscillation tests, numerical simulation and road tests, the first prototype damper was structurally improved and the second prototype damper was made as shown in Fig. 2. Another improvement was nonlinearization of damping force characteristics. The second EMD system was proved by the oscillation tests and road tests that it achieves the nonlinear damping force characteristics for automobiles, and has suitable ability for automobiles dampers.

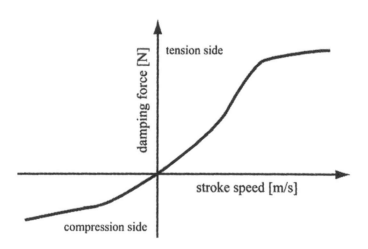

Fig. 1. Properties of Damper for Automobiles.

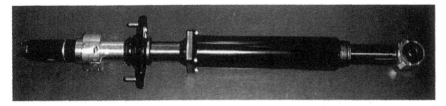

Fig. 2. The Second Prototype Electromagnetic Damper.

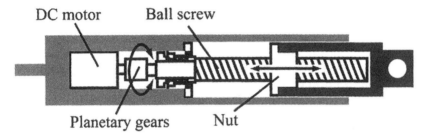

Fig. 3. The Concept of Electromagnetic Damper.

2   PROPOSED ELECTROMAGNETIC DAMPER SYSTEM

2.1   BASIC CONCEPT

The EMD consists of a Direct Current (DC) motor, planetary gears and the ball screw mechanism. The concept of the EMD is shown in Fig. 3. It converts transverse motion of vibration between wheel and car body into rotating motion of a DC motor. The DC motor generating induced electromotive force, the EMD uses this electromagnetic force as damping force.

The following is an equation of motion of the EMD when the DC motor is connected to a variable resistance. Descriptions of symbols are shown in Table 1. Supposing that the stroke speed of damper is $\dot{z}$ m/s, the damping force of the EMD can be written as following equations if the inductance in the circuit and friction can be supposed to be ignored:

$$f_d = -c_{em}\dot{z} - I_d\ddot{z} \qquad (1)$$

where

$$c_{em} = \left(\frac{2\pi}{l}\alpha\right)^2 \times \frac{k_e \cdot k_t}{R}, \quad I_d = \left(\frac{2\pi}{l}\alpha\right)^2 J_m + \left(\frac{2\pi}{l}\right)^2 J_b \tag{2}$$

Equation (2) shows that the damping coefficient can be variable by resistance of circuit. Equation (1) shows that damping force has the hysteresis due to the second inertia term. Then, to make the influence of hysteresis on the damping property small, the equivalent inertial inertia had better be as small as possible.

2.2 Improvement of Hardware

The first prototype actual dampers were made and examined for the basic characteristics of the proposed damper system by numerical simulations and oscillation tests with a shaker. From these results, it was confirmed that it is possible to use as automobile dampers, but it was also found that the equivalent inertial inertia and the structural backlash of the mechanism are slightly large. Improving these two points, the second prototype dampers were designed and manufactured. From Equation (2), the first term has a large effect on the equivalent inertial inertia because of the planetary gear ratio α is greater than 1. So, mainly making a conversion rate small, the equivalent inertial inertia was reduced by 68 %. Figure 4 shows results of shaking experiments of the second prototype EMD with a sinusoid stroke input in the condition of the maximum stroke speed of 0.2 m/s and the stroke of ± 25 mm. The hysteresis band of the second prototype is small enough to be said that damping force is in proportion to stroke speed.

Table 1. Description of Symbols.

Symbol	Description	Unit
I_d	Equivalent inertial inertia	kg m^2
J_m	Moment of inertia of motor's rotor	kg m^2
J_b	Moment of inertia of ball screw	kg m^2
c_{em}	Damping coefficient of the EMD	Ns/m
e_m	Induced voltage of motor	V
f_d	Damping force of the EMD	N
i	Electric current in closed circuit	A
k_e	Voltage constant	V s/rad
k_t	Torque constant	N m/A
l	Lead of the ball screw	m/rev
z	Stroke of EMD	m
R	Variable resistance of the circuit	Ω
α	Planetary gear ratio	—

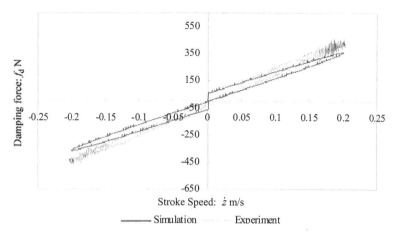

Fig. 4. Characteristics of Stroke Speed: $\dot{z}$ and Damping Force: f_d.

2.3 DEVELOPMENT OF POWER ELECTRONICS CIRCUIT FOR NONLINEAR DAMPING FORCE CHARACTERISTICS

To improve ride comfort and satisfy compatibility between driving stability and isolation, general dampers of suspension for automobiles have nonlinear damping force characteristics. So the proposed EMD system should have the similar damping force characteristics. This section describes the way of realizing the nonlinear damping force characteristics in the proposed system without sensors and external power source.

2.3.1 Concept

The induced voltage of the DC motor in the EMD can be written as the following equation with the stroke speed:

$$e_m = \frac{l}{2\pi\alpha k_e} \cdot \dot{z} \tag{3}$$

The damping force of the EMD is written using the circuit current as follows:

$$f_d = -\frac{2\pi}{l}\alpha \cdot k_t i - I_d \ddot{z} \tag{4}$$

If the equivalent inertial inertia is small enough, the second inertia term in Equation (4) is supposed to be disregarded compared with the first term. So, the damping force is in proportion to the circuit current. In Equation (3), the induced voltage is in proportion to the stroke speed. If the current and voltage have a nonlinear relationship, the damping force and the stroke speed also have the similar relationship.

2.3.2   Proposed Power Electronic Circuits

To get nonlinear relationship of the current and voltage, the authors propose power electronic circuits composed of a MOS-FET, a variable shunt regulator and two variable resistances, which are called variable current restriction circuits, as shown in Fig. 5 (a). A volt-ampere characteristic of this circuit is showed in Fig. 5 (b), where V_{REF} means reference voltage of the variable shunt regulator. This circuit can adjust a saturation point within a gray region in Fig. 5 (b) by changing values of two variable resistances.

Then, several units of the variable current restriction circuits are connected as shown in Fig. 6 (a). This multistage variable current restriction circuit (multistage circuit) has saturation points of the number of units, so that the current and voltage get a nonlinear relationship as shown in Fig. 6 (b). The obtained damping force and the stroke speed of damper, i.e., damping force characteristics, also have the similar nonlinear relationship that is suitable for the automobile damper.

To confirm whether the multistage circuit works well, the property of the circuit composed of four units is measured on five conditions. The four units circuit is powered by DC power supply. Figure 7 shows measurement results that the multistage circuit can tune its volt-ampere characteristics like the nonlinear damping ones.

The actual EMD system with multistage circuit and an EMD was developed to fit for the test vehicle. This system needs no external power source for circuits because the EMD regenerate power. Moreover, the high controllability and tuning of nonlinear damping characteristics were achieved.

3   OSCILLATION TEST WITH THE EMD SYSTEM

The oscillation tests with a shaker were made to examine the property of the EMD system that is composed of the EMD and the multistage variable current restriction circuit. This chapter describes an example of experimental results in oscillation tests.

3.1   METHOD OF OSCILLATION TESTS

The multistage circuit used in these tests has four units, and it is set on three conditions. Three conditions of multistage circuit are different from the number and position of the saturation points. The sinusoid stroke input is set as the maximum stroke speed of 0.3 m/s and the stroke of $\pm$ 25 mm.

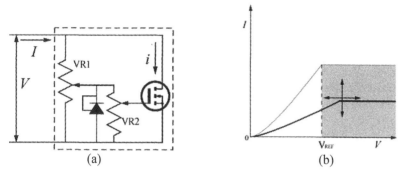

(a)   (b)

Fig. 5. Diagram and Volt-Ampere Characteristic of Variable Current Restriction Circuit

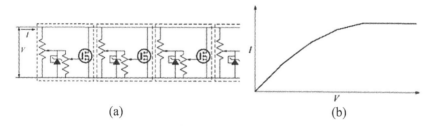

(a)   (b)

Fig. 6. Diagram and Volt-Ampere Characteristic of Multistage Variable Current Restriction Circuit.

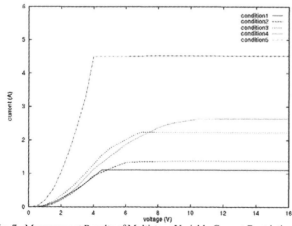

Fig. 7. Measurement Results of Multistage Variable Current Restriction Circuit.

3.2   EXPERIMENTAL RESULTS

Experimental results are shown in Fig. 8 and Fig. 9. Figure 8 shows that the multistage circuit works accurately in spite of the power supply of the induced voltage of EMD. It was found that nonlinear damping characteristics were achieved. For examination of the accrual operating condition, the effect of rubber mount was also examined. Figure 9 shows comparison of the experimental result with rubber mount for that of without rubber mount. It was found that the rubber mount was able to eliminate the small oscillation with high frequency caused by ball-screw mechanism of the proposed EMD.

4   ROAD TESTS WITH THE EMD SYSTEM

The road tests on the proving ground with the EMD system were made to examine the actual property of the EMD system for automobiles. This chapter describes an example of experimental results in road tests

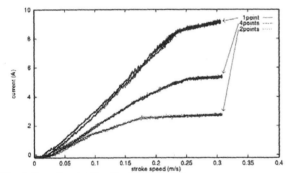

Fig. 8. Experimental Results of Shaker Tests.

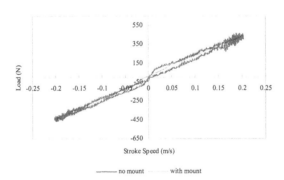

Fig. 9. Effect of Rubber Mount.

4.1   METHOD OF ROAD TESTS

The test car is one of a passenger car, and it is equipped with the EMD system in the rear suspensions. (Fig. 10) To give a disturbance into both wheels of the test car, protrusions of the 18 mm height and the 30 mm width are set on the road face. The shape of protrusions is shown in Fig. 11. The properties of the EMD are set on three conditions as follows:

1.   Nonlinear damping force characteristics with multistage circuit set soft.
2.   Nonlinear damping force characteristics with multistage circuit set hard.
3.   Linear damping force characteristics with a fixed resistance (comparative data).

Moreover, the damping force properties of tension side and compression side are different on both nonlinear conditions.

On the above conditions, the test car stepped over protrusions at a speed of 20 km/h, and the stroke speed and damping force of the EMDs were measured.

4.2   EXPERIMENTAL RESULTS

Experimental results are shown in Fig. 12. These graphs are plotted using the time history data of the stroke speed and damping force. In Fig. 12, both nonlinear conditions show desirable damping force characteristics to automobiles. The nonlinear damping force characteristics are achieved for desired damping level, i.e., soft and hard. So, it was found that the proposed and developed EMD system has suitable ability for automobile dampers from these experiments using new damper system.

Fig. 10. Attached Electromagnetic Damper.

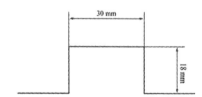

Fig. 11. Shape of Protrusion on Road Face.

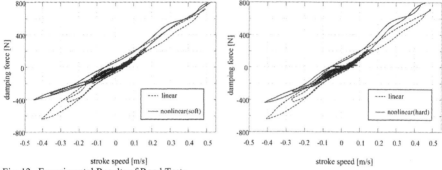

Fig. 12. Experimental Results of Road Tests.

5   CONCLUSIONS

For the automobile damper, the electromagnetic damper that uses the force of electrical motor as damping force has been proposed and improved for passive suspension system. The actual damper has been designed and manufactured. The power electronic circuits for the EMD to achieve nonlinear damping force characteristics have also been proposed and developed. These circuits need no external power source because the EMD regenerate power in circuits. It has an ability to tune any desired nonlinear damping force property. From experiments using a test car equipped with the EMD system, it has been clarified that the proposed and developed EMD system has suitable ability for automobile dampers.

REFERENCES

1.   Nakano, K., Suda, Y. and Nakadai, S.: Self-Powered Active Vibration Control Using Regenerated Vibration Energy, Journal of Robotics and Mechatoronics, Vol.11 No.4, 1999, pp.310-314.
2.   Nakano, K., Suda, Y., and Nakadai, S.: Self-powered active vibration control using a single electric actuator, Journal of Sound and Vibration, 260, 2003-2, pp.213-235.
3.   Nakano, K., Suda, Y., Nakadai, S., Tsunashima, H., and Washizu, T.: Self-powered active control applied to a truck cab suspension, JSAE Review, Vol.20, 1999, pp.511-516.
4.   Nakano, K., Suda, Y. and Nakadai, S.: Anti-Rolling System for Ships with Self-Powered Active Control, JSME International Journal, Series C, Vol.44, No.3, 2001-9, pp. 587-593.
5.   Suda, Y., Suematsu, K., Nakano, K., and Shiiba, T.: Study on Electromagnetic Suspension for Automobiles –Simulation and Experiments of Performance. Proceedings of the 5th International Symposium on Advanced Vehicle Control (AVEC'2000), August 2000, pp. 699–704.
6.   Suda, Y.: Proposal of Electromagnetic Damper for Automobiles –Possibility of Future Automotive Suspension-, Proceeding of the 4th Int. Symposium on Next Generation Vehicle Technology, Nov. 30th, 2001, Gwangju, Korea, pp.1-7.

Electric-power recovery for the mechanical damper in a vehicle-suspension system using artificial intelligent control

ESSAMUDIN A. EBRAHIM[1], N. HAMMAD[2] AND A. M.A. ABU-EL-NOUR[2]

SUMMARY

The suspension elements must be mounted between the vehicle body and wheel axle to minimize undesirable shake for passengers. Parts of mechanical shake energy are transferred to heat energy by such mechanical damper elements in the vehicle-suspension system. So, the new part of this paper is how to utilize or recover this dissipated energy as electrical energy instead of heat energy. So, a new suggested suspension system will be introduced. In this proposed technique, an electric linear generator is used- as an electrical damper- in parallel to the spring instead of the mechanical damper element to recover this dissipated power. This generator is exposed to dynamic forces as equal as to the mechanical damper through the minimized shake. Dependently, the generator transfers undesirable randomly mechanical shake energy in the suspension systems to electric power instead of heat energy due to friction. But, the main problem here is the randomization of such generated power. So, this paper offers an electronic intelligent power card to get the generated electric power in the useful form in different vehicle electrical circuits.

1. INTRODUCTION

A vehicle exists to carry someone or something from place to another. Constructive in that statement is that in doing so their shell is as little shake as possible of who or what is being carried. When the load in-animate, the less the shake the less in the way of special delivery will be necessary. Some of the vehicle-engine power is consumed in driving this shake through the traverse vehicle motion on randomly road roughness.

Vehicle ride models consist of two or more masses, connected to each other and to ground by suspension elements, which produce forces, depend on their relative positions and velocities. Inputs to model take the form of wheels random input displacements, which is generated by the vehicle traversing over a ground profile [1]. The conventional passive suspension system is the basic system used in

[1] *Address corresponding to:* Department of Power Electronics and Energy Conversion, Electronics Research Institute, 12622 Dokki, Cairo, Egypt, Tel: (202) 33 10 551; Fax: (202) 33 51 631; E-mail: essamudin@yalla.com , essamudin@yahoo.com or essamudin@hotmail.com

[2] Address corresponding to: Automotive & Tractor Department, Faculty of Engineering, Mataria Branch, *Helwan University, Cairo, Egypt*

most vehicles. It depends on the various types of spring and damper elements. The characteristics of the suspension elements could be linear or nonlinear. The spring element in the suspension system has two main functions. The first one is carrying on the spring mass and the second is storing the kinetic energy of vehicle masses as potential energy or transmit the latter to masses [2].

The suspension elements must mounted between the vehicle body and wheel axle to minimize this undesirable shake. Parts of mechanical shake energy are transferred to heat energy by such mechanical damper elements in the vehicle-suspension system. So, the new part of this paper is how to utilize or recover this dissipated   energy as electrical energy instead of heat energy.

This paper consists of the following sections: section 1 is introduction to the paper, section 2 describes a general idea about the proposed overall block diagram of the system, section 3 is the mathematical model of the suspension system including the mechanical damper, section 4 describes the proposed electrical linear generator used to recover the power dissipated and use it as an electrical damper, section 5 is the power electronics part used to adapt this random-generated power and transfer it to be utilized, section 6 is the simulation results for the proposed system, section 7 is the conclusion for the paper.

2.   THE OVERALL PROPOSED SYSTEM

The overall proposed system is illustrated in figure 1. The system contains the mathematical model of a mechanical suspension system, the proposed linear generator, rectification, and filtration for smoothing dc output, buck-boost converter and the load. The proposed intelligent control technique is not included in this block diagram and will be explained in detail in the next sections. Each block of the proposed overall system will be explained in details through the next sections.

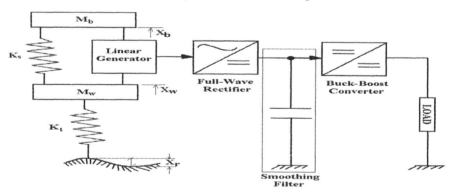

Fig. 1 The overall block diagram of the proposed system.

3. MATHEMATICAL MODEL OF SUSPENSION SYSTEM

Figure (1) describes the mathematical representation for the suspension model. Two main equations express and describe the mathematical model of the mechanical suspension system. These equations are:

$$M_b X_b'' - K_s(X_w - X_b) - C_d(X_w' - X_b') = 0 \tag{1}$$

$$M_w X_w'' + K_s(X_w - X_b) + C_d(X_w' - X_b') - K_t(X_r - X_w) = 0 \tag{2}$$

From Newton's Low of Force Balance:

$$F_d = C_d(X_w' - X_b') \tag{3}$$

Where F_d is the damper force, and the relative speed of the damper can be obtained from the following relation:

$$v = (SWV) = (X_w' - X_b') \tag{4}$$

Then, we can calculate the dissipated power in a heat form from the following equation:

$$P_d = Dissipated - power = F_d(SWV) \tag{5}$$

Where:

M_b, M_w are the masses of the vehicle body and wheel respectively,

X_b, X_w are the linear displacements of vehicle body and wheel respectively,

X_b', X_w' are the linear velocities of vehicle body and wheel respectively,

X_b'', X_w'' are the linear accelerations of vehicle body and wheel respectively,

X_r is the linear displacement of the road roughness

K_s is the suspension spring stiffness constant, C_d is the damper constant

K_t is the tire stiffness constant

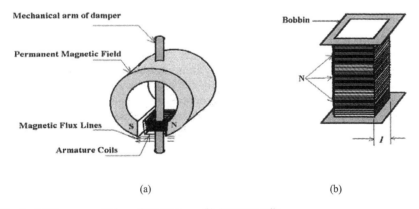

(a)   (b)

Fig. 2. (a) The proposed Linear Generator   (b) Armature coils.

4. THE PROPOSED LINEAR GENERATOR

The proposed linear generator consists of the main two parts: the stator and a linear armature as shown in figure (2-a). The stator is a cylindrical shape and fits on the all space around the suspension system to make as a permanent magnet having a pair of poles (i.e., N-S). The magnetic field strength depends on the size and type of material of this magnet. In this study, we suppose that the magnetic field density (B=0.5 Tesla) is sufficient and suitable for this generator to recover the dissipated-heat power of the damper. The linear armature –shown in figure (2-b) consists of a fibber bobbin is fitted tightly on the mechanical-damper arm with screws to insure moving the armature with the same velocity of the damper. The length of the coil (ι) is the main parameter affects on the induced e.m.f in the armature. Increasing this length- as expressed in eqn. (6), dependently the induced voltage will also increases.

But, we are limited by the space and damper size, etc. The value of the voltage is inversely proportional to conductor length. So, we select the value of length is equal to (ι=0.15 m). Also, the second parameter affects on the induced voltage is the number of coil-turns (N). So we select the value of N equal to (N=50 turns). A copper coil is used with a cross–section conductor diameter ranging from (3-6 mm). This cross section is sufficient to carry the maximum expected induced current. The main induced voltage; current power- formulas are expressed as the following equations [3]:

From Faraday's Low of induction and Neumann's Formulation, the induced e.m.f can be calculated from the following equation:

$$e = B.l.v.N \qquad \text{(Volts)} \qquad (6)$$

Also, from Biot-Savart Force Relationship, induced current can be computed

$$F = B.l.i \qquad \text{(N)} \qquad (7)$$

The relation between current and voltage gives the induced power as follows:

$$P = e.i \qquad \text{(Watts)} \qquad (8)$$

5. RECTIFICATION, FILTERATION AND BUCK-BOOST CONVERTER

The induced generated voltage is a random ac voltage with random amplitude and frequency. So, to utilize this voltage in vehicle electric applications, we must rectify this dc voltage. In this proposed system, an uncontrolled single-phase bridge rectifier is used with four diodes. The output of the bridge is a dc random voltage with variable value, so, a smoothing filter is proposed with large capacitance value as shown in figure (1). The value of the proposed smoothing capacitor is (1200 μf).

The main purpose of this section is transforming this random output induced-generated voltage to a constant dc battery-charging voltage (i.e., V_o=15 V). Because of this induced voltage is a random value – more less or greater than the reference voltage- we must use a buck-boost converter as shown in figure (3) with an intelligent controller that be proposed as an artificial neural network (ANN) controller.

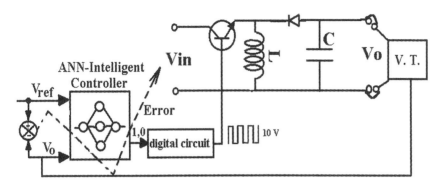

Fig.3. Buck-Boost Converter with intelligent ANN-Controller

5.1. BUCK- BOOST CONVERTER

The buck-Boost converter is a combination of a Buck and Boost converters. This type was explained severally in literature. So, we will focus on the basics of this type. With continuous conduction for the Buck-Boost converter, the voltage of the node connected between the transistor and reactor V_x is equal to the input voltage V_{in} when transistor is on and the output voltage V_o is equal to V_x when the transistor is off. For zero net current change over an inductor, the average voltage across the inductor is zero.

$$V_{in}t_{on} + V_o t_{off} = 0 \tag{9}$$

Which gives the voltage ratio

$$\frac{V_o}{V_{in}} = -\frac{D}{(1-D)} \tag{10}$$

And the corresponding current

$$\frac{I_o}{I_{in}} = -\frac{(1-D)}{D} \tag{11}$$

Since the duty ratio D is between 0 and 1 the output voltage can vary between less or greater than the input voltage in magnitude. The negative sign indicates a reversing of the output voltage.

Figure (4) illustrates the strategy operation for this converter and shows the output waveform for both load voltage and inductor current.

There are two voltages controlling the inductor current. One is the rectified sinusoidal input voltage, which is applied to one side of the inductor, and the other one is the voltage across the boost switch (transistor), which varies between zero and Vo due to switching process and is applied to the other side of the inductor. The voltage that contributes to the ripple is the voltage across the switch. To consider the worst case for ripple, the duty cycle of the switch is taken to be 50% for this analysis [4]. Based on the above assumption and considering that;

V_0 is the output voltage, F_{sw} is the switching frequency
I_{1r} is the rms value of the fundamental component of current ripple at 50% duty cycle. The following equation can be written:

$$(V_o/2)(4/\pi\sqrt{2}) = (2\pi F_{sw}L)I_{ir} \tag{12}$$

Where;

$(V_o/2)(4/\pi\sqrt{2})$ is the rms value of the fundamental component of the switched voltage for a 50% duty cycle.

$(2\pi F_{sw}L)$ is the reactance impedance of L

Considering that the maximum permitted amount of ripple is 5% , therefore, $I_{1r} = .05I$, (where I is the rms value at 100Hz the inductor current)
Then;

$$L = (1.433)(V_s)/(F_{sw})(I) \tag{13}$$

To find the value of C_b for the desired output voltage, it can be determined from the following equation [5]:

$$C = \frac{(4 - r^2)^{\frac{1}{2}}}{2rK_v\omega} \tag{14}$$

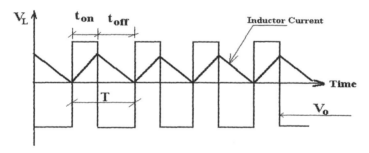

Fig. 4. Waveforms for buck-boost converter

5.2. ARTIFICIAL INTELLIGENT CONTROL

The proposed ANN controller is used as an intelligent controller with three-layer technique. The input layer has two inputs (the reference target output voltage and the actual output voltage, the hidden layer has four hidden neurons and the output layer has one digital output signal that used to drive gate of the IGBT. A feed-forward back-propagation technique is used with delta modulation.

6. SIMULATION RESULTS

The overall proposed system is designed and mathematically simulated with the aid of Matlab/Simulink software package [6] and the data of used vehicle is as:

M_b=330 Kg, M_w=50 Kg, C_d=0.539, C_s=3000, K_s=20000, K_t=20000, and the road roughness displacement is represented and determined according to the graph of figure (5). Figures (6,7) represent the displacement for body and wheel respectively. Figure (8) represents the relative damper velocity and figures (9,10) describe force and power induced respectively. Figures (11,12) demonstrate the induced voltage and current in armature. The rectified voltage is shown in figure (13) and finally, figure (14) is the target output dc voltage recovered to use in electric loads and battery charging in a vehicle.

7. CONCLUSION

The dissipated heat energy induced from the mechanical damper –in vehicle suspension system- was recovered using an electric linear generator as an electric damper. The paper introduced an analytical study for the amount of this power and factors affecting on this power. The main factor is a road roughness displacement that affect directly on this power. A complete design for the proposed system-including ANN as an artificial intelligent controller- was introduced to over come the randomization of this induced power. Also, the paper included some simulation results that can be used as a guide when this proposed technique is implemented experimentally.

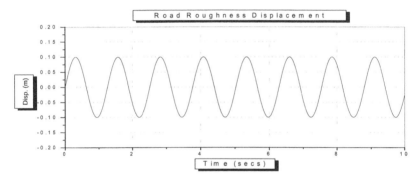

Fig. 5. Road roughness variation

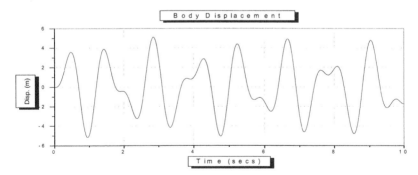

Fig. 6. Body displacement variation with time

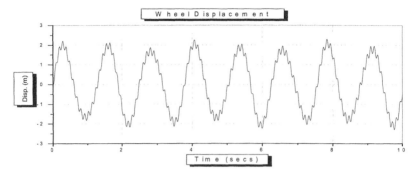

Fig.7. Wheel displacement against time

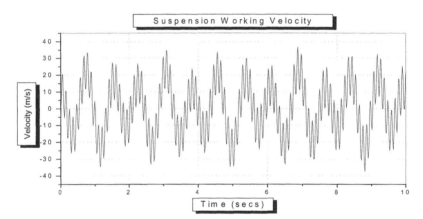

Fig.8. Suspension working velocity

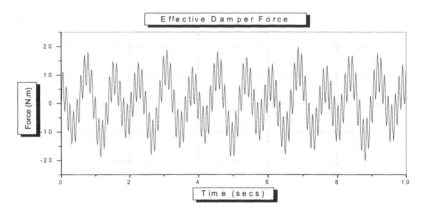

Fig. 9. Effective damper force

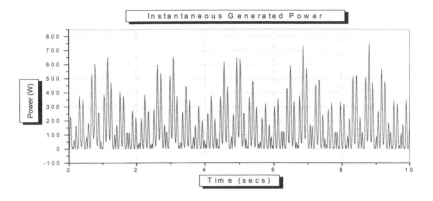

Fig. 10. Induced instantaneous generated power

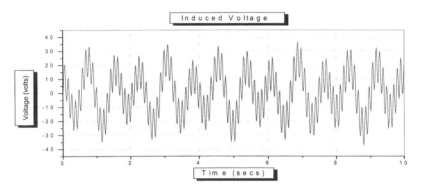

Fig. 11. Induced instantaneous voltage

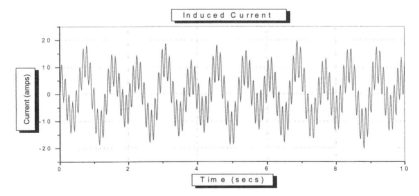

Fig.12. Induced instantaneous current

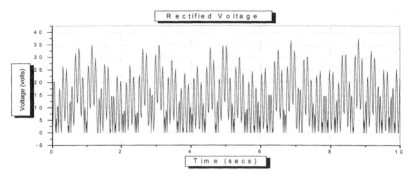

Fig. 13. Rectified output voltage

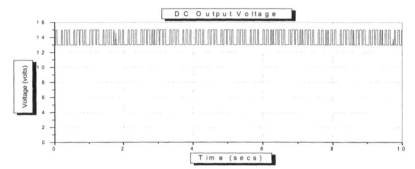

Fig. 14. Output DC charging voltage

REFERENCES

1. Abou-el.Nour, A.M.A.: Recent Developments in Vehicle Systems and Their Effect on Ride Quality. Review paper, Egyptian Scientific Committee for Develops the university Staff Member, 2001.
2. Cebon, D. and Newland, D. E.: The artificial generation of road surface topography by the inverse FFT method. Proc. 8ᵗʰ IAVSD Symp. On Dynamics of Vehicles on Roads and on Railway Tracks, Cambridge MA, 1983, pp 1-12.
3. Ribbens, W. B.: Understanding Automotive Electronics. Fifth Edition, Butterworth-Heinmann Inc., USA, 1998.
4. siebert, A., Tredson, A., and Ebner, S.: AC to DC power conversion now and in the future. IEEE Trans. On Industry Appl., Vol. 38, No. 4, July/Aug. 2002.
5. Abou-el.Nour, A.M.A, Hammad, N., and Ebrahim, E. A.: Electric Utilization of Vehicle-Damper Dissipated Energy. Al-Azhar Engineering 7ᵗʰ International Conference (AEIC 2003), Cairo 7-10 April 2003.
6. Ong, C.M.: Dynamic Simulation of Electric Machinery Using Matlab/Simulink. Text Book, Printice-Hall Inc. and A Simon & Schuster Company, Upper Sadle River, NJ 07458, USA, 1998.

Analytical study on effects of form in transition curve

KEIGO MIYAGAKI[*], MASAHITO ADACHI[*] and YOSHIHIKO SATO[+],

SUMMARY

In Japan straight slope cant and third order parabola for plane curve has been used as the transition curve on existing narrow gauge lines, but the sinuous curve has been used as the transition curve on Shinkansen. In raising the speed of Tokaido Shinkansen to 270km/h the speed could not be raised on several curves because the transition curve could not be elongated due to the obstacles such as turnout. To clarify the effect of transition form on the vehicle response, following analyses have been performed. (1) The frequency characteristics of transition form and their effects on vehicle response. (2) The time dependent response of vehicle to transition from. (3) A new proposal to the transition form and its application to an existing curve.

1. INTRODUCTION

The form in transition curve is important so as to get good comfort entering into the curve or getting out of the curve. In long time the length was discussed from the standpoint of the slope of cant depending on the study in UK[1]. In Japan, when the Tokaido Shinkansen was built, the sinuous slope cant and corresponding plane form (hereafter, called as "sinuous curve or the sinuous") was adopted instead of straight cant and corresponding third order parabola for plane curve ("straight curve or the straight") so as to avoid the shock due to the cant angle at the beginning and end of transition. As for the length of the transition, the experiences on the comfort of tilting train in 1980's gave new tolerances on rolling velocity and roll acceleration of car body[2]. It permits the use of shorter transition.

In realizing 270km/h operation on Tokaido Shinkansen in 1992, the curves were ameliorated by raising the cant from 150mm to 180mm in R3000 and from 180mm to 200mm in R2500 permitting 110mm cant deficiency.

*Central Japan Railway Company & + Railway Track System Institute
Address correspondence to: Railway Track System Institute-RTSI, 1-11-8, Kurosunadai, Inage-ku, Chiba-shi, Chiba-ken 263-0041 Japan. Tel:+81-43-246-3883, Fax:+81-43-246-3922
E-mail:satoy@sa2.so-net.ne.jp,

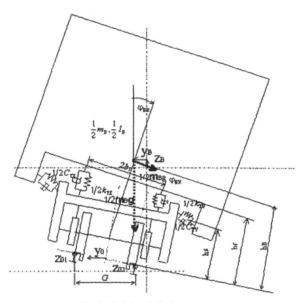

Fig. 1 Model and displacement

Even with those the speed of the latter curve has been restricted to 255km/h. Also, the speed of the curves having no space for elongation of the transition was held under 270km/h.

To raise the speed on such curves, the effects of form in transition curves are analyzed here.

2. ANALYTICAL STUDY ON FORMS OF TRANSITION CURVE

2.1 MODEL

Considering the main effect for the movement of rolling stock on curve depending on secondary suspension, just the car body is considered for the analysis as shown in Fig. 1. The body could move vertically, laterally and rotationally. It is moved by the positions of rails and by the gravity on curve with cant.

2.2 ANALYZED RESULTS

Analyzed results are given as followings for the parameters in Fig. 1 in Laplace Transform.

$$Z_B = \frac{c_{TZ}s + k_{TZ}}{\frac{1}{2}m_Bs^2 + c_{TZ}s + k_{TZ}} \cdot \frac{Z_{01} + Z_{02}}{2} \tag{1}$$

$$Y_B = \frac{\left\{\frac{1}{2}I_Bs^2 + b_1^2(c_{TZ}s+k_{TZ}) - \frac{1}{2}mg(h_B-h_r)\right\}(c_{TY}s+k_{TY})Y_0}{\left\{\frac{1}{2}I_Bs^2 + b_1^2(c_{TZ}s+k_{TZ}) + (h_B-h_S)^2(c_{TY}s+k_{TY})\right\}\left\{\frac{1}{2}m_Bs^2 + c_{TY}s+k_{TY}\right\} - (h_B-h_S)(c_{TY}s+k_{TY})\left\{(h_B-h_S)(c_{TY}s+k_{TY}) + \frac{1}{2}m_Bg\right\}}$$

$$-\frac{\left\{h_S\left(\frac{1}{2}I_Bs^2 - \frac{1}{2}mg(h_B-h_r)\right)(c_{TY}s+k_{TY}) + \left(h_S(c_{TY}s+k_{TY}) + \frac{1}{2}m_Bg\right)b_1^2(c_{TZ}s+k_{TZ})\right\}\frac{Z_{01}-Z_{02}}{G}}{}$$

(2)

$$\Phi_{BX} = \frac{\frac{1}{2}m_Bs^2(h_B-h_S)(c_{TY}s+k_{TY})Y_0 + \left\{\left(\frac{1}{2}m_Bs^2 + c_{TY}s+k_{TY}\right)b_1^2(c_{TZ}s+k_{TZ}) - \frac{1}{2}m_Bs^2h_S(h_B-h_S)(c_{TY}s+k_{TY})\right\}\frac{Z_{01}-Z_{02}}{G}}{\left\{\frac{1}{2}I_Bs^2 + b_1^2(c_{TZ}s+k_{TZ}) + (h_B-h_S)^2(c_{TY}s+k_{TY})\right\}\left\{\frac{1}{2}m_Bs^2 + c_{TY}s+k_{TY}\right\} - (h_B-h_S)(c_{TY}s+k_{TY})\left\{(h_B-h_S)(c_{TY}s+k_{TY}) + \frac{1}{2}m_Bg\right\}}$$

(3)

The transition curve is expressed in Laplace Transform as follows;
Straight curve

Plane form:
$$L\{y\} = -\frac{V^3}{RL}\frac{1}{s^4}(1-e^{-\frac{L}{V}s})$$
(4)

Cant:
$$L\{z_{01} - z_{02}\} = \frac{C}{L}V\frac{1}{s^2}(1-e^{-\frac{L}{V}s})$$
(5)

Sinuous curve
Plane form:

$$L\{y\} = -\frac{1}{2R}\left\{\frac{V^2}{s^3} - \frac{L^2}{\pi^2}\left[\frac{1}{s} - \frac{s}{s^2+\left(\pi\frac{V}{L}\right)^2}\right]\right\}\left(1+e^{-\frac{L}{V}s}\right)$$
(6)

Cant:
$$L\{z_{01} - z_{02}\} = \frac{C}{2}\left(\frac{1}{s} - \frac{s}{s^2+\left(\pi\frac{V}{L}\right)^2}\right) + \frac{C}{2}\left(\frac{1}{s} - \frac{s}{s^2+\left(\pi\frac{V}{L}\right)^2}\right)e^{-\frac{L}{V}s}$$
(7)

2.3 Evaluation of transition curve

The effects of transition curve are given in Fig. 2 for the parameters, the length of 230m, the speed of 270km/h and the radius of 3000m for circular curve. The lateral form of circular curve is assumed as a parabola. In Fig. 2 (a) and (b), the amplitudes of the straight are smaller than those of the sinuous on lower frequency than the natural frequency of car roll of about 0.4Hz, but it is reversed above this frequency though the amplitude is small.

If the length of the sinuous is increased to have the same gradient in cant with that of the straight, the amplitude of the sinuous is smaller than that of the straight even under the natural frequency. If the length of the sinuous is

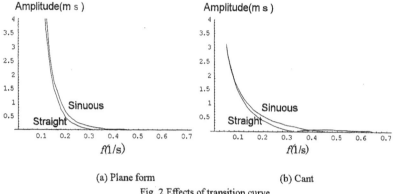

(a) Plane form (b) Cant

Fig. 2 Effects of transition curve

increased to have the same shift with that of the straight, the amplitudes of both transients are nearly same.

2.4 Tilting velocity of car body

The tilting velocities of car body are calculated for the parameters of 700 series car.

The frequency response functions for rolling velocity are given as shown in Fig. 3. In Fig. 3(a) the function for plane form increases up to the natural frequency of car body acceleratedly and, passing it, the function gradually increased. In Fig. 3(b) the function for cant has the peak at the natural frequency of rolling.

The responses to the transition curves are given in Fig. 4. That for cant is given for the hypothetical value of 0.289 (Equilibrium cant). Under the natural frequency the responses of the straight are smaller than those of the sinuous, but above it the peaks of the straight are larger than those of the sinuous. Further, comparing (a) and (b), the effect of cant is about five times larger than that of plane curve. It suggests that the balance of setting values of actual cant and the cant deficiency is important.

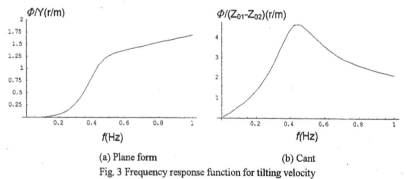

(a) Plane form (b) Cant

Fig. 3 Frequency response function for tilting velocity

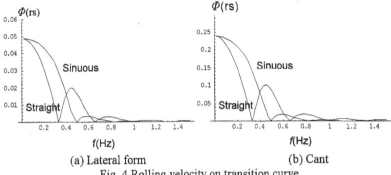

(a) Lateral form   (b) Cant

Fig. 4 Rolling velocity on transition curve

3. ANALYSES BY VAMPIRE

VAMPIRE is the simulator provided by former BR Research (presently, AEA Technology, Rail). The responses to transition curves of the straight and the sinuous run by 700series car with the speed of 260km/h are given in Fig. 5.

As results followings are given.

- Rolling velocity on the straight varies as a trapezoid having two peaks, and that of the sinuous has a single peak of bell form being good for riding quality. The value of the former is smaller than that of the latter.

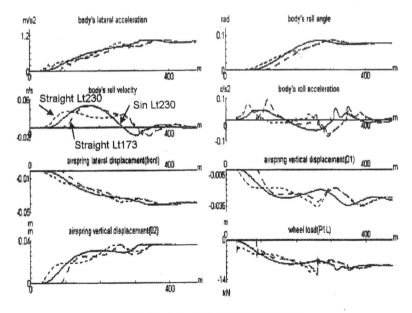

Fig. 5 Simulation by VAMPIRE (260km/h)

- The peaks are smaller than the tolerance of 0.087rad/s (5degree/s) by Koyanagi
- The peaks of rolling acceleration on the straight are larger than that on the sinuous, but much smaller than the tolerance of 0.26rad/s^2 (15degree/s^2).
- Wheel load is changed stepwise on the straight due to the theoretical bent angle, but actually it is mitigated by the roundness brought by rail rigidity.

4. CASE STUDY

4.1 Situation

The case in the station as shown in Fig. 6 is of R3000 with C=160mm run with the speed of 260km/h. The position of both ends are restricted with turnouts. It is hoped to raise the speed to 270km/h. The parameters of the curve are given as followings referring Fig. .7.

The length of transition, 230m, corresponds to $L=0.0055CmV$. The projection of transition curve on X axis, X, shift, intersectional angle, abscissa of circular curve center, total curve length and the projection of transition curve of straight slope having same shift are given as

$$X = L - 0.0226685 \frac{L^3}{R^2} = 229.969m \qquad (8)$$

Shift: $F = 0.02367914868 \frac{X^2}{R}$

$\qquad = 0.417431m$

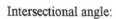

Intersectional angle:

$$\varphi = \theta_0 + \theta_R = \frac{X}{2R} + \frac{L_R}{R}$$
$$= 0.0746616 rad = 4.27779383282 deg$$

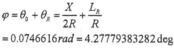

Abscissa of circular curve center

$\qquad X_{RC} = R \sin\varphi = 223.792m$

Total curve length:

$\qquad L_{tc} = 230*2 + 214 = 674m$

Length of straight slope transition curve having same shift:

$\qquad X_{CL} = \sqrt{24RF} = 173.364m$

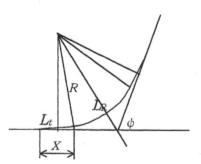

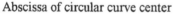

4.2 Study on ameliorated curves

The Equilibrium Cant (EC) for 270km/h running is given in Fig. 8. At R3000 the EC is 290. That is, if the cant deficiency of 110mm is permitted, 180mm cant is enough for R3000. If R2800 is used, the EC is 310mm. That is, the necessary cant is 200mm.

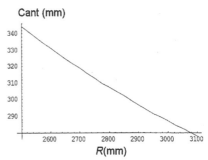

Cant (mm)

Fig. 8 Equilibrium cant

With these, following 4 cases are checked.

Case 1　R=3000m,
　　　　C_m=180mm, Straight slope, Lt=245m, L_{tc}=693m
Case 2　R=2800m, C_m=200mm, Sine slope, Lt=335m, L_{tc}=753m
Case 3　R=3000m, C_m=180m, Sine slope, Lt=245m, L_{tc}=693m
Case 4　R=2800m, C_m=200mm, Straight slope, L_t=270m, L_{tc}=688m

Cases 1, 3 and 4 depends on L=0.005C_mV, but Case 2, L=0.0062C_mV. In these, Case 3 is ideal case and Case 4 has the shortest total length. The slews for these are given as shown in Fig. 9. It suggests that the slew of Case 1 is too large, Case 4, fairly large and Case 2 and 3, executable.

4.3 Study with VAMPIRE

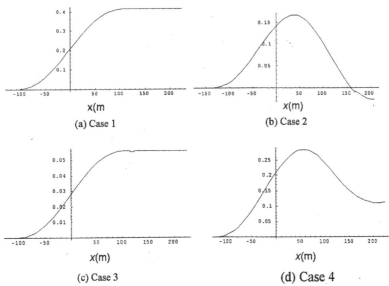

(a) Case 1

(b) Case 2

(c) Case 3

(d) Case 4

Fig. 9 Slews of Cases

The responses of 700series vehicle are simulated with use of VAMPIRE as shown in Fig. 5 for the Cases 1-4. The following results are given.

(1) Case 1 and 3

Roll velocity is very different between the Cases. The sinuous has the peak at the center of transition, but the straight has the variation of trapezoid having two peaks at the beginning and end of the transition. The value in the sinuous is 1.5 times of that in the straight.

(2) Case 2

The peak of roll velocity is much decreased in reverse of length of the transition.

(3) Case 4

Compared with Case 3, the continuation of roll velocity is long, but the maximum values are nearly same.

4.4 Check for realization of necessary transition

To check the possibility of realizing the necessary length in the total length of the case, the relation between the necessary length, $L=0.005C_mV$ ($V=270$km/h with cant defficiency of $C_d=110$mm) and the given length of the transition[3] are calculated changing the radius of circular curve for sinuous and the straight transitions. Here the smaller curves than 2777m having larager cant than 200m are not possible.

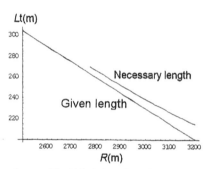

Fig. 10 Relation of R and Lt

The given lengths of transition are not so different by the form. They do not attain the necessary length as shown in Fig. 10.

5. PROPOSAL OF NEW TRANSITION FORM

5.1 Aim

To ameliorate the response of vehicle to the form of transition curve, a new transition curve is proposed. The new curve is considered to have the intermediate characteristics between the sinuous having smooth variation in roll velocity without having sharp rolling acceleration at the beginning and end of the transition and the straight not having so high peak in rolling velocity at the mid of transition curve as the sinuous.

5.2 Form and characteristics

To realize this, a fifth order expression for the cant is used. It is given as

$$C(x) = \left[-8(3-2\alpha) \cdot \left(\frac{x}{Lt}\right)^5 + 20(3-2\alpha) \cdot \left(\frac{x}{Lt}\right)^4 - 2(25-16\alpha) \cdot \left(\frac{x}{Lt}\right)^3 + (15-8\alpha) \cdot \left(\frac{x}{Lt}\right)^2 \right] \cdot Cm \quad (9)$$

where α is the steepst slope in cant.

The corresponding lateral form is given as seventh order expression. The form of cant is given in Fig. 6. The fifth order expression of $\alpha = 1.25$ is in the intermediate position between the straight and the sinuous.

Including the new transition curves, simulations for following condition were performed with VAMPIRE as shown in Fig. 7.

Case 1 Sinuous:
         R=3000m,
         C_m=160mm,
         L=230m,
         V=230km/h
         (Present)

Case 2 5th Order:
         R=2950m,
         C_m=175mm,
         L=230m,
         V=265km/h
         (Proposed)

Case 3 Sinuous:
         R=3000m,
         C_m=180mm,
         L=305m,

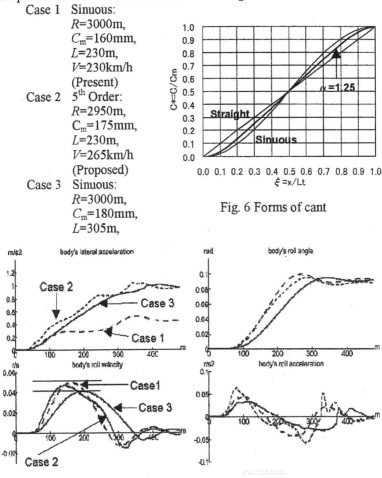

Fig. 6 Forms of cant

Fig. 7 Simulation by VAMPIRE for new transition

V=270km/h (Standard for 270km/h)

As results, the lateral acceleration of proposed case 2 is as same as that of case 3 (Standard). The roll velocity is larger than that of case 3, but smaller than that of case 1 (Present). Though the roll acceleration is larger than that of present case 2, it is much smaller than the tolerance of 0.26rad/s^2.

6. CONCLUDING REMARKS

Through the study, followings are concluded.

(1) The effect of the straight transition on rolling velocity of car body is smaller than that of the sinuous on frequency under natural one of car, but larger above the natural one.

(2) The roll velocity by cant is about five times of that by plane curve suggesting the importance of balance between actual cant and cant deficiency.

(3) In the simulation, the sinuous has smooth variation in rolling velocity, but the straight is smaller than that of the sinuous.

(4) The effect of bent angle in cant could be small by rail rigidity.

(5) The relation between the necessary length of the transition and the possible length of it with restricted ends is discussed.

(6) Although the rolling acceleration for the straight is large at the beginning and end of transition, it is much smaller than the tolerance.

(7) The effectiveness of newly proposed transition form is demonstrated.

REFERENCES

1) Loach, J.C.L. & Maycock, M.G. (1952). Recent developments in railway curve design. Proc., Inst. of Civil Engrs., 503-373.

2) Koyanagi, S. (1985). Ride quality evaluation of a pendulum car. Quarterly Report of RTRI 26-3, 89-92.

3) Kufver, B. (2000). Optimisation of horizontal alignments for railways. TRITA-FKT Report 2000;47, ISSN 1103-470X, ISRN KTH/FKT/D –00/47–SE, Doctorial Thesis, Railway Technology, Department of Vehicle Engineering Royal Institute of Technology, Sweden.

Vehicle System Dynamics Supplement 41 (2004), p.667-676 © Taylor & Francis Ltd.

Dynamic train/track interaction including model for track settlement evolvement

Andreas Lundqvist* AND Tore Dahlberg*

SUMMARY

A ballasted railway track exposed to train traffic will degenerate. Track alignment and track level will deteriorate. Settlements of the track (loss of track level and alignment) require maintenance; the track is lifted and aligned, and new ballast material is injected under the sleepers. This paper presents a computer model by which the dynamic train/track interaction can be simulated and the permanent deformation of the track (i.e. track settlements) can be calculated. The voided sleepers ("hanging" sleepers) phenomenon will also be discussed.

Keywords: Track settlement, train/track interaction, railway ballast, numerical model.

1. INTRODUCTION

A majority of railway tracks rest on ballast. Low speed trains (LST) running with speeds around 200 km/h or less have been operating for a long time on such tracks without any major problem. These trains have transported passengers and freight. High speed trains (HST), running 300 km/h and more, have been operating on certain segments, but these trains are used for passenger transport only. In recent years the experience from HST traffic has revealed that even with lower axle loads, HST have caused track settlement problems at certain sections and the settlements have burdened railway companies with extensive maintenance work. The settlements are caused by dynamic effects of the HST traffic. The dynamic effects lead to degradation of the ballast and underlying layers. Railway companies have restricted high speed tracks to passenger trains only, because heavier loads from freight trains are more likely to impose deformations and wear on the tracks, and this may impede the operation of the high speed trains. This calls for deeper investigations of the problem so that high speed lines can be used with today's passenger traffic but with reduced maintenance costs, or alternatively, with a mixture of passenger and freight traffic. It is envisaged that in a near future the HST traffic will play a major role and be competitive to air

*Address correspondance to: Department of Mechanical Engineering IKP, Linkoping University, SE 581 83 Linkoping, Sweden

traffic in solving the transportation demands of the society.

2. TRACK SETTLEMENT

Ballasted railway tracks will settle as a result of permanent deformation in the ballast and underlying soil. The settlement is caused by the repeated traffic loading, and the severity of the settlement depends on the quality and the behaviour of the ballast, the sub-ballast, and the subgrade. As soon as the track geometry starts to degenerate, the variations of the dynamic train/track interaction forces increase, and this speeds up the track deterioration process. Track settlement occurs in two major phases:

- Directly after tamping the settlement is relatively fast until the gaps between the ballast particles have been reduced and the ballast is consolidated.
- The second phase of settlement is slower and there is a more or less linear relationship between settlement and time (or load).

The second phase of settlement is caused by several basic mechanisms of ballast and subgrade behaviour:

- continued volume reduction, i.e. densification caused by particle rearrangement produced by repeated train loading,
- sub-ballast and/or subgrade penetration into ballast voids. This causes the ballast to sink into the sub-ballast and subgrade and the track level will change accordingly,
- volume reduction caused by particle breakdown from train loading or environmental factors; i.e., ballast particles may fracture due to the loading,
- volume reduction caused by abrasive wear. A particle may diminish in volume due to abrasive wear at points in contact with other particles, i.e., originally cornered stones become rounded and then they occupy less space,
- inelastic recovery on unloading (due to micro-slip between ballast particles),
- movement of ballast and subgrade particles. This causes the sleepers to sink into the ballast and subgrade,
- lateral, and possibly also longitudinal (in the rail direction), movement of sleepers causing the ballast beneath the sleepers to be pushed away, and the sleepers will sink deeper into the ballast.

Concerning the volume reduction or densification caused by particle rearrangement produced by repeated train loading, it could be mentioned that the train load also may have an opposite effect. Due to the elastic foundation, the train load will lift the track (rails and sleepers) in front of and behind the loading point, thus reducing or eliminating the preload (the dead load) caused by the rails and sleepers on the ballast. On the same time, due to the dynamic high-frequency train/track interaction forces, waves will propagate from the wheel/rail contact patches, either through the ballast and subgrade or through the track structure, to the region with the unloaded ballast. These waves will normally propagate faster than the train, giving vibrations in the unloaded ballast. This, in turn, may cause a rearrangement of the ballast particles so that the density decreases. As a result, this may cause a lift, at least temporarily, of the track.

3. TRAIN/TRACK MODELS

The dynamics of the compound train and track system plays an important role when investigating vehicle and track dynamics. The dynamic train/track interaction forces will give rise to vibrations that lead to track deterioration, such as track settlements, railhead corrugation growth, damage to track components (railpads, sleepers, ballast), and so on. Low-frequency (less than 20 Hz) motion of the train is crucial for assessment of safety and riding quality. High-frequency vibrations cause discomfort to passengers and emit noise and vibration to the surroundings.

Many different methods for analysis of a structure subjected to moving loads have been presented in the literature. A survey of railway track dynamics and modelling of the train/track interaction is given in Dahlberg [1]. Basically, the methods can be divided into two groups: methods for calculations in the frequency domain and methods for calculations in the time domain.

The improvement of computer capacity and use of numerical methods have made it possible to make more detailed models of the track structure. The subground may be modelled in a rational manner. Using the finite element method non-linear behaviour of track components such as railpads and ballast can be modelled. Also, Hertzian contact stiffness and other phenomena like loss of contact and recovered contact between wheel and rail as well as between sleeper and ballast may be included without large difficulties. Literature on finite element modelling of railway vehicle/track dynamic interaction may be found in e.g. Knothe and Grassie [2], Popp et al. [3], Nielsen [4], Oscarsson [5], and Andersson [6]. A textbook dealing with vibration of solids and structures under moving loads is Fryba [7].

4. TRACK SETTLEMENT MODELS

Not many numerical models to simulate track settlements can be found in the literature. In a research programme in Germany, aiming at better understanding the dynamic interaction of vehicle and track and the long-term behaviour of the components of the entire system, see [8], track settlement and deterioration of ballast and subgrade materials were examined. In that program a numerical model that describes repeated pas-

669

sages of a vehicle over a rail-track-subsoil system has been developed, see Augustin
et. al. [9]. The model consists of a rail (a Timoshenko beam), railpads (elastic springs),
sleepers (rigid bodies), ballast (hypoelastic material with intergranular strain), and the
subsoil (an elastic material with 3-D wave propagation and energy radiation). The
numerical simulations with this model shows that by reducing the speed of the vehicle the mean as well as the differential settlement of the track bed reduces considerably. Clearly the acceleration forces play a decisive role in the process. Also in Japan,
United Kingdom, France, USA, etc, research on track settlements is performed, see
Dahlberg [10].

In many track models the track stiffness and damping and the track mass are discretized. Then the mass of the track (sleepers, ballast etc.) is modelled by use of rigid
masses and the track stiffness and damping are modelled by springs and dampers. In
a model proposed by Dahlberg [10] it is assumed that also the track settlement can
be discretized. Thus, the settlement of the track is collected in a settlement element in
the track model. The model contains one rail (symmetry used), railpads, rigid sleepers,
ballast stiffness (spring elements), ballast damping (damper elements), and an element
beneath the ballast stiffness to take care of the permanent deformation of the ballast.
In this element the track settlement was accumulated. Using this model it was possible
to simulate so-called differential settlement. A local track settlement was achieved at
some sleepers that were supported by "bad" ballast material.

In a recent project, EUROBALT II, the research on track settlement modelling continued. A computer program for simulation of the dynamic train/track interaction was
further developed. This work was reported by Mansson [11]. In the FEM model developed, focus was on the performance of the ballast bed. Several finite element models
were created and investigated to ensure that a numerical model, suitable for dynamic
train/track interaction calculations, could be established. Further developments of this
track model, including track settlement, are reported in this paper. This development
is performed within a new European project: the SUPERTRACK project (Sustained
Performance of Railway Track). In the new model, the discrete springs, dampers, and
settlement elements have been replaced by a continuum.

4.1. Track model

The new finite element model that has been set up is meant to simulate both the short-
term behaviour of the dynamic train/track system when a wheel passes and the long-
term behaviour of the track due to many wheel passages. A settlement model simulating the permanent ballast and subground deformations is included in this train/track
model.

The model is built-up using the preprocessor TrueGrid [12], and the train/track interaction problem is solved by the commercial finite element program LS-DYNA [13],
which has an effective explicit solver. The time step is automatically made small so

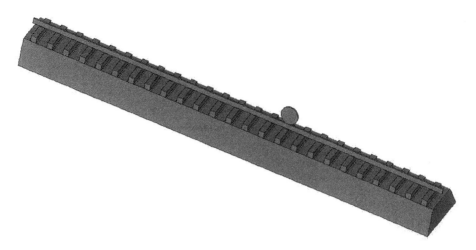

Figure. 1. Train/track model with ballast bed of elastic-plastic material.

that high frequency variations are well represented. According to Gardien *et. al.* [14], LS-DYNA is at least 30 times faster than an implicit solver for investigation of large wave propagation problems. Although an implicit solver does not need a small time step for stability, still a small time step has to be chosen to represent high frequency waves well. With an explicit solver the solution always reaches convergence, which is not always the case with an implicit solver. To define the time step in the explicit solver, the speed of sound c is used. One has $c = \sqrt{E/\rho}$, where E is the the stiffness and ρ the density of any track material. The time step Δt is determined with respect to the shortest element as $\Delta t = l_e/c$, where l_e is the shortest element length.

The model used in this paper consists of 3-D solid elements and it is composed of one rail (symmetry with respect to the centre line of the track is assumed), wheel, railpads, sleepers and ballast bed. The ballast bed is modelled as a continuum with elastic or elastic-plastic material properties. A more realistic constitutive law of the ballast material will also be developed in the SUPERTRACK project and incorporated in the numerical model, see Figure 1. Both wheel and sleepers are modelled as rigid. The railpads are modelled with a predefined rubber material. The loading of the track comes from the moving wheel, which is loaded by a constant force: the dead load of the car body. The wheel mass and half of the axle mass are included, which means that the inertia force from the unsprung mass, i.e. the wheel and the axle, is taken into account.

One problem with modelling the ballast bed and the surrounding soil material is that they are often infinitely large whereas the numerical model is finite. To avoid re-

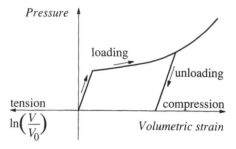

Figure. 2. Pressure versus volumetric strain for soil and crushable foam model. The volumetric strain is given as the natural logarithm of the relative volume V/V_0 (figure redrawn from [15]).

flections from the boundaries of the finite model, also the finite element model should be infinitely large. For the limited model used here, non-reflecting boundary conditions where used to absorb pressure and shear waves at the boundaries.

4.2. Material models

The elastic-plastic material model used for the ballast is predefined in LS-DYNA (MAT_SOIL_AND_FOAM). According to the User's Manual [15] this material behaves like a fluid. It should be used only in situations "when soils and foams are confined in the structure". When loading the material it behaves like a linear elastic, deformation hardening material see Figure 2. Also the "unloading path" is shown in Figure 2, where the slope of the unloading curve gives the bulk modulus. If this option is used in the calculations, and if the existing stress level in the ballast becomes larger than the yield strength of the material, then the relationship between pressure and strain follows the path indicated in the figure. After the ballast has been unloaded, permanent plastic deformations will remain.

The deviatoric perfectly plastic yield function ϕ is described in terms of the second invariant J_2, pressure p, and constants a_0, a_1, and a_2 as:

$$\phi = J_2 - \left[a_0 + a_1 p + a_2 p^2\right] \tag{1}$$

On the yield surface one has $J_2 = \sigma_y^2/3$ where σ_y is the uniaxial yield strength, i.e.,

$$\sigma_y = [3(a_0 + a_1 p + a_2 p^2)]^{1/2} \tag{2}$$

To eliminate the pressure dependency of the yield strength, set $a_1 = a_2 = 0$ and $a_0 = \sigma_y^2/3$. This approach is useful when a von Mises type elastic-plastic model is desired for use with tabulated volumetric data [15].

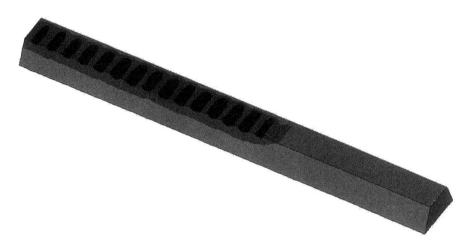

Figure. 3. "Footprints" of sleepers in ballast, i.e. permanent deformation of the ballast, after wheel passage (dark areas).

If the track bed is composed of several layers with different function and characteristics, it may be necessary to use different material models for the different layers.

Validation of this train/track/settlement model will continue with data collected within the SUPERTRACK project from test sites selected at conventional and high speed lines in Spain and France. These data include geometrical information and superstructure characteristics, geotechnical data, maintenance and traffic information, as well as data describing responses of the track during a train passage, such as stresses, displacements, accelerations etc.

As an example, this material model was used for the ballast layer in Figure 1. In the calculated results presented in Figure 3 the yield limit of the ballast material was selected so that the stress beneath the sleepers exceeds the yield limit of the ballast material when the wheel passes. Figure 3 shows the ballast layer when the wheel has passed part of the track model. The dark areas in the figure show the "footprints" of the sleepers, i.e., they show the permanent deformation of the ballast bed.

5. NUMERICAL RESULTS

This section describes a simulation where voided (hanging) sleepers appear after a few load cycles. A voided sleeper is obtained when a gap is formed between the sleeper and the ballast stones, i.e., the sleeper is hanging in the rail without contact between

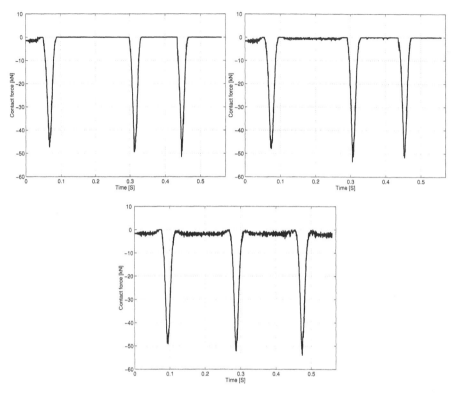

Figure. 4. Contact forces between sleeper and ballast for three different sleepers. (a) Top left figure for sleeper 11, (b) top right figure for sleeper 12, and (c) lower figure for sleeper 15.

the sleeper and the ballast.

The numerical values used in the simulation are as follows: The wheel mass and half of the axle mass is 750 kg, the dead load of the car body is applied to the wheel as a constant force of 100 kN (thus giving a static axle load of 107.5 kN), the track model used has a length of 30 sleeper spans (30 sleepers), the sleeper spacing is 0.6 m, and UIC60 rail is used. The model is divided into tree sections with 10 sleepers in each section. The first and the third part rest on "good" elastic ballast with a modulus of elasticity $E = 100$ MPa and Poisson's ratio $\nu = 0.1$. The second part, which lies between the first and third, rests on "bad" ballast material where settlement will occur (material model SOIL_AND_FOAM as described above). The shear modulus is set to 45.45 MPa and the bulk modulus is 41.67 MPa, $a_0 = 1.2 * 10^{-3}$ MPa^2, $a_1 = a_2 = 0$ for the material used. The ballast density is 2500 kg/m^3 and the sleeper mass is 125 kg.

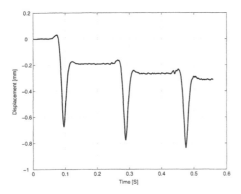

Figure. 5. Displacement curve for sleeper 15. The settlement increases for every load cycle, but the settlement rate decreases.

The wheel starts in the first section and travels over the ballast bed three times with a speed of 90 m/s. The contact forces between the sleepers and the ballast in section two of the model are given in Figure 4. It is seen that the contact force between sleeper 11 (the first sleeper in section two) is zero after the first wheel passage (the force goes to zero at time $t = 0.09$ sec). This is the case also for sleepers 19 and 20. Sleeper 12 gets a lower (static) sleeper/ballast contact force after the first wheel passage, and the contact disappears after the second passage. Sleepers 13 to 18 never lose contact during these cycles, but the permanent deformation (settlement) increases during the loading, see Figure 5. The settlement rate after each wheel passage is, however, decreasing. Figure 5 also shows how the rail and sleeper rise a small amount just before and after the wheel has passed (this can also be seen in Figure 4(c)).

6. CONCLUDING REMARKS

This paper shows that even with a very simple material model, track settlements can be simulated and the settlements can cause voided sleepers to appear.

LS-DYNA has been used to investigate the dynamic train/track problem and non-reflecting boundary conditions were used to absorb shear and pressure waves.

Future work will involve introduction of a bogie with a bogie frame. A more realistic description of the ballast and subgrade materials will be implemented in the model and parametric studies regarding track dynamics and train/track interaction will be performed.

ACKNOWLEDGEMENTS

This work was performed within the European FP5 project SUPERTRACK (Sustained Performance of Railway Track); contract No G1RD-CT-2002-00777.

REFERENCES

1. Dahlberg, T.: Railway track dynamics - a survey. Submitted for international publication, 2003.
2. Knothe, K. L. and Grassie, S. L.: Modelling of railway track and vehicle-track interaction at high frequencies, *Vehicle System Dynamics*, Vol 22 1993, 209-262.
3. Popp, K., Kruse, H. and Kaiser, I.: Vehicle-track dynamics in the mid-frequency range. *Vehicle System Dynamics*, Vol 31(5-6) 1999, 423-464.
4. Nielsen, J.C.O.: *Train/track interaction: coupling of moving and stationary dynamic systems*. PhD Thesis. Dept of Solid Mechanics, Chalmers University of Technology, Gothenburg, Sweden, ISBN 91-7032-891-9, 1993.
5. Oscarsson, J.: *Dynamic train-track interaction: linear and non-linear track models with property scatter*. PhD thesis, Dept of Solid Mechanics, Chalmers University of Technology, Gothenburg, Sweden, ISSN 0346-718X, 2001.
6. Andersson, C.: it Modelling and simulation of train-track interaction including wear prediction. PhD Thesis. Dept of Applied Mechanics, Chalmers University of Technology, Gothenburg, Sweden, ISSN 0346-718X, 2003.
7. Fryba, L.: *Vibration of solids and structures under moving loads*, 3rd edition, Thomas Telford, London, 1999, ISBN 0-7277-2741-9.
8. Popp, K. and Schiehlen, W.: *System dynamics and long-term behaviour of railway vehicles, track and subgrade*. Springer Verlag, Berlin, 2003, ISBN 3-540-43892-0.
9. Augustin, S., Gudehus, G., Huber, G. and Schnemann, A.: Numerical model and laboratory tests on settlement of ballast track. In Popp K and Schiehlen W (editors): *System dynamics and long-term behaviour of railway vehicles, track and subgrade*. Springer Verlag, Berlin, 2003, ISBN 3-540-43892-0, 317-336.
10. Dahlberg, T.: Some railroad settlement models - a critical review, *Proceedings of the Institution of Mechanical Engineers, Part F, Journal of Rail and Rapid Transit*, Vol 215(F4) 2001, 289-300.
11. Mansson, F.: *Numerical modelling of railroad track settlements - a preliminary investigation*. Report LiTH-IKP-R-1144, Solid Mechanics, IKP, Linkoping university, Linkoping, Sweden, 2001.
12. *Truegrid manual*, version 2.1.0, XYZ Scientific Applications, Inc., 2001.
13. Hallquist, J.O.: *LS-DYNA Theoretical Manual*, Livermore Software Technology Corporation, 2876 Waverly Way, Livermore, California 94550, 1998.
14. Gardien, W., Stuit, H. and de Boer, F.: Finite element wave propagation, *Proceedings of Second European LS-DYNA Users Conference*, Gothenburg, Sweden, 1999, 33-42.
15. Hallquist, J.O.: *LS-DYNA Keyword User's Manual*, Version 970, Livermore Software Technology Corporation, 2876 Waverly Way, Livermore, California 94550, 2003.

Vehicle System Dynamics Supplement 41 (2004), p.677-686 © Taylor & Francis Ltd.

Numerical simulation and field experiment of high-speed train-track-bridge system dynamics

W.M. ZHAI [*1], C.B. CAI[1] AND K.Y. WANG[1]

SUMMARY

A numerical simulation technique is developed to analyse the train-track-bridge system dynamic responses when a train traverses a bridge. The basic idea of the simulation is to consider the train, the track and the bridge as a whole interactive system. The modelling aspect of the train-tack-bridge system is presented. A large-scale software, TTBSIM, is designed to simulate the dynamic responses of the complete train-track-bridge system so as to assess the running safety and the ride comfort of trains traversing bridges in the design stage. Two high-speed running experiments were carried out on the Qinhuangdao-Shenyang line, which were used to validate the TTBSIM technique. Results show good correlation between simulation and measurement. TTBSIM software has provided a dynamic assessment tool for the design of bridges on new high-speed railways.

1. INTRODUCTION

In recent years, train speed has been greatly raised than before on the existing main lines in Chinese railways. In 2002, the first special railway line for passenger transport has been built up in China, which is Qinhuangdao-Shenyang line (the Q-S line). The design speed of the whole line is 200 km/h. There is a 66.8 km long high-speed test section from Shanhaiguan to Suizhong, whose design speed of the infrastructure is up to 300 km/h. Beijing-Shanghai high-speed railway line is under planning and its design speed will reach at 300km/h.

With the rapid increase of train operating speed, it becomes very important to investigate the dynamic behaviour of the train-track-bridge system so as to assess the running safety and the ride comfort of trains traversing bridges in the design stage. This is especially important to the design of large bridges on new high-speed railways. In 2000, the Chinese Ministry of Railways (MOR) organized a research group and set up a key project to develop a numerical simulation technique to analyse and then to evaluate the dynamic performance of the train-track-bridge system under high-speed operation. The aim of this study is to provide a unified assessment tool for the design of bridges on new high-speed lines or fast-speed lines. This paper will present the progress of this research, focusing on the basic principle, the dynamic model, the simulation method and the software.

Another part of the paper will present some dynamic measurement results of the most recent high-speed experiments in China so as to verify the numerical simulation.

[1] *Address correspondence to*: Train & Track Research Institute, Southwest Jiaotong University, Chengdu 610031, China. Tel.: +86-28-87601843; Fax: +86-28-87609007; E-mail: wmzhai@home.swjtu.edu.cn

2.   NUMERICAL SIMULATION TECHNIQUE

2.1 BASIC PRINCIPLE

The coupling vibration of a train and a bridge is not a new topic, on which many studies have been done, e.g. [1-3]. However, conventional studies are usually based on the assumption that the displacement of a wheel is always equal to that of the bridge beam under the wheel. They didn't consider the dynamic effect of track structures on the bridge. They didn't consider the wheel/rail interface, which is simplified as the interface of the wheels and the bridge beam. In this study, the basic idea to simulate the dynamic process of a train passing over a bridge is that the train, the track and the bridge are considered as a whole interactive system. The train is coupled with the track structure both in vertical and in lateral by means of the wheel/rail spatial contact relationship. The track structure is coupled with the bridge deck with the ballast springs and dampers.

Dynamic models of train-track-bridge systems are established on the basis of the above idea for different cases. As an example, Fig.1 gives a model of a conventional-bogied high-speed train running on a simply supported bridge with the non-ballasted slab track.

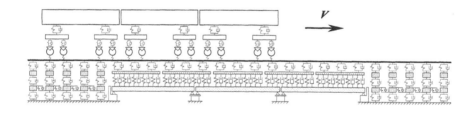

Fig.1. An example of train-track-bridge system dynamic model (with slab track).

In the train model, power cars and a full complement of trailer vehicles should be included. Each vehicle is modelled as a three-dimensional multi-body system. Fig.2 shows a lateral vehicle model.

In the track model, the rail is described as a Bernoulli-Euler beam discretely supported at each fastening point. The slabs are simply modelled as continuous elastically supported plates with free ends in the vertical plane. In order to consider the dynamic effect of train entering into or leaving from the bridge [4], general track model [5] is taken into account at the two ends of the bridge.

The bridge system is modelled with the finite element method according to the specific structure of the bridge. Different types of elements are adopted for different structures of bridges. Experience has shown that five spans are usually sufficient to simulate the dynamic problem of a train traversing a bridge. Too many spans lead to too much computational time without any obvious change in results.

The detailed wheel/rail spatially contact geometry is considered [6], in which the normal forces are described by the non-linear Hertzian contact theory, and the tangential forces are calculated by the Shen-Hedrick-Elkins model [7].

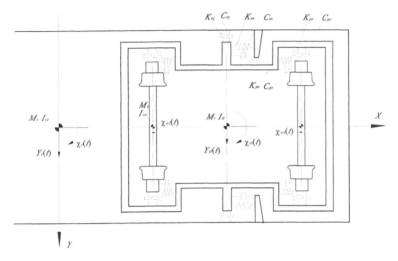

Fig.2 A lateral high-speed vehicle model.

Track random irregularities are used as the excitation of the train-track-bridge systems, which have been proved to have very important influence on the system dynamic performance [8,9]. For new high-speed lines at the design stage, however, the track geometry cannot be obtained from measurements. To represent realistic conditions for the high-speed railways, it is reasonable to use existing high-speed track geometry, e.g. the German high-speed track spectrum.

2.2 SIMULATION SOFTWARE

A large-scale software, TTBSIM, is designed to simulate the train-track-bridge system dynamic responses. Due to the very high degrees of freedom of the train-track-bridge dynamic system, a fast explicit time integration method is employed to numerically solve dynamic responses of such a large system [10].

TTBSIM software mainly consists of five modules as shown in Fig.3. In the train module, 3-demensional dynamic models of typical passenger cars as well as locomotives with four or six axles have been established. In the track module, dynamic models for conventional ballasted tracks and for non-ballasted tracks such as the slab track are included. In the wheel/rail interaction module, the procedures to determine the contact forces and the creep forces at each wheel-rail contact point are well performed. Database of track irregularities is also included in the wheel/rail interaction module. At present, typical track geometry spectra such as AAR standard, the German high-speed track spectrum and the Chinese mainline spectrum have been included in the track irregularity database. Track irregularity can also be input into the TTBSIM software by use of measured data. In the bridge module, eleven types of elements are designed, including the spatial beam element, the spatial rod element, the plate element and the other special combined elements such as the T beam section elements.

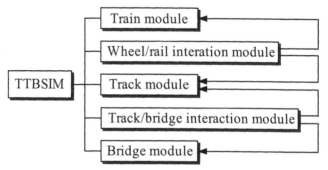

Fig.3 TTBSIM system structure.

The TTBSIM software can produce the responses of the complete train-track-bridge system. The principal results are vertical and lateral accelerations of car bodies, vertical and lateral wheel/rail forces, track dynamic responses, displacements and accelerations of the bridge deck.

3.   FIELD EXPERIMENT

Under the organization of the Chinese Ministry of Railways, we have finished field running experiments twice for two types of high-speed trains on the Q-S line in September and December of 2002, respectively.

3.1 THE FIRST FIELD EXPERIMENT

In the first experiment, the tested train is a power-distributed high-speed train, named "Pioneer". The tested bridges are Shihe bridge and Gouhe bridge, which locate on the high-speed test section of the Q-S line. The Shihe bridge is a conventional ballasted track bridge, whereas the Gouhe bridge is a new slab track bridge. These two bridges are simply supported box-shape beam bridges with the span length of 24 m. The experimental train speed ranges from 180 km/h to 250 km/h.

The dynamic responses of the whole train-track-bridge system were measured, including the carbody accelerations and the ride comfort indices, the wheel/rail forces and the derailment coefficients (Q/P), the rail-pad forces, the displacements and the accelerations of rails and sleepers (or slabs), and the displacements and accelerations of bridges. Fig.4 shows a picture of an instrumented slab track section on a bridge.

Fig.4 An instrumented slab track on a bridge.

3.2 The Second Field Experiment

In this experiment, the tested train was changed into a newly designed high-speed train, named "Chinese Star", whose design speed is 270 km/h. The axle loads of the power car and the trailer vehicles are 195 kN and 130 kN, respectively. The same measurement items as in the first experiment were made when the train passed over the Gouhe bridge, the Shihe bridge, Shuanghe bridge and Xingyan bridge. The latter two bridges locate on a curve section of the Q-S line. The Shuanghe bridge is a simply supported box-shape beam bridges laid with the slab track. The span length of the bridge is 24 m at the test place. The curvature of the line is 4000 m and the super elevation is 115mm. The Xingyan bridge is a conventional ballasted track bridge, used for comparison purpose. The test speed of the train traversing the Gouhe bridge and the Shihe bridge was in the range of 160~315 km/h, whereas the passing speed of the train over the Shuanghe bridge and the Xingyan bridge was from 160 km/h to 230 km/h. Fig.5 shows the high-speed test train "Chinese Star" running on the Shuanghe bridge.

Fig.4 An instrumented slab track on a bridge.

3.3 FIELD EXPERIMENT RESULTS

Detailed measurement results were reported in [11]. Some examples of the measured results are shown in Fig.6~Fig.8. Fig.6 shows a time history of the bridge deck acceleration obtained in the first experiment when the Pioneer test train traversed the Shihe bridge with a speed of 200 km/h. Fig.7 shows the responses of the wheel/rail lateral forces measured from the rail in the second experiment when the Chinese Star test train passed through the Shuanghe bridge with a speed of 200 km/h. Fig.8 gives the time history of a rail pad force in the same experimental condition as in Fig.7.

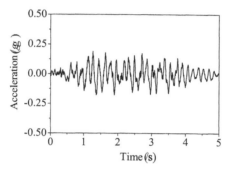

Fig.6 Measured acceleration of bridge deck at the middle of span.

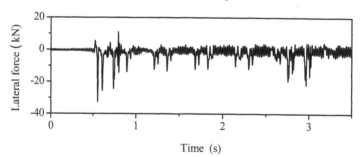

Fig.7 Measured wheel/rail lateral force from instrumented rail.

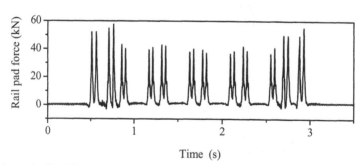

Fig.8 Measured rail pad force response.

4.   VALIDATION OF THE SIMULATIONS

The above high-speed running experiments were used to validate the simulation software TTBSIM. The validation procedure was set by the Chinese MOR. Once the test plan was made, it was required to undertake detailed simulation for the tested trains traversing the tested bridges with the planed passing speeds. The simulation results should be reported to the MOR before the real test measurements started. After the experiments were finished, the comparison would be made between the simulated and the measured results. Some major results will be given below.

4.1 COMPARISON OF THE RIDE COMFORT

The vertical and lateral carbody accelerations of a trailer vehicle are compared in the whole experiment speed range in Fig.9 and Fig.10 when the Pioneer train passed through the Shihe bridge. Fig.11 and Fig.12 compare the ride comfort indices of the power car under the same condition as above.

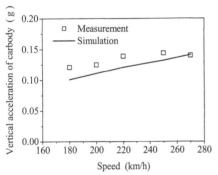

Fig.9 Comparison of carbody vertical accelerations in the trailer vehicle.

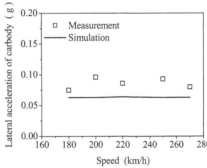

Fig.10 Comparison of carbody lateral accelerations in the trailer vehicle.

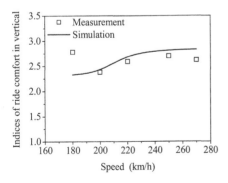

Fig.11 Comparison of indices of vertical ride comfort in the power car.

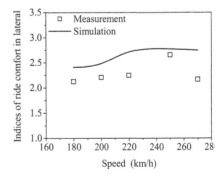

Fig.12 Comparison of indices of lateral ride comfort in the power car.

4.2 COMPARISON OF THE RUNNING SAFETY

Fig.13 and Fig.14 give the comparison of the simulated and the measured derailment coefficient (the Q/P ratio of lateral to vertical wheel/rail force) and the reduction rate of wheel load, which are popularly used for assessing the running safety of a train in Chinese Railways. The running condition was that the Chinese Star train traversed the Shuanghe bridge.

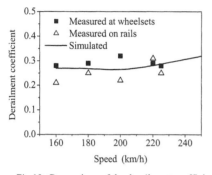
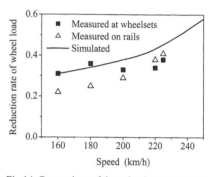

Fig.13 Comparison of the derailment coefficients of power car.

Fig.14 Comparison of the reduction rates of wheel load of power car.

4.3 COMPARISON OF THE BRIDGE AND THE TRACK VIBRATIONS

Table 1 lists the measured maximum values of displacements and accelerations at the middle of span of three tested bridges during the passage of the Chinese Star train with different speeds. The calculated results are also listed in Table 1 for comparison. Fig.15 is an example of calculated acceleration of bridge deck at the middle of span when the Chinese Star train traverses the Gouhe bridge with a speed of 250 km/h. Fig.16 shows the calculated response of a rail pad force. Comparison of the measured and the calculated rail pad forces are shown in Fig.17. A comparison of the slab accelerations is given in Fig.18.

Table 1 Comparison of peak values of displacements and accelerations of bridge decks at the middle of span

Name of bridge	Guohe		Shihe		Shuanghe	
Maximum train speed	300 km/h		315 km/h		225 km/h	
Dynamic index	Displacement (mm)	Acceleration (g)	Displacement (mm)	Acceleration (g)	Displacement (mm)	Acceleration (g)
Measurement	0.94	0.23	1.52	0.24	1.43	0.25
Simulation	1.11	0.19	1.28	0.19	0.92	0.11

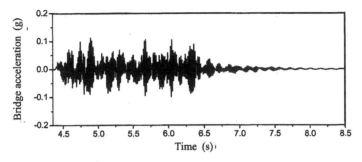

Fig.15 Calculated acceleration response of bridge deck at the middle of span.

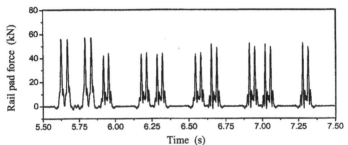

Fig.16 Calculated time response of a rail pad force.

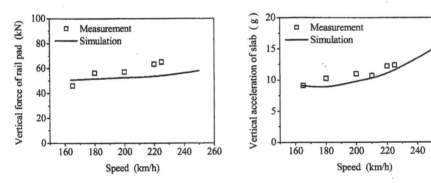

Fig.17 Comparison of the rail pad forces.   Fig.18 Comparison of the slab accelerations.

5. CONCLUSIONS

A numerical simulation technique has been developed to analyse the train-track-bridge system dynamic responses and to assess the running safety and the ride comfort of trains traversing bridges. Unlike the conventional studies, the complete train-track-bridge system is taken into account.

A large-scale software, TTBSIM, has been designed to simulate the dynamic responses of the train-track-bridge system, which is capable of dealing with different types of trains, different track structures, different bridges, and different track geometry spectra.

The simulation method has been verified by two field experiments in which different high-speed trains run over different bridges with different track structures. Good agreement between simulation and measurement has been achieved in the validation although some indices show somewhat difference in some cases.

We believe that the simulation technique described in this paper has provided a dynamic assessment platform for the design of bridges on new high-speed railways, especially on the Beijing-Shanghai high-speed line.

ACKNOWLEDGEMENTS

This work was supported by National Natural Science Foundation of China (NSFC) under Grant 50178061, and also by the Chinese Ministry of Railways under Grant 2000G049-D and 2001G040. The research group of the TTBSIM software consists of the researchers from Southwest Jiaotong University, Chinese Academy of Railway Sciences, Northern Jiaotong University, and Central South University, China.

REFERENCES

1. Chu, K.H., Garg, V.K. and Wiriyachai, A.: Dynamic interaction of railway train and bridges. Vehicle System Dynamics, 1980, 9(4): 207-236.

2. Dahlberg, T.: Vehicle-bridge interaction. Vehicle System Dynamics, 1984, 13(1): 187-206.

3. Diana, G. and Cheli, F.: Dynamic interaction of railway systems with large bridges. Vehicle System Dynamics, 1989, 18(1): 71-106.

4. Zhai, W.M. and True, H. Vehicle-track dynamics on a ramp and on the bridge: simulation and measurements. Vehicle System Dynamics, 1999, 33 (Supplement), 604-615.

5. Zhai, W.M. and Sun, X. A detailed model for investigating vertical interactions between railway vehicle and track. Vehicle System Dynamics, 1994, 23 (Supplement), 603-615.

6. Chen, G. and Zhai, W.M.: A new wheel/rail spatially dynamic coupling model and its verification. Vehicle System Dynamics, accepted for publication.

7. Shen, Z.Y., Hedrick, J.K. and Elkins, J.A.: A comparison of alternative creep force models for rail vehicle dynamic analysis. Proc. 8th IAVSD Symposium, MIT, Cambridge, 1983, 591-605.

8. Rawlings, L., Evans, J. and Clark, G.: Simulation and validation of the dynamic interactions between trains and a new bowstring bridge. Vehicle System Dynamics, 2002, 37 (Supplement), 279-289.

9. Zhai, W.M. and Cai, C.B.: Train/track/bridge dynamic interactions: Simulation and applications. Vehicle System Dynamics, 2002, 37 (Supplement), 653-665.

10. Zhai, W.M.: Two simple fast integration methods for large-scale dynamic problems in engineering. International Journal for Numerical Methods in Engineering, 1996, 39 (24), 4199-4214.

11. Jiang, C. and Zhai, W. M.: A synthetically experimental study of non-ballasted tracks on bridges on Qinhuangdao-Shenyang line (in Chinese). Research Report of China Academy of Railway Science, Beijing, June 2003.

MULTIPLE PANTOGRAPH OPERATION
-EFFECTS OF SECTION OVERLAPS

Pia Harèll*, Lars Drugge* and Marten Reijm†

SUMMARY

The traffic situation of the future goes towards flexibility of train configurations and increasing speed. There is a need for the ability to adjust the number of passenger cars as well as the energy consumption for trains to run efficiently, leading to multiple pantographs and short pantograph distances. Limitations due to this were studied on an existing catenary system. A model of the pantograph catenary system was developed in a finite element program. To verify the correctness of the model, comparison with full scale experiments was done. Results from the simulations indicates that trailing pantographs suffers from high dynamic effects within the section overlap. Effects of changes in design of the section overlap was studied. The results show that it is possible to get lower dynamic effects in the section overlap, even lower than within an ordinary span.

1 INTRODUCTION

To enable for high speed trains to operate freely between countries, heavy demands are put upon safe and uninterrupted traffic. The current collection system is a limiting factor for the speed of a train. In regions where a large amount of people travel by train every day, specially constructed catenary systems giving nearly constant vertical flexibility are built for high speed trains to operate on. This system is called compound system and was developed in the 60's. Similar results can be achieved with high tension catenaries with stitch wires [1]. In regions with low population density the building of new complex catenary systems is often too expensive. There is however a need for the ability to adjust the number of passenger cars for trains to run efficiently with low energy consumption. This can be done by the use of self supporting cars, EMUs (Electric Multiple Unit), with a pantograph at each car for current collection. The usage of EMUs leads to multiple pantographs and short pantograph distances. The dynamic effects due to the fact that trailing pantograph enters a system already

---

*Div. of Vehicle Dynamics, KTH, SE-100 44 Stockholm, Sweden, e-mail: pia@fkt.kth.se, lasse@fkt.kth.se
†Swedish National Rail Administration, e-mail: marten.reijm@banverket.se

in motion increases with speed. There is also a possibility of resonance in some ranges of speed [2, 3], where the motion of trailing pantographs become larger. Damping in the catenary will also have higher effect on trailing pantographs than leading [4]. The speed range in which resonance occurs depends not only on the pantograph spacing but also on the configuration of the catenary. The dynamic force fluctuation should be kept low to ensure reliable contact.

The catenary system is built in sections, each section being approximately 1200 m. In each end of a section the wire is lead to a support pole where a tensile force is induced in the line. This gives an overlap of sections, see figure 1, which causes an irregularity in the system that can lead to increased dynamic effects. Optimised overlaps with good interaction characteristics especially at high speeds have been developed through experimentation [5]. Good results were obtained for a single pantograph with a five span overlap with DB's RE250 catenary up to 400km/h. The contact wires where raised at the middle of the overlap span by 4 cm compared with the nominal contact wire height.

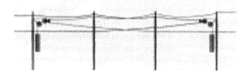

Figure 1: Section overlap

This work focus on multiple pantograph operation and its limitations. The study is based on an existing catenary system that was originally built for a train configuration with only one pantograph or pantograph distances of more than 150 m. Newly introduced intercity trains operating in Sweden with multiple pantographs are now limited to a train speed of 200 km/h.

1.1 Current Collection Quality

High contact forces has the advantage to give continuity of contact, but it increases the wear and the risk of damaging the carbon collector strips of the pantograph as well as the contact wire. It may also lead to severe oscillations. Low contact forces can lead to loss of contact, which results in arcing and voltage dips, leading to wear of both the carbon collector strips and contact wire. Most countries has stated that the loss of contact at normal speed should be less than 1%. Statistical analysis of the quality of the contact force is being used as opposed to loss of contact since use of loss of contact as a measurement gives no indication of the robustness of the system [6]. Here, the range of contact force is evaluated from following criteria;

$$F_{min} = F_m - 3\sigma > 0 \tag{1}$$

and

$$F_{max} = F_m + 3\sigma < 200N \tag{2}$$

where σ is the standard deviation and F_m the mean force. F_{min} and F_{max} are known as the statistical minimum and maximum forces.

2 FINITE ELEMENT MODEL

A two dimensional finite element model of the pantograph-catenary system was developed, neglecting the zigzag of the catenary. That is, the pantograph is assumed to be excited at midpoint. The FE-program ANSYS was used for the analysis. The model of the catenary system analysed is based on a stitch catenary system called SYT15/15, used by the Swedish National Rail Administration (the system has similarities with German RE250).

This system has a design speed of 250 km/h. Both the catenary and contact wire have a tensile force of 15 kN and the stitch wires 2.8 kN. The cross sections are 70, 120 and 35 mm^2 respectively and the material is copper. One span is 60 m and the system height is 1550 mm at the support poles. The catenary wire is connected directly to the support poles, while the contact wire is linked to the supports via steady arms. There are seven droppers in each span and the distance between them are 8.3 m. The model of the analysed system consists of two sections of 12 spans each and a five span overlap. Transition from one section to the next occurs at the central span where the contact wire is lifted 0.15 m at the support poles and the lift is parabolic.

The catenary wire, the stitch wire and the droppers are built up from link elements, the droppers being nonlinear, and the contact wire from beam elements. The model of the pantographs has two degrees of freedom and is built up from spring-damper elements, structural mass elements and a combination of spring-slider and damper in parallel in series with gap. Contact is defined between the upper mass of the pantographs and the contact wire.

To verify the correctness of the model, comparison with full scale experiment was done for a stitch catenary system called SYT7.0/9.8. This system has a tensile force in the catenary and contact wire of 7 and 9.8 kN respectively. The details of this system is not to be explained in this paper. Measurements were performed according to EN 50317 by Schunk Nordiska AB Sweden. Figure 2 shows contact force variation for the third pantograph at 200 km/h with a pantograph spacing of 54 m. Contact force statistics at a pantograph speed of

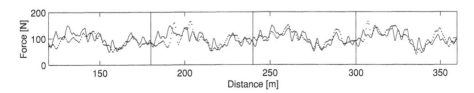

Figure 2: Comparison between simulated (-) and measured ($\cdots$) values of contact force for the third pantograph at 200 km/h

160, 180 and 200 km/h can be seen in table 1. Results from static and dynamic analyses for SYT15/15 are presented in chapter 3.

Table 1: Comparison with full scale experiments for the third pantograph.

	Velocity [km/h]	F_m [N]	σ [N]
Measured	161	104	15
Simulated	160	100	14
Measured	182	98	21
Simulated	180	98	20
Measured	201	100	25
Simulated	200	99	23

3 RESULTS

3.1 Static Analysis

Results from a static analysis for five spans are presented in figure 3. A section overlap can be seen in the middle of each picture, between 540 and 600 m. The

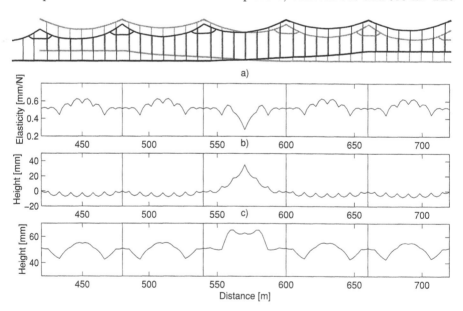

Figure 3: Static characteristics for SYT15/15. *a) catenary elasticity b) catenary height at static equilibrium c) pantograph height for 100 N uplift force*

upper picture (a) shows the system elasticity for a static uplift force of 100 N, the middle one (b) illustrates the height of the contact wire without static load and in the lower (c) the pantograph height for a uplift force of 100 N can be seen. The vertical lines illustrates the position of the support poles. The elasticity of the system is relatively constant within each span and lower at the section overlap. The lower elasticity is due to the fact that the sections crosses here and the pantograph will therefore operate on both lines. This results in half the elasticity at midpoint compared to an ordinary span. The difference in elasticity within the overlap is higher than both before and after. The elasticity plays an important role for the dynamic condition of the system [1] and should preferably be constant within a span. The contact wire has a parabolic lift which gives that the lift at midpoint is 37.5 mm compared to the nominal contact wire height. For a static load of 100 N, which corresponds to the pantograph uplift force at 240 km/h, the pantograph height within the section overlap is higher than in the rest of the system. To obtain a level pantograph height an uplift force of 160 N is required. Since the trains have to operate on new as well as old catenary systems there are restrictions on the pantograph mean uplift force to not exceed 110 N.

3.2 Dynamic Analysis

Figure 4 illustrates results from a dynamic analysis with three Schunk WBL 88 pantographs with a distance of 54 m, travelling at a speed of 240 km/h. The

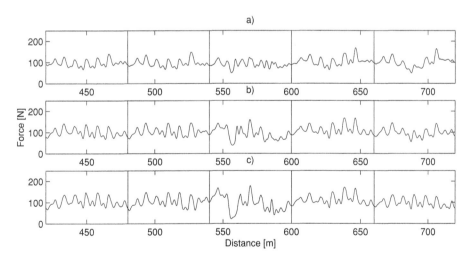

Figure 4: Contact force variation as function of travelled distance at 240 km/h. a) first pantograph b) second pantograph c) third pantograph

pantograph uplift force in the simulations consists of a static part of 70 N and an aerodynamic, velocity dependent, part. The total uplift force at 240 km/h is 98.8 N. The section overlap is between 540 and 600 m. It can be seen that there is a rise in dynamic effects within the overlap for the trailing pantographs. Both second and third pantograph suffers severe dynamic effects, with almost twice the force variation compared to an ordinary span. Dynamic analyses was performed for speeds from 160 to 260 km/h with an increment of 20 km/h. The resulting forces before and in the section overlap can be seen separately in figure 5. The curves represent the statistical maximum and minimum force for each pantograph. According to the evaluation criteria presented in chapter 1.1 these results indicates that there is a risk for contact loss and too high uplift forces in the section overlap at 240 km/h for third pantograph.

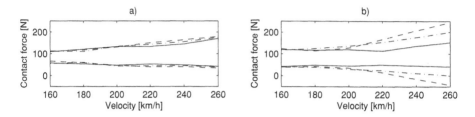

Figure 5: Statistical minimum and maximum contact forces as function of velocity. *a) before section overlap b) within section overlap (- first pantograph, -.- second pantograph, −− third pantograph)*

4   PARAMETER STUDY

To evaluate how changes in the design of the section overlap affects the dynamic behaviour, a parameter study was performed. The section where the pantographs operates on both contact wires was chosen. This was accomplished

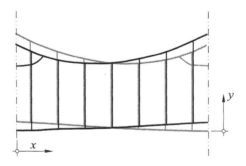

Figure 6: Parameter definition

by changing the position from where the lift of the contact wire starts (x) as well as changing the lift of the contact wire at the support (y), see figure 6. Analysis was performed for a start of lift of contact wire at 0, 10, 20 and 30 m from the support pole and a lift of contact wire at the other support of 70, 100, 150 and 200 mm. Results from static analyses for an overlap when changing the lift of contact wire and start of lift can be seen in figures 7 a) and b) respectively. Static characteristics for an ordinary span is included in the figure for comparison. The upper picture shows the system elasticity, the middle one

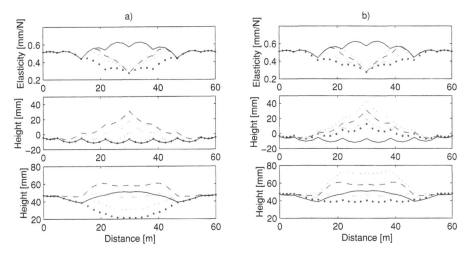

Figure 7: Static characteristics within a section overlap. a) *different start position of lift (− ordinary span, −− 0 m, −.− 10 m, .. 20 m, .. 30 m b) different lift of contact wire at the support (− ordinary span, .. 70 mm, −.− 100 mm, , −− 150 mm, .. 200 mm))*

illustrates the height of the contact wire and in the lower the pantograph height can be seen.

Moving the start of lift towards midspan results in that the pantograph is in contact with both lines for a longer distance and gives a smoother transition in elasticity. This also lowers the static pantograph height. By changing the start of lift to 10 m a more level pantograph height is established while the elasticity remains as for existing system (0 m). Reducing the lift gives a smoother transition in elasticity. An increase in lift will have opposite effect. The static pantograph height is affected in the same direction as the change in lift. A change of lift to 100 mm gives both a smooth elasticity curve and a level static pantograph height within the overlap. Results from dynamic analyses can be seen in table 2 and 3. The travelling speed is 240 km/h. Table 2 shows results from changing the start of lift of contact wire. Contact force standard deviation for different values of lift of contact wire can be seen in table 3. The effect from altering the start of lift was not large, only the third pantograph showed on decreased dynamic effects for a change to 30 m. For that case the first pantograph performed worse. Changing the lift of contact wire to 100 mm gives a consid-

Table 2: Contact force standard deviation within section overlap when changing the start of lift of contact wire (x).

Start [m]	σ, 1:st [N]	σ, 2:nd [N]	σ, 3:rd [N]
0	15.0	27.4	37.1
10	16.7	28.5	35.9
20	21.3	27.0	30.0
30	24.3	26.6	26.3

Table 3: Contact force standard deviation within section overlap when changing the lift of contact wire at the support (y).

Lift [mm]	σ, 1:st [N]	σ, 2:nd [N]	σ, 3:rd [N]
50	15.4	22.1	19.8
100	12.7	17.2	18.2
150	15.0	27.4	37.1
200	20.5	40.3	49.5

erable change, lowering the standard deviation to less than half for the third pantograph in the section overlap. Figure 8 shows the contact force variation at 240 km/h for this system configuration. The position of the section overlap is between 540 and 600 m. Results from dynamic analyses at speeds of 160-260 km/h with a lift of contact wire of 100 mm is shown in figure 9. Figure 10 shows results from dynamic analyses with the existing and new system configuration for two pantographs at 240 km/h with a pantograph distance ranging from 30 to 180 m. The increment is 15 m.

5 DISCUSSION AND CONCLUSIONS

Static analyses showed that the system has a relatively constant elasticity within each span but lower in the section overlap. With a static load of 100 N the pantograph height at the overlap is higher than in the rest of the system, see figure 3.

Dynamic analyses was performed for speeds from 160 to 260 km/h. The results revealed that both the second and third pantograph suffers severe dynamic effects within the section overlap for increasing speed, see figure 5. According to the evaluation criteria presented in chapter 1.1 these results indicates that there is a risk for contact loss and too high uplift forces within the section overlap for the second and third pantograph. Problems with current collection quality on the system of today does not appear for a single pantograph but for multiple pantograph operation.

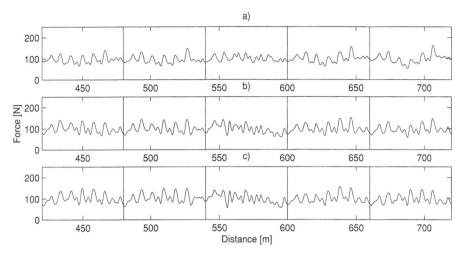

Figure 8: Contact force as function of travelled distance at 240 km/h with a contact wire lift of 100 mm at the support. a) first pantograph b) second pantograph c) third pantograph

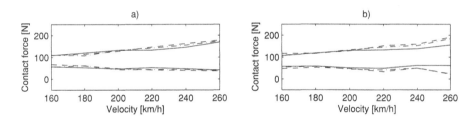

Figure 9: Statistical minimum and maximum contact forces as function of velocity with a lift of contact wire at the support of 100 mm. a) before section overlap b) within section overlap; - first pantograph, -.- second pantograph, −− third pantograph

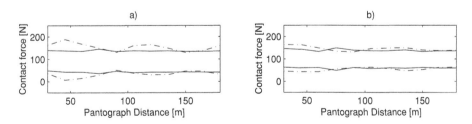

Figure 10: Statistical minimum and maximum contact forces within section overlap as function of pantograph distance. a) 150 mm lift of contact wire at support b) 100 mm lift of contact wire at the support (-first pantograph -.-second pantograph)

In order to evaluate how changes in the design of the section overlap affects the dynamic behaviour, a parameter study was performed. The influence of the part where the pantographs operates on both contact lines was investigated. Both static and dynamic analyses was performed. Changing the start of lift did not affect the results as much as when modifying the lift of contact wire at the support. The system was proven to be very sensitive to change of contact wire lift, see table 2. A change of lift to 200 mm gives a fast transition from the highest to the lowest elasticity as well as having a static pantograph height which is much higher than at the poles. This is also the configuration which gives the highest value of contact force standard deviation. A level static pantograph height as well as a smooth elasticity transition within the overlap was accomplished by a change of lift to 100 mm, see figure 7. This configuration gives the lowest contact force standard deviation, a significant decrease of the dynamic effects was accomplished. The contact force variation for this system configuration can be seen in figure 9, the figure shows results from before and within a section overlap. Analyses for different pantograph distances showed on robustness of the new system configuration, see figure 10. The new system configuration suffers less dynamic effects within the section overlap, in some cases even lower than within an ordinary section.

ACKNOWLEDGEMENTS

The gratefully acknowledged research funding has been provided by The Swedish National Rail Administration.

REFERENCES

1. Bartholomae, H. "High- Performance Overhead Contact System", European Railway Review, pp. 46–52, Issue 3, 2001
2. Manabe, K. "Measure to Multiple Pantographs of High Speed Train Operation", Japanese Railway Engineering, pp. 11–14, March, 1989
3. Manabe, K. and Fujii, Y. "Overhead System Resonance with Multi-Pantographs and Countermeasures", QR of RTRI, vol.3, No.4, pp.175–180, Nov., 1989
4. Drugge, L. "Modelling and Simulation of Pantograph-Catenary Dynamics", Doctoral Thesis, Department of Mechanical Engineering, Luleå University of Technology, January, 2000
5. Bauer, K.-H., Kießling, F. and Seifert, R. "Effects of design parameters of Overhead Contact Lines on High-Speed Operation - a Comparison of Theory and Testing ", Translation of article from "Electrische Bahnen", No.10, pp. 269–279, 1989
6. Poetsch, G., Evans, J., Meisinger, R., Kortum, W., Baldauf, W., Veitl, A. and Wallaschek, J. "Pantograph/Catenary Dynamics and Control ", Vehicle System Dynamics, pp. 159–195, (28)1997

Vehicle System Dynamics Supplement 41 (2004), p.697-706

Pantograph-CATENARY DYNAMIC INTERACTION IN THE MEDIUM-HIGH FREQUENCY RANGE

STEFANO BRUNI*, GIUSEPPE BUCCA*, ANDREA COLLINA*, ALAN FACCHINETTI*, STEFANO MELZI*

SUMMARY

The paper addresses the topics of current collection quality and wear in the pantograph-catenary system, with special emphasis on the effect of high frequency dynamic effects. By means of a suitable mathematical model of pantograph-catenary interaction, it is shown that the percentage of contact losses cannot be easily related to the value of performance indexes typically used by railway administrations, due to the effect of high frequency pantograph catenary interaction.

As to the topic of catenary wear, the results of laboratory tests are presented and used to derive a wear model as function of contact force, current flow and pantograph speed. This model is then used for the numerical simulation of wear progress in the line. The results obtained appear to be in good agreement with line measurements, and open the possibility to use mathematical models to optimise line design and pantograph service conditions in order to optimise the wear of the catenary-collector couple.

1 INTRODUCTION

The behaviour of the system composed by a pantograph travelling along the overhead equipment (OHE) involves dynamic phenomena spanning a wide frequency range: the low frequency behaviour is dominated by the span passing frequency (up to 1.5 Hz) and its multiples, while the dropper passing frequency typically falls in the $10 \div 15$ Hz frequency range for high speed applications. Higher frequencies, up to 100 Hz, are then related to the excitation due to the wire irregularity and to the transients taking place when the pantograph approaches a dropper or a registration arm. This latter field of frequency requires particular attention, because the high frequency fluctations of the contact force are strictly related to contact losses and to the presence of continuous sparking, with important implications on the current collection quality as well as on the wear rate of the contact wire and of the collector strip.

While the low frequency dynamic interaction between pantograph and catenary can be now considered a solved problem, thanks to the contributions of several authors (among which [1, 2, 3, 4, 5]), many important issues related to the high frequency interaction require further investigation. A contribution towards the numerical simulation of the high-frequency pantograph-catenary interaction was provided in [6], based on a detailed schematisation of catenary and

---

*Department of Mechanical Engineering, Politecnico di Milano, Via La Masa 34 20158 Milano Italy

pantograph dynamic properties, together with a suitable model of pantograph-catenary contact.

Aim of this paper is therefore to investigate the relationships between contact force fluctuations, contact loss and wear, further developing the work presented in [6]. Moreover, the paper reports some of the first results obtained on a laboratory test rig for the study of pantograph-catenary contact and wear (see section 4), which recently entered into service at Politecnico di Milano.

2   MATHEMATICAL MODEL OF PANTOGRAPH-CATENARY INTERACTION

The mathematical model used for the numerical simulation of pantograph-catenary interaction includes a three-dimensional finite element model of the overhead equipment (OHE) and a multi-body model for the pantograph. As far as the OHE model is concerned, use is made of high order tensioned beam elements, thereby including the non negligible effects of wire bending stiffness and allowing the model to minimise the disturbance effects generated by the pantograph sliding on a discretised structure and to reproduce the dynamic behaviour of the catenary in a wide frequency range. The registration arms are also modelled in detail, because they represent a singularity along the span which may lead to wear concentration. Another feature of the OHE model is the representation of the droppers as non-linear elements, based on the experimental characterisation of the force-deformation relationship, allowing to consider the very important effect of dropper slackening.

The pantograph model represents the articulated frame as a rigid body or lumped parameter system, and the collector heads as flexible elements, using a modal superposition approach: this last feature allows to reproduce the dynamic behaviour of the pantograph system in the $0 \div 350$ Hz frequency range, as demonstrated by a validation against laboratory measurements reported in [6].

The model of pantograph-catenary contact is based on a penalty method approach, and was tuned to allow a correct representation of the mono-lateral sliding contact in the $0 \div 100$ Hz frequency range, and at the same time to avoid numerical problems associated to a stiff formulation of the equations of motion.

For full details on the formulation of the model, including validation against semi-anaytical solutions and experimental data, the reader is referred to [6].

3   RELATIONSHIP BETWEEN CONTACT FORCE AND CONTACT LOSSES

Contact losses between the current collectors and the contact wire may occur due to dynamic contact force disturbances caused by wire irregularities and discontinuities, or by local changes of the wire impedance, such as at registration

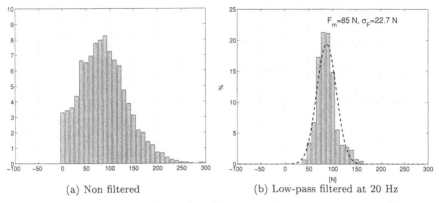

(a) Non filtered   (b) Low-pass filtered at 20 Hz

Figure 1: Distribution of simulated pantograph-catenary contact force

arms or even at the droppers. This phenomenon should be avoided, especially in high speed service, in order to ensure an appropriate quality of current collection and to avoid accelerated wear of the catenary and of the collectors.

To assess the quality of current collection, the railway administrations have estabished criteria which are generally formulated in terms of a performance index I_c, defined as:

$$I_c = F_m - 3\sigma_F \tag{1}$$

where F_m and σ_F are respectively the mean value and standard deviation of the contact force. The reason for introducing this index relies on the assumption of considering the contact force a gaussian random process, so that I_c would represent the minimum value of the contact force, and no contact loss should occur for positive values of this index. Nevertheless, this approach has some important drawbacks: first of all, the bandwidth of the contact force measure is typically limited to 20 Hz or slightly higher, due to the limitations of the present measuring set-ups, leading to a severe underestimation of the actual standard deviation σ_F. Moreover, the statistical distribution of the contact force may be highly non-simmetric and non-gaussian, due to the occurrence of contact losses.

The mathematical model of pantograph-catenary interaction described in section 2 was therefore applied to better understand the validity of the performace index defined by equation (1) to represent the quality of current collection. As a first analysis, in figure 1 the statistical distribution of the contact force as directly obtained from the simulation (pass-band 100 Hz approximately) is compared to that of the same contact force low-pass filtered at 20 Hz to simulate the effect of line measure. As expected, low-pass filtering results in a strong modification of the shape and standard deviation of the statistical distribution, so that the filtered signal is not representative of the statistical properties of the actual contact force. It is particularly important to observe that a relatively high percentage of contact losses (approximately 3%, as indicated by the percentage corresponding to 0 N in figure 1a) takes place, despite the fact that the performance index I_c computed on the filtered signal is greater than zero.

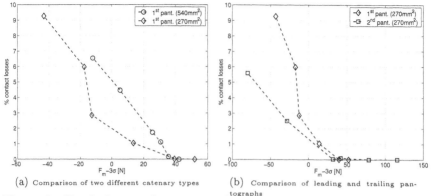

(a) Comparison of two different catenary types

(b) Comparison of leading and trailing pantographs

Figure 2: Simulation results: contact losses percentage vs $F_m - 3\sigma_F$ index, calculated on the filtered (20 Hz) contact force

Figure 2 shows the relationship between the performance index computed on the low-pass filtered contact force and the contact loss percentage, comparing the results obtained for two different catenary types (referred to as "270 mm^2" and "540 mm^2") and for the leading and trailing pantograph on the same catenary, in the case of a double pantograph train configuration, with pantograph distance of 293 m.

The results reported in figure 2 show that very different percentages of contact loss may occur corresponding to the same value of the performance index, depending on the type of catenary considered, or even for the leading / trailing pantograph on the same catenary. Therefore, it can be concluded that a limit placed on the performance index (1) computed on the filtered contact force is purely empyrical, and strongly depends on the particular combination of OHE and pantograph pair considered: this explains why the limit to this index is set to very different values in different countries (e.g. 0 N in U.K., 30 N in Germany, 40 N in Italy).

4 WEAR OF THE CONTACT WIRE
- LABORATORY EXPERIMENTS

4.1 Description of the test rig

In order to investigate wear formation on the contact strip and on the contact wire, a specific test bench was designed and built at the Department of Mechanical Engineering of Politecnico di Milano. This facility, represented in figure 3, allows the full-scale testing of a collector or of the complete pantograph system, at speeds up to 250 km/h under the passage of electrical current up to 1200 A. The main element of the test bench is a 4.3 m diameter disk, rotating around a vertical axis and driven by an a.c. 90 kW motor through a belt transmission. A contact wire with a 120 mm^2 section (same as used in real lines) is elastically

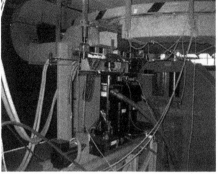

Figure 3: General view of the test bench and detail of the actuators

suspended to the outer circumference. The reason for introducing this flexible connection is to reproduce with more accuracy the contact condition taking place in the line, where the contact wire is highly flexible due to the deformation of the overhead equipment.

The collector head is mounted on a platform moving along the radial direction of the disk following a triangular wave signal: the period of the movement is synchronised with the test speed in order to reproduce the effect of the stagger. The connection between the collector and the platform incorporates a suspension set taken from a Dornier DSA 350S pantograph.

It is important to point out that the other existing test benches of this kind (e.g. [7, 8]) use a short specimen of the contact strip material, rigidly connected to a fixed reference frame and sliding against a rigid contact wire. On the other hand, in this test bench the actual current collector is used, sliding against a flexible contact wire, thereby reproducing with higher accuracy the actual conditions of line service.

The contact force between the collector and the contact wire is applied by means of an hydraulic actuator mounted on the moving platform. This actuator can be used to apply either a constant or time varying contact force, which could in principle be derived from numerical simulation.

The electric circuit of the test bench is used to induce current passage between the wire and the collector. This circuit is made up of a three-phase full-bridge rectifier with capacitor, whose output feeds an IGBT inverter which controls the load current of the output inductor. The commutation frequency of the inverter is suitably high to ensure low current harmonic content. The performance of the inverter can be summarised as follows: up to 1.2 kA d.c. current, 0.5 kA a.c.-16.66 Hz and 0.35 kA a.c.-50 Hz so that all the currents of the European catenaries can be emulated by the test bench.

The measuring set-up includes sensors for measuring the sliding speed, normal and tangential contact force at strip-wire interface; contact strip temperature, current flow through contact and voltage. This last allows to estimate the occurrence of electric arcs and therefore to quantify the percentage of contact loss occurrence. Moreover, the thickness of the contact wire is measured by a laser-optical device, allowing for continuous measurement during the test, while the wear of the collector is measured by weighting the collector before and after the test.

Copper based, metalised carbon and carbon+copper contact strips were tested at speeds up to 200 km/h, current intensities up to 1200 A and static loads around 100 N. In the next subsection, the results of a series of tests performed using a single type of copper based contact strip is used to derive a model for wear in the contact wire.

4.2 A model for the wear of the contact wire

The wear model defined in this section quantifies the "worn area" A (worn volume divided by the distance travelled) as function of the contact force F_c, current flow i and speed V in the following form:

$$A = k_1 \left(1 + \frac{i}{i_0}\right)^{-\alpha} \left(\frac{F_c}{F_0}\right)^{\beta} \frac{V}{V_0} \frac{F_c}{H} + k_2 \frac{R(F_c)i^2}{H\,V} \tag{2}$$

where F_0, i_0 and V_0 are reference values for the contact force, current and speed used to express some terms of equation (2) in a non-dimensional form, H is the hardness of the contact wire material and $R(F_c)$ is the equivalent electrical resistance at the contact, function of the contact force. Finally, parameters α, β, k_1 and k_2 have to be identified from experimental tests, as explained below.

In equation (2) the total wear is expressed as the sum of a mechanical contribution and an electrical contribution. In accordance with experiments, the mechanical contribution to wear (first term at right hand) is assumed to depend non-linearly from the contact force, and also to be function of the current flow, to take into account the effect of "current lubrication" [9] which may lead in some cases to lower wear rates for higher values of current flow. This effect can be explained assuming a local fusion of the material at the contact interface produced by the current flow, which causes a reduction of the friction coefficient and also leads to a redistribution of the material on the contact surfaces rather than to abrasion.

The electrical contribution to wear is defined under the assumption of proportionality between the worn volume and the electrical power dissipated in the contact. The contact resistance R to be used in the expression of electric wear was measured on the experimental test stand for different values of the contact force F_c, by measuring the current flow and voltage drop across the contact. The results of this measure are shown in figure 4: despite a rather high dispersion of these data, the trend of the measurements can be interpolated by the expression:

$$R(F_c) = \frac{k_R}{\sqrt{F_c}} \tag{3}$$

with $k_{FR} = 4.97 \ 10^{-2} \ [\Omega\,N^{\frac{1}{2}}]$.

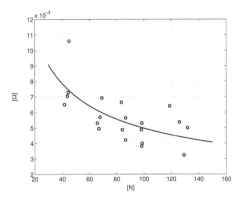

Figure 4: Contact resistance as a function of the contact force

It can then be observed that a decrease in the contact force leads to a higher contact resistance, and therefore to a higher electrical wear, as also observed by [9]. In case of full contact loss, mechanical wear reduces to zero, while the electrical contribution is still computed according to eq. (2), but using the resistance of the air gap between the two conductors. This leads to a high level of wear, corresponding to the effect of electric arcs.

In order to identify the parameters of equation 2, a series of tests were performed using a copper-based contact strip type. Testing conditions included zero current tests at three different levels of contact force (30, 60 and 90 N), and tests at the maximum contact force level (90 N), combined with two different levels of current flow (500 and 750 A). Each test consisted in running the bench for approximately 500 km at a constant value of contact force (except for the small fluctuations produced by the flexibility of the contact wire ring) and of current flow (when applied) and to measure the amount of worn material in 12 sections along the contact wire. The average of the values measured in these sectons was then taken as representative of the test.

The results of wear tests are summarised in figure 5 in terms of specific wear rate (SWR), that is the worn area per unit contact force. The tests performed at zero current flow show SWR values increasing with the contact force, supporting the assumption of non-linear dependence of wear upon the contact load. The test at 500 A current flow shows a wear rate slightly lower than at 0 A, for the same contact load: this can be explained by the effect of current lubrication, while at 1000 A the increase of the electrical wear leads to a higher value of global wear. Nevertheless, some more tests at different values of current flow would be useful to better clarify the dependency of wear rate upon current flow.

All tests reported in figure 5 were performed at the same speed of 160 km/h. Until now, one single test was performed at 90 N contact load, zero current and 200 km/h, producing an SWR of 0.320 mm^3/km N (not reported in figure 5), to be compared with the figure of 0.267 mm^3/km N obtained in the corresponding test at 160 km/h. This result alone obviously does not allow to draw a conclusion

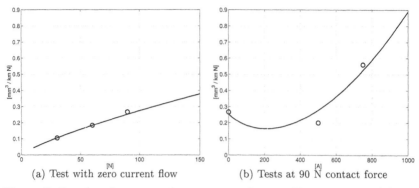

(a) Test with zero current flow   (b) Tests at 90 N contact force

Figure 5: Results of tests on the wear test rig: specific wear rate of the contact wire as a function of the contact force and of the current flow

about the effect of speed, but seems in good accordance with the assumption of proportionality between speed and SWR made in formula (2).

The experimental results reported in figures 4 and 5 were used to identify the parameters of the wear model (2): the values of SWR obtained after parameter identification are drawn in solid line in figure 5(a) for zero current and different values of contact force, and in figure 5(b) for different values of current flow and contact force 90 N.

5 NUMERICAL SIMULATION OF WEAR IN THE CONTACT WIRE

Following the work described in [10], the numerical simuation of wear progress in the contact wire corresponding to real service conditions was performed by combining the mathematical model of pantoghraph-catenary interaction described in section 2 with the wear model described in section 4.2.

Starting from an initial wire irregularity, the simulation of pantograph-catenary interaction provides the time history of the contact force F_c in each collector strip, while the time history of the current flow i is obtained by solving the state equations of a simplified electrical model of the pantograph-catenary system, as reported in [10]. Once known the values of contact force and current flow, formula (2) is used to define the single contribution to wear of each collector strip, and then all contributions are summed considering the spatial shift between the different contact strips, allowing to define the modification to the contact wire profile produced by the train passage.

This modified profile is then fed back to the mathematical model of pantograp catenary interaction, to produce modified time histories of the contact force and current flow. As the amount of wear produced by one single train passage is very small, the same numerical simulation is considered representative of several train passages, so that the entire service life of the line can be simulated by a reasonable number of iterations of the procedure described above.

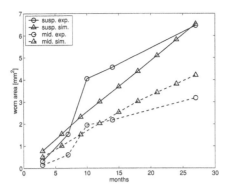

Figure 6: Comparison of simulated and measured wear progress under suspension and at midspan

In order to assess the validity of the procedure, 28 months of service life were simulated for an OHE installed in the Italian high speed line between Florence and Rome, in a short test section, and compared to measurements performed on the test section. The catenary considered in this test case is composed by a 150 mm^2 contact wire and a 120 mm^2 messenger wire, operated under 3kV DC, electric power supply. Though the line is operated using different pantograph types, one single pantograph type was considered in the simulations, to simplify the calculations. The pantograph type is ATR90 (used in Italian high speed trains), in single arrangement, travelling at the speed of 250 km/h, with a static load of 150 N and a total current flow through the two contact strips of 1200 A. The assumption of 125 train passages per day was made in these calculations.

The results of wear progress simulation under suspension and at midspan (averaged over 4 consecutive spans) are compared in figure 6 to the available line measurements, showing good agreement. In particular, the mathematical model reproduces the higher wear rates occurring under suspension with respect to the midspan position. According to the numerical results, the reason for the accelerated wear of the wire under suspension is a contact force unloading which takes place under the suspension due to the presence of the registration arm: according to the wear model (2), this produces mechanical wear, but high electric wear due to the increase of contact resistance, and to huge arch-induced wear if a contact loss takes place. It is important to observe that this effect can be simulated only using a suitable model for high-frequency pantograph-OHE interaction, as described in section 2 and in [6].

Numerical and experimental results appear to be also in rather good agreement from the quantitative point of view, suggesting that the wear model identified on the laboratory test rig can be used to predict line wear. Nevertheless, this point should be treated with great caution, considering that many simplifying assumptions were made in the numerical calculations as described above.

6   CONCLUDING REMARKS

The paper discussed the need of accurate modelling of high-frequency interaction between pantograph and catenary to investigate current collection quality and catenary wear. The performance index currently used to assess current collection quality seems to be rather inadequate to capture the actual relationship between contact force and contact losses, and its use is mainly based on empyrical reasons.

As far as catenary wear is concerned, some first results obtained on a laboratory test rig were presented, allowing for the identification of a wear model which can be used to simulate wear progress in the contact wire during real service. The results of wear progress simulation are in good agreement with line measurements, and suggest the use of this simulation tool to optimise catenary design and service conditions with respect to wear.

ACKNOWLEDGEMENTS

Part of this work received financial support within the European project WIRE, funded by the European Commission under the CRAFT Contract G3ST-CT2001-50127.

REFERENCES

1. Diana G., Bruni S., Collina A., Fossati F. and Resta F., "High speed railways: pantograph and overhead lines modelling and simulation", Computer in Railways VI, Southampton, WIT Press 1997, pp.847-856
2. Nordstrom C. J. and True H., "Dynamics of an electrical overhead line system and moving pantograph", Suppl. to Vehicle System Dynamics, Vol. 29, 1998, pp.104-113
3. Poetsch G. et al., "Pantograph/Catenary Dynamics and Control", Vehicle System Dynamics, Vol. 28(n. 2-3) 1998, pp.159-195
4. M. Reinbold, U. Deckart, "FAMOS - Ein Programm zur Sim ulation von Oberleitungen und Stromabnehmern", ZEV+DET Glas. Ann. 120 (1996), Nr.6, Juni.
5. T. Schulze, W. Baldauf, G. Poetsch, "Validated simulation tools for reliable investigation of the pantograph-catenary interaction", WCRR 2001, Kln, 25-29 November 2001
6. Collina A., Bruni S., "Numerical simulation of pantograph-overhead equipment interaction", Vehicle System Dynamics, Vol. 38(n. 4) 2002, pp.261-291
7. K. Becker, U Resch, A. Rukwied, B.-W. Zweig, "Lebensdauermodellierung von Oberleitungen", Elektrische Bahnen, 94, 1996, eb 11/96
8. Nagasawa H., Kato K., "Wear mechanism of copper alloy wire sliding against iron-base strip under electric current", Wear, 216, 1998, pp. 179-183
9. Klapas D., Benson F. A., Hackam R., Evison P. R., "Wear in simulated railway overhead current collection systems", Wear, 126, 1988, pp. 167-190
10. Collina A., Melzi S., Facchinetti A., "On the Prediction of Wear of Contact Wire in OHE Lines: A Proposed Model", XVII IAVSD Symposium, Copenhagen, Denmark, 20-24 August 2001

A numerical-experimental approach to evaluate the aerodynamic effects on rail vehicle dynamics

FEDERICO CHELI, ROBERTO CORRADI, GIORGIO DIANA
AND GISELLA TOMASINI[1]

SUMMARY

This paper describes the methodology proposed to define the aerodynamic loads acting on a rail carbody, for given turbulent wind conditions and to compute the corresponding vehicle's response. Experimental data of wind tunnel tests, together with numerical simulation results are shown. Finally, the set up numerical model is adopted to estimate critical wind curves.

1   INTRODUCTION

The dynamic response of high speed trains subjected to transversal cross-wind is one of the most critical problems connected with running safety. The main risks associated with cross-wind are vehicle overturning and track instability under the lateral forces exerted by the train.

The European Transportation Authorities are particularly concerned with this problem, since a "Technical Specification for Interoperability" is under definition, covering also this specific theme.

Two approaches can be followed to analyse the problem: wind-tunnel/full-scale experimental tests and numerical models. In this paper a methodology that integrates the two approaches is presented. Starting from the experimental data gathered in wind-tunnel tests and from the numerical model for the simulation of rail-vehicle dynamics, the response of a vehicle subjected to real turbulent wind is computed. The final objective is to define the limit safety conditions of the vehicle, in terms of critical wind curves.

The aerodynamic loads acting on the moving train are calculated according to a new algorithm based on the vehicle admittance, measured in wind-tunnel tests. This algorithm accounts for the actual turbulent wind speed distribution as a function of space and time, for the vehicle's geometry and for its operating conditions.

---

[1] *Address correspondence to*: Politecnico di Milano, Dipartimento di Meccanica, Milano, Italy.
Tel.: +39.02.2399.8480; Fax: +39.02.2399.8492; E-mail: gisella.tomasini@mecc.polimi.it

2   WIND TUNNEL TESTS

An extensive experimental campaign has been carried out over the last two years in the new Politecnico di Milano wind tunnel, in order to measure the aerodynamic forces acting on rail vehicles subjected to cross wind (Fig.1). The wind tunnel tests allowed not only to collect comprehensive experimental information on the aerodynamic behaviour of trains, but also to obtain all the necessary data for setting up the numerical model for the calculation of the aerodynamic forces. Three different 1:20 scale train models, each composed by 3 vehicles, were analysed: two high speed trains (ETR 480 and ETR 500) and one interregional train equipped with UIC-Z1 coach. Figure 1 shows the ETR480 train scale model in the 14m×4m test section of the Politecnico di Milano wind tunnel.

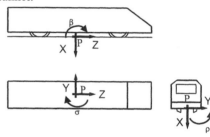

Fig.1. Wind-tunnel tests in turbulent flow conditions.

Fig.2. Reference system for the aerodynamic forces.

The tests have been carried out with both low and high turbulent wind, with different angles of attack and in different configurations: with train on viaduct or embankment, in upwind or downwind conditions, with or without wind fences.

In the wind tunnel tests, all the aerodynamic forces and moments acting on the car body of both the first and the second vehicle, have been measured through a six-components dynamometric balance specifically designed for this application ([1], [2]). Moreover, the wind tunnel has been monitored by means of both Pitot and hot-wire anemometers, in order to evaluate the actual turbulent wind characteristics (mean speed, turbulence intensity, Power Spectral Density, integral length scales) during the experimental tests.

The measured aerodynamic forces allowed to define the static aerodynamic coefficients as a function of the angle of attack, for all the analysed configurations. These non-dimensional coefficients are defined as follows:

$$C_{Fi} = \frac{F_i}{\frac{1}{2}\rho U^2 A} \qquad C_{Mi} = \frac{M_i}{\frac{1}{2}\rho U^2 A h} \tag{1}$$

where F_i (i=x,y,z) are the aerodynamic force components in the train's reference system (Fig.2) and M_i (i=x,y,z) are the corresponding moments. In equation (1), A stands for the lateral surface of the carbody (in the XZ plane, Fig.2), ρ is the air density, $\overline{U^2}$ is the mean square value of the wind speed and h is the vehicle's height.

As an example, Fig.3 shows the modulus of the measured aerodynamic coefficient C_{Fy} of the lateral force F_y acting on the first vehicle of the ETR 500 Italian train, as a function of the wind angle of attack. The coefficient refers to the train on viaduct, in low turbulence flow and in both upwind and downwind conditions.

The proposed wind tunnel tests ([1]) allowed to point out that the aerodynamic coefficients of the first vehicle show a characteristic transition in correspondence with a critical angle of attack, which varies with the test configuration (50° for the configuration considered in figure 3): this is a general characteristic which has been observed for all the first vehicles of the considered trains. As also noticed by other authors ([3], [4]), this trend can be justified considering that, depending on the angle of attack, a transition from slender body to bluff body behaviour occurs. One of the most important consequences is that the maximum lateral force on the first vehicle is reached for an angle of attack lower than 90°, while the C_{Fy} of the second vehicle shows an always increasing trend from 0° to 90°.

Moreover, the comparison between upwind and downwind conditions (fig. 3), shows that in the downwind configuration, the lateral force coefficient reaches a maximum value which is lower than that measured in the upwind configuration. As a matter of fact, depending on the wind direction, the wind itself comes up against the free track on the right side of the train (downwind configuration) or directly against the train (upwind configuration): this obviously modifies the wind flow across the viaduct, with a consequent strong influence on the lateral force acting on the vehicle. This example points out that, when defining the aerodynamic forces with the numerical algorithm, it is important to account for the specific operating conditions.

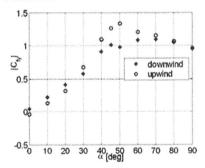

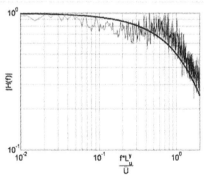

Fig. 3. ETR 500, 1st vehicle on viaduct: comparison between the modulus of the aerodynamic coefficient C_{Fy} in upwind and downwind configuration.

Fig. 4. Vehicle admittance function: experimental data and corresponding interpolating function.

The second objective of the wind-tunnel tests was to identify the vehicle's admittance function ([5], Fig.4), whose meaning and role in the simulation program will be described in the next paragraph.

3   AERODYNAMIC FORCES COMPUTATION AND VEHICLE DYNAMICS SIMULATION

The overall numerical procedure to define the safety conditions of a rail vehicle subjected to cross wind can be divided into three macro-steps (Fig.5):

1) turbulent wind definition, that is wind speed as a function of time and position $u(t,x)$; since turbulent wind is a stationary random process, for given mean wind speed and terrain type, the generated input file must satisfy the statistical properties of the atmospheric boundary layer: wind speed PSD, turbulence intensity, integral length scales (see [6] and [5] for a detailed description of the numerical algorithm for wind speed generation);

2) calculation of aerodynamic loads that act on the rail vehicle subjected to cross wind, by means of the aerodynamic coefficients and the admittance function, both experimentally evaluated;

3) rail vehicle dynamic simulation, to define the limit safety conditions of the vehicle in terms of critical wind curves (CWC).

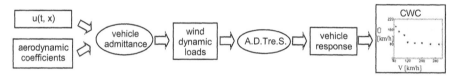

Fig. 5. Numerical-experimental procedure flowchart.

As far as point 2) is concerned, a new numerical algorithm, based on the vehicle admittance function, for the calculation of the aerodynamic loads has been set up. The admittance function allows to account for the spatial correlation of wind pressures at any two points on the carbody's surface: the pressure distribution depends on vehicle's geometry, on the operating conditions (train on viaduct or embankment, in upwind/downwind configurations, …) and on the turbulent wind speed as a function of space and time. The adopted approach is based on the corrected quasi-steady theory, in the frequency domain, that lies in applying the quasi-steady theory and in correcting it with the admittance function ([7], [5]): thus it represents a modifying adjustment of the ideal case of a vehicle enveloped by turbulent wind with full spatial correlation.

Starting from the wind speed function and for a given train speed, the wind velocity relative to the vehicle is defined. In particular, the vehicle experiences a wind profile on its body that varies as a function of the space-time distribution of the

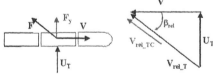

Fig. 6. Corrected relative train-wind velocity V_{rel_TC} and related angle of attack β_{rel} for the train running at speed V.

turbulent wind and of train speed. Referring to the time-depending absolute wind speed $U_T(t)$ evaluated in a reference point moving with the vehicle itself, the 'corrected' train-wind relative velocity is defined as:

$$V_{rel\_TC}^2(t) = V^2 + U_{TC}{}^2(t) \tag{2}$$

where V is the absolute train velocity and $U_{TC}(t)$ is the absolute wind velocity $U_T(t)$, 'corrected' in the frequency domain, by means of the vehicle admittance function $H(f)$:

$$U_{TC}{}^2(f) = H(f)\, U_T{}^2(f) \tag{3}$$

In equation (3), $U_T{}^2(f)$ is the generic harmonic component of the square absolute speed and $U_{TC}{}^2(f)$ is the corresponding component of the "corrected" velocity.

The time history of the lateral aerodynamic force[2] acting on the vehicle is calculated with the following expression:

$$F_y(t) = \frac{1}{2}\rho\, A\, C_{Fy}\left(\beta_{rel}(t)\right) V_{rel\_TC}^2(t) \tag{4}$$

where C_{Fy} is the static aerodynamic coefficient (measured in static wind tunnel tests) evaluated in correspondence of the actual β_{rel} angle of attack and V_{rel_TC} is 'corrected' train-wind relative velocity defined in (2) (see Fig.6).

The square modulus $|H(f)|^2$ of the admittance function is experimentally identified in static model tests with turbulent flow, as the non-dimensional ratio between the PSD of the measured force F and the PSD of the aerodynamic force calculated according to the measured wind speed in the reference point and the static aerodynamic coefficients (quasi-steady theory). Fig.4 shows an example of measured $|H(f)|$, for the considered ETR500 train, as a function of the non-dimensional frequency ($\bar{U}$ and $L_u{}^y$ are respectively the mean wind speed and the integral length scale of the longitudinal wind speed component in the transversal direction). Also the corresponding interpolating function is reported. It is possible to observe that the admittance function tends to 1 when the frequency approaches zero. When the spatial correlation decreases, in correspondence with high non dimensional frequencies, this function slowly decreases up to zero.

As far as point 3) is concerned, in order to simulate the dynamic response of a rail vehicle subjected to cross wind, the algorithm for the aerodynamic loads calculation, has been implemented in A.D.Tre.S., a rail vehicle dynamic simulation program [8], developed by the Mechanical Engineering Department of Politecnico di Milano. A.D.Tre.S. is suitable for simulating the motion of a railway vehicle running on tangent track and curve, also taking into account the deformation of track and substructure and the rail and wheels surface corrugation. The vehicle model is composed of a car body, two bogies and four wheelsets, introduced as rigid or flexible bodies, by means of modal superposition approach [8]. Car body, bogies and

---

[2] Same considerations can be done, obviously, for all the aerodynamic force components.

wheelsets are linked one to each other by means of the elastic and damping elements reproducing the primary and secondary suspensions. The simulation code accounts for the motion of a train-track system in both vertical and lateral direction, while the longitudinal motion of the vehicle is assumed to take place with constant velocity. The vehicle equations of motion are written with respect to a moving reference system, travelling with constant speed along the ideal track centreline.

The contact model is based on a preliminary geometrical analysis on the measured wheel and rail profiles. Contact parameters such as rolling radius and contact angle are reported in table form, as functions of wheel-rail lateral displacement. The geometrical analysis allows also to determine the number of the potential contact points for a given wheel-rail lateral displacement. Lateral and longitudinal creepages are computed and tangential and longitudinal forces at each active contact are obtained according to Shen, Hedrick & Elkins formulation [8].

The equations of motion are integrated in time domain, by means of a numerical step by step procedure [8]. In the present analysis, the simplifying assumptions of "infinitely rigid" track and vehicle rigid bodies were made.

4   NUMERICAL SIMULATIONS AND CRITICAL WIND CURVES

In order to evaluate the *'critical wind curve'* for each running condition, the numerical simulation results have been processed to calculate the limit safety conditions of the vehicle: the *'critical wind curve'* represents the combination of train speed V and wind mean speed $\bar{U}$ which leads to the overcoming of the given safety limits.

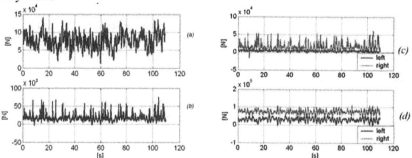

Fig. 7 – ETR500, 2nd vehicle on viaduct, upwind condition, turbulence intensity I_u=17%, L_u^x=90 m, tangent track with irregularity - V=200 km/h , $\bar{U}$ =110 km/h: (a) lateral wind force F_y, (b) ripage force, (c) guidance forces, (d) vertical forces on the front wheelset of the first bogie.

Figure 7 shows an example of input and output of a numerical simulation. The simulation has been performed with the second vehicle of the ETR500 train, running in tangent track on viaduct, in upwind condition, with the following characteristics: train speed V=200 km/h, mean wind speed $\bar{U}$=110 km/h, turbulence intensity I_u=17% and integral length scale L_u^x=90 m ([6],[5]). Also an experimentally measured track irregularity was considered. More in detail, Figure 7a shows the time history of the input lateral force F_y, while Figures 7b, 7c and 7d represent respectively the ripage force, the guidance forces and the right and the left vertical forces on the front wheelset of the first bogie. Starting from the numerical simulation results, it is possible to evaluate the safety indexes of the vehicle and, as a consequence, the *'critical wind curve'*. In particular, four indexes have been monitored to identify the limit vehicle conditions, as proposed in the draft European "Technical Specification of Interoperability":

- Prudhomme criterium;
- Nadal's coefficient Y/Q;
- overturning coefficient η_1, defined as the ratio between the difference of vertical forces on the left and right side wheels of a bogie and the overall vertical load on the four wheels:

$$\eta_1 = \frac{\left| \sum_{Bogie} \left(Q_{left} - Q_{right} \right) \right|}{\sum_{Bogie} \left(Q_{left} + Q_{right} \right)} \leq \eta_{1\_lim} = 0.9 \tag{5}$$

- unloading wheel coefficient η_2,, defined as:

$$\eta_2 = \sum_{Upwind} Q \ / \sum_{Upwind} Q_0 \geq \eta_{2\_lim} = 0.1 \tag{6}$$

where Q and Q_0 are respectively the sum of the dynamic and the static vertical loads on the upwind side wheels of each bogie.

Nadal's coefficient and Prudhomme limit are evaluated as indicated in the UIC 518 Code, while η_1 and η_2 coefficients are both filtered by a 2 Hz low pass filter.

Figure 8 represents the time histories of safety indexes, corresponding to the numerical simulation results shown in Fig.7. It is possible to observe that the two coefficients η_1 ed η_2 give similar information and, in the considered simulation conditions, both these coefficients overcome the established safety limits in various points. On the contrary, the 99.85% percentile of the ripage forces on the rear wheelset of the first bogie (the most loaded) reaches about 96% of Prudhomme limit (49.35 kN) and the Nadal's coefficient is quite far from its limit (max value 0.8).

In general, for all the numerical simulations carried out, the derailment index is the most insensitive to cross wind effect and, especially for the second vehicle, the simulations performed showed that the overturning/unloading indexes are the most restricting safety indexes. This result is a result of the lower mass of the second vehicle. Figure 9 shows the *'critical wind curve'* of the first and the second vehicle of the ETR500 train in tangent track, with medium turbulence intensity and taking into account different levels of track irregularity, as a function of the train speed V. Figure 9 shows that the two *critical wind curves* have different slopes: while, at low train speeds, the first vehicle has a *critical wind velocity* higher than the second vehicle, at high speeds it is possible to observe an opposite trend.

In general, the limit wind speed, for each train velocity, corresponds to the overcoming of either the Prudhomme limit or one of the overturning/unloading indexes (η_1, η_2). From the numerical simulations carried out, it is possible to observe that, as said before, the second vehicle is more influenced by the overcoming/unloading indexes, while the critical wind velocity for the first vehicle is mainly limited by the Prodhomme criterium, especially at high train velocities.

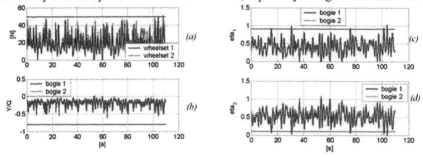

Fig. 8 – ETR500, 2nd vehicle on viaduct, upwind condition, turbulence intensity I_u=17%, L_u^x =90 m, tangent track with irregularity - V=200 km/h , $\bar{U}$ =110 km/h: (a) ripage forces (b) Nadal's coefficient, (c) overturning coefficient, (d) unloading coefficient.

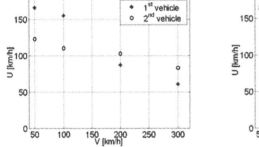

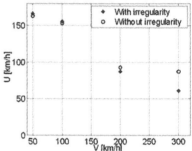

Fig. 9. ETR500, 1st and 2nd vehicle on viaduct, upwind conditions, tangent track with irregularity, medium turbulence intensity, critical wind curves as a function of train speed V.

Fig. 10. ETR500, 1st vehicle on viaduct, upwind conditions, tangent track, medium turbulence intensity, critical wind curves as a function of train speed V, with and without track irregularity.

In order to point out the effects of the track irregularity, the same simulations have also been performed without accounting for the geometrical irregularity and the comparison between the *critical wind curves* with and without irregularity, for the first vehicle of the ETR500 train, in tangent track are shown in Figure 10. It possible to observe that the main differences arise at high train speeds and, obviously, by taking into account the track irregularity, the limit wind velocity is lower.

The effect of the wind turbulence intensity on the *critical wind curve* is shown in Figure 11. The simulation features are: first vehicle of the ETR500 on viaduct, upwind configuration, tangent track, no track irregularity, three different levels of irregularity (I_U=10%, I_U=17% and I_U=30%). For the three different wind turbulence levels, the corresponding *critical wind curves* show the same trend as a function of the train speed but with a different mean value, depending on the turbulence intensity. Figure 12 shows the limit wind velocity for the first vehicle of the ETR500 train running on viaduct in curve, as a function of the cant deficiency and for three train velocities: in this case, the simulations have been carried out by taking into account the track irregularity and with a mean wind turbulence level (I_u=17%, L_u^x =90 m), in upwind conditions.

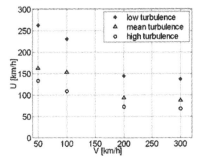

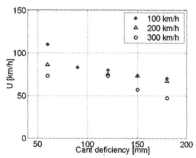

Fig.11. ETR500, 1st vehicle on viaduct, upwind conditions, tangent track without irregularity: critical wind curves as a function of train speed V considering high, medium and low turbulence level.

Fig. 12 - ETR500, 1st vehicle on viaduct, upwind conditions, curve with irregularity, medium turbulence intensity: critical wind curves as a function of train speed V and cant deficiency.

The critical wind curves for different values of train velocities show different mean values and different trends as a function of the cant deficiency: moreover, for all the considered train speed, the limit wind velocity is governed by the Prudhomme criterium. As shown in Figure 10, especially for high train speeds, the lateral forces on the track, which are regulated by the Prudhomme criterium, are strongly influenced not only by the wind loads but also by the specific considered track irregularities conditions.

In conclusion, the limit conditions reported in Figure 12 are related to track instability risk while the safety indexes associated with vehicle overturning and wheel unloading are far below their limit values.

5 CONCLUDING REMARKS

The numerical-experimental methodology set up for the evaluation of the limit safety conditions of a rail vehicle subjected to cross wind has been illustrated. The experimental data collected through wind-tunnel tests have allowed to measure the aerodynamic coefficients in different operating conditions and to identify the vehicle's admittance function, which is the basis of the proposed numerical algorithm for the computation of the aerodynamic forces acting on moving trains. By implementing it into a computer program for the simulation of rail-vehicle dynamics it is possible to study the response of a rail vehicle subjected to cross wind and to make an estimation of the vehicle safety in terms of the maximum allowable wind speed for fixed operating conditions. The presented critical wind curves can not be considered exhaustive for the full comprehension of such a complex problem: however, the proposed numerical simulation procedure has proved to be a useful tool to get a deeper insight in the problem. Moreover, one of the most significant aspects of the problem is to evaluate which are the most suitable safety indexes in order to access the actual safety conditions of a rail vehicle: from the performed analyses, the Prudhomme limit and the overturning/unloading coefficients seem to be the most significant indexes for the evaluation of the rail vehicle safety. Nevertheless, with reference to the overturning/unloading limits, it is important to point out that the overcoming of these indexes does not correspond to the actual vehicle overturning: the adoption of these indexes certainly leads to an extremely conservative estimation of the safety limit with respect to the real overturning mechanism. Finally, a critical analysis has to be done also about the Prudhomme criterium: the moving average on 2m long track section (as indicated in the UIC 518 Code) induces a different filtering effect as a function of the train speed. Keeping in mind these remarks, it is possible to justify some low values of the limit wind velocities shown in the critical wind curves presented in this paper.

ACKNOWLEDGEMENTS

The research has been developed with the financial support of TRENITALIA Gruppo FS, which also provided the physical train models for the wind tunnel experimental tests. Finally, the authors wish to acknowledge Dr. R. Cheli, Dr. G. Mancini and Dr. R. Roberti (Trenitalia S.p.A – UTMR – Direzione Ricerca, Ingegneria e Costruzioni) for their valuable contribution.

REFERENCES

1. Bocciolone, M., Cheli, F., Corradi, R., Diana, G. and Tomasini, G.: Wind Tunnel Tests For The Identification Of The Aerodynamic Forces On Rail Vehicles. Proceedings of 11th ICWE – International Conference on Wind Engineering, June 2003.
2. Corradi, R. et al.: On the aerodynamic loads on rail vehicles: recent wind tunnel experimental activities. To be published on Journal of Wind Engineering.
3. Baker, C.J.: Ground Vehicles In High Cross Winds Part 1 Steady Aerodynamic Forces. Journal of fluids and structures, vol. 5, 69-90 –1991.
4. Baker, C.J.: Ground Vehicles In High Cross Winds Part 2 Unsteady Aerodynamic Forces. Journal of fluids and structures, vol. 5, 91-111 –1991.
5. Bocciolone, M., Cheli, F., Diana, G. and Zasso A.: Suspension Bridge Response To Turbulent Wind: Comparison Of A New Numerical Simulation Method Results With Full Scale Data. Proceedings of 10th International Conference Wind Engineering, June 1999, pp. 871-878.
6. Simiu, E. and Scanlan, R.: Wind Effects On Structures. Wiley-Interscience Publication, New York, (1986).
7. Cheli, R., Mancini, G., Roberti, R., Diana, G., Cheli, F., Corradi, R. and Tomasini, G.: Cross-Wind Aerodynamic Forces On Rail Vehicles – Wind Tunnel Experimental Tests And Numerical Dynamic Analysis. Proceedings of the 2003 Word Congress on Railway Research, September 2003.
8. Bruni, S., Collina, A., Dian,a G. and Vanolo, P.: Lateral Dynamics Of A Railway Vehicle In Tangent Track And Curve: Tests And Simulations. Proceedings of 16-th IAVSD Symposium, Vol. 33 pp 464-477 (1999).

Interaction between Vehicle Vibration and Aerodynamic Force on High-Speed Train Running in Tunnel

KOJI NAKADE[*,1], MASAHIRO SUZUKI[*] AND HIROSHI FUJIMOTO[*]

SUMMARY

This paper describes the investigation of the interaction between vehicle vibration and flow-filed around it for the lateral vehicle vibration of high-speed train running through a tunnel. From the wind tunnel experiment of a excited vehicle vibration model, we found that there was little influence of the interaction when yawing vibration of the 3rd car was added.

Keywords: high-speed train, riding comfort, vehicle vibration, aerodynamic force, flow-induced vibration, wind tunnel experiment.

1. INTRODUCTION

As the maximum train speed increases and the weight of car body decreases, vibration of train becomes a subject of discussion especially in relation to riding comfort. It becomes dominant when a train runs into a tunnel section, in particular. As there are a number of tunnels in the railways in Japan, the riding quality in tunnel is very important from the viewpoint of passenger. Thus, we have extensively been investigating this phenomenon by performing running tests, wind tunnel experiments and computer simulations. The results of running tests are summarized as follows.

The vibration amplitude of trains and the aerodynamic force in tunnel sections are more noticeable than in open sections, and their typical frequency is about 2Hz for the 14th car in a 16-car train set running at 300km/h train speed (Fig. 1). Regarding the vehicle vibration mode, yawing vibration is more prominent than other modes of vibration [1, 2]. There is no correlation between vehicle vibration and track irregularities in tunnel sections [3].

When we study this phenomenon of a train set, the amplitude of vehicle vibration gradually increases from the head toward the tail of the train set [1]. According to the summarized data obtained for several trains, pressure fluctuation develops along the whole length of the train (the typical train length was 400m with 16cars) irrespective

[1] Address correspondance to: Koji Nakade, Railway Technical Research Institute, 2-8-38 Hikari-cho, Kokubunji-shi, 185-8540, Tokyo, Japan.
[*] Railway Technical Research Institute, Japan.

of the train type (Fig. 2). The pressure fluctuation pattern is divided into three regions in a whole train set, which are "developing region," "steady region" and "separating region." From the head to the 6th car (at about 150m from the head), the amplitude of pressure fluctuation increases and its peak frequency decreases (developing region). From the 6th car to the tail car, it remains constant in the steady state (steady region). Finally, a drastic increases in the amplitude of pressure fluctuation is observed at the nose of tail car (separating region). The peak frequencies of vehicle vibration and pressure fluctuation are different in the "developing region" (for example: vehicle: 2Hz, pressure: 3Hz), while they have the same peak frequency (for example: 2Hz) in the "steady region" and "separating region."

In general, the problem of strong interference between the vibration of car body and the flow around it possibly occurs due to the resonance phenomenon when the natural frequency of the car body is close to that of vortex formation, for example, the flow induced self-excitation vibration on long-size vibrating bodies which contact a small gap flow [5]. The results of a running test showed that the dominant frequency of car body vibration was nearly identical with that of the aerodynamic force (Fig. 1, 2). Then, we studied the possibility of interaction between aerodynamic force and vehicle vibration. At the first stage of this study, Ueki et al.[2] analyzed observed data and performed CFD with moving boundaries to calculate the interaction between flow field (two-dimensional flow) and vehicle dynamics (a middle car). Following this study, we analyzed the flow field around a vehicle model, which was excitedly vibrated in the yawing direction in a wind tunnel experiment. In this experiment, we focused on whether a feedback fluid force existed when a car was vibrating.

In this paper, we describe the wind tunnel experiment in Chapter 2. We summarize the results of the study in Chapter 3.

2. WIND TUNNEL EXPERIMENT

2.1. Apparatus and method

A 1/40 scale vehicle model within a tunnel is installed in a wind tunnel (Fig. 3). The model consists of a half-length car as the head car and two full-length cars as middle and tail cars. The shapes of vehicle model and tunnel imitate those of a Shinkansen car and a double-track tunnel. The tail car can be vibrated in a harmonic wave form in the lateral and yawing directions by two electromagnetic vibrators installed at the front and rear bogies, respectively. We measure pressure at eight points on the side surface of the tail car (at four points on each side), and two lateral displacements of the tail car at the position of each bogie with a laser displacement sensor.

Non-dimensional parameters used in this experiment are the reduced velocity $(U/(fD))$ and the dimensionless amplitude (y/D). U is the wind velocity in the wind

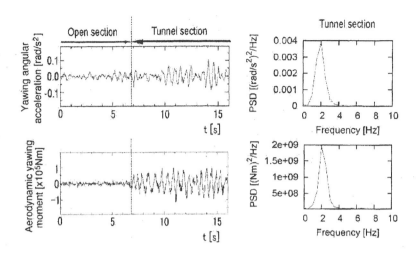

Fig. 1. Time history of yawing angular acceleration and aerodynamic yawing moment of a train entering a tunnel (train speed: 300km/h)

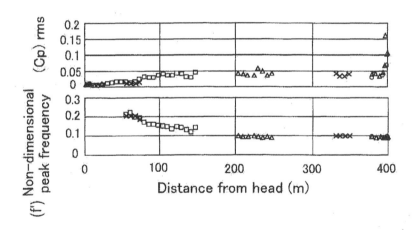

Fig. 2. Development of pressure fluctuation on the whole train set (f' indicates the non-dimensional frequency based on train speed and train width) [4]

Table 1. Parameter values in the wind tunnel experiment estimated from those in a field test using the similarity law

	Field test	Wind tunnel experiment
	$U = 270$km/h (75m/s)	$U = 20$m/s
	$f = 2 \sim 3$Hz	$f = 20 \sim 30$Hz
	$D = 1$	$D = 1/40$
	$y = 10$mm	$y = 0.25$mm

Table 2. Conditions of wind tunnel experiment

	Wind (U =20m/s)	Vibration (yawing mode, $f = 30$Hz, $y = 0.5$mm)
Case (a)	×	○
Case (b)	○	×
Case (c)	○	○

Note. "y=0.5mm" corresponds to the gap of stopper in lateral direction, which means the maximum vibratory displacement. And, a yawing mode is dominant in field test.

tunnel experiment or the speed of a real train; f is the frequency of vibration, y is the amplitude and D is reference length (for example, height of train). From the values in a field test, we estimate those in the wind tunnel experiment using the similarity law (Table 1).

We measure pressure in the following three cases to investigate the effect of vehicle vibration on the flow-field around the train (Table 2). The case (a) is under the condition that only train vibration is given without an inflow wind; the case (b) is under the condition that an inflow wind is given without train vibration and the case (c) is under the condition that train vibration and an inflow wind are given.

We first observe the pressure fluctuation on the surface of the tail car. Then, we calculate the yawing moment coefficient (C_M). C_M is defined as

$$C_M = \frac{F_M}{\frac{1}{2}\rho U^2 AD}$$

where F_M, ρ, U, A and D are the yawing moment of air pressure working on the car body, density of air, wind velocity or train speed, side surface area of train and half length of bogie, respectively. The yawing moment is estimated by integrating the pressure at eight points excluding the nose of tail car.

2.2. Results of experiment

Comparison with running test data to verify the experiment The air pressure wave forms in the running test (tail car in a 16-car train set) and in the experiment of Case

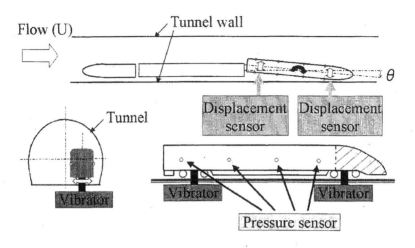

Fig. 3. Schematic pressure measurement with excited vibration of train in tunnel

(b) (Table 2) are compared. The position where air pressure is observed within a measurement car in the running test is identical with that in the experiment. The pressure fluctuation in the running test moves from the former car to the tail car and develops on the train surface facing to the tunnel wall, while keeping the same wave pattern. The pressure observed in the experiment generates a large fluctuation at the tunnel wall side rather than at the center side, and propagates from the upper to the down stream side, and consequently has a agreement with it in the running test (Fig. 4). The wave forms observed in the experiment show the same tendency as those in the running test. Consequently, the wind tunnel experiment is considered to qualitatively represent the phenomenon of real train.

Effect of excited vibration to aerodynamic force The time history of C_M and the power spectral densities of F_M generated under the three conditions (Case (a), (b) and (c)) are shown in Fig. 5 together with the yawing angle (θ) with thin lines. C_M of Case (a) is small although the amplitude of vibration is larger in the experiment than in the field test. C_M of Case (c) approximately equals to the sum of Case (a) and Case (b). Consequently, it is considered that the C_M generated by wind and by forced vibration are independent. We also studied the phase between C_M and θ, and confirmed that C_M had done positive work for the train every time. This means that C_M generated by vibration works as a resistance to the vibration.

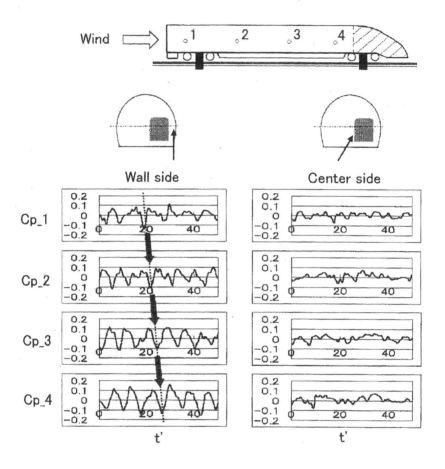

Fig. 4. Time history of Pressure coefficient observed in the wind tunnel experiment

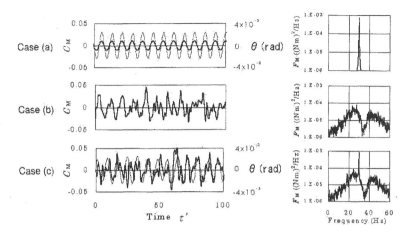

Fig. 5. Time history of yawing moment coefficient (C_M, thick line) and yawing angle (θ, thin line)(Left),
Power spectral density of yawing moment (F_M) (Right)

3. CONCLUSIONS

In order to study the interaction between aerodynamic force and vehicle vibration for high-speed trains running through a tunnel, we performed the wind tunnel experiment of a excited vehicle vibration model.

The wind tunnel experiment shows that there is little interaction for the 3rd vehicle vibrated in the yawing mode.

REFERENCES

1. Fujimoto, H. and Miyamoto, M.: The Vibration of the Tail Car in a Coupled Train (in Japanese). *JSME* Series C 87(359) (1987), pp. 2110–2114.
2. Ueki, K., Nakade.K and Fujimoto, H.: Lateral Vibration of Middle Cars of Shinkansen Train in Tunnel Section, *Proc. 16th IAVSD Symposium* (1999), pp. 749–761.
3. Takai, H.: Maintenance of Long-Wave Track Irregularity on Shinkansen (in Japanese). *RTRI Report* 3(4) (1989), pp. 13–20.
4. Suzuki, M.: Unsteady Aerodynamic Force Acting on High Speed Trains in Tunnel. *Quarterly Report of RTRI* 42(2) (2001), pp. 89–93.
5. Inada, F. and Hayama, S.: A Study on Leakage-flow-Induced Vibrations. Part 1:Fluid-Dynamic Forces and Moments Acting on the Walls of a Narrow Tapered Passage. *J. of Fluids and Structures* 4(4), pp.395–412.

Vehicle System Dynamics Supplement 41 (2004), p. 734-743 © Taylor & Francis Ltd.

Fatigue relevant road surface statistics

KLAS BOGSJÖ[1] AND ANDERS FORSÉN[2]

SUMMARY

Road roughness is a major source of vehicle fatigue. To improve the understanding of vehicle durability, statistical methods are applied to characterise measured road profiles. Different statistical road models are used to generate corresponding synthetic road profiles. Vehicle fatigue is assessed utilising a simple quarter-vehicle model in combination with the Palmgren-Miner damage hypothesis, Basquin's relation and Rainflow counting. Several road realisations (Monte-Carlo simulation) provide an estimate of the expected fatigue damage. The results suggest that actual roads cause more damage than synthetic Gaussian roads, possibly due to occasional road transients (bumps and holes), causing large loads on the vehicle. Thus, a road model being the sum of Gaussian 'noise' and transient events is suggested.

1. INTRODUCTION

Road roughness is a major source of vehicle fatigue. Statistical analysis provides compact description of measured roads and offers the possibility to generate synthetic 'statistically equivalent' roads.

The overall target is to find a statistical, parametric road profile description, with as few parameters as possible, which can be used to generate synthetic road profiles for test and simulation purposes. Expected vehicle fatigue should differ less than, say, 10%, between actual and synthetic roads characterised by the same parametric description (a *very* ambitious target).

The study consists of statistical analysis of measured road profiles, Monte-Carlo simulation of road profiles and assessments of fatigue damage induced in vehicles.

The measured road profile data used in this study is highpass filtered prior to analysis, to remove longwave disturbances (hills), which are irrelevant to road-induced vehicle fatigue, and to remove spurious results caused by profile measurement system 'drift'.

---

[1] *Address correspondence to:* Department of Mathematical Statistics, Lund Institute of Technology, Box 118, SE-221 00 Lund, SWEDEN. E-mail: klas@maths.lth.se
[2] Scania, SE-151 87 Södertälje, SWEDEN.

2. ROAD PROFILE CHARACTERISATION

Reporting of road profile measurements is standardised by ISO [1]. Wavelengths between 0.1 and 100 m are considered relevant to road-induced vibrations.

Figure 1: Road profile sample.

Data from three road measurements are utilised. Profiles from left and right wheel-path are designated 'Road 1L', 'Road 2R' etc. Profiles are characterised by probability distribution and spectrum ("smoothed spectrum" according to ISO8608).

2.1. DISTRIBUTION OF ROAD ELEVATION

2.1.1. Upcrossing intensity
The upcrossing intensity is proportional to the probability density function if the derivative at all locations x, is independent of the value of the process (i.e. the profile height) at the same location x. This is the case for stationary Gaussian processes [2].

2.1.2. Empirical road level distributions

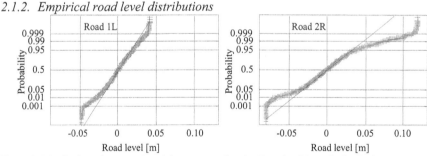

Figure 2: Distribution of measured road data compared to the Gaussian (Normal) distribution.

Figure 2 shows that the height distribution of Road 1L and Road 2R is very close to Gaussian for values above the smallest 2% and below the highest 2%. However, the highest and lowest values differ clearly from the Gaussian distribution, especially for Road 2 (and Road 3).

2.1.3. Non-Gaussian distribution
Gaussian distributions are convenient in statistical analysis, but real roads are not Gaussian, as shown in Figure 2. One way to handle this problem is to transform the actual distribution to a Gaussian, perform the statistical analysis on the

transformed data, and finally apply an inverse transformation. With this method, it is possible to design a synthetic road with statistical properties similar to the measured road's.

The transformation is estimated by applying a smoothing process to the actual (measured) road profile's distribution, an example is shown in Figure 3.

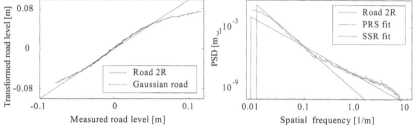

Figure 3: Smoothed transformation.   Figure 4: Estimated and fitted spectra.

2.2. SPECTRAL ANALYSIS

Four different ways to parameterize the measured road's spectra are investigated.

2.2.1. *PRS: Gaussian distribution, broad-band spectrum fit*

ISO8608 [1] suggests a straight line fit in log-log scale (Figure 4) to the power spectrum generated from the measured road profile. The resulting parametric road spectrum (PRS) is described by:

$$R_{PRS}(n) = \begin{cases} A \cdot \left(\dfrac{n}{n_0}\right)^{-w} & 0.01 < n \le 10 \\ 0 & \text{otherwise} \end{cases} \tag{1}$$

where n_0 is 0.1 m^{-1} and A indicates degree of road unevenness. Parameter values are determined by least-square fit over the spatial frequency range $0.011 - 2.83$ m^{-1}. For the three measured roads utilised in this study A varies around 10^{-5} and w around 2-3.

2.2.2. *SSR: Gaussian distribution, resonance-band spectrum fit*

An important range to fit correctly to the road spectrum is the frequency range containing the lowest resonance frequency of the vehicle, i.e. 1-2 Hz. When the vehicle travels at constant velocity v m/s, the corresponding spatial frequency range becomes: $\dfrac{1}{v} < n \le \dfrac{2}{v}$. $\tag{2}$

The shifted spatial range (SSR) spectrum is defined by least square fit in this range,

$$R_{SSR}(n) = \begin{cases} A \cdot \left(\dfrac{n}{n_0}\right)^{-w} & n_{start} < n \le 10 \\ 0 & \text{otherwise} \end{cases} \tag{3}$$

$$n_{start} = \max \left\{ \begin{array}{l} \left(10^{1-w} - \dfrac{1-w}{An_0^w} \sigma^2 \right)^{\frac{1}{1-w}} \\ 0.01 \end{array} \right. \tag{4}$$

followed by adjustment of the spectrums lower frequency limit to preserve the measured profile's variance σ^2, cf Figure 4.

2.2.3. DSE: Gaussian distribution, direct spectrum estimate

In order to be able to produce synthetic roads with similar power spectrum as real roads, the power spectrum of the actual road is smoothed and utilised for generation of synthetic road profiles. This method to generate synthetic roads with a power spectrum given by a Direct Spectrum Estimation is labelled DSE.

2.2.4. TrDSE: Transformed Gaussian distribution, direct spectrum estimate

The procedure outlined in chapter 2.1.2 is applied, an empirical transformation function is estimated from the measured road profile and utilised to produce a transformed road profile. This profile is then analysed and its spectral parameters are calculated with the DSE method described above. The road is described by:

1/ A transformation function (Figure 3).
2/ The direct spectrum estimate of the transformed road (Figure 4).

2.3. SAMPLING AND SIGNAL LENGTH

The highest spatial frequency of interest in the present study is 10 m^{-1}. (The tires are assumed to smooth disturbances shorter than 0.1 m, i.e. spatial frequencies higher than 10 m^{-1}.) Theoretically, the sample rate has to be at least twice the highest frequency in the signal to avoid aliasing. Thus, with appropriate filtering prior to digitalisation, a sampling frequency of 20 m^{-1} should be adequate, but proper description of transient's shape may well require higher sample rates.

Road 1 and road 2 are measured with a sample distance of 0.05 m, just about adequate for random (non-transient) road profiles. Unfortunately the sample distance in road 3 measurement is twice the theoretical minimum, 0.1 m.

To get a smooth and detailed response spectrum, even if the vehicle has narrow (undamped) resonance peaks and travels at high velocity, a fine spatial frequency resolution, maybe $4 \cdot 10^{-4}$ m^{-1}, is desirable. This sets the signal length requirement to (at least) 2.5 km. This is fulfilled by the studied roads, the measured distance on roads 1 and 2 being 5 and 5.8 km, while the measurement on road 3 covers 25 km.

Longer measurements reduce statistical uncertainty and scatter; longer measurements also increase the probability that several rare events (large transients) are included, thus enabling statistical analysis of the extreme events. An old MIRA investigation [3] concludes that in order to get stable statistics of road induced loads in vehicles, the measured distance should be at least 100 miles (160 km). However, this is not fulfilled by the three studied road samples, which are 5-25 km long.

3.   SYNTHETIC ROAD REALIZATION

A stationary zero-mean Gaussian process is uniquely defined by its spectrum. Thus, when the spectrum is known, any number of statistically equivalent realizations may be created. (The realization method is briefly described in the appendix).

The six measured and analysed road profiles provide parameter values to five different stochastic models of each road profile:

- PRS: Gaussian with parameters from a 'broad-band' evaluation of the real road.
- SSR: Gaussian with parameters from a Shifted Spatial Range evaluation.
- DSE: Gaussian with a direct spectrum estimation of the real road.
- Tr.DSE: Non-Gaussian, being the inverse transformed realization of a Gaussian process with a direct spectrum estimate of the transformed roads spectrum.
- SSRq: Non-Gaussian road, q artificial holes per km added to the SSR model.

A MATLAB toolbox, WAFO [4], is utilised to create 80 realizations of each road and model, in total $80 \cdot 6 \cdot = 2400$ synthetic road profiles. An actual, measured road profile is compared to corresponding realizations of synthetic roads in figure 5.

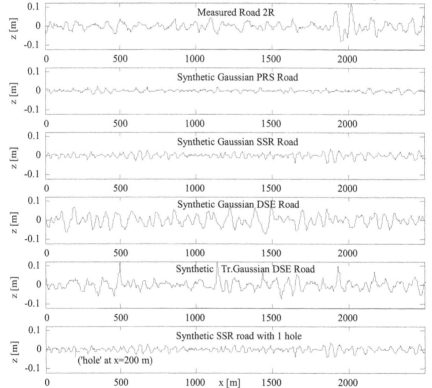

Figure 5: Measured road and synthetic roads.

Realizations of synthetic road profiles are given the same length as the measured profiles, 5.8, 5 and 25 km, respectively. Figure 5 is limited to the first 2.5 km to improve readability.

The Gaussian PRS and SSR road realizations appear different from the measured road. The Gaussian DSE and transformed Gaussian DSE roads are more similar to the measured profile, but still give a different impression, although they have the same variance, similar spectrum and similar upcrossing intensity as the actual road. Furthermore, the variance of the derivative of these processes also agrees with the measured roads'.

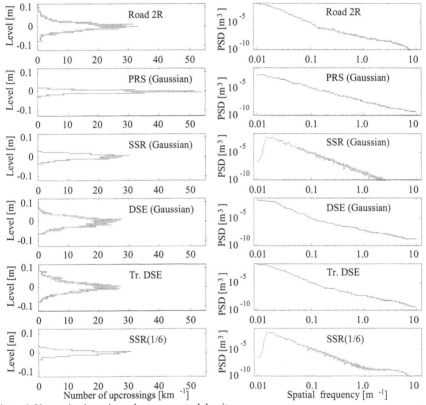

Figure 6: Upcrossing intensity and power spectral density.

Level crossings and PSD:s in Figure 6 result from analysis of one realization. Thus, they may differ somewhat from the theoretical distribution and parametrically described spectra of the corresponding road model.

4. VEHICLE FATIGUE ASSESSMENT

Fatigue damage is assessed by studying a quarter-car model travelling at constant velocity on (actual or synthetic) road profiles. This very simple vehicle model cannot be expected to predict loads on a physical vehicle exactly, but it will high-light the most important road characteristics as far as fatigue loading is concerned; it might be viewed as a 'fatigue load filter'. The utilised quarter-car model includes one non-linearity: it may lose road contact; otherwise it's linear with parameter values modelling a heavy truck.

The total force acting on the sprung mass is rainflow-counted and the resulting load cycles evaluated with Palmgren-Miners linear damage accumulation hypothesis. Fatigue strength is described by Basquin's relation, i.e. $s^\beta N = $ constant, where s is load cycle amplitude, β fatigue exponent and N number of cycles to failure. For vehicle components, β is usually in the range $3 - 8$, making it most important to describe load cycles with large amplitude accurately.

5. DURABILITY SIMULATION RESULTS

Quarter vehicle simulations are performed with three velocities, 15, 19 and 23 m/s. Fatigue damage is calculated for each of the resulting $3 \cdot (2400 + 6) = 7218$ load sequences, using 6 fatigue exponents, $\beta = 3, 4, \ldots, 8$. The 'Monte-Carlo' simulation result thus comprises 43308 fatigue damage values. All results are normalised with the fatigue damage indicated for the corresponding measured road profile.

The stochastic road modelling makes every realization of a road profile different, although it is based on the same model and parameter values. Naturally, the calculated fatigue damage will also vary from one road realization to the next. The *mean* result from a number of realizations provides an estimate of the *expected* fatigue damage on the studied stochastic road. Averaging all results from each input combination reduces the simulation output to 540 values and enables evaluation of the road models' performance as a function of physical road, fatigue exponent and vehicle velocity.

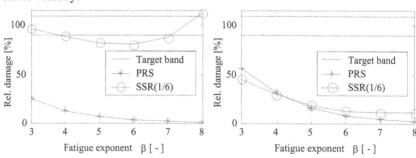

Figure 7 Relative damage, Road1R, v = 23 m/s.   Figure 8 Relative damage, Road1R, v = 15 m/s

Figure 7 and 8 illustrate a typical result, the PRS model underestimates the fatigue damage in all cases, the SSR(1/6) model performs quite well at 23 m/s on road 1R, but it is less satisfactory at 15 m/s. Detailed analysis indicates that this is due to a vehicle resonance, which occurs at 15 m/s on the measured road profile, but is absent at 23 m/s.

Figure 9 summarises the results of the study, the 108 mean relative damage values obtained for each road model are grouped according to calculated damage:

- less than half the damage indicated on the measured road (unsatisfactory)
- 50 – 90 % of damage indicated on the measured road
- 90 – 110 % of damage indicated on the measured road (on target)
- 110 – 200 % of damage indicated on the measured road
- more than twice the damage indicated on the measured road (unsatisfactory)

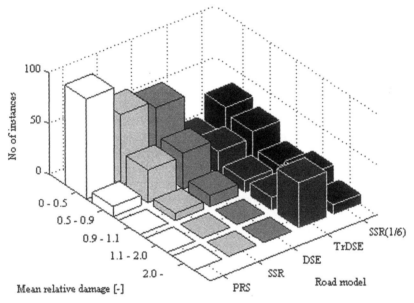

Figure 9: Monte-Carlo simulation results

Figure 9 shows that none of the investigated road model fulfils the ambitious target: mean relative damage 90 - 110%, irrespective of velocity and fatigue exponent. The purely Gaussian road models underestimate the imposed fatigue damage in almost every case, while non-Gaussian TrDSE model frequently overestimate the fatigue damage, often overshooting the target grossly (relative damage 10^5 in several cases). Usually the result deteriorates with increasing fatigue exponent β, as illustrated in figures 7 and 8.

6. CONCLUSIONS

- Gaussian models PRS, SSR and DSE give non-conservative fatigue estimates.
- The PRS model coincides with the method suggested in ISO8608 [1].
- The non-Gaussian TrDSE model produces very conservative fatigue estimates.
- Standard stochastic analysis, (transformed) Gaussian models, is insufficient.
- A prototype model SSRq (Gaussian 'noise' with added transients) is promising.
- It is quite difficult to create a synthetic road that causes the correct amount of fatigue damage, probably due to fatigue's sensitivity to occasional large load cycles, i.e. occasional transients in the road profile.

7. DISCUSSION AND COMMENTS

Fairly good roads, on which a 'normal' driver keeps essentially constant velocity, are investigated. Bad and inhomogeneous roads, where the driver adapts his driving to the varying conditions, may require a more sophisticated approach.

Sampling distances in road profile measurements should be quite small to catch road transients, preferably in the range 0.005 – 0.01 m, but conventional sampling distances are 0.05 – 0.1 m. This makes it difficult to investigate shape, size and frequency of transient events ('holes' and 'bumps') in actual roads.

Synthetic roads with manually added 'holes' display an interesting behaviour, especially when β is large, as illustrated in Figure 7. This suggests a way to deal with the problem of occasional road transients and corresponding large load cycles.

The unsatisfactory performance of the transformed Gaussian DSE road model might be due to shortcomings in the transformation function, Figure 5 indicates that the TrDSE model produces too many large peaks. Another possible cause of the difficulties is that the transformation procedure assumes (strictly) stationary data. If this assumption is not fulfilled, the transformation function should be altered along the road. Longer measurements increase the likelihood of non-stationary data, thus increasing the modelling difficulty, but longer measurements also improve the chance of creating a realistic model, by decreasing the statistical uncertainty and forcing the analyst to consider the physical reality of non-stationary roads.

ACKNOWLEDGEMENT

The support and encouragement offered by professors Georg Lindgren and Igor Rychlik, Mathematical Statistics, Lund Institute of Technology, is gratefully acknowledged.

REFERENCES

1. ISO8608. *Mechanical Vibration - Road Surface Profiles - Reporting of Measured Data.* International Organization for Standardization, Geneva, 1995.
2. Rice, S.O.: *The mathematical analysis of random noise.* Bell Syst Techn. J., Vol 23 (1944), pp 282-332.
3. Drury, C.G. and Overton, J.A.: *Vehicle service loads. Part I: A preliminary study of stress level counting.* MIRA report No 1964/8.
4. WAFO, a MATLAB Toolbox for Analysis of Random Waves and Loads. http://www.maths.lth.se/matstat/wafo/, Lund University, 2000.
5. Wittig, L.E. and Sinha, A.K.: *Simulation of multicorrelated random processes using the FFT algorithm.* Journal of Acoustical Society of America, Vol. 58, No 3 (1975), pp 630-634.
6. Hudspeth, R.T. and Borgman, L.E.: *Efficient FFT simulation of Digital Time Sequences.* Journal of the Engineering Mechanics Division, ASCE, Vol 105, No EM2 (1979), pp 223-235.

APPENDIX: SYNTHETIC ROAD REALIZATION

Two kinds of distributions are used to create synthetic roads, Gaussian and transformed Gaussian. To create a transformed Gaussian processes, a transformation function is applied to the generated Gaussian data. All calculations are performed with MATLAB and the WAFO toolbox [4]. The method to generate realizations of a Gaussian road with given spectrum is described briefly below, see [5] and [6] for a detailed explanation.

To create a Gaussian road realisation with K samples, a set of K independent Gaussian random numbers $\zeta_k = \xi_k + i\eta_k$ is created, such that $E(\xi_k) = E(\eta_k) = 0$ and $E(\xi_k^2) = E(\eta_k^2) = 0.5$. Next, a vector $Z = [Z_1, ..., Z_N]$ is defined, where

$$Z_k = \sqrt{\frac{K}{2h}} a_k \zeta_k, \quad k = 0, ..., \frac{K}{2} - 1. \tag{A1}$$

The second half of vector Z is found from the property $Z_{K-k} = Z_k^*$, where * denotes complex conjugate. h is the sample distance in the realisation and a_k is given by:

$$a_k = \int_{n_k - 1/(2Kh)}^{n_k + 1/(2Kh)} R(n) dn \tag{A2}$$

where $R(n)$ is the one-sided spectrum. The realisation of the Gaussian process is obtained by taking the inverse FFT of Z:

$$z(x_j) = \frac{1}{K} \sum_{k=0}^{K-1} Z_k e^{\frac{i2\pi kj}{K}}, \quad j = 0...K-1, \quad x_j = jh. \tag{A3}$$

Vehicle System Dynamics Supplement 41 (2004), p. 734-743   © Taylor & Francis Ltd.

Rough road simulation with tire model RMOD-K and *FTire*

A. RIEPL, W. REINALTER, G. FRUHMANN[1]

MAGNA STEYR Fahrzeugtechnik, GRAZ / AUSTRIA

SUMMARY

This paper deals with the modeling steps – vehicle model, tire model, road surface description and vehicle control – to realize a virtual rough road ride. The calculation results achieved by means of the physical tire models RMOD-K and *FTire* are compared to six measurement rides, i.e. that the measurement variation range occurring during real rough road rides is taken into consideration. The load path analysis also includes the suspension strut force and the engine mount force. Finally, a few user-relevant comments are made regarding the tire models.

1. INTRODUCTION

In this paper, the stochastic, road-excited vehicle vibrations occurring during rough road rides and causing strength-relevant component loads are analyzed. The results of the simulation carried out with the physical tire models RMOD-K [1] and *FTire* [2] are compared to six measurement rides performed with one and the same vehicle on the MAGNA STEYR Fahrzeugtechnik test track. Based on the calculated and measured quantities

- longitudinal, lateral and vertical wheel forces $F_{Wi,x}$, $F_{Wi,y}$, $F_{Wi,z}$
- suspension strut force $F_{Si,z}$
- engine mount force $F_{Ej,z}$

the simulation quality is analyzed in consideration of the measurement variation range – a very significant aspect – and the effect of the tires.

Generating the vehicle model, however, requires detailed consideration of the components damper, rubber/hydro mount, jounce and rebound stop and their parameterization.

---

[1] Department of MBS-Simulation MAGNA STEYR Fahrzeugtechnik, A-8041 Graz
Tel.: +43 (0)316 404-5224, Fax: +43 (0)316 404-2356
E-Mail: anton.riepl@magnasteyr.com

2. GENERATING OF COMPLETE-VEHICLE MODEL

This chapter deals with the essential modeling steps to realize a virtual rough road ride in the software environment ADAMS/Car [3].

2.1 VEHICLE MODEL

The setup of the hybrid vehicle model shown in Figure 1a is object-oriented and consists of the modules

- front axle: double-wishbone axle
- steering system with speed-sensitive power assistance
- rear axle: multi-link suspension
- powertrain: 4WD with constant torque distribution in the transfer case, flexible exhaust system
- rigid body: The local elasticities of the body in the damper domes are taken into consideration by applying additional elasticities.
- tire model: RMOD-K or *FTire*, see Chapter 2.2

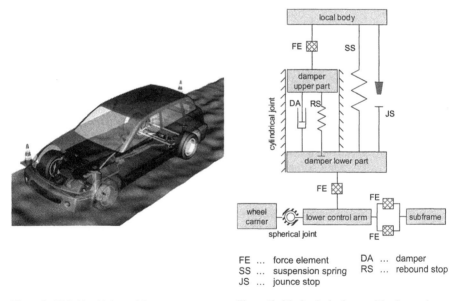

FE ... force element	DA ... damper
SS ... suspension spring	RS ... rebound stop
JS ... jounce stop	

Figure 1a: Hybrid vehicle model.   Figure 1b: Mechanical scheme of the front axle.

Figure 1b shows the mechanical scheme of the double-wishbone axle. A similar modeling depth and parameterization was selected for the rear multi-link suspension.

On the rough road track of MAGNA STEYR Fahrzeugtechnik, the vertical wheel forces dominate as against the longitudinal and lateral wheel forces, i.e. that the vertical transfer behavior of the wheel suspension is of utmost importance. In Figure 1b, the wheel carrier is connected to the lower control arm by means of a spherical joint. The lower control arm is connected to the subframe (fixed to the body) by means of two three-dimensional force elements (FE). The damper lower part is elastically mounted to the control arm by means of a force element. The damper upper part is connected with the damper lower part by means of a cylindrical joint. Between the two parts, two one-dimensional force laws act as damper (DA) and rebound stop (RS). The damper upper part is connected to the body (locally) by means of a three-dimensional force element. The suspension spring and the jounce stop (JS) act between the local body and damper lower part, see Figure 1b.

The stabilizer consists of two rigid bodies coupled together by a revolute joint. A linear torsion spring represents the torsional stiffness of the stabilizer. The wheel carrier is connected by means of a coupling rod.

The damper behavior is approximated by a non-linear, speed-sensitive characteristic curve derived from a VDA damper measurement at test speeds up to ±2.0 m/s. The static gas force of the damper was taken into consideration, the friction forces and temperature effects in the damper were neglected. Due to the ride specifications and the extreme wheel load variations, these effects are of minor significance.

The extremely non-linear behavior of jounce and rebound stop is represented by a characteristic curve in dependence of the respective relative displacement. The input data for the stops are derived from rig test measurements of the relevant component. When determining the damage-relevant component loads, the exact definition of the application points for jounce and rebound stop is of utmost significance.

The force elements in Figure 1b are the equivalent to the bushing elements in ADAMS/Car. The characteristic curves of the wheel suspension mount are derived from test rig measurements and a FEM-simulation for the local additional body stiffness and the additional torsional stiffness of the stabilizer, respectively.

Figure 2 shows the mechanical scheme for the (spatially) elastically supported 4x4-powertrain. The rigid engine/transmission unit (incl. the transfer case and the front axle drive) is three-point mounted to the rigid body. Both engine mounts are hydro mounts [4], the transmission mount is a rubber mount. Since during driving on the MAGNA STEYR Fahrzeugtechnik rough road the eigenforms of the exhaust system are excited and significantly effect the vibration behavior of the engine/transmission unit, it is necessary to integrate this influence into the vehicle model. The front part of the twin pipe FEM-exhaust system with a degree of freedom of 63 is mounted to the engine/transmission unit by means of four fixed joints. The rear part is mounted to the body by means of bushing elements.

The rear axle differential is supported by three rubber mounts attached to the rigid subframe. The latter is connected to the body by four rubber mounts. The driveshaft to the rear axle consists of two parts being interconnected by a translational joint with friction in longitudinal direction. The driveshaft itself is elastically attached to the body by means of an intermediate mount, see Figure 2. Owing to the resulting friction force, the drive torque required to meet the driving specifications is high enough to keep the length compensation constantly closed. Apart from considering the elastic exhaust system, this is another significant effect to simulate the proper vibration behavior of the engine/transmission unit.

The engine torque T_{Engine}, which is saved in the form of a characteristic field as function of speed and throttle position, acts on the flywheel that is rotatably supported in the engine/transmission unit.

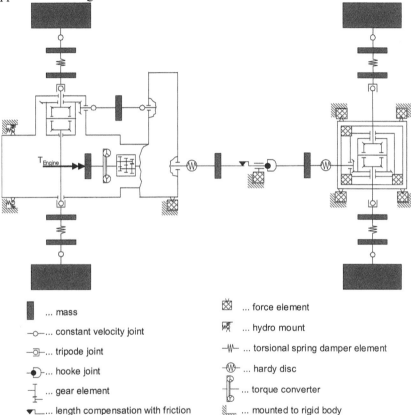

Figure 2: Mechanical scheme of the (spatially) elastically supported powertrain (shown without exhaust system).

Further torque is transferred by the torque converter that is closed in such a case. The transfer case is modeled as open differential and distributes the input torque in a constant ratio to the two outputs. Apart from the kinematic connections, the automatic transmission and the transfer case also take the characteristics of the masses into consideration. The same applies to the axle differentials supported in the axle transmissions. From there the drive torque is transferred via the drive shafts to the rigid body "rim", which the tire model RMOD-K or *FTire* is coupled to. The drive shafts are modeled as torsional dual-mass vibration elements.

Modeling of the powertrain is rather sophisticated due to the fact that comfort analyses are part of a complete-vehicle development carried out with one and the same vehicle model. For these analyses, the kinematic support of the individual shafts relative to each other plays an important role.

2.2 TIRE MODELS RMOD-K AND FTIRE

The tire model RMOD-K [1] denotes a model family. For this work, RMOD-K 31 (Version 6.0) was used. This is a structure-dynamic model consisting of four circular rings with mass points that are also coupled in lateral direction. Because multi-track contact and pressure distribution across the belt width are taken into consideration, version '31' of the RMOD-K family is suitable for rough road rides. To reduce calculation time, the road surface data are provided by a road interface developed by MAGNA STEYR Fahrzeugtechnik. The tire model RMOD-K is integrated parallelly to the vehicle model in ADAMS/Car. RMOD-K provides the rim forces and rim torques that are transferred to ADAMS/Car via gforce-elements at defined times.

In case of *FTire* [2] (built May 19, 2003) the tire belt is modeled as an extensible and flexible ring considering the bending stiffnesses and elastically mounted to the rim by distributed, partially dynamical stiffnesses in radial, tangential and lateral direction. The degrees of freedom of this ring are such that rim in-plane as well as out-of-plane motions are possible. The ring is numerically approximated by a finite number of rigid 'chain-links', the belt elements, which are coupled by stiff springs and by bending stiffnesses both in-plane and out-of-plane. To every belt element, a number of mass-less 'tread blocks' is assigned. These blocks carry non-linear stiffness and damping properties in radial, tangential and lateral direction. The radial deflections of the blocks depend on road profile, locus and orientation of the assigned belt elements. Tangential and lateral deflections are determined by the sliding velocity on the ground and the local values of the sliding coefficient. *FTire*, too, provides the vehicle model in ADAMS/Car with the rim forces and rim torques at pre-defined time steps. Data from the road surface are communicated by the same road interface used for RMOD-K. For *FTire*, an implicit integration process is used, which works parallel to the ADAMS/Car integrator.

2.3 ROAD SURFACE DESCRIPTION

By means of the power spectral density (PSD), road surfaces can be assigned to a track describing the proportion of individual wave lengths in the overall profile.

Figure 3 shows the PSD-analyses for the used virtual rough road, a highway and the reference values of ISO 8608. The power spectral density of the virtual rough road generated on the basis of measured data [4] was almost over the whole relevant wavelength range more than hundred times higher than that of the highway. Above all wavelengths between 0.3 and 2 m are very pronounced on a rough road and – at vehicle speeds between 20 and 25 km/h – excite the frequency range from 2.8 to 23 Hz.

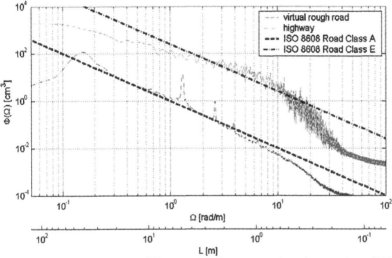

Figure 3: Power spectral density of different road surfaces compared to reference values of ISO 8608 [5].

2.4 VEHICLE CONTROL

The driver model is responsible for keeping to the pre-specified speed (longitudinal control) and for following the pre-specified course (lateral control). The constant, pre-specified vehicle speed of 22 km/h is controlled by the throttle position which influences the engine torque acting on the flywheel.

The two-level model of Donges [6] is applied for the lateral control of the vehicle. The first level, the anticipatory open loop control, represents the driver of the vehicle who follows the course of the road. The steering angle remains the same. The second level, the compensatory closed loop control, compares the reference position of the vehicle's center of gravity to the position estimated by the driver at a certain distance. The difference of these two values is the input value for the control. The design of the driver control and the subsequent integration into ADAMS/Car is referred to in [7].

3. LOAD PATH ANALYSIS, COMMENTS TO RMOD-K AND *FTire*

The hybrid vehicle model introduced in Chapter 2.1 and the tire models RMOD-K or *FTire* were used for virtual rides on the 460 m long, concreted MAGNA STEYR Fahrzeugtechnik rough road. In compliance to the specification, the vehicle speed was 22 km/h. The calculation results are – as already mentioned above – compared to six measurement rides on the real rough road and with a suitable vehicle fitted with four wheel force transducers and further sensors. This provides information regarding the measurement variation to be expected, which is of great interest when generating and parameterizing a model.

Due to the stochastic road surface excitation, the results are processed on the basis of the level crossing method according to DIN 45667. The method is able to characterize the load peaks and their cycles, i.e. the load shape, without any further characteristic quantities.

The following load path analysis begins with vertical, longitudinal and lateral forces occuring in the center of the left front wheel, see Figure 4, 5 and 6. The reference value for the diagrams is the vertical wheel load $F_{W1,z0}$ of wheel 1 at vehicle standstill. The right-oriented, cartesian coordinate system is wheel-fixed and does not rotate. The x-axis points rearwards, the z-axis upwards.

Figure 4 shows the standardized, vertical wheel forces of the individual measurements as against the calculations with tire models RMOD-K and *FTire*. Within the jounce range (level crossings greater than one), simulation and measurements correspond rather well. The difference between the results achieved with RMOD-K and *FTire* can be considered to be neglectable. Within the rebound range, there are differences between the two simulation results. The RMOD-K-result, however, remains within the variation range of the measurements nearly over the whole cycle range.

Figure 5 shows the standardized longitudinal wheel forces. At a cycle value smaller than ten, both tire models deviate from the measurement results. The difference between the calculation results is very small. A detailed analysis showed that the small deviations can be attributed to a few but relatively steep obstacles on the virtual road surface.

The results regarding the lateral transfer behavior of RMOD-K and *FTire* are compared to the measurement results – Figure 6. The complete-vehicle simulation with RMOD-K corresponds to the measured forces. Due to the simpler modeling of *FTire*-model in lateral direction, there are some deviations which, however, are of no significance. The influence of the vehicle`s lateral control – due to the slightly different trajectories – must also be taken into consideration. The comparison of the load spectrum forms clearly showed that the vertical loads dominate, see Figures 4, 5 and 6.

The suspension strut measuring point left front is the next position of the load path analysis, refer to Figure 1b. The standardized suspension strut force shown in Figure 7 is a sum of forces acting on the damper lower part and mainly consisting of the components spring, damper, jounce stop and rebound stop. The z-component points into the direction of action of the damper. Both times, jouncing corresponds

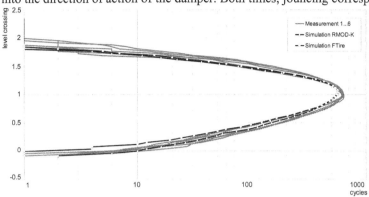

Figure 4: Standardized, vertical wheel force ($F_{W1,z} / F_{W1,z0}$) of left front wheel.

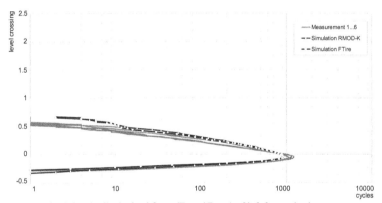

Figure 5: Standardized, longitudinal wheel force ($F_{W1,x} / F_{W1,z0}$) of left front wheel.

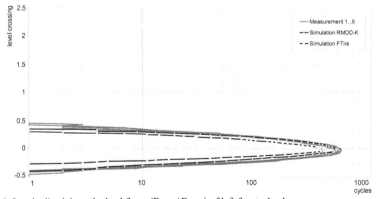

Figure 6: Standardized, lateral wheel force ($F_{W1,y} / F_{W1,z0}$) of left front wheel.

rather well. The results for rebound with *FTire*, however, lie at the lower variation range limit, those of RMOD-K already slightly beyond the lower limit. The vertical wheel forces differ, too – see Figure 4.

Analysis of the left engine mount force, which is transferred to the body, refers to the mount preload $\left| F_{E1,z0} \right|$ and is shown in Figure 8. The majority of the calculated mount forces correspond well to the measured ones. However, if high peak loads occur, there is a distinct difference between measured and calculated results. The difference between simulated and measured load spectra of this analysis point can be attributed among other things to the simplified description of the engine mount – linear hydro mount model [4].

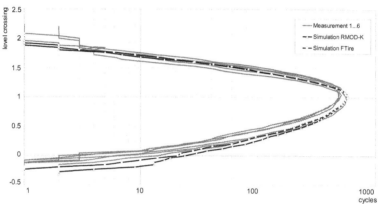

Figure 7: Standardized suspension strut force ($F_{S1,z}$ / $F_{S1,z0}$) left front.

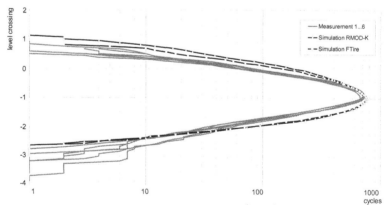

Figure 8: Standardized vertical force in left engine mount ($F_{E1,z}$ / $\left| F_{E1,z0} \right|$).

Conclusion: Both tire models are suitable for virtual rough road rides. Although the models RMOD-K and *FTire* are quite different, the excitation caused by this road surface showed only minor differences in the results. Apart from the evaluation with regard to the quality of the results, a few more criteria are relevant to an efficient simulation when developing a vehicle:

- A detailed measurement specification for parameterizing the tire model, [2].
- User-friendly and stable integration of the tire model into the software environment.
- Calculation time efficiency: The simulation technique on hand is very suitable for relative data, e.g. the effects of a modified axle concept, changed spring/damper tuning [4], etc. Usually the long calculation times in the course of complete-vehicle simulation are caused by the physical tire model. In this respect, *FTire* is considerably better than RMOD-K.
- Documentation, hotline

By way of conclusion it can be stated that for recent virtual rough road tests on MAGNA STEYR Fahrzeugtechnik`s proving ground tire model *FTire* has been used. Further and more comprehensive evaluation results for the current tire models will certainly be achieved at the 3rd International Tyre Colloquium [8].

LITERATURE

[1]   Oertel Ch., Eichler M., Fandre A., „Modellsystem zur Simulation des Reifenverhaltens bei Überrollen kurzwelliger Bodenunebenheiten", gedas-usermanual, 1999.

[2]   Homepage FTire, http://www.ftire.com.

[3]   Homepage MSC, http://www.mscsoftware.com.

[4]   Riepl A., Fruhmann G., Reinalter W., "Rough Road Simulation – A simulation tool to reduce the development risks", 6th International Symposium on Advanced Vehicle Control, Hiroshima 2002.

[5]   ISO 8608: Mechanical vibration – Road surface profiles – Reporting of measured data. International Standard (ISO) 1995.

[6]   Donges E., „Experimentelle Untersuchung und regelungstechnische Modellierung des Lenkverhaltens von Kraftfahrern bei simulierten Straßenfahrten", Dissertation TH Darmstadt 1977.

[7]   Rauscher A. „Entwurf und Simulation eines Fahrerreglers für Kurvenfahrt auf ebener Strecke", Diplomarbeit TU Wien 2002.

[8]   Homepage 3rd International Tyre Colloquium, http://tmvda.tuwien.ac.at.

Vehicle System Dynamics Supplement 41 (2004), p.744-755 © Taylor & Francis Ltd.

Identification of transient road obstacle distributions and their impact on vehicle durability and driver comfort

F. ÖIJER[*+] AND S. EDLUND[*]

SUMMARY

This paper presents methods that enable identification of the distribution of transient obstacles in a set of public roads. The roads are considered as non-stationary random profiles with superimposed transients of varying depths or heights and lengths.

The transient obstacles are detected from both road profiles and vehicle responses by use of wavelet analysis.

The impact of the identified distributions on vehicle durability and driver comfort shows a clear potential for differentiating the designs for different road conditions.

1. INTRODUCTION

The excitation of a road vehicle can be described by environmental parameters, such as, road profile, road curvature, topology, and driver influenced parameters, such as, speed changes and manoeuvring. Of these, the most important parameter for the durability of most components and driver comfort is the road profile.

For computer simulations, it is desirable to have synthetic descriptions of the road surfaces. Real road profiles cannot be described solely by random noise [1], but to correctly describe road profiles, combinations of random noise roads and transient obstacles (potholes, bumps, etc.) should be used to get a balanced excitation of all components [2].

Distributions of stationary random roads have previously been established for different markets and applications [3-4]. This paper is concerned with identification of distributions of transient road contents. To identify the transients, methods based on wavelet analysis are used.

The influence of transients on vehicle damage and driver comfort is exemplified by using different road profiles and road descriptions as input for vehicle simulations. The necessary responses are calculated with finite element models of complete vehicles [5].

---

[*] Department of Chassis Strategies & Vehicle Analysis, Volvo 3P, S-405 08 Gothenburg
[+] Corresponding author: E-mail: fredrik_oijer@vtc.volvo.se, Phone: +46 31 666710

2. IDENTIFICATION OF TRANSIENT ROAD OBSTACLES

The most reliable method to find transient road obstacles is to analyse measured road profiles directly. But as the number of measured roads is limited, one has also to rely on estimating transients from measured vehicle responses, which typically are more numerous. The estimation of transients from vehicle responses will inevitably be more uncertain, but has one advantage in that the vehicle speed when passing the transient event may be identified.

2.1. IDENTIFICATION FROM MEASURED ROAD PROFILES

Wavelets can successfully be used for de-noising of signals, i.e. removing noise from a measurement and retaining the true signal [6]. As it here is believed that roads can be seen as combinations of random roads and transient obstacles, de-noising of a road profile should leave only the transients.

The de-noising procedure is shown in Figure 1 for an excerpt of a measured road profile. The first step is to calculate the wavelet coefficients for the signal. The analysing wavelet used here is the Daubechies-4. In the second step, the noise on each resolution level is estimated from the median absolute wavelet coefficient value [7]. Based on that estimate the coefficients are shrunk towards zero, thus minimizing the noise. The third step is to inverse transform the modified coefficients, leaving a noise-free signal. The wavelet analysis has been performed with WaveLab [8].

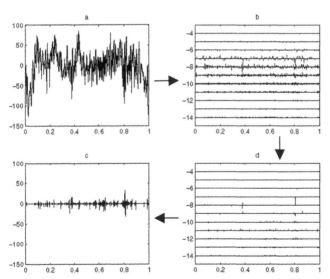

Fig. 1. (a): Part of measured road profile in northern Sweden. (b): Wavelet coefficients for road profile. (c): Reconstruction of de-noised road profile. (d): De-noised wavelet coefficients for road profile.

From the de-noised signal, transient obstacles can be identified and sorted according to type and size. Four different categories are used: potholes, bumps, and if full potholes and bumps cannot be identified: up-steps (half bump) and down-steps (half pothole). In each category the transients are sorted according to depth (or height) and length. Examples of identified pothole distributions sorted according to depth are shown in Figure 2 for a set of US highways [9] and two Swedish roads.

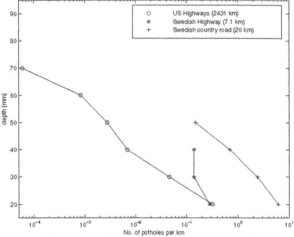

Fig. 2. Identified pothole distributions for different roads. Note that the Swedish measurements are over a very limited distance.

The measured US highways and the Swedish highway are considered to be of mainly good quality, while the measured Swedish country road is considered to be very poor. There is a clear difference between the good roads and the poor in terms of the number of potholes in the roads. This can be utilised as a potential for differentiation of the design with respect to weight and cost.

2.2. IDENTIFICATION FROM MEASURED FRONT AXLE ACCELERATIONS

Estimation of transient road profiles from vehicle responses is a difficult subject. A common approach is to select a set of relevant responses, and run a simulation model over a transient obstacle and calculate the responses. The process is iterated by changing the obstacle shape in an intelligent way, until a certain level of accuracy is achieved [10]. This method is usually very time consuming, and has to be repeated for each event of interest. Moreover, as the tyres are likely to be in the air during part of the event, the exact shape of the obstacle can never be determined, no matter which vehicle responses are used.

In this work the focus is more on the distributions of the transients, rather than their exact shape, so a different approach is used. A simulation model of the vehicle used in measurement is used to calculate the responses for a limited number of idealised obstacles. A few shapes were tested, before it was decided to give the obstacles elliptic shape (see Fig. 3). The lengths of the obstacles are varied between 0.8 – 2 m, and the depth/height between 20 – 120 mm. The obstacles were analysed for speeds varying from 20 km/h to 120 km/h in steps of 10 km/h, which covers the speeds found in the measurements. In all, 42 analyses have to be performed per obstacle type and speed. The run time for each such analysis is about 15 seconds.

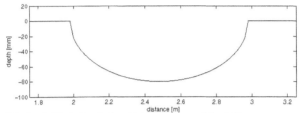

Fig. 3. Example of idealised elliptical pothole with length 1 m and depth 80 mm.

Together with vehicle speed, the only responses used in the identification of the obstacles are the vertical front axle accelerations on left and right hand sides. Mainly this is because they were the only responses available on both sides of the vehicle for all measurements. The measured accelerations are wavelet transformed and de-noised in the same way as the road profile, so that only the transient acceleration responses are left. The transient accelerations are then compared to the ones calculated for the idealised obstacles and if a match is found, the road obstacle is said to be approximately determined.

To test the method, a vehicle model was used to calculate the front axle accelerations for one of the measured road profiles, so that the resulting pothole distributions found by the two methods could be compared. The distributions are plotted in Figure 4. The pothole depth seems to compare quite well to the potholes found directly from the road profile, and on the whole the results are good enough for practical use.

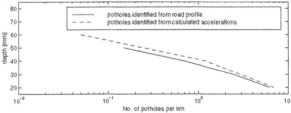

Fig. 4. Comparison of pothole distributions determined from road profile and accelerations for one road.

Four different sets of measured accelerations have been analysed with the above method. Three of the sets were measured on roads that by the drivers were described as good, the fourth on poor roads. The pothole distributions for the roads are plotted in Figure 5. If compared with the distributions determined from the road profiles directly, it can be seen that the distributions found for good roads compare quite well for both methods. The same is true for the poor roads. Even though the described method calculates the road surface indirectly and approximately it is believed that the information gained by this method is of great value when trying to determine distributions of transient obstacles.

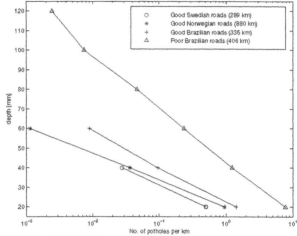

Fig. 5. Pothole distributions identified from measured accelerations.

However, all transient accelerations will not be identified as coming from one of the idealised obstacles. These accelerations are put in a residual and their maximum and minimum amplitudes are matched to the accelerations from the idealised obstacles regardless of the shape of the transient acceleration. Identified and residual potholes are plotted in Figure 6 for one good and one poor road.

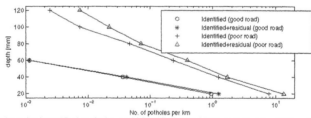

Fig. 6. Distributions for identified potholes and identified + residual potholes for different roads.

For the good road the residual is very small, i.e. most transients are properly identified. For the poor road, however, the number of residual transients is of same size as the number of identified. Part of the reason for the size of the residual, may be that interaction between transient obstacles becomes a factor on poor roads. Another reason is that the so-called up-steps and down-steps have not been included among the idealised obstacles, but only potholes and bumps. If the steps are included the residual will most likely decrease. Other ways to minimise the residual may be use more basic shapes and to include more responses in the analysis, or perhaps to use different methods altogether, for example neural nets and/or fuzzy logic. In the end it comes down to having a balance between computational efficiency and quality of the achieved results.

3. VEHICLE MODEL

The vehicle model in this investigation is based on finite elements with certain extensions to allow for more correct dynamic behaviour.

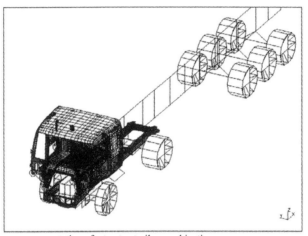

Fig. 7. Finite element representation of a tractor-trailer combination.

The model consists of a flexible frame modelled by shell elements, and cab, engine and axle suspensions modelled with springs, dampers and beams.

In this work, the tyres have been modelled as point-contacts with the possibility to loose contact with the road. As the focus of this work is on potholes and bumps, the longitudinal tyre model is of great importance. The implemented model is similar, but not identical, to the one presented in [11]. The models are similar in that they both pre-process the road profile by rolling a flexible tyre over the surface at crawl speed, prior to the actual response analysis. They are different in the way the parameters are determined, which is done more empirically here. The tyres have linearised slip models in both longitudinal and lateral directions.

Typical correlation for this type of model traversing a pothole is plotted in Figure 8.

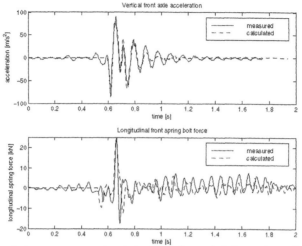

Fig. 8. Correlation between measurement and simulation for vertical front axle acceleration (top) and longitudinal front spring bolt force (bottom).

4. SIMULATION RESULTS

The vehicle simulations have been performed using modal transient analysis in MSC.Nastran [12].

4.1. VEHICLE DURABILITY

The influence of transients is exemplified using a 1.1 km section of a freeway. The road profile was measured on four different occasions, twice before it was resurfaced and twice afterwards. The number of identified transient obstacles in the road profiles is listed in Table 1.

Table 1. Number of identified transient obstacles in measured road profile

Date of road profile measurement	No. of transients
1991-05-11	23
1992-04-14 (before resurfacing)	24
1993-04-06 (after resurfacing)	1
2000-11-10	10

In Figure 9, the calculated longitudinal front spring bolt force is plotted for the four different road profiles. Even though the general vibration level is higher prior to resurfacing, most of the damage is caused by the transients. This becomes quite clear if the two uppermost signals in Figure 9 are studied. The damage ratio calculated for a slope of 5 in the S-N-curve before and after resurfacing is about 7000 for this force. Similar ratios can be seen for other component forces.

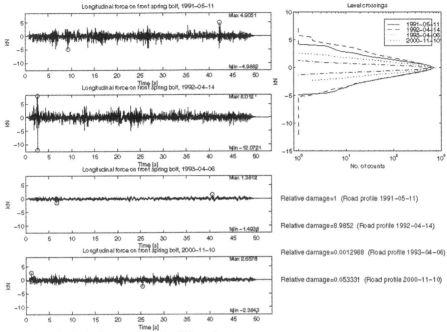

Fig. 9. Longitudinal cab mount force for different inputs.

To give a clearer view of the influence the transients have on vehicle responses, the road measurement giving most damage was analysed in more detail. Two alternative descriptions of it were generated, one random noise road with the same severity as the real road profile minus the identified transients, and one with the identified transients included in the form of idealised obstacles also included. Calculated damage results are presented in Table 2. The results show quite clearly that random input alone is incapable of representing road surfaces for durability assessments, but with a limited number of transient shapes included results are much improved.

Table 2. Calculated damage for different descriptions of road measurement 1992-04-14.

Road description	measured road	Synthetic roads	
		generated noise	generated noise + idealised transient obstacles
vertical front axle acceleration	1.0	0.008	0.36
vertical fuel tank bracket force	1.0	0.17	1.38
vertical battery box force	1.0	0.09	0.71
longitudinal front spring force	1.0	0.026	0.5
longitudinal rear engine force	1.0	0.37	4.3
vertical rear engine force	1.0	0.1	2.4
lateral front cab force	1.0	0.09	2.9

4.2. DRIVER COMFORT

Truck driver comfort is to a large degree defined by the cab accelerations. The same roads that were used to exemplify influence on vehicle damage is used here. Also, two different suspension systems have been used for rear axle and cab to highlight what type of improvements can be expected by using softer suspensions. It was found that the differences in responses before and after resurfacing are much greater than can possibly be achieved by changing to a softer suspension type. The softer suspension reduces accelerations with about 6 - 20 %, which is a clearly noticeable difference. However, the resurfaced road reduces the accelerations by a factor three or more. Calculated accelerations weighted according to ISO 2631 [13] are summarized in Table 3.

Table 3. Cab floor accelerations weighted according to ISO 2631.

Date of road profile measurement	Suspension types: front axle/rear axle /cab	Vertical acc. [m/s²]		Longitudinal acc. [m/s²]	
		RMS	Max	RMS	Max
1992-04-14 (before resurfacing)	steel/steel/steel	0.45	4.6	0.47	4.7
1992-04-14 (before resurfacing)	steel/air/air	0.41	3.8	0.37	3.2
1993-04-06 (after resurfacing)	steel/steel/steel	0.15	0.71	0.11	0.66

The synthetic descriptions of the road surface prior to resurfacing have been analysed for comfort as well. As can be expected, the transient obstacles do not contribute much to the RMS value, but only the maximum response values (see Table 4).

Table 4. Cab floor accelerations weighted according to ISO 2631 for different road descriptions of road measurement 1992-04-14.

Road type	Vertical acc. [m/s²]		Longitudinal acc. [m/s²]	
	RMS	Max	RMS	Max
measured road profile	0.45	4.6	0.47	4.7
generated noise road	0.35	1.5	0.39	1.6
generated noise + idealised transients	0.50	4.2	0.52	4.3

5. CONCLUSIONS

Transient obstacles in roads do have a significant impact on vehicle damage and driver comfort, and it is therefore necessary to know about their distributions when designing road vehicles.

The transient response measurements using wavelet analysis.

To minimise vehicle events can be identified from both road profiles directly and indirectly from vehicle damage and driver fatigue caused by vibrations, soft road-friendly suspensions may be used, but the importance of proper maintenance to keep road surfaces from deteriorating should not be underestimated.

REFERENCES

1.   *Bruscella, B., Rouillard, V. and Sek, M.: Analysis of Road Surface Profiles, J. Transportation Eng., Vol. 125, No. 1, pp. 55-59, 1999.*

2.   *Öijer, F. and Edlund, S.: Complete Vehicle Durability Assessments Using Discrete Sets of Random Roads and Transient Obstacles Based on Q-distributions. True, H. (Ed.): The Dynamics of Vehicles on Roads and Tracks, Proc. 17th IAVSD Symposium, Copenhagen, Denmark, 2001, Swets & Zeitlinger, Lisse, 2003.*

3.   *Aurell, J. and Edlund, S.: Operating Severity Distribution a Base for Vehicle Optimization. Anderson, R. (Ed.): The Dynamics of Vehicles on Roads and Tracks, Proc. 11th IAVSD Symposium, Kingston, Canada, 1989. Swets & Zeitlinger, Lisse, 1989.*

4.   *Aurell, J. and Edlund, S.: Prediction of Vibration Environment in Real Operations Described by Q-distributions. Segel, L. (Ed.): The Dynamics of Vehicles on Roads and Tracks, Proc. 14th IAVSD Symposium, Ann Arbor, USA, 1995. Swets & Zeitlinger, Lisse, 1996.*

5.   *Öijer, F.: FE-based Vehicle Analysis of Heavy Trucks; Part II: Prediction of Force Histories for Fatigue Life Analysis, Proc. 2nd MSC Worldwide Automotive Conference, Dearborn, USA, 2000.*

6.   *Donoho, D.: Non-linear Wavelet Methods for Recovery of Signals, Densities, and Spectra from Indirect and Noisy Data, Proc. of Symposia in Applied Mathematics, Volume 47, 1993.*

7.   *Donoho, D. and Johnstone, I.: Ideal Spatial Adaptation via Wavelet Shrinkage, Biometrika, 81:425-455, Dec. 1994.*

8.   *Buckheit, J. et al: WaveLab Reference Manual, Version 0.700, Stanford University, USA, Dec. 1995.*

9.   *Long Term Pavement Performance IMS Data, Release 13.1, Federal Highway Administration, USA, 2002.*

10.   *Nordberg, P.: A Novel Approach to the Identification of Automotive Road Loads, Thesis for the Degree of Licentiate of Engineering, Dept. of Applied Mechanics, Chalmers University of Technology, Gothenburg, Sweden, 2002.*

11.   *Zegelaar, P. and Pacejka, H.: The In-plane Dynamics of Tyres on Uneven Roads. Segel, L. (Ed.): The Dynamics of Vehicles on Roads and Tracks, Proc. 14th IAVSD Symposium, Ann Arbor, USA, 1995. Swets & Zeitlinger, Lisse, 1996.*

12.   *MSC.Nastran Advanced Dynamic Analysis User's Guide, Version 70, MSC.Software Corporation, Los Angeles, USA, 1997.*

13.   *ISO 2631-1:1997(E), Mechanical Vibration and Shock – Evaluation of Human Exposure to Whole-body Vibration, Part I: General Requirements*

Analysis of The Improvement of Running Performance by Secondary Suspension

MASAHITO ADACHI

SUMMARY

The main purpose of this study is to investigate how to improve the running performance by the secondary suspension of the railway vehicle with the aim of high speed curving. In this paper the analysis and the simulations about ride comfort on curves and running stability are conducted. On ride comfort on curves, it is confirmed that how much the stationary lateral acceleration is reduced by new method of adding the auxiliary lateral springs between the car body and the bogie without changing the lateral and the vertical vibrations of the car body. On running stability, it is confirmed that how effective the yawing rotational stiffness and the lateral damping between the bogie and the car body are in running stability, and it is concluded that the lateral damper can heighten running stability in case of the bogie without the yaw damper and also the yawing rotational stiffness can be smaller.

1   INTRODUCTION

In the railway, in order to shorten the achievement time to the destination, it is necessary to improve running speed on curve track. Therefore, it is important to reduce the stationary lateral acceleration of a car body on curve track at high speed, and to realize both running stability at high speed (as yaw dampers between a car body and bogie are failed) and curving performance on sharp curve. Recently tilting cars or steering bogies have been introduced by limited express trains of (narrow-gauge) conventional lines, but these are not always appropriate methods in the high-speed vehicle because vehicle's weight increases. However, actually, it is difficult to find other methods.

In this study, with the aim of improving ride comfort on curves and running stability, the analysis and the simulation on the curving behavior and hunting motion are made in cases of newly adding the lateral auxiliary springs to the secondary suspension and changing the characteristics of the secondary suspension.

Address correspondence to: Running Dynamics Group, Technology Research and Development Department, Central Japan Railway Company, Komaki-shi, Aichi, 485-0801, JAPAN.
Tel : +81-568-47-5371; Fax : +81-568-47-5364;E-mail: adachi@jr-central.co.jp

2 ANALYSIS CONDITION

2.1 Analysis model

Two kinds of analysis are carried out in this study. One is the theoretical study according to the simplified cross section model of a railway vehicle as shown in Fig.1 and Fig.2, and another one is the time domain simulation according to one vehicle model as shown in Fig.3. The vehicle's characteristics are based on the series 700 of the Shinkansen vehicle.

The auxiliary springs are placed on the side beam of the bogie as shown in Fig.1 and Fig.2. The lateral stiffness of secondary suspension with the auxiliary springs is equal to the conventional secondary suspension. That is, the lateral stiffness of the conventional secondary suspension is shared by both air springs and auxiliary springs.

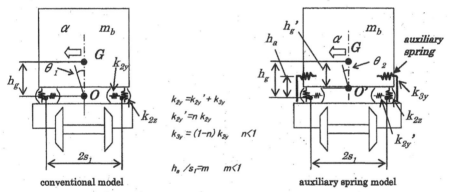

$$k_{2y} = k_{2y}' + k_{3y}$$
$$k_{2y}' = n \, k_{2y}$$
$$k_{3y} = (1-n) \, k_{2y} \quad n<1$$

$$h_a \, / s_1 = m \quad m<1$$

conventional model   auxiliary spring model

Fig.1. Cross section model on ride comfort on curves

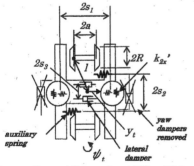

λ : taper of the wheel tread

f : creep coefficient in a wheelset

Fig.2. Cross section model on running stability at high speed

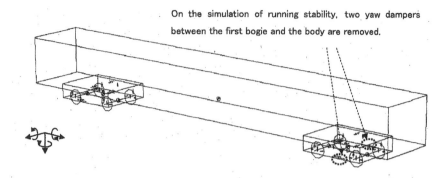

On the simulation of running stability, two yaw dampers between the first bogie and the body are removed.

Fig.3. One vehicle model on time domain simulation

Simulations are conducted with the analysis software VAMPIRE. One vehicle body, two bogies and four wheelsets are modeled as rigid bodies and have six degrees of freedom each except for the wheelsets (wheelsets have five degrees of freedom except for pitching). Hence, there are 38 degrees of freedom. Creep forces are determined according to the non-linear theory of Kalker on the basis of the contact data of the wheel/rail profile of the Shinkansen.

2.2   Running condition on simulation

The analysis on the stationary lateral acceleration is made on condition that the vehicle is passing the curve with radius 3000m and cant 160mm at a speed of 230 km/h. The analysis on the side thrust on sharp curve is made on condition that the vehicle is passing the curve with radius 400m and cant 160mm at a speed of 90 km/h.

On the analysis of running stability, alignment irregularity of 3 sine waves is set on the beginning of the straight line. The single swing width and wavelength are respectively 3mm and 83.3m.

3   THEORETICAL STUDY BY CROSS SECTION MODEL

3.1   Improvement of ride comfort on curves

The car body tilt angle can be shown by the following equations (1) and (2) when lateral acceleration α is given to the car body center of gravity, as shown in Fig.1.

The equations (1) and (2) are respectively in case of the conventional suspension and the auxiliary suspension. In these equations, only a balance of the rolling moment is considered and that of lateral translation is not considered.

$$\theta_1 = \frac{m_b \cdot a_y \cdot h_g}{2 \cdot k_{2z} \cdot s_1^2} \tag{1}$$

$$\theta_2 = \frac{m_b \cdot a_y \cdot h_g{}'}{2 \cdot \left(k_{2z} \cdot s_1^2 + \dfrac{k_{2y}{}' k_{3y}}{k_{2y}{}' + k_{3y}} \cdot h_a^2 \right)} \qquad h_g{}' = h_g - \frac{k_{3y}}{k_{2y}{}' + k_{3y}} \cdot h_a \le h_g \tag{2}$$

The ratio of equations (1) and (2) is shown as

$$p = \frac{\theta_1}{\theta_2} = \left\{ 1 + n \cdot (1-n) \cdot \frac{k_{2y}}{k_{2z}} \cdot m^2 \right\} \cdot \left\{ \frac{h_g}{h_g - (1-n) \cdot m \cdot s_1} \right\} \ge 1 \tag{3}$$

Equation (3) shows that, by appropriately choosing the stiffness and the height of auxiliary springs (i.e. n and m as shown in Fig.1), it is possible to decrease the car body tilt angle. And the ratio p also shows the increasing ratio of the rolling rotational stiffness between the car body and the bogie. For example, in case of $n=0.4$ and $m=0.2$, p will become 1.24. It means that the rolling rotational stiffness increases by approximately 20%.

3.2 Improvement of running stability at high speed

The hunting motion of the bogie can be shown as the following equations (4) and (5) [1], which are based on the cross section model of Fig.2. The conditions of the model are as followings: the yaw damper is removed from the bogie, the wheelsets are rigidly combined with the bogie frame, and then the car body is fixed in the plane.

$$m_t \cdot \ddot{y}_t + 2 \cdot \left(c + \frac{f}{v} \right) \cdot \dot{y}_t + 2 \cdot \left(k_{2y}{}' + k_{3y} \right) \cdot yt = 2 \cdot f \cdot \psi_t \tag{4}$$

$$m_t \cdot i^2 \cdot \ddot{\psi}_t + 2 \cdot \left(c \cdot s_3^2 + \frac{f \cdot l^2}{v} \right) \cdot \dot{\psi}_t + 2 \cdot \left(k_{2x}{}' s_1^2 + k_{3y} \cdot s_2^2 \right) \cdot \psi_t = -2 \cdot f \cdot \frac{a \cdot \lambda}{R} \cdot y \tag{5}$$

By substituting $y_t = Y_t e^{pt}$ and $\psi_t = \Psi_t e^{pt}$ into equations (4) and (5), and rearranging the expression, the following fourth characteristic equation is obtained.

$$A_4 \cdot P^4 + A_3 \cdot P^3 + A_2 \cdot P^2 + A_1 \cdot P + A_0 = 0 \tag{6}$$

The characters of A_4-A_0 in equation (6) are as follows.

$$A_4 = m_t^2 \cdot i^2 \tag{7}$$

$$A_3 = 2 \cdot \left\{ m_t \cdot i^2 \cdot \left(c + \frac{f}{v} \right) + m_t \cdot \left(c \cdot s_3^2 + \frac{f \cdot l^2}{v} \right) \right\} \tag{8}$$

$$A_2 = 2 \cdot \left\{ m_t \cdot i^2 \cdot \left(k_{2y}' + k_{3y} \right) + 2 \cdot \left(c \cdot s_3^2 + \frac{f \cdot l^2}{v} \right) \cdot \left(c + \frac{f}{v} \right) \right\} \tag{9}$$

$$A_1 = 4 \cdot \left\{ \left(c \cdot s_3^2 + \frac{f \cdot l^2}{v} \right) \cdot \left(k_{2y}' + k_{3y} \right) + \left(k_{2x}' s_1^2 + k_{3y} \cdot s_2^2 \right) \cdot \left(c + \frac{f}{v} \right) \right\} \tag{10}$$

$$A_0 = 4 \cdot \left\{ \left(k_{2x}' s_1^2 + k_{3y} \cdot s_2^2 \right) \cdot \left(k_{2y}' + k_{3y} \right) + f^2 \cdot \frac{a \cdot \lambda}{R} \right\} \tag{11}$$

The stability of the hunting motion can be examined by the Routh-Hurwitz stability discrimination formula as follows:

$$\Delta = A_1 \cdot A_2 \cdot A_3 - A_0 \cdot A_3^2 - A_1^2 \cdot A_4 (\geq 0 : stable) \tag{12}$$

In order to heighten the stability in equation (12), it is most effective to increase A_2. Hence, increasing the values of c or s_3 of the lateral damper is considered one of valid ways of heightening the running stability. However, increasing the value of c affect lateral riding comfort, therefore it is proper to increase only the value of s_3.

And also, increasing A_1 has an effect on the stability. However, A_1 is included in both the first and the third term of the equation (12), therefore A_1 is not as effective as A_2. And then, the main parameters to increase A_1 are the lateral stiffness ($k'_{2y} + k_3$) and the yawing rotational stiffness ($k'_{2x}s_1^2 + k_3 s_2^2$), but the way of increasing these parameters is not appropriate, because it will be disadvantageous for the lateral riding comfort and the side thrust on sharp curve.

4 DYNAMIC SIMULATION OF ONE VEHICLE MODEL BY VAMPIRE

4.1 Simulation of ride comfort on curves

The effect of auxiliary springs (n=0.4 and m=0.2) on the stationary lateral acceleration and the roll angle (to the ground) is shown in Fig.4. It can be read off the figure that it is possible to reduce about 8% of the stationary lateral acceleration

(= tilting coefficient) of the car body. And this is correspondent with the increase of about 30% of the rolling rotational stiffness between the car body and the bogie. This result is larger than that of section 3.1. (However, in this simulation, leveling valves between the car body and the bogie are not modeled, hence the absolute values in the simulation are different from the measured values.)

Fig.5 shows the comparison between the cases of auxiliary suspension and conventional suspension on the lateral and vertical acceleration of car body. The track geometry used in the simulation of Fig.5 is the same as that of Fig.4. But, in the simulation of Fig.5, the four kinds of track irregularities (cross level, lateral, vertical and gauge variation) are superimposed on the track geometry. This result shows two suspensions are much the same in both vibrations of the car body. Hence, it is concluded that the auxiliary suspension can make the rolling rotational stiffness between the car body and the bogie increase without changing vibration characteristics in vertical and lateral direction.

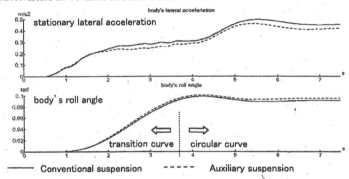

Fig.4. The comparison on ride comfort on curves. (radius 3000m,cant 160mm,230 km/h)

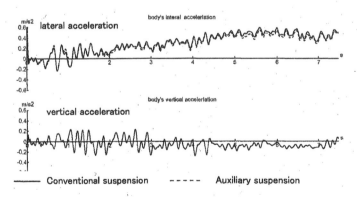

Fig.5. The comparison on the car body's accelerations between two suspensions.

4.2   Improvement of both running stability and curving performance

In this section, for the purpose of decreasing the side thrust on sharp curve, the analysis of improvement of running stability is made on condition that the yawing rotational stiffness between bogie and car body is a half of conventional one. And also, in the vehicle model which is shown in Fig.3, yaw dampers between the first bogie and car body are removed in consideration of being failed.

The simulations are conducted in such four cases as follows:

A : Conventional vehicle (except for the longitudinal position of lateral dampers=0).

B : The yawing rotational stiffness between a bogie and a car body= a half of case A.

C : The yawing rotational stiffness between a bogie and a car body= a half of case A. The longitudinal distance between the two lateral dampers is expanded to the half of the wheel base.

D : Vehicle with the auxiliary suspension(n=0.4 and m=0.2). The vertical position of the lateral dampers is equal to the auxiliary springs. The yawing rotational stiffness between a bogie and a car body and the longitudinal position of lateral dampers are equal to those of case C

Fig.6 shows lateral displacements of the first wheelset on simulation in order to examine the hunting motion of the bogie. V_{lim} expresses the divergence speed of the hunting motion (= hunting speed) in case A.

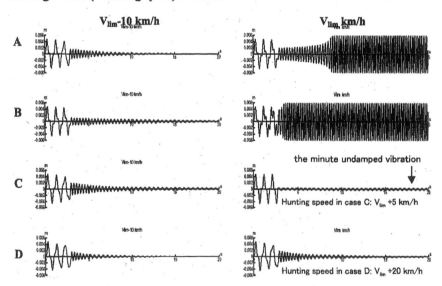

Fig.6.   Lateral displacements of the first wheelset in hunting motion.

The hunting speed in case B is lower than case A, but both of them are almost same. That is, the reduction to half of the yawing rotational stiffness between the bogie and the car body has only a little effect on running stability.

In case C, the hunting speed is a little higher than that in case A and case B, namely V_{lim} +5 km/h. Hence, the increase of longitudinal distance between the two lateral dampers is effective in heightening hunting speed. But, in spite of heightening the hunting speed, the minute undamped vibration is generated at lower speed than the hunting speed, as shown in Fig.6 (The speed at which the minute undamped vibration is generated becomes lower, as the longitudinal distance between the two lateral dampers becomes longer.). According to frequency analysis as shown in Fig.7, it is found that frequency of this vibration is about 2Hz and less than a half of hunting motion frequency. Moreover, the result of investigating the motion of car body is as shown in Fig.8. As a result, in case C, it is found that the minute undamped rolling vibration of the car body is also generated.

In case D, high running stability can be achieved, namely hunting speed is $V_{lim}+20$. And the minute undamped vibration is not generated as shown in Fig.8.

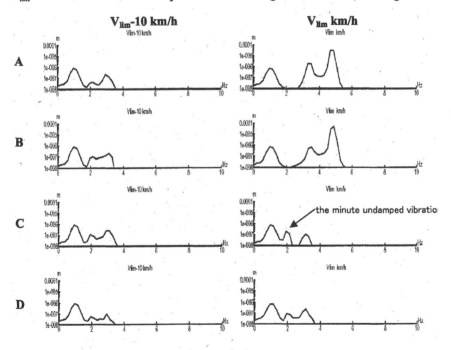

Fig.7. Frequency analysis of lateral displacements of the first wheelset in Fig.6.

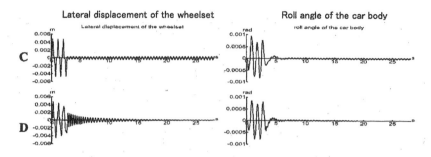

Fig.8. Relation between lateral displacement of the wheelset and roll angle of car body at V_{lim} km/h.

The result of comparison between case C and case D makes clear that the minute undamped vibration is caused by the rolling moment which is added to the car body by the lateral dampers in yawing motion of the bogie. Because the vertical distance between the lateral dampers and the center of gravity of the car body is shorter than that in case C. And then, the hunting speed can be heightened in case D in comparison with case C. The reason for this is considered that the lateral-vibration interaction between the bogie and the car body decreases.

On the other hand, Fig.4.3 shows the comparison of static side thrust on sharp curve between case A and case D. The static side thrust in case D (= that in case B and case C) decreases approximately 5% in comparison with case A. Hence, the decrease of yawing rotational stiffness between the bogie and the car body is effective in reducing the static side thrust on sharp curve, without deteriorating running stability almost.

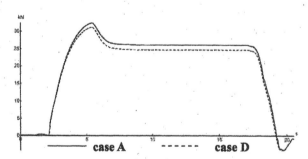

Fig.9. The comparison of static side thrust on sharp curve(radius 400m,cant 160mm) at a speed of 90 km/h

5   CONCLUSION

The above analysis and simulation conclude as follows:

(1)Adding the lateral auxiliary springs between the bogie and the car body makes it possible to reduce the stationary lateral acceleration (tilt coefficient) of car body in high-speed curving, without affecting lateral and vertical vibration of the car body (=without changing the lateral and vertical stiffness between the bogie and the car body).

(2)Improving the running stability by the lateral auxiliary springs is impractical. Because, increasing the stiffness in yaw and lateral direction between the bogie and the car body is less effective than the yaw damping force by the lateral damper and is disadvantageous for the lateral riding comfort and the side thrust on sharp curve.

(3)The reduction to the yawing rotational stiffness between the bogie and the car body has only a little effect on running stability. Hence, reducing the yawing rotational stiffness between the bogie and the car body is one of effective ways of decreasing static side thrust on sharp curve.

(4)In case of the bogie without the yaw damper between the bogie and the car body, expanding the longitudinal distance between the lateral dampers and reducing the vertical distance between the lateral damper and the center of gravity of the car body, is advantageous for heightening the hunting speed. However, it is important to choose the position and the damping of the lateral dampers appropriately, because there is possibility that the minute undamped vibration of the bogie and car body by the lateral damping force will be generated and make the running stability decrease.

(5)Heightening both positions of the auxiliary springs and the lateral dampers are advantageous for the vehicle motion, i.e. respectively ride comfort on curves and running stability. Hence, if the auxiliary spring and the lateral damper should be unified as one device, the improvement on curving performance and running stability could be realize by setting the position and characteristics (stiffness and damping) of the devices appropriately.

6   REFERENCES

1) Matsudaira,T. Arai,S. and Yokose,K.(1965). "Combined Effects of Frictional and Elastic Moments Against truck turning upon Hunting of Truck," RAILWAY TECHNICAL RESERCH REPORT, RTRI, NO.512.

Ride Quality Improvement of AGT Vehicle by Active Secondary Suspension

HITOSHI TSUNASHIMA[1] and SHUNJI MORICHIKA[2]

SUMMARY

This paper describes the design of an active suspension system for Automated Guideway Transit (AGT) vehicles. The modeling of the AGT vehicles with a steering system and a single axle bogies system is carried out first . The vehicle model includes lateral transition, roll and yaw motion of the vehicle body, front bogie and rear bogie, respectively. Actuator dynamics that produce lateral forces is also considered. H-infinity control theory is used to design the reduced order controller for an active suspension, which has an advantage over designing frequency dependant controllers and on the robustness of the controller. The designed controller is evaluated in a simulation study with a AGT vehicle model from the view point of improving ride quality.

1 INTRODUCTION

The Automated Guideway Transit (AGT) system was developed and commercialized in the 1970s [1], adding to the many new transport systems now in operation domestically to meet the increasing demand for intermediate capacity transport between mass-transit railways and small-capacity transport buses. In Japan, the AGT system is recognized as a medium capacity transportation system that is elevated and run on an exclusive guideway using rubber tires.

The AGT system has a lateral guidance system that can be separated into a steering system and a single-axle bogie system. The steering system is designed so that the guide wheels can sense its lateral displacement relative to the guide rail and steer the running wheels according to the amount of lateral displacement as shown in Fig. 1(a). The lateral force required to control the motion of the vehicle largely depends on the cornering force produced by the side-slip of the running wheels.

The single-axle bogie system is designed to restrict a vehicle's movement by the guideway through the four guide wheels provided at the front and rear of the running wheels. Under this system, the guide wheels directly produce the lateral force required to guide the vehicle as shown in Fig. 1(b).

The dynamic characteristics of the steering system vehicle were studied in the research and development stages of the AGT system [2, 3]. As for the single-axle bogie

[1]*Address correspondence to*: Department of Mechanical Engineering, College of Industrial Technology, Nihon University, 1-2-1 Izumi-cho, Narashino-shi, Chiba 275-8575, Japan; Phone: +81-47-474-2339, Fax: +81-47-474-2349, E-mail: tsuna@cit.nihon-u.ac.jp
[2]Graduate school of Nihon University

system, studies were conducted for railway vehicles [4] but no detailed analysis was made of AGT vehicles. The present study aims to discuss the effectiveness of active secondary suspensions [5] for AGT vehicles with two types of mechanical guidance systems [6].

2 DYNAMIC MODEL OF AGT VEHICLE WITH ACTIVE SECONDARY SUSPENSION

The dynamic model for the AGT vehicle consists of the body, front and rear bogie with rubber tires and mechanical guidance system as shown in Fig.1. The typical symbols used in equations for the vehicle motion are shown in Fig. 1

In this section, a dynamical model for an AGT with a single-axle bogie is described in order to apply an active secondary suspension. The lateral transition, roll and yaw motions of the vehicle body, front bogie and rear bogie can be expressed as follows:

Vehicle body:

$$m_B \ddot{y}_B = -K_L \Delta y_F - C_L \Delta \dot{y}_F - K_{AL} \Delta y_F - K_L \Delta y_R - C_L \Delta \dot{y}_R - K_{AL} \Delta y_R + F_F + F_R \tag{1}$$

$$I_{BX} \ddot{\phi}_B = -h_k K_L \Delta y_F - h_c C_L \Delta \dot{y}_F - h_a K_{AL} \Delta y_F - h_k K_L \Delta y_R - h_c C_L \Delta \dot{y}_R - h_a K_{AL} \Delta y_R$$
$$-(K_\phi + 2l_a^2 K_{AH}) \Delta \phi_F - (K_\phi + 2l_a^2 K_{AH}) \Delta \phi_R$$
$$-2(l_a^2 C_A + l_d^2 C_h) \Delta \dot{\phi}_F - 2(l_a^2 C_A + l_d^2 C_h) \Delta \dot{\phi}_R + h_{ac} F_F + h_{ac} F_R \tag{2}$$

$$I_{BZ} \ddot{\psi}_B = -l_{TB}(K_L \Delta y_F + C_L \Delta \dot{y}_F + K_{AL} \Delta y_F) + l_{TB}(K_L \Delta y_R + C_L \Delta \dot{y}_R + K_{AL} \Delta y_R)$$
$$-C_\psi \Delta \dot{\psi}_F - C_\psi \Delta \dot{\psi}_R - l_{TB} F_F + l_{TB} F_R \tag{3}$$

where $\Delta y_F = y_B - y_{FT} + l_{TB} \psi_B - h_k \phi_B + (h_k + h_{TB}) \phi_{FT}$,
$\Delta y_R = y_B - y_{RT} - l_{TB} \psi_B - h_a \phi_B + (h_k + h_{TB}) \phi_{RT}$, $\Delta \phi_F = \phi_B - \phi_{FT}$, $\Delta \phi_R = \phi_B - \phi_{RT}$,
$\Delta \psi_F = \psi_B - \psi_{FT}$, $\Delta \psi_R = \psi_B - \psi_{RT}$, F_F, F_R are actuator forces for the front and rear bogies respectively. Here, we assumed that the actuator has first order lag characteristics (cut off frequency: 10Hz).

Front Bogie:

$$m_T \ddot{y}_{FT} = -K_g \Delta y_{FF} - K_g \Delta y_{FR} + 2K_t \beta_F + K_L \Delta y_F + C_L \Delta \dot{y}_F + K_{AL} \Delta y_F - F_F \tag{4}$$

$$I_{TX} \ddot{\phi}_{FT} = -h_g K_g \Delta y_{FF} - h_g K_g \Delta y_{FR} + (h_k + h_{TB}) K_L \Delta y_F + (h_c + h_{TB}) C_L \Delta \dot{y}_F + (h_a + h_{TB}) K_{AL} \Delta y_F$$
$$+2h_t K_t \beta_F + (K_\phi + 2l_a^2 K_{AH}) \Delta \phi_F + 2(l_a^2 C_A + l_d^2 C_h) \Delta \dot{\phi}_F - 2l_b^2 K_{TZ} \dot{\phi}_{FT} + (h_{ac} + h_{TB}) F_F \tag{5}$$

$$I_{TZ} \ddot{\psi}_{FT} = -2t_P K_t \beta_F + C_\psi \Delta \dot{\psi}_F - l_g K_g \Delta y_{FF} + l_g K_g \Delta y_{FR} \tag{6}$$

where $\Delta y_{FF} = y_{FT} + l_g \psi_{FT} + h_g \phi_{FT} - y_{wFFg}$, $\Delta y_{FR} = y_{FT} - l_g \psi_{FT} + h_g \phi_{FT} - y_{wFRg}$,
$\beta_F = \psi_{FT} - \dot{y}_{FT}/V - l_{TB} \dot{\psi}_{FT}/V$

Rear Bogie:

$$m_T \ddot{y}_{RT} = -K_g \Delta y_{RF} - K_g \Delta y_{RR} + 2K_t \beta_R + K_L \Delta y_R + C_L \Delta \dot{y}_R + K_{AL} \Delta y_R - F_R \tag{7}$$

$$I_{TX} \ddot{\phi}_{RT} = -h_g K_g \Delta y_{RF} - h_g K_g \Delta y_{RR} + (h_k + h_{TB}) K_L \Delta y_R + (h_c + h_{TB}) C_L \Delta \dot{y}_R + (h_a + h_{TB}) K_{AL} \Delta y_R$$
$$+ 2h_t K_t \beta_R + (K_\phi + 2l_a^2 K_{AH}) \Delta \phi_F + 2(l_a^2 C_A + l_d^2 C_h) \Delta \dot{\phi}_F - 2l_b^2 K_{TZ} \phi_{FT} + (h_{ac} + h_{TB}) F_F \tag{8}$$

$$I_{TZ} \ddot{\psi}_{RT} = -2t_P K_t \beta_R + C_\psi \Delta \dot{\psi}_R - l_g K_g \Delta y_{RF} + l_g K_g \Delta y_{RR} \tag{9}$$

where $\Delta y_{RF} = y_{RT} + l_g \psi_{RT} + h_g \phi_{RT} - y_{wRFg}$, $\Delta y_{RR} = y_{RT} - l_g \psi_{RT} + h_g \phi_{RT} - y_{wRRg}$,
$\beta_R = \psi_{RT} - \dot{y}_{RT}/V - l_{TB} \dot{\psi}_{RT}/V$

For the steering system, the lateral motion and the yaw motion of bogie should be replaced with a suitable steering equation [6].

3 CONTROLLER DESIGN

3.1 H-infinity control approach

An H infinity control approach is employed for designing the controller. Figure 2 shows the structure diagram of the H-infinity control system.

The generalized plant can be described by the following equation;

$$\begin{bmatrix} z \\ y \end{bmatrix} = G \begin{bmatrix} w \\ u \end{bmatrix} = \begin{bmatrix} G_{11} & G_{12} \\ G_{21} & G_{22} \end{bmatrix} \begin{bmatrix} w \\ u \end{bmatrix} \tag{10}$$

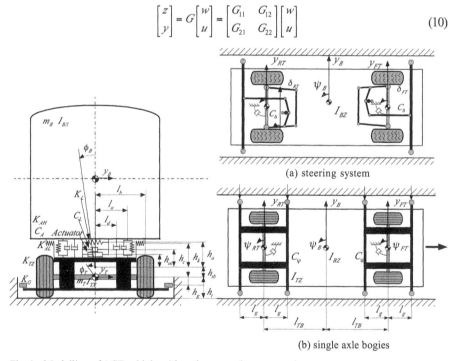

(a) steering system

(b) single axle bogies

Fig. 1. Modelling of AGT vehicle with active secondary suspension.

The transfer function between the external input and the output for the system can be described by the following equation.

$$G_{zw} = G_{11} + G_{12}K\left(I - G_{22}K\right)^{-1}G_{21}.$$ (11)

The internally stable controller, $K(s)$, that satisfies

$$\left\|W(s)G_{zw}(s)\right\|_{\infty} < 1$$ (12)

is obtained.

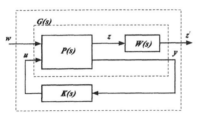

Fig. 2. Stracture of control system for H-infinity control.

3.2 Lateral and roll controller design

The active suspension system is designed by using a reduced order model, where the lateral transition and roll motions are considered for a half vehicle model. The reduced order vehicle equations for the lateral and roll motions are given by

$$m_B\ddot{y}_B/2 = -K_L(y_B - y_T - h_k\phi_B) - 2K_{AL}(y_B - y_T - h_a\phi_B) - C_L(\dot{y}_B - \dot{y}_T - h_c\dot{\phi}_B) + A_s u_L,$$ (13)

$$I_{BX}\ddot{\phi}_B/2 = -h_k K_L(y_B - y_T - h_k\phi_B) - 2h_a K_{AL}(y_B - y_T - h_a\phi_B)$$
$$-h_c C_L(\dot{y}_B - \dot{y}_T - h_c\dot{\phi}_B) - 2K_{AH}l_a^2\phi_B - 2C_Al_a^2\dot{\phi}_B - 2C_Hl_d^2\dot{\phi}_B + h_{ac}F_L.$$ (14)

The state space model for the reduced order model can be expressed by

$$\dot{x}_L = A_L x_L + B_{L1}w_L + B_{L2}u_L,$$ (15)
$$z_L = C_{L1}x_L + D_{L11}w_L + D_{L12}u_L,$$ (16)
$$y_L = C_{L2}x_L + D_{L21}w_L + D_{L22}u_L,$$ (17)

where state variable vector, control output vector and output vector are,
$$x_L = \begin{bmatrix} \dot{y}_B & y_B - y_T & \dot{\phi}_B & \phi_B \end{bmatrix}^T, \quad z_L = \begin{bmatrix} \ddot{y}_B & y_B - y_T & \ddot{\phi}_B & u_{Lp} \end{bmatrix}^T, \quad y_L = \begin{bmatrix} \ddot{y}_B & y_B - y_T & \ddot{\phi}_B \end{bmatrix}^T,$$
$$w_L = \dot{y}_F$$
The weighing functions for the lateral and roll controller are designed as,

$$W_L(s) = \mathrm{diag}(W_1(s), W_2(s))$$ (18)

where

$$W_1(s) = \frac{10\pi}{s + 10\pi}\mathrm{diag}(5.0, 1.0, 0.5), \quad W_2(s) = \frac{s}{s + 14\pi} \times 10^{-6.3}$$

3.3 Yaw controller design

An reduced order vehicle equation for the yaw motions is given by

$$I_{BZ}\ddot{\psi}_B = -l_{TB}C_L(l_{TB}\dot{\psi}_B - \dot{y}_{FT}) - l_{TB}C_L(l_{TB}\dot{\psi}_B + \dot{y}_{RT})$$
$$-l_{TB}(K_L + 2K_{AL})(l_{TB}\psi_B - y_{FT}) - l_{TB}(K_L + 2K_{AL})(l_{TB}\psi_B + y_{RT}) + 2l_{TB}F_\psi . \quad (19)$$

The state space model for the reduced order model can be expressed by

$$\dot{x}_\psi = A_\psi x_\psi + B_{\psi 1} w_\psi + B_{\psi 2} u_\psi , \quad (20)$$
$$z_\psi = C_{\psi 1} x_\psi + D_{\psi 11} w_\psi + D_{\psi 12} u_\psi , \quad (21)$$
$$y_\psi = C_{\psi 2} x_\psi + D_{\psi 21} w_\psi + D_{\psi 22} u_\psi . \quad (22)$$

where state variable vector, control output vector and output vector are,
$x_\psi = [l_{TB}\psi_B \quad l_{TB}\dot{\psi}_B - (y_{FT} - y_{RT})/2]^T$, $z_\psi = [l_{TB}\ddot{\psi}_B \quad l_{TB}\dot{\psi}_B - (y_{FT} - y_{RT})/2]^T$, $y_\psi = l_{TB}\ddot{\psi}_B$,
$w_\psi = (\dot{y}_{FT} - \dot{y}_{RT})/2$.

The weighing functions for the yaw controller are designed as,

$$W_\psi(s) = \text{diag}(W_1(s), W_2(s)) , \quad (23)$$

where

$$W_1(s) = \frac{10\pi}{s + 10\pi}\text{diag}(5.0, 1.0) , \quad W_2(s) = \frac{s}{s + 14\pi} \times 10^{-7}$$

The required control forces calculated from the two types of controllers are mixed and this is then used as a control force command for each actuator as shown in Fig. 3.

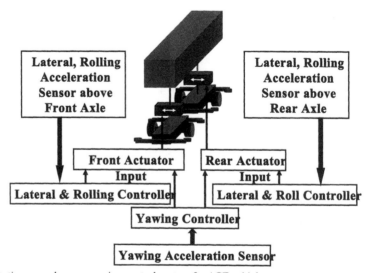

Fig. 3. Active secondary suspension control system for AGT vehicle.

4 SIMULATION

A running simulation with a random roughness input from the guide-rail is carried out to show the effectiveness of the designed active suspension.

4.1 Generation of guide rail irregularity

In this simulation example, the external forces are applied from the guide rail to the front and rear guide wheels. The guide-rail irregularities are generated from PSD of guide-rail irregularity. The generated irregularity of the guide-rail is shown in Fig. 4. In this

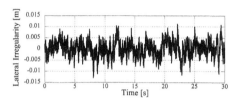

Fig. 4. Generated guide rail ireregularity used for simulation.

simulation example, irregularity from the left and right rail is assumed to be same.

4.2 Simulation condition

Typical vehicle parameters used in this simulation are listed in Table 1. The vehicle motions are calculated for a speed of 80[km/h]. For the consideration of the wear on the guide wheels, a 3[mm] clearance between guide wheel and guide rail is set for both the left and right side of the guideway.

Table 1. Vehicle parameters

Symbol	Description	Value
m_B	Body Mass	8325.06[kg]
m_T	Truck Mass	1435[kg]
I_{BX}	Body Roll Moment of Inertia	17102[kgm^2]
I_{TX}	Truck Roll Moment of Inertia	1300[kgm^2]
I_{BZ}	Body Yaw Moment of Inertia	113604[kgm^2]
I_{TZ}	Truck Yaw Moment of Inertia	5000[kgm^2]
I_{TS}	Steering Yaw Moment of Inertia	100[kgm^2]
C_ψ	Yaw Damper	4776[Ns/m]
C_L	Damping Coefficient	1.6×10^3[Ns/m]
C_A	Vertical Damper	7.84×10^3[Ns/m]
K_{AL}	Stiffness of Lateral Air Spring	300×10^3[N/m]
K_L	Stiffness of Side Stopper Rubber	300×10^3[N/m]
K_G	Stiffness of Gude Wheel	154×10^3[N/m]
K_{AH}	Stiffness of Vertical Air Spring	232×10^3[N/m]
K_{TZ}	Vertical Stiffness of Vertical Tire	450×10^3[N/m]
K_T	Cornering Coefficient	300×10^3[N/rad]
tp	Pneumatic Trail	0.03[m]
$G_g G_r$	Steering Gain	1.3

769

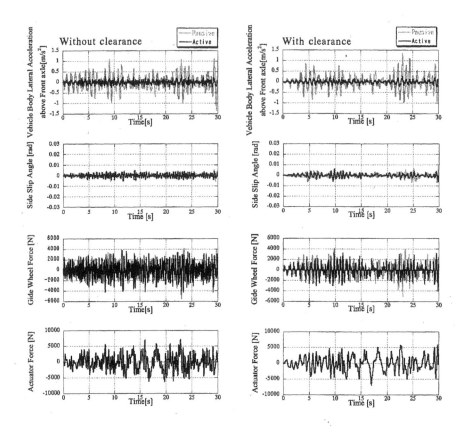

Fig. 5: Simulation results of single-axle bogies

4.3 Simulation results

Figures 5 and 6 show the time history for the lateral acceleration of the vehicle body above the front axle, side-slip angle of tire, guide wheel force and actuator force of active suspension for the single-axle bogie system and steering system respectively. The left side of the figure illustrates the case without any guideway clearance, whereas the right side of the figure illustrates the case with guideway clearance. In both cases, we can see that the designed active suspension system can suppress the lateral acceleration with a suitable actuator force. It should be noted that the guideway clearance does not make much difference in the single-axle bogie system when the passive suspension is used for an AGT vehicle.

It can be seen from Fig. 6 that the active suspension system works well when there is no guideway clearance. However, performance of the active suspension deteriorates when there is clearance between the guide-rail and the guide wheels. We can also see that a larger actuator force is required for the steering system when there is clearance between the guide-rail and the guide wheels.

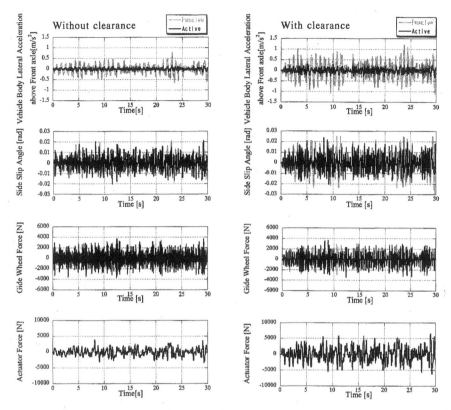

Fig. 6: Simulation results of steering system

4.4 Ride quality evaluation

The ride comfort level can be computed by using the following equations.

$$L_T = 20\log(\overline{a}_w/a_{ref}) \qquad (24)$$

$$\overline{a}_w^{\,2} = \int_{0.5}^{80} S(f)\,df \qquad (25)$$

where $S(f)$ is the frequency weighted power spectral density of acceleration, and $a_{ref} = 10^{-5}\,[\mathrm{m/s^2}]$.

Figure 7 and 8 show the power spectral density (PSD) for the lateral acceleration above the front axle. The ride comfort level can be computed from Eqs. (24) and (25) with Figs. 7 and 8. Comparison of the calculated ride comfort level is shown in Fig. 9. It can be seen that ride quality improvement is achieved from the active secondary suspension. Figure 9 indicates that there is a slight difference in ride quality between the AGT vehicle with a single-axle bogie system than one with a steering system when there is no clearance between the guide-rail hand the guide wheels. We can see that the ride quality for an AGT vehicle with a single-axle bogie is not affected by the

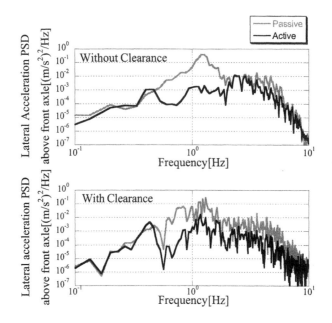

Fig. 7. Lateral Acceralation PSD above front axle (single-axle bogie).

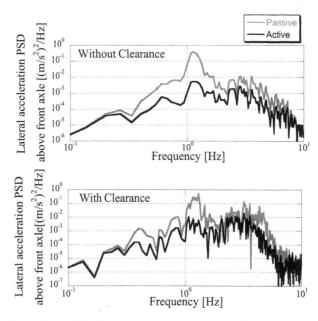

Fig. 8. Lateral Acceralation PSD above front axle (steering system).

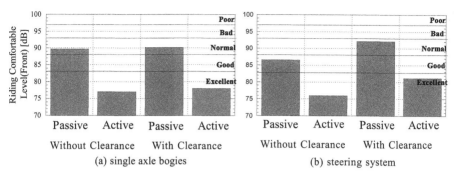

Fig. 9. Ride comfort level of vehicle body above front axle.

clearance. However the ride quality for an AGT vehicle with a steering system is affected by the clearance.

5 CONCLUSIONS

An active secondary suspension system for AGT vehicles with two types of mechanical guidance systems was discussed. The nonlinear equations for the AGT vehicles are derived and used to examine the reduced order independent controller, which is equipped in the front and rear bogie. An H infinity control approach is employed for designing the controller. Numerical simulation was carried out to show the performance of the active secondary suspension.

It is shown that the obtained controller has good ride quality improvement by suppressing the lateral vibration effectively . It is also shown that the ride quality of an AGT vehicle with a single-axle bogie is not affected by the clearance, however the ride quality of an AGT vehicle with a steering system is affected by the clearance.

REFERENCES

1. Shladover, S.: Review of the State of Development of Advanced Vehicle Control System (AVCS), *Vehicle System Dynamics* Vol.24 No. 6-7 (1995), pp.551-595.
2. Shladover, S.: Steering controller design for Automated Guideway Transit vehicles, *Transaction of the ASME Journal of Dynamic System, Measurement and Control* Vol.100 (1978).
3. Abe, M.: Computer simulation study on lateral dynamics of Automated Guideway Transit Vehicle, *Modeling and Simulation in Engineering IMAC*S (1983), pp.145-153.
4. Arrus, P., de Pater, A.D. and Meyers, P.: The Stationary Motion of One-Axle Vehicle Along a Circular Curve with Real Rail and Wheel Profiles, *Vehicle System Dynamics* Vol.37 No. 1 (2002), pp.29-58.
5. Goodall, R.: Active Railway Suspension: Implementation Status and Technological Trends, *Vehicle System Dynamics* Vol.28 No.2-3 (1997), pp.87-117.
6. Tsunashima, H: Dynamics of Automated Guideway Transit Vehicle with Single-Axle Bogies, *Vehicle System Dynamics* Vol.39 No. 5 (2003), pp.365-397.

Vehicle System Dynamics Supplement 41 (2004), p.774-783   © Taylor & Francis Ltd.

Dynamic Response Analysis of Railway Passenger Car With Flexible Carbody Model Based on Semi-Active Suspensions

Pingbo Wu, Jing Zeng, Huanyun Dai[1]

SUMMARY

A multi-body system dynamical model including elastic deformation effect of structure is put forward based on the test results of principal modes of passenger carbody. The influence of the carbody elastic vibration on ride comfort is analyzed numerically in detail. In order to settle the vehicle vibration problems including the elastic deformation effect due to high speed, the method of semi-active suspensions is utilized. The dynamic responses of the high speed passenger car with passive, semi-active On/Off damper are studied and compared. The numerical simulation results for the passenger car are validated with a roller test rig.

1   INTRODUCTION

In recent years Chinese railway has developed very fast and the speed of trains has been raised extensively. On the main railway lines the speed of trains has reached 140~160 km/h and the traveling time of the passengers is greatly shorten. A high attention is also paid to the development of high speed passenger trains by China. Up to now the maximum test speed of the Chinese high speed trains in site has reached 320km/h. The lightening design of the vehicles is very important for the running of high speed trains. The lightening of carbody of railway vehicle with high speed can effectively reduce the fierce actions between the wheels and rails, manufacturing cost, and saves energy. In order to lighten railway vehicle systems, for an example, stainless steel and aluminum alloy have been widely used in the structure of the carbody. But the carbody lightening causes the reduction of stiffness of the structure, lowers the natural frequencies, and easily leads to resonance vibration. The aggravated vibration of carbody affects the ride comfort of passengers seriously. So the study on vibration problems of railway vehicle systems caused by carbody lightening is very important.

---

[1] *Address correspondence to*: National Traction Power Laboratory, Southwest Jiaotong University, Chengdu, Sichuan 610031, P. R. of China.
Tel.:(86)28-87600882; Fax:(86)28-87600868;E-mail:wupingbo@263.net

To consider elastic deformation effect of structure in the dynamical analysis of vehicle systems is an effective way to study this vibration problem. In reference [1], an aggregation of three pieces of flexible beams is used as a carbody model to establish the multi-body system dynamical model. The analysis result in ref. [1] proves that sufficient carbody stiffness ensures good ride comfort for passengers when a railway vehicle moves at higher speed. In reference [2], the elastic deformation effect of structure is taken into consideration in the detailed analysis of the dynamic response for a high speed train with three-car articulated type, in which five pieces of flexible beams are used to model the carbody. Extensive comparisons between measurements and corresponding numerical simulations were carried out by Carlbom P, and the carbody modes which make main contributions to the structure vibration were also predicted. His main conclusions from the study are that the structure flexibility of the carbody must be taken into account when one predicts the vertical vibration comfort [3-4]. There are many other papers published on automobiles, aviation, etc., and the method of flexible body is used in the multi-body system dynamical analysis [5-9].

In order to improve the ride quality of the passenger car at different speeds, the semi-active suspension is proved to be an economical and efficient way and a lot of work have been done by many researchers in this respect[10-20].

In this paper, a high speed passenger car dynamical model with flexible carbody model based on semi-active suspensions is set up. The influence of the carbody elastic vibration on ride comfort is analyzed numerically in detail. The dynamic responses of the high speed passenger car with passive and semi-active On/Off damper are studied and compared.

2 DYNAMICAL MODEL

When deriving the dynamic equations of the multibody system, normally 6 generalized coordinates including 3 translational and 3 rotational for a body is used. The generalized coordinates are $q_i = [x, y, z, \psi, \theta, \varphi]^T$, $q = [q_1^T, ..q_n^T]^T$. By use of the method of Lagrangian multiplier, the system dynamic equation is [8]

$$
\begin{cases}
\dfrac{d}{dt}\left(\dfrac{\partial T}{\partial \dot{q}}\right)^T - \left(\dfrac{\partial T}{\partial q}\right) + \phi_q^T + \theta_q^T \mu - Q = 0 \\
\\
\phi(q,t) = 0 \\
\\
\theta(q,\dot{q},t) = 0
\end{cases}
\tag{1}
$$

where the energy $T = [M \cdot v \cdot v + \omega \cdot I \cdot \omega]/2$, $\varphi(q,t) = 0$ is the integral constraint and $\theta(q,\dot{q},t) = 0$ is the non-integral constraint. q and Q are the generalized coordinate matrix and general force matrix. μ is the Lagrange multiplier vector corresponding to the non-integral constraint. M and I are the mass and moment of inertia matrices, and v and ω are the generalized velocity and angular velocity vectors respectively.

In order to consider the effect of the flexible body in the system dynamic equation, the flexible body is taken as a set composed of nodes of the finite element model with each node having small linear deformation corresponding to the local axis system which can have large translational and rotational movements. The linear local movement of each node can be approximately taken as the linear summation of the vibration mode shapes [8].

The position of the i th node is

$$r_i = x + A(s_i + \phi_i q) \tag{2}$$

where x is the position vector from the origin of global coordinate axes to the local axes, A is the directional cosine matrix, s_i is the position of the ith node in the local axes before deformation, ϕ_i the mode shape of the ith node, and q the modal amplitude vector.

Let the Euler angle representing the directions, the general coordinates of the motion are

$$\xi = \left\{ \begin{array}{c} x(x,y,z) \\ \Psi(\psi,\theta,\phi) \\ q(q_{j,j} = 1, m) \end{array} \right\} \tag{3}$$

where x,y and z denote the position of the local coordinate axes corresponding to the global axes, ψ, θ, ϕ are the Euler angles of the local coordinate axes corresponding to the origin of the global axes, and q_j the jth modal amplitude.

In the same way, the velocity of the ith node can be obtained, then the kinetic energy and potential energy of the system can be expressed as

$$T = \frac{1}{2} \sum_{i=1}^{N} m_i v_i^{\ T} v_i = \frac{1}{2} \xi^T M(\xi) \dot{\xi} \tag{4}$$

$$\nabla = \frac{1}{2} \xi^T K(\xi) \dot{\xi} \tag{5}$$

Substituting the above expressions into the Lagrangian's Equation, the dynamic equations considering the flexible bodies can be obtained as

$$M\ddot{\xi} + \dot{M}\dot{\xi} - \frac{1}{2}\left[\frac{\partial M}{\partial \xi}\dot{\xi}\right]\dot{\xi} + K\xi + f_g + D\dot{\xi} + \left[\frac{\partial \Psi}{\partial \xi}\right]^T \lambda = Q \qquad (6)$$

In the equation, K indicates the modal stiffness of the flexible body which relies on the deformation of the body, and D denotes the damping matrix of the flexible body. f_g is the gravitational force, λ is the Lagrangian Multiplier of the constrain equation, Ψ and Q are the external forces.

Using the flexible body modeling theory, the model taking into account the elastic effect of carbody is set up. The schematic diagram of the vehicle system is shown as Fig 1. The coordinate system of each body in the vehicle system is chosen to have its origin at the center of gravity of the body and moves along the track center line at a constant forward speed V in the x direction. The y axis is parallel to the track plane and directs to the left of the track and the z axis is normal to the plane of the track and directs upward. The total number of degrees of freedom of the vehicle system is 31. The dynamic characteristics of the equipped carbody are measured in the site [9]. The mode shapes, frequencies and damping ratios of the carbody from the site measurement are used in the establishment of multi-body system dynamic model for the passenger car. The carbody is treated as a flexible body. In the model the first vertical bending mode, the first lateral bending mode and the first torsion mode are take into account for the carbody, the wheelset and bogie frame are assumed to be rigid bodies. The nonlinearitis arising due to the suspension system and wheel/rail interaction are also considered in the model[21].

In the dynamic model, the method of semi-active suspension is utilized with On/Off damping control law used for the secondary lateral suspensions. The On/Off damping control law is as below

$$c_{syj}(t) = \begin{cases} \min(c_h, c_{sy}\dfrac{\dot{x}_{cj}(t)}{\Delta\dot{x}_{cj}(t)}), & \dot{x}_{cj}(t)\Delta\dot{x}_{cj}(t) > 0 \\ \\ c_l, & \dot{x}_{cj}(t)\Delta\dot{x}_{cj}(t) \le 0 \end{cases} \qquad (7)$$

where c_h, c_l are the higher and lower limit values of the semi-active damper. In ideal case it should have $c_l = 0$. c_{sy} is the basic damping value of the damper. For the case of passive suspension, we can chose $c_h = c_l = c_{sy}$ in eq.(7). $\dot{x}_{cj}, \Delta\dot{x}_{chj}$ indicate the absolute velocity of carbody suspension point and the relative velocity between the carbody and bogie frame suspension points, which can be obtained by the integration of the relevant accelerations. For the carbody, the semi-active damper only supplies damping forces to do negative work which can reduce the carbody vibration and improve the passenger ride comfort. But for the bogie frame, the semi-active damping forces may do negative work or positive work.

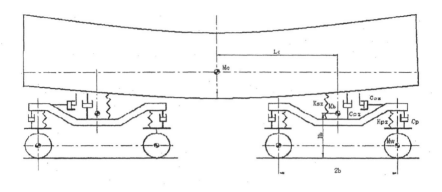

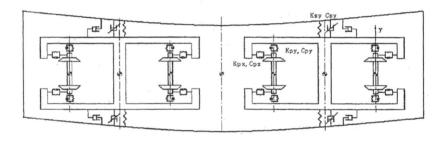

Fig.1. The Dynamic model of passenger car with semi-active and flexible carbody

3 SIMULATION AND TEST RESULTS

The dynamic analysis of the high speed passenger car is carried out by using rigid multi-body system and flexible multi-body system respectively. The influence of the carbody elastic vibration on ride comfort at different speeds is investigated in detail. The numerical simulation is carried out in the time domain with speed from 50km/h to 330km/h. The data at two points on the carbody floor in the places of carbody center and bogie center with 1m offset laterally are calculated and the ride index calculation is carried out according to the Chinese National Standard 95J01-M.

The ride performance of the high speed passenger car with rigid carbody model and flexible carbody model are shown in Fig.2 to Fig.5 respectively. It can be known from Fig.2 that the lateral ride index on the carbody floor of the bogie center based on the flexible model is larger than that based on the rigid model. When the speed increases, the difference of the ride index becomes larger. Fig.3 shows that the vertical ride indices on the carbody floor of the bogie center obtained with the two models nearly have no difference. Fig.4 and Fig.5 illustrate the lateral and vertical ride indices on the floor of the carbody center respectively. It is obvious that the indices obtained with the flexible model are larger than those based on the rigid body model, and the higher the speed is and the larger the difference is.

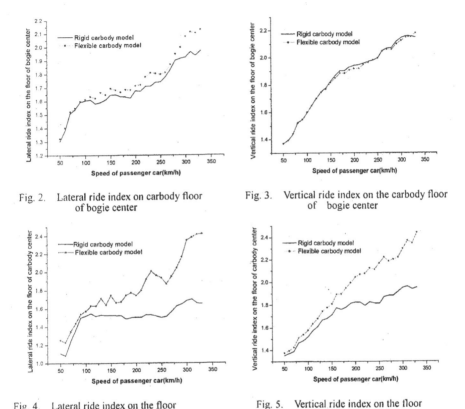

Fig. 2. Lateral ride index on carbody floor of bogie center

Fig. 3. Vertical ride index on the carbody floor of   bogie center

Fig. 4. Lateral ride index on the floor of carbody center

Fig. 5. Vertical ride index on the floor of carbody center

Fig.6 to Fig.9 show the lateral and vertical acceleration frequency spectrum based on the rigid body and flexible body respectively. The simulation speed is 300km/h and the data are based on the point of the carbody center. These results indicate that the influence of the carbody elastic vibration on the response is significant. In order to improve the dynamic performance of high speed passenger car, the feasibility of the carbody lightning scheme should be considered carefully, and the method of semi-active suspensions should be utilized to improve such vibration problems.

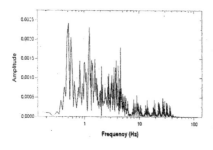

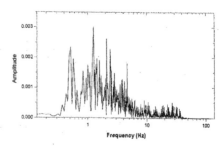

Fig. 6. Lateral acceleration frequency spectrum on carbody center (Rigid carbody, v=300km/h)

Fig. 7. Vertical acceleration frequency spectrum on carbody center(Rigid carbody, v=300km/h)

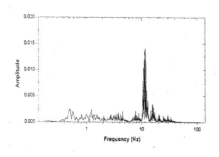

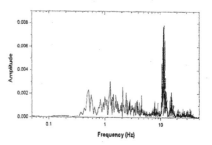

Fig. 8. Lateral acceleration frequency spectrum on carbody center (Flexible carbody, v=300km/h)

Fig. 9. Vertical acceleration frequency spectrum on carbody center(Flexible carbody, v=300km/h)

The ride performance of the high speed passenger car with lateral semi-active damper and passive damper are shown in Fig.10 to Fig.13. It can be seen from Fig.11 that using semi-active damper, the lateral ride index on the carbody floor of the bogie center can be reduced greatly compared with the passive damper. When the speed increases, the difference of the ride index becomes larger. Fig.11 shows that the vertical ride indices on the carbody floor of the bogie center based on these two models have not much difference. Fig.12~13 illustrate the lateral and vertical ride indices on the floor of the carbody center respectively. It is obvious that using lateral semi-active damper can greatly reduce the lateral ride index on the carbody floor of the carbody center.

Fig.14~15 show the passenger car on roller test rig. The simulation results for the passenger car ride comfort are validated by the roller rig test. By using the lateral semi-active damper, the lateral ride index on carbody floor of bogie center will reduce from 2.030 to 1.896 at speed of 200km/h and from 2.227 to 2.058 at speed of 300km/h under large track random irregularity inputs. But under the case of small track random irregularity inputs, the lateral ride index will reduce from 1.641 to 1.457 at speed of 160km/h, and reduce from 1.928 to 1.750 at speed of 300km/h. The test results show that the vertical ride indices with semi-active or passive dampers have not much difference. Therefore it can be concluded that the use of lateral semi-active damper can greatly improve the lateral dynamic performance which validates the numerical simulations.

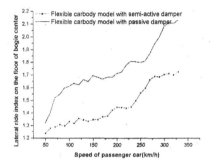

Fig. 10. Lateral ride index on carbody floor of bogie center

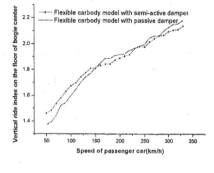

Fig. 11. Vertical ride index on the carbody floor of bogie center

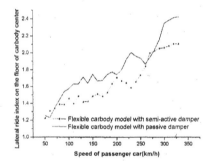

Fig. 12. Lateral ride index on the floor of carbody center

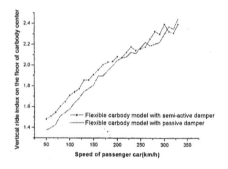

Fig. 13. Vertical ride index on the floor of carbody center

Fig. 14. High Speed Passenger car with semi-active on roller rig

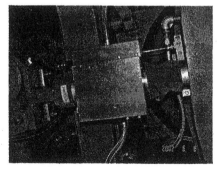

Fig. 15. Semi-active later damper on Truck

5. CONCLUSIONS

(1) The dynamic model of the vehicle system with flexible carbody based on the carbody modal test results is established. This model can precisely describe the actual vehicle system and make the numerical simulation results more believable. The work of the paper can provide the theoretical basis for the feasibility of the optimum design of the carbody structure.

(2) The simulation results show that the ride indices on the carbody floor in the place of bogie center between the flexible carbody model and rigid carbody model have not much difference, but in the place of carbody center there is a big difference. The flexible carbody model will increase the ride index of the carbody floor in the place of carbody center, the higher the speed it is, the larger the difference is.

(3) The numerical and roller rig test results indicate that the use of semi-active damper can greatly improve the ride comfort of the passenger car at different speeds.

6. ACKNOWLEDGMENT

This Research is supported by the Trans-Century Excellent Talent Foundation and the Key Project of Science and Technology (No. 01150) of the Chinese Ministry of Education.

REFERENCES

1. Linwukangwen: Theory analysis of elastic vibration of carbody. *Rly. Techn Res. Inst.* ,No.1(1990), pp42-48.

2. Li S. L., Wang W. D.: Respond Analysis and Modelling of the Articulated Car Body Considering the Effect of Elasticity for High Speed Train. *China Railway Science*,Vol(18)(1997), pp77-86.

3. Carlbom P.: Combining MBS with FEM for rail vehicle dynamics analysis. *Euromech Colloquium* 404, IDMEC/IST, Lisbon,1999.

4. Carlbom P.: Structure Flexibility in a rail vehicle car body – dynamic simulations and measurement. *KTH. Stockholm*, ISRN/KTH/FKT/FR-98/37-SE,1998.

5. Eickhoff, Evans J.R., Minnis A.J.: A Review of Modeling Methods for Railway Vehicle Suspension Components. *Vehicle System Dynamics* Vol(24) (1995), pp469-496.

6. Fanti G., Berti G., Vianello A.: Dynamic analysis of rail vehicles by means of models of equivalent finite elements. *Ingegneria Ferroviara*, Vol.43(5) (1998), pp257-263.

7. Shawn P. M., Steven C. P.: Modeling Vehicle Suspension Structural Compliance at Ford Motor Company Using a Coupling Of ADAMS. *Theory and MSC/NASTRAN*. SAE Paper, 1994.

8. Zhong X., Yang R.Q., Xu Z. F., Gao J. H.: Dynamic Modeling of Multi-Flexible System---Theory and Application. *Mechanical Science and Technology*, Vol(21) (2002), pp387-389.

9. Wu Y.: Test model analysis for passenger coach body. *Journal of the China Railway Society*, Vol16(4)(1994), pp6-14.

10. Suda, Y., Anderson, R. J.: Improvement of Dynamic Performance of Trucks with Longitudinally Unsymmetric Structures by Semi-Active Control for Rail Vehicles. *JSME Int Journal, Series C*, 37-3(1994), pp542-548.

11. Gordon T. J., Palkovics L., Pilbeam C. Sharp R.S.: Second Generation Approaches to Active and Semi-Active Suspension Control System Design. *13th IAVSD Symposium, Chengdu ,China,* Vol.23(1994), pp158-171.

12. Hedrick J.K.: Railway Vehicle Active Suspensions. *Vehicle System Dynamics*, Vol.26(1996), pp381-393.

13. Wilson D.A., Sharp, R.S., Hasson, S.A.: The Application of Linear Optimal Control Theory to the Design of Active Suspensions. *Vehicle System Dynamics*, Vol.15(1986), pp105-118.

14. Goodall R.: Active Railway Suspension: Implementation Status and Technological Trends. *Vehicle System Dynamics*. Vol.28(1997), pp87-118.

15. Dai H.Y., Zhang H.Q., Zhang W.H.: Robust Performance Analysis of Active Suspension with Model Uncertainty Using Structured Singular Value, μ Approach. *15th IAVSD Symposium, Budapest, Hungary, Suppl. To Vehicle System Dynamics*, Vol.29(1998), pp634-647.

16. Zeng J., Zhang W.H., Dai H.Y.: Hunting Instability Analysis and H_∞ Controlled for Active Stabilizer Design for High Speed Railway Passenger Car. *15th IAVSD Symposium, Budapest, Hungary, Suppl. To Vehicle System Dynamics*, Vol.29(1998), pp655-668.

17. Goodall R. J., Pearson J. T., Pratt I.: Design of Complex Controllers for Active Secondary Suspensions On Railway Vehicles. *Vehicle System Dynamics*, Vol.25(1996), pp217-228.

18. Hrovat D., Barak, P., Rabins M.: An Approach Toward the Optimal Semi-Active Suspension. *Trans, ASME,J. Dyn. Sys. Meas. Cont.*, Vol.110(1988), pp691-705.

19. Hac A.: Stochastic Optimal Control Of Vehicles With Elastic Body and Active Suspensions. *Trans. ASME. J. Dyn. Sys. Meas. Cont.*, Vol.108(1986), pp106-110.

20. Hac A.: Control of Suspensions for Vehicles with Flexible Bodies –Part II : Semi-Active Suspension. *Trans. ASME, J. Dyn. Sys. Meas. Cont.*, Vol.118(1996), pp518-525.

21. Shen Z. Y., Li Z. L.:A Fast Non-state Creep Forcee Model Based on the Simplified Theory. *Wear,* 191(1996), pp242-244.

Vehicle System Dynamics Supplement 41 (2004), p. 784-790 © Taylor & Francis Ltd.

Research on Detection of Braking Reactions in Emergency Situations

Yoji Seto[1], Kouki Minegishi[2], Zhengrong Yang[3] and Takashi Kobayashi[4]

SUMMARY

In this research, an investigation was made of detectable indexes and methods of measuring them among a number of candidate physiological responses that precede a driver's emergency braking action. Evaluation methods that could be used in detecting emergency situations were identified on the basis of the results obtained.

Measurements were made of the electrical activity of the frontal muscle, ocular muscle, brachial biceps, femoral rectus and tibial anterior as well as the operation of the accelerator and brake pedals, gripping force on the steering wheel and pressure on the footrest as potential index candidates. In experiments conducted with an actual vehicle on a proving ground course, test subjects wore a face-mounted display in order to reproduce emergency situations involving potential rear-end collisions. Thirty subjects, including ten older drivers, took part in the experiments. The results showed that the tibial anterior was the fastest to show electrical activity among the muscle responses involved in emergency braking.

We have started research on a braking system for reducing collision speed, including devices that detect the driver's reactions. The system judges whether to apply braking force based on the driver's reactions and operations, in addition to the distance to an object ahead of the vehicle and the characteristic of vehicle deceleration. This enables the system to apply braking force while giving utmost priority to the driver's intentions.

1 INTRODUCTION

Research and development work on safety devices that utilize sensor technologies to detect the circumstances around a vehicle and driver's behavior has been actively pursued in recent years with the aim of assisting drivers with the execution of driving tasks.

[1] Vehicle and Transportation Research Laboratory, Nissan Research Center, Nissan Motor Co., Ltd., 1, Natsushima-cho, Yokosuka-shi, Kanagawa 237-8523, Japan. Phone: +81-468-67-5154, Fax: +81-468-65-5699; E-mail:y-seto@mail.nissan.co.jp
[2] Research Prototype and Test Department, Nissan Research Center, Nissan Motor Co., Ltd., 1, Natsushima-cho, Yokosuka-shi, Kanagawa 237-8523, Japan.
[3,4] Safety and IT Research Division, Japan Automobile Research Institute, Inc., 2530, Karima, Tsukuba-shi, Ibaraki 305-0822, Japan.

The present research probed detectable indexes of physiological responses preceding a driver's emergency braking action and examined methods of measuring such indexes. The following items were examined in this connection and evaluation methods that could be used in the detection procedure were identified on the basis of the results.

1. Selection of physiological and operational responses that precede braking action

2. Identification of detectable indexes of physiological and operational responses and the order of reactions

This paper describes the research procedure, results obtained and an application to a driving support system.

2 EVALUATION INDEXES

A preliminary study was made of possible indexes of the physiological responses involved in braking actions in order to determine which indexes attention should be focused on. Based on the results, it was decided to use the following physiological indexes in this research.

2.1 Physiological Indexes Related to Information Acquisition

Most of the information needed for driving is obtained visually. In response to an external visual stimulus, the brain sends commands to activate the muscles around the eyes, resulting in changes in facial expression. The frontal muscle and the ocular muscle move the muscles around the eyes when the eyes are opened wider in response to an external stimulus such as an emergency-driving situation. These two muscles were used as physiological indexes related to information acquisition [1].

2.2 Physiological Indexes Related to Operational Responses

In the course of depressing the brake pedal quickly with considerable force in emergency braking situations, it is thought that drivers pull the steering wheel toward them or press against it with greater force than they normally do. It is obvious that the brachial biceps in particular react quickly to produce the force for executing such action.

The tibial anterior acts to bend the foot backward as drivers release the accelerator pedal so that they can depress the brake pedal, and the femoral rectus then acts when drivers raise their thigh to move their foot to the brake pedal. The actions of these muscles were made clear by the results of the preliminary study.

Accordingly, the brachial biceps, femoral rectus and tibial anterior were selected as physiological indexes directly related to the action of pulling or pressing against the steering wheel and pedal operations [2].

Measurements were also made of the gripping force and tension on the steering wheel and the pressure directly applied to the brake pedal and footrest when braking. Examples of the measurements made in the experiments are shown in Fig. 1.

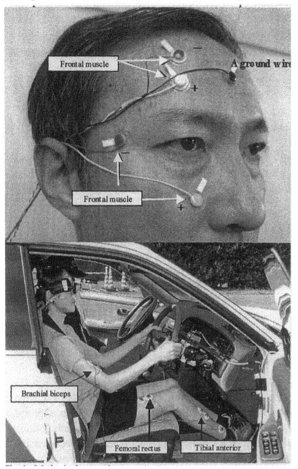

Fig. 1. Method of measuring muscle electrical activity

3 EXPERIMENTAL PROCEDURE

Driving tests that simulated rear-end collisions, representing a typical accident pattern in Japan, were conducted on a proving ground course at the Japan Automobile Research Institute (JARI). The test vehicle used was an ordinary passenger car with a 2.0-liter engine. The test subjects wore a face-mounted display incorporating a CCD camera, as shown in Fig. 2. The subjects watched images presented on the display as they drove the test course.

In the experiment that simulated a rear-end collision, the subjects first followed a preceding vehicle at a speed of approximately 40 km/h while maintaining a constant headway distance. At the point when they became accustomed to driving, the display switched from an image of the surrounding scenery to a prerecorded image of an emergency braking situation. This created the illusion that the vehicle in front, which they had been following up to that point, had stopped suddenly and they needed to execute emergency braking. Each subject's physiological and operational responses were recorded at that moment along with information on vehicle behavior. A typical test scene is shown in Fig. 3.The test subjects were 30 ordinary drivers.

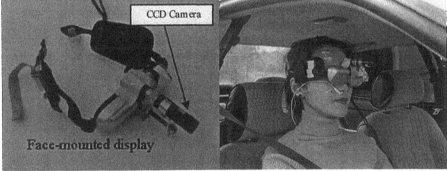

Fig. 2. Usage of face-mounted display

Fig. 3. Test vehicle

4 EXPERIMENTAL RESULTS

Figure 4 shows the orders of physiological and operational responses plotted on a time line. These plots are the average values of all subjects. An examination of the order of responses indicates that the electrical activities constituted the main responses that preceded the operation of the accelerator pedal, brake pedal and steering action. The order of responses was as follows. The tibial anterior showed the first response as the subjects released the accelerator pedal. Electrical activity of the frontal muscle, ocular muscle and femoral rectus then occurred at approximately the same time. That was followed by the electrical activity of the brachial biceps, the action of lifting the foot off the accelerator pedal and depressing the brake pedal.

In terms of the order of operational responses, the subjects first lifted their foot off the accelerator pedal, then applied pressure to the footrest, gripped the steering wheel, applied tension to the steering wheel and depressed the brake pedal. In other words, in the interval between taking their foot off the accelerator pedal and depressing the brake pedal, the subjects either applied pressure to the footrest with their left foot or applied force to the steering wheel in some form.

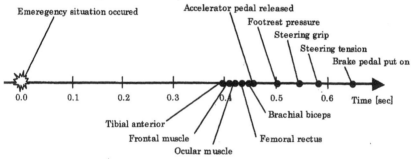

Fig. 4. Order of physiological responses and operational responses

5 APPLICATION TO A BRAKING SYSTEM FOR REDUCING COLLISION SPEED

We initiated a study concerning the application the detectable indexes and related measurement methods to a system that applies braking force just before a collision occurs with a forward object, with the aim of reducing the collision speed. To ensure that the system applies braking force only in emergency situations, its activation is based on the following judgment procedure. If the system judges that the driver is going to depress the brake pedal and that there is a very high possibility of a collision with a forward object even if the driver executes braking action, it applies braking force.

Figure 5 shows the configuration of this braking system. Based on the signals for both physiological and operational responses, the system judges whether the driver is going to depress the brake pedal or not. Based on the distance to a forward object detected with a radar sensor, it judges if there is high possibility of a collision or not.

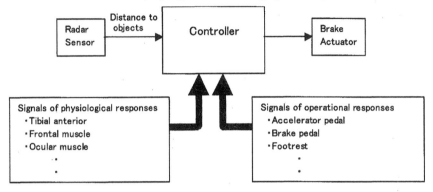

Fig. 5 Configuration of braking system

The earlier the system applies braking force, the more the collision speed is reduced. Accordingly, if Equation (1) is satisfied, the system applies braking force. The first term on the right side of the equation is the distance needed for deceleration and the second term is the distance that the vehicle travels prior to the driver's braking action.

$$d_r \leq \frac{v_r^2}{2a} + t_{brk} v_r \tag{1}$$

where

d_r : distance to a forward object

v_r : relative velocity the forward object

a : assumed deceleration resulting from the driver's emergency braking action

t_{brk} : estimated time gap between the physiological index of action and driver's braking action

We have started to research the above-mentioned braking system as one system that is based on the driver's reactions.

6 CONCLUSION

An investigation was made of detectable indexes of physiological and operational responses that precede a driver's emergency braking action and ways of measuring such indexes. Various indexes of the physiological and operational responses involved in braking action were examined and evaluated in experiments conducted with a test vehicle. The findings obtained are summarized below.

1. The order of physiological response time and operational responses in relation to an emergency braking situation while driving was identified.

2. The tibial anterior was the fastest to show electrical activity among the muscle responses involved in emergency braking. In terms of the responses involved in or related to driving operations, it was found that lifting the foot off the accelerator pedal was followed by the application of pressure to the footrest.

Research has been initiated on a braking system that incorporates devices to detect the driver's reactions with the aim of reducing collision speeds.

It should be noted that the findings of this research are based on an analysis of experimental data obtained in tests conducted with a limited number of subjects. In order to gain a clearer understanding of the details of physiological and operational responses, it will be necessary to conduct validation testing with a larger number of subjects in the future.

REFERENCES

1. Nakamura, R., et al., Clinical Kinesiology, Ishiyaku Publishing, Inc. 1979.

2. Asoh, T., Automotive Research, 7-5, 1985.

Non-Linear Dynamic Techniques v. Equivalent Conicity Methods For Rail Vehicle Stability Assessment

R.M.GOODALL [1], S.D.IWNICKI [2]

SUMMARY

This paper is to record a special discussion meeting held on 26th August 2003 during the 18th IAVSD Symposium in Atsugi, Kanagawa, Japan, with the particular objective of answering the question "Should we use the term Equivalent Conicity?". The subject is introduced, the subsequent discussion is documented, and some final conclusions and recommendations are given.

1. INTRODUCTION

Chair: *Professor Roger Goodall, Loughborough University, UK*

There have been discussions in several of the early sessions at the Symposium about the validity of the use of the parameter equivalent conicity (λ) [1] in assessing railway vehicle stability. This special session has been convened to allow open discussion and for a possible decision on the appropriateness of the technique. Professor Hans True will provide an overview of non-linear dynamics techniques, there will then be a period for discussion and at the end we will try and come to some general conclusions.

Professor Hans True, The Technical University of Denmark

I'd like to start with a true story. The Danish State Railways (DSB) wanted to lease cars to run at 160 km/hr. The manufacturer (Bombardier) and DSB had tested the cars on all lines, where they should be used, with up to 175 km/hr with no problems. In places the gauge on the route was very narrow – down to 1425 mm but this had not caused problems in the tests. The track authority refused to accept the vehicles because the equivalent conicity was too high according to the UIC to run at over 140 km/hr.

Initially several simulations using SIMPACK and our own routines were performed with various vehicle models passing from a section with 1435 mm gauge onto a section with 1427 mm gauge and then back again. The speeds were chosen to be close to but under the critical speed (the critical speed as defined by the saddle-node bifurcation point in diagram 2). As expected the vehicles started to hunt

---

[1] *Department of Electronic and Electrical Engineering Loughborough University, Leicestershire, LE11 3TU, UK Email: R.M.Goodall@lboro.ac.uk*

[2] *Department of Engineering and Technology, Manchester Metropolitan University, Manchester M1 5GD UK, S.D.Iwnicki@mmu.ac.uk*

immediately after having entered the 1427 mm section and stopped hunting soon after having left this section. Then simulations were performed with a verified model of the actual car in Bombardier's plant in Görlitz on their computer, and the results of the simulations showed that the cars run stably - even on 1427 mm gauge track - up to 216 km/hr. When these results were presented to the Danish State Railway Inspectorate the cars were allowed to run at 160 km/hr, but the Danish State Railway Inspectorate took their revenge: From the following week the maximum speed of all locomotive types that were earlier permitted to run faster had their maximum speed limited to 140 km/hr.

There is ABSOLUTELY NO THEORETICAL BASIS for the use of the concept 'equivalent conicity' in rail-wheel contact problems. The real conicity on the other hand is mathematically well defined on a sound scientific basis and should therefore ALWAYS be used instead of the popular 'equivalent conicity'. Neither can alone be used to estimate the dynamics of a railway vehicle, since the dynamical behaviour depends strongly also on the design of the vehicle - especially its suspension system. As a 'rule of thumb', if anybody wants to use that phrase, the real conicity is as good, and it is better because the real conicity is scientifically sound. Therefore the concept 'equivalent conicity' ought to be abandoned before more harm is done by its abuse.

I am worried that these issues could spread to other railways.

Not all problems are non-linear but the wheel-rail contact certainly is. Non-linear problems have different characteristics to linear problems:

- The principle of superposition does not hold;
- There can be more than one equilibrium position for a fixed set of parameter values.

So the first task in the solution of a non-linear problem is to find all the possible solutions. This can be done easily by using modern multibody simulation tools. Then the stability of all the solutions must be found. This can be done by linearising around each equilibrium solution and then by checking that all the real parts of the eigenvalue are negative.

Figure 1 shows an example of two asymptotically stable spiral points in the phase plane separated by a saddle point. It is easily seen that the asymptotes to the saddle point that describe the motion approaching the saddle split the phase plane into two distinct domains called domains of attraction. One consists of all initial states that will approach one attractor (a spiral point) and the other domain consists of all initial states that will approach the other attractor. This dynamical problem is a stationary problem, where the three equilibrium solutions are fixpoints.

In railway dynamics we deal with a periodic solution describing hunting coexisting in a certain speed interval with a stationary solution - a fixpoint. Let us now make the following supposition: Rotate the phase plane around a vertical axis through the right hand side spiral point. The left hand side spiral point will describe a circle in the now three-dimensional phase space. Since it is a closed curve and the points on it are stable, it can be considered a stable limit cycle. The saddle point will in the same way become an unstable limit cycle in the three dimensional phase space. The asymptotes mentioned above will with an appropriate little transformation connect into surfaces that will divide the three dimensional phase space into two domains of attraction: One for the stable fixpoint on the axis and one for the stable limit cycle. This is NOT the phase space for railway dynamical problems, but it may serve as a simple three-dimensional analogy in order to demonstrate the structure of the phase space in the speed range in railway dynamics in which a stable fixpoint coexists with a stable limit cycle.

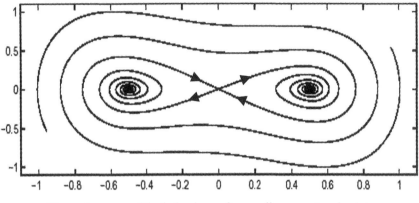

Figure 1.   The behaviour of a non-linear system in state space

Consider the bifurcation diagram shown in Figure 2. It can be seen that there exist two asymptotically stable solutions for all speeds between the critical speed and the Hopf bifurcation point at velocity = 1: A periodic (hunting) motion characterized by a non-zero amplitude in the diagram and the non-periodic trivial solution. The last solution is the desired one. As soon as there is a coexistence of asymptotically stable solutions in a nonlinear dynamical problem, then each stable solution has its own domain of attraction in phase space as seen in figure 1. The domain of attraction consists of all states from where the solution trajectories will approach its stable solution. These domains are finite and therefore finite disturbances exist that will take a trajectory away from a given domain of attraction and put it into another domain of attraction. Related to the railway hunting problem and the critical speed it means: For all speeds where there exists a stable periodic solution in addition to the 'straight' or non-periodic motion, there will be a finite probability that a finite disturbance will push the motion onto the periodic one, which in reality has a large domain of attraction. Therefore it is the lowest value of the speed for which a periodic solution exists that determines the critical speed!

There is a finite probability that a disturbance will take the solution to the asymptotically stable periodic solution. This has been seen in practical tests in France, Poland, USA and Denmark.

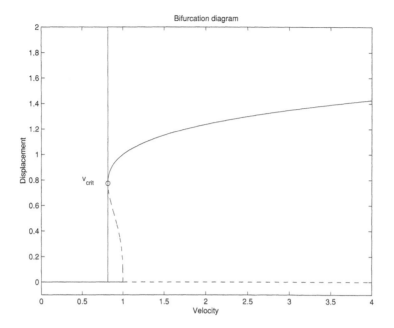

Figure 2.   The bifurcation diagram

The following method may be used to calculate the critical speed:

Start a numerical simulation for the system at a speed slightly above the Hopf bifurcation point (the initial bifurcation) and with a small perturbation. The solution will approach the asymptotically stable periodic solution (the hunting). Repeat the simulation with a slightly reduced speed and with initial conditions taken from the end of the previous step. Repeat the procedure with decreasing speeds until the solution will no longer be periodic but has returned to a stable non-periodic motion.

The lowest speed at which a periodic oscillation is found in this way is strictly spoken a lower bound for the real critical speed, but in reality it has proven to be a very accurate value under normal service conditions. Note that the simulation results only are valid for the parameter values other than the speed that are used in the simulation. In order to find the 'most critical speed' more simulations with other parameter combinations like the coefficient of adhesion, gauge and so on must be performed.

"Each truth passes through three stages before it is recognized: First it is ridiculed, then it is vigorously opposed and finally it is regarded as obvious" Arthur Schopenhauer 1788-1860.

2. DISCUSSION

Mr. Jeremy Evans, AEA Technology Rail, UK

Professor True has demonstrated the mathematical case but as an engineer I need to know if a vehicle will run smoothly down the track whereas a mathematician might want to know if there is a stable solution.

British Rail worked hard to find a range of practical values of conicity (equivalent conicity as wheel profiles are not conical) and found that if vehicles were designed to be stable for λ between 0.05 and 0.4 then they proved to be stable on the whole of the UK network. The rule adopted was that for 5% damping and 10% overspeed the vehicle must be shown to be stable for this range of conicity.

As a case study the class 365 vehicle was showing hunting behaviour with worn P8 wheel profiles when running on good quality, straight track. The profiles were found to be very hollow and the equivalent conicity was over 1.0. A limit of 0.4 was set for the equivalent conicity to force the track owner to grind the rail profiles to restore the conicity.

We took 50 matched wheel and rail pairs with equivalent conicity varying from 0.2 to 0.48. We ran non-linear stability tests with an initial disturbance and found that for λ up to 0.38 there was no hunting observed but at values of λ over 0.45 there was always hunting.

So, although mathematically the concept of equivalent conicity may be indefensible, in practical terms the concept is useful to railway engineers.

Prof. Mats Berg, KTH Stockholm, Sweden

When defining the operating conditions for evaluation of ride stability ride comfort it is quite convenient to use the parameter equivalent conicity since it summarises several other parameters such as wheel profile, rail profile, rail inclination wheel width, flange thickness and track gauge (including the variation in track gauge). Although a linearisation must be carried out to arrive at a value for λ, the use of this method is a reasonable approach as it significantly reduces the number of vehicle and track measurements and simulations required. The idea of defining this procedure in standards is good but the limits should only be recommendations.

Prof. Yuri Romen VNIZT, Moscow, Russia

I am not a mathematician but an engineer. I would like to start my comments with a story:

Two people came to a wise man, one said 'Grey is almost like black', the wise man said 'Yes, you are right'. The other man said 'Grey is very similar to white', the wise man said 'Yes you are right'. A passer by heard this and said 'But they can't both be right' and the wise man said 'Yes, you are right too'.

The discussion here is about the definition of equivalent conicity. The value of equivalent conicity depends on where the contact is. In the 1950s we didn't understand the concept of a limit cycle but anyway we had to know the range of parameters that would give stability. Yesterday there was a paper that showed that even in curves, if the irregularities are great and the rail is worn, there can be problems with instability as flange contact is reached and conicity becomes high.

Equivalent conicity has a right to exist but you must specify the railway conditions. In natural experiments we have seen examples where a new train can be unstable in the morning and then stable in the afternoon. Gauge can vary, creep forces can vary and we must define the conditions that are important. We should not use the UIC definition of stability.

Dr. Ingemar Persson, DEsolver, Sweden

Effective conicity is a useful technique for engineers who must be aware of the limitations of the method, but it should not be written into standards.

Professor Stefano Bruni, Politecnico di Milano, Italy

Equivalent conicity is useful when we want to compare different wheel/rail profiles, but if we want to get quantitative agreement with experiments, or to predict accurately the behaviour of a given vehicle on a given track, we need to use a more detailed approach.

We should not forget that track irregularities mean that we move across the rolling radius difference plot, so that the amplitude of the wheelset motion will influence its stability.

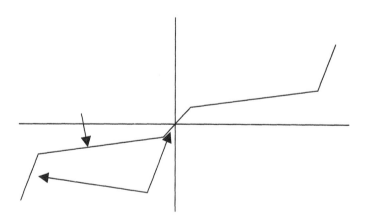

Figure 3.   The rolling radius difference plot

So we in Milan are using the following method. We perform non-linear simulations in the time domain with real profiles and track irregularities, for increasing vehicle speed values, and we plot the RMS lateral acceleration of the bogie against the running speed, as shown in Figure 4. We usually see that there is a jump in the lateral acceleration and we take this value as the critical speed.

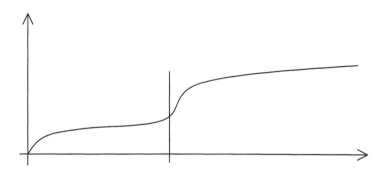

Figure 4.   Lateral acceleration against speed

This, of course, is not useful when writing a standard but it is useful in practical work.

Apart from the problem of computing the stability threshold using a mathematical model, we also need to find a reliable way of detecting instability during line tests. A typical way to do that is to look at the measurement of the lateral axlebox acceleration, searching for 6 consecutive cycles with amplitude higher than 8 m/s^2, this could however also be the result of the stable vehicle response to a cyclic track irregularity. I think we should make an effort to find out more reliable methods and have them incorporated into the standards. A possible solution could be to use methods like autocorrelation / random decrement techniques in conjunction with parameter estimation algorithms such as the methods by Prony or Ibrahim to quantify the amount of damping in the system.

Dr. TX Mei, Leeds University, UK

I feel that we are discussing this issue too much as a mathematical question but the fact is that the wheel-rail contact law (as we know it) is not really a precise science.

Professor Hans True, The Technical University of Denmark

If there is track excitation we can see chaotic motion. It is a highly complicated problem and is outside the linear range. We have to prove values by using non-linear simulation.

If you use maths you must use the mathematical laws.

Dr. Anna Orlova, St. Petersburg University of the Means of Communication, Russia

For most general applications the equivalent conicity is suitable for understanding the qualitative behaviour of the system.

Professor Hans True, The Technical University of Denmark

Ok, if you don't tell anyone! But you can use the real conicity to obtain the same degree of understanding.

Chair: Professor Roger Goodall, Loughborough University, UK

This type of discussion is just what we should be doing at the IAVSD symposia. We have not exactly answered the question that we asked ourselves, but we have had a useful exchange of ideas.

Professor True has proposed that we should take the calculation of equivalent conicity out of the current UIC standard and perhaps that is something that we all need to think about.

Thank you all for attending.

3.   CONCLUSIONS AND RECOMMENDATIONS

The following conclusions have been prepared by the authors to reflect the overall results of the discussion meeting:

i)   There is no question that wheel-rail contact with profiled wheels creates a non-linear problem and that the actual critical speed will usually be lower than that predicted by simple linear theory.

ii)   Despite the above, the general view emerging from the discussion is that the concept of equivalent conicity is a useful way of capturing certain aspects of the non-linearity and is therefore a practical engineering approach.

iii)   The concept of equivalent conicity should only be used by vehicle dynamicists who understand the non-linear issues and are therefore properly aware of the context in which it is being used.

The following recommendations are made:

a)   It is inappropriate to include equivalent conicity as a mandatory part of railway standards. If the equivalent conicity stability assessment procedure is defined in standards, any limits based upon the approach should be recommendations only.

b)   A recommended alternative would be to require stability to be demonstrated for the specified conditions and to allow the vehicle or track designer or operator to prove that stable running will be achieved by appropriate simulation or testing.

c)   A special session should be arranged at the next IAVSD Symposium in 2005 in order to pursue this discussion further. Particular practical examples can be used to provide objective comparisons of the stability predictions between the rigorous non-linear dynamics approach and the linearised equivalent conicity method.

ACKNOWLEDGEMENT

The authors are grateful to all those people who attended the discussion meeting, and in particular to Professor Hans True, not only because his scientific challenge of the concept of equivalent conicity stimulated the discussion, but also for preparing and presenting the introduction to non-linear dynamics. Thanks also to Jerry Evans for leading the case for the defence and to all those who contributed to the vibrant discussion.

REFERENCES

1.   Wickens A H, "Fundamentals of rail vehicle dynamics – guidance and stability", Swets and Zeitlinger, 2003.

BIBLIOGRAPHY

True H. "Does a Critical Speed for Railroad Vehicles exist?", *RTD-Vol. 7, Proc. of the 1994 ASME/IEEE Joint Railroad Conference*, Chicago Ill., March 22--24, 1994, American Society of Mechanical Engineers, pp125-131

Ahmadian M. and Yang, "Hopf Bifurcation and Hunting Behavior in a Rail Wheelset with Flange Contact", *Nonlinear Dynamics*, V15, 1998, pp15-30

Jensen J.C., Slivsgaard E.C. and True H., "Mathematical Simulation of the Dynamics of the Danish IC3 Train", *Proc. 15th IAVSD Symposium on Vehicle System Dynamics, The Dynamics of Vehicles on Roads and Tracks,* Swets & Zeitlinger,1997,Lisse, pp760-765

True, H., "Some Recent Developments in Nonlinear Railway Vehicle Dynamics", *Proceedings of the 1st European Nonlinear Oscillations Conference*, Hamburg, August 16-20 1993, Akademie Verlag,1994, pp129-148

True H., Jensen J.C. and Slivsgaard E., "Nonlinear Railway Vehicle Dynamics and Chaos", *Proc. 5th Miniconf. on Vehicle System Dynamics, Identification and Anomalies*, Budapest, Nov. 11-13, 1996, Budapest, Hungary, pp51-60

True H., Jensen J.C. and Slivsgaard E., "Systemdynamik von Schienenfahrzeugen", *Der Eisenbahningenieur*, V50(2) 1999, pp37-42

Y. Marumo and T. Katayama

Energy flow method for the analysis of effects of structural flexibility on motorcycle stability

Z. Nagorski, J. Piotrowski and T. Szolc

Parametric analysis of the thermal trace caused by medium frequency wheel-rail contact excitations

A. Johansson and C. Andersson

Out-of-round railway wheels - study of formation of long periodic defects by combining analyses of wear and dynamic train-track interaction

P.M. Belotserkovskiy

Two-axle bogie parametric oscillations due to sleeper spacing and rail corrugation

G. Bogomaz, H. Kovtun, O. Markova, V. Maliy and V. Raznosilin

Determination of derailment stability coefficient values for development of on-line diagnostic system for passenger train motion safety

J.R. Evans and D.H. Gilbert

Real-time dynamic simulation as a tool in track geometry monitoring

Y. Boronenko, A. Orlova, V. Bubnov and Y. Romen

Development of higher capacity freight bogies with low track forces

S. Wagner and C. Cole

Modeling train-wagon interaction on curves

N. Asanuma and M. Abe

Yaw rate control by sensing tire forces to stabilize vehicle motion with independent ABS wheel control

L. Nouveliere and A. Chaibet and S. Mammar

Low speed automated and shared longitudinal control using a second order sliding modes

K. Ramji, V.H. Saran., V.K. Goel and K. Deep

Optimum design of suspension system for three-wheeled motor vehicle– using random search optimization technique –

H. Furusho, M. Shimakage and R. Shirato

A lane recognition method using the tangent vectors of white lane markers

H. Soma and D. Haino

Optimum avoidance strategy in automatic collision avoidance systems

Z. Tian, T. Kamada and M. Nagai

Study on a vehicle behavior under earthquake using driving simulator

W. Zhang, Y. Liu and G. Mei

A study on dynamic stress of catenary

M. Nagai, H. Yoshida and T. Sueki

A new tilting control for railway vehicle using variable link mechanism

K. Tanifuji, H. Soma and N. Ishizaka

Curving behavior of actively steered rail vehicles on sharp curves

J. Li, R.M. Goodall, T.X. Mei and H. Li

Steering controllers for rail vehicles with independently driven wheel motors

M. Pennati, M. Gobbi and G. Mastinu

A dummy for measuring the ride comfort of vehicles

M. Valasek, Z. Sika and M. Stefan

Nonlinear identification for improvement of vehicle suspension control

S. Toyama and M. Yokoyama

Sliding mode control for semi-active suspensions under actuator uncertainties

I. Youn and S. Lee

Discrete time preview suspension control for the real time simulation of a tracked vehicle

Milton Keynes UK
Ingram Content Group UK Ltd.
UKHW050304161024
449569UK00033B/133